CLASSICAL MATHEMATICS FROM AL-KHWĀRIZMĪ TO DESCARTES

This book follows the development of classical mathematics and the relation between work done in the Arab and Islamic worlds and that undertaken by the likes of Descartes and Fermat.

'Early modern' mathematics is a term widely used to refer to the mathematics which developed in the West during the sixteenth and seventeenth century. For many historians and philosophers this is the watershed which marks a radical departure from 'classical mathematics,' to more modern mathematics; heralding the arrival of algebra, geometrical algebra, and the mathematics of the continuous. In this book, Roshdi Rashed demonstrates that 'early modern' mathematics is actually far more composite than previously assumed, with each branch having different traceable origins which span the millennium. Going back to the beginning of these parts, the aim of this book is to identify the concepts and practices of key figures in their development, thereby presenting a fuller reality of these mathematics.

This book will be of interest to students and scholars specialising in Islamic science and mathematics, as well as to those with an interest in the more general history of science and mathematics and the transmission of ideas and culture.

Roshdi Rashed is one of the most eminent authorities on Arabic mathematics and the exact sciences. A historian and philosopher of mathematics and science and a highly celebrated epistemologist, he is currently Emeritus Research Director (distinguished class) at the Centre National de la Recherche Scientifique (CNRS) in Paris, and is the former Director of the Centre for History of Medieval Science and Philosophy at the University of Paris (Denis Diderot, Paris VII). He also holds an Honorary Professorship at the University of Tokyo and an Emeritus Professorship at the University of Mansourah in Egypt.

Michael H. Shank is Professor of the History of Science at the University of Wisconsin-Madison, where he teaches surveys of the history of science from antiquity to Newton. His research interests focus on, and often stray beyond, the late-medieval Viennese astronomical and natural philosophical traditions.

CULTURE AND CIVILIZATION IN THE MIDDLE EAST

General Editor: Ian Richard Netton

Professor of Islamic Studies, University of Exeter

This series studies the Middle East through the twin foci of its diverse cultures and civilisations. Comprising original monographs as well as scholarly surveys, it covers topics in the fields of Middle Eastern literature, archaeology, law, history, philosophy, science, folklore, art, architecture and language. While there is a plurality of views, the series presents serious scholarship in a lucid and stimulating fashion.

PREVIOUSLY PUBLISHED BY CURZON

THE ORIGINS OF ISLAMIC LAW
The Qur'an, the Muwatta' and Madinan Amal
Yasin Dutton

A JEWISH ARCHIVE FROM OLD CAIRO
The History of Cambridge University's Genizah Collection
Stefan Reif

THE FORMATIVE PERIOD OF TWELVER SHI'ISM
Hadith as Discourse Between Qum and Baghdad
Andrew J. Newman

QUR'AN TRANSLATION
Discourse, Texture and Exegesis
Hussein Abdul-Raof

CHRISTIANS IN AL-ANDALUS 711–1000
Ann Rosemary Christys

FOLKLORE AND FOLKLIFE IN THE UNITED ARAB EMIRATES
Sayyid Hamid Hurriez

THE FORMATION OF HANBALISM
Piety into Power
Nimrod Hurvitz

ARABIC LITERATURE
An Overview
Pierre Cachia

STRUCTURE AND MEANING IN MEDIEVAL ARABIC AND PERSIAN LYRIC POETRY
Orient Pearls
Julie Scott Meisami

MUSLIMS AND CHRISTIANS IN NORMAN SICILY
Arabic-Speakers and the End of Islam
Alexander Metcalfe

MODERN ARAB HISTORIOGRAPHY
Historical Discourse and the Nation-State
Youssef Choueiri

THE PHILOSOPHICAL POETICS OF ALFARABI, AVICENNA AND AVERROES
The Aristotelian Reception
Salim Kemal

PUBLISHED BY ROUTLEDGE

1. THE EPISTEMOLOGY OF IBN KHALDUN
Zaid Ahmad

2. THE HANBALI SCHOOL OF LAW AND IBN TAYMIYYAH
Conflict or Concilation
Abdul Hakim I Al-Matroudi

3. ARABIC RHETORIC
A Pragmatic Analysis
Hussein Abdul-Raof

4. ARAB REPRESENTATIONS OF THE OCCIDENT
East-West Encounters in Arabic Fiction
Rasheed El-Enany

5. GOD AND HUMANS IN ISLAMIC THOUGHT
Abd al-Jabbār, Ibn Sīnā and al-Ghazālī
Maha Elkaisy-Friemuth

6. ORIGINAL ISLAM
Malik and the madhhab of Madina
Yasin Dutton

7. AL-GHAZALI AND THE QUR'AN
One Book, Many Meanings
Martin Whittingham

8. BIRTH OF THE PROPHET MUHAMMAD
Devotional Piety in Sunni Islam
Marion Holmes Katz

9. SPACE AND MUSLIM URBAN LIFE
At the Limits of the Labyrinth of Fez
Simon O'Meara

10. ISLAM AND SCIENCE
The Intellectual Career of Nizam al-Din al-Nizaburi
Robert G. Morrison

11. IBN 'ARABÎ – TIME AND COSMOLOGY
Mohamed Haj Yousef

12. THE STATUS OF WOMEN IN ISLAMIC LAW AND SOCIETY
Annotated translation of al-Tāhir al-Haddād's *Imra'tunā fi 'l-sharī'a wa'l-mujtama'*, with an introduction
Ronak Husni and Daniel L. Newman

13. ISLAM AND THE BAHA'I FAITH
A Comparative Study of Muhammad 'Abduh and 'Abdul-Baha 'Abbas
Oliver Scharbrodt

14. COMTE DE GOBINEAU AND ORIENTALISM
Selected Eastern Writings
Translated by Daniel O'Donoghue Edited by Geoffrey Nash

15. EARLY ISLAMIC SPAIN
The History of Ibn al-Qūṭīya
David James

16. GERMAN ORIENTALISM
The Study of the Middle East and Islam from 1800 to 1945
Ursula Wokoeck

17. MULLĀ ṢADRĀ AND METAPHYSICS
Modulation of Being
Sajjad H. Rizvi

18. SCHOOLS OF QUR'ANIC EXEGESIS
Genesis and Development
Hussein Abdul-Raof

19. AL-GHAZALI, AVERROES AND THE INTERPRETATION OF THE QUR'AN
Common Sense and Philosophy in Islam
Avital Wohlman, translated by David Burrell

20. EASTERN CHRISTIANITY IN THE MODERN MIDDLE EAST
Edited by Anthony O'Mahony and Emma Loosley

21. ISLAMIC REFORM AND ARAB NATIONALISM
Expanding the Crescent from the Mediterranean to the Indian Ocean (1880s–1930s)
Amal N. Ghazal

22. ISLAMIC ETHICS
Divine Command Theory in Arabo-Islamic Thought
Mariam al-Attar

23. MUSLIM FORTRESSES IN THE LEVANT
Between Crusaders and Mongols
Kate Raphael

24. BEING HUMAN IN ISLAM
The Impact of the Evolutionary Worldview
Damian Howard

25. THE UAE AND FOREIGN POLICY
Foreign Aid, Identities and Interests
Khalid S. Almezaini

26. A HISTORY OF EARLY AL-ANDALUS
The Akhbar Majmu'a
David James

27. INSPIRED KNOWLEDGE IN ISLAMIC THOUGHT
Al-Ghazali's Theory of Mystical Cognition and its Avicennian Foundation
Alexander Treiger

28. SHI'I THEOLOGY IN IRAN
The Challenge of Religious Experience
Ori Goldberg

29. FOUNDING FIGURES AND COMMENTATORS IN ARABIC MATHEMATICS
A History of Arabic Sciences and Mathematics, Volume 1
Roshdi Rashed

30. THE MUSLIM CONQUEST OF IBERIA
Medieval Arabic Narratives
Nicola Clarke

31. ANGELS IN ISLAM
Jalal al-Din al-Suyuti's *al-Haba'ik fi akhbar al-mala'ik*
Stephen Burge

32. THEOLOGICAL APPROACHES TO QUR'ANIC EXEGESIS
A Practical Comparative-Contrastive Analysis
Hussein Abdul-Raof

33. IBN AL-HAYTHAM AND ANALYTICAL MATHEMATICS
A History of Arabic Sciences and Mathematics, Volume 2
Roshdi Rashed

34. GHAZALI'S POLITICS IN CONTEXT
Yazeed Said

35. ORIENTALISM REVISITED
Art, Land and Voyage
Edited by Ian Richard Netton

36. IBN AL-HAYTHAM'S THEORY OF CONICS, GEOMETRICAL CONSTRUCTIONS AND PRACTICAL GEOMETRY
A History of Arabic Sciences and Mathematics, Volume 3
Roshdi Rashed, translated by J. V. Field

37. SCIENCE AND RELIGION IN MAMLUK EGYPT
Ibn al-Nafis, Pulmonary Transit and Bodily Resurrection
Nahyan Fancy

38. THE QUR'AN IN CHRISTIAN-MUSLIM DIALOGUE
Historical and Modern Interpretations
C Jonn Block

39. ARABIC AND HEBREW LOVE POEMS IN AL-ANDALUS
Shari L. Lowin

40. RELIGIOUS SCHOLARS AND THE UMAYYADS
Piety-Minded Supporters of the Marwanid Caliphate
Steven C. Judd

41. SKEPTICISM IN CLASSICAL ISLAM
Moments of Confusion
Paul L. Heck

42. FREE WILL AND PREDESTINATION IN ISLAMIC THOUGHT
Theoretical Compromises in the Works of Avicenna al-Ghāzālī and Ibn 'Arabī
Maria de Cillis

43. IBN AL-HAYTHAM, NEW SPHERICAL GEOMETRY AND ASTRONOMY
A History of Arabic Sciences and Mathematics, Volume 4
Roshdi Rashed, translated by J.V. Field

44. CLASSICAL MATHEMATICS FROM AL-KHWĀRIZMĪ TO DESCARTES
Roshdi Rashed, translated by Michael H. Shank

CLASSICAL MATHEMATICS FROM AL-KHWĀRIZMĪ TO DESCARTES

Roshdi Rashed

Translated by Michael H. Shank

Routledge
Taylor & Francis Group
LONDON AND NEW YORK

مركز دراسات الوحدة العربية
CENTRE FOR ARAB UNITY STUDIES

First published 2015 by Routledge

2 Park Square, Milton Park, Abingdon, Oxfordshire OX14 4RN
52 Vanderbilt Avenue, New York, NY 10017

Routledge is an imprint of the Taylor & Francis Group, an informa business

First issued in paperback 2019

British Library Cataloguing in Publication Data
A catalogue record for this book is available from the British Library

Library of Congress Cataloging in Publication Data
A catalog record for this book has been requested

ISBN: 978-0-415-83388-2 (hbk)
ISBN: 978-0-367-86761-4 (pbk)

Typeset in Times New Roman
by Cenveo Publisher Services

This book was prepared from press-ready files supplied by the editor

CONTENTS

II. ARITHMETIC

PART II: GEOMETRY

PART III

APPLICATION OF MATHEMATICS: ASTRONOMY AND OPTICS

FOREWORD

Classical Mathematics from al-Khwārizmī to Descartes includes two new chapters – one on the transmission of Greek heritage into Arabic and the other on Descartes's mathematics – that did not appear in the original French of *D'al-Khwārizmī à Descartes*. Conversely, I have omitted here the chapter on burning mirrors ('Les miroirs ardents, anaclastique et dioptrique'), a subject to which I devoted an entire book, which is now available in English.[1]

The English translation of the present work by Professor Michael Shank has benefited greatly from both his competence in the history and philosophy of science and his refined bilingualism. I mention this to express my profound gratitude for his hard work and for apposite comments that improved the text.

I warmly thank Aline Auger (Centre National de la Recherche Scientifique), who focused her competence and flawless attention to detail on assembling the index and preparing the book for the press.

I am also grateful to Kathryn Rylance and Joe Whiting at Routledge for the care that they gave to this project at every stage of its development.

Last but not least, Dr Khair El-Din Haseeb has spared no effort in bringing this historical research to audiences beyond the original francophone one. I hereby offer him my friendly gratitude.

Roshdi RASHED

[1] *Geometry and Dioptrics in Classical Islam*, London, al-Furqān, 2005.

TRANSLATOR'S NOTE

In agreeing to undertake this translation, I hoped to learn much. The project has exceeded my expectations in many ways. Most directly, it has given me a new appreciation for the exceptional range and importance of Professor Rashed's contributions to the histories of mathematics and of the mathematical sciences, from manuscript discoveries to pointed analyses. Most strikingly, his scholarship offers a panoramic and up-to-date view of Arabic mathematics, of its historical and conceptual connections with Greek, medieval Latin, and early-modern mathematics. If this book has taught me the most about new research in the remarkable range of Arabic mathematics, its expositions have also given me a new appreciation for the significance of both Apollonius and Fermat. Not least, looking to the future, the book also makes clear that much remains to be done in the history of Arabic mathematics and mathematical sciences. May it serve to inspire young scholars when it hints that exciting discoveries are likely to reward the patient reader of Arabic manuscript holdings.

In translating, I have striven for clarity above all. When this could be achieved by retaining the gallic flavour of the original, I have done so, sometimes in the case of mathematical terminology.

Whenever they were readily available, I have used standard English translations of primary sources. Since every translation is also an interpretation, I have in some instances (usually noted) modified an existing translation to make it consistent with Professor Rashed's interpretation of the source.

Many thanks to Professor Rashed for his confidence, and to Carol Troyer-Shank for her able assistance contributions. Without her managerial acumen, and editorial and clerical skills, I doubt that I could have completed the translation.

Michael H. SHANK

PREFACE

When historians become concerned with organizing and illuminating the various stages of mathematical thought, it is not unusual for them to isolate mathematics at the dawn of the modern age by distinguishing it from ancient and medieval mathematics. The expressions 'early modern mathematics' or 'mathematics in the classical age' are the ones customarily used to designate the mathematics that developed during the 16th and 17th centuries in Western Europe – to the exclusion of every other territory. For many historians, and even more so for philosophers such as Edmund Husserl, it is in this era that mathematicians carried out a radical revolution by breaking with ancient and medieval ways of thinking in order to forge new ones that heralded modern times. According to these historians and philosophers, it is precisely in this era that one encounters the beginning of algebra, the eruption of geometrical algebra, and the emergence of the mathematics of the continuous. It is thus easy for them to take the next step: these new contributions, which were born in the 16th and 17th centuries, allegedly mark this radical revolution and seal the unity of classical mathematics. On this account, classical mathematics would therefore be characterized by the simultaneous presence of the aforementioned chapters, among others.

If one examines the components of this 'early modern mathematics', however, one soon notices that these chapters are far from being contemporaneous: each has its own history, and the inventions or discoveries are by no means simultaneous. More generally, the global landscape of mathematics in the 16th and 17th centuries appears as a composite, an edifice constituted by different elements with origins traceable to many different dates. Indeed, some chapters, such as plane geometry, the geometry of the conic sections, and the geometry of the sphere, go back a millennium. They take us back to Euclid, Apollonius, and Menelaus. Without Apollonius, for example, how can one understand anything about Mydorge, Descartes, Fermat, Desargues, and Wallis, to mention only a few? These ancient geometers are effectively fixtures in the landscape of 'early modern mathematics', a point that obviously undermines the thesis of a 'revolution'. If one now wishes to consider algebra, the books of al-Khwārizmī, Abū Kāmil, and Fibonacci are prerequisites for anyone who wishes to deal with the authors of the 16th and 17th centuries. In algebraic geometry, finally, the surest way of seeing novelty where it does not exist and overlooking it where it does is to forget all about the works of al-Khayyām and Sharaf al-Dīn al-Ṭūsī.

One can easily multiply similar examples from the study of projections, the theory of parallels, spherical geometry, trigonometry, Diophantine analysis, number theory, combinatorial analysis, infinitesimal geometry, etc. And the same analysis holds also for such mathematical sciences as astronomy, optics, and statics. In each of these cases, one can see that a global look is necessarily deceptive, unsuited to depicting the landscape of early modern mathematics, and especially to tracing the history of its various components. Now that new results have accumulated in the history of Arabic and Latin science, one can be even less satisfied than before with the sketchy and simplistic frame that hems in early modern mathematics. It is therefore necessary to return this mathematics to a horizon that is truly its own by shattering the old tripartite periodization inherited from political history – ancient, medieval, modern – in order to conceive of a new one that will be faithful to the facts. Indeed, the very expression 'early modern mathematics' does not even have the same conceptual or temporal extension depending on the chapter of mathematics under consideration. As far as plane geometry is concerned, for example, the differences between the practice of a 3rd-century BC mathematician and that of a 17th-century mathematician are reduced to a few nuances of style that are insufficient to distinguish them from each other. And in order to understand the algebraic geometry of the early modern period, it will be necessary to integrate into it some of the work carried out in the 11th to the 12th centuries. The same goes for other chapters of mathematics and for the mathematical sciences.

To sketch a picture that conforms better to the reality of these mathematics, I have thought it indispensable to go back to the beginnings of these various chapters in order to identify the concepts and the practices that these mathematicians put to work, and to understand some of their main developments and corrections. For several decades, this effort has been the goal of my research in the history of ancient, medieval, and classical (early modern) mathematics. The studies collected here represent only a few of these results. Although carried out in a variety of circumstances, they all nevertheless grow out of the same intention and follow the same method.

With competence and rigour, Aline Auger (CNRS), has prepared the manuscript for print and compiled the index.

Roshdi RASHED, Bourg-la-Reine – 30 March 2011

Introduction

Problems of method

– 1 –

THE HISTORY OF SCIENCE: BETWEEN EPISTEMOLOGY AND HISTORY

Never before has the history of science, in all of its forms and specialities, prospered as much as it has in the twentieth century, and especially its last half. This unprecedented prosperity is evident everywhere: in the new domains conquered, in the number of valuable works published, in the number of teaching and research positions created, in that of institutions founded, of specialised journals launched, and of collections published... In other words, we can say without exaggeration that the accomplishments of the last five decades outweigh everything we owe to the last two centuries, which invites us, quite naturally, to ask ourselves without complacency what our discipline has achieved and what remains to be accomplished. Such an examination is all the more necessary because the prevailing impression is one of incessant and increasing dispersion, and because the profession of 'historian of science' is progressing more rapidly than the discipline itself: a singular situation, the consequences of which are uncertain and unpredictable at the very least. Before we begin this questioning, however, it is appropriate to recall the principal achievements of this past century; those that, since the middle of the last century, have sketched the landscape of the history of science. These contributions are distributed among several themes: methods, new fields investigated, and new relations established.

At the end of the 19th century, and especially in the first decades of the 20th, historians of science discovered the full importance of textual research, and the necessity of tracing the textual tradition of every scientific writing they examined. To a large extent, this new duty was a product of the development of the historical and philological disciplines, which were themselves influenced by the German school of philology. This research on textual traditions brought into the history of science a host of auxiliary disciplines and historical techniques – paleography, codicology, philology, etc. – and in the end became part of the permanent patrimony of the discipline, represented in the past by Hultsch, Tannery, Heiberg..., and attested today by the work on the translation of Archimedes by William of

Moerbeke and those on the work of Newton, Leibniz, Euler and, more recently, by the studies devoted to the writings of Einstein, among others.

And yet – symmetrically, as it were – these achievements, to which one must add much other accumulated wealth to be discussed below, did not take long to raise the problem of the gap between the history and the prehistory of the sciences, which in turn gave rise to many other issues concerning scientific change. Such are the questions raised by the famous methodological debate that began in the sixties. In this debate, which was salutary, the goals went far beyond the questions raised. The historians who participated in it wished, in fact, to break with purely descriptive history, spontaneous history, the 'history-novel' of scientists and their facts, and with history as the eclectic sum of people and facts. These were the first attempts to reflect on the discipline as such. In the life sciences, it was Georges Canguilhem who led the reflection; in astronomy, mechanics, and physics, it was, among many others, Gaston Bachelard, Alexandre Koyré, and above all Thomas Kuhn.

This debate interested those sociologists who, whether Weberians or Marxists, wished to give the history of science the social dimension it was lacking, by returning to institutions or social behaviors. In any case, this undertaking of methodological reflection, which in essence could not help but remain unfinished, made it possible to begin the first genuine effort to elucidate the discipline.

In addition to these methodological achievements, we must mention all the new fields to which we have laid claim. In the first place, the goal was to extend the history of sciences further back in time, thanks to the integration of Egypt and of Babylon. It was the work of Thureau-Dangin and Neugebauer, in particular, that allowed this extension, at the same time as it incited historians to rethink the notion of 'origin', and to situate Greek science differently. The other field was the rectification of the representation of medieval science: research on medieval Latin science was renewed, and scientific contributions which had until then been held to be peripheral were somewhat better integrated. Such was the case, in diverse degrees and in different modalities, with science in Arabic, Chinese, and Sanskrit. Yet the view of these sciences as peripheral was to be long-lived, since it has not yet been abandoned. This task was accomplished by entire schools, some of which are associated with the names of Joseph Needham, Adolf Youschkevitch, Pierre Duhem, Anneliese Maier, and Marshall Clagett.

Other fields were soon added to these newly-claimed domains, which opened the field of investigation of the discipline still further. The history of social sciences, the history of the diffusion of science from the centers of production towards the periphery, the history of institutions and of the great scientific laboratories, the history of applications, etc., today belong to the history of science, as is shown by the work presented at the various colloquia.

In these conditions, we will no doubt see the modification of the relations between the history of science and much other historical research, such as the history of philosophy and of technology.

In view of such diversity, not to say dispersion, we cannot avoid raising the question of the discipline itself, which was brought up in the course of the great methodological debate, only in order to be subsequently forgotten. Today, thanks to the work accomplished during the preceding decades – Charles Gillespie's *Dictionary of Scientific Biography (D.S.B.)*, and Robert Cohen's immense collection of Boston Studies in the Philosophy of Science, and the 'Storia della scienza' of the *Enciclopedia italiana*, among others – the question may be formulated as follows: what is this discipline that, throughout its existence, and particularly since the 18th century when it was born as an independent discipline – deals both with epistemology and with history? Whether we think of Condorcet, in his *Sketch* or in his *Academic Eulogies*; or of August Comte and the role of the history of science in his *Course of Positive Philosophy*; or whether, closer to our time, we bring up Joseph Needham, for instance: is the history of science really a discipline? and what is its true place between epistemology and social history?

With the first part of the question – is it really a discipline? – we dispense quickly. As it presents itself today in the works of those who claim allegiance to it, the history of science is a *domain of activity*, and not by any means a discipline. Indeed, it lacks the unifying principle that could provide it with the means and the power to exclude; for a domain of activity does not exclude, but is indefinitely distended by successive additions. It is a heading designated by a label, not a discipline characterised by a genuine definition. Thus, in the history of science, the various doctrines are juxtaposed and opposed on the basis of dogmatic and exclusive options, and even by *petitio principii*. According to some, the history of science presents itself as a history of ideas, or a history of mentalities. For others, by contrast, it is a history of scientific concepts – their formation, their development, and their rectification. For still others who were originally historians, concepts and their nature have little importance, and the history of science is the history of a cultural production, in the same sense as the histories of painting or religion. Let us also mention those for whom it is a kind of social psychology of scientific actors, and those who turn the history of science into an empirical sociology, such as has been developed particularly in the United States after the Second World War: a sociology of groups, laboratories, and institutions. This list is by no means closed, and this diversity continues to increase, not because of some internal necessity of research in the history of science, but rather as an effect of the successive importation of the views and methods from the social disciplines, and the trends that succeed one another therein.

This growing multiplicity has all the appearance of blind progress, which might spare us the examination of the second part of the question: what is the place of the history of science between epistemology and social history? But if we thus leave this question in the shadows, it will force us, whether we like it or not, to make up our minds about the object of the history of science. The difficulty – and it is considerable – is to be able to say in what sense the historian of science practices history, without formulating an arbitrary choice, and without imposing a methodology, be it empirical or transcendental. It was in order to avoid these shoals, upon which the methodological debate had run aground, that it seemed to me appropriate to start, in accordance with a well-known phrase, 'from the things themselves'; that is, from works of science and the traditions within which they are integrated.

It will easily be granted that every work of science belongs to at least one tradition, and often to many (whether or not they are known to us) relatively to which it takes on its meaning. This means that we cannot understand anything about individual creations, however revolutionary they may be, unless we insert them within the traditions that witnessed their birth. If, by 'work of science', we understand a result established in accordance with the precise norms of proof and consigned within a text, or realised within an object or an instrument, we shall for the moment give the word 'tradition' the vague meaning attributed to this term, which has the advantage of not isolating the work of science from the community to which the scientist who conceives it belongs. Let us begin by considering this notion of tradition.

Whatever their allegiance, historians of science are quite willing to admit that one of their essential tasks is the reconstitution of scientific traditions. Yet the paths they follow in order to reach this goal are divergent and branching. In fact, part of the methodological debate in the history of science refers to this diversity of conceptions of tradition and its nature. At first glance, the undertaking may seem easy and almost immediate: do not traditions most frequently present themselves under names, titles, institutions, and networks that ensure the exchange of information and human beings between poles, centers, places, and forms of learning? In this case, traditions would be immediately recognizable: we would speak of the tradition of the Euclidean theory of numbers, of the Japanese Wasan, of the tradition of the Italian algebraic school of the 16th century, of the British quantum physics of the 1920s, or of Bourbakian mathematics. To be sure, there are some exceptions, but they confirm the rule: I am thinking, for instance, of the Alexandrian tradition – or traditions – which finds its summit in the work of Diophantus, and of which, nevertheless, we are completely ignorant. How could we fail to be tempted to describe such easily identifiable facts as men, titles, and institutions? In fact, it is this tendency that dominates a large part of historical writings, which present themselves under different names: history of ideas, social history of science, etc.

Nevertheless, if we are not satisfied with a simple empirical description, the status of a tradition is not easy either to delimit or to establish. How are we to isolate a tradition, to assign to it a beginning and an end, and to trace its borders, without making an arbitrary cut in the indefinitely mobile totality of living history? Who can establish the unity of a tradition that is constantly evolving through time? Why is it constituted, and why does it cease? What system of rules might its existence obey?

There is, it seems, no *a priori* answer to these questions.

With simple description, the historian is only at the beginning of his labour. No sooner has he hitched himself to the task of reconstitution than the illusion is dissipated. The apparent simplicity evaporates, and all the empirical data – names, titles, etc. – proves impotent to delimit a tradition while dominating all its ramifications. Let us try to be more precise, by describing the principal stages in a work of the history of science. At the first stage, the historian must restore a work of science – a mathematical theorem, a physical result, an astronomical observation, a biochemical experiment, etc. – in all its materiality. He must examine inscriptions, tablets, papyri, manuscript texts, and printed texts; he must redo experiments, and re-fabricate objects, if necessary... All these procedures contribute, in the first instance, to the reconstitution of the textual tradition, as well as of the technological tradition... ; in a word, of the 'objectal' tradition (relative to the notion of object in general). Although in many cases, this research is not independent of the very contents of the work of science, it requires competencies other than scientific knowledge: those that deal with the various historical disciplines, such as archaeology, codicology, paleography, philology, the history of technology, etc.

This level of analysis is indispensable, but it is not sufficient: in such a reconstitution, we are still far from having exhausted the work of science. All that is known to us at this stage is its textual and technical authenticity, the networks along which it circulates, and the social context in which it was conceived and composed. All these elements are no doubt important, but they do not enlighten us about its place within the science to which it belongs. Even more seriously, at this stage we would not be in a position to perceive the cleavages that may mark the work of one and the same scientist. In order to consolidate these remarks, let us consider the example of the arithmetical work of Fermat.

Paul Tannery and Charles Henry have reconstituted the textual tradition of this work, as well as the networks of exchange that developed around them, and investigations of the work's social context could be still further refined and multiplied. Yet Fermat's place within arithmetic still remains to be determined. Is his work that of an algebraist, for instance in the tradition of Viète, or a specialist in the theory of numbers? Or else, does such work later belong to algebraic geometry, as André Weil maintains? Finally, was it simply the first arithmetical theory? I have shown elsewhere that Fermat's work does not have only one aspect, and that around the 1640s a line of cleavage splits it in two. One part of his arithmetical work does indeed belong to the tradition of the algebraists, whereas another part has to do with integral Diophantine analysis. Not a single *mathesis* but two *matheseis* – two conceptual traditions – are thus necessary to shed light on Fermat's arithmetical work. The one goes back *via* Bachet de Méziriac to the algebraists, whereas the other, in the wake of the work of mathematicians like al-Khāzin, taken up in Fibonacci's *Liber quadratorum*, renews the theory of numbers, thanks to the invention, for the first time, of an arithmetical method of demonstration: the 'infinite descent'. If therefore we wish to situate Fermat's arithmetical work historically, we must shift to another level of analysis, and direct our attention this time to the reconstitution of the *conceptual tradition*. The case of Fermat is far from being rare; in fact, it even seems to be the most common scenario, especially for scientists who change the trajectory of their science. To take only a few old examples from French science, this is what Descartes did in algebraic geometry, with his seminal distinction between 'geometrical curves' and 'mechanical curves'. It is also what Ampère did in physics, when he gave up explaining electromagnetism by magnetism, in order to choose the opposite path; and what Fresnel did when, against the dominant conception, he defended the necessity of transverse vibrations; that is, vibrations that are perpendicular to the ray. As a historian, the historian of science thus cannot allow himself the luxury of failing to reconstitute this conceptual tradition – or traditions – in other words, of carrying out this epistemological work.

Along this path, other obstacles inevitably arise, which have their essential origin in the dialectic between increasing multiplicity and fundamental stability. After studying numerous traditions, one general result imposes itself on us: a work of science of a certain stature cannot be explained in terms of one single conceptual tradition, not even of the one to which this work has contributed the most. Moreover, a conceptual tradition of some importance is distinguished by a certain stability, in spite of the diversity of authors and of contributions. Two somewhat paradoxical necessities seem to dominate the progress of a conceptual tradition: to exhaust all the logical possibilities inscribed within a given type of rationality, on the one hand; and, on the other, to reform this rationality and its means, in order to account for the new facts inexplicable within its framework. By way of example, suffice it to reflect upon the Archimedean tradition in infinitesimal mathematics; the Euclidean tradition in the theory of parallels, etc. To such obstacles, however, we must also add the question of the scientific 'style' which, behind such multiplicity and beyond the variety of forms and the transformations that mould a tradition, distinguish it and place a seal upon its identity. This 'style' reflects not only the dominant rationality, but also rhetorical procedures of exposition, such as the type of language used, symbolism, graphical representations, etc. The difficulty consists in isolating this 'style', which task is indispensable for placing an individual or collective work of science in perspective, and thereby expressing its meaning. This phenomenological procedure seems inevitable if we wish to invest the tradition with its role as an ordering principle, bringing out the interconnection of the works out of which it is woven.

These two terms – 'objectal' tradition (to which textual tradition belongs) and conceptual tradition – seem to translate concretely the question of the place of the history of science between social history and epistemology. As an element of 'objectal' tradition, the work of science is a material and cultural product, a product of men in a specific place and time. As Karl Marx would have advised, it is incumbent upon the historian to seek out the social and material conditions of this production. As a part of the conceptual tradition, however, this work also calls for an analysis of its conceptual structure, able to bring out its meaning, which will allow us to delimit the very notion of tradition. It might, of course, turn out that such a translation of our initial question may impoverish it. It does, however, seem susceptible of protecting us against two hazards: the reduction of history to pure epistemological analysis, such as it is for many of our eminent contemporaries; or else, even more, to a philosophy of history like that of Auguste Comte. The second risk is its assimilation to the history of any cultural domain, a practice that is current among historians. Yet the difficulty remains intact if we do not further specify what we mean by a conceptual tradition, to which a work of science belongs. Does this last question have the same meaning for all the scientific disciplines? Does the work of science belong to one conceptual tradition, or to several? These questions, among many others, emerge immediately, and they lead us necessarily to question ourselves about this notion of a work of science, and to ask ourselves what distinguishes it from all other social production of cultural works.

In order to answer this question, it is not uncommon for the philosopher to invoke a conception of certainty and of proof. Although this is perfectly legitimate, we will here abandon this path, which – however wrongly – might appear dogmatic. It is also common for the historian to ask the opinion of the scientist who is the object of his study, about the distinctive features of a work of science. There can then seem to be a historical answer to this epistemic question, whereas in fact it receives only an ideological one. Finally, it may happen that when he is confronted with this question, the reflective historian of science may propose two types of distinction, historical and epistemic in nature. The first separates two modes of knowledge: in order to define a work of science, he distinguishes it from a work of proto-science. The second distinction, which is much less powerful, isolates several forms of a work of science, and helps us to understand this cumulative march, necessary and universal, as much as those of its characteristics that are proper to science. The example most frequently invoked is that of Galileo in mechanics. With regard to the second, suffice it to recall the numerous examples that illustrate it: that of Lebesgue in the theory of integration, of Kolmogorov in the theory of probabilities, etc. It is clear that in both cases, such distinctions are intended to account for the emergence of new forms in the work of science; but whereas the former is 'creationist', and concentrates absolutely on initial forms, the second is 'transformationist' or evolutionary, and deals with new forms on the basis of the old ones. Let us therefore pause on this first distinction, the importance of which is fundamental for our purposes.

The distinction between proto-scientific and scientific presents itself as an exclusive one that dominates the entire history of science. This opposition is always understood as simultaneously historical and logical. The proto-scientific always precedes the scientific, both logically and historically; and the radical break between the two is presumed to have been essentially accomplished in the seventeenth century. As a result, this opposition supposedly allows us to distinguish a work of science from all others that claim to deal with the same object; this is the most commonly accepted doctrine. When we look more closely, we shall not hesitate to grant a grain of truth to this distinction, even if the relations between proto-scientific and scientific are much more varied and complex, both logically and historically, and even if the seventeenth century did not have the exact role attributed to it. Let us begin by subtracting, as it were, mathematics from this exclusive opposition, for a contingent reason: nothing proto-mathematical has come down to us, and the only evidence we have of proto-mathematics already belongs to mathematics: indivisibles; considerations about the notion of limits in the eighteenth century; objective and subjective doctrines of probability prior to axiomatic theory, etc. In the other scientific disciplines, the term 'proto-scientific' seems to cover at least four different types of knowledge. Aristotle's physics is proto-scientific, as is the social contractualism of the eighteenth century; the social Darwinism of the following century, and the social physics of Quetelet; the optics of Euclid; the marginalism of a Jevons, a Walras or a Pareto; as well as the ballistic model of a Tartaglia, Condorcet's *homo suffragans*, the economists' *homo bernoulliensis*, etc.

These examples flagrantly reveal the variety of statuses of the 'proto-scientific', since the realities that this term designates cannot be jumbled together under the same denomination, either *de iure* or *de facto*. Thus, Aristotelian physics, like social contractualism, is proto-scientific in the sense of a systematic doctrine of lived experience that one intends to be coherent: that of displacement or that of voting in an assembly. Social Darwinism and social physics are proto-scientific if we understand by this a science which is annexed to a domain other than that of its origin. Euclid's optics, and the marginalist contributions, are proto-scientific in the sense of a 'pure' knowledge, produced by the application – direct, as it were – of mathematics to doctrines of lived experience, whether direct vision or the distribution of goods. Finally, the models of Tartaglia in ballistics, Condorcet in social science, or Von Neumann in economics, are proto-scientific in the sense of an indirect application of mathematics, with the help of analogies to a third discipline that is mathematized or considered to be such, to a doctrine of lived experience.

We can easily see that proto-scientific knowledge is not only multiple, but also, for the most part, linked to other sciences that deal with objects other than its own. Two consequences therefore impose themselves: the criteria of a work of science necessarily differ from all the criteria of such proto-scientific knowledge; moreover, the notion of tradition splits apart, both from the point of view of diachrony and from the point of view of synchrony. Let us begin by examining the question of criteria, since they forbid us from treating the object of science not only like that of a proto-science, or a pre-science, but also like the object of every other cultural production. We have seen that proto-scientific knowledge is always linked to an experience that is lived, and therefore particular. Yet we must not be deceived: the doctrine or philosophy thus elaborated is not restricted to expressing the contents of this experience directly, and it does not proceed by placing in brute correspondence a concept and an event, or a proposition and a datum, but a proposition and another proposition. That is to say, it proceeds by placing in correspondence two relations between concepts. It is in this sense that we may say that the data of immediate experience are mediated. The linguistic task of systematisation and the denominations that we always encounter in the authors of such doctrines are the instruments of this mediation.

In other words, the data of lived experience constitute only a starting point, and mediation is required in order to achieve the constitution of a doctrine. In this regard, recall that the Aristotelian doctrine of motion is by no means made up of propositions directly linked to the sensible experience of the movement of displacement, but only to those which concern the correspondence of 'the act of what is potentially such' with propositions relative to 'determined natures' and to the cosmological order. Likewise, the social doctrine of Jean-Jacques Rousseau does not concern the lived practice of suffrage, but links a conception of the social contract to that of suffrage as the declaration of the general will. In the last analysis, it is thanks to this mediation, and to the transcendence it insures with regard to the data, that we introduce the other criterion: coherence, which philosophers hold must be severe. This coherence refers, moreover, simultaneously to logical consistency and to architectonic action.

To this mediation and search for logical consistency and architectonic perfection, we should add another criterion, with regard to which this doctrine of lived experience may progress: the successive amendments, intended to exhaust the data of a particular experience in an ever-more-coherent exposition. One thinks of the amendments carried out by the partisans of *impetus* for the Aristotelian doctrine of motion. To sum up: mediation, transcendence, logical consistency and architectonic action, progress by successive amendment: such are the criteria of knowledge produced by phenomenology in order to provide a framework for events – the doctrine of Aristotle or of Rousseau, for example – or by confiscation, that is, the restriction of a phenomenology initially intended for another universe than that for which the explanation is undertaken – as in social physics or social Darwinism.

The first type of application of mathematics to this doctrine of experience consists in the will directly and completely to replace its notions by mathematical structures; this is the example of Euclid's optics, and of

Walrasian marginalism. In this case, mathematics is only a language. The second type of application, by contrast, subordinates the substitution of mathematical structures to the intercession of a third discipline, which is dominated by mathematical knowledge, or else considered to be so. Analogical correspondences between the two disciplines are the means to mathematize the doctrine of experience itself: this is the method of models.

Proto-scientific knowledge is thus multiple; moreover, not all its instances are of equal value. Their goals, their explanatory powers, their syntactic and technical controls differ, even if all of them have as their point of departure one of the doctrines of lived experience, subject to the criteria we set forth previously. Such instances of knowledge thus cannot have the same relations with future science. It is true, as has often been affirmed, that science is made in opposition to these instances of knowledge, by breaking with them; but this break does not have the same significance in every case. Even if, at the deepest level, it always takes place against the above-mentioned doctrine of lived experience and the criteria of its operation, its paths subsequently never cease diverging. Thus, Ibn al-Haytham's geometrical optics broke with all the doctrines of his predecessors, it is in so far as he separated the conditions of the propagation of light from those of vision, in order to consider, in the first instance, only material entities – 'the smallest parts of light' – which retain only those properties which may be geometrically and experimentally controlled, thus abandoning all sensible qualities other than the energetic ones. Yet this profound break, which allowed the introduction of a new category of proof in optics, and more generally in physics – experimental proof – did not take place in the same way with Euclid's optics and with Aristotle's doctrine of vision. Likewise in mechanics: Galileo was the first scientist able to separate, within the doctrines of motion, what falls under the domain of kinematics from what falls under dynamics, in order to consider only the relations between the positions of material entities within time. These entities now took on only those properties susceptible of being controlled geometrically and experimentally, thus excluding all sensible qualities other than those of resistance to movement. This profound break was carried out not so much with regard to Aristotelian doctrine as to the doctrine of *impetus*, the doctrines of the Calculators of Oxford and Paris, or the models of al-Qūhī and Tartaglia.

This diversity of relations between proto-science and future science obliges the epistemologist not only to differentiate among the conceptual traditions of the various proto-scientific instances of knowledge, but also, and more importantly, it gives him the means to order and to hierarchize them. It is this possibility that is the privilege of works of proto-science, compared to the other cultural works that present themselves to the historian. In other words, future science dictates a principle of order, and – to use a metaphor – a notion of distance that helps to situate proto-scientific knowledge. Yet this privilege of works of proto-science is not affirmed at the historian's expense: quite the contrary, it operates in his favor, for the distinction between these conceptual traditions allows him better to spot, within an often formless mass, the textual and technical traditions which underlie them. Thus, he is in a position to ask all the social-historical questions necessary to understand their formation, their development, and the interaction of the various social and ideological factors that have insured the constancy of their formulation.

The break with the doctrines of lived experience and, at the same time, with the criteria of their elaboration takes place thanks to a conception of an object that contains an operative and judicatory norm. Not only is the knowledge produced invested with an accumulative power, but it can effectively realize this accumulation only by means of a constant rectification of its comprehension. It is in these acts of rectification that new forms appear. This is why in scientific knowledge, if we think only by ready-made concepts, we may say that continuities and discontinuities are inscribed one within the other. Such discontinuities are sometimes called 'revolutions', and they designate the passage from one theory to another: from the mechanics of Galileo and of Newton to Special Relativity; from the latter, from electrodynamics, and from continuist thermodynamics to quantum theory. We have here the emergence of new forms of the same science, which redefine their object each time without, however, replacing it with a different one, as was the case for proto-scientific knowledge. In this discontinuous succession of forms, the old form presents itself as an approximate case of the new one, yet one that is expressible in the language of the latter. It is, so to speak, the new form that explains the old form, and specifies its conditions of validity. The former includes the latter as an approximate case. The emergence of new forms no longer cancels out the old ones; it rectifies and integrates them. In these conditions, the notion of conceptual tradition is profoundly modified. The best proof of this is the style of its death. In pre-science, conceptual traditions are assassinated; here, they die by the exhaustion of their own possibilities. This difference – fundamental, in my view – seems to show that the questions and problems that presided over its birth are internal to science; or, at the very least, that it has been possible to make them adopt its language completely. Thus, each tradition can speak the language of the other, and all can be translated into the language of distant successors. In optics, the language of the Alhazanian tradition may be translated into that of the Newtonian tradition, which is impossible for Euclidean optics; and the language of the first two traditions may later be translated into that of Fresnel, etc. Such translation is not only in the diachrony of victorious science, but it is also valid in synchrony. Let us mention the example of two rival contemporary traditions: that of fluxions begun by Newton and that of differential calculus, founded by Leibniz. Despite the famous controversy and the distance that separates their styles – the former is geometrical, the latter algorithmic – each can speak the other's language, and both are translatable into the standard language of analysis. This fundamental feature is proper not only to mathematics, but also common to all scientific knowledge, even that which, according to Bachelard's expression, are phenomeno-technical.

In science, thanks to a certain epistemological closure that characterises it, the notion of conceptual tradition is liberated to a much greater degree than in the pre-science of the corresponding 'objectal' tradition. The role of exogenous elements not only becomes minimal, but above all it is controlled at the time of the constitution of theoretical models and the demonstration of their validity. Linguistic and technical surveillance protects us against hidden Gods.

This independence by no means diminishes the role of the 'objectal' tradition – quite the contrary. If the conceptual tradition precisely indicates the temporal and human constituents of the 'objectal' tradition, the latter, in order to be established, requires the undertaking of work which would allow us to understand the formation of the community of scientists, their modes of learning, their choice of which sectors to develop and with which rhythm, etc.; in other words, all the material and social elements that have established the framework of the conceptual tradition. These elements may illuminate the rhythms and the diffusion of the conceptual tradition, but they cannot shed light on systems of concepts and the proofs of their validity. To be sure, the choices of investment and allocation of resources, the training of scientists and the multiplicity of competencies, the stratification of their community, social ideologies as well as scientific ideologies, among many other factors, may explain controversies when the facts are imperfectly established and proofs not rigorously carried out. They also shed light on the conflicts of interpretation which almost always accompany the passage to application, the unequal development of different disciplines, etc.; but they cannot inform us about the constitution of valid theoretical models. This last task is, so it seems, the particular duty of the history of science; it is this task that it must define if it wants to constitute itself as a genuine discipline. Work on the 'objectal' tradition, which the historian of sciences can probably not do without, falls under other specialities, subject to other criteria, which range from archaeology to social psychology, *via* codicology and economics, among others. The differences between the objectal tradition and the conceptual tradition lie not only in objects and methods, but they are much more deeply rooted in the very nature of their necessity. Moreover, it is perhaps here that resides the source of all the conflicts and controversies, or, if one prefers a ready-made expression, the reason for the chasm between 'internalists' and 'externalists', or between the adepts of 'social history' and historians of science. In a word, the objectal tradition deals, with our actions: with psychological, social, and historical events, with things that are here and now; in brief, with contingent facts. The formation of an academy, the functioning of a great research centre, the organisation of a laboratory, the modes of transmission of knowledge, the material support of texts, the allocation of resources, the scientist's social affiliations, his psychological profile, etc., are all so many contingent facts. Even if psychology, sociology, economics, and so on, can identify a kind of necessity in them, there is none in their relations to the facts of science. Conversely, it is the necessity characterizing them that makes these facts recognizable. Such is the case for a mathematical theorem, a physical law, etc. It is for this reason, moreover, that an objectal fact is not liable to be true or false, whereas for the conceptual fact its character of necessity is also a criterion of truth. We can thus understand that every globalising vocation is condemned in advance to theoretical defeat. Thus, today, the flourishing temptation to extend social history to the conceptual tradition resembles like two grains of sand the ambition to extend psychology to logic: not long ago, the latter resulted in the famous 'psychologism' which unleashed the fury of philosophers like Kant, Husserl, and Cavaillès; the former did not fail to result in 'historicism'. What is more, the thesis of the extension of social history cannot itself be defended, for it in turn will be of the order of contingency, and the vicious circle will be closed. Besides, if we wanted this thesis of social history to be possible, we would have to evacuate from science both its truth-value and the distinction between true and false. Conversely, to extend conceptual history to the objectal tradition leads to a 'pure history', or a philosophy of history that is no longer a history of science. Yet the whole problem of the history of science, the one to which all of its difficulty is reduced, consists in this: the production of the facts of science, quite determinate as the production of human beings and the results of their actions, transcends, as an effect, the contingent conditions of its advent; and this production transcends these conditions in order to distinguish itself from them by its necessary character. In a nutshell, the whole problem consists in knowing how necessity emerges from contingency. The historian of science then reveals himself to be what he has always tried to be: neither a 'science critic', in the sense of an art critic; nor a historian, in the sense in which we understand a specialist in social history; nor a philosopher, like the philosophers of science; but simply a phenomenologist of conceptual structures, of their genesis and of their affiliations, in the midst of conceptual traditions that are always in transformation. Today more than ever, this self-awareness seems to me necessary if we want the history of science to be constituted as a genuine discipline instead of being merely a domain of activity. Also today, we must construct a new discipline, as necessary as it is legitimate, simultaneously with the history of science, but independently from it: that of social research on the sciences. Such independence is the guarantee that both the history of science and social research on the sciences may be formed as true disciplines, which deal with the cultural phenomenon of science.

THE TRANSMISSION OF GREEK HERITAGE INTO ARABIC

Historians of Arabic science and philosophy, whatever their allegiance, all acknowledge the importance of the translation of the Greek heritage into Arabic. They know that if they ignore this fact, they will understand nothing about the emergence and development of these disciplines in Arabic, and thereafter, in Latin. This is not at all surprising: to measure the impact of the Greek patrimony, one could either become familiar with the actual evolution of these domains in Islamic science, or simply rely on the testimony of ancient historians and biobibliographers themselves (*e.g.*, Ibn Iṣhāq al-Nadīm).[1]

Historians of Greek science and philosophy, also provide evidence, however indirect, for the significance of the Greek heritage that was translated into Arabic. Indeed, if these historians ignore the Arabic versions of Greek writings, they are condemned to missing a considerable part of their quarry, and to depriving themselves of a precious means of understanding it. Indeed, some of these writings, for which the Greek text is lost either in part or as a whole, exist only in a single Arabic translation. The commentaries of Arabic scientists, along with the progress they made in the disciplines pertinent to these writings, are a powerful tool for reaching a better understanding of the latter and situating them in the history of the discipline. One need only think of Diocles, Apollonius, Ptolemy, Diophantus, Alexander of Aphrodisias, among many others.

Everyone recognizes the exceptional breadth of this phenomenon of scientific and philosophical transmission and its importance for the history

[1] Ibn Iṣhāq al-Nadīm, *Kitāb al-fihrist*, ed. R. Tajaddud, Teheran, 1971. See particularly the seventh chapter, pp. 299–360 and pp. 417–25. English translation by B. Dodge, *The Fihrist of al-Nadim*, 2 vols, New York, 1970. One of the earliest studies, now a classic, is Max Meyerhof's 'Von Alexandrien nach Bagdad. Ein Beitrag zur Geschichte des philosophischen und medizinischen Unterrichts bei den Arabern', *Sitzungsberichte der Berliner Akademie der Wissenschaften*, Philologisch-historische Klasse, 1930, pp. 389–429.

of science and of philosophy. And yet it is far from getting the attention it deserves. Many texts have yet to be properly edited and, in order to give a satisfactory account of them, many studies must yet be undertaken. Even more importantly, a change of perspective is necessary to put this research back on a more fruitful path. This change, which is now underway, must occur in both the method and the very concept of the object of research. To study the transmission of the Greek patrimony into Arabic solely from a philological perspective, which is by far the most frequent approach, is the surest way of missing the most essential points: the motivation for the translation, its extension, and the new forms that it constantly adopted. To examine this transmission solely with the goal of restoring Greek writings that were definitively or temporarily lost is to miss the very evolution of the phenomenon. To be sure, these studies are completely legitimate and sometimes locally important. As soon as they are generalized, however, they become the trees that hide the forest: they are taken to be the means of describing the evolution of the translation movement from Greek into Arabic. Recent research on this phenomenon has tried to correct this perspective.[2] It is this correction that we attempt to present and examine here.

1. TRANSMISSION AND TRANSLATION: SETTING UP THE PROBLEM

1. *Towards a new approach*

It therefore seems urgent to abandon the dominant conception of transmission and translation. To this end, we must remind ourselves of two well-known elementary facts. First, it is crucial to remember that the new Muslim state stretched over the major part of the Hellenistic world. The two regimes therefore involved the same peoples, but the latter underwent a more or less massive change of language and religion. These peoples therefore inherited a conglomeration of know-how, technical objects, and institutions; together, these formed the elements of a social and economic

[2] R. Rashed, 'Problems of the Transmission of Greek Scientific Thought into Arabic: Examples from Mathematics and Optics', *History of Science* 27, 1989, pp. 199–209; repr. in *Optique et Mathématiques: Recherches sur l'histoire de la pensée scientifique en arabe*, Variorum reprints, Aldershot, 1992, I. See also D. Gutas, *Greek Thought, Arabic Culture. The Graeco-Arabic Translation Movement in Baghdad and Early ʿAbbāsid Society (2nd–4th/8th–10th centuries)*, London/New York, 1998; J. L. Kraemer, *Humanism in the Renaissance of Islam: The Cultural Revival during the Buyid Age*, 2nd edn., Leiden, 1992.

patrimony that is as relevant to the history of techniques as to the history of institutions. At the heart of this heritage, however, also lies a body of dormant texts, as it were, and a body of elementary teaching, notably in theology, astrology, alchemy, or medicine. The second fact is the following: to this heritage are added other contributions, arising from different horizons, especially Persian, Sanskrit, and Syriac. To ignore these facts is to neglect the important role of practices, know-how, technical objects, and institutions in the circulation of knowledge. Such an omission quickly reduces the issue of transmission only to matters of translation, and the only portion of the Greek heritage to survive would then be its bookish part. In this case, in other words, one runs the risk of missing everything that happened in the wake of the means mentioned earlier: elementary geometry, logistics, agronomy, hydrostatics, metrology, etc. – in short, all the branches that will later become parts of full-fledged disciplines, or simply of practical geometry. Even though this heritage can probably not single-handedly explain the emergence and development of the sciences and philosophy in the new Islamic culture, it nevertheless remains among the latter's important elements.

Conversely, it is not unusual for the act of *translating* to be presented as passive, related to teaching, and always set in the same register, regardless of the context in which it is carried out. On this account, the act is that of a *translator*, often a physician, who knows Greek and who renders into Arabic – haphazardly and based on chance encounters – Greek writings pertinent to various disciplines that do not always fall within his competence. Translation from Greek is therefore cast as a matter of chance, with no good reason for either the choice of books, or the opportunity to translate them. In short, according to this often implicit account, translation focused on what it stumbled upon, as best it could, and had a scholastic motivation, insofar as the translated texts allegedly were intended solely for teaching. Finally, the register remained the same, since the act of translation required nothing more than knowledge of Greek (or Syriac).

This picture, of transmission first and of translation second, has led to a doctrine that one encounters here and there, notably among modern biobibliographers.[3] If one believes its proponents, translation to acquire Greek sciences and philosophy is the first stage of a 'law', according to which three stages succeed each other logically and historically: assimilating the acquired knowledge is the second stage, followed by creative production in the third stage. This rather naïve doctrine presents the same picture as the

[3] See for example in F. Sezgin, *Geschichte des arabischen Schrifttums*, vol. V, Leiden, 1974, pp. 25 ff.

preceding one, and sees the translation effort as merely a desire for acculturation. Unfortunately, both this doctrine and the picture that underlies it collide head-on with several facts, only two of which we mention here.

It has not been sufficiently emphasized that translation was concurrent with innovation. A few examples illustrate the point: optics and catoptrics with al-Kindī, the geometry of the conics with al-Ḥasan ibn Mūsā and his student Thābit ibn Qurra (d. 901), and number theory, also with the latter. Once it is properly grasped, this concurrence raises the long-forgotten question of the intimate relationship between translation and research, or of the very form of the translation in relation to its audience, as we shall see.

2. *Cultural transmission, scientific transmission*

The second fact pertains to the hypothesis – widely conceded but rarely discussed – of a water-tight continuity between the scientific and philosophical research in Antiquity and Late Antiquity and the research that developed in Arabic. Now this continuity is in effect only point by point and seems both fragile and paradoxical. First, the institutional point of view will raise the question of the Arabization of the administration and of the tools of power, that is, the *Dīwāns*.[4] In earlier work, we have shown that the Arabization and evolution of the *Dīwān* made possible the translation of a 'logistics' and the beginning of research into it, which in turn contributed to the conception of a non-Hellenistic discipline: algebra, with al-Khwārizmī. We have also shown how the culture of the *Dīwān*, which was necessary for the creation of a document-intensive bureaucracy, created a specific social stratum. It was the latter's linguistic and literary needs, as well as the demands of – among other disciplines – logistics, algebra, and geometry that stimulated both translation and inventive research.[5] At this level, one can truly discern a certain continuity, as also appears in such other sectors as architecture, farming techniques, etc. But matters are rather different when one considers scientific and philosophical research. This

[4] R. Rashed, 'Les recommencements de l'algèbre aux XIe et XIIe siècles', in J. E. Murdoch and E. D. Sylla (eds), *The Cultural Context of Medieval Learning*, Dordrecht, 1975, pp. 33–60; repr. in R. Rashed, *The Development of Arabic Mathematics: Between Arithmetic and Algebra*, Boston Studies in the Philosophy of Science 146, Dordrecht, 1994, pp. 34–84, particularly pp. 53 ff.

[5] Abū al-Wafā' al-Būzjānī's book *On the Needs for Scientific Arithmetical Science for Kuttāb (writers, secretaries, civil servants in administrative offices), 'Ummāl (prefects, tax collectors) and Others* belongs to this tradition. Cf. an edition of Abū al-Wafā''s book in A. S. Saidan, *'Ilm al-Ḥisāb al-'Arabī*, Amman, 1971, vol. 1.

kind of research was becoming rare and disappearing in both Alexandria and Byzantium.[6] In Arabic from the 9th century on, however, one witnesses a genuine scientific and philosophical renaissance, the foundations of which – linguistic and historical as well as philosophical and theological – had been solidly laid in the 8th century.

For the new scientific community, in short, Alexandria, Byzantium, and the other cities of the *oikoumene* constituted a 'dormant library' that preserved a wealth of manuscripts from Antiquity and Late Antiquity. All the historical evidence points in this direction.[7] However, the absence of continuity at this level raises two closely related questions, only one of which matters here. How indeed can one make sense of the renewal that occurred by leapfrogging over the centuries in order to return to Apollonius and Aristotle, for example? What are the relations between this renaissance and the transmission of the Greek heritage, and notably its translation? Indeed, only in light of this scientific and philosophical renaissance does the question of translation makes sense: this very link that seems to offer the best means of grasping it.

The transmission of the Greek heritage into Arabic followed mainly, but not uniquely, two paths that were interrelated despite their unequal significance and their different natures. Although it is well known to historians of society and culture, the first, which we mentioned earlier, has barely been explored: it is the one associated with technical subjects, crafts, and institutions. These are techniques, organisations, and ideologies that the ancient citizens and inhabitants of the Greek-speaking Mediterranean had established to ensure both their material and social existence. This path is that of the transmission of the *Dīwān* translated into Arabic under

[6] J. F. Haldon, *Byzantium in the Seventh Century*; *The Transformation of a Culture*, Cambridge, 1990; H. D. Saffrey, 'Le chrétien Jean Philopon et la survivance de l'école d'Alexandrie au VIe siècle', *Revue des études grecques*, 1954; L. G. Westerink, *Anonymous Prolegomena to Platonic Philosophy*, Amsterdam, 1962.

[7] R. Rashed, *Les Mathématiques infinitésimales du IXe au XIe siècle*. Vol. I: Fondateurs et commentateurs: *Banū Mūsā, Thābit ibn Qurra, Ibn Sinān, al-Khāzin, al-Qūhī, Ibn al-Samḥ, Ibn Hūd*, London, al-Furqān, 1996, p. 142; English translation: *Founding Figures and Commentators in Arabic Mathematics*. A History of Arabic Sciences and Mathematics, vol. 1, Culture and Civilization in the Middle East, London, Centre for Arab Unity Studies, Routledge, 2012. See also T. M. Green, *The City of the Moon*, Leiden, 1992. Al-Masʿūdī's description in *Murūj al-dhahab* (*Les prairies d'or)*, ed. C. Barbier de Meynard and M. Pavet de Courteille, revised by C. Pellat, Section XI Historical Studies, Beirut, 1966, vol. 2, §§ 1389–1398, pp. 391–6, shows that the traces of Hellenism in Ḥarrān around the 3rd century after the Hegira are essentially religious.

Hishām ibn Abī al-Malik (r. 724–743);[8] it is also the path taken by the procedures of practical geometry, logistics, and disciplines such as medicine, alchemy, astrology, agronomy, the military arts, or architecture. To this category also belong the treatises on elementary logic and theology, deemed necessary to religious teaching in the context of the Nestorian and Jacobite monasteries.[9] On this first path, which was in effect completely natural, since the Hellenized populations had followed it for a millennium, also circulated translations of scientific texts, as we shall see below. The second path, which is much clearer and better known, is that of learned translation, pertaining to the philosophical and scientific writings of Antiquity and Late Antiquity. By its extent, this path stands out sharply from all prior translation efforts, including those in the Latin and Syriac worlds.[10] It would be unrealistic to think that these two paths were impermeable and mutually exclusive. Indeed, several clues prove the contrary, and future research will no doubt find intermediate routes that will help us circumscribe better the social phenomenon of the transmission of heritage and translation. For now, it suffices to highlight one incontestable general characteristic: this translation movement went hand in hand with the unification, the Arabization, and the Islamization of the Muslim empire and its administration.

3. *Scholarly transmission: one myth and several truths*

However that may be, if one can trust the legend, the second path was officially 'opened' by one of the grand Caliph al-Maʾmūn's dreams. As the legend goes, the Caliph had conversed with Aristotle in a dream. After having recounted this episode, the ancient biobibliographer al-Nadīm writes:

[8] Al-Nadīm, *al-Fihrist*, p. 303.

[9] One might consider the iconic figure of Patriarch Timothy, who collaborated on the translation of Aristotle's *Topics* from Syriac into Arabic, which was commissioned by the Caliph al-Mahdi. Cf. S. P. Brock, 'Two Letters of the Patriarch Timothy from the Late Eighth Century on Translations from Greek', *Arabic Sciences and Philosophy*, 9, 1999, pp. 233–46. See also J. van Ess, *Theologie und Gesellschaft im 2. und 3. Jahrhundert Hidschra. Eine Geschichte des religiösen Denkens im frühen Islam*, Bd. III, Berlin/New York, 1922, pp. 22–8.

[10] H. Hugonnard-Roche, 'Les traductions du grec en syriaque et du syriaque en arabe', in J. Hamesse and M. Fattori (eds), *Rencontres de cultures dans la philosophie médiévale*, Louvain-la-Neuve, 1990, pp. 131–47.

This dream was one of the surest reasons for rendering the books (into Arabic). Al-Maʾmūn then wrote to the king of the Romans to ask for permission to send what he (al-Maʾmūn) would choose from the ancient sciences stored and preserved in the land of the Romans. After some hesitation, the king accepted. Al-Maʾmūn then sent a group, including al-Ḥajjāj ibn Maṭar, Ibn al-Biṭrīq, Salmān who was associated with the House of Wisdom, and others. They took what they had chosen. When they brought these to al-Maʾmūn, he ordered them to translate the texts. It is said that Yūḥannā ibn Māsawayh was among those who went to the land of the Romans.[11]

Al-Nadīm then notes that many others imitated the imperial model. Thus al-Ma'mūn's protégés – the Banū Mūsā – had sent to the 'land of the Romans the famous translator Ḥunayn ibn Isḥāq (d. 873)', who returned 'with precious books and unique works on philosophy, geometry, music, arithmetic and medicine'.[12] According to another account, it seems that Muḥammad (d. 873), the eldest of the Banū Mūsā, was part of an expedition to the Byzantine empire.[13] Several other historical sources allude to emissaries sent to Byzantium, to Alexandria, to the monasteries in the interior of the ancient Hellenistic world, in order to search for Greek manuscripts on science and philosophy throughout the entire 9th century and even later.

Legendary though it may be, al-Maʾmūn's dream expresses, among historians and biobibliographers of the time, the clear awareness that this translation movement differed qualitatively from its predecessors. It is this difference that we must try to grasp.

3.1. *The rebirth of research*

The ancient historians were aware that the translation movement had begun well before the reign of al-Maʾmūn (813–833). More precisely, one can discern before this period two stages within a first phase. Several fragments of information recounted by biobibliographers tell us that translators had already been hired under the Umayyads. So it happened that the grandson of the dynasty's founder, Khālid ibn Yazīd (d. after 704), had asked a certain Stephanus, to translate books of alchemy from the Coptic and the Greek. Al-Nadīm offers the following commentary on this testimonial: this is 'the first translation from one language to another in

[11] Al-Nadīm, *al-Fihrist*, pp. 303–4.

[12] *Ibid.*

[13] Ibn Khallikān, *Wafayāt al-aʿyān*, ed. I. ʿAbbās, 8 vols, Beirut, 1978, vol. 1, p. 313.

Islam'.[14] This testimonial has recently been questioned,[15] but at least it has the merit of noting that ancient historians already ascribed to this period a definite interest in translation and attributed a specific role to Khālid ibn Yazīd. The same al-Nadīm reports another testimonial that corroborates the first: at this time, during the reign of the Caliph Hishām ibn ʿAbd al-Malik, the *Dīwān* was translated from Greek into Arabic. Also in this very same period, in the reign of the latter's father, and on the advice of the same Khālid ibn Yazīd, coins were first minted in Arabic and no longer in Greek (see the accounts of Ibn al-Athīr and al-Nuwayrī).[16] Yet another witness of the same type claims that, at the end of the very same 7th century, Māsarjawayh translated into Arabic a medical compendium by Ahrun.[17]

These past vestiges – the aptest expression in this case – indicate that, simultaneously with the movement for the Arabization of the *Dīwāns* in particular, that is, of the administration and of its texts, several translations took place, thanks to the initiative of individuals and in response to immediate practical needs. Other vestiges, of indeterminate date, but very likely belonging to the era between this period and the beginnings of the ʿAbbāsid dynasty, testify to the existence of translations, most notably in astronomy: for example, the translation of Theon of Alexandria's *Introduction* to the *Almagest*, which al-Nadīm characterized as an 'ancient translation' (*naql qadīm*).

With the beginning of the ʿAbbāsid dynasty, Arabization which had advanced considerably, continued to progress. To this, one should add a policy of great construction projects on account of the displacement of the centre of the Empire and increasing urbanisation. The translation effort could thus only speed up and spread farther. One name exemplifies this movement, that of the second ʿAbbāsid caliph al-Manṣūr (r. 754–775).

Ancient historians[18] all agree in emphasizing al-Manṣūr's personal interest in astrology. Thus when he decided to found the new capital, Baghdad, he commissioned astrologers to compute the astrological chart and to determine the most propitious moment to begin construction. It is on this occasion that the names of Abū Sahl ibn Nawbakht, Ibrāhīm al-Fazārī,

[14] Al-Nadīm, *al-Fihrist*, p. 303. See also p. 419.

[15] M. Ullmann, 'Khālid ibn Yazīd und die Alchemie. Eine Legende', *Der Islam*, 55, 1978, pp. 181–218.

[16] Al-Nuwayrī, *Nihāyat al-arab fī funūn al-adab*, ed. M. al-Ḥīnī, Cairo, 1984, pp. 223–4; Ibn al-Athīr, *al-Kāmil fī al-tārīkh*, ed. C. J. Tornberg under the title *Ibn El Athiri Chronicon quod perfectissimum inscribitur*, 12 vols, Leiden, 1851–71; repr. 13 vols, Beirut, 1965–67.

[17] Al-Nadīm, *al-Fihrist*, p. 355.

[18] Al-Masʿūdī, *Muruj al-dhahab*, Beirut, 1991, vol. 4, p. 333.

and Māshā'allāh surface. From many different provinces, the caliph also brought workers, craftsmen, jurists, and geometers:[19] all of these guilds were necessary to bring to fruition this colossal project. Let us pause briefly on this information. Abū Sahl ibn Nawbakht was not only an astrologer; but also a *mutakallim,* that is, a theologian-philosopher. According to al-Nadīm's report, he left an autograph of a kind of legendary history of the sciences, the epistemological and historical origins of which he traced to Babylonian-Persian astrology.[20] Was this doctrine intended to justify the practice of astrology in which the Caliph himself believed? However, even this practice required a true knowledge of astronomy and particularly of the composition of the *zījs*. As for al-Fazārī (second half of 8th century), he was not only an astronomer but also a mathematician. He not only composed and edited a *zīj*, but also wrote on astronomical instruments – astrolabes and sundials, which required a solid knowledge of stereographical projections. It therefore seems possible that this group of astrologer-astronomers, accompanied by other geometers, would have carried out all the necessary surveys for the foundation of Baghdad, as well as calculating its astrological chart.

New needs appear that implicitly encourage a particular kind of research: the composition of *zījs* and the exact representation of spheres in a plane. Although the disappearance of texts drastically limits our knowledge of sources that would have allowed us to evaluate this incipient research, clues remain that alert us to the new intellectual environment. Al-Manṣūr allegedly received an Indian delegation that included an astronomer, in the presence of al-Fazārī to whom he would have given an Indian *zīj*. On this account al-Fazārī together with Ya'qūb ibn Ṭāriq would have taken on the adaptation into Arabic. The history may perhaps be uncertain, but it nevertheless captures the picture that people had of the period.[21] Another, equally late account (from 330/941) by a certain al-Akhbārī, and recorded by historian al-Mas'ūdī, also draws attention to al-Manṣūr's interest in astrology, and also to the presence around him of Abū Sahl ibn

[19] As al-Nuwayrī says in *Nihāyat al-arab fī funūn al-adab*, vol. 22, p. 90: 'He (al-Manṣūr) wrote to every country to ask that they send artisans and masons and he decreed that eminent and just men, honest and educated in jurisprudence and geometry, should be chosen'.

[20] Al-Nadīm, *al-Fihrist*, pp. 299–300.

[21] For a similar case, see al-Jāḥiẓ, *Kitāb al-bayān wa-al-tabyīn*, ed. 'A. Hārūn, 4 vols, Cairo, n.d., vol. 1, pp. 88–93. French translation of the passage in M. Aouad and M. Rashed, 'L'exégèse de la *Rhétorique* d'Aristote: Recherches sur quelques commentateurs grecs, arabes et byzantins', *Medioevo* 23, 1997, pp. 43–189, at pp. 89–91.

Nawbakht, al-Fazārī, and ʿAlī ibn ʿĪsā – the astrolabe expert, who was considerably younger. It is reported that al-Manṣūr is 'the first Caliph for whom books in foreign languages were translated into the Arabic language.'[22] Al-Akhbārī then names several translated titles, including the *Almagest*, the *Elements*, Nicomachus of Gerasa's *Introduction to Arithmetic*. He further writes that for him were translated 'all the various Ancients in Greek, Byzantine, Pahlavi, Persian and Syriac and that these were diffused among individuals who examined them and made efforts to master them.'[23]

Regardless of the historical value granted late evidence, it does indeed reflect the opinion of those who followed the period of al-Manṣūr. Thanks to the rulers' initiative, translations were undertaken; but in the background, research was beginning that required the translation of specific works. The accelerated Arabization demanded the constitution of a new library on the scale of an empire extending from the Indus to the Atlantic. As to the books that al-Akhbārī mentions, what should one make of them? Nothing contradicts the accuracy of the information about the *Almagest*, which is indeed corroborated by a passage from al-Nadīm, according to which al-Manṣūr's vizier, Khālid ibn Barmak, would have ordered a first translation, which proved unsatisfactory and was later corrected at his request.[24] This may be the translation in question. By contrast, Nicomachus's *Introduction to Arithmetic* was first translated from Syriac by Ḥabīb ibn Bihrīz. But the latter 'translated several books for al-Ma'mūn',[25] that is, at least 40 years later – which is possible but improbable. As to Euclid's *Elements*, it presupposes a translation that antedates al-Ḥajjāj's first translation. Since no other information corroborates its existence, the question remains open.

The intervention of political power to request translations from the Greek and other languages; the constitution in Arabic of a library in the scale of the new world, which is, at least, a consequence of the continuous arabization of the State and of the culture for more than a century and a half; and finally the response to research needs: such are the imperatives to which translation movement was responding at the end of its first phase and at the beginning of the second. Several ancient translations, unknown until very recently, may well belong to this intermediary phase. Thus we know that al-Kindī had access to a translation of Archimedes' *The*

[22] Al-Masʿūdī, *Murūj al-dhahab*, Beirut, 1991, vol. 4, p. 333.
[23] *Ibid.*
[24] Al-Nadīm, *al-Fihrist*, p. 327.
[25] *Ibid.*, p. 304.

Measurement of the Circle, different from the one that will later be made from a Greek manuscript.[26] The selfsame al-Kindī knew about a translation of Euclid's *Optics* that differs from the one that has reached us and that most probably was produced before the latter. Finally, we have just found an ancient translation of the beginning of Anthemius of Tralles's *Mechanical Paradoxes*.[27]

The diversity of these translated texts is striking: Euclid's *Optics*, Archimedes' *Measurement of the Circle*, Anthemius of Tralles. To these, one could also add several other treatises. In the current state of our knowledge, however, these involve relatively short texts that nevertheless seem to be linked with research, as we shall see. As to the translations themselves, they are literal and use a terminology that will be thoroughly recast in the second phase of the translation movement (see below).

3.2. *Institution and profession: the age of the Academies*

This movement accelerates as it enters the second phase, in which the translation becomes at once *institution* and *profession*. Al-Maʾmūn's dream not only characterizes this phase, but also draws its meaning from it.

Even at its apogee at the beginning of the ʿAbbāsid dynasty, the first phase of the translation movement cannot be confused with the one that will soon follow, whether one considers the number of translations, the diversity of the translated writings, or the increased technicality and specialization of the translators. At this point, translation becomes both a scientific profession and an institution. There are several reasons for this transformation, which begins at the time of al-Maʾmūn and increases even more under his successors. One reason that is often overlooked is the change in the encyclopaedia of knowledge. Between the middle of the 8th century and the middle of the 9th, several disciplines emerge that are directly tied to the new society, its ideology, and its organisation. These involve, for example, the various areas of research stimulated by the need to treat the sacred texts and their interpretation. One thus sees the emergence of a full spectrum of linguistic disciplines ranging from ethnolinguistics to lexicography based on genuine phonological research; and thanks to combinatorial thinking (al-Khālil ibn Aḥmad), they included grammar and philology.[28] Consider also the development of *kalām*, the

[26] R. Rashed, 'Al-Kindī's Commentary on Archimedes' *The Measurement of the Circle*', *Arabic Sciences and Philosophy*, 3, 1993, pp. 7–53.

[27] R. Rashed, *Les Catoptriciens grecs, I: Les miroirs ardents*, edited, translated with commentary, Collection des Universités de France, Paris, 2000, pp. 343–59.

[28] See below 'Algebra and linguistics'.

philosophico-theological science, with its many schools and their ramifications.[29] One can also mention the various domains of history and the birth of methods for the critical analysis of testimony; the development of hermeneutical studies, notably that of the Koran; the various logico-juridical sciences necessary for research into Islamic law, etc. To this, one should also add algebra itself, as well as other disciplines born from practice and from the administration of the Empire. This encyclopaedia of knowledge is therefore far removed from that of Late Antiquity: somewhat later, al-Fārābī will offer a sketch of its contents in his *Enumeration of the Sciences*.[30]

Whereas this new encyclopaedia echoes the disciplines and their diversity as well as of the culture of the time, it also points to an advance that becomes noticeable upon reading the books of the *ṭabaqāt* (classes of scholars) and of the ancient biobibliographies: a growing specialisation. A scholar belongs not only primarily to a profession, sometimes two interconnected ones (for example, *mutakallim,* philosopher-theologian, and jurist), but within the same profession, he belongs to one school or the other: Kūfā and Baṣra for example, for grammarians; Baṣra and Baghdad for theologian-philosophers.[31] All of these new disciplines and the specialists in them, the number of which is constantly growing, created both a demand and an audience. The philosopher-theologian wanted to have more and better knowledge in philosophy, logic, and even statics and physics.[32] The religious requirements – determining the direction of Mecca and the time of prayer in such a vast empire – demanded new knowledge in astronomy, just as the progress of the medical sciences was now necessary to meet the needs of medicine in the urban centres. Such positions as

[29] R. M. Frank, 'The Science of Kalām', *Arabic Sciences and Philosophy*, 2.1, 1992, pp. 7–37; J. van Ess, *Frühe Muʿtazilitische Häresiographie*, Wiesbaden, 1971. For an insight into the extremely wide-ranging branches of thought, cf. Shahrastani, *Livre des religions et des sectes*, French translation with introduction and notes by D. Gimaret and G. Monnot, Louvain/Paris, 1986.

[30] Al-Fārābī, *Iḥsāʾ al-ʿulūm*, ed. ʿU. Amīn, 3rd ed., Cairo, 1968.

[31] From the outset these differences have been perceived as constituent parts. In the two fields mentioned, see Abū Saʿīd al-Anbārī, *Al-Inṣāf fī masāʾil al-khilaf bayn al-naḥwiyyīn al-baṣriyyīn wa-al-kūfiyyīn*, Beirut, 1987 and Abū Rashīd al-Nīsābūrī, *Al-masāʾil fī al-Khilāf bayn al-Baṣriyyīn wa-al-Baghdādiyyīn*, ed. M. Ziyāda and R. al-Sayyid, Beirut, 1979.

[32] For example, Abū al-Hudhayl and his nephew al-Naẓẓām, Cf. M. A. Abū Rīda, *Ibrāhīm b. Sayyār al-Naẓẓām wa-arāʾuhu al-kalāmiyya al-falsafiyya*, Cairo, 1946; A. Dhanani, *The Physical Theory of Kalam. Atoms, Space and Void in Basrian Muʿtazili Cosmology*, Leiden, 1994.

administrator of the *Dīwān* and private secretary (which became a genuine profession[33]) required a rather extensive general culture. In short, these individuals all constituted a large audience for disciplines and a culture that had to be translated, most notably from the Greek and Persian. Among the translated books, one thus finds works with a cultural aim that focus on such subjects as the moral opinions of the philosophers[34] or the interpretation of dreams.[35]

This second phase of the translation movement soon witnessed the institutionalization of both translation and the Greek heritage. An abundance of facts and anecdotes shows that caliphs, viziers, princes, the wealthy, and even some scholars founded libraries and observatories and encouraged translation and research.[36] It has not been sufficiently emphasized, however, that these new institutions included not only individuals, as had been the case before, but also groups and teams that were often rivals and competitors.[37] All of these means served to integrate the Greek heritage into the new scientific regime. By way of example, the House of Wisdom (*Bayt al-Ḥikma*) founded by al-Maʾmūn in Baghdad, included astronomers such as Yaḥyā ibn Abī Manṣūr, translators such as al-Ḥajjāj ibn Maṭar (who translated Euclid's *Elements* and Ptolemy's *Almagest*), and mathematicians such as al-Khwārizmī. Later, in another group linked with this same House, the three Banū Mūsā brothers, mathematicians and astronomer who encouraged and financed translation; the translator of Apollonius, Hilāl ibn Hilāl al-Ḥimṣī, and the translator and mathematician, Thābit ibn Qurra. We also know that translators and scholars formed groups around the Banū Mūsā, around al-Kindī and Ḥunayn ibn Isḥāq, etc. Finally, the mosque, the observatory, and the hospital were so many venues and institutions in which other groups of specialists worked.

[33] See for example Ibn Qutayba, *Adab al-kātib*, ed. A. Fāʿūr, Beirut, 1988; al-Jahshayyai, *Kitāb al-wuzarāʾ wa-al-kuttāb*, Beirut, n.d.

[34] See *Testament de Platon pour l'éducation des jeunes*, ed. L. Cheikho, *Traités philosophiques anciens*, Beirut, 1911, for Ḥunayn ibn Isḥāq's translation of Plato.

[35] For example, Ḥunayn ibn Isḥāq's translation of the *Book of Dreams* by Artemidorus of Ephesus (see the critical edition by T. Fahd, Damascus, 1964).

[36] M. G. Balty-Guesdon, 'Le Bayt al-Ḥikma de Bagdad', *Arabica*, 39, 1992, pp. 131–50; Y. Eche, *Les bibliothèques arabes publiques et semi-publiques en Mésopotamie, en Syrie et en Égypte au Moyen-Âge*, Damascus, 1967.

[37] Ancient bibliographers report conflicts, for example, between al-Kindī and his collaborators on the one hand, and the Banū Mūsā and their group, on the other.

The way translation is organized during this period brings to the fore two interrelated characteristics with a very special significance. First of all, carried out on a grand scale, translation is not limited only to writings with a practical goal. Second, on more than one occasion, works will be retranslated – not only those from the first phase, but even some from the beginning of the second phase. Thus Euclid's *Elements* was translated three times; the *Almagest* at least three times, etc. These retranslations correspond to a change in the criteria of the act of translating. In short, translation had become the act of individuals belonging to competing groups and schools. The criteria of translation had changed. The translator, no longer what he was during the first phase, had now acquired a two-fold training in languages and in science and philosophy. But before explaining this evolution and examining who was translating, and how and why, one must first note that translation followed neither a didactic order (from the easiest to the most difficult books), nor the chronological order in the series of Greek authors. Certainly, no pre-established plan governed the translation; one should not therefore believe that it occurred haphazardly, with the discovery of books to be translated. On the contrary, several witnesses from this period suggest that scholars chose the work to be translated before they searched for the manuscripts needed for the translation. Thus Ḥunayn ibn Isḥāq had decided to translate Galen's *On Demonstration* before he set out in search of manuscripts;[38] likewise when the Banū Mūsā wanted to trans-

[38] In an autobiographical note on his search for a manuscript of Galen's *Demonstration*, Ḥunayn ibn Isḥāq states: 'He [Galen] composed this work in 15 books. His aim was to show the way of demonstrating what is demonstrable necessarily. This is Aristotle's goal in Book IV of his *Organon*. To date, none of our contemporaries has seen a complete set of this work; but Jibril took great care about his search, as did I in my quest for this work in Mesopotamia, throughout Syria, Palestine, and Egypt, until I reached Alexandria. I found nothing from it, except about half of it in Damascus. But these books were incomplete and out of order. Jibril, however, had found books of this work, which were not those that I found.' From this account, one can infer that searches took place not only in Byzantium, but throughout the whole of the ancient empire; that scholars went to Alexandria, among other places, in search for Greek manuscripts; that the manuscript of such an important work as this was 'simply' found in Damascus; that the translators themselves travelled independently of the great manuscript expeditions, such as the one ordered by al-Maʾmūn; and finally that our knowledge of translations of Greek into Syriac and Arabic is still inadequate. These conclusions are confirmed by another account that is worth recounting. Yaḥyā (Yūḥannā) ibn al-Biṭrīq, a member of the famous expedition sent by the Caliph al-Maʾmūn to Byzantium in search of Greek manuscripts, tells how he received the order from the Caliph to look for the manuscript of the *Secret of Secrets*: the translator Yūḥannā b. al-Biṭrīq said: 'I left unvisited not one of these temples in which the philosophers had hidden their secrets; and not one of the

(*Cont. on next page*)

late Apollonius's *Conics*.[39] All of these characteristics of the translation movement's second phase reveal a phenomenon that has too long gone unnoticed: the tight connections between a massive translation effort and active and innovative research. These links in particular interest us here.

3.3. *An ideal type of translator: Ḥunayn ibn Isḥāq's journey*

Before examining these links, let us consider briefly the training of this new generation of translators, the very ones who will transmit the essential core of the Greek philosophical and scientific heritage throughout the 9th century and especially during its second half. In contrast to the majority of their predecessors, these translators were neither enlightened amateurs familiar with an ancient language nor specialists, physicians or alchemists – capable of giving an approximate Arabic translation of one of the books in their discipline. Henceforth we confront genuine professionals of both the languages and the sciences. The ideal type of the latter, so to speak, is the famous Ḥunayn ibn Isḥāq.[40] The narrative of his biography that has reached us is of considerable interest: a colourful literary piece that, whether true or legendary, certainly highlights the ideal characteristics of the new profession. (Everything nevertheless suggests that this ideal trajectory has more to it than vague similarities with historical reality.) A Christian (Nestorian) Arab born in 808 to a pharmacist father in Hira, Ḥunayn's journey begins in Baṣra, where we see him perfect his Arabic. He thus knew that the language of translation was not the vernacular used in daily life. This choice

---

(*Cont.*) great men among the ascetics, made wise by the knowledge thereof, whom I thought might have in his possession the object of my search, until I came to the temple that Asclepius had built for himself. There I met a devout and pious ascetic eminent in knowledge and possessed of a penetrating intelligence. I made him aware of my goodwill towards him, I stayed as his guest, and I used guile to get him to put into my safekeeping books which were in his temple, and among them I found the book I had been looking for, the object of my quest and of my covetousness' (ed. A. Badawi, *Fontes Graecae doctrinarum politicarum islamicarum*, Cairo, 1954, p. 69).

39 *Les Mathématiques infinitésimales du IXe au XIe siècle*, vol. III: *Ibn al-Haytham. Théorie des coniques, constructions géométriques et géométrie pratique*, London, al-Furqān, 2000, Chap. I; English translation: *Ibn al-Haytham's Theory of Conics, Geometrical Constructions and Practical Geometry*. A History of Arabic Sciences and Mathematics, vol. 3, Culture and Civilization in the Middle East, London, Centre for Arab Unity Studies, Routledge, 2013.

40 See G. Bergsträsser, *Ḥunain ibn Isḥāq über die syrischen und arabischen Galen-Übersetzungen*, Abhandlungen für die Kunde der Morgenlandes, XVII, 2, Leipzig, 1925. See also G. C. Anawati and A. Z. Iskandar, 'Ḥunayn ibn Isḥāq', *Dictionary of Scientific Biography*, 1978, vol. 15, Suppl. I, pp. 230–48.

underlies the legend according to which he met one of the greatest Arabic linguists: al-Khālil ibn Aḥmad.[41] In the story of his life, this legend plays a role that is at once seminal and emblematic: for Arabic, he would have had al-Khālil ibn Aḥmad himself as a patron. We meet him next in Baghdad, the stage of his proper scientific training; he also studies medicine under the tutelage of one of the great physicians of the time: Yūḥannā ibn Māsawayh. It is here that the hero meets his destiny. Driven away from his circle by Ibn Māsawayh, Ḥunayn sets out again on the road of his training; this is the third stage. He travels to one of the centres of Hellenism in order to perfect his Greek – the biographers are not clear whether it is in the Byzantine empire or in Alexandria. Several years later, he reappears in Baghdad, reciting by heart Homer's poetry;[42] at the emblematic level, such a mastery of Greek naturally constitutes the counterpart to al-Khālil's earlier tutelage in Arabic.

Here are then three distinct stages, each necessary for the training of the new type of translator: henceforth he will be a *translator-scholar* who masters Greek, Arabic, Syriac, and science. These stiff requirements echo two facts: the translated science is still a living science. As we shall see below, the point of translating was not to reconstitute the history of a science, but rather to pursue lively research and to carry out a living practice. One of the tasks of the translator – this is the second fact – is henceforth to build Arabic as a scientific language. This involves linguistic research in the proper sense of the term, which requires a training similar to the one Ḥunayn ibn Isḥāq ostensibly received.

Ḥunayn spends the rest of his life translating Greek medical books as well as several books of philosophy, some of which were required by the medical curriculum. It is while carrying out these translations, the quality of which meets unanimous approbation, that he started researching scientific Arabic. The census of his translations includes 129 books, roughly two thirds in Syriac and one third in Arabic. The clear preponderance in favour of Syriac reflects the make-up of the medical community of his day, and also the structure of the demand. The positions of court physicians were still largely filled by physicians of Syriac origin: it is from them that most of the requests for translation come, to meet the needs of practice or

[41] 'He [Ḥunayn ibn Isḥāq] stayed some time and his master in Arabic was al-Khalīl ibn Aḥmad' (Ibn Abī Uṣaybiʿa, *ʿUyun al-anbāʾ fī ṭabaqāt al-atibbāʾ*, ed. N. Riḍā, Beirut, 1965, pp. 257 and 262).

[42] *Ibid.* p. 258.

research.[43] And in fact, among the commissioners named in the historical sources, one finds Bakhtīshūʿ ibn Jibrāʾīl, Salmawayh, Dāʾūd, Yūḥannā ibn Māsawayh, who are all Syriac physicians; and the Banū Mūsā, highly cultured mathematicians. In addition to his medical practice, Ḥunayn ibn Isḥāq also wrote several medical works of his own, and some books of Arabic grammar and lexicography.[44] To understand this enormous production, let us remember a second feature that characterizes it: the organisation of translation and research into genuine working teams. Around Ḥunayn, one finds an entire school, the members of which include his son Isḥāq, his nephew Ḥubaysh and ʿĪsā ibn Yaḥyā, as well as the copyists al-Aḥwal and al-Azraq.[45]

As one can see, this new type of translator is distinguished not only by the requirements of the linguistic and scientific training for which he is responsible, but also by the new tasks he has taken on: research into both scientific Arabic and in science. As the century progresses, a gradual and latent transformation starts to confirm what was embryonic at the time of Ḥunayn ibn Isḥāq: the transformation of the *translator-scientist* into *scientist-translator*. This is the distance that separates Ḥunayn from Thābit ibn Qurra (d. 901).

3.4. *Third phase: from translator-scientist to scientist-translator*

Thābit ibn Qurra is one of the great mathematicians not only of Islam, but also of all time. He began life as a moneychanger. His mother tongue was Syriac, and he perfected his Greek and Arabic enough to translate astronomy, mathematics, and philosophy. In any case, it is on account of these talents and his linguistic knowledge that Muḥammad ibn Mūsā, returning from an expedition in search of manuscripts in the Byzantine Empire, 'discovers' him in his native town of Ḥarrān (or in one of the surrounding villages, Kafr Tūtha) and takes him along to Baghdad. Welcomed by Muḥammad ibn Mūsā into his own home, Thābit receives mathematical training under the guidance of the three Mūsā brothers and especially the youngest, al-Ḥasan, the mathematician of genius. Once his training is complete, Thābit ibn Qurra translates a considerable number of

[43] Cf. H. Hugonnard-Roche, 'L'intermédiaire syriaque dans la transmission de la philosophie grecque à l'arabe: le cas de l'Organon d'Aristote', *Arabic Sciences and Philosophy* 1, 1991, pp. 187–209.

[44] Ibn Abī Uṣaybiʿa mentions among his writings a book on grammar (*Kitāb fī al-naḥw)* and a book on the classification of names of simple medicines (*Kitāb fī asmāʾ al-adwiyya al-mufrada ʿalā ḥurūf al-muʿjam)*, (*ʿUyūn al-anbāʾ*, ed. N. Riḍā, p. 273).

[45] *Ibid.*, pp. 260, 270.

Greek mathematical treatises, including Archimedes' *On the Sphere and Cylinder*, the three last books (lost in Greek) of Apollonius's *Conics* and Nicomachus of Gerasa's *Introduction to Arithmetic*. He also revises many translations – Euclid's *Elements* and Ptolemy's *Almagest*, among others. Finally, Thābit ibn Qurra composes numerous works in astronomy and mathematics, which are so important that they practically dwarf his crucial work as translator.

Between the translator-scientist in the pattern of Ḥunayn and the scientist-translator in the mould of Thābit falls an entire intermediate category. It is composed of eminent translators whose scientific training is as wide-ranging as it is solid, notably Ḥunayn's own son, Isḥāq ibn Ḥunayn (d. 911) and Qusṭā ibn Lūqā (d. beginning 10th century), among many others. Nonetheless, two traits accompany this new phase of translation: a noticeable change in both training requirements and the criteria used in translating, and a major strengthening of the links between scientific and philosophical research and translation. As we have noted with Thābit, all of these factors generated previously unknown activities: the revision of older translations or of those done by a non-specialist in the subject.

2. TRANSLATION AND RESEARCH: A DIALECTIC WITH MANY FORMS

If we ignore scientific and philosophical research, we will not understand the first thing about the translation movement from Greek into Arabic. It is this research that illuminates the selection of books to be translated and that drives its evolution. This statement does not derive from anything I have postulated; nor is it a Gestalt-like grasp of the act of translation. On the contrary, it is nothing more, but also nothing less, than history. We will therefore draw several examples from various domains, in order to illustrate and to clarify as much as possible this dialectic between research and translation. The variety of the contexts and, of course, the limits of my own competence have determined my choice: mainly optics, geometry, and arithmetic.

1. *Coexisting and overtaking: optics and catoptrics*

Let us begin empirically, by listing the titles of the main Greek works on optics and catoptrics translated into Arabic, along with the names of their translators.

1. Euclid's *Optics* was translated at least twice into Arabic, once before the middle of the 9th century. Al-Kindī wrote a critical commentary on it, on the basis of his own research in optics.[46]

2. Ptolemy's *Optics*; the Greek text is lost, as is the Arabic translation, which was very probably not done before the end of the 9th century. Only the Emir Eugenius of Sicily's Latin translation survives.[47] The surviving documents suggest that this work, and particularly the fifth book on refraction, entered the discussion rather late in the development of optics, that is, during the 10th century (notably in the research by al-ʿAlāʾ ibn Sahl, among others).

3. The *Catoptrica* attributed to Euclid. I have shown that traces of this work exist in Arabic, notably in a 9th-century book written by Qusṭā ibn Lūqā.[48]

4. Diocles's *Burning Mirrors*: only two propositions from it were cited by Eutocius.[49] The Greek original is lost; only an Arabic version survives – a relatively precocious one, judging from its vocabulary.[50]

5. Anthemius of Tralles's *Burning Mirrors* (*Mechanical Paradoxes*). The extant Greek text is incomplete. It was translated at least twice, perhaps three times, into Arabic; the first time, before the middle of the 9th century, the second, rather later. At least one of the Arabic versions seems complete.[51]

6 *Burning Mirrors and Abridged Conics*. This is an Arabic translation of a lost Greek work, written – according to the Arabic transcription – by a certain 'Dtrūms', who has yet to be identified.[52]

7. The *Bobbio Fragment* on burning mirrors, a text that has left no trace in Arabic.[53]

To this list, one can add a few titles of lesser importance, such as Hero of Alexandria's *Catoptrica*, fragments of which have survived in an ancient Arabic translation.[54]

[46] R. Rashed, 'Le commentaire par al-Kindī de l'*Optique* d'Euclide: un traité jusqu'ici inconnu', *Arabic Sciences and Philosophy*, 7.1, 1997, pp. 9–57.

[47] A. Lejeune, *L'Optique de Claude Ptolémée dans la version latine d'après l'arabe de l'émir Eugène de Sicile*, Louvain, 1956.

[48] R. Rashed, *Œuvres philosophiques et scientifiques d'al Kindī*. Vol. I: *L'optique et la catoptrique*, Leiden, 1996, Appendix II, pp. 541–645.

[49] R. Rashed, *Les Catoptriciens grecs*, 1st part.

[50] *Ibid.*, p. 21.

[51] *Ibid.*, Appendix, pp. 343–59.

[52] *Ibid.*, chap. II, pp. 155–213.

[53] *Ibid.*, chap. IV, pp. 272 ff.

[54] Various writings in this tradition survive in Arabic.

This is the sum total of the texts in optics and catoptrics. Several conclusions are immediately obvious. The essential core of Greek works was known and translated into Arabic, sometimes more than once. This is what I mean when I describe translation as large scale and multiple – moreover, several treatises were translated into Arabic before the middle of the 9th century. Finally, these treatises were not only studied, but also subjected to scientific criticism beginning in the middle of the same century. Al-Kindī, for example, offers a detailed and thorough critique both of Euclid's *Optics* and Anthemius of Tralles' book.[55] One should not be under the impression that the translations occurred in the order given above; rather translation followed the order of research. Before coming back to this point, however, we might start by noting the difference between the two phases of translation, in order to lift out their criteria. The example of Anthemius of Tralles offers a good example.

It is certain that the first translation of Anthemius' *Mechanical Paradoxes* took place before the middle of the 9th century, at the very time when Arabic research on burning mirrors seems to get underway. The efforts of al-Kindī and Qusṭā ibn Lūqā in this area leave no room for doubt about this point. A detailed examination of this translation shows both that it is literal and that it uses an archaic vocabulary, that is one that al-Kindī himself has already abandoned. The second translation took advantage of the research that was underway, not only by choosing a more precise and appropriate lexicon, but also by improving the syntax to produce a more readable text.[56]

[55] R. Rashed, *L'optique et la catoptrique d'al-Kindī*.

[56] In order to illustrate this situation, let us take a single example.

Anthemius wrote: τοῦ Η σημείου μεταξὺ τῆς τε χειμερινῆς ἀκτῖνος καὶ τῆς ἰσημερινῆς νοουμένου ὡσανεὶ κατὰ τὴν διχοτομίαν τῆς ὑπὸ ΕΒΓ γωνίας καὶ ἐκβληθείσης τῆς ΗΖ ὡς ἐπὶ τὸ Θ σημεῖον (*Les Catoptriciens grecs*, p. 350, 10–13).

The first translation reads as follows:

wa-li-yufʿal ʿalāmat H *wāsiṭatan bayna al-shuʿāʿ al-shatwī wa-shuʿāʿ al-istiwāʾ, ka-annahā qāṭiʿa wasaṭ zāwiyat* EBC. *wa-li-yukhraj khaṭṭ* HG *ilā ʿalāmat* I.

[…] Let the sign *H* be made intermediate between the winter solstitial ray and the equinoctial ray, as if it cut the middle of the angle *EBΓ*. Let the straight line *HZ* be drawn to point Θ (*ibid.*, p. 287, 7–9).

The translator renders the Greek νοεῖσθαι by the verb *faʿala* (to make), a very awkward translation to say the least. Under the rather improbable assumption that he might have wished to avoid a form of the verb *wahama* (to imagine, to conceive of), he could have chosen *jaʿala* or *kana*. Note moreover his usage of *ʿalāma* (sign) to render the Greek σημεῖον. Although one still occasionally encounters this usage in the 9th century, it is already becoming much rarer.

(*Cont. on next page*)

The difference between these two types of translation was already noticed at the time, even though its historic dimension was not. Indeed it was not a coincidence that, both in the 9th century and later, scholars raised the matter of different styles of translation: al-Kindī debates it; his

---

(*Cont.*) The second translation of this same phrase reads:

> *wa-li-takun nuqṭat* H *fī al-wasaṭ fīmā bayna khaṭṭay* BE, BC *ʿalā niṣf zāwiyat* EBC. *wa-nukhrij* HG *ilā nuqṭat I*.

> Let the point *H* be in the middle, between the two straight lines *BE* (the solstitial line of winter) and the straight line *BΓ* (the equinoctial line), on half of the angle *EBΓ*. Extend *HZ* to point Θ.

Both the lexicon and the syntax of the second translation are better suited to Arabic and to the language of geometrical optics.

Let us pursue this example a little farther. The Greek text continues as follows: ἐὰν τοίνυν κατὰ τὴν θέσιν τῆς HZ εὐθείας νοήσωμεν ἐπίπεδον ἔσοπτρον, ἡ BZE ἀκτὶς προσπίπτουσα πρὸς τὸ HZΘ ἔσοπτρον λέγω ὅτι ἀνακλασθήσεται ἐπὶ τὸ A σημεῖον (*ibid*., p. 350, 14–17), which the translator renders as follows:

> *fa-matā mā naḥnu tawahamnā mirʾa dhāt saṭḥ mustawin fī mawḍiʿ khaṭṭ* HG *al-mustaqīm mawuqiʿan li-al-shuʿāʿ alladhī dalāʾiluhu* BGE *ʿalā mirʾat* GHI, *azʿumu annahu yuʿṭafu rājiʿan ilā mawḍiʿ A*.

> When we then imagine a mirror having a plane surface in the position of straight line *HZ*, the locus for the ray whose signs are *BZE* on mirror *ZHΘ*, I maintain that it is reflected onto position *A*.

Note that the expression *fa-matā mā naḥnu tawahhamnā* is redundant and that its syntax seems non-Arabic; it would have been better to write *fa-matā tawahhamnā*. Likewise, it is more correct to use the preposition *ʿalā* instead of *fī*. The remainder is no better; after *al-mustaqīm*, the passage should read:

> *wa-kānat mirʾat* GHI *mawqīʿan li-shuʿāʿ dalāʾiluhu* BGE; *fa-aqūlu innahu yanʿakisu ilā mawḍiʿ A*.

Note that, in this case, literalism has a negative effect. Conversely, the use of *dalāʾiluhu BGE* is archaic; later translators will abandon it. Likewise, *ʿaṭafa rājiʿan* as a translation of 'to reflect' begins to disappear in the 9th century. Finally, the use of *azamu* (I opine, I claim) instead of *aqūlu* (I say) to translate λέγω does not appear in translations after the middle of the 9th century.

Turning now to the second translation of the second sentence, we read:

> *fa-in tawahamnā saṭḥan mirāʾiyan mawḍūʿan ʿalā mawḍiʿ khaṭṭ* HGI, *fa-innahu yakūn shuʿāʿ* BGE, *idhā waqaʿa ʿalā mirʾat* HGI, *yarjaʿu ilā nuqṭat* A.

> If we imagine a mirror surface located in the position of the straight line *HZΘ*, then if the ray *BZE* falls on the mirror *HZΘ*, it returns to point *A*.

Although this less literal translation does not convey exactly the letter of the Greek text (assuming that we are dealing here with the very same text, which is by no means certain), it does render the meaning in an Arabic that is correct from the vantage points of both vocabulary and syntax.

contemporary, the learned philosopher al-Jāḥiẓ,[57] also treats it. To illustrate this point, one need only cite al-Kindī's words to a correspondent who did not understand Ptolemy's description of an instrument in Book V of the *Almagest*:

> You asked me, o brother showered with compliments, to describe the instrument that Ptolemy mentioned at the beginning of Book V of the *Almagest*, when you began to have doubts about the description that he gives of this instrument and its usage. Your doubts pertain not to a defect in his exposition, but to the difficulty in the organization of his words, for this man with his sophisticated language is far removed from observing most people's customary use of words; for this reason, access to the order of his words is difficult, even though their significations are clear to those who undertake to translate his books from Greek into Arabic, since the difficulty in the order of his words has become the reason why the translator has difficulty understanding them. Thus, for fear both of presenting their own opinions instead of the meaning of his words, and of being led into error about their true essence, they [the translators] have striven to present the same order in Arabic, and they have successively substituted word-for-word the Arabic equivalent.
>
> Translators have thought constantly about what they have gained from this book, and have sought, and lifted out, its significations in order to eliminate errors. Not all of those who translated some of his books succeeded, but only those who were most competent and sufficiently skilled in Greek to avoid missing two things at once: knowledge of the significations of the book and the precision of its words. Indeed, whoever sets out to interpret the meaning of what he has translated without understanding that meaning, brings about two things at once – he loses the significations and he loses the words. In this respect, this is prejudicial to whomever examines their translation with the aim of truly grasping something about the views of the book's author.
>
> Conversely, if the translator describes the word such as it is, even if it is difficult to understand it, he then elicits understanding of the thought of the book's author, even if one attains this only with great effort.[58]

In the language of the period, this crucial text tells us what translation from Greek into Arabic involved, and it alludes to the two main styles we

[57] *Kitāb al-Ḥayawān*, ed. ʿA. Hārūn, vol. 1, pp. 75 ff. Cf. Abū Ḥayyān al-Tawḥīdī, *Kitāb al-imtāʿ wa-al-muʾānasa* ed. A. Amin and A. al-Zayn, reprod. Būlāq, n.d., pp. 112, 115–16, 121. See also M. Mahdi, *Language and Logic in Classical Islam*, in G. E. von Grunbaum, *Logic in Ethical Islamic Culture*, Wiesbaden, 1970, pp. 51–3.

[58] Al-Kindī, *Risāla fī dhāt al-ḥalaq*, ms. Paris, Bibliothèque Nationale, no. 2544, fols 56–60.

have just mentioned. In addition to the lexical difficulty, it is syntactic difficulty that predominates. These two difficulties characterize specialized language (in this case, astronomy). In fact, there are three styles of translation: that of the translators who proceed word for word, thus running the risk of losing the meaning; that of the translator-scientists who try first to grasp the meaning of the concepts; and finally, among the latter, only those who are 'competent and skilled in Greek' and who succeed in eliminating errors. Absent this distinction, which characterizes Ḥunayn, Isḥāq, etc., al-Kindī prefers the word-for-word style.

The historical significance of these reflections is clear even though al-Kindī, who was himself in constant contact with both types of translator, does not mention it. In fact, if the first translation was often undertaken in response to the needs of incipient research, the second was usually linked to much more advanced research. For catoptrics, al-Kindī and his contemporary Ibn Lūqā are good examples of this point. Since he had access to the first Arabic version of *The Mechanical Paradoxes*, al-Kindī wrote a whole treatise on burning mirrors.[59] One finds therein not only a critique of the many weakness in Anthemius's text, but a mass of new results. Ibn Lūqā was also doing research in catoptrics[60] and composed a treatise on burning mirrors. It is at this time that most of the Greek treatises on mirrors were translated into Arabic, as a rigorous study of the vocabulary demonstrates. The research progress that al-Kindī and others who followed him achieved will have a rather paradoxical result: on the one hand, it stimulates a new and better translation of *The Mechanical Paradoxes*, which will be used by such successors of al-Kindī as Ibn ʿĪsā (a second-tier author);[61] on the other hand, it reduces the role of the translated Greek texts to one of mere historical significance. The likes of ʿUṭārid and Ibn ʿĪsā were still interested in them at the beginning of the 10th century. In the work of Ibn Sahl, his contemporaries, and his successors at the end of that century, however, the Greek texts are little more than a rather pale memory.

Burning mirrors, however, represent only one chapter of Hellenistic optics. The latter also includes optics proper: that is, the geometrical study of perspective and of optical illusions; catoptrics, that is, the geometrical study of the reflection of visual rays on mirrors; meteorological optics, which examines such atmospheric phenomena as the halo and the rainbow... These are the chapters that al-Fārābī mentions in his *Enumeration of*

[59] R. Rashed, *L'optique et la catoptrique d'al-Kindī*.
[60] *Ibid*., see Appendix II.
[61] *Ibid*., Appendix III, pp. 647–701.

the Sciences. To these geometrical chapters, one should also add the doctrines on vision that mark the works of the physicians and the writings of the philosophers. In each of these domains, the transmission of the Greek heritage occurred along the lines of the model previously analyzed for burning mirrors.

Historical research is not yet in a position to tell us which optical concepts medical practice had transmitted before the end of the 8th century. Around this time, however, and throughout the first half of the 9th century, one encounters ophthalmological research among such physicians as Jibrā'īl ibn Bakhtīshū' (d. 828/9)[62] and Yūḥannā ibn Māsawayh after him. In any case, the topic garnered sufficient interest for Ḥunayn ibn Isḥāq to write for the medical community a compendium in which he presents the contents of Galen's writing on the anatomy and physiology of the eye.[63] Ḥunayn also translates the pseudo-Galenic treatise *On the Anatomy of the Eye*.[64] Did ophthalmological research and practice stimulate the study of optics and catoptrics? This seems plausible, even though it is too early to answer the question. In any event, it is around this time that the main Greek works in optics and catoptrics were translated – Euclid, Theon, Hero – (Ptolemy's *Optics* was in all likelihood not translated until the end of the century). Thus, whereas Arabic optics is the sole heir of Greek optics, its history is from the outset that of the correction and critique of the latter.

It is a significant fact that in the middle of the 9th century Euclid's *Optics* was not only available, but already the focus of corrections. We now know that, at this date, Euclid's *Optics* existed in not one, but two translations. The first, which exists in several manuscripts, often departs from the text of the two Greek versions that we know today – notably in passages as fundamental as the preliminary definitions. Two 13th-century mathematicians, Naṣīr al-Dīn al-Ṭūsī and Ibn Abī Jarrāda, will comment on this Arabic translation. The second translation is at least as old as the first, since al-Kindī was using it in the middle of the 9th century. The identification of this version has profoundly changed our ideas about the textual history of Euclid's *Optics,* which Heiberg had outlined and which was recently a

[62] Al-Nadīm, *al-Fihrist*, pp. 354–5.

[63] See his two books *Daghal al-'ayn* (The Disorder of the Eye) and *Fī ma'rifat miḥnat al-kaḥḥalin* (On the Knowledge of the Proof of Occulists). Cf. M. Meyerhof and C. Prufer, 'Die Augenheilkunde des Juhana ben Masawaih', *Der Islam* 6, 1915, pp. 217–56 and especially M. Meyerhof, *The Book of the Ten Treatises on the Eye Ascribed to Ḥunain ibn Isḥāq (809–877 AD)*, Cairo, Imprimerie Nationale, 1928, pp. 11–12.

[64] M. Meyerhof, *The Book of the Ten Treatises on the Eye*, pp. 18 ff.; P. Sbath and M. Meyerhof, *Livre des questions sur l'œil de Honaïn ibn Isḥaq*, in *Mémoires présentés à l'Institut d'Égypte*, Cairo, 1938, vol. 36.

matter of controversy. In brief, Heiberg distinguished between the *Optica genuina* (Vienna, phil. gr. 103) and the version he labeled 'of Theon', the oldest manuscript of which is Vat. gr. 204. It is only recently, however, that this thesis was invalidated: the texts that Heiberg attributed to Theon (Vat. gr. 204) are now confirmed as Euclid's, and the *Optica genuina* should be considered a later development. It was recently thought if we take into account the two Arabic versions, we are able to show that there were not only two, but four, pairwise independent textual traditions of Euclid's *Optics*. This confirms that no one tradition preserves *the* sole correct version of Euclid's text.

As far as we know, al-Kindī wrote the very first critical commentary of Euclid's *Optics*. The title of his book unambiguously expresses his intention: *Correction of the Errors and Difficulties Owing to Euclid in his Book Called Optics*. But this book is preceded by another of al-Kindī's works, *On the Diversity of Perspectives*. Although the Arabic is lost, this text was translated into Latin as *Liber de causis diversitatum aspectus (De aspectibus)*. The first quarter of the book seeks to justify the rectilinear propagation of light rays by means of geometrical considerations about shadows and the passage of light through slits. Al-Kindī was thus developing points from the epilogue to the second version of Euclid's *Optics*, which Heiberg had attributed to Theon. For now, it matters little whether this attribution is well founded. What matters here is that by the middle of the 9th century, at least this prologue, if not the whole version, was known in Arabic. In the second part of *De aspectibus*, al-Kindī takes up the main doctrines of vision known since Antiquity, adopting in the end extramission theory with a few modifications. This discussion has at the least the advantage of showing that al-Kindī knew about his predecessors' theories of vision. In the last part of *De aspectibus*, he studies the phenomenon of reflection and establishes the equality of the angles that the incident and reflected rays form on either side of the perpendicular to the mirror at the point of incidence. His demonstration is not only geometrical but also experimental. 'Experimental verification' takes place in a traditional language that bears traces of the prologue to the *Optics* attributed to Theon and that Ibn al-Haytham will fundamentally rethink at the beginning of the 11th century.

The point of this quick overview of *De aspectibus* is to display both the kind of research carried out in optics in the middle of the century and the gap that separates it from Euclidean optics in the strict sense, which formed the background for the reception of the latter. In fact, it is after he composes *De aspectibus* that al-Kindī writes his critical commentary on Euclid's *Optics*. The chronology of al-Kindī's optical writings is thus clear: his critical commentary on Euclid follows his own contribution to optics.

This order explains, at least in part, the meaning that his critical commentary embodies. In light of his own results, al-Kindī then examines one after the other of Euclid's definitions and proposition. He integrates criticisms that he had already levelled at Euclid during the composition of his own book, corrects what appears incorrect to him, proposes other demonstrations that he judges better, and, drawing on his own resources, tries to bring out the underlying ideas.

Like his works on burning mirrors, al-Kindī's optical writings are a perfect example of the conjunction of research with the translation of the Greek heritage. They also show that, without a meticulous study of the Arabic versions, it is impossible to reconstitute either the conceptual or the textual tradition of the *Optics*.

The 9th century holds other examples besides that of al-Kindī. One of his collaborators and colleagues, Qusṭā ibn Lūqā, is also interested in optics and catoptrics. Later, in the 870s, he too writes a book called *On the Causes of the Diversity of Perspective Found in Mirrors*.[65] This catoptric research shows that Ibn Lūqā knew not only Euclid's *Optics*, but also the *Catoptrica* attributed to him. In Chapter 10, Ibn Lūqā seems to use the first proposition of the latter work; likewise, in Chapter 22, one can identify traces of Propositions 7, 16, and 19 from the same book. In the following chapter, one finds traces of Proposition 21; in Chapter 28, he uses Proposition 5. Although these similarities do not prove that Ibn Lūqā had in hand an Arabic version of the *Catoptrica*, they nevertheless strongly suggest that he had access to a still unknown source that had picked up several propositions from this book.

The second important treatise that we have inherited from Greek optics is Ptolemy's. Unfortunately, we are at a loss when it comes to dating and setting the context of this translation, which is no longer extant. One scholar believed that he could recognize in al-Kindī's *De aspectibus* several passages 'that are clearly inspired by expositions found in the version of Eugenius'[66] (*i.e*, the Latin translation from Arabic). On this account, al-Kindī would thus be a *terminus ante quem* for the Arabic translation. This seems incorrect, as I have shown elsewhere, that the prologue to the version attributed to Theon[67] suffices to explain what appears in *De aspectibus*. The first evidence that we currently have for the

[65] R. Rashed, *L'optique et la catoptrique d'al-Kindī*, Appendix II.

[66] A. Lejeune, *L'Optique de Claude Ptolémée*, p. 29.

[67] 'Le commentaire par al-Kindī de l'*Optique* d'Euclide: un traité jusqu'ici inconnu'.

Arabic translation of Ptolemy's *Optics* is that of ʿAlāʾ ibn Sahl – rather late, towards the end of the 10th century.[68]

Since it is therefore necessary to proceed by means of conjecture, it seems that this translation was carried out at the end of the 9th century or at the beginning of the 10th. I believe that the translation of this book became necessary when research on refraction developed in both optics and catoptrics for lenses, as the works of Ibn Sahl pointedly attest. It is no coincidence that Book V of Ptolemy caught his attention. As long as we do not know the date of the translation, however, every statement, including my own, remains pure conjecture. However that may be, it remains the case that the progress of Arabic optics with Ibn Sahl and Ibn al-Haytham (d. after 1040) demoted these translations to items of mere historical interest and often could not prevent their fall into oblivion.

By using geometrical optics, we have seen how the various phases of the translation movement are interrelated. Although they are easy to spot, these phases multiply and overlap. We also notice a particular kind of translation, which might be called *in medias res* insofar as it is directly tied to research and follows its evolution. Anthemius and Euclid are translated *in phase* with the research of al-Kindī, Qusṭā ibn Lūqā, and others. Progress of this domain of study in turn stimulates efforts to take up anew the translation of these very same writings. Everything indicates that the translation of Ptolemy's work also follows upon the emergence of the study of refraction.

2. Translation and recursive reading: the case of Diophantus

We now turn to another type of translation, distinct from the preceding insofar as the translation does not go hand-in-hand with research, but follows it after a certain interval. In short, translation is stimulated in order to enrich research that is already well-developed, active, and prosperous. In this situation, translation is akin to the masterful recovery of an ancient text, which will be reactivated and in effect reinterpreted with a meaning that was not originally its own. In cases of this sort, there of course was neither a revision nor a second translation. Diophantus's *Arithmetic* is a perfect illustration of this type of translation.

[68] R. Rashed, *Géometrie et dioptrique au Xe siècle: Ibn Sahl, al-Qūhī et Ibn al-Haytham*, Paris, 1993; English version: *Geometry and Dioptrics in Classical Islam*, London, 2005.

Probably, but far from certainly, in the 2nd century, Diophantus of Alexandria wrote an arithmetical summa in 13 books that was probably modelled on Euclid's *Elements*. In this work, Diophantus clearly states his intention in the preface to the first book: to construct an arithmetic theory, ἀριθμητικὴ θεωρία. The elements of this theory are the integers considered as pluralities of units, μονάδων πλῆθος, and the fractional parts as fractions of magnitudes. The constituent elements of the theory are not only present 'in person', but also as species of numbers. The term εἶδος, translated into Arabic as *naw*ʿ and later into Latin as *species*, is in no sense reducible to the meaning of 'power of the unknown'. In the *Arithmetic*, this concept covers equally and without distinction the indeterminate plurality and the power of a number with any plurality whatsoever, that is, provisionally indeterminate. This last number is called the 'unstated number', ἄλογος ἀριθμός. To understand better this concept of *species,* one must remember that Diophantus refers to three *species:* that of the linear number, that of the plane number, and that of the solid number. The species generate all the others, which at the limit take their names. Thus the square-square, the square-square-square, and the square-cube are all squares. The cubo-cubo-cube is a cube. In other words, the species thus generated can only become such by composition, and the power of each is necessarily a multiple of 2 or 3. In the *Arithmetic*, there is no 7th power, for example, nor a 5th power in the statement of the problems. All told, the concept of polynomial is missing. At the same time, the composition of Diophantus's work thus suddenly becomes clearer: the point is to combine these species with one another, under certain conditions, using the operations of elementary arithmetic. To solve a problem is to try to continue in every case 'until there remains one species on one side and the other'.[69] Contrary to what one often reads, Diophantus's *Arithmetic* is not a book of algebra, but a genuine treatise of arithmetic in which one tries to find, for example, two square numbers the sum of which is a given square.

The second point that demands an explanation pertains to a work written at the time of the Caliph al-Maʾmūn, between 813 and 833: al-Khwārizmī's *Algebra,* the book that first conceived of algebra as an autonomous discipline. Having defined basic terms and operations, al-Khwārizmī studies algebraic equations of the first and second degree, together with associated binomials and trinomials, the application of algebraic procedures to numbers and to geometrical magnitudes, then concludes his book with indeterminate problems of the first degree. He

[69] See our edition, *Diophante: Les Arithmétiques*, *Livres V*, *VI*, *VII*, Collection des Universités de France, Paris, 1984, p. 103.

frames these problems in algebraic terms and solves them by means of these concepts. Al-Khwārizmī's successors, notably Abū Kāmil, carried out further research on 'indeterminate analysis as an integral part of algebra'.[70]

It was while he was doing research on indeterminate analysis as a chapter of algebra that Qusṭā ibn Lūqā translated seven books of the *Arithmetic*. The first three correspond to the first three of the Greek version. The following four are lost in Greek, and for Books 4, 5 and 6 of the Greek version they seem never to have been translated into Arabic.

These two preliminary explanations allow us better to formulate the problem of the translation of the *Arithmetic*. We confront, on the one hand, algebra as a new discipline that is definitely not Hellenistic, for having been established only half a century earlier, and one chapter of which focuses on indeterminate analysis; and, on the other hand, the *Arithmetic* – which treats problems that, after being translated into the terminology of this new discipline, will belong to the latter's domain. This interpretation was not accessible to the first randomly chosen translator. But Qusṭā ibn Lūqā, who did translate the *Arithmetic*, understood the utility of Diophantus's book for research in this new discipline, and in particular for indeterminate analysis. It was he who anachronistically gave the *Arithmetic* its first algebraic reading. It is not hard to imagine the consequences of this reading, not only for research, but also for translation. Before examining these effects, let us introduce the translator. Qusṭā ibn Lūqā was a Greek Christian from Baalbek, who, according to al-Nadīm, was a good translator who had mastered Greek, Syriac and Arabic.[71] Also according to the ancient biobibliographers, he was called to Baghdad, the capital, to participate in the translation of the Greek patrimony around the 860s. He thus belonged to the generation of slightly later translators who, by virtue of both heritage and training, possessed an elaborate and polished terminology in many different areas of learning. Included among these lexica, it is important to emphasize that of algebra. Qusṭā ibn Lūqā also belonged to the professional category of translator-scientist, fully conversant with the various scientific disciplines, and therefore in full possession of the means of understanding the meaning of the works they were translating. The titles of the works that Qusṭā translated and that have come down to us belong to a wide range of competencies. Among them are the little astronomical treatises: Autolycus's *On Risings and Settings*; *On Habitations*, *On Days and*

[70] R. Rashed, 'Combinatorial Analysis, Numerical Analysis, Diophantine Analysis and Number Theory', in Rashed (ed.) *Encyclopedia of the History of Arabic Science*, 3 vols, London/New York, Routledge, 1996, vol. 2, p. 376–417.

[71] Al-Nadīm, *al-Fihrist*, p. 304.

Nights, and the *Sphaerica*, all by Theodosius; Aristarchus's *On the Sizes and Distances of the Sun and Moon*; Hero of Alexandria's *Barulkos*; Archimedes' *On the Sphere and Cylinder*; Alexander of Aphrodisias's commentary of *On Generation and Corruption*; as well as a part of his commentary on the *Physics*. And last but not least, Hypsicles's Books XIV and XV, additions to Euclid's *Elements*. Qusṭā must therefore have translated mainly books on mathematics and philosophy, fields to which he had contributed books of his own.

It is thus with all of the competence expected of the translator-scientist that Qusṭā ibn Lūqā took on the *Arithmetic* in the 860s. His translation stands out by its blatantly algebraic appearance. It is as if this translator-scientist took Diophantus to be al-Khwārizmī's successor and to speak the same language. Indeed he drew on al-Khwārizmī's lexicon to translate into Arabic both mathematical entities and operations. This lexical choice reflects Ibn Lūqā's interpretative bias: the *Arithmetic* is a work of algebra. This bias will have a very long life; it reappears in the work of Thomas Heath[72] and can still be found today.

Ibn Lūqā's choice transpires already in his translation of the title of the work. Instead of *Arithmetical Problems*, προβλήματα ἀρίθμητικά, which can be found in the colophons of some books (*al-Masā'il al-'adadiyya*), he translates the title as *The Art of Algebra*, *Fī ṣina'at al-jabr*. The fundamental terms are also translated by those that algebraists used, despite the irreducible semantic difference separating them. Take, for example, the expression ἄλογος ἀριθμός, a key concept in Diophantus's arithmetical theory, which designates the provisionally indeterminate number that will necessarily be indeterminate at the end of the solution. This concept puzzles such modern translators as Ver Eecke, who renders it in French as *arithme*. Ibn Lūqā translated it with the word 'thing' (*cosa*, *res*), that is, 'the unknown' of the algebraists. The successive powers of this entity δύναμις, κύβος, etc., are also rendered by the algebraic terms *māl* (square), *ka'b* (cube), etc. What is more, Ibn Lūqā translates the term πλευρά by the Arabic *jidhr*, 'root of the square', thereby distinguishing Diophantus's usage from his own.

The operations have also been made algebraic. Thus, Diophantus formulates the first operation with the phrase προσθεῖναι τὰ λείποντα εἴδη ἐν ἀμφοτέροις τοῖς μέρεσιν, 'to add the *species* subtracted on one side and the other from the two members'. Ibn Lūqā translated the phrase by a single noun: *al-jabr*, the very word from which the discipline takes its

[72] See T. L. Heath, *Diophantus of Alexandria: A Study in the History of Greek Algebra*, Cambridge, 1885.

name. Likewise, when Diophantus writes: ἀ φελεῖν τὰ ὅμοια ἀπὸ τῶν ὁμοίων, 'to subtract the similar from the similar', Ibn Lūqā also translates this formula by a single word, which is used by algebraists to describe this operation: *al-muqābala*. Further investigation leads one to the conclusion that this algebraic choice was deliberate and systematic.

This choice could not, however, cover all of Diophantus's vocabulary: Ibn Lūqā also had to invent new terms and expressions, if only to translate the very terms that Diophantus used to designate certain methods of resolution. He thus coined his own terminology when he translated ἡ διπλῆ ἰσότης, a concept of which Diophantus was particularly fond, by *al-musāwāt al-muthanna*, the 'double equality', replicating with this expression the Greek mathematician's own semantic procedure. Finally, to translate the originally philosophical expressions that Diophantus used, such as γένος, εἶδος, οἰ κεῖον, φύσις, μέθοδος. Ibn Lūqā, who was himself a translator of minor philosophical treatises, borrows from philosophy its already-canonical lexicon.

Diophantus's *Arithmetic* was therefore translated in light of al-Khwārizmī's *Algebra*. This translation clearly stands out by comparison with the writings on *Burning Mirrors* and *Optics*; it is also distinct from the translations of Euclid's *Elements* or Ptolemy's *Almagest*. The remaining question is to discover what were the reasons for this translation, and what were the reasons that motivated the translator's choice. We will then perhaps be in a better position to understand the transmission of this portion of the Greek patrimony.

To answer this question, we need to examine the subsequent life of this translation. In Arabic, the first research on indeterminate analysis (nowadays called Diophantine analysis) in all likelihood began immediately after al-Khwārizmī. As we have already pointed out, al-Khwārizmī takes on some indeterminate problems in the last part of his *Algebra*. Nothing, however, indicates that he is interested in indeterminate equations for their own sake; in any case, indeterminate analysis does not appear in his work. That said, the place of this analysis in Abū Kāmil's later book (*c*. 870) tells a different story. Its high level of understanding, the allusions to other mathematicians working in this field after al-Khwārizmī and whose writings are now lost, and finally the references to their own terminologies leave no room for doubt: Abū Kāmil was neither the first nor the only successor of al-Khwārizmī to work actively on indeterminate equations. It was therefore during one half-century that the context attuned to interest in Diophantus's *Arithmetic* took shape. Conversely, when read in light of the new algebra, which was precisely Ibn Lūqā's reading, the *Arithmetic* immediately found its place among the work in progress on indeterminate analysis. It even went so far as to give a genuine thrust to the development of this chapter, which will later get its own proper name: *fī al-istiqrāʾ*. One can also see that the impact of the *Arithmetic* on the Arab algebraists was more of an extension than an innovation.

3. *Translation as a vehicle of research: the Apollonius project*

We have just examined two translation types: translations contemporaneous with, and in the same domain as, research; and translation that follows research by a certain interval and eventually integrates the translated work into a tradition that was initially different from it. There were also three styles of translation – by an amateur, a professional, or a scientist translator, with the latter gradually predominating as the century advances. But these types and these styles were not the only ones: translation is sometimes stimulated not by one single research activity, but by a whole variety of them, some of which do not properly pertain to the field of the translated work. In such a case, the work is translated in order to carry out the research relevant to it, as well as to the research of other disciplines that are either fully established or in the process of formation. The translation of Apollonius's *Conics* is a perfect example of such a pattern.

We should not forget that the study of conic sections represents the most advanced part of Hellenic research in geometry. Apollonius's *Conics* had been considered the most difficult mathematical work from the patrimony of antiquity. It includes the sum total of the knowledge of conic curves produced by geometry after Euclid, Aristaeus the Elder, etc., which Apollonius magisterially enriched, in the last three books in particular. His treatise will remain the most complete on this subject, at least until the 18th century. Although composed of eight books, only seven survive. Indeed the eighth disappeared relatively early, perhaps before Pappus[73] in the 4th century. The first seven all survived in Arabic translation. The extant Greek text, consisting of only the first four books, is the one Eutocius edited in the 6th century.

Recall also that, beginning in the second half of the 9th century, mathematicians are treating problems that require an appeal to conic sections: some problems were raised in astronomy and optics (parabolic, ellipsoidal, and conical mirrors); the determination of the areas and volumes of curved surfaces and solids, etc. To appreciate this, one need only skim the titles of the works by al-Kindī, al-Marwarūdhī, al-Farghānī and the Banū Mūsā.

[73] R. Rashed, *Les mathématiques infinitésimales*, vol. III, Chap. I.

Thus al-Farghānī drew on conic sections in order to provide the first demonstrative exposition of the theory of stereographical projections, which the astrolabe requires. Even more significant is a research direction that begins to dawn during this time and that will grow ever stronger as the century progresses: beginning with the Banū Mūsā, researchers are interested *simultaneously* in the geometry of the conic sections and in the measurement of curved surfaces and volumes. So it was that al-Ḥasan, the youngest of the three Banū Mūsā brothers, wrote a treatise of enormous importance on the generation of elliptical sections and the measurement of their areas.[74] Al-Ḥasan then conceived of a theory of the ellipse and of elliptical sections using the bifocal method, that is, a path different from Apollonius's. He thus considers the properties of the ellipse as a plane section of a cylinder, as well as the different kinds of elliptical sections. His own brothers report that al-Ḥasan wrote his book without truly knowing Apollonius's *Conics*. He only had a faulty copy, which he could neither translate nor understand. If proof were needed, the path that he took clearly substantiates this testimony.

It is therefore easy to understand the interest that the *Conics* generated; not only did the desire to see the work translated come from several directions, but the need to study this chapter of geometry also became an urgent priority. The Banū Mūsā thus set out to find a translatable copy of Apollonius's work. It was only after al-Ḥasan's death that his brother Aḥmad found in Damascus a copy of Eutocius's edition of the first four books. This key would unlock the eventual translation of the seven books. Such a task exceeded the skills of an ordinary translator. It was a team that tackled it until a scientist-translator eventually took it up. It was Thābit ibn Qurra who had translated the three last books, which are the most difficult, and, according to Apollonius, the most original. It was very probably also Thābit who collaborated with the two living Banū Mūsā brothers – Aḥmad and Muḥammad – on the revisions and translation of the whole work.

Without a doubt, this was teamwork, since such other translators as Hilāl ibn Hilāl al-Ḥimṣī also participated in the translation.[75] As noted above, however, this undertaking remains the product of scientist-translators: Thābit ibn Qurra and the Banū Mūsā, acting at the very least as revisors. Overseen by creative scientists of the highest order, this translation cannot be confused with either an ordinary translation or one by a translator-scientist. To be sure, like the latter, this translation of Apollonius renders into Arabic a Greek work that was perfectly understood and mas-

[74] *Ibid.*, vol. I.
[75] *Ibid.*, vol. III.

tered. In addition, however, it takes on a clear heuristic value. The translation of the scientist-translators constitutes genuine means for discovery and the reconfiguration of knowledge. If it plays this role, it is because, of all translations, it is the one most intimately connected with research. To illustrate this new role, let us return to Thābit ibn Qurra, and begin with his book *On the Sections of the Cylinder and its Lateral Surface*.[76]

Having access to the seven translated books of the *Conics*, Thābit in a sense goes over the book of al-Ḥasan ibn Mūsā we have just mentioned. Apollonius's *Conics* thus serves as a model for a new theory of the cylinder and of its plane sections. His master's book will give him means that he will further develop himself, namely, geometric projections and transformations. Indeed, Thābit ibn Qurra takes the first step in this direction by treating the area of the cylinder as a conical surface and the cylinder as a cone whose summit is cast off to infinity in a given direction. He begins by defining the cylindrical surface, then the cylinder, just as Apollonius in the *Conics* had first defined the conical surface, then the cone. Thābit also follows the order of Apollonius's definitions: axis, generator, base, right or oblique cylinder. The analogy of his procedure is further confirmed when one examines the first propositions of Thābit's book.[77] The *Conics* serves as his model for developing his new theory of the cylinder; and it is to meet the needs of the latter that he develops the study of geometric transformations.

Reliance on the translation of the *Conics* is thus built into the research that al-Ḥasan ibn Mūsā and his student Thābit ibn Qurra carry out. But, as we have said, this is not the only line of research. Thābit and his contemporaries also take up geometrical constructions by means of conic sections, namely the two means and the trisection of the angle. Astronomer-mathematicians, such as al-Farghānī, also draw on conic sections to study projections, in order to give a rigorous foundation to the plates of the astrolabe.

One might easily think that the translation of the *Conics* constitutes a unique example on account of its high geometrical level. Important as its sophistication is, this is not the main point. After all, Thābit ibn Qurra also translated Nicomachus of Gerasa's *Introduction to Arithmetic*, a lower-level neo-Pythagorean book of arithmetic.[78] Here, too, all signs indicate that this translation also fit into the research of this scientist-translator. Starting from one of Nicomachus's descriptive statements, Thābit elabo-

[76] *Ibid.*, vol. I, pp. 458–673.
[77] *Ibid.*, vol. I.
[78] Ed. W. Kutsch, Beirut, 1958.

rates the first theory of amicable numbers along with his famous theorem.[79] Thābit establishes his new theory based not only on Nicomachus, but also on the arithmetic books of Euclid's *Elements*. Such a research program could not have taken place without an extensive scientific culture; indeed, the latter was growing ever richer and more extensive as the scientific community and its institutions grew. One of the means for this growth was precisely translation.

The translation activities of the Greek patrimony were growing not only in extent but also – thanks to research – in understanding. Thus the criteria of good translation developed continuously, which explains the equally massive movement of retranslation and of revision focused on the translated works. In other words, retranslating and revising became two of the hallmarks of the translation movement that turned the Greek patrimony into Arabic. Thus Euclid's *Elements* was translated three times, and the last version was also revised. Likewise for the *Almagest* and for certain works by Archimedes, some works in optics, etc. Revision became a kind of standard, beginning from the time when al-Kindī revised some translations done by Qusṭā ibn Lūqā, and Thābit ibn Qurra those of Isḥāq ibn Ḥunayn.

4. *Ancient evidence of the translation-research dialectic: the case of the Almagest*

In spite of its highly technical character, the translation of the *Conics* fairly reflects the general state of affairs. This example illustrates concretely the reasons that underlie the act of translation, those that stimulated retranslation, and finally those that initiated revisions of the translations. If we overlook a few differences attributable to the very nature of the discipline and its objects, as well as to the high standard of proof required in it, the situation is analogous to that found not only in the other mathematical sciences, but also in alchemy or in medicine. The case is similar for astronomy, for example, and so is the context for the translation of the most important work of ancient astronomy – the *Almagest*. Ibn al-Ṣalāḥ, the learned scholar of the 12th century, gives us an invaluable account when he writes:

> There were five versions of the *Almagest*, in various languages and translations: a Syriac version that had been translated from the Greek; a second ver-

[79] F. Woepcke, 'Notice sur une théorie ajoutée par Thābit ben Qorrah à l'arithmétique speculative des grecs', *Journal Asiatique* IV, 2, 1852, pp. 420–29.

sion that al-Ḥasan b. Quraysh translated from Greek into Arabic for al-Ma'mūn; a third version that al-Ḥajjāj b. Yūsuf b. Maṭar and Halyā b. Sarjūn translated from Greek into Arabic, also for al-Ma'mūn; a fourth version from Greek to Arabic by Isḥāq b. Ḥunayn, for Abū al-Ṣaqr b. Bulbul – we have Isḥāq's original autograph; and a fifth version that is Thābit ibn Qurra's revision of Isḥāq b. Ḥunayn's translation.[80]

In approximately one half-century, then, there appear at least three translations of the *Almagest*, plus one revision by one of the most prestigious mathematicians and astronomers of the period. With a few exceptions, the Greek astronomical library was translated into Arabic during the 9th century. Equally significant is the fact that two translations of the *Almagest* were produced in the two decades of al-Ma'mūn's reign. The only way to understand this remarkable fact is to take research into consideration. We owe an account of the latter to Ḥabash al-Ḥāsib, an eminent astronomer of the day.

Ḥabash begins by describing the state of astronomical research before al-Ma'mūn. He mentions that some astronomers established principles and claimed to have reached a high level of scientific knowledge about the Sun, the Moon, and the stars, but without having 'proposed about this a clear demonstration of it and a true deduction'.[81] Ḥabash says nothing about the identity of these astronomers or their works. This, in his view, is how matters stood until al-Ma'mūn. Thereafter, astronomers set out to verify and to compare the various astronomical tables already translated into Arabic, namely the Indian astronomic table (*zīj al-Sindhind*), the astronomical table of Brahmagupta (*zīj al-Arkand*), the Persian astronomical table (*zīj al-shāh*), the 'Greek cannon', that is, Ptolemy's *Handy Tables*, and 'other *zījs*'. This verification of the results from the various astronomical tables led to the following result: 'each one of these *zījs* is sometimes correct and sometimes strays from the path of truth'.[82]

Who took on the first step in this research? Ḥabash does not say. We know, however, that this activity had begun much earlier, with al-Fazarī and Ya'qūb ibn Ṭāriq, and that at the time of al-Ma'mūn the names are plentiful. However that may be, it was after this verification and in the wake of this negative result that al-Ma'mūn ordered Yaḥyā ibn Abī Manṣūr

[80] R. Morelon, 'Eastern Arabic Astronomy between the Eighth and the Eleventh Centuries', in R. Rashed (ed.), *Encyclopedia of the History of Arabic Science*, vol. 2, p. 22 (Arabic text p. 155, 12–18 of P. Kunitzsch's edition: Ibn al-Ṣalāḥ, *Zur Kritik der Koordinatenüberlieferung im Sternkatalog des Almagest*, Göttingen, 1975).

[81] Ḥabash al-Ḥāsib, *al-Zīj al-Dimashqī*, ms. Berlin 5750, fol. 70r.

[82] *Ibid*., fol. 70r.

al-Ḥāsib to 'return to the fundamentals of astronomical tables and to gather the astronomers and scientists of his time to collaborate on research about the foundations of this science, with the intention of correcting it, since Ptolemy proved that it is possible to grasp what astronomers try to know'.[83] The mathematician and astronomer Yaḥyā ibn Abī Manṣūr al-Ḥāsib followed al-Ma'mūn's orders. He and his colleagues chose the *Almagest* as their foundational book, and in Baghdad made observations of the movements of the Sun and the Moon at various times. After Yaḥyā ibn Abī Manṣūr's death, al-Ma'mūn directed another astronomer, Khālid ibn 'Abd al-Malik al-Marwarūdhī, this time in Damascus, to undertake the first historically recorded continuous year-long observation of the movements of the Sun and the Moon.[84] During this period of active research in astronomy, Ptolemy's *Almagest* was translated twice. A similar analysis pertains to the other disciplines linked with astronomy, such as research on sundials and the translation of Diodorus's *Analemma*, or spherical geometry, and the translation of Menelaus's famous work.

This analysis makes it possible for us not only to understand the various phases of the translation movement, but also to glimpse its end, which occurs when the results and methods of the new research overtakes the inherited science. This does not happen at the same moment in every discipline but, for a good number of them, the translation movement ends at the turn of the century.

3. PROSPECTIVE CONCLUSION

Despite the fundamental historical research undertaken in this domain,[85] the question of the transmission of the Greek philosophical patrimony into Arabic remains unanswered. With a few exceptions, the work to date has taken place under one of the following rubrics: philological, philologico-archaeological, or historical. Philological research is devoted to the problems of lexicon and syntax that Arabic translation raises. Philologico-archaeological study tries to identify real or virtual Greek

[83] *Ibid.*, fol. 70^{r}.

[84] *Ibid.*, fol.70^{v}.

[85] See M. Steinschneider, *Die Arabischen Übersetzungen aus dem Griechischen*, Graz, 1889; 'A. Badawi, *Al-Turāth al-yūnānī fī al-ḥaḍāra al-islāmiyya*, Cairo, 1946; *id.*, *La Transmission de la philosophie grecque au monde arabe*, Paris, 1968; R. Walzer, 'Arabische Übersetzungen aus dem Griechischen', *Miscellanea Medievalia*, 9, 1962, pp. 179–195.

works behind the Arabic text, based on the postulate that the diffusion of a word reflects that of the concept. For its part, historical research examines the impact of the translated text on the philosophers of classical Islam. Although philology and history are necessary, the studies of translation that belong to these disciplines do not absolve the scholar from trying to discover either what was at stake in the translation, or the reasons for it, or the translator's choices. These tasks will apparently require us to look beyond the 'first philosopher among the Arabs', al-Kindī, in order to encounter the theologian-philosophers (*al-mutakallimūn)* who were his predecessors or his contemporaries. As the examples of, among many others, Abū Sahl ibn Nawbahkt, Abū al-Hudhayl and al-Naẓẓām[86] show, the *mutakallimūn* were customers, albeit highly critical ones, of metaphysics, physics, biology, and logic in order to develop their own deliberately rational discourse. In the writings of the Aristotelian neo-Platonic tradition, al-Kindī himself found the discipline that not only provides a foundation of a rational discourse acceptable to all, but also is susceptible to argumentation of a mathematical sort.[87] In all likelihood, the context of the theologian-philosophers will someday give us the key to understanding the reasons motivating the first steps of this massive movement of philosophical translation, and the choice that, in this context, privileged the writings from the Aristotelian tradition of Neoplatonism.

[86] M. A. Abū Rīda, *Ibrāhīm b. Sayyār al-Naẓẓām wa-arā'uhu al-kalāmiyya al-falsafiyya*, Cairo, 1946; R. M. Frank, 'The Science of Kalām'; J. van Ess, *Theologie und Gesellschaft im 2. und 3. Jahrhundert Hidschra*.

[87] R. Rashed, 'Al-Kindī's commentary on Archimedes' *The Measurement of the Circle*'.

– 3 –

READING ANCIENT MATHEMATICAL TEXTS: THE FIFTH BOOK OF APOLLONIUS'S *CONICS*

How ought one read an ancient mathematical work? What are the means best suited to interpreting it? Both historians and philosophers of mathematics have constantly grappled with these questions. Even a historian who is so indifferent to mathematical facts that he treats them as he would a painting or a theological text (*i.e.*, as a sociological fact) cannot avoid this question. At the very least, he will have to classify mathematical works, if not organize them into a hierarchy.

The answers to these questions are nevertheless far from being immediate and simple: one need only recall the debates and controversies that opposed various historians when dealing with these queries. In order to circumscribe the problem, let us distinguish two tasks that are implicit in every interpretation of a mathematical work. The first is also the duty of every historian who reads a philosophical work: the need to bring out the structure both of its networks of signification and of the argumentation even as one draws out the author's intuitions. It is by examining the articulation of these structures that the historian can reconstruct the work and insert it into the tradition(s) to which it belongs. But whereas the historian of philosophy can restrict himself to this interpretative task without passing judgment on the truth of its doctrinal elements, matters are completely different for the historian of mathematics. Indeed, everyone knows that it is in examining the truth of its mathematical facts (theories and theorems) and the validity of its demonstrations that one truly reads the work. But this task runs into many difficulties when the work that one reads is ancient, that is, the product – sometimes preserved in a dead language – of a mathematical rationality that is no longer our own, and of a society and culture that have long ago disappeared, such as those of ancient Egypt, ancient Mesopotamia, ancient Greece, and the like. It is true that, when reading, the historian will take advantage of certain characteristics that make a mathematical text distinctive. Indeed, whatever its nature, this text is translatable into other mathematical languages.

This possibility is itself a consequence of an even more fundamental property: once established or proven, a mathematical theory or theorem remains such for all time and everywhere. In no instancc has a theorem been rejected after having been proven. This same theorem can thus be stated in languages other than the one in which it was originally formulated. Now this possibility of translation is at once theoretical and historical.

Indeed, theoretically, one can state the same fact in several languages. Thus the plane in hyperbolic geometry can be defined axiomatically, as it is in Lobatchevsky, and the latter's plane geometry was thus elaborated as in Euclid's *Elements*. One can also consider a portion of a plane in hyperbolic geometry as the surface of a pseudosphere, where the geodesic curves play the role of straight lines. One can also take the interior of a circle as a plane in hyperbolic geometry. This possibility of translation is moreover at the origin of the concept of model in logic.

But it often happens that these multiple translations are historical readings in which one can, moreover, see one of the main vectors in the development of mathematics. One takes up again the ancient mathematical facts in another language, in a *mathesis* that is different from theirs. So it was that the mathematicians of the 10th and 11th centuries, notably Alhazen, then those of the 17th century, such as Fermat, read certain works of Archimedes, and later this same work of Archimedes was translated into the language of integral sums. One can also call to mind Diophantus's *Arithmetic* read in the language of classical algebra and, more recently, the works of Euler and Lagrange on the theory of quadratic equations which were later rethought by Kummer, Dedekind, and Kronecker in the language of field of algebraic numbers.

There are many examples of this theoretical and historical plurality of readings, each of which finds a new richness in the mathematical object. For this reason, the historian of mathematics finds himself in a slightly paradoxical situation, one in which the stability of the mathematical fact is opposed to the variety of the *matheseis* into which this very fact is integrated. To return to the reading of Archimedes by his successors, the organization of his ontology is not the same in Archimedes or in Alhazen or in Fermat, no more than are the methods, the languages, and the power of extension of mathematical thought.

Take a simple example, that of Diophantus's famous problem (II.8), 'To divide a given square into two squares'.[1]

[1] Here is the text: 'To divide 16 into two squares. Let us assume that the first square is the square of *arithmos* (x^2). The second square will be then 16 units minus a

(*Cont. on next page*)

Following the invention of algebra, Arab mathematicians read this problem as an indeterminate second-degree equation of two variables. $x^2 + y^2 = a^2$. Others, who had developed the entire Diophantine analysis, saw in it an arithmetic problem – a numerical right triangle. One can also read it as a problem of rational parametrization of the circle ($x = ut$; $y = ut - a$; from which $x = \frac{2au}{1+u^2}$ and $y = a \cdot \frac{u^2-1}{1+u^2}$; the point $(0, -a)$ is rational). It is with respect to this problem that Fermat in the margin formulated the impossibility of decomposing an nth power into a sum of two n^{th} powers when $n \geq 3$. This remark is at the origin of Fermat's famous last theorem, which was only proven in 1994.

The mathematical fact is true regardless of the reading of it, but the *mathesis* is different each time.

The questions thus read: Can the historian indifferently choose any given reading? Or is there one reading which, being better than the others, allows one to situate a mathematical work in history? Or, finally, is it necessary to multiply the readings? And if so, which ones?

The most common temptation is to read the work in light of the works of the author's predecessors. This is precisely what happened when one tried to read Diophantus's *Arithmetic* in the language of logistics and of the arithmetic of his predecessors; or to read Descartes's *Géométrie* only in the language of the late-medieval cossists or of Christopher Clavius. But this unique reading runs the risk of missing what the work contains by way of new forms, and what the successors will not cease to bring to light and enrich. Thus the history of a mathematical work is also the history of the later mathematicians' exploitation of it. Moreover, the historian who would restrict himself to examining the predecessors would run another risk, that of drifting into a search for origins that are frequently buried in limbo. The

---

(*Cont.*) square of *arithmos*. 16 units minus a square of *arithmos* must therefore equal a square. I make up the square of any indeterminate quantity of *arithmoi* minus a number of units equal to the root of 16 units. Let it be two *arithmoi* minus 4 units. The square will then be 4 squares of *arithmos* plus 16 units minus 16 *arithmoi*. This is equal to 16 units minus a square of *arithmos*. Add to both sides the quantity subtracted, and subtract the quantity of the same nature, then 5 squares of *arithmos* are equal to 16 *arithmoi*, and the indeterminate number will be $\frac{16}{5}$. The first square will be $\frac{256}{25}$, the second one $\frac{144}{25}$, and their sum $\frac{400}{25}$; that is, 16 units, and each of them is a square' (*Apollonii Pergaei quae graece exstant cum commentariis antiquis, edidit et latine interpretatus est I. L. Heiberg*, 2 vols, Leipzig, 1891, 1893; repr. Stuttgart, 1974).

discussion underway for almost two centuries on the origins of al-Khwārizmī's *Algebra* illustrates well the limits of this type of research.[2]

Another temptation can overcome historians, especially when their knowledge of the predecessors is uncertain or fragmentary, namely the temptation to comment on the statements of the author by drawing on other phrases borrowed from his own text. This type of commentary, which is necessarily limited, in fact runs the risk of becoming nothing more than a dull paraphrase of the mathematical transcription of the text. What is more, this reading is even more poorly equipped than the preceding one to uncover new truths masquerading in ancient garb. It is not rare for such readings, which tout their 'fidelity' to the text, to end up betraying its mathematical content. It will therefore be necessary to elaborate a genuine strategy for the reading of ancient works.

Let's remember, first of all, that a good number of these works suffered serious accidents during the course of their transmission, and that our knowledge of their authors and predecessors is poor and sketchy. Such is the case of Apollonius and his *Conics*, with Menelaus and his *Spherics*, with Diophantus and his *Arithmetic*, and of many other Alexandrian and Arabic mathematicians. It may also happen that one must wait for centuries and for the help of a different kind of mathematics in order to begin reading and exploiting the work. Only in the 9th century did one begin to read Diophantus. Apollonius had to wait until the 10th century for the reactivation of research on the geometry of the *Conics*, and so on.

It is therefore necessary, first of all, to begin by rigorously establishing the texts of the author and of his mathematical successors. At this stage, it is advisable to seek the help of another mathematics, from which one can borrow tools sufficient to actualizing all the mathematical information in the work one is reading. In other words, starting from another mathematics, it is a matter of elaborating a model that allows one to go farther in one's understanding of the text. It even happens that this model, sometimes conceived on the basis of recent mathematics, plays a revelatory role in unveiling methods that underlie the work in question. This model thus has an instrumental and heuristic role. Thus, to read Diophantus's *Arithmetic*, some have proposed a model forged from the concepts of algebraic geometry applied to the field of rational numbers. Such a model is on the surface ahistorical. In other cases, the mathematics of the model, in this case still different from that of the work one is reading, is nevertheless inscribed

[2] R. Rashed, *Al-Khwārizmī: Le commencement de l'algèbre*, Paris, Librairie A. Blanchard, 2007; English translation: *Al-Khwārizmī: The Beginnings of Algebra*, History of Science and Philosophy in Classical Islam, London, Saqi Books, 2009.

in the posterity of the latter. As we will see, this is the case for Apollonius's *Conics*, and for algebraic-analytic mathematics.

If the reliance on models for the interpretation of an ancient work seems indispensable to me, it is because the work evinces a blurry relationship of identity and difference with later mathematics, whether this link is theoretical or historical. It is a truism that this model is not the object. The model and the interpreted work pertain, as I have said, to two different *matheseis* or conceptual traditions.

But this instrumental and heuristic use of models runs the risk of a double displeasure. First of all it will displease those who do not distinguish the model from its objects. Indeed, some eminent mathematicians did not hesitate to find in Diophantus's *Arithmetic* not only algebra, but also the very notions of algebraic geometry and its methods (the methods of the chord and of the tangent). This attitude nevertheless has nothing to do with the procedure that consists in making a brutal regression, without any model at all, in order to discover in the ancient text concepts and procedures that required several more centuries to conceive them. This is the procedure that Jean Dieudonné followed when he wrote about Apollonius's *Conics*:

> [In the study of normals to conics], the evolutes of conics are completely characterized and studied. Apollonius's theorems translate immediately in our notation into the equation of the evolute that only the underdeveloped state of Greek algebra prevents him from writing.[3]

It is one thing to rely on a model developed on the basis of another mathematics. It is a very different thing to project its concepts and methods onto a work conceived in a different mathematics.

The recourse to models will also displease historians who are upset by the reflection of recent mathematical concepts on the tarnished mirrors of ancient times; in this, they will see an anachronistic procedure.

Note that the model is not unique. One can elaborate several of them, beginning from different mathematics. The *Arithmetic* of Diophantus, for example, can be accommodated to an algebraic model, an arithmetic model, a geometric model, even though Diophantus was no more an algebraist than a geometer. The same can be said of the fifth book of Apollonius's *Conics*, as we shall see. The whole problem is therefore to

[3] J. Dieudonné, *Cours de géométrie algébrique*, I: *Aperçu historique sur le développement de la géométrie algébrique*, Paris, PUF, 1974, p. 17; quoted from the English translation, *History of Algebraic Geometry: An Outline of the History of Development of Algebraic Geometry*, transl. J. D. Sally, New York, Chapman and Hall, 1985, p. 2.

find a model that is, so to speak, minimal, capable of gathering all the information contained in the text and of exhaustively explaining all of the mathematical facts found in it.

Finally, one must confront the model with the mathematics of the author and of his period in order to remove from it all notions extraneous to the context of the work. Thus, in the case of the *Arithmetic*, once one has removed the concepts of algebraic geometry, there remain a small number of algorithms (corresponding in particular to the method of the chord and the method of the tangent) which give an account of all the problems that the author considers. The model has thus allowed one to identify a small number of methods and to illuminate Diophantus's procedures, about which it has been claimed since Hankel that they were only the chance examination of a succession of problems.[4] It is precisely this confrontation that is the test of truth and that allows one to judge the pertinence of the model.

To illustrate briefly this historical research and this strategy, I will discuss the fifth book of Apollonius's *Conics*.

The fifth book is certainly one of the high points of ancient and classical mathematics. If one compares it to all the others books in Apollonius's treatise, it is without a doubt the most important and most difficult. Its difficulty is all the greater because Apollonius's analysis is missing. Whereas this book is fundamentally the most analytical of the seven books of the *Conics*, its style of composition is purely synthetic. Understandably, it is not easy to make a systematic commentary on it. Of course, such a commentary requires first a genuine critical edition as well as a rigorous translation, which I believe I have accomplished. One can then attempt a commentary that requires multiple angles of attack. The first is certainly not a reading of the contribution of Apollonius in those of his predecessors but only, and insofar as the documents allow it, a precise mapping of the questions that his predecessors raised and the way in which Apollonius picks up on them. One approaches the second reading by means of an algebraic-analytic model, the elaboration of which occurs a millennium after Apollonius and the development of which will continue for several more centuries. Although alien to the mathematics of the *Conics*, this algebraic-analytic mathematics nevertheless finds one of its historical roots in Apollonius's book. One can see here all of the complexity of the relations of identification and of difference. A third reading, using the theory of the singularities of differentiable applications, even if it is stripped of every

[4] H. Hankel, *Zur Geschichte der Mathematik in Altertum und Mittelalter*, 1st ed., Leipzig, 1874; reprod. Hildesheim, Georg Olms, 1965, pp. 164–5.

historical dimension, nevertheless helps one to appreciate all of the richness of the objects that the Alexandrian mathematician considered. Of course, within the framework of a brief study, one can only sketch the broad strokes of these readings. In this rapid sketch, I will pause on a single curve – the parabola.

What were Apollonius's intentions when he was writing the fifth book? What was his project? Apollonius's explanations are miserly. In the prologue to the first book of the *Conics*, there is just one small phrase in which he says that the Book V deals 'more fully with maxima and minima',[5] that is, with the extremal lines that one can draw from a given point to the points of the curve. To understand this allusion, our only source is the prologue to Book V.

In his cover letter when sending Book V to Attalus, Apollonius sketches a rapid chronology of the research that he hopes to undertake in it, and explains his own contribution. Regrettably, both the chronology and the explanation are very brief, indeed rather allusive. He writes first to Attalus:

> In this book are found propositions on maximum and minimum lines.[6]

Thus is the territory designated. He continues:

> You must know that our predecessors and contemporaries scarcely took up the investigation of minimal lines and they showed thereby which straight lines touch the sections, and also the the reciprocal of this proposition; that is, what happens to the straight lines that touch the sections, such that, if this happens, the straight lines are tangent.

From this historical reflection, we learn that the predecessors and contemporaries took an interest only in minimum lines with the unique goal of determining the tangents to conic sections. One can thus see that, for the ancients, this was a continuation of Euclid's study of tangents to the circle. Even though our information about the predecessors and contemporaries of Apollonius is incomplete and full of lacunae, one can see in it an allusion to the second book of Archimedes's *On Floating Bodies*, and to mathematicians in the tradition of Conon and of Dositheus, who treated the parabolic mirror, several of whom Diocles mentions. The latter indeed draws on two properties of the parabola that pertain to the tangent and to the normal:

[5] *Apollonius of Perga. Conics, Books I–III*, translation by R. R. Catesby Taliaferro, New Revised Edition, Sante Fe New Mexico, Dana Densmore Editor, 2000, p. 2.

[6] See our edition, translation, and commentary on Book V of *Coniques*, *Apollonius: Les Coniques*, Tome 3: *Livre V*, Berlin/New York, Walter de Gruyter, 2008, p. 223, 5.

the vertex of the parabola is the middle of the subtangent; the subnormal is equal to half of the *latus rectum*.[7] Apollonius himself takes up this study in Propositions 27 to 33, and informs us thereby about the type of research that his predecessors and contemporaries pursued. In fact, the point is to study the orthogonality of the minimum straight line ending at point *A* of the conic section, beginning from a point *B* in the concavity of the curve and from the tangent in *A*. Recall, for example, Proposition 27:

> The straight line drawn from the extremity of one of the minimum straight lines that we mentioned and which is tangent to the section, is perpendicular to the minimum straight line.[8]

Given a parabola of axis *BΓ*, the tangent to the extremity *A* of a minimum straight line is perpendicular to this straight line.

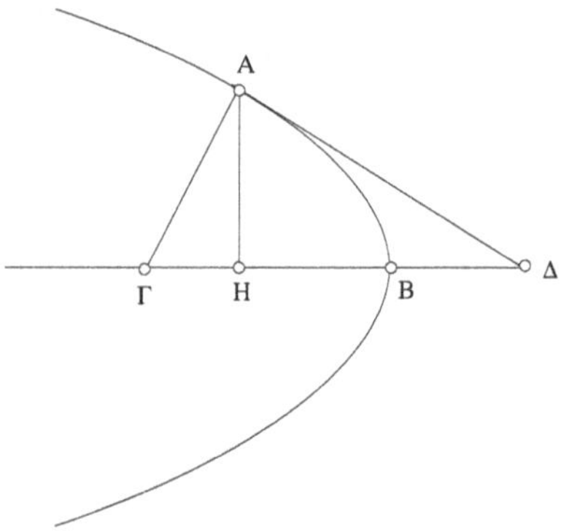

Fig. 1

If we can believe Apollonius, his predecessors and contemporaries had determined that the tangents were perpendicular to the minimum straight lines originating from the axis, a study that Apollonius, as we shall see, integrates into a larger field.

Let us listen to more of Apollonius.

> For our part, we have shown these things in the first book without using, in order to demonstrate this, what pertains to minimal lines; and we wanted to arrange matters such that they would be located near the place where we explained the generation of the three sections, in order to show thereby that, for each of the three sections, there can be an infinite number of these

[7] R. Rashed, *Les Catoptriciens grecs*. I: *Les miroirs ardents*, Collection des Universités de France, Paris, Les Belles Lettres, 2000, pp. 103 ff.

[8] See our edition, *Apollonius*: *Les Coniques*, Tome 3: *Livre V*, p. 313, 17–18.

> tangent straight lines, and on account of what happens and what is necessary to them, like that which happened for the first diameters.[9]

Apollonius thus explains that, if like his predecessors, he did not draw on the minimum straight lines when searching for tangents, it is because, unlike them, he wanted to elaborate a theory of tangents to conical curves as he had done for the diameters, and in relation to the diameters rather than the orthogonals. Thus, in order to give an account of their infinite number and of their necessary properties, it is not enough to study the tangents using the orthogonals; but one must proceed as one does for the diameters, by studying them on their own. Contrary to the studies carried out by Euclid in Book III of the *Elements* or by Archimedes in *On Spirals*, Apollonius's research is devoted not to the tangent to the curve – the circle or the spiral – but to the tangent to a class of curves, namely, conic sections. In this new study, one must moreover treat several themes of research other than tangent or orthogonal, such as (for example) tangent and ordinate, tangent and diameter, tangent and asymptote, different methods for determining the tangents, etc. Now, this unprecedented extension of both the domain and the themes seems to have required a more detailed elucidation of the concept of tangent as well as an elaboration of the theory that encompasses it. This task was all the more necessary because Apollonius was beginning deliberately to examine the properties of the tangent for an entire class of conic sections before returning to his study of a sub-class: central conics on the one hand, and parabolas on the other. It is in his first book, where he treats the generation of conic sections that, as we have shown,[10] Apollonius lays the foundation for this theory of tangents, to which he devotes a dozen propositions.

Having elaborated a theory of the tangent in which the minimum straight lines play no foundational role, Apollonius returns to the matter to undertake a systematic study of them. But this study very naturally requires a study of their alter-ego: the maximum straight lines. This is precisely the theme associated with Book V: extremal straight lines. Let's listen again to Apollonius writing to Attalus:

> As to the propositions in which we discussed the minimal lines, we have distinguished and isolated them, separately, after a lengthy inquiry; and we have gathered everything stated about them together with what was stated earlier about the maximal lines, because we have seen that those who study this science need this in order to know the determination and analysis of

[9] *Apollonius: Les Coniques*, Tome 3: *Livre V*, p. 223, 10–14.

[10] See R. Rashed, *Apollonius: Les Coniques*, Tome 1.1: *Livre I*, Berlin/New York, Walter de Gruyter, 2007.

problems, as well as their synthesis, in addition to what pertains to them in themselves: this is one of the things to which study aspires.[11]

The statements are transparent, and the goal is clear: the fifth book is a treatise devoted entirely to extremal lines, both for the intrinsic interest of these mathematical objects and for the utility they have in *diorismoi* and the analysis and synthesis of the problems. The main point is to study the distance from a given point on the plane to a variable point describing one conic section or the other. In particular, one should determine if there are solutions for each of the three conic sections and, if the latter exist, how many. Such is the goal of the fifth book. In the course of this study, however, a subgroup of propositions will appear that expressly concerns the normal: Propositions 27 to 33. In addition to this group, one constantly encounters this notion of orthogonal as soon as one examines the progression of the fifth book in a little more detail.

To grasp the place of the study of the normal in the fifth book and before undertaking a detailed commentary, let us recall that the study of the distance from a given point E in the plane of the conic section to a variable point M on the section, namely the distance $l = EM$, takes place in two phases. The three sections are related to their axis which, in the case of the ellipse, may be the large or the small axis. The vertex Γ of the section is taken as origin on this axis. Apollonius considers in general the points M on one half of the section separated from this axis.

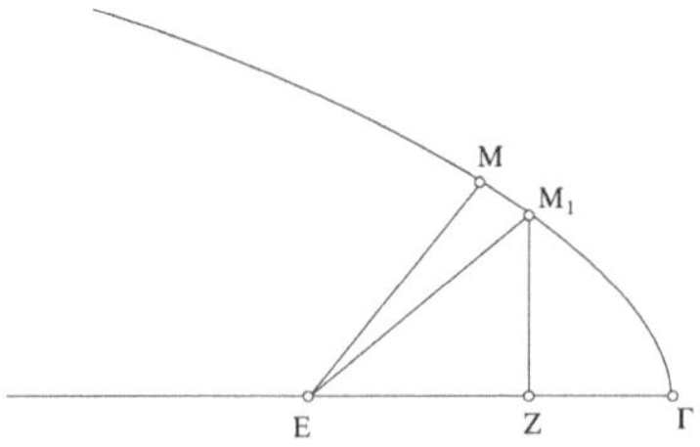

Fig. 2

I: *The point* E *is taken as a point on the axis of the section*, which is a straight half-line internal to the section in the case of the parabola or the hyperbola, and a line segment in the case of the ellipse. In Propositions 1 to 43, Apollonius studies the variation of l as a function of the abscissa of a variable point M, and shows the existence of a straight line EM_1 of a minimal length l_1 (if E is taken on the minor axis of the ellipse, it will be a

[11] *Apollonius*: *Les Coniques*, Tome 3: *Livre V*, pp. 223, 15–225, 3.

maximum straight line). Apollonius shows in particular that in every case a minimum (or maximum) straight line is *normal* to the section.

The exposition takes place in the following way:

1. The propositions of the first group (1–3) are lemmas.

2. Group (4–11): Apollonius shows that every point E of the axis is associated with a minimal distance $l = EM$, or a maximal distance if E is on the minor axis of the ellipse. For the three conics, Proposition 12 is a corollary of Proposition 7.

3. Group (13–25): he shows that every point M of the arc under consideration is associated with a *unique* point E on the axis, for which the distance l is minimal. In group (16 to 32) he shows that this distance is maximal.

4. Group (27–29): he shows that every straight line EM that yields a minimum (or maximum) distance is orthogonal to the tangent at point M; that is to say, that it is *normal* to the section.

5. Group (31–33): he shows reciprocally that with every point M on the arc he considers, the normal cuts the axis at E, and $l = EM$ is the minimal distance associated with point E (the distance is maximal if E is on the minor axis of the ellipse).

6. Group (35–37): he studies the angle that the normal makes with the axis.

7. Group (38–40): he studies the position of the point of intersection of the two minimum (or maximum) straight lines; in Proposition 38, for any conic section; in the two other propositions, for the ellipse.

8. Group (41–43): he studies the conditions under which a minimum straight line cuts the section again.

9. In the two remaining propositions (12 and 34), he offers some remarks on the distances.

Such is the structure of the first part of Book V, which consists of forty-three propositions. Nine propositions directly concern the normal.

II: *The point* E *is not taken on the axis*. This point E and the part of the section considered are on one side and the other of the axis. Apollonius then studies the straight lines passing through E and which are the supports of the minimum straight lines examined in I. He discusses in this part (Propositions 44 to 63) the existence and number of such straight lines. The central part consists of Propositions (51, 52) and (62, 63). The group (44, 49, 51, 58, 62) is devoted to the parabola, whereas group (45 to 48, 50, 52 to 57, 59, 63) treats the ellipse, and group (45 to 50, 52 to 61 and 63) the hyperbola.

III: The third group (64 to 77) is devoted to the study of the *variation* of the distance l when the point M describes the section under consideration; Apollonius brings into play the results of the proceeding discussions and notably Propositions 51 and 52.

The group (64, 67, 72 plus Lemma 68) is devoted to the parabola; the group (65, 67, 72 and the Lemma 69) treats the hyperbola, and the group (66, 73 to 75 and the Lemmas 70 and 71) of the ellipse.

The remaining Propositions 76 and 77 are particular cases.

In these groups one encounters, as in Propositions 73, 74, 75, results pertaining to the normals.

The fifth book is thus organized in these three parts. At this stage in the discussion, one could say that the examination of normals is naturally essential during the course of this study of the distances, but without being the focus all by itself; it represents an important part of this study without the latter being the goal. A more detailed description will show the meaning and consequences of this conclusion.

We now turn to the commentaries.

First reading:

From Apollonius's own remarks, it transpires, on the one hand, that his predecessors and contemporaries were raising questions about minima and maxima, and on the other hand, that it is precisely this reflection that he intends to take up in greater detail. Apollonius names neither his predecessors nor his contemporaries, and he does not expose the results of their research. We know, however, from other sources that two mathematical techniques for the examination of the problems of solids were flourishing at the time: on the one hand, intercalation; and on the other, the intersection of conics. Thus, in his treatise *On Spirals* (Propositions 5, 7, 8, 9, in particular), Archimedes reduces the most difficult propositions to intercalations. On the other hand, according to the testimony of Apollonius himself, we know that scholars in the circle of Conon of Alexandria used the intersection of conics to study the problems of solids. But we know also that some problems, such as the two means, were studied by means of both techniques. Now, many propositions of Book V can be reduced to *neusis*, under different forms. There, Apollonius treats the diorism of the intersection of the conic section with a circle with a given centre and a variable radius. He also gives there the intercalation of a given straight line directed to a given point between a conic and one of its axes. The topic seems to have been treated by his predecessors, if we can believe his prologue. One knows moreover that Archimedes, in Book II of *On Floating Bodies*, draws on normals to the parabola. We also just saw that the testimony of Apollonius, according to which his predecessors had determined that the tangents were perpendicular to the shortest line issuing from the axis, proceeding as it were as Descartes would do later, in the second book of his *Géométrie*.

The determination of the minimum straight lines issuing from the points of one of the axes of a conic section was therefore a well-known problem at the time of Apollonius. Not only does the mathematician take it up, but he considers a problem that is more general and that requires other means: the straight lines issuing from any given point in order to study how it reduces to the problem of determining a solid locus. Let us take up this problem.[12]

Since a conic is determined by its axis, its vertex and its *latus rectum*, one tries to draw normals to it from a fixed point P. One requires that P and the feet of these normals be in the opposite half-planes facing the axis of the conic. Let us restrict ourselves to the case of the parabola.

If PM is normal to parabola $\mathscr{P}$ and cuts the axis at Q, we know that sub-normal QZ (Z being the orthogonal projection of M on the axis) is equal to the semi-*latus rectum* p. The search for them resembles a *neusis*: to insert between the axis and $\mathscr{P}$, a straight line QM directed towards P and whose projection QZ is equal to a given straight line. The point P being in the inferior half-plane, one begins by determining the locus of points M of the superior half-plane such that, if PM cuts the axis at Q, the projection QZ from QM be equal to p.

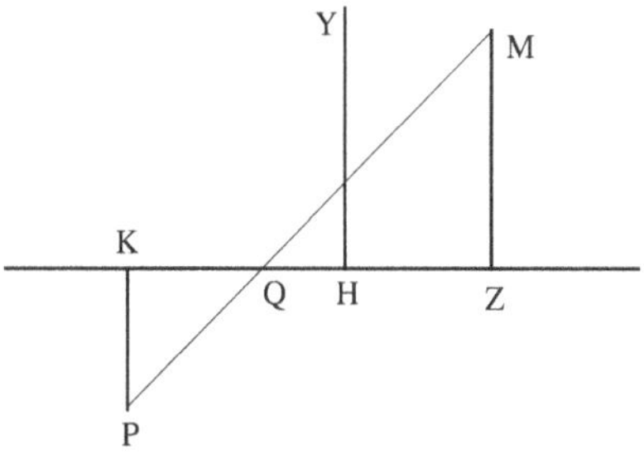

Fig. 3

[12] Cf. also H.-G. Zeuthen, *Histoire des mathématiques dans l'Antiquité et le Moyen Âge*, Paris, Gauthier-Villars, 1902, pp. 178–82 and J. Itard, 'L'angle de contingence chez Borelli: commentaire du livre V des *Coniques* d'Apollonius', *Archives internationales d'histoire des sciences*, no. 56–57, 1961; repr. in Jean Itard, *Essais d'histoire des mathématiques*, collected and introduced by R. Rashed, Paris, Librairie A. Blanchard, 1984, pp. 112–38, at pp. 118–24.

The triangles PQK (K being the projection of P on the axis) and MQZ are similar, therefore

$$\frac{QZ}{ZM}=\frac{KQ}{KP}=\frac{KZ}{ZM+KP}=\frac{KH+HZ}{ZM+KP},$$

where one has introduced the point H of line KZ such that $KH = p = QZ$, therefore $HZ = KQ$.

Thus

$$QZ\cdot(ZM+KP)=ZM\cdot(KH+HZ),$$

therefore

$$KH\cdot KP=QZ\cdot KP=ZM\cdot HZ$$

if one eliminates the equal expressions $QZ \cdot ZM = ZM \cdot KH$. This relation means that the point M belongs to the hyperbola $\mathscr{H}_P$ with asymptotes HZ, HY that passes through point P. More precisely, M is on the branch of $\mathscr{H}_P$ in the superior half-plane and P is on the other branch.

Thus the feet of the normals at $\mathscr{P}$ issuing from point P are the points of intersection of $\mathscr{P}$ with $\mathscr{H}_P$. Thus

$$\frac{HZ}{KP}=\frac{KH}{ZM},$$

therefore

$$\frac{HZ^2}{KP^2}=\frac{KH^2}{ZM^2}=\frac{KH^2}{2KH\cdot AZ}=\frac{KH}{2AZ}$$

where A is the vertex of the parabola and where one has taken into account the *symptoma* of the parabola $ZM^2 = 2p \cdot AZ$. Since the square KP^2 and the segment $\frac{KH}{2}$ are known, the determination of Z is pertinent to the lemma of Archimedes for Book II, Proposition 4 of *On the Sphere and Cylinder*: To divide the given straight line AH at point Z such that the ratio of the square of HZ to the given square KP^2 be equal to the ratio of the given segment $\frac{KH}{2}$ to AZ.

The diorism of Proposition 51 can easily be reconstructed. Given that point K is fixed, let us find a position of point P on the perpendicular KP to the axis such that hyperbola $\mathscr{H}_P$ is tangent to $\mathscr{P}$ at a point B that is projected on the axis at E. One knows that the tangent to $\mathscr{P}$ at B meets the axis at a point F such that $EA = AF$. If this straight line is also tangent to

$\mathcal{H}_P$, E is the middle of HF, therefore $HE = EF = 2EA$ and E is therefore located at one third of AH starting from A. This determines point E, and therefore point B and point G where PB meets the axis, since $GE = p$. Thus one finally has P at the point where BG meets KP. Apollonius determines $KP = \Lambda$ starting from BE by means of the proportion $\frac{\Lambda}{BE} = \frac{KG}{GE} = \frac{HE}{KH}$. One has

$$\frac{\Lambda^2}{HE^2} = \frac{EB^2}{KH^2} = \frac{2AE}{KH}$$

where $AE = \frac{1}{3}AH$ and $HE = \frac{2}{3}AH$; thus $\Lambda^2 \cdot KH = \frac{8}{27}AH^3$, which determines K as a function of AH.

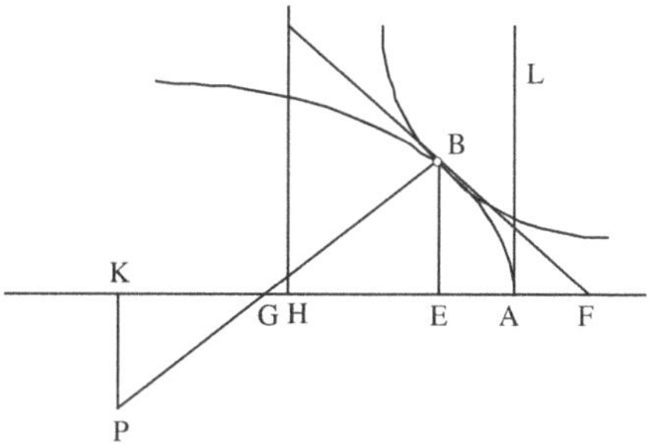

Fig. 4

Let us consider now a point P' of KP such that $KP' > \Lambda$. One thus has $KH \cdot KP' > KH \cdot KP$. As a consequence, for every point M of $\mathcal{H}_{P'}$, $ZM \cdot HZ > EB \cdot HE$. Thus the superior branch of the hyperbola $\mathcal{H}_{P'}$ is above that of $\mathcal{H}_P$ and this branch is therefore completely exterior to $\mathcal{P}$; there is therefore no normal originating from P' with a foot in the superior half-plane.

If on the contrary $KP' < \Lambda$, then $KH \cdot KP' < KH \cdot KP$ and $ZM \cdot HZ < EB \cdot HE$ for every point M of $\mathcal{H}_{P'}$. Consequently, the superior branch of $\mathcal{H}_{P'}$ is below that of $\mathcal{H}_P$ and it meets $\mathcal{P}$ at two points separated by B. There are therefore two normals originating in P' with feet in the superior half-plane.

Note therefore that the points of intersection of this hyperbola and of the given parabola will be the feet of the normals originating from $P(x_0, y_0)$.

The entire problem consists in determining the points P whose hyperbolae are tangent to the parabola. The locus of these points is a curve (a semi-cubic parabola) that divides the plane into two regions such that, from the points of one, two normals can be drawn, and from the points of the other, a single normal. It is while searching for the conditions under which these hyperbolae are tangent to the parabola that one determines the ordinate of a point of this curve, its abscissa being known.

We now know that this curve is the evolute of the parabola. But nothing allows us to say that Apollonius or anyone else before Huygens thought of this curve.

1. In this study, Apollonius is as close as can be to the definition of the evolute of the parabola $\mathscr{P}$, since to every abscissa x of a point P of the axis, he links a reference length Λ that is the ordinate of the point of the abscissa x on the evolute. It is nevertheless clear that he in no way considers this curve, and that in his treatise his consideration of the ordinates and the abscissas of points only has meaning for the points of a conic.

2. We have already noted that in the case of the parabola the construction of normals issuing from P reduces to Archimedes' problem of division. The commentary of Eutocius on *The Sphere and the Cylinder* contains a construction of this division, which Eutocius restored from a corrupt text that he attributes to Archimedes. This construction, applied to the problem that concerns us, determines the point Z as the projection on the axis AH of the intersection of a parabola of vertex A with axis AL perpendicular to AH and a right side $\frac{KP^2}{AH}$, with a hyperbola with asymptotes AH and AL passing through the point B such that HB is perpendicular to AH and equal to $\frac{KH}{2}$. These two curves are different from $\mathscr{P}$ and from $\mathscr{H}_P$. The hyperbola depends only on K whereas the parabola depends on the position of P on KP, $\mathscr{H}_P$ depends on the position of P, and $\mathscr{P}$ is fixed.

In the commentary of Eutocius, the diorism that allows one to determine under which conditions the two curves meet is also obtained by the determination of the position that guarantees the contact of the two curves. The hyperbola is fixed and one makes the parabola vary by changing the value of KP. As in the case of Apollonius, the properties of the tangents to the conics allow one to conclude that the contact occurs when on AH, Z is one third of the way from A. One also observes a kinship between the tradition that goes back to Archimedes and the research of Apollonius.

3. One can express the idea that underlies the diorism of Archimedes or of Apollonius by saying that the properties of intersection of a fixed conic $\mathscr{C}$ with a mobile conic $\mathscr{H}_P$ change only when $\mathscr{H}_P$ becomes tangent to $\mathscr{C}$, that is, when the transversality of the two curves is lost. One is close to the intuition according to which transversality is a stable property.

Let us add that the set of points P for which $\mathcal{H}_P$ and $\mathcal{C}$ are not transversal is the evolute of $\mathcal{C}$, therefore a *closed rare* set. This is a perfectly simple and elementary case of the famous transversality theorem of Thom.

4. Apollonius's study of the determination of extremal straight lines originating from a given point is presented in a completely static manner in the sense that he compares the lengths of segments according to different positions of the extremity on the conic. Apollonius does not yet consider at this stage the continuous variation of a straight line joining a fixed point of the axis to a mobile point on the conic. Starting from Proposition 64 of the same book, however, he studies the continuous variation of the distance of a point E on the plane to a variable point M on the conic. One cannot overstate the innovative character of this study in Hellenistic mathematics.[13]

In the 11th century, Ibn al-Haytham further develops research into continuous variation. Among other things, he studies the asymptotic behavior of quantities such as segments, but also ratios of segments or of arcs of circles, using infinitesimal notions; his astronomical concerns (the apparent motion of a planet on the celestial sphere) are probably not alien to this work.[14]

It is entirely plausible that Apollonius had, in one way or the other, carried out an analysis of this sort. Such a commentary is satisfactory since, on the one hand, it relies on no concept unknown to Apollonius and, on the other hand, takes into account his relations with the mathematicians of his day. Nevertheless, this commentary does not sufficiently clarify for us the link between the concepts elaborated by Apollonius and the mathematical rationality that suffuses them and with which Apollonius had to deal. If therefore one wants to grasp the true reasons for his research in the fifth book, and to elucidate all the mathematical facts present in it in order to understand what makes this book what it effectively is, it will be necessary to define this link and to follow its genesis. Now this explanatory task cannot be carried out properly in the author's own mathematics. It will therefore be necessary to draw upon a model elaborated on the basis of another mathematics, at the risk of having to return to the text in order to

[13] See our edition, translation and commentary of Book V of the *Conics*.

[14] R. Rashed, *Les mathématiques infinitésimales du IX*^{e} *au XI*^{e} *siècle*. Vol. V: *Ibn al-Haytham: Astronomie, géométrie sphérique et trigonométrie*, London, al-Furqān, 2006; English transl. *Astronomy and Spherical Geometry: The Novel Legacy of Ibn al-Haytham*, London, Centre for Arab Unity Studies, Routledge, 2014.

appreciate the power that this model has of exhausting the information that it carries. One can trace the source of the first model to an algebraic-analytical mathematics, stimulated by the reading of Apollonius's *Conics* by al-Khayyām, Sharaf al-Dīn al-Ṭūsī, Descartes, Fermat, etc.

Second reading:

One expects that this model will describe the evolution of research in the course of the fifth book, will illuminate the links between the different themes broached by the mathematician, and will draw out the reasons for the mathematical facts he establishes. It is in this manner that the themes emerge around which the book is organized: the extremal distance of a variable point of a conic curve to a given point on the plane, which can be on the axis or outside the axis; a theory of normals and a study of the variation of a geometrical quantity: the distance between the point given on the plane to the points on the curve. Now, if the study of minimum and maximum straight lines circumscribes the domain of research, the study of normals and of the variation of distances, stands out as an area of research that is as fertile as it is innovative. It is moreover with this research that Apollonius's study distinguishes itself from that of his predecessors and contemporaries. Let us consider quickly and partially the example of the parabola.

Apollonius begins by studying the length l – the distance – of a given point E (x_0, y_0) to a point M (x, y) that describes the parabola $\mathscr{P}$ related to the rectangular system $(\Gamma x, \Gamma y)$ formed by the axis and the tangent to the vertex with $y^2 = 2\,px$. He considers the following case:

• E is given on the axis $x_0 > 0, y_0 = 0$ and M on the semi-parabola with a positive ordinate; on the axis Γx, one takes the point Z such that $\Gamma Z = p$.

$$l^2 = EM^2 = (x - x_0)^2 + y^2 = x^2 - 2x\,x_0 + x_0^2 + 2px,$$

$$l^2 = f(x) = x^2 - 2x(x_0 - p) + x_0^2,$$

$$f'(x) = 2x - 2(x_0 - p).$$

The derivative $f'(x)$ is positive or zero if $x \geq x_0 - p$, which always occurs when $x_0 \leq p$, that is when point E is between Γ and Z. In this case $f(x)$ always increases and its minimum is obtained for $x = 0$, that is $M = \Gamma$. The minimal value l_0 of EM is then $E\Gamma = x_0$ (if $x_0 = p$, one obtains $l_0 = p$).

If on the contrary $x_0 > p$, that is, if E is beyond Z on the axis Γx of the parabola, $f'(x) < 0$ for $0 \leq x < x_0 - p$ and f decreases in this interval. In this case, $f(x)$ has a minimum for $x = x_0 - p$, of value:

$$x^2 - 2x(x_0 - p) + x_0^2 = x_0^2 - (x_0 - p)^2 = 2x_0 p - p^2$$

and the minimum value of EM is $EM_0 = l_0 = \sqrt{2x_0 p - p^2}$ where M_0 is the point of the parabola of abscissa $x_0 - p$. Its projection H_0 on the axis is such that $EH_0 = p$, and one therefore sees that EM_0 is normal to the parabola.

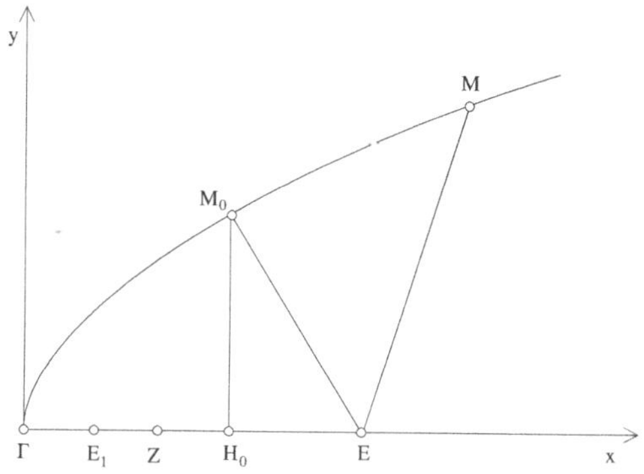

Fig. 5

For every other point M, one has

$$EM^2 = l^2 = l_0^2 + (x + p - x_0)^2$$

$$EM^2 = EM_0^2 + (x_M - x_{M_0})^2.$$

Conclusion: To every point E on the axis Γx with the abscissa $x_0 \geq p$ corresponds, on the semi-parabola with a positive ordinate, a point M_0 of abscissa $x_{M_0} = x_0 - p$ such that the length $EM_0 = l_0$ be the minimal length of all the segments EM.

Conversely, to every point M of abscissa x_M on the semi-parabola, there corresponds, on axis Γx, a point E of the abscissa $x_0 = x_M + p$. Thus, through every point M, there passes one and only one straight line on which the axis isolates a minimum straight line.

Next, Apollonius studies the angle that the minimum straight line makes with the axis.

In triangle MEH, angle $M\hat{H}E$ is right, therefore $M\hat{E}\Gamma < 90°$. Assuming $M\hat{E}H = \alpha$, one has $\tan\alpha = \frac{y_M}{p}$. When M describes the parabola, y_M increases from 0 to ∞, therefore $\tan\alpha$ increases from 0 to $+\infty$ and α increases from 0 to $\frac{\pi}{2}$.

Immediate consequence: Two minimum straight lines, issuing from two points M and M_1 with positive ordinates, intersect at a point O with a negative ordinate.

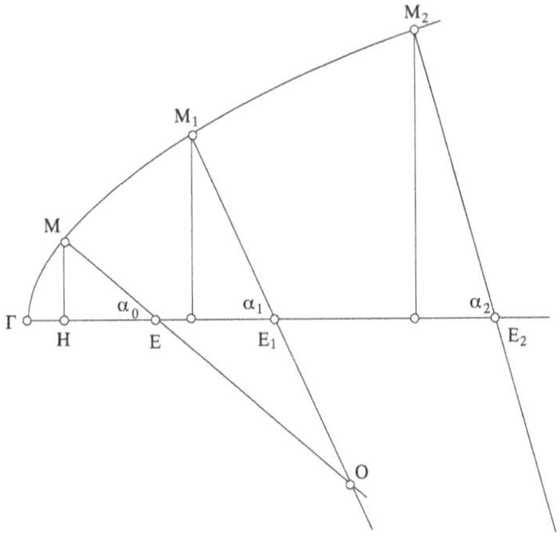

Fig. 6

The beginning of this commentary shows that, for Apollonius, the study of normals is intimately connected to that of minimum straight lines.

The study of the number of minimum straight lines that meet in a given point reduces to the study of normals to the parabola that pass through that point.

$$y^2 = 2px \Rightarrow 2yy' = 2p \Rightarrow y' = \frac{p}{y}.$$

$\frac{p}{y}$ is the slope of the tangent at the point of the ordinate y, therefore $-\frac{y}{p}$ is the slope of the normal.

Apollonius studies the normals that pass through O (x_0, y_0) and that meet the parabola at point (x, y). One has $y - y_0 = m(x - x_0)$, where $m = -\frac{x}{p}$ is the slope; whence $m^2p^2 = 2px$ (equation of the parabola) and $x = \frac{m^2p}{2}$; and, substituting in the first equation, one has

(*) $f(m) = pm^3 + 2m(p - x_0) + 2y_0 = 0$.

• If $y_0 = 0$, $m = 0$ is the solution for every value x_0; $m = 0$ gives the straight line $O\Gamma$ which is normal to the vertex Γ.

If $pm^2 + 2(p - x_0) = 0$, two cases emerge:

$x_0 \leq p$, in which case the only root of equation (*) is $m = 0$; the only normal is $O\Gamma$.

$x_0 > p$; $m^2 = \frac{2(x_0 - p)}{p}$, whose two opposite roots m' and m'' that yield two normals at two symmetrical points in relation to the axis Γx.

• If $y_0 \neq 0$, one can suppose that $y_0 < 0$; the equation (*) is written

(**) $f(m) = m^3 + 2m\frac{(p - x_0)}{p} + \frac{2y_0}{p} = 0$.

The study of the number of roots of the equation ($x > 0$, $y < 0$) is deduced from the sign of $27y_0^2 - \frac{8}{p}(x_0 - p)^3$.

One is thus led to study the equation $y^2 = \frac{8}{27p}(x - p)^3$.

Recall, however, that Apollonius had defined a 'reference length' k in two of the most important propositions of the fifth book, such as $k^2 = \frac{8(x_0 - p)^3}{27}$. Thus, we can now understand the origins of this 'reference length', which Apollonius gives without explanation.

The other particularly important theme is the study of the variation of the distance EM when M describes the semi-parabola. One demonstrates that, if the points M' and M'' correspond to two normals originating from point E, then the distance EM increases when M traces the arc $\Gamma M'$, decreases when M traces the arc $M'M''$ and increases indefinitely when M moves away indefinitely.

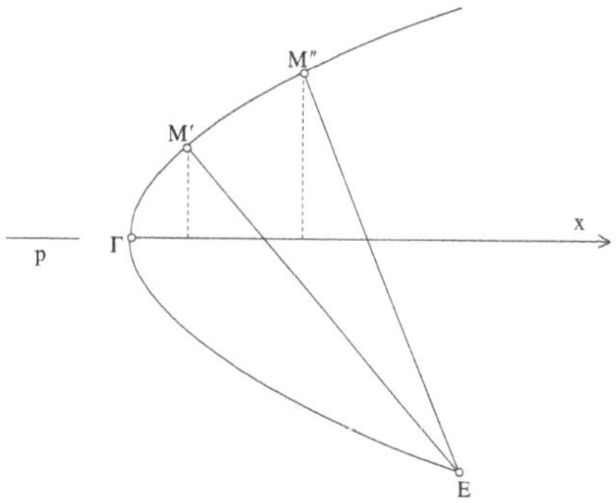

Fig. 7

This reading gathers all of the information contained in Book V and gives an account of all the mathematical facts established in it. Once one has eliminated all of the notions that are alien to Apollonius's geometry, it sheds light on all of the facts that he established.

Third reading:

One can also read the research of Apollonius with the help of singularities with differential applications. In this case, when the text of Apollonius is no longer one of the sources of this theory, this reading nevertheless allows one to uncover a potentiality of his research, one that could only be actualized in another mathematics. At issue is a theory of evolutes (envelopes of the normals to the parabola in this case), which allows one to unveil *a posteriori* the organization of the fifth book and to reveal the rational linkage of the propositions that compose it.

Given a parabola $\mathscr{P}$ of equation

(1) $$y^2 = 2\,px$$

and a point E on the plane, with coordinates (ξ, η); one seeks the extremal values of the distance $d\,(E, M)$ when M traces $\mathscr{P}$.

One can parametrize $\mathscr{P}$ in order to express this distance as a function of one variable; by assuming that $x = uy$ in equation (1), one finds

(2) $$x = 2\,pu^2, y = 2\,pu.$$

The square of the distance is written

(3) $$d(E,M)^2 = f(u;\xi,\eta) = 4p^2u^2(1+u)^2 - 4pu(\xi u + \eta) + \xi^2 + \eta^2.$$

The problem thus reduces to studying the values of u for which this function passes through one *extremum*, also called *critical values*; they are also given by zeroing the derivative

(4) $$f'(u;\xi,\eta) = 4p\left[4pu^3 - 2u(\xi - p) - \eta\right],$$

which goes to zero together with the polynomial (5).

(5) $$P(u;\xi,\eta) = 4pu^3 - 2u(\xi - p) - \eta = 0\,.$$

One notes that, with u fixed, the polynomial P is linear at (ξ, η); the equation $P(u;\xi,\eta) = 0$ thus defines a straight line N on the plane of E. This straight line passes through the point M defined by equation (2); its slope is equal to $-2u$. Since the slope of the tangent to $\mathcal{P}$ at M is $\frac{u}{2}$, one sees that N is one normal to $\mathcal{P}$ at M (this is the foundation of algebraic-analytic study).

Equation (5) is of the third degree, of the form $\alpha^3 + \alpha a + b = 0$. It thus admits one or three finite solutions according to whether the discriminant

$$\Delta(\xi,\eta) = 8(\xi - \eta)^3 - 27p\eta^2 \tag{6}$$

is positive or negative. The limiting case, $\Delta(\xi,\eta) = 0$ defines a curve $\mathcal{Q}$ with the equation.

$$(\xi - \eta)^3 = \frac{27p\eta^2}{8}. \tag{7}$$

This curve is a semi-cubic parabola, and it divides the plane into two regions: the interior, $\{(\xi,\eta) | \Delta(\xi,\eta) > 0\}$ and the exterior $\{(\xi,\eta) | \Delta(\xi,\eta) < 0\}$.

One shows that, when E is inside $\mathcal{Q}$, P has three finite roots; and when E is outside $\mathcal{Q}$, there is only one root.

The curve $\mathcal{Q}$ is decomposed into a regular part X and a cusp Z at which the two arcs terminate: $\mathcal{Q} = X \cup Z$.

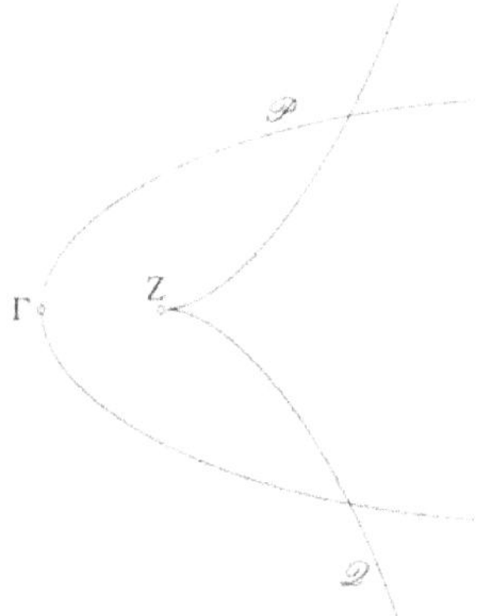

Fig. 8

Without going into more detail, let us merely recall that the point of view adopted here leads to an examination of the dependence of the solutions u of the equation $f'(u;\xi,\eta) = 0$ in relation to $(\xi, \eta) = E$. One knows, for example, that outside the curve $\mathcal{Q}$, these solutions are differential functions of (ξ, η) that depend regularly on E; in other words, the feet of the normals drawn from E depend regularly on E. Likewise, along the curve $\mathscr{Q}$, the double root u of P depends regularly on E.

The possible readings of an ancient mathematical work, and especially one that is as foundational as the *Conics*, are many, different, and in no way mutually exclusive. The first consists in trying to understand the work in the context of the research of its time, in relation to which, but also against which, it constituted itself. The job is neither easy nor definitive. Indeed scientific works are neither uniform, nor of a single piece, and those of ancient mathematicians also have their protrusions, their cleavages, and their stratifications. As ancient contributions, they are, moreover scarred by time and the vicissitudes of transmission and translation. Thus the reconstruction of such a work is never more than a temporary arrangement, governed by the criteria of the period. It is always perfectible, subject to progress in our knowledge of the facts, or to a better understanding of mathematical results and methods. The goal of the historian who undertakes such a reconstruction is twofold: to understand the intentions of the mathematical author; and to grasp the rational linkage of the concepts that he brought to bear in his work in order to grasp the mathematical reality at which he aims.

One can also – this is the second route – forge a model capable of bringing out the latent structure of a work as well as what it contains in potentiality and will be exploited by later mathematicians. It is true that, by taking this approach, one tries to examine the work intrinsically, outside of diachronicity. This time, one is not at all interested in what it owes to the works of predecessors and contemporaries. No doubt this approach will provoke some to denounce this route as well as the elaborated 'model' by accusing them of anachronism. But this would be to ignore the function – instrumental and heuristic – assigned to the model. And in fact, the example of Book V has shown us how the models have brought to light the themes of research – the theory of normals and the study of variation, among others – the networks of relations and the argumentative procedures. Moreover, these models are neither arbitrary nor exclusive, nor are they the only possible ones. All of them concern the same mathematical reality that Apollonius studied but, in every case, with a different recasting of the ontology. And each time new strata of meaning come to light in this mathematical reality that Apollonius struggled to grasp already two millennia ago. The contribution of Apollonius is thus in no way a first approximation of one of the models invoked; it is itself a model elaborated from a particular organization of the ontology.

To read an ancient text, the historian of mathematics has no choice but to mobilize all of these methods and to give up all pretences to a chimerical fidelity to it. He must convince himself once and for all that sole subjection to the words alone does not necessarily guarantee faithfulness to the concepts, and that the history of mathematics is always incomplete, always in need of reconstruction, always at the mercy of future mathematical advances.

– 4 –

THE FOUNDING ACTS AND MAIN CONTOURS OF ARABIC MATHEMATICS

The classical mathematics is neither homogeneous nor seamless. Some of its chapters go back as far as Greek mathematics; consider, for example, plane geometry, the geometry of conic sections, and spherical geometry. Other chapters are rooted in Arabic mathematics, as the algebraic disciplines and the study of geometrical transformations testify. Yet others develop in 17th century Europe, such as the infinitesimal calculus. Nevertheless, one can safely say that the distinctive characteristic of classical mathematics is that it is 'algebraic and analytic'.

The problem is to determine as exactly as possible when and how this distinctive trait revealed itself, that is, when this algebraic-analytic rationality was born and how it developed. Several founding acts were necessary for the emergence and constitution of this new rationality. I limit myself here to the founding acts of this rationality that one encounters in Arabic mathematics. Some others follow in Italian mathematics (the introduction of imaginary numbers), others with Viète and Descartes (the invention of a genuine symbolism), others yet in the same era that I call 'the liberation of the infinite'.

Recent historiography has made it possible to establish, definitively it seems, a fact, the reality of which one has always suspected: for at least five centuries, it was mainly in the lands of Islam and in Arabic that intense and fruitful mathematical research took place. Between the 8th and the 14th centuries, this research was carried out by personalities such al-Khwārizmī, Thābit ibn Qurra, Ibn al-Haytham, al-Khayyām, etc., but also by dynasties of scientists that constituted genuine teams and in the midst of schools of thought. Recall the dynasties of the Banū Mūsā, and the Banū Karnīb; the Niẓāmiyya of Baghdad, the schools of Marāgha, Samarqand, and many others. As one might expect, this high-level research was cumulative, diversified, and sometimes revolutionary. It was certainly cumulative: not only because it never stopped enriching the heritage of ancient mathematicians, mainly Greek and Hellenistic, with new results in all the domains that the latter cultivated; but also because it did not take long to organize itself into traditions to which every new generation continued to add its own discoveries to the acquisitions of predecessors. This high-level research was also diversified, for it witnessed the birth of many chapters unknown to the ancients, the elaboration of which remodelled both the organization and the extent of the encyclopedia of the mathematical sciences. Consider, for example, algebra, algebraic geometry, combinatory analysis, the whole of Diophantus's analysis, trigonometry, and many other fields. Finally, it was sometimes revolutionary, violating ancient prohibitions and inventing new procedures: treating irrational quantities arithmetically, changing the criteria of permissible geometric constructions, demonstrating geometrically algebraic algorithms as well as those of quadratic interpolations, explicitly introducing movement into geometry, etc.

To paint this scientific activity, or at least some of its chapters, is to write its history with all requisite precision, to show when it started, what were the conditions that made its beginning possible, what was its regime, what obstacles appeared on its journey, and when it stopped. This epistemological history, which tries to combine the history of concepts with the history of texts, is the only one that, in my view, allows us to answer the foregoing questions. I would like to pause briefly at the beginnings of this research activity, that is, at the ideas and the concepts that founded it before examining the renewal of these ideas and of these concepts, that is, the new beginning of this same activity.

My goal is not to rewrite the history of this research movement, but rather to try to describe it phenomenologically, as it were, to grasp what these beginnings signify, and what their reach was. This attempt will help us to situate this mathematical research in relation to both the ancients and the moderns, that is, in relation to Archimedes, Apollonius, Menelaus, etc., on the one hand, and Descartes, Fermat, Cavalieri, etc., on the other.

1. It has not been emphasized sufficiently that mathematical research in Arabic begins in a rather paradoxical way. Even as Euclid's *Elements* and Ptolemy's *Almagest* were being translated, Arabic mathematicians took advantage of the situation to consummate their first break with the Hellenistic tradition. In other words, even as they were falling in line behind Hellenistic mathematics, they were distancing themselves from it. It was al-Khwārizmī who brought about this first break while he was putting to use the translation of the *Elements* – al-Khwārizmī, whose colleague in the Academy of Baghdad, al-Ḥajjāj, was the translator of Euclid and of Ptolemy. This event occurred when al-Khwārizmī in the first third of the 9th century conceived of a new discipline: algebra. This was a founding act in

several respects – by what he assumes, by the nature of the object he proposes, by the language that he forges, and by the new mathematical possibilities that he generates.[1]

Let us not forget that in this first book of algebra al-Khwārizmī wants to reach a goal that is precise and clearly formulated as such: to elaborate a theory of equations soluble by roots on which one can then rely indifferently for arithmetic and geometrical problems, a theory that is thus applicable to calculation, to commercial exchanges, to inheritances, to land surveying, etc.[2]

Contrary to his predecessors and to his mathematical contemporaries in all languages who began by setting themselves problems in order to formulate them as equations, al-Khwārizmī starts with equations, the theory that allows one to obtain and classify them. As to the problems that can be solved with them, they arc arithmetic as well as geometrical and infinite in number. Thus, in the very first part of his book, al-Khwārizmī first defines in his theory the primitive terms which, because of the requirement of resolution by roots and because of his know-how in this area, could pertain only to equations of the first two degrees. These primitive terms are the unknown – the 'thing' (*al-shay'*; later *cosa*) –, its square, rational positive numbers, the laws of elementary arithmetic, and equality. The main concepts that he introduces next are the equation of the first degree, the equation of the second degree, related binomials and trinomials, the normal form, the algorithmic solution and the geometrical proof of the algorithm. In each case, al-Khwārizmī seeks to establish on the same geometrical basis that the algorithm is justified and that it leads to the result. It is during this presentation that he takes care to justify the arithmetic treatment of quadratic irrational quantities, and to show geometrically the validity of this treatment. This brief reminder shows clearly that the novelty of al-Khwārizmī's project is theoretical, not technical. On the technical level, in fact, his book does not reach the level of Diophantus's *Arithmetic*. The theoretical innovation of al-Khwārizmī's algebra is that he does not draw on the notion of equation while he is solving problems, but that it is a primitive notion, originating from primitive terms, the combination of which must yield all possible equations.

[1] By 'founding act', I in no way mean symbolic gesture, or the manifestation of some kind of subjectivity, but a genuine project whose elements gradually make their appearance in the process of its effective realization.

[2] R. Rashed, *Al-Khwārizmī: Le commencement de l'algèbre*, Paris, Librairie A. Blanchard, 2007. English transl. *Al-Khwārizmī: The Beginnings of Algebra*, History of Science and Philosophy in Classical Islam, London, Saqi Books, 2009.

But this conceptual break with Babylonian, Greek, Indian, or other traditions, has even deeper roots that draw on a new mathematical ontology and on a new epistemology. The central concept of algebra – the unknown or 'thing' (*cosa*) – indeed, does not designate a particular entity, as was the case in the traditional ontology, but an object that can be indifferently numerical or geometrical. In other words, the object of the new discipline is neither the geometrical figure nor the rational numbers; and the properties that this discipline is supposed to study are no more those of measure than those of position and form. Its object is new and is not defined negatively: the mathematical entity finds itself invested with a new meaning, that of an entity that is sufficiently general to allow several determinations, geometrical as well as arithmetical. This original indetermination is itself heavily laden with logical possibilities, destined to be realized more and more as a new means of studying one or the other perspective of the object are discovered. In other words, the algebraic object that al-Khwārizmī conceives, cannot be obtained by abstraction from particular entities; nor can one reach it by close imitation of a form or an idea. The new ontology is neither Aristotelian nor Platonic; it is, so to speak, formal, in all likelihood the first ever encountered in the history of mathematics. Its impact first on mathematics and then on philosophy will be considerable, witness the mathematician al-Karajī and the philosopher al-Fārābī, for example.

It is because algebra is conceived as a science that one can call it an epistemic novelty. Although apodictic like every other mathematical science, it shares with art the feature of having its goals outside of itself, because it also seeks to solve arithmetic and geometrical problems. Thus algebra does not fit the Aristotelian-Euclidean schema.

Finally, this new apodictic discipline is also algorithmic. In itself, the algorithm of solution admittedly must be an object of geometrical demonstration. Indeed, if one conceives of the solution as a simple decision procedure, the procedure must be justified in a different mathematical language, that of geometry. Al-Khwārizmī thus breaks with all preceding and contemporary traditions of algorithmic mathematics.

If such a *sui generis* conception of mathematical science was possible, it was thanks to a formal and combinatory choice that made it possible to establish an *a priori* classification of equations. This choice unfolds along the following stages: first, to determine a finite set of discrete elements, the number, the unknown ('thing', *cosa)* and the square of the unknown; second, from these elements, to rely on a combinatorics to obtain *a priori* all the possible equations; third, thanks to the theory, to isolate from these possible equations the cases that correspond to the criteria of the latter.

Thus from the eighteen equations he obtains, al-Khwārizmī selects the six canonical ones, thereby avoiding redundancy and repetition.

This *a priori* classification of possible equations, which, along with the other characteristics we have just enumerated, presents the configuration of this beginning of algebra and characterizes a domain of classical mathematics that will continue to grow. Thus did algebra begin, thanks to a break with the style of Hellenistic mathematics and of the other mathematics known at the time.

But one recognizes a genuine incipient from, on the one hand, the conceptual and textual tradition that it inaugurates; on the other, the new ruptures that it in turn provokes. In the flesh, the tradition appears in the names of the mathematicians and the titles of their works. The successors of al-Khwārizmī soon multiplied: they set out first to extend algebraic calculation much farther than the latter had done; second, to integrate Diophantine rational analysis into algebra; and, finally, to formulate the proofs of algorithms more rigorously, in the language of Euclidean geometry. Let's mention, among other mathematicians linked to these tasks, the names of Ibn Turk, Thābit ibn Qurra, Sinān ibn al-Fatḥ, and especially Abū Kāmil. To the latter we are indebted for the first treatise of algebra containing a chapter devoted to rational Diophantine analysis. This book is also known for its translations into Latin and Hebrew translation, and for Fibonacci's borrowings from it.

As to the other breaks (alternatively, the other 'new beginnings') provoked by this new conception of a mathematical science, embryonic forms of them exist in the new possibilities that the founding ideas and acts contain. With the algebra of al-Khwārizmī, it indeed became possible to apply one discipline to another: arithmetic to algebra, algebra to arithmetic, geometry to algebra, algebra to geometry, and algebra to trigonometry... Every one of these applications resulted in the creation of new chapters in mathematics and, with the same stroke, redrew the map of the mathematical continent.

Thus the application of arithmetic to algebra made it possible to conceive the algebra of polynomials in the ancient sense, that is, the algebra of the elements of the ring $Q[x,\frac{1}{x}]$. We named this 'new beginning' of algebra for the act that made its realization possible, that is, the 'arithmetization' of algebra undertaken by al-Karajī (end of the 10th century) and his successors, such as al-Samaw'al ibn Yaḥyā. This arithmetization led to an unprecedented development of abstract algebraic calculation, extended to irrational quantities, the species of which multiplied to infinity. Combinatorial analysis figures among the other means forged to this end. It is precisely according to this model of polynomial calculation that decimal calculation was reinterpreted and decimal fractions were invented. The difficulty that still persisted for al-Karajī was to apply Euclid's divisibility algorithm to polynomials. But the only invertible elements of the ring $Q[x,\frac{1}{x}]$ are the monomials. Al-Karajī thus divided a polynomial by a monomial, not a polynomial. It was to overcome this obstacle that his 12th century successor, al-Samaw'al, conceived the idea of continuous division and thus by approximation (limited development).

In this algebra, research on rational Diophantine analysis delved deeper yet by introducing a new classification according to forms (linear, quadratic, cubic).

It is also in relation to this algebra, but against it as well, that mathematicians such as al-Khujandī, al-Khāzin, Abū al-Jūd, al-Sijzī, etc., conceived and developed complete Diophantine analysis. They often started with the study of numerical right triangles, before setting themselves many other problems, notably the theorem of Fermat for $n = 3, 4$. This choice of numerical right triangles and of similar problems can be explained both by the limitation of domain of the solution to integers and especially by the new requirements that these mathematicians imposed: to demonstrate in the Euclidean style and to justify in Euclidean terms the algorithms of solution of Diophantine equations. At the same time, one notices an important inflection in this domain: from now on, one searches for purely arithmetic proofs, notably by using congruences.

The application of algebra to number theory made it possible not only to renew the proofs in domains that had already been cultivated, such as the theory of amicable numbers (Kamāl al-Dīn al-Fārisī), but also to conquer new domains: the study of elementary arithmetical functions, the sum and number of divisors.

It was also in relation to algebra and, more specifically, to the development of abstract algebraic calculation that they established a discipline that Hellenistic mathematics had never conceived: combinatorial analysis. It is precisely for the sake of algebraic calculation that al-Karajī established the binomial formula and Pascal's triangle. The explicitly combinatorial interpretation is found among many mathematicians, particularly Naṣīr al-Dīn al-Ṭūsī. Applied to diverse branches of linguistics, number theory or proportion theory, as well as philosophy, this combinatorial analysis was explicitly based on two ideas that, for most domains, characterized the thought forms of the era: to classify *a priori* all possible forms or all the elements of a finite set of discrete possibilities. Scholars were taking the same steps in domains as diverse as lexicography, prosody, cryptanalysis, mathematics, etc.

Even as the arithmetization of algebra was being pursued, however, the foundations of another program were being laid: the geometrization of that very same algebra. Two acts presided over the elaboration of this new program. The first, carried out by several mathematicians of the 10th century beginning with al-Māhānī, consisted in translating problems of solid geometry into cubic equations. The second was repeated several times during the 10th century: to solve the cubic equation by means of the intersection of several conic curves (al-Qūhī, Abū al-Jūd, etc.). This last act is moreover the echo of a doubly negative predicament: they could not succeed in solving the cubic equation by means of roots and in addition they did not have the means of justifying the algorithm used for the solution of certain forms of cubic equation and of biquadratic equation, since these solutions are not constructible with a straight edge and compass. One thus witnesses several advances, all headed in the same direction. The first is that of al-Qūhī who, for the cubics, conceives a theory equivalent to the application of areas for plane equations. As for al-Khayyām, he elaborates the first geometrical theory of cubic equations from a classification of all possible forms, which imposes the absence of equalization to zero. Next, he develops a classification as a function of the curves that are involved in the solution of equations. Finally Sharaf al-Dīn al-Ṭūsī, scarcely half a century after al-Khayyām, gives an analytical inflection to the theory, on account of the new requirement of demonstrating the existence of positive roots. These last two figures bring us face to face with the first research into elementary algebraic geometry.

2. Scarcely three decades after al-Khwārizmī, the mathematicians of Baghdad threw themselves into other conquests, this time armed with a broader knowledge of the Greek heritage. Indeed, the research they had undertaken had breathed life into an entire translation movement devoted to many Greek books, for example, the *Conics* of Apollonius, *On the Cutting-off of a Ratio* (by the same author), *The Measurement of the Circle* and *On the Sphere and the Cylinder* by Archimedes, Book 8 of Pappus's *Collection*, and the *Spherics* of Theodosius and of Menelaus. Beginning in the 9th century, one can identify three interrelated research traditions: infinitesimal geometry, the geometry of conic sections, and spherical geometry. It remains to be seen which new founding acts distinguished each of these traditions and led the heirs of Hellenistic mathematics to develop these chapters. As I see it, two expressions suffice to designate these acts: first, point-wise transformations; second, continuous movement. Let us examine them in that order.

In the works of Archimedes and Apollonius, the use of point-wise transformations crops up in some proofs. In *On Conoids and Spheroids*, Archimedes draws on an orthogonal affinity. Apollonius apparently uses several transformations, notably in *Plane Loci*. That said, Archimedes' book was never translated into Arabic; as to that of Apollonius, we know only what Pappus says about it. Neither Apollonius's book nor the remarks of Pappus reached the Baghdad mathematicians. The only exception is Book VI of the *Conics*, in which Apollonius follows a proto-transformational procedure, insofar as he tries to determine the conditions under which two conic sections are superimposable – that is, homothetic or similar – by means of *symptomata* (properties characteristic of each of the three conic sections) without, however, taking an interest in the very nature of these point-wise transformations. We therefore have every reason to believe that point-wise transformations as such were not part of the heritage received through translation. This historical fact is confirmed in the mathematicians' conception of point-wise transformations: among the moderns, they are not merely used in demonstrations but also surface more and more as elements in the concept of the geometrical object. As one gradually advances from the mid-9th century, one no longer studies only the figures, but also their transformations and the relations that unite them. This is precisely what one begins to notice among the three Banū Mūsā brothers, their student and collaborator Thābit ibn Qurra, the astronomer-mathematician al-Farghānī (in his book *al-Kāmil*), and many others. Their successors devoted themselves so heavily to this domain that, by the end of the 10th century, al-Sijzī borrows from Thābit ibn Qurra a generic term (*al-naql*) to name the object 'point-wise transformation'. Several decades later, in order to provide a foundation for this object, Ibn al-Haytham invents an entirely new geometrical discipline devoted mainly to studying the invariable elements of a figure when all the others change; he calls this discipline 'the knowns'.

Precisely this is the second founding act of Arabic mathematics: the introduction of point-wise transformations both into the concept of the geometrical object and into demonstration. Contrary to the act that founded algebra, this one is the achievement not of one individual, but of several simultaneously in different chapters: infinitesimal geometry and 'the science of projection', as the ancient mathematicians and biobibliographers called it. Notably traceable to the Banū Mūsā, especially the youngest of the three brothers, al-Ḥasan, their student Thābit ibn Qurra, and the astronomer-mathematician al-Farghānī, this act was continuously extended, affirmed, and deepened in the following centuries, as one sees in the 9th century with al-Bīrūnī and Ibn al-Haytham. But it is impossible to understand anything about the way it sprang up without bearing in mind the geometry of conic sections, beginning with the translation of Apollonius's *Conics*.

Let us turn first to al-Farghānī and to the 'science of projection'.

One of the first disciplines born from this founding act is the 'science of projection' (*ʿilm al-tasṭīḥ*), which first appears around the mid-9th century. Nothing surprising here: in this century astronomy flourished as it had not done since the second century. The astronomers did not restrict themselves to translating Greek works and a few Sanskrit texts, but they examined critically the theories and the calculations that they found in them. Specifically with this flowering of research in astronomy, the chapter about projections will separate itself from astronomy proper to become a chapter of geometry, even if its main area of application remains astronomy or, more precisely, astronomical instruments. Al-Farghānī played an essential role in this shift: he demanded that the procedures that astronomers used for the precise representation of the sphere be established on solid geometrical foundations. As far as we know, he was the first to require such a condition. These procedures enter into both the drawing of geographical maps and the construction of such astronomical instruments as the astrolabe.

To establish these procedures on solid geometrical foundations is to try to make apodictic the knowledge of the domains that are in play. Everything points to the following fact: if al-Farghānī was able to raise this question and to conceive of such a project, it is because of his very recent knowledge of Apollonius's *Conics*. Although Apollonius himself does not treat projections in this treatise, nevertheless in *Conics,* Book I, Propositions 4 and 5, he answers a specific question – about the nature of the intersection of the conic surface with a plane. It is with the help of these propositions that al-Farghānī proves the following:

Given a circle of diameter AG, the tangent to the circle at G, and any chord BC. The projections from pole A of points B and C on the tangent are I and K respectively.

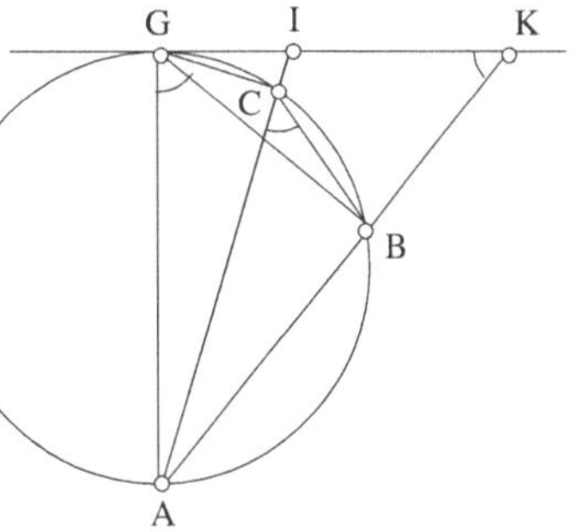

Fig. 9

Now $A\hat{G}B = A\hat{C}B$ (they are inscribed angles) and $A\hat{G}B = A\hat{K}G$ (they have the same complement $B\hat{A}G$); therefore

(1) $A\hat{C}B = A\hat{K}G$

and

(2) $C\hat{B}A = K\hat{I}A$.

Stated in a different language, one can interpret these results as follows: given GB and CG, the heights of triangles GAK and GAI, respectively, one has $AG^2 = AB \cdot AK = AC \cdot AI$; therefore, in the inversion τ of pole A and of the power AG^2, one has

$$I = \tau(C) \text{ and } K = \tau(B).$$

And, according to (1) and (2), points B, C, I, K belong to a circle that is invariant under the inversion τ.

Although al-Farghānī did not formulate the concept of inversion, he nevertheless recognized the transformation of a circle into a straight line and saw that the extremities of the chord and of the segment are on a circle that remains *invariant* under this transformation.

This lemma in fact amounts to the following: the conic projection from pole A of a chord onto the tangent to the diametrically opposed point is a segment of the tangent such that the extremities of the chord and of the segment are on a circle invariant under the inversion τ with the same pole A, which transforms the given circle into the tangent line. By means of this lemma, al-Farghānī establishes that the projection of a sphere with point A as a pole onto a plane tangent to the point diametrically opposed to it, or on a plane parallel to that plane, is a stereographic projection.

We cannot trace here the history of this chapter beyond al-Farghānī to Ibn ʿIrāq and al-Bīrūnī. Nevertheless, in order to appreciate how far al-Farghānī had advanced, let us simply recall a few themes studied by al-Qūhī and Ibn Sahl during the second half of the 10th century. These two mathematicians studied the conical projection from a point on the axis or off the axis of the sphere, and the cylindrical projection of a direction parallel or nonparallel to the axis of the sphere. They obtained many important new results, which were formerly attributed to later mathematicians. What is more, in this chapter, mathematicians such as al-Qūhī, Ibn Sahl, Ibn ʿIrāq, etc., invented new procedures of demonstration and hammered out a new language. Take, for example, al-Qūhī when he proves the following property: with every circle traced on the sphere and whose plane does not contain the pole, the stereographic projection associates a circle in the plane of projection, and inversely. To do so, he uses *Conics*, Proposition I.5, in which Apollonius studies the section by a plane of a cone with a circular base, for the case in which the plane of the base and the secant plane are anti-parallel. Al-Qūhī resorts to the technique of rabatment in order to make constructions in plane geometry possible. Thus the proofs are composed of comparisons of ratios, of proportions, and of rabatment; in other words, of techniques that are traditional as well as non-traditional, that is, projective.

As to the language, it is mixed: the vocabulary of proportion theory mingles with terms that henceforth designate projective concepts.

The 10th century thus already had access to the beginning of a new chapter in geometry, one destined to be enriched by successive generations for whom Ptolemy's *Planisphere* would at best be nothing more than a very distant ancestor.

3. Another chapter that testifies to developments in the use of point-wise transformations concerns the drawing of conic curves. It is likely that this problem is as ancient as the geometrical study of sundials and burning mirrors. Yet we know for sure that research on the drawing of conical curves was renewed by several occurrences: the reactivation of research in astronomy, on the one hand, and on burning mirrors and eventually lenses, on the other; and finally, beginning in the 9th century, the theory of cubic equations. Never before the 9th century had the search for effective procedures for tracing these curves been so intense, manifold, and continuous. Beginning in the 10th century, entire chapters and even treatises were devoted to this question, which was at the time thematicized. Many famous mathematicians took an interest in it, such as Ibrāhīm ibn Sinān, al-Khāzin, al-Qūhī, Ibn Sahl, al-Sijzī, al-Bīrūnī, etc. Methods of tracing by means of points were proposed, then methods of continuous tracing for which one invents mechanical instruments, such as the famous perfect compass, or other optical procedures. Whether theoretical or technical, all of these methods rest on one or the other affine point-wise transformation, and sometimes even on a projective transformation, as is the case in Ibn Sinān.

This research brought to the fore a question that, to date, had never been formulated: can one obtain conical curves starting from the circle, that is, with a transformation of the circle, and thus trace these curves by using that transformation? This question, which is already implicit in the writings of Ibn Sinān and Abū al-Wafā' al-Būzjānī, is formulated explicitly by the mathematician al-Sijzī in the second half of the 10th century.

Here too, everything begins with the youngest of the Banū Mūsā, al-Ḥasan, and his student Thābit ibn Qurra. Before he had access to a translation of Apollonius's *Conics* al-Ḥasan had studied the ellipse and its properties as the plane section of a cylinder, as well as the different varieties of elliptical sections. Contrary to Apollonius, he proceeds by means of the bifocal method and designates the ellipse with the significant expression 'the elongated circular figure'.[3] He then shows that this figure can be obtained from a circle by means of an orthogonal affinity, which is a contraction (or a dilation, as the case may be) according as the ratio of the major axis to the minor axis is less than 1 (or greater than 1, as the case may be). His student Thābit ibn Qurra, for his part, starts with a profound knowledge of Apollonius's *Conics*. He begins by demonstrating the following proposition: the plane sections of two cylinders with a circular base that have the same axis and the same height are homothetic, the center of the homothesis being their common center situated on the axis and the homothetic ratio being the ratio of the diameters of the base circles.[4] Next, Thābit demonstrates the proposition that al-Ḥasan had established. By using points, one can therefore rigorously trace the ellipse starting from the circle. But what about the parabola or the hyperbola? The grandson of Thābit ibn Qurra, Ibrāhīm ibn Sinān, soon raises the question and traces any point of the parabolic section by means of a circle. As to the hyperbola, Ibrāhīm ibn Sinān traces it by means of a circle and of a projective transformation aimed at transforming the circle into a hyperbola whose *latus rectum* is equal to the transverse diameter.

If Ibrāhīm ibn Sinān raises in general the problem of drawing conics by means of points starting from the circle, this is neither contingent nor circumstantial. The title of his book alone is a program: *On Drawing the Three Conic Sections*. Moreover, in order to succeed in conceiving this project and in formulating the question, it was necessary to be interested much more than before in the study of geometrical transformations. One need only examine such works by Ibn Sinān as *The Sundials*, or *The Anthology of Problems*, in order to note his frequent usage of transformations.[5] This interest only grew after Ibn Sinān. In the second half of the century, al-Sijzī wrote a treatise with a title that reflects his intention

[3] R. Rashed, *Les Mathématiques infinitésimales du IXe au XIe siècle*, vol. I: *Fondateurs et commentateurs*: *Banū Mūsā, Thābit ibn Qurra, Ibn Sinān, al-Khāzin, al-Qūhī, Ibn al-Samḥ, Ibn Hūd*, London, al-Furqān, 1996, Chap. I; English transl. *Founding Figures and Commentators in Arabic Mathematics*. A History of Arabic Sciences and Mathematics, vol. 1, Culture and Civilization in the Middle East, London, Centre for Arab Unity Studies, Routledge, 2012, Chap. I, p. 8.

[4] *Ibid.*, Chap. II.

[5] See the edition, translation and commentary on these treatises in R. Rashed and H. Bellosta, *Ibrāhīm ibn Sinān. Logique et géométrie au Xe siècle*, Leiden, E.J. Brill, 2000.

perfectly: in *All Figures are from the Circle*,[6] he explicitly takes up and in effect generalizes the book of his predecessor.

During the 10th century, however, tracing these curves by means of points was no longer satisfactory, notably because of the construction of geometrical problems on the one hand, and the solution of cubic equations by means of the intersection of conic curves on the other. Thereafter, it was necessary to be convinced of the continuity of curves in order to discuss the existence of points of intersection. Added to these theoretical reasons for dissatisfaction were technical ones: the construction of patterns for parabolic and elliptical burning mirrors, plano-convex and biconvex lenses, as well as the construction of astrolabes and sundials. Two contemporaries, Ibn Sahl and al-Qūhī, invented instruments to make a continuous drawing, and an entire tradition of mathematicians focused on the geometrical study of these instruments, and of the continuous drawing of curves. Al-Sijzī belonged to it and wrote a treatise on the perfect compass.[7] Everything was in place for the conception of the first treatise devoted entirely to methods of drawing by means of points and of continuous drawing of conic curves; such is the focus of al-Sijzī's book on *The Description of Conic Sections*.[8]

But to allow procedures for continuous drawing is to allow the notion of movement in geometry. Moreover, since point-wise transformations are now involved in demonstrations, transformations and continuous movements therefore become the foundation for a new chapter in the geometry of the conic sections. Mathematicians like Kamāl al-Dīn ibn Yūnus (1156–1248) and such students of his as Athīr al-Dīn al-Abharī (d. 1265) will continue to enrich this endeavour until the second half of the 13th century).[9]

4. Point-wise transformations, or a continuous movement associated with point-wise transformations, characterized the founding acts of the new chapters as well as the renewal of the ancient chapters of geometry, beginning in the mid-9th century. This is what we just saw with the 'science of projection' and 'drawing conic curves'. Three other chapters also illustrate this: that devoted to geometrical constructions; the one on the theory of parallels; and finally the chapter that treats infinitesimal geometry.

[6] Edition, translation and commentary in R. Rashed, *Œuvre mathématique d'al-Sijzī*, vol. I: *Géométrie des coniques et théorie des nombres au Xe siècle*, Les Cahiers du Mideo, 3, Louvain-Paris, Éditions Peeters, 2004.

[7] See R. Rashed, *Geometry and Dioptrics in Classical Islam*, London, al-Furqān, 2005, Chap. V.

[8] Edition, translation and commentary in R. Rashed, *Œuvre mathématique d'al-Sijzī*.

[9] See R. Rashed, *Geometry and Dioptrics in Classical Islam*, Chap. V.

Especially after the translations of Diocles's *Burning Mirrors*,[10] Eutocius's *Commentary*, and Archimedes' *Sphere and Cylinder*, the mathematicians of this era inherited the construction of several solid problems: the two means, the trisection of the angle, Archimedes' division of the line (*Sphere and Cylinder*, II.4). To these problems, they added many others, and they multiplied the constructions in particular. We know for example that many mathematicians took up the trisection of the angle and the construction of the regular heptagon. Unlike the ancients, however, they did not hesitate to modify the very criterion of the admissibility of construction for a solid problem. As a construction procedure, transcendental curves were banished, retaining only conic curves. The latter construction became admissible on the same terms as the use of the straight edge and compass for plane problems. It is important to emphasize that, if this new criterion was introduced, it was in answer to the practice of the algebraists who were beginning to translate solid problems into cubic equations, as attested in the works of al-Māhānī at the end of the 9th century. And it was precisely this new criterion that made it possible to gather into one chapter the dispersed studies of particular examples. But these studies drew on transformations and notably on similarity. Consider for example the problem of the regular heptagon. One begins by constructing the triangles, the ratio of whose angles are of one type or the other: (1, 2, 4), (1, 5, 1), (1, 3, 3), (2, 3, 2), before transforming it in order to inscribe it in the circle.[11] In short it was thanks to conic sections and to transformations that this new chapter became constituted as such.

Applications of the theory of conic section by geometers and algebraists led to several developments at the heart of this very theory. The 10th century mathematicians thus examined more closely the properties of the harmonic division of the conics. Ibn Sahl even wrote a treatise on this topic. His younger contemporary, al-Sijzī, studied a new theme: plane sections and the classification of them. No one went beyond them before Fermat and his successors in the 17th century. No less important were the contributions of the mathematicians of the 11th century: Ibn al-Haytham in

[10] *Les Catoptriciens grecs*. I: *Les miroirs ardents*, edition, translation and commentary by R. Rashed, Collection des Universités de France, Paris, Les Belles Lettres, 2000.

[11] R. Rashed, *Les Mathématiques infinitésimales du IXe au XIe siècle*, vol. III: *Ibn al-Haytham. Théorie des coniques, constructions géométriques et géométrie pratique*, London, al-Furqān, 2000; English translation: *Ibn al-Haytham's Theory of Conics, Geometrical Constructions and Practical Geometry*. A History of Arabic Sciences and Mathematics, vol. 3, Culture and Civilization in the Middle East, London, Centre for Arab Unity Studies, Routledge, 2013.

Egypt and ʿAbd al-Raḥmān ibn Sayyid in Andalusia. Both men were concerned with the generalization of the following classic problem: given two quantities, find two other quantities such that the four are in continuous proportion. This problem translates into a cubic equation that the algebraists solved with the intersection of two conic curves. According to al-Khayyām, Ibn al-Haytham generalized this problem for four quantities between two given quantities, which leads to a fifth degree equation solved by the intersection of a conic and a cubic. Everything thus indicates that Ibn al-Haytham had in hand a method analogous to that of Fermat in his *Dissertatio tripartita*. But the true generalization is that of ʿAbd al-Raḥmān ibn Sayyid, who, according to Ibn Bājja, wrote a work on the theory of conics in which he studied the intersection of a nonplane surface with a conic surface, that is, the general case of skew curves. Ibn Bājja recalls that Ibn Sayyid thus resolved the problem of two mean proportions 'for as many straight lines as one wishes between two straight lines, in continuous proportion, and in this way he divided the angle according to any given proportion'. Let us not forget that one will have to wait for Jacques Bernouilli to raise this problem a second time. To these developments, one must add the study of the optical properties of conic sections, which al-Kindī took up and which Ibn Sahl and Ibn al-Haytham generalized.

Without a doubt, the theory of parallels is one of the fundamental chapters of the geometry of this era. The mathematicians who devoted themselves to it include Thābit ibn Qurra, al-Khāzin, Ibn al-Haytham, al-Khayyām, Naṣīr al-Dīn al-Ṭūsī, among others. In two successive treatises that became the foundation for later research, Thābit ibn Qurra intentionally introduced the concept of continuous movement in order to define that of equidistance between parallels.[12]

Finally, transformations were used heavily in a vast domain that one can call 'infinitesimal geometry', or 'infinitesimal algebra'. It includes the measure of areas of curved surfaces and of the volumes of curved solids; isoperimetric and isoepiphanic problems; the solid angle; the study of variations of functional expressions such as trigonometric functions.

Let us pause at the most traditional example on this list, that of the measure of areas and of volumes. Since research in this domain ended after Archimedes, not until al-Kindī and the Banū Mūsā in the 9th century does one witness the beginning of a return to it, probably the effect of a unique first meeting between the traditions of Archimedes and of Apollonius. This meeting occurred not in a vacuum, but in a milieu informed by the algebra of al-Khwārizmī and his successors. The Banū Mūsā and their student

[12] See below, 'Thābit ibn Qurra on Euclid's Fifth Postulate'.

Thābit ibn Qurra, who had been the first to arrange the meeting of these two traditions, stimulated research along two roads that would continue to branch and advance: on the one hand, a much more substantial and systematic arithmetization than had occurred before; on the other, a more deliberate and more frequent use of point-wise transformations. To illustrate these perspectives quickly, recall that, in order to determine the area of a portion of a parabola, Thābit ibn Qurra begins by presenting twenty-one lemmas, eleven of which are arithmetic. These arithmetical lemmas pertained to the summation of many arithmetical progressions. Next, by using the arithmetic lemmas, he proves four lemmas on the sequencing of segments, on which he draws to study the requisite majoration. Starting from these lemmas, Thābit ibn Qurra took up to calculate the area of a parabolic portion.

He would follow this arithmetically based procedure when calculating the volume of the paraboloid of revolution. Later, Ibn al-Haytham would use it to determine the volume of the parabaloid generated by the rotation of a parabola about its ordinate.[13] He too begins with arithmetic lemmas in which he calculates the sum of the progressions of integer powers, that is the sums $\sum_{k=1}^{n} k^i$ for $i = 1, 2, 3, 4$; and he reaches a general rule thanks to a slightly archaic complete induction. He goes on to prove the following double inequality.

$$\frac{8}{15}n(n+1)^4 \le \sum_{k=1}^{n}\left[(n+1)^2 - k^2\right]^2 .$$

But al-Ḥasan ibn Mūsā and Thābit ibn Qurra draw on transformations in the demonstration of other areas of different species of plane sections of a right cylinder and an oblique cylinder, the area of an ellipse and the area of elliptical segments.

In this undertaking, the mathematicians draw mainly upon orthogonal affinities, homothesis, and the composition of these transformations; and they show that this composition preserves the areas.

This approach based on point-wise transformations will be followed by the successors of the Banū Mūsā and of Thābit ibn Qurra, who will try to reduce the number of lemmas. Thus Ibrāhīm ibn Sinān (909–946) in a brief treatise takes up the measure of a parabola.[14] Ibn Sinān's central idea,

[13] *Les Mathématiques infinitésimales du IXe au XIe siècle*, vol. II: *Ibn al-Haytham*, London, al-Furqān, 1993; English translation: *Ibn al-Haytham and Analytical Mathematics*. A History of Arabic Sciences and Mathematics, vol. 2, Culture and Civilization in the Middle East, London, Centre for Arab Unity Studies, Routledge, 2012.

[14] See edition, translation and commentary in *Les Mathématiques infinitésimales*, vol. I.

which he insisted on proving first, is the following: the affine transformation leaves the proportionality of the areas invariant. He then needs only two lemmas of a single proposition to complete his study.[15]

From the middle of the 9th century and throughout the 10th century, point-wise transformations and continuous movement were among the main foundational elements of the various chapters of geometry. Failing to study these developments will ensure that one will understand nothing about the constitutive acts of these chapters and about the development of geometry in this era. The presence and ever-more-frequent intervention of these concepts could not help but force new questions on mathematicians and face them with new tasks. How ought one legitimate the role of the one or the other? How could one allow the notion of movement in both statements and demonstrations when such a notion had never been defined? These two related questions raised a third of equal importance: If henceforth one focuses on the relations among figures, their transformations and their movements, is it not necessary to rethink the concepts of place and space? It is indeed impossible to leave aside this question of place; and it is equally impossible to retain the Aristotelian *topos* of place-envelope. These questions first emerge at the end of the 10th century, as some works of al-Sijzī attest, later becoming sources of reflection and intervention in Ibn al-Haytham. Let us pause very briefly to consider only the question of continuous movement.

To banish *de jure* the consideration of movement of the elements of geometry: such was the attitude of the Platonic geometers, dictated by the Theory of Forms. It was also the attitude of Aristotelian geometers on account of their doctrine of abstraction. But might it not be that the real reason for this attitude was less their ontological commitments than their very modest need for the notion of movement in a geometry that focused essentially on the study of figures? Even when this need is felt ever so slightly, it is not rare to find that, although avoided *de jure*, movement is slipped back in surreptiously or unintentionally. Is this not in fact Euclid's position in the *Elements*? He avoids movement, but introduces it in disguise when he resorts to superposition. Indeed the latter cannot occur without a displacement, even if the latter is the effect of an intellectual vision. We know moreover that, when he defines the sphere, Euclid opens the door to movement in spite of himself, as it were. Nevertheless, the injunction against movement remained in force for a long time. Recall that

[15] See below, 'The Archimedeans and problems with infinitesimals', pp. 500 ff.

al-Khayyām criticized Ibn al-Haytham's use of movement when the latter attempted to prove the fifth postulate.[16]

It is another matter altogether to resort to movement *de facto* without worrying about the legitimacy of using it. To keep silent about the issue of legitimacy is not to contradict the preceding opinion: hence the success of this practical – if not pragmatic – use of movement and of transformations by both the ancient geometers and those of the 9th and 10th centuries. In any event, this is the dominant position among the ancient geometers who focused on curves, whether transcendental or algebraic. Later, this position will also be that of Archimedes in *On Conoids and Spheroids* as well as *On Spirals*; that of Apollonius in the *Conics*, etc. This usage became even more intense during the 9th and 10th centuries.

It is, however, something else again to introduce movement into the primitive terms of geometry, a step that betrays a positive attitude towards movement and its role in definitions and demonstrations. But such an approach demands that one rearrange the concepts of geometry (at least some of them), that one rethink them in terms of movement, and that one forge anew the concept of geometrical 'place'. Now, a reworking of this sort evidently requires new foundations, another discipline, and a different method. To my knowledge, Ibn al-Haytham was the first to have attempted such a reworking by conceiving of a discipline of 'knowns', by elaborating an *ars analytica* (analytical art), and by reformulating the notion of 'geometrical place'.[17] To encounter similar attempts after him, one must wait until the second half of the 17th century, and especially the *analysis situs* of Leibniz.

Other foundational acts take place in other geometrical disciplines: the combination of spherical geometry and of trigonometry to get rid of the theorem of Menelaus, for example. Yet others occur even later: the invention of symbolism by Descartes, the introduction of imaginary numbers, exact representation. It is also the case for many other branches of mathematics: decimal arithmetic, the study of the classification of mathematical propositions (Ibn Sinān, al-Samaw'al), the theme of analysis and synthesis, and more generally the philosophy of mathematics. In a word, in all of

[16] R. Rashed and B. Vahabzadeh, *Al-Khayyām mathématicien*, Paris, Librairie A. Blanchard, 1999; English translation: *Omar Khayyam. The Mathematician*, Persian Heritage Series no. 40, New York, Bibliotheca Persica Press, 2000.

[17] See R. Rashed, *Les Mathématiques infinitésimales du IXe au XIe siècle*, vol. IV: *Méthodes géométriques, transformations ponctuelles et philosophie des mathématiques*, London, al-Furqān, 2002.

these chapters one can detect the main founding acts of classical mathematics. For other founding acts to take place, one must wait for the Italian algebraic school of the 16th century, the *Géométrie* of Descartes, the Diophantine analysis of Fermat, as well as the infinitesimal geometry of the 17th century. The latter, along with those identified above, mark the beginnings of modern mathematics and thus participate in the genesis of the new rationality.

Part I

Algebra

– 1 –

ALGEBRA AND ITS UNIFYING ROLE

The second half of the 7th century witnessed the constitution of the Islamic empire; the 8th century, its institutional and cultural consolidation. Throughout this century, many disciplines emerged that were directly tied to the new society and its ideology, with numerous attempts to found and to develop hermeneutic, Quranic, linguistic, juridical, historical, and theological disciplines. To gauge the breadth of this unprecedented intellectual movement, one need only think of the schools of Kūfā and Baṣra during the second half of the 8th century. One cannot overemphasize the importance of the research that developed at that time in these disciplines. It made possible the creation of the means necessary for the acquisition of other sciences, notably for the reception and the integration of the mathematical sciences and of the philosophical disciplines, for example. It also created the demand for them and molded their context. This same research, which was pursued, was amplified, and branched out in the 9th century, led to two fundamental epistemic results: a new classification of the sciences and a new conception of the encyclopedia of knowledge. Each of these, evidenced first in the domain of scientific research itself, would also be reflected in the awareness that the philosophers had of them. To appreciate this fact, one need only read carefully al-Fārābī's *Enumeration of the Sciences*, for example. Algebra, which was constituted at the beginning of the 9th century, played a leading role both in this new classification and in the conception of the new encyclopedia. But the event of algebra's birth itself around 820 cannot be understood without the light provided by the intellectual context of the 8th century, and without the demand induced by research into the disciplines mentioned above: indeed it was this research that gave algebra its domains of application, and effectively its social justification, as it were.

To recount this 8th-century intellectual – some might say epistemological – context would require a thick book and, to avoid being arbitrary, a wide range of competences. For now, let us do no more than note a few characteristics derived from only two examples. We borrow the first from the founder of Arabic prosody and lexicography, al-Khalīl ibn Aḥmad, and his successors. Ibn Aḥmad elaborated the elements of combinatory analysis necessary to found prosody and to establish the first Arabic dictionary. His method consists in elaborating *a priori* all possible combinations. Thus, in lexicography, the part that is phonetically realised in accordance with the rules of phonology he elaborated constitutes the real language. But this approach, which we have treated elsewhere,[1] was itself possible only thanks to a positive, theologically neutral, conception of language and a new conception of the linguistic object. The body of the language includes both the 'divine language' (the Quran), and the language of the 'pagans', which the Arabs spoke before Islam, and which we know thanks to poetry and to ethnolinguistic study, which were both thriving at the time. Even more importantly, however, the new linguistic object is a combination of phonemes independent of all phonetic material. Resulting from this formal association are possible words – that is, words with neither phonetic nor semantic value.

Thus, in this discipline, as in prosody and other activities of the time, the arrangements and combinations of elements devoid of signification produced certain knowledge by themselves. In the intellectual context of the 8th century, then, such a concept of the object of knowledge was admissible, and the algebraists adapted themselves perfectly to it. After all, al-Khalīl himself was a mathematician and wrote a book on computation, which has not survived. But the approach of al-Khalīl reflects yet another characteristic of the intellectual landscape of his day: there are no air-tight boundaries between science and art, not only in the respective practice of each, but especially from the theoretical point of view. A particular science – *e.g.*, lexicography – has an end beyond itself, which does not prevent it from being apodictic. It clearly aims to rationalize a practice without, however, ceasing to be a theory. In turn, an art can be conceptualized as the technical means of solving a theoretical problem: the composition of the language, to stay with our lexicographic example. The algebraists adapt themselves to this type of relationship as well.

This example borrowed from the linguists shows at the very least that on the eve of Algebra's birth, concepts both of apodictic knowledge and of its object had already been elaborated which perfectly suited the algebraists. From this point of view, the intellectual context of the 8th century favored the advent of algebra and made its integration into the midst of the disciplines of the time as natural as it was easy.

[1] R. Rashed, *The Development of Arabic Mathematics: Between Arithmetic and Algebra*, Boston Studies in Philosophy of Science, vol. 156, Dordrecht, Kluwer, 1994, pp. 253 ff.

The second example shows us that, by providing domains for the application of algebra, the scholars of the 8th century participated in the construction of one of its chapters. Likewise for jurisprudence, which in this same period witnesses the birth of three main schools: Abū Ḥanīfa (died in 767), founder of the school that carries his name, and his student Abū Yūsuf (died in 798); Mālik ibn Anas, the head of the school of Medina (died in 795); and finally al-Shāfiʿī, the founder of the third school, died in 820. These eminent jurists and their students also wrote on legal theory; it is in this respect that their successors compared them to Aristotle in logic. What interests us in particular here is the development in these schools during the 8th century of works in civil law, notably pertinent to wills and to inheritances according to the edicts of the Quran. These rules were often very complex and, to be exact, they required the invention of a genuine calculus. From the outset, this calculus was of an algebraic type. The student of Abū Ḥanīfa and of Mālik ibn Anas, Muḥammad ibn Ḥasan al-Shaybānī (749–803) himself composed a work called *Ḥisāb al-waṣāya* (*Calculation of Wills*). We shall see below that an entire chapter, constituting five-twelfths of the book of al-Khwārizmī, is developed precisely to this type of calculation. As it develops, this chapter will lead to an algebraic discipline, under the same title or entitled, *ḥisāb al-farāʾiḍ* (*Calculation of Obligations*). It will be taught as a field of research for algebraists and jurists after the creation of schools of law such as the Niẓāmiyya.

1. THE BEGINNING OF ALGEBRA: AL-KHWĀRIZMĪ

It is in this context that Muḥammad ibn Mūsā al-Khwārizmī writes a book entitled *Kitāb al-jabr wa-al-muqābala* (*Book of Algebra and al-Muqābala*).[2] The appearance of this work, under the reign of Caliph al-Maʾmūn (between 813 and 833), constitutes a turning point in the history of mathematics. For the first time, the term 'algebra' appears in a title to designate a distinct discipline with its own technical vocabulary.

The term *al-jabr* (algebra) is indeed an Arabic term, a name for the action of the verb (a *maṣdar*, according to the grammarians), the root of which has the general meaning of rectifying or correcting something using some form of constraint, such as setting a broken bone, for example. It is thus an ordinary-language term that can take on multiple meanings. It was

[2] R. Rashed, *Al-Khwārizmī: Le commencement de l'algèbre*, Paris, Librairie A. Blanchard, 2007; English transl. *Al-Khwārizmī: The Beginnings of Algebra*, History of Science and Philosophy in Classical Islam, London, Saqi Books, 2009.

devoid of technical meaning before al-Khwārizmī for the first time gave it a two-fold technical signification. When associated with the word *al-muqābala*, it designates both a discipline and an operation. The successors of al-Khwārizmī will quickly give pride of place to the first word, 'algebra', to name the discipline, and derive from this single word the name of the professional, 'algebraist'. This usage already appears in Thābit ibn Qurra (826–901). But this word also designates an operation: that of 'restoring' an equation, that is, adding to its two members the subtracted terms. For example, in

$$x^2 + c - bx = d \qquad \text{where } c > d,$$

the operation 'algebra' consists in adding bx to each side,

$$x^2 + c = bx + d,$$

and the operation *al-muqābala*, that is, 'opposition' or 'reduction' amounts to

$$x^2 + (c - d) = bx.$$

The point of the two related operations is to bring the equation back to one of the canonical types that al-Khwārizmī defined *a priori*.

Thus, between 813 and 833, this Baghdad mathematician and astronomer, a distinguished member of the 'House of Wisdom', would write a book with this technical terminology in the title. The event was crucial, and was recognized as such by both ancient and modern historians. The mathematics community of the time and of subsequent centuries did not miss its importance either. Thus Abū Kāmil (*c*. 830–*c*. 900) wrote about al-Khwārizmī:

> He who first composed a book of algebra and *al-muqābala*; he who initiated it and invented all the foundations found in it.[3]

The same Abū Kāmil wrote another book, no longer extant, but cited by the biobibliographer Ḥajjī Khalīfa:

> I established in my second book (*al-waṣāya bi-al-jabr*) the proof of Muḥammad ibn Mūsā al-Khwārizmī's authority and priority in algebra and *al-muqābala*, and I answered what that rash fellow, Ibn Barza, attributes to ʿAbd al-Ḥāmid, whom he mentions as his grandfather.[4]

[3] Abū Kāmil, ms. Istanbul, Beyazit Library, Kara Mustafa, 379, fol. 2ʳ; see R. Rashed, *Abū Kāmil: Algèbre et analyse diophantienne*, Scientia Graeco-Arabica, vol. 9, Berlin, Walter De Gruyter, 2012, p. 245, 2–3.

[4] Ḥajjī Khalīfa, *Kashf al-ẓunūn*, ed. Yatkaya, Istanbul, 1943, vol. 2, pp. 1407–8.

One could cite many more testimonials of this sort. Sinān ibn al-Fatḥ, in the introduction of his *Opuscule*, mentions only al-Khwārizmī and credits him with the algebra: 'Muḥammad ibn Mūsā al-Khwārizmī composed a book that he called algebra and *al-muqābala*'.

This book of al-Khwārizmī remained a constant source of inspiration and an object of commentaries by mathematicians, not only in Arabic and Persian, but also in Latin and the languages of Western Europe. But this event reveals a paradox. Contrasting starkly with the novelty of al-Khwārizmī's book in terms of conception, vocabulary, and organization is the simplicity of the mathematical techniques he deploys, when they are compared to the famous mathematical writings of Euclid and Diophantus, for example. But this technical simplicity is precisely a function of the new mathematical conception of al-Khwārizmī. To the superficial observer, most of al-Khwārizmī's ideas are found in one or the other of his predecessors. His book thus appears to be a worthy work, to be sure, but not distinguished by its novelty. Although one element of his project occurs twenty-five centuries earlier among the Babylonians, another in Euclid's *Elements*, and a third in Diophantus's *Arithmetic*, no work before al-Khwārizmī's reconstituted these elements and organized them in this specific way in order to create a discipline that no one had ever conceived. But what are these elements? and what is this organization? Al-Khwārizmī's goal is clear and unprecedented: to elaborate a theory of equations soluble by radicals, to which one can indifferently reduce both arithmetic and geometrical problems, and therefore to be able to use them in calculation, commercial transactions, inheritances, land surveying, etc. The new discipline is deliberately an applied one, as al-Khwārizmī states in his introduction. The domain of application may include problems in arithmetic as well as geometry, and the book is organized with this intention.

In the first part of his book, al-Khwārizmī begins by defining the primitive terms of this theory, which, because of the requirement of solution by radicals and because of his competence in this domain, only concerns equations of the first two degrees. These definitions pertain to the unknown (designated indifferently by *radical* or by *thing/cosa*), the square of the unknown, rational positive numbers, the laws of arithmetic $\pm$, $\times/\div$, $\sqrt{}$, equality. The main concepts that al-Khwārizmī introduces next are the equations of the first and second degrees, the binomials and trinomials associated with them, the normal form, algorithmic solutions, and the demonstration of the solution formula. In al-Khwārizmī's book, the concept of equation emerges to designate an infinite class of problems, and not, as was the case for the Babylonians for example, in the midst of solving this or that specific problem. Conversely, the equations themselves are presented not in the midst of solutions to problems, as the Babylonians or Diophantus had done, but at the outset, beginning from primitive terms, the combinations of which must yield all possible forms. Thus, once al-Khwārizmī has introduced the primitive terms, he immediately gives the following six types:

$$ax^2 = bx,$$
$$ax^2 = c,$$
$$bx = c,$$
$$ax^2 + bx = c,$$
$$ax^2 + c = bx,$$
$$ax^2 = bx + c.$$

He then introduces the concept of normal form, and requires that each of the preceding equations be reduced to the normal form corresponding to it. From this, there follows in particular, for trinomial equations,

$$(1) \qquad x^2 + px = q, \quad x^2 = px + q, \quad x^2 + q = px.$$

Al-Khwārizmī then moves on to the determination of algorithmic formulas of the solutions. He then treats each case, and obtains formulas equivalent to the following expressions:

$$x = \left[\left(\frac{p}{2}\right)^2 + q\right]^{\frac{1}{2}} - \frac{p}{2}$$

$$x = \frac{p}{2} + \left[\left(\frac{p}{2}\right)^2 + q\right]^{\frac{1}{2}}$$

$$x = \frac{p}{2} \pm \left[\left(\frac{p}{2}\right)^2 - q\right]^{\frac{1}{2}} \text{ if } \left(\frac{p}{2}\right)^2 > q;$$

and in this last case, he specifies:

– if $\left(\frac{p}{2}\right)^2 = q$, 'then the root of the square [*māl*] is equal to half of the number of the roots, exactly, without excess or shortfall';[5]

– if $\left(\frac{p}{2}\right)^2 < q$, 'then the problem is impossible.[6]

Al-Khwārizmī demonstrates also the different formulas, not algebraically but by using the notion of equality of areas. He was probably inspired by his very recent knowledge of Euclid's *Elements*, translated by his

[5] *Al-Jabr wa-al-muqābala*, ed. R. Rashed in *Al-Khwārizmī: Le commencement de l'algèbre*, p. 107, 3–4; English transl., p. 106.

[6] *Ibid*., p. 107, 2; English transl., p. 106.

colleague in the House of Wisdom, al-Ḥajjāj ibn Maṭar. Al-Khwārizmī presents each of these demonstrations as the 'cause' (*'illa*) of the solution. So al-Khwārizmī not only requires that each case be demonstrated, but he sometimes proposes two demonstrations for one and the same type of equation. Such a requirement highlights beautifully how far al-Khwārizmī has come: it separates him not only from the Babylonians, to be sure, but also, on account of its systematic aspect, from Diophantus.

For the equation $x^2 + px = q$, for example, he takes two line segments $AB = AC = x$ and next takes $CD = BE = p/2$. If the sum of the areas *ABMC*, *BENM*, *DCMP* is equal to q, then the area of the square *AEOD* is equal to $(p/2)^2 + q$, whence[7]

$$x = \left[\left(\frac{p}{2}\right)^2 + q\right]^{\frac{1}{2}} - \frac{p}{2}.$$

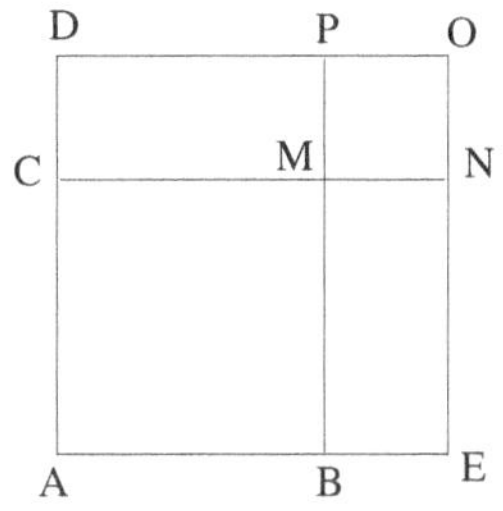

Fig. 10

With al-Khwārizmī, the concepts of the new discipline – namely 'the thing (*al-shay'*)' or '*cosa*', the unknown – do not designate a specific entity, but an object that can be indifferently numerical or geometrical; conversely, the algorithms of the solution must themselves be an object of demonstration. This is where the main elements of al-Khwārizmī's contributions are located. From now on, every problem treated in algebra, whether it be arithmetical or geometrical, must, according to him, be reduced to an equation with one variable and a rational positive coefficient of the second degree at least. Algebraic operations (transposition and reduction) are then applied so that the equation is written in normal form, which makes possible the idea of solution as a simple decision procedure, an algorithm for each class of problems. The formula of the solution is then justified mathematically with the help of a proto-geometrical demonstration, so that al-Khwārizmī can then write that everything pertinent to alge-

[7] *Ibid.*, pp. 100–4.

bra ‘necessarily leads you to one of the six procedures that I described in the introduction of my book’.[8]

Next, al-Khwārizmī turns briefly to several properties associated with the application of the elementary laws of arithmetic to the simplest algebraic expression. He thus studies products of the type

$$(a \pm bx)\,(c \pm dx) \qquad \text{where } a, b, c, d \in \mathbf{Q}_+.$$

Rudimentary though it may seem, this inquiry nevertheless represents the first effort devoted to algebraic calculation as such, for the elements of this calculation become the object of relatively autonomous chapters. In the following chapters, al-Khwārizmī goes on to apply the theory he has elaborated in order to solve numerical and geometrical problems, before finally treating problems of inheritance by algebraic means.

Thus algebra presents itself as a type of arithmetic, an arithmetic of unknowns, as some of al-Khwārizmī’s successors will call it. This arithmetic is more general than ‘logistics’ (since, thanks to its concepts, it makes possible a more rigorous solution of problems in the latter), but also more general than a metrical geometry. The new discipline is in fact a theory of linear and quadratic equations with a single unknown and soluble by radicals, and a theory of algebraic calculation on related expressions, with as yet no concept of polynomial. Finally, note that this new discipline was constituted by drawing not at all on symbolism, but only on terms from natural language. It was nevertheless well understood that the combined terms – the ‘thing/*cosa*’ and its square – are conceived without any specific content and in effect function as symbols. One early reader, an anonymous lawyer ignorant of symbolism, conceived a combinatorics of legal judgments. He then represented each of them by one letter of the alphabet:

> We designate the examples (the judgments) by letters devoid of signification such that they indicate by themselves, and not by reason of some existential matter. They are, moreover, concise and analogous to the ‘things’ and the squares (*māl*) that calculators use to determine unknowns.[9]

[8] *Ibid*., p. 121, 7–8; English transl., p. 120.

[9] Ms. Florence, Laurenziana, Or. 428, fol. 3^{v}:

من هاهنا نجعل الأمثلة بالحروف الخالية عن معنى حتى إن أنتجت، أنتجت بنفسها لا لمادة وجودية، وهي مع ذلك أخصر، ووازتها الأشياء والأموال التي يستعملها الحاسب في استخراج المجهولات.

2. AL-KHWĀRIZMĪ'S SUCCESSORS: GEOMETRICAL INTERPRETATION AND DEVELOPMENT OF ALGEBRAIC CALCULATION

To grasp better al-Khwārizmī's idea of the new discipline and of its fruitfulness, it is not enough to compare his book to ancient mathematical writings; it is also necessary to examine his impact on his contemporaries and his successors, the research that he succeeded in stimulating, and the traditions that he engendered. Only then will his true historical stature become clear. In our opinion, an essential characteristic of his book is that it immediately stimulated a current of algebraic research. Al-Nadīm, the biobibliographer of the 10th century, already draws up a long list of al-Khwārizmī's contemporaries and successors who carried on his research. Particularly notable among many others are Ibn Turk, Sind ibn ʿAlī, Abū Ḥanīfa al-Daynūrī, al-Ṣaydanānī, Thābit ibn Qurra, Abū Kāmil, Sinān ibn al-Fatḥ, al-Ḥubūbī, Abū al-Wafāʾ al-Būzjānī. Although many of their writings have disappeared, enough have survived to reconstruct the main outlines of this tradition. Given the limited scope of this chapter, however, we cannot take up the analysis of each of these contributions. We will try only to lift out the main axes of the development of algebra in the wake of al-Khwārizmī.

At the time of al-Khwārizmī and immediately thereafter, one witnesses essentially the extension of research that the latter had already pioneered: the theory of quadratic equations, algebraic calculation, and the application of algebra to problems of inheritance, distribution, etc. The research into the theory of equations itself took several routes. The first was the one al-Khwārizmī himself had already opened up, but now with improvements in his proto-geometrical demonstrations. This is the route followed by Ibn Turk,[10] who without adding anything new, takes up a tighter discussion of the proof. More important is the route taken a little later by Thābit ibn Qurra. He returns to Euclid's *Elements*, both to set the demonstrations of al-Khwārizmī on a firmer geometrical footing and to translate second-degree equations geometrically. Ibn Qurra is, moreover, one of the first to draw a clear distinction between algebraic and geometrical methods: he tries to prove that both lead to the same result, that is, to a geometric interpretation of algebraic procedures. Ibn Qurra begins by showing that the equation $x^2 + px = q$ can be solved by means of Proposition II.6 of the *Elements*. At the end of his demonstrations, he writes: 'this approach

[10] Aydin Sayili, *Logical Necessities in Mixed Equations by ʿAbd al-Ḥāmid ibn Turk and the Algebra of his Time*, Ankara, 1962, pp. 145 ff.

accords with that of the algebraist (*aṣḥāb al-jabr*)'.[11] He starts over again for $x^2 + q = px$ and $x^2 = px + q$, drawing on *Elements*, II.5 and II.6, respectively. In each case, he shows the correspondence with the algebraic solutions and writes: 'The path for this problem is that for the two preceding ones, in that the method of resolution by means of geometry accords with the method of resolution by means of algebra'.[12]

To our knowledge, Abū Kāmil and Thābit ibn Qurra were the first to make al-Khwārizmī consistent with Euclid, and thereby to give a geometrical interpretation of al-Khwārizmī's algebra and an algebraic interpretation of the geometry of *Elements*, Book II. If the historians' expression 'the geometrical algebra of the Greeks' makes any sense at all, it is only after al-Khwārizmī, with Abū Kāmil and Thābit ibn Qurra. Here is an example of Ibn Qurra's research that is important for both algebra and its history. Let's go back to the example of al-Khwārizmī's first equation, $x^2 + px = q$. Ibn Qurra posits that x^2 is equal to a square *ABCD*. He next considers a unit u, which is a unit of length if the unknown is a geometrical magnitude, but equal to 1 if it is a number. He posits that $BE = p \cdot u$. Clearly $AB = x \cdot u$. Ibn Qurra then proceeds by means of the following calculation. One has $AB \cdot BE = px = DE$, therefore $CE = x^2 + px = q$, a known number. The product of *EA* and *AB* is known, and the segment *BE* is known; the problem thus reduces to a geometrical problem:

> The segment *BE* is known; one adds *AB* to it, and the product of *EA* and *AB* is known. But Proposition 6 of Book II of the *Elements* shows that, if one divides the segment *BE* into two halves at point *F*, the product of *EA* and *AB* plus the square of *BF* is equal to the square of *AF*. But the product *EA* and *AB* is known and the square of *BF* is known; the square of *AF* is thus known and *AF* is known. If one subtracts from it *BF*, which is known, the remainder *AB*, which is the root, is known.[13]

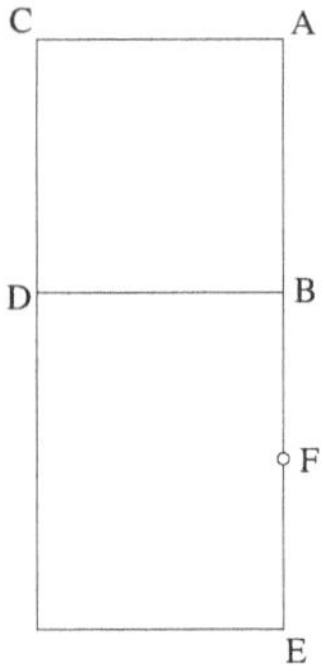

Fig. 11

Ibn Qurra then establishes the correspondence between the method of the algebraists and that of the geometers. He writes:

[11] *Fī taṣḥīḥ masā'il al-jabr bi-al-barāhīn al-handasiyya*, ed. R. Rashed, 'Résolution géométrique des équations du second degré', pp. 153–69, in *Thābit ibn Qurra. Science and Philosophy in Ninth-Century Baghdad*, Scientia Graeco-Arabica, vol. 4, Berlin, Walter de Gruyter, 2009, at p. 163, 13.

[12] *Ibid.*, p. 169, 3–4.

[13] *Ibid.*, p. 163, 3–11.

This procedure agrees with the algebraists' procedure for solving this problem. Indeed when the latter (the algebraists) take half of the number of the roots $\left(\frac{p}{2}\right)$, it is as when we take half of segment *BE*; when they multiply it by itself $\left[\left(\frac{p}{2}\right)^2\right]$, it is as when we take the square of half the segment *BE*; when they add the number to the product, it is as when we add the product of *EA* and *AB* to obtain from all of this the square of the sum of *AB* and of half the segment <*BE*>; when they take the root of the sum, it is as when we say that the sum of *AB* and of half the segment <*BE*> is known, given that the sum is a known square; when they subtract half from the number of the roots, it is as when we subtract half of *BE*. They obtain the remainder, which is the size of the root, and when they subtract from this half the size of the root, it is as when we subtract the segment *BF* to obtain the remainder, just as we obtained *AB*.[14]

If we quoted this text despite its length, it is to show that that Ibn Qurra's double interpretation – geometric for al-Khwārizmī, algebraic for Euclid – was both intentional and rigorous.

Mathematicians will follow an analogous route to establish a geometrical theory of equations, but they will do so by appealing to the Euclidean procedure of the application of areas, that is, to Book VI of the *Elements*. One of them writes:

It has been shown that the procedure that led to the determination of the sides of unknown squares in each of the three trinomial equations is the procedure presented by Euclid at the end of Book VI of the *Elements*, and which consists in applying to a given straight line a parallelogram that exceeds the whole parallelogram, or falls short of it, by a square. Indeed, the side of the exceeding square is the side of the unknown square in the first and second trinomials [$x^2 + q = px$, $x^2 + px = q$]; and in the third trinomial, it is the sum of the straight line to which the parallelogram is applied and of the side of the exceeding square.[15]

As we shall see, however, Ibn Qurra's geometrical translation of al-Khwārizmī's equations turns out to be particularly important for the development of the theory of algebraic equations. At almost the same time, another slightly different translation takes place that will also be fundamental for the development of the same theory: the translation of geometrical problems into algebraic terms. Indeed al-Māhānī, a contemporary of Ibn Qurra's, begins to translate into algebraic equations not only some

[14] *Ibid*., pp. 163, 13–165, 3.

[15] Anonymous manuscript no. 5325 Āstān Quds, Meshhed, fol. 24^{r-v}, falsely attributed to Abū Kāmil, copied in 581/1185.

biquadratic problems from Book X of the *Elements*, but also a solid problem – that given in Archimedes' *Sphere and Cylinder* – into a cubic equation (see below).

Another direction in the development of the theory of equations was also pursued at the time, namely, research on the general form of the quadratic equations

$$ax^{2n} + x^n = c, \qquad ax^{2n} + c = bx^n, \qquad ax^{2n} = bx^n + c,$$

as one encounters it in Abū Kāmil and Sinān ibn al-Fatḥ, among others.

Furthermore, after al-Khwārizmī, one sees an expansion of algebraic calculation. This is perhaps both the main research theme and the most widely shared among the algebraists who followed him. They began by extending the very terms of algebra to the 6th and 8th powers of the unknown, as one finds, respectively, in Sinān ibn al-Fatḥ and in Abū Kāmil. Sinān ibn al-Fatḥ, moreover, defines these powers multiplicatively,[16] in contrast to Abū Kāmil, who provides an additive definition. But it is the algebraic work of the latter that leaves its mark on both the epoch and the history of algebra. Beyond the expansion of algebraic calculation, he integrates into his book a new chapter of algebra: indeterminate rational analysis. After having taken up the theory of equations with demonstrations more solid than those of his predecessor, he studies in much greater depth and extent the arithmetical operations on binomials and trinomials, in each case demonstrating the result he obtained. He announces and justifies the sign rule, establishes the rules for calculation with fractions, before moving on to systems of linear equations with several unknowns and to equations with irrational coefficients, such as

$$\left(x^2+\frac{1}{\sqrt{2}}x\right)^2=4x^2, \quad \frac{\sqrt{10}x}{\left(2+\sqrt{3}\right)}=x-10, \quad \left(3+\sqrt{\frac{1}{2}x}\right)\left(2+\sqrt{\frac{1}{3}x}\right)=20, \text{ etc.}$$

Abū Kāmil integrates into his algebra auxiliary numerical methods, some of which purportedly appeared in a lost book of al-Khwārizmī, such as

$$\sum_{k=1}^{n} k, \quad \sum_{k=1}^{n} k^2, \quad \sum_{k=1}^{n} 2k.$$

[16] See R. Rashed, *The Development of Arabic Mathematics*, p. 20, n. 8.

Next, Abū Kāmil studies numerous problems that are reducible to second-degree equations.

We thus notice that the research of al-Khwārizmī's successors, and first and foremost of Abū Kāmil, contributed both the theory of equations and to the extension of algebraic calculation to the field of rational numbers, and to quadratic irrational numbers. Abū Kāmil's research on indeterminate analysis made a considerable impact on the development of this discipline but also gave it a new meaning and a new status. Originating in algebra, this analysis from now on constitutes a chapter in every treatise that seeks to cover the discipline.

3. THE ARITHMETIZATION OF ALGEBRA: AL-KARAJĪ AND HIS SUCCESSORS

We will understand nothing about the history of algebra if we do not emphasize the contributions of two currents of research that developed during the period just considered. The first focuses on the study of irrational quantities, whether the result of reading Book X of the *Elements* or in some sense independently. Among the many other mathematicians who took part in this research, one can name al-Māhānī, Sulaymān ibn ʿIṣma, al-Khāzin, al-Ahwāzī, Yūḥannā ibn Yūsuf, al-Hāshimī. It goes without saying that we cannot review these various contributions here. We merely want to emphasize that in this work, they actively developed calculation on irrational quantities and sometimes even began to read parts of Book X of Euclid's *Elements* in light of al-Khwārizmī's *Algebra*. To take only one example, consider al-Māhānī who in the 9th century sought the square root of five apotomes. Thus, to extract the square root of the first apotome, al-Māhānī proposes that 'we proceed by the method of algebra and *al-muqābala*',[17] that is, to posit $a = x + y$ and $b = 4\,xy$, and to obtain the equation $x^2 + b/4 = ax$. One then determines the largest positive root x_0, one deduces y_0, and one obtains

$$\sqrt{a-\sqrt{b}} = \sqrt{x_0} - \sqrt{y_0}\,.$$

Recall that, in the language of Book X of the *Elements*, $a + \sqrt{b}$ is the first binomial, with a and b rational, $a > \sqrt{b}$, $\sqrt{b}$ is irrational and $\frac{\sqrt{a^2-b}}{a}$ is rational, and the conjugate straight line $a - \sqrt{b}$ is the first apotome.

[17] *Tafsīr al-maqāla al-ʿāshira min kitāb Uqlīdis*, ms. Paris, BN, 2457, fol. 182r.

Al-Māhānī starts over again in this way for the next four apotomes; for the second apotome ($\sqrt{b} - a$), for example, with $b = 45$ and $a = 5$, one obtains the equation

$$x^4 + \frac{625}{16} = \frac{65}{2}x^2.$$

For these apotomes, as for the four that remain, al-Māhānī applies the same method to reduce the problem to one of the equations of al-Khwārizmī. Thus consider an apotome A, the square root of which one seeks to extract. Apotome A can be written, $A = a_1 - a_2$ with $a_1, a_2 > 0$, a_1 and a_2 incommensurable in length but commensurable in square, $a_1 \geq a_2$. Al-Māhānī writes $\sqrt{A}$ in the form of an apotome. Let x and y such that $x + y = a_1$ and $xy = \left(\frac{a_2}{2}\right)^2$; x and y are chosen such that $\sqrt{A} = \sqrt{y} - \sqrt{x}$.

One has $y = a_1 - x$, $xy = x(a_1 - x)\left(\frac{a_2}{2}\right)^2$, whence

$$x = \frac{a_1 - \sqrt{a_1^2 - a_2^2}}{2}$$

assuming $y > x$

$$y = \frac{a_1 + \sqrt{a_1^2 - a_2^2}}{2}$$

and $\sqrt{A} = \sqrt{y} - \sqrt{x}$.

One verifies that $A = (\sqrt{y} - \sqrt{x})^2 = a_1 - a_2$.

Now, these mathematicians have not only extended algebraic calculation to irrational quantities (algebraic extensions of **Q**), but they also made it possible to confirm the generality of the algebraic tool.

The second current of research was stimulated by the translation of Diophantus's *Arithmetic* into Arabic, and notably by an algebraic reading of it. Indeed, around 870, Ibn Lūqā translated seven books of Diophantus's *Arithmetic*, significantly with the title *The Art of Algebra*.[18] The translator drew upon al-Khwārizmī's language to render the Greek of Diophantus, thereby inflecting the content of the book towards the new discipline. Now, even if this *Arithmetic* is not an algebraic work in the sense of al-Khwārizmī, it nevertheless contains techniques of algebraic calculation that are powerful for the era: substitutions, eliminations, changes of variables, etc. It became the object of commentaries by mathematicians such as Ibn Lūqā, who translated the work in the 9th century, and Abū al-Wafā' al-

[18] R. Rashed, *L'Art de l'Algèbre de Diophante*, Cairo, National Library, 1975.

Būzjānī, a century later, but these texts are unfortunately lost. We know nevertheless, that al-Būzjānī wanted to demonstrate the Diophantine solutions. This same Abū al-Wafā᾽, in a text that has survived, proves the formula of the binomial, which is used often in the *Arithmetic*, for $n = 2, 3$.[19]

However that may be, this progress of algebraic calculation – whether by its extension to other domains or by the sheer mass of technical results it obtained – led to the renewal of the discipline itself. A century and a half after al-Khwārizmī, the Baghdad mathematician al-Karajī conceived another research project: to apply arithmetic to algebra, that is, to study systematically the application of the laws of arithmetic and of some of its algorithms to algebraic expressions and to polynomials in particular. It is precisely this calculation on algebraic expressions of the form

$$f(x) = \sum_{k=-m}^{n} a_k x^k \qquad m, n \in \mathbf{Z}_+$$

which became the main object of algebra. To be sure, the theory of algebraic equations is still present, but it occupies only a modest place among the concerns of the algebraists. One can therefore understand that books of algebra undergo modification not only in their content, but also in their organization.

Al-Karajī devotes several treatises to this new project, notably *al-Fakhrī* and *al-Badī*ʿ. These books will be studied anew, developed, and commented upon by mathematicians until the 17th century. In other words, in the research on arithmetic algebra, the work of al-Karajī occupies a central place for centuries, whereas the book of al-Khwārizmī becomes a historically important exposition, on which only mathematicians of the second tier will comment. Without getting into the history of six centuries of algebra here, let us illustrate the impact of al-Karajī's work by turning to one of his 12th-century successors. Al-Samaw'al (d. 1174) integrates into his algebra, *al-Bāhir*, the main writings of al-Karajī, and notably the two works cited above. He begins by defining very generally the concept of algebraic power and, thanks to the definition $x^0 = 1$, he states the rule equivalent to $x^m x^n = x^{m+n}$, where $m, n \in \mathbf{Z}$.

Here is what al-Samaw᾽al writes, after having noted in a table the exponents on one side and the other of x^0:

> If the two powers are on one side and the other of unity, starting from one of them we count in the direction of unity the number of elements in the table

[19] *Fī jamʿ aḍlāʿ al-murabbaʿāt wa-al-mukaʿʿabāt wa-akhdh tafāḍulahā*, ms. Meshhed, Āstān Quds 5521.

that separate the other power from unity, and the number is on the side of unity. If the two powers are on the same side of unity, we count in the direction opposite to unity.[20]

There follows his study of the arithmetic operation on monomials and polynomials, notably on the divisibility of polynomials as well as the approximation of fractions by elements of the ring of polynomials. One has, for example,

$$\frac{f(x)}{g(x)}=\frac{20x^2+30x}{6x^2+12}\approx\frac{10}{3}+\frac{5}{x}-\frac{20}{3x^2}-\frac{10}{x^3}+\frac{40}{3x^4}+\frac{20}{x^5}-\frac{80}{3x^6}-\frac{40}{x^7},$$

where al-Samaw'al obtains a kind of limited development of $f(x)/g(x)$, which works only for a sufficiently large x.

Next one encounters the extraction of the square root of a polynomial with rational coefficients. To all of these calculations on polynomials, al-Karajī had devoted a treatise, now lost but fortunately cited by al-Samaw'al, in which he devotes himself to establishing the formula of the binomial expansion and the table of its coefficients (see the image of the arithmetic triangle of al-Karajī, p. 147 below):

$$(a+b)^n=\sum_{k=0}^{n}\binom{n}{k}a^{n-k}b^k \qquad \text{with } n=1,2,3\ldots$$

It is when he demonstrates this formula that one sees in archaic form the complete finite induction as a procedure of mathematical proof. Among the auxiliary means of calculation, al-Samaw'al gives, following Abū Kāmil and al-Karajī, the sums of the different numerical progressions, along with a demonstration of them:

$$\sum_{k=1}^{n}k,\quad \sum_{k=1}^{n}k^2,\quad \left(\sum_{k=1}^{n}k\right)^2,\quad \sum_{k=1}^{n}k(k+1),\ldots$$

Next comes the answer to the following question: How can multiplication, division, addition, subtraction and the extraction of roots be used for irrational quantities?'[21] The answer to this question led al-Karajī and his successors to read Book X of the *Elements* algebraically (deliberately so),

[20] *Al-Bāhir en Algèbre d'as-Samaw'al*, edition, notes and introduction by S. Ahmad and R. Rashed, Damascus, 1972, p. 19.

[21] *Ibid.*, p. 37.

to extend to infinity the monomials and binomials presented in this book, and to propose rules of calculation, among which those of al-Māhānī are explicitly formulated:

$$\left(x^{\frac{1}{n}}\right)^{\frac{1}{m}} = \left(x^{\frac{1}{m}}\right)^{\frac{1}{n}} \quad \text{and} \quad x^{\frac{1}{m}} = \left(x^{n}\right)^{\frac{1}{mn}}$$

along with others like the following:

$$\left(x^{\frac{1}{m}} \pm y^{\frac{1}{m}}\right) = \left[y\left(\left(\frac{x}{y}\right)^{\frac{1}{m}} \pm 1\right)^{m}\right]^{\frac{1}{m}}.$$

One also finds an important chapter on rational Diophantine analysis, and another on the solution of systems of linear equations with several unknowns. Al-Samaw'al gives a system of 210 linear equations with 10 unknowns.

With al-Karajī's work begins the constitution of a new current of research in algebra, a tradition recognizable by the content and organization of each of its works. The latter 'are almost innumerable', to reuse Ibn al-Bannā''s expression in the 13th and 14th centuries.[22] By way of example, one finds among their authors the teachers of al-Samaw'al (al-Shahrazūrī, Ibn Abī Turāb, Ibn al-Khashshāb), al-Samaw'al himself, Ibn al-Khawwām, al-Tanūkhī, Kamāl al-Dīn al-Fārisī, Ibn al-Bannā', and, much later, al-Kāshī, al-Yazdī, etc.

In this stream, the chapter on the theory of algebraic equations proper made some progress, without, however, being central. Like his predecessors, al-Karajī himself considered quadratic equations. Some of his successors, however, tried to study the solution of cubic and fourth-degree equations. Thus, al-Sulamī in the 12th century tackled the cubic equation to find a solution by means of radicals,[23] bearing witness to the interest that the mathematicians of his day took in this problem. At the time, he himself considered two types as possible:

$$x^3 + ax^2 + bx = c \quad \text{and} \quad x^3 + bx = ax^2 + c.$$

[22] *Kitāb fī al-jabr wa-al-muqābala*, ms. Cairo, Dār al-Kutub, Riyāḍa 1, Muṣṭafā Fāḍil, 40, fol. 1.

[23] *Al-muqaddima al-kāfiya fī ḥisāb al-jabr wa-al-muqābala*, Collection Paul Sbāṭ, no. 5, fols 92^{v}–93^{r}.

He nevertheless required that $a^2 = 3b$. Al-Sulamī himself put it this way: 'This type (the first) has two conditions: one is proportionality and the other is that the third of the number of squares (x^2) be equal to the square root of the number of things (= unknowns). If these two conditions are met, then it can be solved'. As he explains next, by 'proportionality', he means that $\frac{1}{x} = \frac{x}{x^2} = \frac{x^2}{x^3} \ldots$; whereas the second condition is the one given above. For the second type, he adds one more condition: 'The things [= unknowns] must be with the cube'; when the things or unknowns are in the same member as the squares, he mentions that there is as yet no known method. Next, he gives a real, positive root for each equation.

$$x = \left(\frac{a^3}{27} + c\right)^{\frac{1}{3}} - \frac{a}{3} \text{ and } x = \left(c - \frac{a^3}{27}\right)^{\frac{1}{3}} + \frac{a}{3}.$$

We can reconstruct al-Sulamī's procedure as follows: using an affine transformation, he reduces the equation to its normal form; but instead of finding the discriminant, he zeroes the coefficient of the first power of the unknown, in order to reduce the problem to that of extracting the cubic root. For the first equation, for example, one takes the affine transformation $x \mapsto y - \frac{a}{3}$ and rewrites the equation as

$$y^3 + py - q = 0$$

with

$$p = b - \frac{a^2}{3} \text{ and } q = c + \frac{a^3}{27} + \left(b\frac{a}{3} - \frac{a^3}{9}\right);$$

given $b = \frac{a^2}{3}$, one gets

$$y^3 = c + \frac{a^3}{27},$$

and therefore y, and then x.

For the second equation, he takes the transformation $x \mapsto y + \frac{a}{3}$. Finally, he considers the equation

$$x^3 = ax^2 + bx + c$$

and recalls that he cannot solve it by means of the transformation $x \mapsto y + \frac{a}{3}$. In this case, he obtains a negative number, that is, $b = -\frac{a^2}{3}$, which he cannot allow.

Al-Sulamī continues with a 4th-degree equation. Here he naturally tries to find the change of variable that reduces it to a quadratic equation, that is, by forming the beginning of a square.

Thus, for the equation

$$x^4 + ax^3 + bx^2 + cx = d,$$

with $a = 2$, $b = 6$, $c = 5$, and $d = 66$, he finds the solution $x = 2$.

Al-Sulamī's method is the following: in effect, he posits $a = 2$ and $b - 1 = c$, that is, $b = 6$ and $c = 5$, and also $b \geq 1$.

Rewritten, the equation reads

$$(x^2 + x)^2 + (b - 1)(x^2 + x) = d.$$

By positing

$$x^2 + x = y,$$

one has

$$y^2 + (b - 1)y = d,$$

a second-degree equation with a positive root y_1.

Next, one solves $x^2 + x = y_1$, which also has a positive root.

If one takes $b = 1$ and $c = 0$, the equation is written as

$$(x^2 + x)^2 = d$$

and the equation

$$x^2 + x = \sqrt{d}$$

has a positive root.

Likewise, if $b = 0$, one has $c = -1$; the equation is rewritten as

$$(x^2 + x)^2 = x^2 + x + d,$$

whence

$$y^2 = y + d$$

and for every d, one has y_1 and one can solve $x^2 + x = y_1$.

If $a \neq 2$, one applies the method once again. The equation is rewritten

$$x^4 + ax^3 + \frac{a^2}{4}x^2 + x^2 + x^2\left(b - \frac{a^2}{4}\right) + cx = d,$$

whence

$$\left(x^2+\frac{a}{2}x\right)^2+\left(b-\frac{a^2}{4}\right)\left(x^2+\frac{a}{2}x\right)=d.$$

After positing

$$\left(x^2+\frac{a}{2}x\right)=y,$$

one has

$$y^2+\left(b-\frac{a^2}{4}\right)y=d,$$

an equation with a positive root; next, one solves

$$\left(x^2+\frac{a}{2}x\right)y_1,$$

another equation with a positive root.

In short, whatever the positive number a may be, if b and c are positive and meet the condition $b-\frac{a^2}{4}=2\frac{c}{a}$, then one can solve the original equation by positing

$$y=x^2+\frac{a}{2}x.$$

In every case, the equation will have a positive root.

Although credited to the 14th-century Italian mathematician Master Dardi of Pisa,[24] such attempts are frequent in the algebraic tradition of al-Karajī (10th–11th century). For example, take the case of the mathematician Ibn al-Bannā', who claims, that for equations that 'are reduced to other degrees (= other than the second), one cannot resolve them by means of algebra except for "cubes are equal to a number"'.[25] Although he implicitly recognizes here the difficulty of solving by means of radicals all cubic equations except $x^3 = a$, he nevertheless takes up the equation

$$(*) \qquad x^4+2x^3=x+30,$$

which he solves as follows: one rewrites the equation as

[24] See W. van Egmond, 'The Algebra of Master Dardi of Pisa', *Historia Mathematica*, 10, 1983, pp. 399–421.

[25] *Kitāb fī al-jabr wa-al-muqābala*, fol. 26v.

$$x^4 + 2x^3 + x^2 = x^2 + x + 30,$$

which is rewritten again

$$(x^2 + x)^2 = x^2 + x + 30;$$

positing that $y = x^2 + x$, one has

$$y^2 = y + 30.$$

By solving this equation, one obtains $y = 6$. One then solves $x^2 + x = 6$, and finds $x = 2$ as the solution of (*) above.

It is still too early to know precisely what contribution the mathematicians in this tradition made to the solution of third- and fourth-degree equations; these pieces of evidence nevertheless show that, contrary to what has been believed to date, some of them tried to go beyond al-Karajī.

4. THE GEOMETRIZATION OF ALGEBRA: AL-KHAYYĀM (1048–1131)

The algebraist arithmeticians were committed to the solution by radicals and wanted to justify the algorithm of the solution. Sometimes it even happened that the same mathematician – Abū Kāmil for example – offered two justifications, one geometrical, the other algebraic. For the cubic equation, not only were the solutions by radicals missing, but also the justification for the algorithm of solution, because the solution was not constructible by means of a straight edge and compass. The mathematicians from this tradition were fully conscious of this fact. As one of them wrote well before 1185:

> Since the unknown that one wants to determine and to know in each of these polynomials is the side of the cube mentioned in each, and since the analysis leads one to apply a known rectangular parallelepiped to a line segment that is known and exceeds or falls short of the whole parallelepiped by a cube; and one can perform the synthesis of this only by means of conic sections.[26]

[26] Ms. Meshhed, Āstān Quds, no. 5325, fol. 25.

Box 1

Here is how al-Khayyām himself recounts this history in his famous algebra treatise:

'As for the Ancients, no statement about these <premises> has come down to us from them: maybe they did not, after a study and an examination, understand them; or they were not compelled by their research to examine them; or what they have said about them has not been translated into our tongue.

And as to the moderns, it is to al-Māhānī, from among them, that the resolution of the premise which Archimedes had taken for granted in the fourth proposition of the second Book of his work *On the Sphere and the Cylinder* suggested itself by means of Algebra. Thus he ended up with an equation between cubes, squares and numbers. However, its solution did not occur to him after he had thought it over for a long time.

Hence he settled the matter by asserting that it was impossible. Until Abū Jaʿfar al-Khāzin appeared and solved it by means of conic sections. And after him, many geometricians needed several species thereof. Thus some of them solved some. But none of them has said anything substantial about the enumeration of their species, and about the study of the forms of each species thereof and their demonstration: except for two species which I shall mention.

And I always aspired intensely – and I still do – to investigate all their species, and to distinguish by means of demonstrations what is possible from what is impossible with respect to the forms of every species; for I knew that they were very urgently needed in the difficulties of the problems.

But I was not in a position to devote myself exclusively to the study of this good nor to meditate on it assiduously, because of my being detained from it by that part of the vicissitudes of fate which had befallen me. For we have been affected by the gradual disappearance of men of science, except for a group, small in number, great in afflictions, whose concern is to seize the opportunities of fate in order to devote themselves in the meantime to the achievement and the ascertainment of a science.'[27]

Box 2

'The ancient mathematicians who did not speak our language drew attention to none of this, or else nothing either came down to us or was translated into our language. And among the moderns who do speak our language, the first who needed a trinomial species of these fourteen species is al-Māhānī the geometer. He solved the lemma that Archimedes used, having considered it as admitted, in Proposition 4 of the second book of his work *On the Sphere and the Cylinder*. This is what I will present.

Archimedes said: the two straight lines *AB* and *BC* have a known magnitude and are in the extension one of the other; and the ratio of *BC* to *CE* is known. Thus *CE* is known, as is demonstrated in the *Data*. Then he said: let us posit that the ratio of *CD* to *CE* is equal to the square of *AB* to the square of *AD*.

[27] R. Rashed and B. Vahabzadeh, *Omar Khayyām: the Mathematician*, Persian Heritage Series no. 40, New York, Bibliotheca Persica Press, 2000, pp. 111–12; Arabic text in *Al-Khayyām mathématicien*, Paris, Librairie Blanchard, 1999, pp. 116–18.

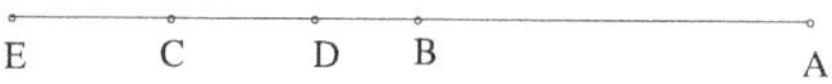

He did not say how to know this, since one necessarily needs conic sections. And, apart from this, he did not introduce anything in this book that was based on <conic> sections. He also considered this as admitted. The fourth proposition concerns the division of a sphere by a plane in a given ratio. But al-Māhānī used the terms of the algebraists in order to make things easier; since analysis led to numbers, to squares and to cubes in an equation, and since he could not solve it by means of conic sections, he settled the matter by saying that is impossible. The solution of one of these species thus remained concealed from this eminent man, despite his eminence and his supremacy in this art until Abū Jaʿfar al-Khāzin appeared and took advantage of a method that he presented in a treatise; whereas Abū Naṣr ibn ʿIrāq, a protégé of the Prince of the Believers from the land of Khwārizm, solved the lemma that Archimedes had used to determine the side of the heptagon in the circle, and that is based on the square that evinces the cited property. He used the terms of the algebraists.

The analysis led to "a cube plus some squares are equal to a number", which he solved by means of <conic> sections. By my life, this man belongs to a superior class in mathematics. This is the problem that left powerless Abū Sahl al-Qūhī, Abū al-Wafāʾ al-Būzjānī, Abū Ḥāmid al-Ṣāghānī, and a group of their colleagues, who were all devoted his Lordship ʿAḍud al-Dawla, in the City of Peace. This problem, as I was saying, is the following: If you divide 10 into 2 parts, the sum of their squares plus the quotient of the greater by the lesser is 72. The analysis led to squares equal to a cube plus some roots and a number. For a long time, these eminent men remained perplexed by this problem until Abū al-Jūd solved it. They deposited it in the library of the Samanid kings. There are thus three species, two trinomials and one quadrinomial, compound equations. Our eminent predecessors solved the only binomial equation, namely "the cube is equal to a number". We have received nothing from them about the ten remaining <equations>, and nothing as detailed as this. If time gives me leave, and if success follows me, I will record these fourteen species with all of their branches and sections by distinguishing the possible cases from the impossible ones (for some of these species require certain conditions in order to be valid) in a treatise that will contain several preliminary lemmas of great utility for the principles of this art.'[28]

Now this appeal to conic sections, which was explicitly intended to solve cubic equations, quickly followed the first algebraic translations of solid problems. We mentioned al-Māhānī and the lemma of Archimedes (see Box 1) in the 9th century. The other problems, such as the trisection of the angle, the two means, and the regular heptagon among others, would also soon be translated into algebraic terms. Conversely, however, faced with the difficulty mentioned earlier and with the added one of solving the cubic equation by radicals, mathematicians such as al-Khāzin, Ibn ʿIrāq, Abū al-Jūd ibn al-Layth, al-Shannī and others, were led to translate this equation into the language of geometry (see Box 2). They were thus in a

[28] *Risāla fī qismat rubʿ al-dāʾira*, in *Al-Khayyām mathématicien*, pp. 254–6.

position to apply to the study of this equation a technique that was then commonly used to examine solid problems, that is, the intersection of conic curves. Precisely herein lies the main reason for geometrizing the theory of algebraic equations. In this instance, contrary to Thābit ibn Qurra, one seeks not to translate geometrically the algebraic equations in order to find the geometrical equivalent of an algebraic solution that has already been discovered, but to determine, by means of geometry, the positive roots of the equation that one could not otherwise obtain. The attempts of al-Khāzin, al-Qūhī, Ibn al-Layth, al-Shannī, al-Bīrūnī, and others are so many partial contributions, until al-Khayyām's conception of the project, namely, the elaboration of a geometrical theory of equations with a degree less than or equal to 3. Al-Khayyām (1048–1131) intends first to go beyond fragmentary research, that is, research tied to this or that specific form of the cubic equation, in order to elaborate a general theory of equations, and to propose at the same time a new method of composition. He then studies all types of third-degree equations, which he classifies formally according to the distribution of the constants, from the first degree, the second degree, and the third degree, among the two members of the equation. For each of these types, al-Khayyām finds a construction with a positive root by means of the intersection of two conics.

For example, to solve the equation 'a cube plus a number are equal to squares', that is

$$x^3 + c = ax^2.$$

It follows that $a > x$. Let $AC = a$. Let H be a segment such that $H^3 = c$. Three cases thus present themselves: $H = AC$, $H > AC$ and $H < AC$.

Al-Khayyām begins by examining these cases.

• If $H = AC$, the problem is impossible.

Three cases present themselves again according to whether the solution x_0 is equal to, greater than, or less than H.

For $x_0 = H$, $ax_0^2 = H^3 = c$, which is impossible.

For $x_0 < H$, $ax_0^2 < c$, whence $ax_0^2 < c < x_0^3 + c$, which is absurd.

For $x_0 > H$, $x_0^3 > ax_0^2$, which is also absurd.

• If $H > AC$, the problem is *a fortiori* impossible.

For $x_0 = H$, $x_0 > a$ and $x_0^3 > ax_0^2$, which is impossible.

For $x_0 < H$, $ax_0^2 < H^3 = c$, since $H > a$, which is also impossible.

For $x_0 > H$, $x_0^3 > ax_0^2$, since $H > a$, which is impossible.

One thus has the necessary condition $H < AC$, that is, $c < a^3$.

Since $H < AC$, take a point B on AC such that $BC = H$, and let us examine three cases of the figure, namely (1) $BC = AB$, (2) $BC > AB$ and (3) $BC < AB$.

First case of figure: $BC = AB \Leftrightarrow c^{\frac{1}{3}} = \frac{a}{2}$.

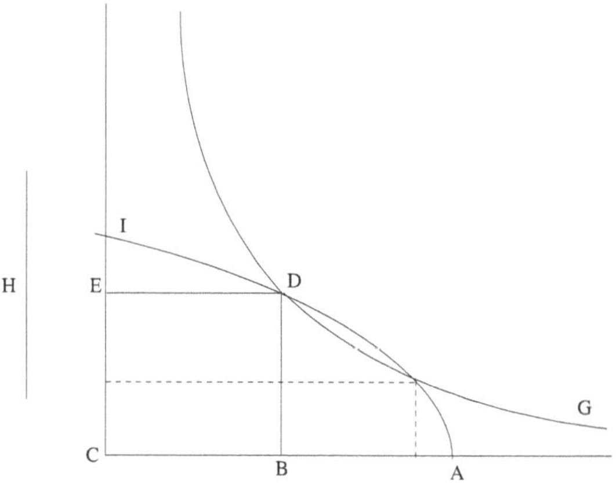

Fig. 12

Complete the square DC and trace the hyperbola $\mathscr{H}$ that passes through D and admits AC and CE as asymptotes. Trace also the parabola $\mathscr{P}$ with summit A, axis AC and *latus rectum* BC. The parabola $\mathscr{P}$ passes through D, for

$$DB^2 = AB \cdot BC,$$

which is the equation of the parabola, therefore $\mathscr{H}$ and $\mathscr{P}$ intersect at D. But $\mathscr{P}$ cuts $\mathscr{H}$ at another point, as al-Khayyām notices but without justifying it.

Second case of figure: $BC > AB \Leftrightarrow c^{\frac{1}{3}} > \frac{a}{2}$.

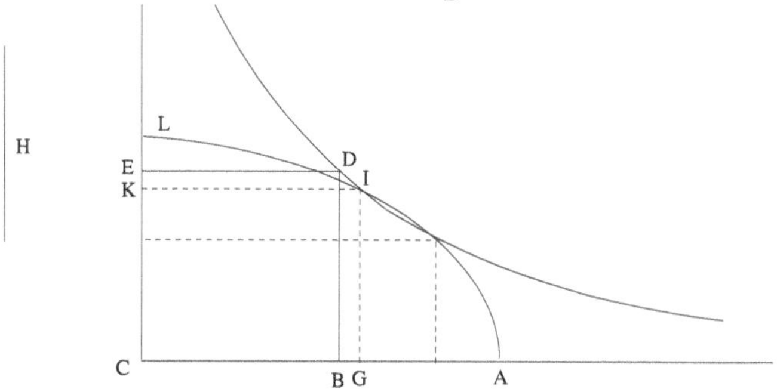

Fig. 13

Let us complete the square DC; let us trace the hyperbola $\mathscr{H}$ passing through D, and the parabola $\mathscr{P}$. Point D is exterior to the parabola for

$$DB^2 > AB \cdot BC.$$

If $\mathscr{P}$ and $\mathscr{H}$ intersect or are tangent at a point $D_1 \neq D$, the projection of D_1 onto AC is necessarily between A and B, and the problem is possible; otherwise, it is not. For the demonstration of this conclusion, see the commentary.

Third case of figure: $BC < AB \Leftrightarrow c^{\frac{1}{3}} < \frac{a}{2}$.

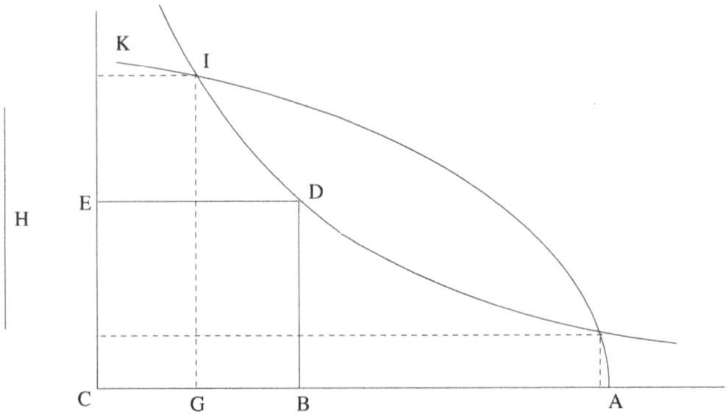

Fig. 14

In this case, point D is interior to parabola $\mathscr{P}$; $\mathscr{H}$ and $\mathscr{P}$ therefore intersect in two points.

In all these cases of figure, let I be one of the two intersections of $\mathscr{H}$ and of $\mathscr{P}$. Let G be the projection of I onto CA in Fig. 13. One has $(IC) = (DC)$, according to the equation of $\mathscr{H}$. Therefore

$$\frac{GC}{BC} = \frac{BC}{IG}.$$

But $I \in \mathscr{P}$, whence $IG^2 = AG \cdot BC$. Therefore

$$\frac{BC}{IG} = \frac{IG}{GA},$$

therefore

$$\frac{GC}{BC} = \frac{BC}{IG} = \frac{IG}{GA},$$

whence

$$\frac{GC^2}{BC^2} = \frac{BC}{GA},$$

whence

$$c = BC^3 = GC^2 \cdot GA,$$

therefore

$$c + GC^3 = GC^2\,(GA + GC) = GC^2 \cdot CA = a \cdot GC^2,$$

therefore GC is the solution.

One proceeds in the same way for the two other cases of figure.

For the third case of figure $\left(c^{\frac{1}{3}} < \frac{a}{2}\right)$, two distinct solutions correspond to two points of intersection of $\mathscr{H}$ and $\mathscr{P}$.

Let us comment on the text of al-Khayyām. We have noted that al-Khayyām demonstrates that if $c^{\frac{1}{3}} \leq \frac{a}{2}$, the problem is always possible. The case $\frac{a}{2} < c^{\frac{1}{3}} < a$ thus warrants discussion. Let us consider the first two cases, that is, $c^{\frac{1}{3}} \geq \frac{a}{2}$. In each case, al-Khayyām has shown that the problem is impossible. In the third case, that is, when $a > c^{\frac{1}{3}}$, he has shown that there are three cases of figure, according to whether $c^{\frac{1}{3}}$ is equal to, greater than, or less than $\frac{a}{2}$.

For the choice of two curves, one reduces the equation to the proportion:

$$\frac{CG^2}{BC^2} = \frac{BC}{GA}.$$

One introduces GI perpendicular to BC such that

$$\frac{CG}{BC} = \frac{BC}{GI},$$

so that

$$\frac{CG^2}{BC^2} = \frac{CG}{GI}$$

and the equation becomes

$$\frac{CG}{GI} = \frac{BC}{GA} \text{ or } \frac{CG}{BC} = \frac{GI}{GA} \text{ or again } \frac{BC}{GI} = \frac{GI}{GA}.$$

The point I such that $CG \cdot GI = BC^2$ and $GI^2 = BC \cdot GA$ lies at the intersection of hyperbola $\mathscr{H}$ with parabola $\mathscr{P}$.

Note, moreover, that this choice of curve is identical to that of the equation $x^3 + ax^2 = c$, which al-Khayyām had studied earlier. The only difference that accounts for the change in the sign of the coefficients, is that the concavity of the hyperbola is turned in the other direction.

In other words, consider the system of axes (CA, CE), that is (Ox, Oy), and let the point $D\ (c^{\frac{1}{3}}, c^{\frac{1}{3}})$ be given. This equation is rewritten as

$$c^{\frac{1}{3}}(a-x) = \frac{c^{\frac{4}{3}}}{x^2},$$

since zero is not a solution.

Let us posit

$$y = \frac{c^{\frac{2}{3}}}{x}, \qquad \text{equation of } \mathscr{H}$$

$$y^2 = c^{\frac{1}{3}}(a-x), \qquad \text{equation of } \mathscr{P}.$$

1) If $c^{\frac{1}{3}} = \frac{a}{2}$, then $D \in \mathscr{P}$ since $c^{\frac{2}{3}} = c^{\frac{1}{3}}\left(a - c^{\frac{1}{3}}\right)$. Al-Khayyām then claims that $\mathscr{P}$ and $\mathscr{H}$ intersect at another point I (x, y). The problem has two solutions. To obtain the second solution, one knows that if $c^{\frac{1}{3}} = \frac{a}{2}$, $x_1 = c^{\frac{1}{3}}$ is a solution. In this case, the second root is equal to $x_2 = \frac{c^{\frac{1}{3}}}{2}\left(1+\sqrt{5}\right)$.

2) If $c^{\frac{1}{3}} > \frac{a}{2}$, the point D is exterior to $\mathscr{P}$, since $c^{\frac{2}{3}} > c^{\frac{1}{3}}\left(a - c^{\frac{1}{3}}\right)$.

If therefore $\mathscr{H}$ and $\mathscr{P}$ have in common two points of intersection or if they are tangent at one point, their abscissas necessarily verify

$$c^{\frac{1}{3}} < x < a,$$

and the problem has one solution; if the two curves are tangent, two solutions if they intersect in two points; and no solutions if they do not intersect. The abscissas of these two points of intersection are on the same side of $BC = c^{\frac{1}{3}}$; let us demonstrate that they are larger. The polynomial $x^3 - ax^2 + c$ reaches a minimum for $x = \frac{2a}{3}$ and its value is $\frac{27c - 4a^3}{27}$; the condition of existence of the roots is therefore $c^{\frac{1}{3}} \leq \frac{a\sqrt[3]{4}}{3}$.

Note that the ordinates y_1, y_2 of the abscissa points $x = \frac{2a}{3}$ on the hyperbola and the parabola, respectively, are determined by $y_1 = \frac{3c^{\frac{2}{3}}}{2a}$ and $y_2^2 = \frac{ac^{\frac{1}{3}}}{3}$; the condition $c^{\frac{1}{3}} \leq \frac{a\sqrt[3]{4}}{3}$ means that $y_1 \leq y_2$, that is, that the point of the hyperbola is interior to the parabola. The value $\frac{2a}{3}$ occurs between the two roots of the equation and one must verify that $c^{\frac{1}{3}} < \frac{2a}{3}$, namely $27c < 8a^3$, a consequence of the condition of possibility.

3) If $c^{\frac{1}{3}} < \frac{a}{2}$, then $\mathscr{H}$ and $\mathscr{P}$ necessarily intersect in two points and the problem has two solutions. Indeed, given $I(x_0, y_0) \in \mathscr{P} \cap \mathscr{H}$, one has

$$\frac{x_0}{c^{\frac{1}{3}}} = \frac{c^{\frac{1}{3}}}{y_0} = \frac{y_0}{a - x_0},$$

whence

$$c = x_0^2(a - x_0).$$

Finally, let us remark that if it is easy to obtain the condition $c^{\frac{1}{3}} < a$ by comparison with the respective 'weights' of the monomials of equation ($x^3 < ax^2 \Rightarrow x < a$, and $c < ax^2 < a^3 \Rightarrow c^{\frac{1}{3}} < a$), it is more difficult to obtain the condition $c^{\frac{1}{3}} < x < a$ without geometrical considerations about the asymptotic behaviors of curves. Since CE is an asymptote to $\mathscr{H}$, all the points of the latter have their abscissas to the right of C. But by construction, the points of $\mathscr{P}$ have their abscissas to the left of A. It is therefore the case that every point belonging to the intersection of $\mathscr{H}$ and $\mathscr{P}$ belong to the projection between C and A, whence $0 < x < a$. From the convexity of $\mathscr{H}$ and the concavity of $\mathscr{P}$, one concludes that they cannot intersect to the left of D. If therefore (x, y) is a point of intersection, one has $c^{\frac{1}{3}} < x$; whence the condition $c^{\frac{1}{3}} < x < a$; it was perhaps in some equivalent fashion that al-Khayyām found this inequality.

Finally this seminal discussion of al-Khayyām the algebraist, that is the conditions he found, represents a step on the road that would establish the existence of the positive root. It is therefore not a matter of finding necessary and sufficient conditions. To complete them, al-Khayyām's successor, Sharaf al-Dīn al-Ṭūsī,[29] took another road by determining first the maximum of the expression $x^2(a - x)$.

To elaborate this new theory, al-Khayyām was forced better to conceive and to formulate the new relations between geometry and algebra. Recall in this regard that the fundamental concept al-Khayyām introduced is that of the unit of measure: suitably defined in relation to that of dimension, it makes possible the application of geometry to algebra. This application now leads al-Khayyām in two directions that may at first glance seem contradictory. Whereas algebra is now identified with the theory of algebraic equations, al-Khayyām seems from now on to transcend, however timidly, the gap between algebra and geometry. The theory of equations is more than ever before the meeting ground of algebra and geometry, and increasingly of analytical reasoning and methods. The concrete translation of this situation is the appearance of treatises fully devoted to the theory of equations, such as that of al-Khayyām, as a matter of fact. Indeed, in contrast to the algebraist arithmeticians, al-Khayyām leaves out of his treatise the chapters devoted to polynomials, to the arithmetic of polynomials, to the study of algebraic irrationals, etc. He shapes on the contrary a new model of composition: he begins with a discussion of the concept of algebraic quantity in order to define the concept of unit of measure; he then

[29] R. Rashed, *Sharaf al-Dīn al-Ṭūsī, Œuvres mathématiques. Algèbre et géométrie au XIIe siècle*, Collection Sciences et philosophie arabes – textes et études, 2 vols, Paris, Les Belles Lettres, 1986, vol. I.

moves on to the necessary lemmas as well as to a formal classification of equations as a function of the number of terms, and finally to examine, in order of increasing difficulty, the binomial equations of the second degree, the binomial equations of the third degree, the trinomial equations of the second degree, the trinomial equations of the third degree, and finally, the equations that contain the inverse of the unknown. In his treatise, al-Khayyām attains two remarkable results, which historians usually attribute to Descartes: on the one hand, a general solution of all third-degree equations by means of the intersection of two conics; on the other, a geometrical calculus made possible by the choice of a unit of length, while remaining, contrary to Descartes, faithful to the rule of homogeneity.

Al-Khayyām does not stop here, however. He also attempts to give an approximate numerical solution to the cubic equation. Thus, in a treatise called *Risāla fī qismat rubʿ al-dāʾira* (*On the Division of the Quadrant of a Circle*),[30] in which he announces his new project on the theory of equations, he reaches an approximate numerical solution by means of trigonometric tables.

5. THE TRANSFORMATION OF THE THEORY OF ALGEBRAIC EQUATIONS: SHARAF AL-DĪN AL-ṬŪSĪ

Until very recently, people believed that the contribution of mathematicians of the time to the theory of algebraic equations was limited to al-Khayyām and his work. This is not so. Not only did the work of al-Khayyām inaugurate a genuine tradition, but what is more, the latter was profoundly transformed scarcely a half century after his death.

According to historical testimony, the student of al-Khayyām, Sharaf al-Dīn al-Masʿūdī,[31] composed a book that treats the theory of equations and the solution of cubic equations. But if this book was once written, it certainly never reached us. Two generations after al-Khayyām, we encounter one of the most important works in this current: the treatise by Sharaf al-Dīn al-Ṭūsī on *The Equations*.[32] Now by comparison with al-Khayyām's treatise, that of al-Ṭūsī (*c*. 1170) introduces very important innovations. Unlike his predecessor's, al-Ṭūsī's procedure is no longer global and algebraic, but local and analytical. This change, which is of

[30] See our edition, translation, and commentary in Rashed and Vahabzadeh, *Al-Khayyām mathématicien*, Partie I; English transl., pp. 97 ff.

[31] R. Rashed, 'Résolution des équations numériques et algèbre: Sharaf al-Dīn al-Ṭūsī – Viète', *Archive for History of Exact Sciences*, 12.3, 1974, pp. 244–90.

[32] R. Rashed, *Sharaf al-Dīn al-Ṭūsī, Œuvres mathématiques*.

peculiar importance in the history of classical mathematics, deserves a longer examination.

The treatise of al-Ṭūsī opens with a study of two conical curves, which he will later use: the parabola and the hyperbola; adding to them the circle, assumed as known, exhausts all the curves to which he will appeal. Indeed al-Ṭūsī seems to assume that his reader is familiar with the equation of the circle, obtained from the power of a point in relation to the latter, and he uses this preliminary part to establish this equation of the parabola, and the equation of the equilateral hyperbola, in relation to two systems of axes.

Next comes a classification of equations of degree less than or equal to three. Contrary to al-Khayyām, al-Ṭūsī chooses a criterion of classification that is not intrinsic, but extrinsic. Whereas al-Khayyām, as noted, organizes his presentation according to the number of monomials that form the equation, al-Ṭūsī's criterion for the succession of equations is the existence of positive solutions or the lack thereof; in other words, the equations are organized according to whether they admit 'impossible cases' or not. It is therefore easy to understand that the *Treatise* is composed of only two parts, corresponding to the preceding alternative. In the first part, al-Ṭūsī treats the solution of twenty equations; in each case, he proceeds to the geometrical construction of the roots, the determination of the discriminant only for the quadratic equations, and finally the numerical solution by means of the so-called method of Ruffini-Horner. He reserves the application of this method for polynomial equations, and no longer only for the extraction of the root of a number.

By now already, we can therefore see the constitutive elements of the 12th-century theory of equations in the tradition of al-Khayyām: the geometrical construction of roots, the numerical solution of equations, and finally the return to solutions by radicals of quadratic equations, rediscovered in this case on the basis of the geometrical construction. In the first part, after having studied the second-degree equations and the equation $x^3 = c$, al-Ṭūsī examines eight third-degree equations. The first seven all have a single positive root. They can have negative roots, which al-Ṭūsī does not recognize. To study each of these equations, he chooses two second-degree curves or, more precisely, two portions of curves. By means of geometrical considerations, he shows that the considered arcs have a point of intersection whose abscissa verifies the proposed equation (they may have other points of intersection). Except for a few details about which he remains silent but which he verifies in the data he chooses, the geometrical properties that al-Ṭūsī describes are characteristic properties, and consequently lead to the equations of the curves he has used. Thanks to his use of the terms 'interior' and 'exterior', al-Ṭūsī draws on the continuity of curves and on their convexity. One can translate thus his procedure for the equation

$$x^3 - bx = c \qquad b, c > 0;$$

he effectively considers the two expressions

$$g(x) = \left[x\left(\frac{c}{b} + x\right)\right]^{\frac{1}{2}} \text{ and } f(x) = \frac{x^2}{\sqrt{b}}$$

and shows that, if there exist α and β such that

$$(f - g)\,(\alpha) > 0 \text{ and } (f - g)\,(\beta) < 0,$$

then there exists $\gamma \in \,]\,\alpha, \beta\,[$ such that $(f - g)\,(\gamma) = 0$.

Upon reading this first part, ones sees that al-Ṭūsī, just like al-Khayyām, mainly studies the geometrical construction of the positive roots of these twenty equations of degree ≤ 3, since the remaining ones will be reduced with the aid of affine translations to one or the other of these equations. In a manner analogous to that of al-Khayyām, he proceeds by means of plane geometrical constructions if the equation, reduced as much as possible, is of the first or second degree, and by means of constructions using two of the three curves mentioned above, if the simplest form of the equation is cubic.

Whereas the first part of the *Treatise* depends narrowly on al-Khayyām's contribution, one nevertheless already sees differences, the consequences of which will not appear until the second part. Indeed, for each equation he studies, al-Ṭūsī demonstrates the existence of the point of intersection of two curves, whereas al-Khayyām truly undertakes this study only for the twentieth equation. Also, al-Ṭūsī has introduced several notions, which he will use heavily in the second part, such as the affine transformations, the distance of a point to a straight line.

The second part of the *Treatise* is devoted to five equations that admit, according to al-Ṭūsī's expression, 'impossible cases', that is, cases for which no positive solution exists. They are the following equations:

(1) $x^3 + c = ax^2$
(2) $x^3 + c = bx$
(3) $x^3 + ax^2 + c = bx$
(4) $x^3 + bx + c = ax^2$
(5) $x^3 + c = ax^2 + bx.$

Contrary to al-Khayyām, al-Ṭūsī could not be satisfied with simply pointing out these 'impossible cases'. Indeed, concerned as he was with proving the existence of points of intersection and therefore the existence of roots, he therefore had to characterize such cases and find the reason behind them. It is precisely the intersection of this technical problem with the question that followed from it that led al-Ṭūsī to break with the tradition of al-Khayyām in order to modify his original project. In order to understand this profound shift, it is necessary to analyze al-Ṭūsī's procedure.

Each of the five equations is written in the form $f(x) = c$; f is a polynomial. To characterize the 'impossible cases', al-Ṭūsī studies the intersection of the curve representative of $y = f(x)$ with the straight line $y = c$. For al-Ṭūsī, this is a portion of the curve, one for which $x > 0$ and $y = f(x) > 0$, obtain simultaneously, and a portion that may not exist. Thus, in equation (1), he sets as a condition $0 < x < a$; in equation (2), $0 < x < \sqrt{b}$; in (3), he gives the condition $0 < x < \sqrt{b}$, which is not sufficient. Al-Ṭūsī is thus forced to examine the relations between the existence of solutions and the position of the constant c in relation to the maximum of the polynomial function. It is at this point that he introduces new concepts, new procedures, and a new language; much more importantly, he defines a new object. He thus begins by formulating the concept of the maximum of an algebraic expression, which he designates by 'the greatest number (*al-ʿadad al-aʿẓam*)'. Let $f(x_0) = c_0$ be the maximum; the latter gives the point (x_0, c_0). Al-Ṭūsī next determines the roots of $f(x) = 0$, that is, the intersection of the curve with the axis of the abscissas; finally, from this he deduces the upper and lower bounds of the roots $f(x) = c$.

For him, then, the entire problem is now to find the value of x that gives the maximum of $f(x)$. Al-Ṭūsī then proceeds to the solution of an equation, that, even if it is written differently, is none other than $f'(x) = 0$, where f' is the polynomial, the derivative of f. But before confronting the central problem of the derivative, let us note the change and the introduction of local analysis. Let us begin by recalling al-Ṭūsī's results. For equation (1), the derivative admits two roots, 0 and $2a/3$, which yield respectively a minimum $f(0) = 0$ and a maximum $f(2a/3) = c_0$. Conversely, the equation $f(x) = 0$ admits a double root $\lambda_1 = 0$ and a positive root $\lambda_2 = a$. Al-Ṭūsī thus concludes that if $c < c_0$ the equation (1) has two positive roots x_1 and x_2 such that $\lambda_1 = 0 < x_1 < x_0 < x_2 < \lambda_2 = a$. Note that there is a third root, x_3, which is negative and al-Ṭūsī does not consider. For equations (2), (3) and (5), his reasoning is similar. In these three cases, the derivative admits two roots with contrary signs. The positive root x_0 gives the maximum $c_0 = f(x_0)$ and the equation $f(x) = 0$ admits three simple roots, one of which is negative, and the others are $\lambda_1 = 0$ and λ_2, whence the preceding conclusion. To illustrate better al-Ṭūsī's procedure, let us summarize his discussion of equation (1). This equation is rewritten:

$$c = x^2(a - x) = f(x).$$

Al-Ṭūsī considers three cases:

• $c > \frac{4a^3}{27}$; the problem is impossible according to al-Ṭūsī (it admits a negative root);

• $c = \frac{4a^3}{27}$; al-Ṭūsī determines the double root $x_0 = \frac{2a}{3}$ (but does not recognize the negative root);

• $c < \frac{4a^3}{27}$; al-Ṭūsī determines the two positive roots, with $0 < x_1 < \frac{2a}{3} < x_2 < a$.

He then studies the maximum of $f(x)$; he shows that

$$(*) \qquad f(x_0) = \sup_{0<x<a} f(x) \qquad \text{with } x_0 = \frac{2a}{3},$$

by proving first

$$\text{a)} \qquad x' > x_0 \Rightarrow f(x') < f(x_0),$$

then

$$\text{b)} \qquad x'' < x_0 \Rightarrow f(x'') < f(x_0);$$

it is from a) and b) that he draws (*) above.

To find $x_0 = \frac{2a}{3}$, al-Ṭūsī solves $f'(x) = 0$. He then calculates

$$f(x_0) = f\left(\frac{2a}{3}\right) = \frac{4a^3}{27},$$

which allows him to justify the three cases he considered earlier. He next determines the two positive roots x_1 and x_2. He posits $x_2 = x_0 + y$; this affine transformation leads him to the equation

$$y^3 + ay^2 = k,$$

with $k = c_0 - c = \frac{4a^3}{27} - c$, an equation that he has already solved in the first part of the *Treatise*. He next justifies this affine transformation. He also proceeds to the affine transformation $x_1 = y + a - x_2$, with y being a positive solution of an equation previously solved in the *Treatise*. Al-Ṭūsī again justifies this last affine transformation and finally shows that $x_1 \neq x_0$ and $x_1 \neq x_2$.

A difficulty arises in equation (4), for the maximum $f(x_0)$ can be negative. Al-Ṭūsī then imposes a necessary condition, in order to consider only the case in which $f(x_0) > 0$, and proceeds as he has done earlier. The equation $f'(x) = 0$ then has two roots x'_0 and x_0 $(x'_0 < x_0)$, to which there correspond respectively a negative minimum and a positive maximum. Al-Ṭūsī considers only the root x_0 and obtains $c_0 = f(x_0)$. Moreover the equation $f(x) = 0$ in this case has three roots: 0, $\lambda_1 > 0$, $\lambda_2 > 0$, with $\lambda_1 < \lambda_2$. From this, al-Ṭūsī deduces that for $c < c_0$, equation (4) has two positive roots x_1 and x_2 such that

$$0 < \lambda_1 < x_1 < x_0 < x_2 < \lambda_2.$$

Box 3

Let's take the example of the numerical solution of the equation $x^3 = bx + N$.

Al-Ṭūsī writes: 'To determine the number that is sought, we place the number in the table and we count its rows by cubic root, no cubic root, cubic root. We place the zeros of the cubic root, we also count <the rows> of the number by root, no root, until we arrive at the homonymous root of the last affected place of cubic root. We next place the number of the roots, and we count its rows by root, no root. The homonymous row of the last affected place of root for this number of roots is the last row of the root of the number of roots. The problem has three cases.

First case: in which the homonymous root of the last affected place of cubic root is higher than <the row> of the last part of the number of roots, as when we say: a number with the form 3 2 7 6 7 0 3 8 plus nine hundred sixty-three roots equal a cube. We count from the homonymous root of the last affected place of cubic root to the last row of the number of roots, and we count the same number beginning from the last affected place of cubic root in this direction; and where we reach that point, we place the last part of the number of roots reduced to a third; one then obtains this figure:

3 2 7 6 7 0 3 8.

3 2 1

Since the homonymous root of the last affected place of cubic root is the third affected place of root, it is in the row of the tens of thousands, which is higher than the last row of the number of roots, which is in <the row> of the hundreds. We count starting from the row of the homonymous root of the last affected place of cubic root until the hundreds, and we count with this number also from the row of the last affected place of cubic root; one reaches the tens of thousands; we place the last part of the third of the number of roots in this row and next we place the cubic root that is sought, which is three, in place of the last zero. We subtract its cube from what is under it, we multiply it by the rows of the third of the number of roots and we add three times the product to the number. We place the square of the sought number parallel to it under the number according to this figure:

3
6 0 5 5 9 3 8
3 2 1
9

We subtract a third of the number of roots from the square of the sought number. We eliminate the third of the number of roots, and this figure then remains:

3
6 0 5 5 9 3 8
8 9 6 7 9

We shift the upper line by two rows and the lower line by one row; we place the second sought number, two, and we subtract its cube from the number; we multiply it by the sought number, we add the product to the lower line, we multiply it by the lower line, and we subtract three times each product of the number; we add the square of the second sought number to the lower line, we multiply it by the first sought number, we add the product to the lower line, and we shift the upper line by two rows and the lower line by one row. We place another sought number, which is one; we subtract its cube from the number, we multiply it by the first sought number and by the second, we add the result to the lower line, we multiply it by the lower line, and we subtract three times the product of this number. The upper line is then the figure 3 2 1 which is the root we sought.

Second case: in which the last row of the number of roots is higher than the homonymous root of the last affected place of cubic root, as when we say: roots in number equal to 1 0 2 0 2 1 plus a number of the form 3 2 7 4 2 0 equal a cube.

We count the number of roots by root, no root, and we add to the number two rows by having two zeros precede it; we seek the highest affected place of root corresponding to the number of roots; we then place the zeros of the cubic root; then we seek the homonymous cubic root of this affected place of root (the highest). We shift the row of the number of roots parallel to this root so that it will be parallel to the cubic root homonymous to it. We place the other rows of the number of roots, in order; one then has the figure:

0 0 3 2 7 4 2 0
1 0 2 0 2 1

for the highest place (affected of root) that corresponds to them is the third, and it is in the column of the tens of thousands: its homonym is the third affected place of cubic root which is in the <column> of the thousands of thousand. We shift the row of the tens of thousands by the number of roots so that it will be parallel to the affected place of the third cubic root, and we look for the greatest number such that one can remove its square by the number of roots; it is three; we place it in the third affected place of cubic root; we multiply it by the rows of the number of roots, we add the product to the number and we remove its cube from the number. We reduce the number of roots to a third; it will then begin in the row of the hundreds according to this configuration:

3
3 9 3 3 7 2 0.
3 4 0 0 7

Next we place the square of the sought number parallel to it under the number; one subtracts from it the third of the number of roots and one eliminates the line that is the third of the number of roots; we shift the upper line by two rows and the lower line by one row and we apply the procedure to the end.'[33]

This quick recapitulation shows that the presence of the concept of derivative is neither fortuitous nor secondary, but on the contrary intentional. Moreover, this is not the first time that one encounters the expression of the derivative in the *Treatise*. Al-Ṭūsī had already introduced it to build his method for the numerical solution of equations. Indeed, this method is organized in the following way: al-Ṭūsī determines the first digit of the root as well as its decimal order. The root is then written $x = s_0 + y$, with $s_0 = \sigma_0 10^r$ (r is the decimal order). He next determines the second digit by means of the following equation in y: $f(s_0 + y) = 0$; the algorithm that is named Ruffini-Horner then serves to determine the different terms of the preceding cubic equation in y. The algorithm introduced by al-Ṭūsī is used to dispose the calculations for the smallest possible number of multiplications, and it is nothing but a slight modification of the algorithm of Ruffini-Horner, which he adapts to cubic equations. Al-Ṭūsī then brings out as coefficient of y the value $f'(s_0)$ of the derivative of f in s_0. He obtains the highest possible digit of y, that is, the second digit of the sought root, by taking the integer part of

$$\frac{-f(s_0)}{f'(s_0)};$$

one can recognize here the method called 'of Newton' for the approximate solution of equations. After having determined the second digit, which is the first of y, one applies the same algorithm to the equation in y, in order to find the third digit, and one continues in this fashion until one obtains the root, which is an integer in the cases that al-Ṭūsī considers (see Box 3). But if it were not an integer, one would find the digits after the decimal point for the cases in which the root is not an integer, as attested by the text of al-Iṣfahānī in the 18th century.[34]

If there is no doubt about the presence of the expression of the derivative, it remains that al-Ṭūsī does not explain the route by which he reached such a concept. To understand better the originality of his procedure, let us consider the example of equation (3), which can be rewritten

$$f(x) = x(b - ax - x^2) = c.$$

[33] R. Rashed, *Sharaf al-Dīn al-Ṭūsī, Œuvres mathématiques*, vol. I, pp. 49–52.
[34] *Ibid.*, vol. I, pp. 118 ff.

The fundamental problem is to find the value $x = x_0$ that yields the maximum. It is in explaining the passage from equation (3) to two equations solved earlier by means of affine transformations

$$x \to y = x - x_0 \text{ and } x \to y = x_0 - x,$$

that al-Ṭūsī gives

$$f(x_0) - f(x_0 + y) = 2x_0(x_0 + a)y - (b - x_0^2)y + (3x_0 + a)y^2 + y^3,$$

and

$$f(x_0) - f(x_0 - y) = (b - x_0^2)y - 2x_0(x_0 + a)y + (3x_0 + a)y^2 - y^3.$$

Al-Ṭūsī had to compare $f(x_0)$ to $f(x_0 + y)$ and to $f(x_0 - y)$ by remarking that on $]0, \lambda_2[$, the terms

$$y^2(3x_0 + a + y) \text{ and } y^2(3x_0 + a - y)$$

are positive. Next he was able to deduce from the two equalities that

if $b - x_0^2 \geq 2x_0(x_0 + a)$, one has $f(x_0) > f(x_0 + y)$;
if $2x_0(x_0 + a) \geq b - x_0^2$, one has $f(x_0) > f(x_0 - y)$;

and therefore

$$b - x_0^2 = 2x_0(x_0 + a) \Rightarrow \begin{cases} f(x_0) > f(x_0 + y) \\ f(x_0) > f(x_0 - y) \end{cases}.$$

In other words, if x_0 is the positive root of equation

$$f'(x) = b - 2ax - 3x^2 = 0,$$

then $f(x_0)$ is the maximum of $f(x)$ in the interval considered. Note that the two equalities correspond to the Taylor series with

$$f'(x_0) = b - 2ax_0 - 3x_0^2, \quad \frac{1}{2!}f''(x_0) = -(3x_0 + a), \quad \frac{1}{3!}f'''(x_0) = -1.$$

It would seem that al-Ṭūsī's procedure consists in ordering $f(x_0 + y)$ and $f(x_0 - y)$ according to the powers of y and to show that the maximum occurs when the coefficient of y in this series is 0. The value of x that makes $f(x)$ a maximum is thus the positive root of the equation represented by the equation $f'(x) = 0$. The virtue of the affine transformations $x \rightarrow x_0 \pm y$, with x_0 as the root of $f'(x) = 0$, is to make the y term disappear from the new equation. It is probably from this property that al-Ṭūsī discovered the derivative equation $f'(x) = 0$, perhaps in connection with the consideration of the curve representing f which he never traces in the *Treatise*: for a small y, the main part of the series of $f(x_0 \pm y)$ is in y^2 and does not change signs with y. We have shown that in his search for the maxima and minima of polynomials, al-Ṭūsī's procedure resembles like a sister that of Fermat.[35]

We have thus just seen that the theory of equations is no longer only this one chapter of algebra, but encompasses a much larger domain. Under this theory, the mathematician brings together the geometrical study of equations and their numerical solutions. He poses and solves the problem of the conditions of possibility of each equation, which leads him to study systematically the maximum of a third-degree polynomial by means of a derivative equation. During his numerical solution, he not only applies certain algorithms in which one once again encounters the concept of derivative of a polynomial, but also attempts to justify these algorithms by means of a concept of 'dominant polynomials'. This is clearly very high-level mathematics for the period; let us say simply that here already one touches upon the limits of mathematical research that can be carried out without efficacious symbolism. Indeed al-Ṭūsī carries out all of his research in natural language, without any symbolism at all, except for a certain tabular symbolism, which makes it particularly complicated. A difficulty of this sort indeed emerges as an obstacle not only to the internal progress of the research itself, but also to the transmission of its results. In other words, as soon as the mathematician manipulates analytical notions like those mentioned above, natural language quickly proves to be inadequate to express the concepts and the operations that are applied to them, and limits innovation as well as the diffusion of this mathematical knowledge. Al-Ṭūsī's successors very probably faced the same obstacle until mathematical notation was truly transformed, beginning with Descartes in particular.

But the example of al-Ṭūsī suffices to show that the theory of equations not only was transformed after al-Khayyām, but also continued to diverge ever more from the search for solutions by radicals; it thus ended up by covering a vast domain that included sectors that will later belong to analytical geometry or simply to analysis.

[35] R. Rashed, *Sharaf al-Dīn al-Ṭūsī, Œuvres mathématiques*, vol. I, p. XXVII.

6. THE DESTINY OF THE THEORY OF EQUATIONS

But what was the destiny of this theory of cubic equations that al-Khayyām constituted and that al-Ṭūsī transformed? Future historical research alone will be able to give a true answer to this question. But let's not nourish false hopes. To go beyond al-Khayyām and al-Ṭūsī, it was necessary to conceive not only an efficacious symbolism – not simple abbreviations, as one finds them in al-Qalaṣādī – but especially a new mathematical program that would lead beyond conic curves to tackle the study of algebraic curves by means of equations. In short, the true successors of al-Khayyām and of al-Ṭūsī are Descartes and Fermat.[36] This insight in no way prevents us from following, in the domain of the history of Arabic mathematics, the works from the posterity of al-Khayyām and al-Ṭūsī. The first successor is al-Ṭūsī's very own student: Kamāl al-Dīn ibn Yūnus, the mathematician of Mosul. He takes up again the most difficult equation that al-Khayyām and al-Ṭūsī had examined, the twenty-fifth in their classification:

$$(1) \qquad x^3 + c = ax^2 + bx.$$

Al-Khayyām[37] distinguishes the three cases of $c < ab$, $c = ab$, or $c > ab$ and in the first case distinguishes the greatest positive root, $x^2 > a$. Sharaf al-Dīn al-Ṭūsī, for his part, gives a complete discussion and distinguishes the following cases:

$a = \sqrt{b}$: he shows in this case that the equation $3x^2 - 2ax - a^3 = 0$ has a positive root, $x_0 = a$, and he finds $c_0 = ab$, whence $c < ab$ is a necessary and sufficient condition for equation (1) to have two positive roots, $0 < x_1 < a < x_2 < 2a$. For the two other cases, $a > \sqrt{b}$ and $a < \sqrt{b}$, al-Ṭūsī discusses in a similar manner the necessary and sufficient conditions for the existence of positive roots.

Ibn Yūnus seems to start with the study of al-Khayyām, in order to complete it and to determine the smallest positive root. He proceeds with the help of two methods to determine $x_1 < a$. The second corresponds to that of al-Khayyām: the intersection of an equilateral hyperbola with a

[36] See below, 'Descartes's *Géométrie* and the distinction between geometrical and mechanical curves' and 'Fermat and algebraic geometry'.

[37] R. Rashed and B. Vahabzadeh, *Al-Khayyām mathématicien*, pp. 82–4; English transl., pp. 81–2.

second hyperbola.[38] This is therefore research in the wake of al-Khayyām, destined to complete this study following the same, or a similar, method.

Kamāl al-Dīn ibn Yūnus's student, the astronomer and mathematician Athīr al-Dīn al-Abharī (d. 1262), composed an algebraic treatise that has reached us in an incomplete state, as the scribe himself testifies. In the surviving part, however, al-Abharī applies to the equation $x^3 = a$ al-Ṭūsī's method of numerical solution, and on the same terms as the latter. He is also familiar with the use of affine transformations to reduce one equation to another. Thus, even when he discusses the three canonical types of second-degree equation, he considers $x^2 = px + q$, recalls that $x > p$, posits that $x = y + p$ and returns to the previously studied equation $y^2 + py = q$.

Al-Khilāṭī, another algebraist from this era, reminds the reader that al-Ṭūsī was 'the master of his master' and studied cubic equations, but that Al-Khilāṭī himself remains faithful to the tradition of al-Karajī.[39] Other witnesses from the period mention al-Ṭūsī, but nothing that has come down to us thus far indicates that any other mathematician picked up on al-Ṭūsī's theory.[40]

Much later, 18th-century mathematicians in the school of Iṣfahān, such as al-Iṣfahānī, give in the treatise mentioned above an interesting method for finding the positive root of a cubic equation, one based on the property of the fixed point.[41]

[38] *Ibid.*, pp. 83–4.

[39] *Nūr al-Dalāla fī 'ilm al-jabr wa al-muqābala*, ms. of the University of Teheran, no. 4409, fol. 2.

[40] See Shams al-Dīn al-Mārdīnī, *Niṣāb al-ḥabr fī ḥisāb al-jabr*, ms. Istanbul, Feyzullah, no. 1366, fols 13–14.

[41] See the chapter on algorithmic methods, pp. 387 ff.

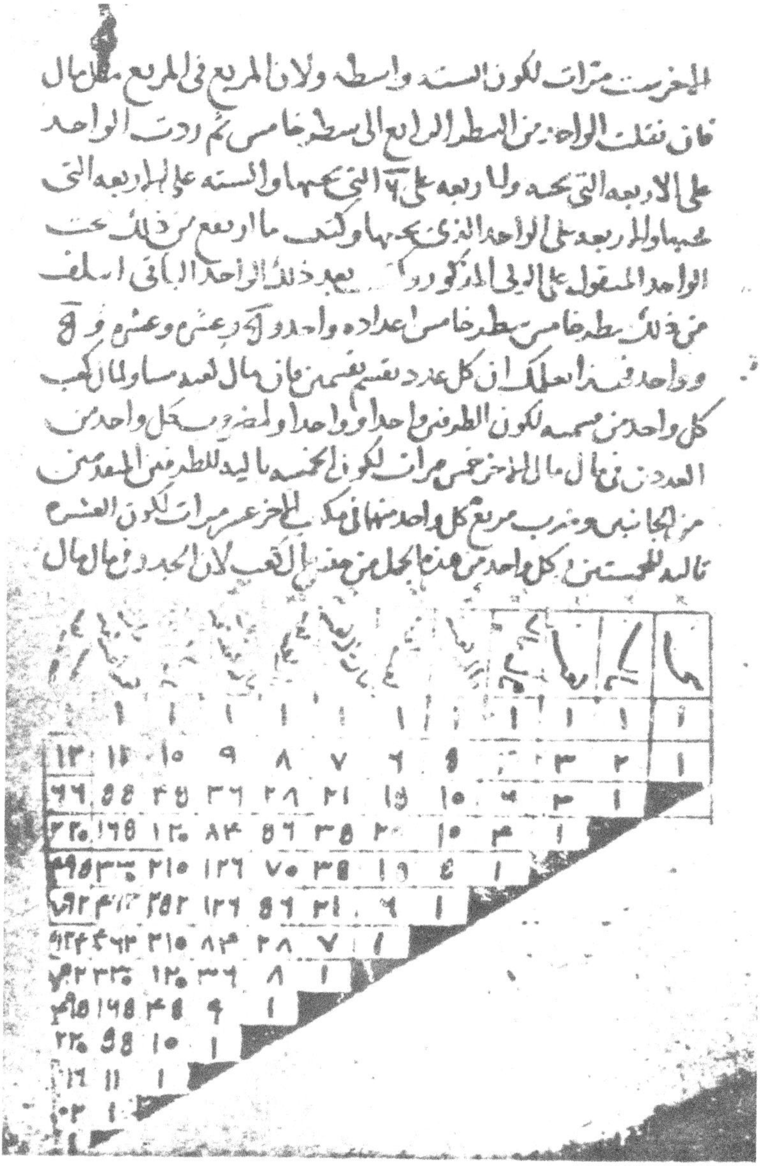

Reproduction of the arithmetic triangle of al-Karajī
in *al-Bāhir* by al-Samawʾal (ed. S. Ahmad and R. Rashed), p. 111.

– 2 –

ALGEBRA AND LINGUISTICS THE BEGINNINGS OF COMBINATORIAL ANALYSIS

Around the middle of the 17th century, the study of combinations presents itself as a full-blown, independent field of research with its own dedicated literature. Recall, for example, the *Abrégé des combinaisons (Summary of Combinatorics)*, which Frenicle composed in this period but published later.[1] Consider also the writings devoted to the calculus of probabilities, particularly Pascal's *Opuscule* on 'Combinationes' (Combinatorics),[2] and later the second part of Jacques Bernoulli's *Ars conjectandi*,[3] as well as many others. Finally recall the *De arte combinatoria* of Leibniz.[4] A quick examination shows that research on combinations originates in the brand new calculus of probabilities – first as a doctrine of chance, then as a genuine calculus of the probable – in arithmetical studies, in studies of the universal language and 'universal characteristic' (Leibniz's *characteristica universalis*), a part of theoretical philosophy. It is true that, much more than any other discipline, the calculus of probabilities revived research on combinations to suit the needs of its own constitution. Yet what one witnesses is a genuine self-conscious insight: the object 'combination' is itself an object of study, independently of the places where it occurs. In this self-consciousness resides the autonomy, as it were, of combinatorial analysis. Would it therefore be legitimate to date the beginning of this chapter to the middle of the 17th century? Certainly one would go badly astray if, prior to this era, one merely collected here and there a few combinations carried out by

[1] Frenicle, 'Abrégé des combinaisons', in *Divers ouvrages de mathématiques et de physique par Messieurs de l'Académie Royale des Sciences*, Paris, Imprimerie Royale, 1693; also in *Mémoires de l'Académie royale des sciences, depuis 1666 jusqu'à 1699*, vol. V, Paris 1729, pp. 87–125.

[2] Pascal, *Œuvres de Blaise Pascal*, Collection des Grands Écrivains de la France, Paris, 1908, vol. III.

[3] Jacques Bernoulli, *Ars conjectandi*, Basel, 1713.

[4] Leibniz, 'Dissertatio de arte combinatoria', in *Die philosophischen Schriften von G.W. Leibniz*, Hildesheim, Georg Olms, 1965.

logicians, linguists, alchemists, etc. On the contrary, one would need to find analogous situations from which the calculus of probability is absent, that is, in which this chapter, a leading factor in the development of combinatorial analysis, has not yet been conceived. Arabic mathematics offers us just such a situation, for, as far as we know, no study of the calculus of probabilities occurs in this mathematics. After noting this exception, however, one observes that linguistic studies, research in algebra and also in arithmetic (studies of the compounding of ratios), and some writings in theoretical philosophy, led mathematicians up to the domain of combinations. Moreover, some mathematicians seem to have devoted proper treatises to combinations. We know at least one, Ibrāhīm al-Ḥalabī, who in the 16th century wrote a book called *Fī istikhrāj ʿiddat al-iḥtimālāt al-tarkibiyya min ayy ʿadad kāna* (*On the Determination of Combinable Eventualities Starting from Any Given Number*),[5] which we had the good fortune of finding.

One last item should be noted: whereas some participate in this research on combinations, as we shall see from the 8th century on, neither the classifiers of the sciences, nor the biobibliographers, neither ancient nor modern, mention them: this is the case of al-Fārābī as well as Ibn al-Akfānī several centuries later. A similar silence prevails among historians of mathematics and of science: it is only recently – four decades ago – that we permitted ourselves to speak of combinatorial analysis in Arabic mathematics.[6] But neither we nor those who followed us have broached the legitimacy of this usage, that is, asked ourselves why this great absence had never ceased to be present. The answer to this question will no doubt allow us a better understanding of the conditions for the formation of the discipline, even though the latter still remained unnamed.

1. LINGUISTICS AND COMBINATORICS

From the 8th century on, one name dominates several chapters of Arabic linguistics: al-Khalīl ibn Aḥmad (718–786). A mathematician, he is the author of a book of arithmetic; a musicologist, he is the founder of Arabic phonology, of prosody, of lexicography, to say nothing of his

[5] Ms. Istanbul, Hamadiye 873; ed. forthcoming.

[6] R. Rashed, 'Algebra and linguistics: combinatorial analysis in the Arabic science', in R. Cohen (ed.), *The Development of Arabic Mathematics: Between Arithmetic and Algebra*, Boston Studies in the Philosophy of Science, Boston, Reidel, 1994, pp. 261–74.

contributions to grammar and to other linguistic disciplines. His case is exceptional and deserves a large book. Here, we consider only his contribution to lexicography.

As far as we know, al-Khalīl was the first to conceive the project of composing not merely a lexicon, of Arabic, but a dictionary. It is to this project that he devoted *Kitāb al-ʿayn*, which is the first known dictionary. It is possible that al-Khalīl wrote only one part, the remainder being the work of his student, al-Layth; it is possible that others also participated. Scholars who debate these questions of attribution nevertheless agree that the project itself is al-Khalīl's, that it is he who shaped the means of realizing it, and that we are indebted to him for one part of this book.

Al-Khalīl's project is both clear and precise: to rationalize the empirical practice of lexicographers and to extend it so that a single book contains the entire vocabulary of Arabic. It is neccssary to find the means of enumerating exhaustively the words of the language and in addition to establish a bi-univocal correspondence between the set of words and the entries of the dictionary. It is in this connection that al-Khalīl elaborates the theory that may be encapsulated thus: the language is a phonetically realized part of the possible language. The words of the latter are obtained by combinations and permutations of letters; the words of the former are those of the possible language, verify the rules of phonetic compatibility, and are effectively used. The lexicographer thus confronts two tasks at once: the first is deliberately and uniquely combinatorial; the second is phonological. These are the two main tasks that interest us here, but al-Khalīl has added to them several others; ethnolinguistics, history, etc.

Al-Khalīl begins by recalling that the roots of Arabic words are at least two-lettered, and at most five-lettered. Indeed if the arrangement r at a time of the 28 letters of the alphabet, with $1 < r \leq 5$, gives us the set of roots, and therefore of words, of the possible language, the language will be formed by only one part of the result, limited by the rules of phonology (that is, the compatibility of the roots' phonemes). To construct a dictionary, one must therefore constitute first the possible language in order to extract from it, according to preceding rules, all the words of the real language. To compose his dictionary, al-Khalīl thus begins by calculating the combination without repetition of the letters of the alphabet taken r at a time, with $r = 2, 3, 4, 5$; and then the number of permutations of each group of r letters. In other words, he calculates

$$A_n^r = r!\binom{n}{r}$$

$n = 28$ and $1 \leq r \leq 5$.

At this stage, it is important to discover if, in order to obtain these results, al-Khalīl proceeded by simple direct enumeration (that is, empirically), or if earlier he had elaborated rules for the computation of combinations and permutations. Only on this condition will we be able to judge if he contributed to combinatorial analysis.

The question is a thorny one, because some of al-Khalīl's writings are lost. Fortunately, there survives the following citation from the later linguist, Hamza al-Aṣfahānī (or Iṣfahānī), transmitted to us in the *Muzhir* of the famous linguist al-Suyūṭī:

> Al-Aṣfahānī writes: 'Al-Khalīl mentioned in his *Kitāb al-ʿayn* that the number of these forms (roots) of the language of the Arabs, those used and those neglected in the four categories – two-lettered, three-lettered, four-lettered, and five-lettered – without repetition, is twelve thousand thousand, three hundred thousand, five thousand, four hundred and twelve.[7]

Al-Aṣfahānī continues by parsing this number in the following way: words that are two-lettered 756; three-lettered 19.656; four-lettered 491.400; five-lettered 11.793.600; the sum is therefore 12.305.412. This citation does not appear in the *Kitāb al-ʿayn* as it has come down to us. Perhaps it appeared in another version, or in another of al-Khalīl's books, now lost. However that may be, this quotation was picked up by his successors, such as al-Suyūṭī and it effectively corresponds to the work accomplished in this last book.

The numbers cited above are exact and cannot be obtained by direct enumeration. Without the shadow of a doubt, al-Khalīl must have proceeded by means of a genuine computation. But which one?

To answer this last question, we appeal to a later author who seems to perpetuate an old tradition: Ibn Khaldūn in his *Prolegomena*. According to the latter, to obtain the two-lettered roots, al-Khalīl combines a first letter with all those that followed, then the second letter with all those that followed, and so on. He then sums the whole 'by a process known among the arithmeticians'.[8] In other words, he combines the first letter with the 27 that follow, then the second letter with the 26 that follow, and so on; whence the sum $\sum_{k=1}^{27} k = 378$, 'which one doubles on account of the inversion

[7] Al-Suyūṭī (ʿAbd al-Raḥmān Jalāl al-Dīn–), *al-Muzhir fī ʿulūm al-lugha wa-anwāʿihā*, ed. Muḥammad Aḥmad Jād al-Mawla, ʿAlī Muḥammad al-Bijāwī, Muḥammad Abū al-Faḍl Ibrāhīm, Cairo, n.d., vol. I, pp. 74–5.

[8] Ibn Khaldūn, *al-Muqaddima*, Cairo, n.d., p. 548.

of the two-lettered words', hence 756 words. In short, he effectively calculates $2!\binom{28}{2}$.

To obtain the triliterals, one uses the same calculation. One considers the biliterals, namely (a, b) and one combines with this couple all the successive letters; likewise for all the other couples. The total number of words will be

$$\frac{1}{2}\sum_{k=1}^{26} k(k+1) = 3.276 \text{ words.}$$

But for every triliteral word, one can obtain six words by permutation, for a total of 6 × 3.276 = 19.656; here, too, he calculates $3!\binom{28}{3}$. Likewise for the quadriliterals:

$$\frac{1}{6}\sum_{k=1}^{25} k(k+1)(k+2) = 20.475;$$

whence 24 × 20.475 = 491.400, al-Khalīl's total, that is, $4!\binom{28}{4}$. Finally, for the quadriliterals:

$$\frac{1}{24}\sum_{k=1}^{24} k(k+1)(k+2)(k+3) = 98.280,$$

whence 120 × 98.280 = 11.793.600, that is $5!\binom{28}{5}$.

The whole problem is to know how al-Khalīl effectively proceeded: by combination and permutation? or by calculating sums of integers? and, in the latter case, did he know the relations between the two? And if so, one must suppose either that he knew how to form the figurate numbers as well as the relations among these numbers and the formula of combinations, or he knew how to compute the sum of powers of natural integers, for the fourth and fifth power. Now such calculation does not occur before Ibn al-Haytham, at the beginning of the 11th century and in a very different mathematical context.

Ibn Khaldūn's presentation seems to leave no doubt about the matter: despite confusion and errors, he suggests that al-Khalīl had proceeded by combinations and permutations. This strong suggestion draws directly on the probable sources of Ibn Khaldūn. Indeed the latter seems to have gathered his information from Ibn al-Bannā᾽ (d. 1321), notably in his book *Raf῾ al-ḥijāb*.[9] In this treatise of arithmetic, Ibn al-Bannā᾽ explicitly links figurate numbers and the combinations used in lexicography, in the tradition of al-Khalīl. He thus sets up for the letters of the alphabet $n = 28$ and for $1 < r \leq 5$

$$\binom{n}{r} = \sum_{k=1}^{n-r+1} F_k^r = \frac{n(n-1)\ldots(n-r+1)}{r!}$$

with F_k^r the kth figurate number of order r.

During his demonstration by archaic recurrence,[10] he uses

$$\binom{n}{r} = \frac{n-r+1}{r}\binom{n}{r-1}$$

and knows the rules that, in our notation, we can express thus:

$$(n)_r = n(n-1)\ldots(n-r+1)$$
$$(n)_n = n!$$

To be sure Ibn al-Bannā᾽ does not study figurate numbers with the generality that one finds in his contemporary al-Fārisī, but the meaning is there for $1 < r \leq 5$. What matters here, however, is the articulation of combinatorial formulas based on knowledge of the arithmetic triangle, the law of its formulation, and the lexicography of al-Khalīl. There is therefore not the slightest doubt about either the combinatorial interpretation of formulas, or the manner of obtaining the sum of natural integers presented by Ibn al-Bannā᾽ and probably borrowed – rather awkwardly – by Ibn Khaldūn.

But Ibn al-Bannā᾽ himself seems to belong to a tradition that articulated the combination of lexicography and knowledge of the arithmetic triangle. Thus Ibn al-Mun῾im (d. 1228) also tried to find all the words of the language, and not only the roots. After recalling the project of al-Khalīl, he wants to extend it to all Arabic words, whose maximum length is no more than 10 letters out of an alphabet of 28 letters. He conceives an original model:

[9] R. Rashed, 'Nombres amiables, parties aliquotes et nombres figurés aux XIIIe et XIVe siècles', *Archive for History of Exact Sciences*, 28, 1983, pp. 107–47; English transl. in *The Development of Arabic Mathematics*, pp. 275–319.

[10] R. Rashed, *The Development of Arabic Mathematics*, pp. 299–303 and 'Matériaux pour l'histoire des nombres amiables et de l'analyse combinatoire', *Journal for the History of Arabic Science*, 6, 1982, pp. 209–78.

> One has access to ten colors of silk. One wishes to make tassels (*sharārīb*) some of which have only one color, others two colors, yet others three colors, and so on, until the last tuft has ten colors; and one would like to know the number of each individual type of tuft.[11]

Ibn al-Mun῾im then gives the arithmetic triangle.

To return to al-Khalīl, it seems that he had obtained combinatorial formulas, at least by induction. Being both a mathematician and a linguist of genius, he was no doubt up to the task.

Note finally that al-Khalīl the lexicographer considers the alphabet of 28 letters. He omits several phonemes; he does not consider the 'hamza (ʾ)' to be a letter that enters into combination. Recall that this phoneme had no fixed form, and is written *alif* as in *badaʾ*, *wāw* as in *yūʾmin*, *yāʾ* as in *yastanbiʾūnaka*, and sometimes without a letter, as in *bināʾ*. This is why linguists such as al-Mubarrad do not consider it a letter. Perhaps for this reason, al-Khalīl eliminates it as he is composing his dictionary. But al-Khalīl the phonologist reintegrates the 'hamza'; he considers it to be a phoneme and thus counts 29 phonemes. Beyond the calculation of combinations, we note al-Khalīl also proceeds by permutation. As cited by his student al-Layth, al-Khalīl writes in *Kitāb al-῾ayn*:

> Know that a biliteral word takes two forms, such as qd, dq; šd, dš; and triliteral words take six forms that one calls *masdūsa*, 'in the form of six', such as ḏrb, ḏbr, brḏ, bḏr, rḏb, rbḏ. The quadriliteral word takes twenty-four forms since its letters, which are four, are multiplied by the number of forms of the triliteral, which is six; one has twenty-four forms [...]. The quinqueliteral word takes one hundred twenty forms for its letters, which are five, are multiplied by the forms of the quadriliterals, which are twenty-four; this makes one hundred twenty forms, of which a small number is used and the larger number is neglected.[12]

This quotation shows us, on the one hand, how al-Khalīl reasons – this is consistent with the preceding interpretation – and, on the other, that he knows the two expressions $P_n = 1, 2, 3 \ldots n$ and $P_n = nP_{n-1}$.

The fact that $n = 2, 3, 4, 5$ takes nothing away from the generality of the expression, just as is the case in the 17th century.

After having obtained the possible language by means of these combinations and permutations, al-Khalīl exploits both his phonological

[11] Cited in Driss Lamrabet, *Introduction à l'histoire des mathématiques maghrébines*, Rabat, 1994, p. 215.

[12] Al-Khalīl ibn Aḥmad, *Kitāb al-῾ayn*, ed. Abdullāh Darwish, Cairo, 1967, vol. I, p. 66.

and his ethnolinguistic knowledge to isolate the real language. Thus, once he has distinguished two levels of analysis – signs and significations – and once he has reconstituted the possible language solely from the level of signs, he moves on to an additional differentiation among sounds: the periodic sound (musical) and the irregular sound (aperiodic, that is between vowels and consonants). The consonants are then arranged in classes according to their points of articulation. Starting with the laryngials and ending with the labials, he lists the following:[13]

1	ʿ	*ḥ*	*h*	*ḫ*	*ġ*
2	*q*	*k*			
3	*ǧ*	*š*	*ḍ*		
4	*ṣ*	*s*	*z*		
5	*ṭ*	*t*	*d*		
6	*ẓ*	*ḏ*	*ṯ*		
7	*r*	*l*	*n*		
8	*f*	*b*	*m*		
9	*ī*	*ū*	*ā*	ʾ	

For some classes, he distinguishes between voiced and unvoiced letters: thus, in the first class, ʿ is voiced whereas *ḥ* is unvoiced; and in the fifth, *d* is voiced but *t* is unvoiced. An examination of al-Khalīl's classification and of his explanations in the *Kitāb al-ʿayn* easily shows that, in light of modern phonetics, he has correctly approached the distribution of sounds into classes according to points of articulation, on the one hand, and the opposition of voiced/unvoiced, on the other. The order of the consonants within each class nevertheless remains rather approximate, and the students of al-Khalīl, *e.g.*, Sībawayh, will take up his analysis in order to improve it.

In this phonological analysis, al-Khalīl will find the necessary conditions for recognizing, among the words of the possible language, those that can be real. It so happens that not all of the words that meet the conditions of reality are necessarily used. It is at this point that ethnolinguistics, knowledge of pre-Islamic literature and of the literature of the first century of Islam, the Qurʾan, etc., enter the picture: these linguistic treasures allow one to distinguish between words that are utilized and those that are neglected (*muhmal*). Nevertheless, one ought not forget that this phonological study allowed al-Khalīl to discover a property of Arabic (and

[13] *Ibid.*, p. 65.

of Semitic languages more generally) that became essential to his lexicographic project. Indeed, he noticed a morphological characteristic of Arabic, that is, the importance of roots in the derivation of its vocabulary and the relatively small number of its roots. As a group of consonants (and consonants only) and as a signified to which a generic signifier is most often attached, the root could not emerge as a theoretical unit of analysis before the preceding distinctions between meaning and signification, on the one hand, and vowel and consonant, on the other. These roots are, moreover, limited forms, restricted to the four forms mentioned above: at most quinqueliterals, but mostly triliterals. This analysis thus allowed al-Khalīl to conceive both his project and the means of bringing it to fruition. Among the latter, we emphasize the possibility of omitting the half-vowels, which would have made the combinatorics much more complex. It also gave him the rules of incompatibility between phonemes within the same root. We cannot here present in detail these rules of incompatibility. Roughly summarized, the first two consonants of the root can belong neither to the same class of localization nor (frequently) to neighboring classes of localization. The last two consonants of the root fall under the same rule but can be similar. The derivation of words from roots occurs by means of finite patterns, which are themselves the object of combinatorics. These finite patterns and their combinations will be recognized as research develops, that is, when Arabic phonology as well as morphology will be considered in their own right and not merely as auxiliaries of lexicography. This will be the contribution of the students and successors of al-Khalīl.

The *Kitāb al-ʿayn* will not only survive al-Khalīl, but also become a model for a very long tradition. In short, every Arabic lexicographer is in a sense a student of al-Khalīl's. To be sure, his heirs corrected the errors he committed as he collected words from the real language, varied the form of the dictionary, and perfected its composition; but the method remains essentially the same. To cite only one example from among al-Khalīl's successors, consider Ibn Durayd. Born in 223/834, less than a half-century after al-Khalīl, Ibn Durayd was also a member of the school of Baṣra; he wrote *al-Jamhara*, in which he proceeds to calculate n^r, for $n = 28$, the number of letters, and $1 < r \leq 5$. He insists on distributing the various classes of forms obtained according to whether or not they contain one or more defective letters – the *wāw*, the *yāʾ* and the *hamza*, in other words, according to a morphological principle. To reinforce the point, let us examine his calculation for $r = 2$. He obtains $n^r = 784$ forms; he removes 28 of them, that is, the forms consisting of the repetition of one and the same letter. There remain $756 = 28 \times 27 = A_n^r$. He emphasizes that these 28 forms are invariant under permutation (*qalb*). Next, he examines the morphology of all forms and finds $600 = 24 \times 25 = A_{25}^2$ forms without any defective letter; 150 forms, each of which contains a defective letter; then 6 forms, each of which contains 2 defective letters; and finally 3 forms each of which contains a defective letter repeated twice. Ibn Durayd carries out the calculation for triliterals, quadriliterals, and quinqueliterals. Like al-Khalīl and explicitly, he considers this combinatorial study 'as a kind of calculation (*bi-ḍarbin min al-ḥisāb*)'. He writes: 'I explain to you what you obtain from biliteral, triliteral, quadriliteral, and quinqueliteral forms, if the High God wills it, by a kind of clear calculation (*ḥisāb*)'.[14]

One can follow this lexicographic tradition for another millennium, through a goodly number of dictionaries such as those of Aḥmad ibn Farīs (*al-Maqāyīs*), Ibn Manẓūr (*Lisān al-ʿArab*), al-Zabīdī (*Taj al-ʿArūs*), etc.

We have just seen that, beginning in the 8th century, lexicographers not only undertook combinatorial studies but were in possession of the elementary forms of this new chapter: $\binom{n}{r}, A_n^r, P_n, n^r, n!$, in our notation. Since al-Khalīl, they were clearly conscious that these procedures and expressions pertained to 'a kind of *ḥisāb* (calculation)'. This is the first title given to this chapter, which consists of a calculation of combinations. The latter gained authoritative recognition on its own, so to speak, when one tried first to resolve *theoretically* this practical problem of composing a dictionary. Language then emerges as a privileged domain as much for elaborating this new calculation as for exercising it. This phenomenon is in a sense associated with the history of elementary combinatorial analysis. Indeed, is not language one of the readily available domains in which discreteness and finitude are both verified? The letters are discrete objects, finite in number. Later, algebraists and number theorists will try precisely to return to language to draw from it at once examples, notations, and methods in order to illustrate the combinatorics that they will conceive, in all likelihood, independently of the linguists.

Nevertheless, lexicography is not the only discipline, the constitution of which required the elaboration of a combinatorics. In prosody, the procedure is analogous and was that of the very same al-Khalīl. He is also credited with one of the first treatises composed in a discipline that was constituted as such in this period: namely, cryptography and cryptoanalysis.[15] Even if these disciplines are not a part of linguistic research,

[14] See al-Suyūṭī, p. 72.

[15] Cf. *Tabaqāt al-naḥawiyyīn wa-al-lughawiyyīn* of al-Zabīdī, ed. Muḥammad Abū al-Faḍl Ibrāhīm, Cairo, 1973.

they are intimately tied to it. It is therefore understandable that numerous linguists, for many centuries, have written works of cryptography or cryptoanalysis. In these disciplines, as in lexicography and prosody, one wants to solve a practical problem theoretically: to invent efficacious algorithms in order to hide from everyone who does not know them the meaning of a text or message. The original name of this discipline is therefore *al-taʿmiya*, from *ʿamiya*, to go completely blind. In any case, by the 9th century at the latest, with al-Kindī, this discipline had not only its own name, but also an entire technical vocabulary. Beginning in this period, an immense specialized literature will thrive until the middle of the 13th century; witness the writings of al-Zabīdī, the author of both the famous Arabic dictionary and a treatise on the foundations of cryptography.[16] Among the names of famous authors, one encounters ʿAlī ibn ʿAdlān (583–661 H./1187–1263), Isḥāq ibn Ibrāhīm ibn Wahb, Ibn Ṭabaṭaba (d. 322 H./934), ʿAlī ibn Muḥammad ibn al-Durayhim (712–762 H./1312–1361), etc.[17] As is easy to understand, these authors drew on transposition, substitution, and permutation to create some of their algorithms. Although they did not introduce new rules to enrich combinatorics, on the one hand, they showed that this tool could be put to use in a domain other than lexicography and prosody, and therefore that it did not depend on any particular field; on the other hand, they contributed to its diffusion among a readership broader than that of the linguists; finally, they displayed the truly combinatorial meaning of such expressions as permutation, transposition, and substitution.

One could cite many more examples, both tied to and independent of the preceding domains, which reveal that both combinatorial knowledge and combinatorial practice had diffused among scholars and philosophers. Thus, the literary scholar Abū al-Ḥayyān al-Tawḥīdī mentions in his *Muqabasāt* the example of his contemporary, the philosopher Yaḥyā ibn ʿAdī, who sought the numbers of the 'logical division' according to which the figures of the phrase *inna al-qāʾim ghayr al-qāʿid*, that is, the number of configurations obtained from a phrase of 14 letters (the hamza is not a letter). This is the model case of a distribution of 14 balls in 2 boxes. Yaḥyā

[16] Al-Zabīdī, *Tāj al-ʿArūs*, ed. ʿAbd al-Satār Aḥmad Farāj, Kuwait, 1965, Introduction, p. XI.

[17] See M. Mrayātī, Yaḥyā Mīr ʿAlam, Ḥassān al-Ṭayyān, *Origins of Arab Cryptography and Cryptoanalysis*, vol. I: *Analysis and Editing of Three Arabic Manuscripts: Al-Kindī, Ibn Adlān, Ibn al-Durayhim* (in Arabic), Damascus, Arab Academy of Damascus publications, 1987 and vol. II.

ibn ʿAdī finds exactly $2^{14} = 16\,384 = \sum_{k=0}^{14}\binom{14}{k}$. Al-Bīrūnī confirms this testimony in *Taḥdīd nihāyāt al-amākin* and himself proceeds to a simple calculation of combinations to treat the lunar eclipse.[18]

Another domain into which combinatorial procedures enter is that in which Thābit ibn Qurra (826–901) studies the compound ratios as well as the sector figure.[19]

2. ALGEBRAIC CALCULATION AND COMBINATORICS

Engaged in this famous movement of 'arithmetization of algebra'[20] and thus of the development of abstract algebraic calculation, mathematicians came to conceive of new techniques such as the development of the binomial of any degree. It was while they were elaborating these techniques that they were led to the table of binomial coefficients, the rule of its formation, and the binomial formula enunciated for integer powers. Indeed al-Samawʾal tells us that, at the end of the 10th century, al-Karajī not only had access to the formulas below (** and *) but had in addition established the binomial theorem by complete finite induction.[21]

** $$\binom{n}{r} = \binom{n-1}{r-1} + \binom{n-1}{r}$$

and

* $$(a+b)^n = \sum_{r=0}^{n}\binom{n}{r}a^{n-r}b^r$$ where n is an integer.

[18] Edward S. Kennedy, *A Commentary upon Bīrūnī's* Kitāb Taḥdīd al-Amākin, *an 11th Century Treatise on Mathematical Geography*, Beirut, American University of Beirut, 1973, pp. 169–70; Arabic edition *Kitāb Taḥdīd nihāyāt al-amākin*, ed. P. Bulgakov, in *Revue de l'Institut des manuscrits arabes*, Cairo, vol. 6, fasc. 1&2, mai-nov. 1962, pp. 101–2; English translation by Jamil Ali, *The Determination of the Coordinates of Cities*, Beirut, 1967, pp. 132–3.

[19] See P. Crozet, 'Thābit ibn Qurra et la composition des rapports', in R. Rashed (ed.), *Thābit ibn Qurra. Science and Philosophy in Ninth-Century Baghdad*, Scientia Graeco-Arabica, vol. 4, Berlin/New York, Walter de Gruyter, 2009, pp. 391–535 and H. Bellosta, 'Le traité de Thābit sur *La figure secteur*', *ibid.*, pp. 335–90.

[20] See above, 'Algebra and its unifying role'.

[21] R. Rashed, *The Development of Arabic Mathematics*, pp. 62–84 and al-Samawʾal, *al-Bāhir en Algèbre*, ed., and commentary by S. Ahmad and R. Rashed, Damascus, Presses de l'Université de Damas, 1972.

Since the end of the 10th century, the expressions of the form (*) surfaced with a few minor variants throughout books of algebra or *ḥisāb*.

It is almost certain that al-Khayyām (1042–1131) had access to them. Does he not write the following?

> And the Indians have methods for determining the sides of squares and cubes based on a restricted induction, that is, on the knowledge of the squares of the nine figures, that is, the square of the unit, of two, of three, and so on – and likewise of their product one by the other – I mean the product of 2 and 3, and so on. And we have written a book to demonstrate the correctness of these methods, and the fact that they fulfill the requirements. And we have spoken abundantly about the kinds thereof, I mean the determination of the sides of the squared-square, of the squared-cube, of the cubed-cube, whatever degree it may reach. And no one did it before us. But these demonstrations arc only numerical demonstrations based on the arithmetical Books of <Euclid's> *Elements*.[22]

Later in the 13th century, the same formulas will reappear with very few changes. Thus Naṣīr al-Dīn al-Ṭūsī in his *Jawāmiʿ al-ḥisāb*[23] gives

$$(a+b)^n - a^n = \sum_{r=1}^{n} \binom{n}{r} a^{n-r} b^r .$$

One finds again the same expression in the 15th century, in al-Kāshī's *Key of Arithmetic*.

But these formulas were well known to al-Zanjānī, al-Fārisī (d. 1319), Ibn Malik al-Dimashqī, al-Yazdī, and Taqī al-Dīn ibn Maʿrūf, among many others. In short, the so-called triangle of Pascal as well as the binomial theorem were common knowledge among Arab mathematicians since the end of the 10th century. It is moreover not rare to find some rules of permutation in the books of *ḥisāb* (calculation), placed after their examination of the elementary laws of arithmetic and their summary of arithmetic progressions, but before the study of the extraction of square, cubic, and higher roots. To be sure, it is one thing to know this triangle, the rule of its

[22] *Risāla fī al-jabr wa-al-muqābala*, in R. Rashed and B. Vahabzadeh, *Al-Khayyām mathématicien*, Paris, A. Blanchard, 1999, pp. 129, 17–131, 2. English version: *Omar Khayyām the Mathematician*, Persian Heritage Series no. 40, New York, Bibliotheca Persica Press, 2000 (without the Arabic texts), pp. 116–17 (emended slightly).

[23] *Jawāmiʿ al-ḥisāb bi-al-takht wa-al-turāb* (Arithmetic Complete, by Board and Dust)', ed. A. S. Saidan, *al-Abhath*, XX.2, June 1967, pp. 91–164, at pp. 145–6; and 3, Oct. 1967, pp. 213–29.

formation, and the binomial theorem as mathematical tools necessary to the algebra of polynomials, the extraction of the nth root of an integer, etc. It is something else again to conceive of them as elements of a new discipline devoted to the partition of a finite set of elements. These tools will belong to this discipline when they are interpreted in combinatorial fashion. Indeed, it is the explicit implementation of such an interpretation that will consecrate the birth of this new discipline. It would nevertheless be far-fetched to believe that the algebraists had not grasped this interpretation rather early on, even if only faintly. Consider for example one of al-Samaw'al's studies: he sets himself ten unknowns, $x_1, \ldots, x_{10}$, and seeks a system of linear equations with six unknowns; he obtains $\binom{10}{6} = 210$ linear equations with 6 unknowns. Next, al-Samaw'al examines the compatibility of these equations and, also by combination, finds 5.040 conditions, if one carries out all of the replacements. After eliminating the repetitions, there remain only 504 for the system to be compatible.

In this study, al-Samaw'al represents the x_i by the number i, which nowadays are called indices. Al-Samaw'al's application of the calculation of combinations thus disposes of the claim that these algebraists ignored everything about this combinatorial interpretation of the rules that they had elicited. We are accordingly convinced that the algebraists had not overlooked this interpretation, but nothing required them to formulate it explicitly. This step will taken shortly, when algebraists engage in new arithmetic research, or when they concern themselves with philosophy. It is precisely in this diversity of fields of application (algebra, arithmetic research, philosophy, linguistics, cryptography, etc.) that combinatorial analysis first appears, before the consciousness of the field's unity emerges, not despite, but rather originating in, the diversity of fields of application. It is at this point that authors such as al-Ḥalabī will write works completely devoted to combinatorial analysis.

3. ARITHMETIC RESEARCH AND COMBINATORICS

Starting certainly with Naṣīr al-Dīn al-Ṭūsī (1201–1273) but very probably before him, one constantly encounters the combinatorial interpretation of the arithmetic triangle and of the law of its formation, as well as the elementary rules of combinatorial analysis (see below). The fact is that, throughout this entire century, and most notably in its last years, this interpretation is bodily present in arithmetic research. Indeed we have shown that, at the end of the 13th century, Kamāl al-Dīn al-Fārisī returns to this interpretation in a memoir on number theory and establishes the usage of the arithmetic triangle for numerical orders, that is, that he obtains the result for which Pascal ordinarily gets credit.[24] Indeed, for figurate numbers (see below), al-Fārisī establishes a relation equivalent to

$$F_p^q = \sum_{k=1}^{p} F_k^{q-1} = \binom{p+q-1}{q},$$

with F_p^q the pth figurate number of order q, $F_1^q = 1$ for every q.

Thus al-Fārisī establishes a relation between the combinations and the figurate numbers of any order. From now on, it is thus possible to refer to the table of figurate numbers in order to know the number of proper divisors of an integer. Here is al-Fārisī's explanation:

> The method for knowing the proper divisors – binary aliquot parts (two-term combinations) or ternary, or others with any given number of sides, on condition that they all be prime and all distinct, consists in seeking in the series of homonymous sums of the number of times by which one combines (*ʿadad al-taʾlīf*) minus one, the number whose row – that is, the first numbers <in the series> of the sums (indices of the columns) – is homonymous of the number of sides minus the number of times by which one combines. This is the number of combinations.[25]

To grasp the meaning of this text, let us suppose that the given integer can be decomposed in n distinct prime factors, and that one seeks the number of aliquot parts (proper divisors) with m elements, where $0 < m < n$. One then seeks in the table the element on the $(m-1)^{\text{th}}$ line and in the $(n-m)^{\text{th}}$ column. One then obtains F_{n-m+1}^{m}, which, according to the preceding expression, is equal to $\binom{n}{m}$.

To demonstrate this proposition, al-Fārisī operates in a fully combinatory manner, with successive applications of the arithmetic triangle, the different 'cells' of which are explicitly interpreted as the combinations of p objects taken k at a time. This combinatorial style of al-Fārisī seems to be a common trait of the period; more precisely: without this style, he could not have established the theorem about the elementary arithmetic functions,

[24] See R. Rashed, 'Matériaux pour l'histoire des nombres amiables et de l'analyse combinatoire', *Archive for History of Exact Sciences*, 28, 1983, pp. 107–47, and 'Nombres amiables, parties aliquotes et nombres figurés aux XIIIe et XIVe siècles', *Journal for the History of Arabic Science*, 6, 1982, pp. 209–78.

[25] R. Rashed, 'Matériaux pour l'histoire des nombres amiables et de l'analyse combinatoire', Proposition 17, p. 251.

about the number of proper divisors of a number. Indeed, in his calculation of the combinations established to determine the number of proper divisors of an integer, al-Fārisī again takes up binomial coefficients, but now to give them a deliberately combinatorial interpretation. Such an act, one of those that founded combinatorial analysis itself, also made it possible for him to conceive of figurate numbers in a sense incomparably more general, as far as we know, than anything one finds among al-Fārisī's predecessors and contemporaries.

4. PHILOSOPHY AND COMBINATORICS

Linguistics, cryptography, and arithmetic research are not the only domains in which one sees combinatorial procedures accompanied by a deliberately combinatorial interpretation. Theoretical philosophy played a particularly important role in the formation of combinatorial analysis. Well before Leibniz and much more effectively than R. Llull, the 13th-century mathematician Naṣīr al-Dīn al-Ṭūsī applies the rules and the formulas that he himself had used in his *Jawāmī' al-ḥisāb* (*Compendium of Calculation*) in order to make Avicenna's ontology speak with precise language. In this book, as we have said, al-Ṭūsī reproduces the arithmetic triangle and the binomial theorem. In the metaphysical treatise entitled, *On the Demonstration of the Mode of Emanation of Things Infinite <in Number> Beginning with a Unique First Principle*, al-Ṭūsī wants to solve the problem of the emanation of a multiplicity from a first principle. The issue, then, is the emanation of Intellects.[26] Al-Ṭūsī represents these Intellects by the letters of the alphabet, and then applies the combinatorial rules. He begins by introducing the following lemma: the number of combinations of n elements taken k at a time is equal to $\sum_{k=1}^{n}\binom{n}{k}$, $k = 1, 2, \ldots, n$; and he uses the equation $\binom{n}{k}=\binom{n}{n-k}$ to calculate this number. He next proceeds to calculate the beings derived row by row by means of the expression

$$(*) \qquad \sum_{k=0}^{m}\binom{m}{k}\binom{n}{p-k} \text{ in which } 1 \leq p \leq 16, m = 4, n = 12, \ldots$$

whose value is the binomial coefficient $\binom{m+n}{p}$.

[26] See the chapter 'Philosophy of mathematics' below.

Al-Ṭūsī's contribution will be recalled in the first known treatise on combinatorial analysis.

5. A TREATISE ON COMBINATORIAL ANALYSIS

Al-Ṭūsī takes the deliberately combinatorial interpretation of the arithmetic triangle and of the binomial theorem completely for granted, a given that expresses in a technical terminology that reappears among his successors, al-Fārisī and Ibn al-Bannā', for example. Everything points to the fact that, at the time of al-Ṭūsī, and perhaps even before him, the formulas that the algebraists established accepted this interpretation when one applied them to the various disciplines, including algebra itself. Now these formulas, notably the triangle and the theorem, are frequently reproduced in the books of *ḥisāb* and algebra. One finds them in the books of *Miftāḥ al-ḥisāb* (*Key of Arithmetic*) by al-Kāshī, in the books of Ibn al-Malik al-Dimashqī (*al-Is'āf al-atamm*), in that of Tāqī al-Dīn ibn Ma'rūf (*Bughyat al-ṭullāb*) who borrows examples from linguistics to illustrate certain rules, in al-Yazdī (*'Uyūn al-ḥisāb*), ...

Everything indicates that, with its massive presence and its frequent applications, the time had come for combinatorial analysis to become more independent, to present itself with a certain autonomy; it was ripe for becoming a topic of composition. Exactly when did this happen? We do not know precisely. Nevertheless, a philosopher mathematician, Ibrāhīm al-Ḥalabī, did write a treatise entitled *Fī istikhrāj 'iddat al-iḥtimālāt al-tarkībiyya min ayy 'adad kāna* (*On the Determination of Combinable Eventualities Starting with Any Given Number*). As far as we now know, this is the first treatise entirely and uniquely devoted to combinatorics. In it, the rules of the latter no longer appear simply as the rules of algebraic calculation. Instead of being simply used in an application that is algebraic, linguistic, philosophical, etc., they are now considered in themselves, in a book entitled *Combinable Eventualities*. The generic designation of this title refers as much to permutation as to arrangements, combinations, etc., in other words, to all of the combinations studied at that time. This treatise gives pride of place to the text of al-Ṭūsī, which is developed and amplified, and plays the role of the method for determining and establishing combinations.

Let us turn quickly to Ibrāhīm al-Ḥalabī's treatise. He begins by raising questions about the different possible methods of combining 'eventualities' (*al-iḥtimālāt al-tarkībiyya*). Al-Ḥalabī's goal is clear: 'to determine the number of combinable eventualities for any given number of objects'.[27] He sidelines the empirical method of enumeration, which offers no general rule, despite its effectiveness for simple cases. This method consists in enumerating, for a set of three elements (a, b, c) for example, the seven 'combinable eventualities' $\{a, b, c, ab, ac, bc, abc\}$. The difficulty is clear for a set with n elements.[28] The second method,[29] offers a general rule, of which al-Ḥalabī is proud. It consists of an expression equivalent to $u_n = 2u_{n-1} + 1$, where u_n is the set of 'combinable eventualities' with n elements. In our language,

$$u_n = \sum_{k=1}^{n} \binom{n}{k}$$

with

$$\binom{n}{k} = \frac{n!}{k!(n-k)!} \qquad \text{for } 1 \le k \le n.$$

This method was perhaps established from the following rule, already known since the end of the 10th century:

$$\binom{n}{k} = \binom{n-1}{k-1} + \binom{n-1}{k}.$$

By summation, one obtains

$$\begin{aligned} u_n &= \sum_{k=1}^{n} \binom{n-1}{k-1} + \sum_{k=1}^{n} \binom{n-1}{k} \\ &= \sum_{k=0}^{n-1} \binom{n-1}{k} + \sum_{k=1}^{n} \binom{n-1}{k} \\ &= 2u_{n-1} + 1 \end{aligned}$$

Al-Ḥalabī also puts aside this method, which requires a complicated calculation, that of all u_i for $1 \le i \le n-1$. To define a better method, al-Ḥalabī starts with the expression

$$\binom{n}{k} = \binom{n}{n-k}$$

[27] *Risālat fī istikhrāj ʿiddat al-iḥtimālāt al-tarkībiyya*, ms. Istanbul, Süleymaniye, Hamidiye 873, fol. 69ᵛ.

[28] *Ibid.*, fol. 70ʳ.

[29] *Ibid.*, fols 70ʳ–71ᵛ.

knowing that

$$\binom{n}{n+r}=0; \qquad \binom{n}{n}=\binom{n}{0}=1.$$

He then defines several 'combinable eventualities', along with the rules of calculation corresponding to them. This is how he obtains:

1) The matter (*al-mādda*)[30] of the eventuality of the kth kind – that is, the combinations without repetition given by the preceding formula;

$$\binom{n}{k}.$$

2) The matter and the form (*majmūʿ al-mādda wa-al-ṣūra*)[31] of the eventualities of the kth kind – that is, arrangements without repetition

$$A_n^k = k!\binom{n}{k} = \frac{n!}{(n-k)!}.$$

3) The form (*al-ṣūra*)[32] of the eventualities of the kth kind: one need only subtract the matter, in 1) above, from the matter and from the form, in 2) above.

$$k!\binom{n}{k} - \binom{n}{k} = \binom{n}{k}(k!-1).$$

4) The form of the eventualities, independent of the kind: that is, the permutations of n objects, namely

$$n! = n\,(n-1)\,\ldots\,2\cdot 1.$$

5) The matter, the form, and the repetition of eventualities of the kth kind,[33] that is, the arrangements with repetition of n objects taken k at a time, namely n^k.

Note that the technical lexicon of the language of combinatorial analysis on which al-Ḥalabī draws in this treatise is a composite of terms already used by al-Ṭūsī (*tarkība*), of terms that are idiosyncratic with al-Ḥalabī, such as *iḥtimālāt* (eventuality), *tikrār* (repetition), but also of borrowings from Aristotelian terminology, such a *mādda* (matter) and *ṣūra*

30 *Ibid.*, fol. 71ᵛ.
31 *Ibid.*, fol. 72ʳ.
32 *Ibid.*, fols 72ᵛ–73ʳ.
33 *Ibid.*, fols 73ᵛ–74ʳ.

(form). These last two terms force him to introduce problems alien to his subject, if not superfluous in this context, and in any case, prejudicial to the clarity of his exposition: he wonders, for example, if one can separate matter and form.

After having set out these rules, al-Ḥalabī writes: 'To determine the material eventualities (*al-iḥtimālāt al-māddiyya*) (that is, the combinations without repetition), there is another method that was mentioned to determine the Accidental Intellects (*al-ʿuqūl al-ʿaraḍiyya*).' It is at this point that he integrates the text of al-Ṭūsī, sometimes in words, sometimes by developing the calculation. Thus, he traces the arithmetic triangle up to 12, and adds the elements of the diagonal, which he calls 'simple combinations' (*al-iḥtimālāt al-basīṭa*), in order to obtain the number 4.095 mentioned by al-Ṭūsī. He calls 'composite combinations' (*al-iḥtimālāt al-murakkaba*)[34] the expression

$$(**) \qquad \left(\sum_{k=1}^{m}\binom{m}{k}\right)\left(\sum_{j=1}^{n}\binom{n}{j}\right) \qquad \text{for } m = 4, n = 12,$$

and shows that the expression above (*) is the sum of the simple combinations and the composite combinations. That is, one obtains

$$(***) \qquad \sum_{p=1}^{m+n}\left(\sum_{k=0}^{m}\binom{m}{k}\binom{n}{p-k}\right) = \sum_{k=1}^{m}\binom{m}{k} + \sum_{j=1}^{n}\binom{n}{j} + \left(\sum_{k=1}^{m}\binom{m}{k}\right)\left(\sum_{j=1}^{n}\binom{n}{j}\right)$$
$$= \sum_{k=1}^{m}\binom{m}{k} + \left(\sum_{k=0}^{m}\binom{m}{k}\right)\left(\sum_{j=1}^{n}\binom{n}{j}\right).$$

When one subtracts 1 from both sides, one gets

$$\sum_{p=0}^{m+n}\left(\sum_{k=0}^{m}\binom{m}{k}\binom{n}{p-k}\right) = \left(\sum_{k=0}^{n}\binom{m}{k}\right)\left(\sum_{j=0}^{m}\binom{n}{j}\right),$$

whence, beginning with the equivalence with formula (*),

$$2^{m+n} = 2^m 2^n.$$

Al-Ḥalabī moves on to other calculations on the data provided by al-Ṭūsī, and delves into reflections on his predecessor's work. These all

[34] *Ibid.*, fol. 81ʳ.

pertain to combinatorial properties. We are now far from the problem of the emanation of multiplicity from the One, of which only a shadowy memory remains: already dim in al-Ṭūsī, the ontological content disappears completely in this treatise on combinatorial analysis, leaving only the methods and results that are necessary or useful for the body of the latter. Whereas the 'axiomatic' cast of Ibn Sīnā's doctrine and an inclination towards a formal ontology initially gave al-Ṭūsī the hope of finding a mathematical solution to this metaphysical problem, this solution found itself eventually integrated into mathematical works, independently of the metaphysical problem that had once given rise to it. This was possible insofar as the entities of the combinatorics may be Intellects or any other objects, on the sole condition that they be separate and as numerous as one might wish, but always finite.

6. ON THE HISTORY OF COMBINATORIAL ANALYSIS

The history of the establishment of combinatorial analysis as an autonomous discipline between al-Khalīl in the 8th century and al-Ḥalabī in the 16th, presents itself as the history of stripping away objects belonging to different domains, removing them from all of their ontological roots, and in the end preserving nothing but formal components: a set of any finite and discrete elements. As we have seen, it is precisely because of, not despite, the multiplicity of these domains that these formal elements could be extricated and gain their independence. One can easily understand, however, that this multiplicity prevents the early history of combinatorial analysis from coinciding with a progressive march that gradually unveils the central properties of a specific entity that is given in advance. Nevertheless, because of their ontological neutrality and their aspect, the letters of the alphabet lend themselves directly to combinatorial study. Whatever one might say, the latter did not appear immediately, and as a kind of Gestalt apprehension, but only when al-Khalīl wanted to resolve theoretically the practical problem of composing a dictionary of Arabic. The procedures he invented and the formulas that were very likely established probably concerned only the formed words and did not yet have this character of generality necessary to the conception of an autonomous discipline dedicated to combinations. It nevertheless remains the case that al-Khalīl and such successors as Ibn Durayd, who were sensitive to the formal character of these procedures and formulas, were able to recognize in them 'a kind of calculation' (*ḍarbun min al-ḥisāb*). As an autonomous discipline, combinatorial analysis was not yet born, but as a type of calculation, its gestation was already advanced. Among cryptographers, it was in the same state.

At the end of the 10th century, the algebraists (al-Karajī and his successors) had for their part established the arithmetic triangle, the rule of its formation, and the binomial theorem. But if the mathematicians knew that they were manipulating combinatorial procedures, they were, according to al-Samaw'al,[35] using them only as auxiliary means for the abstract algebraic calculation that they were then trying to develop.

It is after the end of the 10th century and before the 13th that an act essential to the history of combinatorial analysis took place: the explicit, no longer merely implicit, recognition that the rules applied by linguists, cryptographers, and others were the same as those that the algebraists provided and established, and moreover that one could apply them to a variety of situations: arithmetic research, philosophy, etc. At present, however, one cannot say with complete certainty who made the leap. But it is impossible to ignore the fact that Naṣīr al-Dīn al-Ṭūsī was already moving very comfortably in this new universe of thought, the very one to which belong the works of al-Ṭūsī's successors, such as al-Fārisī and Ibn al-Bannā'.

An additional stage has yet to be crossed, but it does not present any particular difficulties: to speak of these rules and formulas in a combinatorial language, but with no reference whatever to any domain of application. This step inscribes the discipline's act of recognition and forges its autonomy. It is al-Ḥalabī who takes this step, which is more theoretical than technical; the importance of his book lies in this epistemic gesture. By inventing a technical language, albeit a clumsy one, which Leibniz will later find in his *Horizon of Human Doctrine*,[36] al-Ḥalabī asserts once again his will to distinguish a branch of learning: combinatorial analysis. Was al-Ḥalabī the first to do so? To the best of our present knowledge, he is indeed.

The history of Arabic combinatorial analysis stops here. The discipline will be reactivated when it is applied to the vast and new field of probability theory. But for that step, one must wait for Fermat, Pascal, and J. Bernoulli, among others.

[35] Al-Samaw'al, *al-Bāhir en Algèbre*, ed. S. Ahmad and R. Rashed, pp. 104–12.

[36] Leibniz, *De l'horizon de la doctrine humaine (1693)*, ed., French transl. from Latin and postface by M. Fichant, Paris, Vrin, 1991.

– 3 –

THE FIRST CLASSIFICATIONS OF CURVES

1. INTRODUCTION

It is no exaggeration to say that research on curves has given birth to some of the most important inventions in mathematics. Consider the histories of the differential calculus, the calculus of variations, differential geometry, algebraic geometry... The mathematicians of Plato's day had invented curves aimed at solving several problems of geometrical construction raised at the time, notably the quadrature of the circle, the two means, the duplication of the cube. Later, it was on curves that the most advanced works of Hellenistic geometry focused: Archimedes, Zenodorus, Apollonius, Diocles... In 9th-century Baghdad, mathematicians evinced no less interest in curves; on the contrary, it was the study of curves that wove together the main networks among the different mathematical chapters of the period. The intervention of the algebraists of that time (al-Khayyām and Sharaf al-Dīn al-Ṭūsī) opened up other perspectives, which Descartes and Fermat would go on to rethink and transform. It is still in research on curves that mathematicians began to discover the main methods of mathematical analysis. One can complete the panorama by noting that, if one were to forget research on curves, entire chapters of the history and the philosophy of mathematics would drop off the map. Among the themes encountered at the crossroads between effective mathematical research and the philosophical reflection it stimulates, the classification of curves is surely one of the most ancient and the most fruitful. Plato already alluded to it, Aristotle discussed it, Geminus lingered in it, followed by Xenarchus, Sporos, Pappus, Proclus, Simplicius, and probably yet others. When mathematical research revives in 9th-century Baghdad, first-order mathematicians such al-Qūhī and al-Sijzī, among others, take up the theme of the classification of curves on other foundations. Beginning in the 9th century, algebraists, notably al-Khayyām and Sharaf al-Dīn al-Ṭūsī, renew that line of research until it becomes the object of a second transformation, thanks to

Descartes in the 17th century. Descartes and Fermat complete this first period and open a new one that will occupy the efforts of Newton, MacLaurin, Cramer, and their contemporaries and successors. For at least two millennia, then, this problem was a steady preoccupation, even though the number of known curves was rather limited. Indeed until 1637, transcendental curves were limited to the quadratrix and spirals of Archimedes, and plane algebraic curves could be counted on the fingers of one hand. To these, Descartes himself added two others after 1630. Indeed the following were known: a cubic (cissoid), two quartics (the conchoid of Nicomedes and the so-called 'snail of Pascal', to which Descartes had added two cubics: the 'folium' and the 'trident'. In addition, two algebraic space curves were known – that of Archytas and the hippopede of Eudoxus. Now this paradoxical situation is more than mildly interesting. Why indeed exert oneself in classifying such a small number of objects?

Moreover, in the mathematical and philosophical literature from the ancient to the classical period at least until the middle of the 17th century, one encounters several concepts of the curve that coexist or follow one another. These conceptions, of which some are kinematic and others geometrical, reflect the procedures for generating curves, by points, by motion, or by combinations of motions. This is what transpires from the examination of such curves as the quadratrix, the spiral, the helix, the conchoid, etc. The continuity of the curve is ensured by that of the motion(s). For his part, Descartes requires in addition a perfect organic coordination of the motions. But it is one thing to conceive of a way of tracing a curve, it is quite another to define it. For a long time, the term 'curve' was understood as a line that satisfies none of the definitions of the straight line, whether that of Euclid or that of Archimedes, for example.[1] Alongside this negative definition, one encounters in the Archimedean tradition the concept of curve as the limit of a polygon with an infinite number of sides. Such is the conception of Ibn al-Haytham, for example, which one encounters again in Fermat, Pascal, Leibniz... By way of example, consider the Marquis de l'Hospital in his *Traité analytique des sections coniques* (1720):

[1] See for example M. Federspiel, 'Sur la définition euclidienne de la droite', in R. Rashed (ed.), *Mathématiques et philosophie de l'antiquité à l'âge classique*, Études en hommage à Jules Vuillemin, Paris, Éditions du CNRS, 1991, pp. 115–30.

> If one imagines that any curved line whatever is divided into an infinite number of infinitely small arcs [...], it is clear that, by taking the chords of these arcs instead of the arcs themselves, one will see emerging a polygon with an infinite number of sides, each indefinitely small, which one can take to be the curved line, since it will in no way differ from the latter.[2]

This is a well-known concept, which Father Castel formulates in almost identical terms when he writes eight years later in his book of 1728:

> Since every curved line is a polygon with an infinity of infinitely small sides, geometers consider the tangent to be the extension of the small side to which it adjusts itself by touching the curve.[3]

This is still the same concept, but now joined to a definition of the tangent that differs from that of the ancients. These two concepts of the curve are still alive in the 18th century, that is, an era in which the algebraic curve is already defined by an equation. If one restricts oneself to a particular curve, such as a conic, the fundamental property of which had already been demonstrated before Apollonius and which he himself took up in the first book of the *Conics*, the curve is known as a plane section; but this fundamental property does not characterize it, because its reciprocal had not been demonstrated. Not until the algebraists of the 11th and 12th centuries was this particular curve defined by its equation. One sees, therefore, that for centuries people undertook to classify a small number of objects, the generation procedures of which they knew, while still hesitating about their definitions. The latter will become precise later, with the gradual evolution of the infinitesimal calculus, thanks to the elaboration of such new concepts as the curvature, evolutes, involutes. Is it not precisely the case, as Cramer put it, that 'to know its Nature perfectly, one must know in addition how much the curve moves away from this direction [of the tangent at every point]; one must be able to measure its curvature. For a given curve is not equally curved everywhere'.[4] For such a study to see the light of day, no long wait was necessary, since Newton was already engaged

[2] *Traité analytique des sections coniques*, Paris, Montalant, 1720, p. 129.

[3] L.-B. Castel, *Mathématique universelle. Géométrie transcendante*, Paris, P. Simon, 1728, p. 566.

[4] Gabriel Cramer, *Introduction à l'analyse des lignes courbes algébriques*, Geneva, 1750, p. 539.

in it at the end of 1664.[5] Not least, even from a completely modern point of view, it is far from easy to define what a curve is with complete generality.

Finally, notice that the names listed above include mathematicians as well as philosopher-mathematicians and philosophers. This third remarkable characteristic immediately leads us to wonder whether, at least at the beginning of Greek mathematics, this theme was not the result, as Paul Tannery wrote, of 'an unfortunate encroachment of philosophy onto the domain of mathematics';[6] or whether this theme is not rather the reflection of mathematical rationality itself, of its power of discernment among existing objects – curves, in this case – however few they might be, and a reflection of its capacity to conceive an *a priori* classification of objects, the full knowledge of which is a pledge to be redeemed in the future. Indeed, as a matter of fact, is not the classification of mathematical objects itself a mathematical activity? This is the fourth remarkable aspect in the theme of the classification of curves, and of classification in mathematics more generally. I restrict myself here to the first classification of curves, that is, those that were proposed before the complex plane intervenes and before the foundation of the differential calculus, and later, of the theory of functions.

To write even the merest historical sketch of this question thus demands that one multiply the points of view and superimpose the perspectives: the effective study of curves by mathematicians; the attempts at classification; but also the analysis of the consciousness that mathematician-philosophers and philosophers had of it. Indeed this consciousness not only indicates the extent and the limits of the various formulations of the problems of classifications, but also can illuminate the reasons that guided these formulations. These classifications can in fact be empirical, that is, *a posteriori*, according to the form of the curve; they can be 'experimental', so to speak, according to the mode of the generation of the curves; or they can also be *a priori*, corresponding to the mathematical formula that characterizes the curve. These are the modes that interest us here, particularly when one begins to combine the last two: it is indeed a major event in the history of mathematics when the mode of genera-

[5] *The Mathematical Papers of Isaac Newton*, ed. D.T. Whiteside, Cambridge, Cambridge University Press, 1967, vol. I: *1664–1666*, pp. 245–8.

[6] P. Tannery, 'Pour l'histoire des lignes et surfaces courbes dans l'antiquité', in *Mémoires scientifiques*, published by J.-L. Heiberg and H.-G. Zeuthen, Toulouse/Paris, Ed. Privat/Gauthier-Villars, 1912, t. II: Sciences exactes dans l'antiquité, no. 30, pp. 1–47, at p. 37.

tion of the curve and its formula jointly guide its classification. For the conics, such an event first takes place in the 10th century, with al-Qūhī and his successors; next, come the algebraic curves with Descartes. But before reaching this point, it was necessary to travel a long road, the main stages of which we sketch here.

All along this road, as we shall see, what is under consideration is not a random line, *i.e.*, a line traced by chance on the sand or on paper, but only the line obtained thanks to a technical process controlled by the rules of geometry, thus allowing one to make an exact replica of it as often as one wishes. Both the line and the instrument invented to trace it are thus geometrical, a fact of which mathematicians are fully conscious. Listen, for example, to Gabriel Cramer in the middle of the 18th century:

> Every line is *regular* or *irregular*. The irregular lines are those that are described without any certain or known rule. Such is the doodle that a writer makes randomly. These lines are not the object of Geometry: they give it no traction [...]. Regular lines, on the contrary, are those that are described according to a constant law that determines the position of all of their points. There is some uniform property that pertains equally to all of the points of one and the same regular line, and that pertains to them alone. This property constitutes this line's *Nature* or *Essence*.[7]

It nevertheless remains that the characterization of these curves and of the curve in general is intimately tied to the mathematics of the times. It is precisely these objects, in their mathematical horizons, that one wanted to classify.

This horizon was quickly redrawn during the first half of the 18th century. Already towards the middle of this century, mathematicians were occupied with the classification of functions (*e.g.*, Euler's *Introductio in analysin infinitorum*, 1748).[8] Indeed, they were beginning to discuss arbitrary functions

[7] Cramer, *Introduction à l'analyse des lignes courbes algébriques*, pp. 1–2. Note also that Newton in 1667 or 1668 seems to distinguish geometrical curves, mechanical curves, and randomly drawn curves ('curva sive Geometrica sive Mechanica & casu ducta', translated by D. T. Whiteside as 'a curve be it algebraic or transcendental and drawn haphazardly', *The Mathematical Papers of Isaac Newton*, Cambridge, Cambridge University Press, 1968, vol. II: *1667–1670*, pp. 142–3).

[8] Euler expresses this new concept thus: 'Although one can describe several curves mechanically with the continuous motion of a point which reveals to the eye the curve as a whole, we consider them here primarily as the result of functions, since this way of

(*Cont. on next page*)

and their representations by a curve forming a continuous tracing or drawn freehand (Euler, d'Alembert, Lagrange, D. Bernoulli). Up to that point, this type of curve seemed to have escaped the attention of mathematicians; to establish the study of it, one had to wait for many other 19th-century works, beginning with Fourier.

2. SIMPLE CURVES AND MIXED CURVES

The first such classification is that of the *Parmenides* (145 b). With respect to the One, Plato already discussed three possible figures: straight line, circle, and a mixture of straight line and circle. Aristotle takes up this classification, but by pairing the first two curves to two types of motion, on the one hand, and by considering them as the only simple curves, on the other. This is what he clearly affirms when he writes:

> But all movement that is in place, all locomotion, as we term it, is either straight or circular or a combination of these two, which are the only simple movements. And the reason for this is that these two, the straight and the circular line, are the only simple magnitudes.[9]

In his *Commentary on the First Book of Euclid's Elements*, Proclus takes up the preceding points to summarize Aristotle's teaching. He writes:

> Aristotle's opinion is the same as Plato's; for every line, he says, is either straight, or circular, or a mixture of the two. For this reason there are three species of motion – motion in a straight line, motion in a circle, and mixed motion.[10]

According to this commentary, and the citation of Aristotle, one understands that the forms of lines precede – logically at least – the forms of motion

---

(*Cont.*) conceiving them is more analytic, more general, and more suited to calculation. Thus any function of x will yield a straight line or a curved one, whence it follows reciprocally that curved lines can be connected to functions' (From *Introduction à l'analyse infinitésimale*, transl. from Latin into French, with notes and explanations by J. B. Labey, Paris, 1797, t. II, p. 4).

[9] *De Caelo* I 2, 268 b 17–20 (J. L. Stocks translation).

[10] *Procli Diadochi in primum Euclidis Elementorum librum commentarii*, ed. G. Friedlein, Leipzig, Teubner, 1873, p. 104, 21–25; Proclus, *A Commentary on the First Book of Euclid's Elements*, transl. with introduction and notes by G. R. Morrow, Princeton, Princeton University Press, 1970, p. 85.

in place. What one must remember, is the distinction between the first two motions and the motion that is a result of their mixture. Indeed the latter depends on the nature of the first two and cannot be called 'simple'. In *Physics*, VIII, 8, 261 b 28–31, Aristotle writes:

> The motion of everything that is in process of locomotion is either rotatory or rectilinear or a compound of the two (πᾶν μὲν γὰρ κινεῖται τὸ φερόμενον ἢ κύκλῳ ἢ εὐθεῖαν ἢ μικτήν): consequently, if one of the former two is not continuous, that which is composed of them both cannot be continuous either.[11]

No need to tarry longer: Aristotle does not inflect his teaching by stating that he is dissatisfied simply to take up Plato's opinion; rather, he takes pains to link the classification of curves with that of local motion, and to distinguish in his classification – both of curves and of motions – between the entities that are simple and those that are not. According to him, there are only two simple curves, the straight line and the circle, since all others are compounded of these two elementary curves and therefore cannot be simple.

The remaining step is to find out what precisely is meant by 'mixed curves', which amounts to wondering: at the time, of what did the knowledge of curves consist? Tradition assigns the discovery of the quadratrix to Hippias of Elis (*c*. 420 BC), a sophist and contemporary of Plato's.[12] It also credits Plato's friend, Archytas of Tarentum, with the discovery of the eponymous curve. The discovery of conic curves is attributed to Menaechmus, a student of Eudoxus's (*c*. 350 BC). Neither Plato nor Aristotle could therefore be ignorant of these curves, and notably the sophist's quadratrix. On the same grounds as the others mentioned above, it is a mixed curve.

It is in fact Proclus[13] who attributes the discovery of this curve to Hippias of Elis. From his testimony alone, however, it is impossible to know which quadratrix he meant. For his part, Pappus gives a precise description of the curve in the fourth book of his *Collection*,[14] but attributes it to Dinostratus, the

[11] R. P. Hardie and R. K. Gaye translation.

[12] See T. Heath, *A History of Greek Mathematics*, 2 vols, Oxford, 1921; repr. Dover Publications, New York, 1981, vol. I, p. 182. This curve was used for the division of the right angle in any ratio – notably trisection – and for the quadrature of the circle. See also, pp. 226–30.

[13] Proclus, ed. Friedlein, p. 356, 8–11; English transl. G. R. Morrow, p. 277 and more clearly, p. 212 (Proclus, ed. Friedlein, p. 356, 8–11 and, more clearly, p. 272, 7–10).

[14] *Pappi Alexandrini Collectionis quae supersunt e libris manu scriptis edidit latina interpretatione et commentariis instruxit F. Hultsch*, 3 vols, Berlin, 1876–1878; *La*

(*Cont. on next page*)

brother of Menaechmus, and dates it two generations after Hippias. P. Tannery,[15] who maintains that Hippias made the discovery, states that the two curves of Dinostratus and Hippias can be assimilated. For the purposes of our present discussion, however, the priority dispute does not matter. Let us therefore follow Tannery and begin by examining the description of the curve as Pappus reports it:

Let *DAB* be a quarter circle, and *AB* and *AD* be two orthogonal semi-diameters; and let a semi-diameter *AE* rotate uniformly from *AB* to *AD*. Let the straight line *B′Z* move uniformly from *B* along *BA* parallel to *AD* and let it cut the semi-diameter *AE* at *Z*. Let it continue to displace itself uniformly until its foot *B′* arrives at *A*; point *E* thus moves from point *B* to *D*. The locus of the points *Z* is the quadratrix of Dinostratus, the curve with the vertex at *H*.[16]

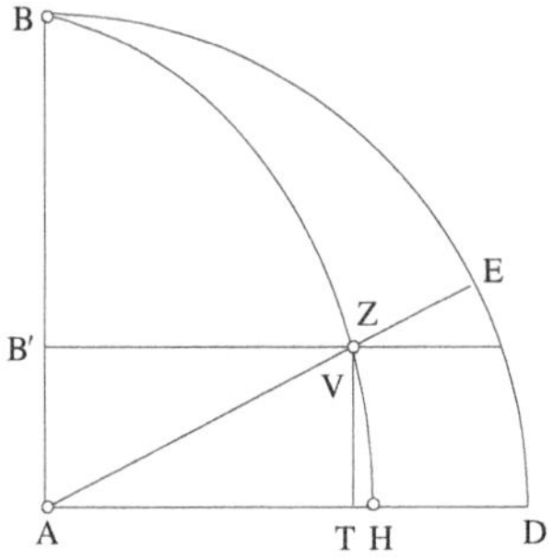

Fig. 15

This description by Pappus stimulates the following remarks:

• From point *B* one can construct, by means of an iterated bisection of the angles, an infinity of points *Z* of the curve, everywhere dense on the arc *BH*;

(*Cont.*) *Collection mathématique*, Work translated for the first time from Greek into French, with an introduction and notes by P. Ver Eecke, 2 vols, Paris/Bruges, 1933; New printing Paris, 1982.

[15] P. Tannery, 'Pour l'histoire des lignes et surfaces courbes dans l'antiquité', in *Mémoires scientifiques*. On the history of the quadratrix, see for example E. Knobloch, 'Sur le rôle de Clavius dans l'histoire des mathématiques', in Ugo Baldini (ed.), *Christoph Clavius e l'attività scientifica dei gesuiti nell' età di Galileo*, Rome, 1995, pp. 35–56, esp. pp. 50–2.

[16] Cf. *La Collection mathématique*, French transl. Ver Eecke, p. 192.

but point H cannot be constructed in this way. This is the case for all quadratrixes.

• One could thus trace this curve by means of points. But the question of the necessity of the continuous tracing is not pertinent to the mathematics of the period.

• The term 'mixed' seems to designate negatively, so to speak, the form of the curve (it is neither straight nor circular); but, especially positively, the method of generation of this curve, which is traced by means of two distinct motions, one circular (a rotation), the other rectilinear (a displacement). It is on this mode of generation that Aristotle's classification rests. In contradistinction to a simple curve – a straight line or a circle – which results from a single uniform motion, a 'mixed' curve is traced by means of two distinct uniform motions.

• Finally, it is clear that the class of 'mixed' curves includes curves of very different kinds, the quadratrix as well as a conic. This will soon cause difficulties, which will surface as soon as the classification is reformed or at least amended. But attempts at reform or rectification cannot be isolated from effective mathematical research. One of the first such attempts seems to be tied precisely to research on the quadratrix.

Still discussing the quadratrix of Dinostratus, Pappus records a criticism addressed by his immediate predecessor, Sporos, pertaining to the use of this curve.[17] It essentially concerns the determination of the vertex H. With good reason, Sporos believes that the mode of construction of the moving point Z is not applied to H, since the two straight lines whose intersection determines the point Z – the radius of the circle and the parallel to AD – are superimposed during the determination of H, for which there is no precise construction. In other words, if the unicity of H is established by Dinostratus's reasoning, or by the continuous decrease of AZ and the continuous increase of AT, point H is not constructed according to the proposed rules.

In other words, if one takes (AD, AB) as a system of axes (x, y), point H is the limit position of Z when $y \to 0$, and one runs into a form of indetermination. The equation of the curve is written

[17] *La Collection mathématique*, French transl. Ver Eecke, pp. 193–4.

$$x = \frac{y}{\tan \dfrac{\pi y}{2A}},$$

where $a = AB$, and AH is the value of x for $y = 0$, which yields the indeterminate form $\frac{0}{0}$; since $\tan \frac{\pi y}{2a} \approx \frac{\pi y}{2a}$; when y tends to 0, one has $AH = \frac{2a}{\pi}$.

Sporos concludes his criticism thus:

> [...] it is not appropriate that, by trusting the reputation of the men who invented it, one admit this line, which is in some fashion too mechanical (μηχανικωτέραν πως οὖσαν).[18]

Following this last phrase is another: 'and useful to mechanicians for many problems', which the editor believes to be an ancient interpolation. The presence of the expression 'too mechanical' to characterize the curve or the reference to 'its usage', whether or not the phrase is interpolated, implies that, in this era already, they were isolating a certain number of curves, including the quadratrix, within the category 'mixed'. Did this distinction rest only on the use of these curves, as the incriminated phrase specifies? or, in part and indirectly, on their mode of generation? The question is important since it elicits criteria of classification. P. Tannery, whom P. Ver Eecke follows,[19] has hypothesized that the phrase indicated 'that set-squares in which a quadratrix replaces the hypotenuse (*équerres en quadratrice*) were used in practice since the time of Hippias and that Sporos insists that the necessarily approximate construction of these instruments does not stand up to comparison with the compass and straightedge';[20] but he offers no historical argument in support of such a hypothesis. J. Itard[21] has followed the eminent historian down this hypothetical road, and tried to imagine such a set-square.

Let us return to the first figure and draw on it a circle of center A and of radius AZ; it meets AB at M and AD at S. Let us draw an arc with the same

[18] *Pappi Alexandrini Collectionis*, ed. Hultsch, vol. I, p. 254, 24.

[19] *La Collection mathématique*, p. 194, n. 3.

[20] P. Tannery, 'Pour l'histoire des lignes et surfaces courbes dans l'antiquité', in *Mémoires scientifiques*, p. 11.

[21] J. Itard, 'La espiral y la cuadratriz en los Griegos', *Ciencia y Tecnologia*, 5, 1955, pp. 53–8.

center that passes through point H and meets AB at N. From the property of the quadratrix, one has

$$\frac{ZT}{AB}=\frac{\widehat{DE}}{\widehat{DB}}=\frac{\widehat{ZS}}{\widehat{MS}}.$$

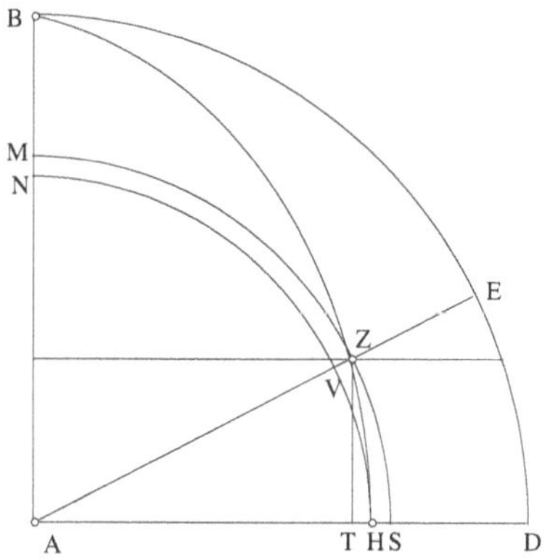

Fig. 16

For sufficiently small arcs, one can approximate arcs with chords. Thus ZT and arc ZS whence $AB \approx \widehat{MS}$. One then has $ZT \approx \widehat{HV}$ and the triangle AZH is approximately equal to sector AVH; one thus has a set-square with a quadratrix. But this set-square will in no way allow a precise determination of point H.

One must nevertheless concede that, according to this conjecture, one could conclude that the term 'mechanical' refers not only to the use, but also to the manufacture, of a tool destined to determine the curve, in the manner of the 'horn-shaped ruler' that Diocles will later use to join the points of a cissoid.[22] But in each of these cases, the adjective 'mechanical' pertains to practice, without any theoretical justification. Nevertheless, with this term, Sporos offers a first amendment to the Aristotelian classification. The second critique aimed at the latter originates in the study of another curve, the cylindrical helix, which will also lead to a correction of Aristotle's classification.

[22] *Les Catoptriciens grecs*. I: *Les miroirs ardents*, ed., transl. and commentary by R. Rashed, Collection des Universités de France, Paris, Les Belles Lettres, 2000.

The long-dominant Aristotelian doctrine was evidently not endorsed by all. If we can believe Proclus,

> Some dispute this classification, denying that there are only two simple lines and saying that there is also a third, namely, the cylindrical helix, which is traced by a point [reading *sêmeion* instead of *sêmeiou* – Friedlein] moving uniformly along a straight line that is moving around the surface of a cylinder. This moving point generates a helix, any part of which coincides homoeomerically with any other, as Apollonius has shown in his treatise *On the Cochlias*. This characteristic belongs to the helix alone.[23]

This idea rests on another, to be explicated later: the homogeneity of the curves in question (the straight line and the circle in the plane, the helix in three dimensions) relative to the group of displacements. Indeed the only homoeomeric curves (with respect to the group of displacements) in the plane are the straight line (translation) and the circle (rotation); and in three dimensions, they are a straight line, the circle, and the helix, which corresponds to a one-parameter group of screw-motions. If one replaces the group of displacements with that of similarities, a new homogeneous curve appears: the logarithmic spiral, studied in the 17th century by Mersenne, Descartes, and Torricelli as well. Proclus did not think it necessary to record either the names or the arguments of these contradictors. Nevertheless, one gathers that this classification of curves became the arena of a lively debate, between not only mathematician-philosophers, but also theologian-philosophers.[24] This debate goes back at least to the first century of our era since – still according to Proclus – Geminus took part in it. We also know from Simplicius that Xenarchus, at the beginning of the millennium, had rejected the Aristotelian classification, and that Alexander of Aphrodisias in turn criticized him two centuries later. In Simplicius's words:

[23] Proclus, ed. Friedlein, p. 104, 26–105, 7; transl. G. R. Morrow, p. 85.

[24] M. Rashed, 'La classification des lignes simples selon Proclus et sa transmission au monde islamique', in C. d'Ancona and G. Serra (eds), *Aristotele e Alessandro di Afrodisia*, Padua, Il Poligrafo, 2002, pp. 257–69. The Neoplatonists, and Pseudo-Dionysius in their wake, related the three constitutive moments of the Triad to the two simple lines and the composite of them. In the domain of the real, these three geometrical entities are the imperfect images of the three divine 'motions'.

But Xenarchus, in answering many things in his work *Against the Quintessence*, also answered the argument that 'the cause of it is that only these magnitudes likewise are simple, namely the straight and the circular'.[25]

Xenarchus opposes this Aristotelian thesis in the following terms:

> for another simple line is the helix on the cylinder, on the grounds that every part of the latter is superposable on every equal part. But if there is a simple magnitude alongside the first two, there could also be a simple motion alongside the other two, and another simple body, alongside the five, that would be moved by this motion.[26]

To defend the simplicity of the helix is to strike at the heart of several points of Aristotelian doctrine: the number of simple lines, the number of simple motions, and the number of simple bodies. Alexander of Aphrodisias, the great commentator of Aristotle, was therefore duty-bound to reply to Xenarchus, to whom he addresses two criticisms. Only the second matters here, which Simplicius records as follows:

> that the helix on the cylinder is not even a simple line, even though it is true that it is generated from two dissimilar motions, circular and rectilinear (ἐκ δύο κινήσεων ἀνομοίων ... κυκλικῆς τε καὶ ἐπ' εὐθείας). Indeed, having drawn a straight line circularly around the surface of the cylinder, and letting any point move regularly (ὁμαλῶς) on this straight line, the cylindrical helix (ἡ κυλινδρικὴ ἕλιξ) is generated, as Xenarchus himself recognizes when he writes as follows: 'let us make a square revolve while keeping immobile only one of its sides, which will be the axis of then cylinder. On the side that is parallel to it and that revolves, let us draw a point such that it covers the length of the line in the time it takes for the parallelogram to return to its starting point. The parallelogram thus produces a cylinder, whereas the point drawn on the line produce a helix, which, according to his statement, is simple (ἁπλῆν) because it is homoeomeric (διότι ὁμοιομερής)'. But even if it is homoeomeric, it is not simple. For the simple line is necessarily also homoeomeric, but the homoeomeric line is not necessarily simple, on account of its not being uniform (μονοειδής); and if <the simple ligne> is produced as the result of a motion, this motion is also uniform and, what is more, one.[27]

[25] Simplicius, *In Aristotelis de caelo commentaria* (C.A.G. 7), ed. Heiberg, p. 13, 22–25.

[26] *Ibid.*, p. 13, 22–28.

[27] *Ibid.*, p. 14, 10–24. See A. Falcon, *Corpi e movimenti, Il* De caelo *di Aristotele e la sua fortuna nel mondo antico*, Elenchos XXXIII, Naples, Bibliopolis, 2001.

In short, the cylindrical helix is homoeomeric, on the same grounds as the straight line and the circle; unlike these, however, it is not a simple line, because two dissimilar motions generate it. Now, according to Proclus, well before Alexander, Geminus in effect had already answered Xenarchus pre-emptively. For Proclus, Geminus gets the credit for having noted this point:

> Hence from these distinctions it may be gathered that the only three lines that are homoeomeric are the straight line, the circle, and the cylindrical helix. Two of them lie in a plane and are simple; one is mixed and lies around a solid. This has been clearly shown by Geminus, who had previously demonstrated that the two lines drawn from a point to a homoeomeric line and making equal angles with it are themselves equal.[28]

These angles are mixed angles, formed by a straight line and a curve: it is therefore hard to see how Geminus could manipulate them and complete his demonstration about them. Such a difficulty cannot be avoided, however, for the simple reason that here Geminus apparently wishes to give a *geometrical meaning* to this rather imprecise notion of homoeomery. To disentangle this difficulty, it would have been necessary to state that, for every homoeomeric curve, there exist points from which one can draw straight lines that are both equal and isogonic. Thus Geminus could have said:

Given the curve ABC, with point B as the middle of the arc AC, and BD as an axis of symmetry. Let us place this arc on itself, A at C and C at A; then B does not move, nor does the middle D of chord AC, which is generally distinct from B. For every point P of the straight line BD, the line segments PA, PC share this property of equality, not only for A and C, but also for every pair of points on the curve that has B as their middle. AC, however, is perpendicular to D at BP.

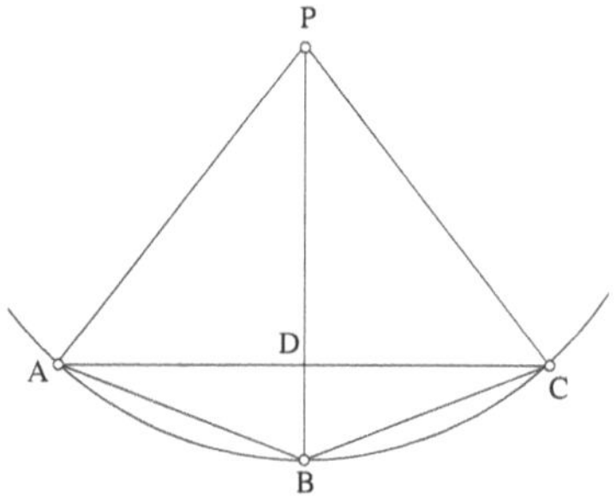

Fig. 17

Thus all chords AC are thus perpendicular to BP at their middle. If this property is true of every point B of the curve, that is, if every point B is the extremity of an axis of symmetry, the curve is homoeomeric. To every point B of the homoeomeric curve (other than a straight line), one thus associates one

[28] Proclus, ed. Friedlein, pp. 112, 19–113, 3; Morrow, pp. 91–2.

and only one straight line *BP*.[29] In the case of the helix, it is the generatrix of the right helicoid that plays an important role in the passages that Pappus devotes to this curve. It is clear that, according to this definition, a homoeomeric curve allows a transitive group of displacements on itself. But an argument of this sort could not be put forward without having defined the concept of symmetry.

Such an interpretation makes it possible to display the 'geometric' nature of this concept of homoeomery without, however, forgetting motion. Indeed, still according to Proclus, the very same Geminus '[...] has rightly declared that, although a simple line can be produced by a plurality of motions, not every such line is mixed, but only one that arises from dissimilar motions'.[30] If therefore the helix is not simple, it is on account of the type of motion that generates it, whereas its homoeomery is, for Geminus at least, answerable, to geometry. For the cylindrical helix, the concept of homoeomery thus seems irreducibly opaque, whereas this is not the case for the circle, which is both homoeomeric and simple – a case in which motion and geometry combine. Geminus's step forward is certainly important: henceforth, the classification of curves depends on the type of motions and on their composition, as well as on a geometric characterization that still remains to be determined.

According to the Proclus citation above, already in Apollonius's *On the Cochlias*, the cylindrical helix stood out as an exceptional curve insofar as it is the only one all of whose parts coincide; in other words, it is homogeneous.[31]

Not only is the Greek original of Apollonius's book lost, but no trace of it, either direct or indirect, has been preserved by the heirs of Hellenistic mathe-

[29] J. Itard, 'La espiral y la cuadratriz en los Griegos'.

[30] Proclus, ed. Friedlein, p. 105, 26–106, 3; transl. Morrow, p. 86.

[31] A displacement of the plane is composed of two symmetries with respect to two straight lines *D* and *D′*. If *D* and *D′* are parallel, the displacement is a translation; if not, it is a rotation around the point common to *D* and *D′*. It follows that the only curves that are homogeneous (with respect to the group of displacements) in the plane are the straight line (translations) and the circle (rotations).

A displacement in three dimensions is composed of two or four symmetries with respect to the planes because it is composed of a rotation (about a point chosen as origin) and of a translation; each of these two transformations reduces to two symmetries. A suitable choice of origin allows one to reduce a given displacement to the composite of a rotation about an axis and of a translation parallel to this axis (screw-motion). The curves that are homogeneous in three dimensions are thus the straight line, the circle, and the helix, which corresponds to a one-parameter group of screw-motion helicoids.

matics. The mathematicians of the 10th century therefore knew nothing about it. Did they hear some echoes, even indirectly, of Proclus's *Commentary*? Whereas this question remains largely open, they certainly had some knowledge of this curve, either from the *Commentary* of Alexander on the first book of the *De caelo*, from which Simplicius draws his knowledge of Xenarchus, or from the studies of Hero of Alexandria and Pappus. The first book of Alexander's *Commentary*, which interests us here, was indeed translated into Arabic; better yet, the 10th-century mathematician al-Khāzin commented on it.[32] Hero's *Mechanics* was also translated into Arabic, and as was book eight of Pappus's *Mathematical Collection*. More importantly, al-Sijzī (second half of the 10th century) personally transcribed this eighth book from a copy made in the 9th century for the famous mathematician brothers, the Banū Mūsā.[33]

In book two of the *Mechanics*, Hero defines the cylindrical helix and explains how to draw it; Pappus will take up Hero's exposition in almost identical terms.[34] Here are Hero's words:

> We claim that the nature of the line drawn on it (the screw) is the following: if one assumes that one of the sides of a cylindrical figure moves on the surface of the cylinder, and if one assumes that any point (starting) at the end of this side that moves on this side by covering its entire length in the same time that the side takes to make one full turn around the entire cylindrical surface, then the line that this point traces on the surface of the cylindrical figure is one turn of a helix, which is called 'helix'.
>
> When we wish to trace this line on the surface of the cylinder, we use the following procedure: on any given plane, we assume two straight lines, one of which is erected perpendicularly on the other, one of the two straight lines being equal to the side of the cylinder and the other equal to the circle of the cylinder, that is, to the circle of its base. If we join the ends of the two straight lines that surround the right angle with a straight line that subtends the right angle, and if we apply the straight line equal to the side of the cylinder onto the side of the cylinder, and the straight line equal to the circle of the base of the cylinder onto the base of the cylinder, then the straight line that subtends the right angle is

[32] See al-Nadīm, *Kitāb al-Fihrist*, ed. R. Tajaddud, Teheran, 1971, p. 311.

[33] Ms. Istanbul, Ahmet III 3457, fol. 34v.

[34] Ms. Istanbul, Ahmet III 3457, fols 26v–27r; *Pappi Alexandrini Collectionis*, ed. Hultsch, pp. 1122, 26–1124, 24.

rolled around the surface of the cylinder, and on the latter one will obtain a turn of a helix.[35]

Challenges to Aristotelian doctrine always originated from its incapacity to distinguish between the various kinds of 'mixed'. The study of mixed curves – now the quadratrix, now the cylindrical helix – was not the only thing feeding this antagonism; from the 3rd century on, both the multiplication of newly discovered mixed curves and the elaboration of a theory of conics gave it a lively stimulus. Even the most rigid Aristotelians could not help but emend the Stagirite's classification. Thus, in his *Commentary on Book I of Euclid*,[36] Simplicius returns to it in his study of the Euclidian definition of the straight line. Following Aristotle, he distinguishes two classes, the simple (the straight line and the circle) and the 'intermediate', that is, the curves composed of the first two. In contrast to Aristotle, however, within the 'intermediates' he distinguishes two sub-classes: the first includes mixed curves that have neither 'order' nor 'regularity' and that geometers do not use; in the second are curves such as the conics, the helicoids, among many others that geometers do use. If, therefore, by 'geometrical' one means the curves that interest geometers, these are the straight line, the circle, and the curves that are governed by an 'order'.[37] How ought one understand these two important concepts that define mixed geometrical curves in opposition to mixed curves that are not geometrical? To this question, Simplicius does not offer the slightest answer. The most likely interpretation, which is also the simplest, brings us back to tradition: 'ordered' and 'regular' are the terms that characterize what can exist in only one way, whatever its mode of construction may be, even if one cannot determine it directly.

The preceding classification is not free of all ambiguity. Despite its Aristotelian partisanship, it does not rely on a doctrine of motion and remains essentially descriptive. Among the 'intermediaries', Simplicius thus numbers the conics as well as the helix: this is justifiable if one is considering order and regularity, but not if one is thinking about the motion required to draw them. Moreover, the term 'intermediary' means, in a first sense, the curves other than the circle and the straight line, and in a second, those that are 'like' (*ka*)

[35] Translated from a slightly emendation of the edition in Héron d'Alexandrie, *Les Mécaniques ou l'élévateur des corps lourds*, Arabic text of Qusṭā ibn Lūqā established and transl. by B. Carra de Vaux, repr. Paris, Les Belles Lettres, 1988, pp. 46–7.

[36] See below Appendix, p. 237.

[37] *Ibid.*

those composed of the straight line and the circle. Now this last meaning surely is not appropriate for conics, which are nevertheless considered to be intermediaries. Finally, there remains the most important point of this classification: the distinction between the curves that geometers use and those that they do not, a distinction that later mathematicians will retain.

3. GEOMETRICAL AND MECHANICAL: THE CHARACTERIZATION OF CONIC SECTIONS

Beginning in the middle of the 9th century, mathematicians were familiar with not only the writings of Aristotle, Alexander, and Simplicius, but also Hero's work and Pappus's borrowings from it. But they elaborated their knowledge in a completely different context, and directed it to other ends. Hero and Pappus had encountered the cylindrical helix in their studies of instruments designed to raise heavy bodies, most specifically the screw and 'the peculiarities of its construction and its usage', as Pappus writes. For the mathematicians of the 9th century, however, the dominant concerns were pure geometry and the classification of curves.

Since we have given a detailed account of this new mathematical landscape elsewhere, we will recall here only a few of its characteristics. Without fear of contradiction, one can state unequivocally that never since Apollonius had the geometry of conics flourished so remarkably. To make the point, one need only allude to the works of the Banū Mūsā, Thābit ibn Qurra, and Ibrāhīm ibn Sinān, among others.[38] Never up to that point had the geometry of conics built so many bridges and woven so many networks among the various mathematical disciplines: the establishment of anaclastics and of the first theory of plano-convex and biconvex lenses; the study of certain projective properties of conics; extensive studies of geometrical constructions by means of conics; the geometrical theory of cubic equations and their solution by the

[38] R. Rashed, *Les Mathématiques infinitésimales du IXe au XIe siècle*, vol. I: *Fondateurs et commentateurs: Banū Mūsā, Thābit ibn Qurra, Ibn Sinān, al-Khāzin, al-Qūhī, Ibn al-Samḥ, Ibn Hūd*, London, al-Furqān, 1996; English transl. *Founding Figures and Commentators in Arabic Mathematics*. A History of Arabic Sciences and Mathematics, vol. 1, Culture and Civilization in the Middle East, London, Centre for Arab Unity Studies, Routledge, 2012. R. Rashed and H. Bellosta, *Ibrāhim ibn Sinān. Logique et géométrie au Xe siècle*, Leiden, E. J. Brill, 2000.

intersection of conics, etc. One of the results of this research, the importance of which needs no demonstration, is a distinction that ultimately became the standard, namely, that between the curves that one can obtain geometrically as plane sections of a cone or cylinder and draw with a continuous motion; and all the others. This continuous motion, which is conceived geometrically and not kinematically, is generated thanks to new instruments designed with this goal in mind; among these, the perfect compass in particular stands out.[39] Thanks to these instruments, the mode of generation of a class of curves and their geometrical *characterization* are now unified. This unification has a specific date, and al-Qūhī is responsible for it.

Al-Qūhī distinguishes the curves traced by the perfect compass – the straight line, circle, and conic sections – by giving them a generic name: 'the *qiyāsiyya* lines', which we translate as 'measurable lines'. At the beginning of his book he writes:

> This is a treatise on the instrument called the perfect compass, and it contains two books. The first deals with the demonstration that it is possible to draw measurable lines *(qiyāsiyya)* by this compass – that is, straight lines, the circumferences of circles, and the perimeters of conic sections, namely parabolas, hyperbolas, ellipses, and opposite sections.[40]

Henceforth the tradition will dedicate the term *qiyāsiyya* to designating these curves and also to excluding the others.

What did al-Qūhī and his successors mean by 'measurable lines'? If one follows the geometrical terminology of the period, the expression denotes lines governed by the theory of proportions. This is precisely what al-Qūhī means: They are curves generated by a single continuous motion – that of the leg of a perfect compass, for example – and to which the theory of proportions can be applied. Such is the case of the straight line and the circle, but also of the conic sections characterized by *symptomata* or by the properties of the focus and the directrix, or later the eccentricity.

Al-Qūhī has thus established a classification of curves: those that are measurable, and the others. By doing so, he has just left behind a distinction

[39] R. Rashed, 'Al-Qūhī et al-Sijzī: sur le compas parfait et le tracé continu des sections coniques', *Arabic Sciences and Philosophy*, 13.1, 2003, pp. 9–44 and *Geometry and Dioptrics in Classical Islam*, Chap. V.

[40] *Fī al-birkār al-tāmm*, ed. R. Rashed in *Geometry and Dioptrics in Classical Islam*, Chap. V, Arabic p. 727, 3–6.

anchored in tradition: the distinction that opposes the straight line, on the one hand, and curves (including the circle), on the other. In addition, he has upset the ancient classification between 'simple' and 'mixed', since the straight line, the circle, and the conics now belong to the same class. Finally, he has opened up a new perspective: every curve is measurable that an instrument can generate in a single continuous motion and that one can study geometrically in terms of the proportion theory. This is al-Qūhī's message to the future. In other words, he has just isolated plane curves of the first and second degree. Now, by distinguishing this class of curves, the very criterion of classification changes: the line of demarcation now runs between 'measurable' and 'non-measurable' curves, according to their mode of generation and according to whether or not they are governed by proportion theory.

'Non-measurable' curves remain, and al-Sijzī, al-Qūhī's younger contemporary, will make progress in this domain. In his *Introduction to Geometry*,[41] a book devoted to the classification of geometrical entities, he distinguishes three types of curve: the 'measurables' we have just discussed; the 'non-measurables' that have an order (*niẓām*) and a regularity (*tartīb*), adopting the terminology that we have already encountered in Simplicius; and the 'non-measurables' without 'order' or 'regularity'.

The first, generated by a single continuous motion, are 'geometrical', that is, one draws upon them in geometry; the second are generated by two continuous motions that are dissimilar. They are no longer 'geometrical' but 'mechanical'. Finally, the third, also generated by two dissimilar continuous motions, are not even mechanical. They have no specific name. The example that al-Sijzī gives is a curve that we know well: the cylindrical helix. In his words:

> As to the cylindrical helix (*al-khaṭṭ al-lawlabī*), which is used in mechanics (*al-ḥiyal*) and not in geometry (for it is not measurable [*ghayr qiyāsī*] but has an order and a regularity), it is generated by the motion of the point following a straight line and following a circle in common with the cylinder.[42]

He continues:

> Here is its figure. Given the cylinder *ABCD* with its two bases *AB* and *CD*. If we imagine that point *A* moves with uniform motions following the straight line

[41] *Al-Madkhal ilā 'ilm al-handasa*, ms. Dublin, Chester Beatty 3652, fols 2^{v}–8^{r}, ed. R. Rashed and P. Crozet in *Œuvre mathématique d'al-Sijzī*, vol. II, forthcoming.

[42] *Al-Madkhal ilā 'ilm al-handasa*, ms. Dublin, Chester Beatty 3652, fol. 4^{r}.

AC and that the cylinder rotates in uniform motion, the line *AEGHID* is generated, which is a cylindrical helix. As to the line that has no order, it therefore has neither limit (*ḥadd*, also translated by 'definition') nor extremity, and it is used in none of the arts; that is why it is neither described nor defined.[43]

Al-Sijzī draws the cylindrical helix, but gives no example of nonmeasurable curves without order or regularity. Perhaps he had in mind such curves as the quadratrix or the spiral.

About the meaning of the distinction between measurable and nonmeasurable curves, there is not the shadow of a doubt. A clear confirmation of the meaning of these terms in al-Sijzī's use of them when he defines angles: the nonmeasurable angles are precisely curvilinear angles and the angle of contingency (the horn-angle), whereas the measurable angles are those one can study by means of proportion theory.[44]

Note that 'measurable', 'nonmeasurable', 'order', and 'regularity' are qualifications of curves and not of motions, and even less of the problems, in distinction to 'plane', 'solid' and 'linear' in Pappus. The same continuous and dissimilar motions generate curves that cannot be qualified as 'mechanical'. The singularity of the cylindrical helix, which Apollonius had already established by affirming that same-length parts of the cylindrical helix coincide homoeomerically, recurs in al-Sijzī, but in a different form. In this case, the wine and the wineskins are both new.

Nowadays we know that the helix traced on the cylinder of revolution is the only skew curve whose radii of curvature and radii of torsion are constants. It remains to be discovered whether Apollonius's formula – 'homoeomeric' – as well as al-Sijzī's concepts – 'nonmeasurable', 'with order and regularity' – were means of expressing qualitatively properties that they had glimpsed without yet having the means of knowing them. In the case of al-Sijzī, one can suggest the following conjecture: as a young contemporary of Ibn Sahl and al-Qūhī, that is, two mathematicians who devoted themselves to the study of tangents and tangent planes, al-Sijzī was in a position to know a distinctive property of this curve: the tangents to it form a constant angle with the axis. This is indeed a property of order and regularity.

In his classification of curves, al-Sijzī proceeds by means of a double reference point: the number and nature of the motions involved in generating the

[43] *Ibid.*
[44] *Ibid.*, fol. 8^r.

curve; and the geometrical properties of the curve itself, 'order and regularity', destined to characterize the curves (or not).

One should also remember that the proposed classification very naturally echoes the mathematical knowledge of the time, reflecting its extent and boundaries. Indeed some of its distinctive characteristics derive from two limitations in the latter. As has already been pointed out, al-Sijzī alludes to no 'mechanical' curve apart from the cylindrical helix: in doing so, he is no exception. Short of being contradicted by a surprising discovery, one can venture that the 10th-century mathematicians, as well as their successors, had little interest in 'mechanical' curves. To explain this fact, however, one cannot restrict oneself to the history of transmission of the Greek geometrical *corpus*. Although it is true that Archimedes' treatise on the spiral, for example, was not translated into Arabic (thereby depriving mathematicians of this curve), his treatise *On Conoids and Spheroids* was not translated either; yet this did not prevent them from reinventing and transcending its content. It is therefore necessary to look elsewhere for an explanation, which will turn out to be the flip-side of a fact that is itself positive.

We have shown that during the 10th century, with mathematicians like al-Qūhī,[45] new requirements have now established themselves as norms: one must provide a genuine proof of existence when it is necessary, and one must also establish construction procedures on solid geometrical foundations. Thus, the mechanical system of Ibn Sahl and the perfect compass of al-Qūhī, which are both intended for the construction of conics, are not just any old instruments; they are themselves shaped by the theory of conics that they embody. The mathematicians thus restricted themselves to the only curves for which they had the means of establishing their existence and of proceeding to their construction. To put the point clearly and succinctly, it is thanks to these very requirements that mathematicians were able to distinguish the class of plane curves of the first two degrees, and turned away from the active study of 'mechanical' curves.

[45] R. Rashed, *Les mathématiques infinitésimales du IXe au XIe siècle*, vol. III: *Ibn al-Haytham. Théorie des coniques, constructions géométriques et géométrie pratique*, London, al-Furqān, 2000 (English translation: *Ibn al-Haytham's Theory of Conics, Geometrical Constructions and Practical Geometry*. A History of Arabic Sciences and Mathematics, vol. 3, Culture and Civilization in the Middle East, London, Centre for Arab Unity Studies, Routledge, 2013); and *Geometry and Dioptrics in Classical Islam*.

The second limitation in the mathematical knowledge of the time is inherent in the same class of plane curves. Having now defined the class of 'measurable curves', why indeed did they stop with the first two degrees, even though they had encountered solid and 'super-solid' problems? The question is a fundamental one. After all, al-Qūhī had generalized the Euclidean method of applying areas to solids[46] and had changed the criteria for permissible constructions: no longer restricted to the straightedge and compass as the Ancients had been, he now allowed constructions by means of conics.[47] In other words, until new evidence to the contrary turns up, no one ever tried to trace a cubic. To do that, it would have been necessary to have the definition of *any* plane curve – not merely those of the first two degrees – by its equation. In other words, one would have needed the establishment of a new chapter in which curves are studied by means of their equations. Not until the end of the 17th century, at least, would this be truly realized. For now, we will be content with stating that, beginning in the middle of the 10th century, it is the very framing of the problem that evolves: its formulation appears increasingly to be that of a geometrical problem. The mode of generation of an entire class of curves is always conceived in terms of motions, but these motions are themselves studied either in the geometrical terms of the perfect compass or other instruments that are themselves incarnations of mathematical theories; or in terms of loci on quadratic surfaces. If the 'geometrization' of the problem took place at the cost of the limitations emphasized above, it in turn generated new technical questions that goaded mathematicians to fine-tune the classification of second-degree curves. This is how al-Sijzī undertook to determine conic sections as plane sections of solids. In three of his writings, he made it his goal to characterize several quadratic surfaces by means of their plane sections, thus classifying them by means of a new concept: 'the rank'. The rank is a number, a function of the number of sections, whether limited or unlimited, associated with the surface. In al-Sijzī, the term 'unlimited' has a double meaning. A curve is unlimited if it goes to infinity, which happens only for parabolas and hyperbolas; conversely, a family of ellipses is 'unlimited' if the diameters of these ellipses are not limited. We have studied al-

[46] R. Rashed, *Les mathématiques infinitésimales*, vol. III, pp. 919–35; English transl. *Ibn al-Haytham's Theory of Conics*, pp. 714–27.

[47] To understand the importance of this extension, recall that Viète allowed only constructions made with a straightedge and compass.

Sijzī's classification elsewhere,[48] and will therefore simply reproduce the table here.

Solids	*Plane Sections*	*Rank*
Sphere	Circle	1
Oval or lenticular solid	Circle, ellipse	2
Cylindrical solid	Circle, unlimited ellipse	3
Parabolic or hyperbolic cupola	Circle, ellipse, parabola or hyperbola	4
Conic solid	Circle, ellipse, parabola, hyperbola, triangle (unlimited curves)	5

Note what is missing from among the solids that al-Sijzī considers here: elliptical, parabolic, and hyperbolic cylinders. But he examined these three solids in another study of their plane sections and gave the following classification:[49]

Solids	*Plane Sections*
Elliptical cylinder	Parallelogram Ellipse Circle
Hyperbolic cylinder	Parallelogram Hyperbola
Parabolic cylinder	Parallelogram Parabola

This table completes the preceding one, and hence the classification of quadratics of revolution. One will have to wait another eight centuries for a consideration of ruled quadrics.

In his writings, al-Sijzī characterizes a certain number of surfaces by their plane sections. He limits himself to cases in which these sections are circles or conics, and he does not mention the straight line when the surface is a plane. No one will take up this problem until Fermat does so in his *Introduction aux*

[48] R. Rashed, *Œuvre mathématique d'al-Sijzī*. Vol. I: *Géométrie des coniques et théorie des nombres au Xe siècle*, Les Cahiers du Mideo, 3, Louvain/Paris, Peeters, 2004, Chap. I.

[49] *On the Properties of Elliptical, Hyperbolic and Parabolic Solids*, in R. Rashed, *Œuvre mathématique d'al-Sijzī*, vol. I.

lieux en surface.[50] Note also that, more generally, this method will play a privileged role in the study of algebraic surfaces at the end of the 19th century.

4. GEOMETRICAL TRANSFORMATION AND THE CLASSIFICATION OF CURVES

Thus far the opposition between geometrical and mechanical is the principal tool in the first classifications of curves. It will remain such for several centuries, as we shall see later, still accompanied by the doctrine of motion that undergirds it. But another equally important constant of these first classifications of curves is that this opposition will be pushed into the background as soon as mathematicians attempt to refine the classification of geometrical curves. In this case, either one tries to translate the concept of motion into operational terms, or one associates with it other concepts that allow the isolation of families of curves in the midst of this category; thus the concept of equation, for example.

We have just alluded to the examples of Ibn Sahl for the conics, and especially of al-Qūhī's perfect compass. The theory of this instrument, developed by al-Qūhī and his successors to draw continuously curves of the first two degrees, made it possible not only operationally to translate the motions (rotation, translation), but also to unify in the language of proportion theory both the procedure for generating curves that involve motions and the characterization of the latter, that is, their equations. It thus became possible to identify this family of geometrical curves. Fine-tuning this classification a little farther, however, puts one in a position to organize the curves of this same family. This is precisely what al-Sijzī accomplished, thanks to his studies of certain quadratic surfaces and to his concept of 'rank'. In parallel with these attempts, however, mathematicians were successfully clearing the way for a further refinement of the classification. The point at issue was to draw on geometrical transformations as a tool for establishing a class of objects. The stunning fertility of this route will become clear later, in this field as well as many others.

To understand how mathematicians from the beginning of the 10th century set out on this new path, it will be useful to recall a few salient facts. Ever since the Banū Mūsā in the middle of the 9th century, the study of geometrical

[50] *Œuvres de Fermat*, published by P. Tannery and C. Henry, Paris, 1896, vol. III, pp. 102–8.

transformations spread to such an extent that the new geometrical research eventually became distinct from the legacy of the Hellenistic period.[51] Conversely, from the beginning of the 10th century, geometers such as Ibn Sinān and al-Khāzin raised the theoretico-technical question of drawing conics both by points and continuously. Finally, an apparently modest result was obtained, with mathematical and philosophical consequences that were not: the circle belongs explicitly and by birthright to the family of the conics. Everything was therefore in place to raise an unprecedented question: can one obtain the three conic sections by means of geometrical transformations of the circle? If so, one will, on the one hand, effectively solve the theoretical-technical problem of drawing conics and, on the other, conceive of the act of classification as an authentically mathematical act. The rather vague concept of continuous motion will be replaced by the more rigorous and operational concept of transformation. Finally, this act of classification will henceforth be a deliberately unifying one in the sense that the family of curves will now be composed of transforms of one of its elements.

To our knowledge, the first mathematician who took this route is Ibrāhīm ibn Sinān (909–946). In the introduction of a treatise that he titled *On the Drawing of the Three Conics*, he wrote:

> When we discovered that the continuous drawing of these three sections by means of the compass or other instruments is difficult, we exerted ourselves to trace many points, the number of which one can increase at will and such that these points will be on one of the three sections. Everything that we determined thus shows how these sections, as well as others, can be generated starting from the circle.[52]

There is no doubt that Ibn Sinān knew that the continuous drawing of the conics is impossible with a straightedge and compass. He therefore tried to invent a method to trace them using these same instruments, but by points. This method is nothing other than the transformation of the circle into these curves. In his treatise, he thus spells out how to obtain the parabola starting from a family of circles by means of affine transformation; the ellipse by

[51] R. Rashed, *Les mathématiques infinitésimales du IXe au XIe siècle*, vol. IV: *Méthodes géométriques, transformations ponctuelles et philosophie des mathématiques*, London, al-Furqān, 2002; R. Rashed and H. Bellosta, *Ibrāhīm ibn Sinān. Logique et géométrie au Xe siècle*.

[52] R. Rashed and H. Bellosta, *Ibrāhīm ibn Sinān. Logique et géométrie au Xe siècle*, p. 264; Arabic p. 265, 14–17.

orthogonal affinity; and the hyperbola by projective transformation. We have studied this treatise elsewhere after having established and translated the text.[53] Let us interpret here Ibn Sinān's study of the case of the hyperbola, using a language other than his own.

Given a circle of radius $r = \frac{AB}{2}$, and the line *EG*, tangent to the circle at *E*. One draws the straight line *GO* in any fixed direction such that *GO* = *GE*. Let *P* and *T* be the orthogonal projections of *E* and *O* respectively. In triangle *CEG*

$$CE^2 = CP \cdot CG,$$

whence

$$CG = \frac{r^2}{CP}.$$

Moreover

$$\frac{GE}{CE} = \frac{PE}{PC},$$

whence

$$GE = r \cdot \frac{PE}{PC}.$$

Letting θ be the angle *ZGO*, one has

$$GT = GO\cos\theta = GE\cos\theta = r\frac{PE}{PC}\cos\theta$$

and

$$TO = GO\sin\theta = GE\sin\theta = r\frac{PE}{PC}\sin\theta.$$

[53] *Ibid.*, Chap. III.

If one notes x, y as the coordinates of E, and X, Y as those of O on the orthogonal axes with origin C, the x-axis being CZ, one thus has

$$X = \frac{r^2}{x} + \frac{ry\cos\theta}{x}, \quad Y = \frac{ry\sin\theta}{x};$$

these are the parametric equations of hyperbola $\mathscr{H}$ of equation $X^2 - \frac{2XY}{\tan\theta} - Y^2 = r^2$ as a function of the parameters x and y that verify $x^2 + y^2 = r^2$. For all values of x, one has

$$\frac{Y}{X} = \frac{y\sin\theta}{r + y\cos\theta} \text{ (equation of straight line } GO\text{).}$$

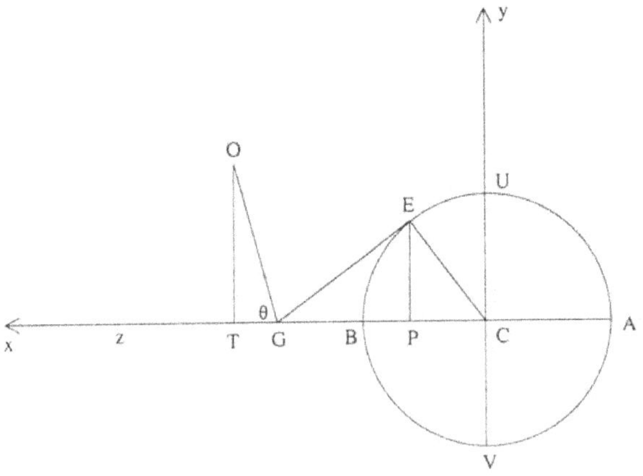

Fig. 18

If $x = 0$, one has $y = r$ where $y = -r$ (point U or point V of the circle); in this case

$$\frac{Y}{X} = \frac{\sin\theta}{1+\cos\theta} \text{ or } \frac{Y}{X} = \frac{\sin\theta}{\cos\theta - 1};$$

one thus has the equations of the asymptotes of the hyperbola $\mathscr{H}$. The straight line with the equation $Y = X\tan\theta$ is the conjugate diameter of the transverse diameter AB.

One half-century after Ibn Sinān, al-Sijzī drafts a treatise with a title that is nothing less than an entire research program – *All Figures Start from the*

Circle – as well as a letter with the same title addressed to a contemporary.[54] No trace of the book remains but we have the letter, in which al-Sijzī takes up briefly the terms of his treatise. Along with another work entitled *On the Description of Conic Sections*,[55] the letter shows that al-Sijzī stands in the tradition of Ibn Sinān. The next step in this same direction will not be taken until Desargues, who will develop fully the idea that all conic curves are obtained by conic projection starting from the circle, and make this proposition the foundation of the theory of conic sections.

5. THE INTERVENTION OF THE ALGEBRAISTS: THE POLYNOMIAL EQUATION AND THE ALGEBRAIC CURVE

To the geometers, the algebraists of this period lent a strong hand, insofar as these curves of the first two degrees were henceforth *characterized* by their equations.

The geometers of the 10th century knew about not only conic sections, but also other non-mechanical curves, such as the conchoid of Nicomedes and also the cissoid of Diocles. What is more, they themselves – notably the Banū Mūsā – invented another curve, the 'conchoid of the circle', which Roberval would later call 'Pascal's snail'. They relied on some of these curves for the construction of solid problems – finding two means and trisecting an angle. But their successors soon turned away from these curves to focus on conic sections in the construction of solid problems, which in the end constituted a full and important chapter in the geometrical research of the era.[56] Thus they limited themselves exclusively to curves admissible to geometry. Nevertheless, alongside the active research in this area, which deliberately turned its back not only on mechanical curves, but also on curves other than conics, al-Khāzin and Abū al-Jūd among others were developing a geometrical theory of algebraic equations. It was al-Khayyām (1048–1131) who in the end elaborated it, followed by Sharaf al-Dīn al-Ṭūsī.

[54] R. Rashed, *Œuvre mathématique d'al-Sijzī*, vol. I, Chap. II and Text no. 6.
[55] *Ibid.*, Text no. 3.
[56] R. Rashed, *Geometry and Dioptrics in Classical Islam*, Chap. IV.

Without retracing here the history of this theory,[57] let us recall two of its results, which are of particular significance for the topic at hand.

For the equations of the third degree, al-Khayyām gives a classification that indicates the two conic curves used to determine a positive root. For example, he gives the following pairs for these quadrinomial equations:

$x^3 + ax^2 + bx = c$	circle and hyperbola
$x^3 + ax^2 + c = bx$	hyperbola and hyperbola
$x^3 + bx + c = ax^2$	circle and hyperbola
$x^3 = ax^2 + bx + c$	hyperbola and hyperbola
$x^3 + ax^2 = bx + c$	hyperbola and hyperbola
$x^3 + bx = ax^2 + c$	circle and hyperbola
$x^3 + c = ax^2 + bx$	hyperbola and hyperbola.

One thus sees that the quadrinomial cubic equations of this class are constructed by means of an equilateral hyperbola and a circle, or of two equilateral hyperbolas.

Henceforth – and this is not the least result of the classification of equations of the first three degrees – it is known that a polynomial equation (here the first three degrees, and also the fourth) is constructed starting from two conics, that is, from their equations. Moreover, it is also known that the *symptoma* (once the reciprocal has been established) is nothing more than a second-degree equation.

This correspondence between polynomial equations and conic curves carries within itself a seminal idea that will later renew the very criterion of classification. Once it is exploited generally, before the invention of instruments suitable to tracing curves of a higher degree, it will isolate a class of curves: algebraic curves. For now, however, this correspondence stops on the threshold of this development. Take, for example, the case in which one encounters the problem raised, and probably solved, by Ibn al-Haytham (d. after 1040), namely, the problem of determining four segments between two segments such that the six be in continuous proportion:

[57] R. Rashed and B. Vahabzadeh, *Al-Khayyām mathématicien*, Paris, Librairie A. Blanchard, 1999.

$$\frac{a}{x} = \frac{x}{y} = \frac{y}{z} = \frac{z}{u} = \frac{u}{b} \qquad \text{(with } a \text{ and } b \text{ given),}$$

which reduces to a fifth degree equation

$$y^5 = a^3b^2$$

and which can be solved by the intersection of a hyperbola $yz = ab$ and of a cubic $y^3 = az^2$.

Al-Khayyām writes:

> This has been demonstrated by Abū ʿAlī ibn al-Haytham; it is very difficult, however, so that we cannot include it in our book.[58]

In other words, since al-Khayyām wished to develop a systematic theory from which he wanted to exclude the fourth degree and, *a fortiori*, higher degrees, he stopped here both in the correspondence between polynomial equation and curve and in the examination of his concept. As a last resort, he no doubt could have included the fourth degree insofar as the intersections of conics were sufficient to give a solution; but his concept of dimension kept him from heading in this direction. For degrees higher than the fourth, the geometrical solution requires that one consider algebraic curves of degree ≥ 3.

Nevertheless, in this same seminal idea is inscribed the origin of Descartes's project and the classification that he establishes between 'geometrical' – that is, algebraic – and 'mechanical' curves.[59]

This intervention of algebra – more precisely incipient algebraic geometry – could only reinforce the classification that al-Qūhī and his contemporaries had conceived and, better yet, made it possible to consolidate explicitly the new basis of the future classification. Indeed, this classification made it possible to establish a direct and explicit relationship between conic curve and equation. To translate an irreducible cubic equation geometrically in order to solve it by the intersection of two conic curves effectively required that the latter be given by their equations. The conic curve is henceforth defined by its equation, which now becomes the main tool of classification.

[58] *Treatise on Algebra*, ed. R. Rashed in *Al-Khayyām mathématicien*, p. 223, 10–11.

[59] See below, 'Descartes's *Géométrie* and the distinction between geometrical and mechanical curves'.

To illustrate this situation, let us take a single example from the preceding list. Given the equation

(1) $$x^3 + ax^2 + c = bx,$$ with a, b real and positive,

which is rewritten as

$$x^2(a+x) = bx - c,$$

whence

$$\frac{b}{x^2} = \frac{a+x}{x - \frac{c}{b}}.$$

Al-Khayyām had begun his study of cubic equations by demonstrating how to find two segments between two segments such that the four were in continuous proportion. Thus one has

$$\frac{\sqrt{b}}{x} = \frac{a+x}{y} = \frac{y}{x - \frac{c}{b}},$$

and therefore

$$xy = \sqrt{b}\left(x - \frac{c}{b}\right),$$ the equation of a hyperbola $\mathscr{H}_1$

and

$$y^2 = \left(x - \frac{c}{b}\right)\left(x+a\right),$$ the equation of a hyperbola $\mathscr{H}_2$.

The intersection of these two hyperbolas gives us once again equation (1). Al-Khayyām shows that the point $\left\{\frac{c}{b}, 0\right\} \in \mathscr{H}_1 \cap \mathscr{H}_2$, and that if this intersection is reduced to a single point, the equation has no solution; but if $\mathscr{H}_1 \cap \mathscr{H}_2 \neq \left\{\frac{c}{b}, 0\right\}$, then there exist two different points of $\left\{\frac{c}{b}, 0\right\}$ belonging to the intersection, which are the solutions of the problem.

In al-Khayyām's discussion, one notices the first use of equations to discuss the intersection points of two conic curves, that is, an attempt, admittedly timid, to learn certain properties of curves by starting with their equations. This orientation finds a confirmation in the research of al-Khayyām's successor, Sharaf al-Dīn al-Ṭūsī.[60]

The latter rewrites equation (1) in the form

$$x\left(b-x^2\right)-ax^2=c \qquad \text{with } 0<x<\sqrt{b}$$

and then considers

$$f(x)=x\left(b-x^2\right)-ax^2;$$

he tries to determine the maximum of this expression. He then posits

$$g(x)=b-3x^2-2ax$$

and determines x_0, the positive root of $g(x)=0$. Next he shows that

$$f(x_0)=\sup_{0<x<\sqrt{b}} f(x),$$

and obtains the following three cases:

$c>f(x_0)$, the problem is impossible;
$c=f(x_0)$, x_0 is a solution;
$c<f(x_0)$, the equation has two positive solutions x_1 and x_2, such that $x_1<x_0<x_2$, which he determines.

Al-Ṭūsī thus goes farther than al-Khayyām, since he derives 'analytic' properties starting from equations: the existence of the maximum.

[60] R. Rashed, *Sharaf al-Dīn al-Ṭūsī, Œuvres mathématiques. Algèbre et Géométrie au XIIe siècle*, Collection Sciences et philosophie arabes – textes et études, 2 vols, Paris, Les Belles Lettres, 1986.

To develop the geometrical calculation necessary for the theory of equations of the first three degrees, al-Khayyām and his successor Sharaf al-Dīn al-Ṭūsī took the following steps, as we have noted. On the one hand, they classified cubic equations with one unknown as a function of the conic sections used to solve them and starting from the property of the two means. On the other hand, they proceeded to characterize these second-degree curves by their equations. Al-Ṭūsī, for his part, took an additional step forward by developing an idea present only in germ in his predecessor's theory: to rediscover certain properties of curves by starting from their equations and determining other analytic properties of algebraic equations. It is precisely to this task that he devoted himself in his research on *maxima*, to which his work on the existence of intersection points of conic curves had led him. But neither al-Khayyām nor al-Ṭūsī had studied an algebraic curve defined by a cubic equation. Not until Descartes would an unprecedented program devoted to the study of algebraic curves emerge.

6. THE CLASSIFICATION OF CURVES AS MECHANICAL AND GEOMETRICAL

Beyond straight lines and conics, only three other algebraic curves were known: the cissoid of Diocles, which is a cubic,[61] the conchoid of Nicomedes, and the so-called 'Pascal's snail,' which is a conchoid of the circle (two quartics). It is even more significant that no algebraist before Descartes thought of defining these cubics or quartics by their equations. Beginning in 1630, Descartes would add two cubics to these three curves: the *folium* and the *trident*.

This was the situation before 1619 when a young Descartes, still equipped with a modest mathematical culture, conceived of this new program for studying algebraic curves. Since we have explained this program elsewhere,[62] only the leading ideas will be mentioned here. Descartes seeks to classify geometrical problems as a function of the curves used to solve them. Thus one has problems that are soluble by means of a straight line and a circle (with a straightedge and compass); those that require the intervention of other curves

[61] R. Rashed, *Les Catoptriciens grecs*.

[62] See below, 'Descartes's *Géométrie* and the distinction between geometrical and mechanical curves'.

different from the circle but generated by a single continuous motion; and yet others solved by curves generated by motions that differ from one another and are not subordinated to one another. The first curves are geometrical, the others are all mechanical, and only the former are allowed in geometry.

According to the other axis of his research program, the challenge is to classify the curves themselves, first by considering the motions by which they are drawn (as we have just seen) and later as a function of the equations that define them. Not until the *Géométrie* did the concept of equation operate explicitly in the classification of curves.[63]

The intention that governs his program is clear: to go beyond the conics in order to isolate a class of curves of which the latter are only a subclass, and on the strict condition of drawing on no concept alien to the ancients; that is, by respecting the rules of proportion theory. What deserves special emphasis here is that, as his research advanced from 1619 to the composition of the *Géométrie* in 1636, Descartes seems to have realized that it would be advantageous to add two concepts: to the concept of movement, which allowed him to exclude mechanical curves from geometry – an eminently positive act – he added another, more effective and more operational, that of the equation of a curve.[64] The latter notion allowed him to circumscribe this new class of geometrical curves, to extend it beyond the conics, and finally to bring about a new classification on which an entire stream of research would rely, already among contemporaries such as Fermat, and later among his successors into the late 18th century.

Let us now turn to the *Géométrie*, and book 2 in particular. As Descartes writes, the second book concerns 'some general statements (*quelque chose en*

[63] See the text cited later (at note 77) in which Descartes explains himself without any ambiguity (A.T. VI, pp. 392–3). See also J. Itard, 'La *Géométrie* de Descartes', *Les Conférences du Palais de la Découverte*, série D, no. 39, 7 January 1956; repr. in J. Itard, *Essais d'histoire des mathématiques*, collected and introduced by R. Rashed, Paris, A. Blanchard, 1984, pp. 269–79, and J. Vuillemin, *Mathématiques et métaphysique chez Descartes*, Paris, P.U.F., 1960, esp. pp. 91–3.

[64] 'In 1618 the curves constructed with compasses are called geometrical; these curves will later be called "organic curves". In 1632 geometrical curves are those which, in relation to one axis – Descartes never used two coordinate axes – one could express an algebraic equation between the abscissas and the ordinates' (J. Itard, 'La *Géométrie* de Descartes', p. 276). See below, 'Descartes's *Géométrie* and the distinction between geometrical and mechanical curves'.

général) about the nature of curved lines'.[65] Both the project and the formulation are new: at stake is nothing less than the construction of a theory of curves. But in order to reach 'some general statements', several routes are possible, one of which, not the least traveled, consists in classifying, mostly *a priori*, objects or operations. It is therefore not surprising that, as his point of departure, Descartes settles on a classification that is already available, and that he then takes up the one that Pappus had proposed for geometrical problems: planes, solids, and linear.[66] Whereas he agrees with Pappus in asserting that plane problems are soluble with straightedge and compass and that solids are soluble by means of conics, Descartes diverges from him when it comes to linear problems, the construction of which requires other curves. Descartes criticizes Pappus's classification at two points, which he articulates as follows:

> But I am astonished that they (= the ancients according to Pappus) did not distinguish different degrees among these more complex curves besides this (those for the construction of linear problems), and I am unable to understand why they called these curves mechanical, rather than geometrical.[67]

Descartes explains that this can be justified neither by the mode of construction of these curves, nor by their complexity. He continues, in his own words:

> But it is appropriate to note that there is a great difference between this method of finding several points in order to trace a curved line passing through them, and the method used for the spiral and similar curves. For in the latter we do not

[65] *La Géométrie*, A.T. VI, p. 387; Descartes, *Discourse on Method, Optics, Geometry, and Meteorology*, English transl. with intro. by P. J. Olscamp, Indianapolis, Bobbs-Merrill Co., 1964, p. 189.

[66] The text of Pappus reads (Book III, Prop. 5): 'The ancients considered three classes of geometric problems, which they called plane, solid, and linear. Those which can be solved by means of straight lines and circumferences of circles are called plane problems, since the lines or curves by which they are solved have their origin in a plane. But problems whose solutions are obtained by the use of one or more of the conic sections are called solid problems, for the surfaces of solid figures (conical surfaces) have to be used. There remains a third class which is called linear because other "lines" than those I have just described, having diverse and more involved origins, are required for their construction. Such lines are the spirals, the quadratrix, the conchoid, and the cissoid, all of which have many important properties'. (*The Geometry of René Descartes*, transl. D. E. Smith and M. Latham, Open Court Publishing, 1925, reprint Dover, 1954, p. 40, n. 59; cf. *La Collection mathématique*, French transl. by P. Ver Eecke, vol. I, pp. 38–9, see also p. 207.)

[67] *La Géométrie*, A.T. VI, p. 388; transl. P. J. Olscamp, p. 190.

find all the points of the required line indiscriminately, but only those which can be determined by some process which is simpler than that which is required for composing the curve. And so, strictly speaking, we do not find any one of its points, that is to say, not any one of those which are so properly points of this curve, that they cannot be found except through it; on the other hand, there is no point on the lines that can be used for the proposed problem which cannot be found among those which can be determined through the method (*façon*) just explained.[68]

This 'great difference' that Descartes mentions refers in effect to the 'method (*façon*)' of tracing every sort of curve, according to whether the procedure conforms or not to proportion theory. Indeed, whereas for the two types of curves one proceeds by means of a continuous motion or motions, for geometrical curves these motions must have the same nature, and consequently must be comparable, or better, commensurable, whereas for the spiral, the quadratrix, and the other mechanical curves, these motions have a different nature and are therefore incommensurable. We know that tracing the spiral, the quadratrix, etc., relies on a rotation and a translation. But according to the doctrine of motion that goes back to Aristotle,[69] and to which Kepler,[70] Viète,[71] as well as Descartes, among others, adhered, these motions do not pertain to the same measure '[...] because the ratios between straight and curved lines are unknown, and even, I believe, unknowable to men, so that we cannot thereby reach any exact and assured conclusions'.[72] This is certainly a sufficient reason to exclude mechanical curves during geometrical constructions. As to geometrical curves, Descartes describes in Book 2 of the

[68] *La Géométrie*, A.T. VI, pp. 411–12; transl. P. J. Olscamp, p. 206.

[69] Aristotle, *Physics*, VII 4, 248 a 10–249 a 29.

[70] J. Kepler, *Paralipomena*, 1604, p. 2^r: 'Per cossam id fieret si etiam a rectis ad curvas esset in cossicis denominationibus'.

[71] According to François Viète, the analyst does not compare the straight line with the curve, because the angle is a certain mean between the straight line and the plane line. This is why the law of homogeneity is repugnant to it. This is what he writes in his *In Artem Analyticen Isagoge*: 'Lineam rectam curvæ non comparat, quia angulus est medium quiddam inter lineam rectam & planam figuram. Repugnare itaque videtur homogeneorum lex' (*Opera mathematica*, recognita Francisci à Schooten, Vorwort und Register von J. E. Hofmann, Hildesheim/New York, Georg Olms, 1970, p. 12, § 28).

[72] *La Géométrie*, A.T. VI, p. 412; transl. P. J. Olscamp, p. 206.

Géométrie an instrument related to the *mesolabon* of Eratosthenes that allows one to draw them continuously and consistently with proportion theory.[73]

According to Descartes, then, the ancients were wrong to include, among mechanical curves, curves that were eminently geometrical, such as the conchoid of Nicomedes and the cissoid of Diocles. They were also wrong for failing to distinguish the various subclasses within the class of complex curves. It is therefore necessary to correct these defects by reworking the classification of curves. This is what Descartes proposes to do.

To rework Pappus's classification, however, is first of all to give a brand new meaning to the opposition between mechanical and geometrical. It is by clearly isolating the class of geometrical curves and by enriching it with new curves that Descartes will conceive this new meaning. He goes about the task in two related stages. He begins with the conception of motion or movement, and writes:

> We must no more exclude complex lines from it (= geometry) than simple ones, provided that we can conceive them as being described by a continuous movement, or by several successive movements of which the latter are completely determined by those which precede.[74]

Note that in 1629 already, in a letter to Mersenne on November 13, he states about the helix and the quadratrix:

> Although one can find an infinity of points through which the helix and the quadratrix pass, yet one cannot find geometrically any of the points that are necessary for the effects desired from the one as much as from the other; and one can trace them in their entirety only by the conjunction of two movements that are not mutually dependent.[75]

In other words, in contrast to geometrical curves, for which one can determine the generic points, for mechanical curves one can only determine an incomplete set of points, even if it is everywhere dense.[76]

[73] Cf. H. Lebesgue, *Leçons sur les constructions géométriques professées au Collège de France en 1940–1941*, preface by M. Paul Montel, Paris, Gauthier-Villars, 1950, pp. 16–18.

[74] *La Géométrie*, A.T. VI, pp. 389–90; transl. P. J. Olscamp, p. 191.

[75] 'Descartes à Mersenne, 13 novembre 1629', A.T. I, p. 71.

[76] This is how the point was made at the time: 'The geometrical curves are those whose nature can be expressed and determined by the ratio of ordinates and abscissas, which are each finite magnitudes; the mechanical curves are those whose nature cannot be

(*Cont. on next page*)

Now, to invoke as a criterion the movement that intervenes in the drawing of curves is not completely unprecedented. Indeed the mathematicians of the 10th–12th centuries appealed to this same criterion. What is new, however, is the second stage; when Descartes – sometime after 1632 defines the geometrical curve of whatever degree as that expressed by a polynomial equation irreducible between the two coordinates. Here are his own words:

> I could give here many other ways of drawing and conceiving curved lines whose complexity would increase, by degrees, to infinity. But in order to understand together all curves that are present in nature, and to classify them by order into certain types, I know of nothing better than to say that all points of those curves which can be called 'geometrical' – that is, which fall under some precise and exact measure – necessarily have a certain relation to all the points of a straight line; and this relation can be expressed by a single equation for all the points. And when [no term of] this equation is higher than the rectangle of two undeterminate quantities, or else of the square of a single unknown quantity, the curved line is of the first and simplest class, which comprises only the circle, parabola, hyperbola, and ellipse. But when one or both of the equation's two unknown quantities (for there must be two, in order to explain the relation between two points) reaches the third or fourth degree, the curve is of the second class; and when the equation reaches the fifth or sixth degree, the curve is of the third class; and so on for the others, to infinity.[77]

From now on, it is possible to refine the classification of curves and to give a foundation to the famous 'geometrical calculus'. As to mechanical curves, Descartes in the *Géométrie* adds nothing and is content to give examples: spirals and the quadratrix of Dinostratus. Leibniz and his students will define these curves by a differential equation connecting not the abscissa and

(*Cont.*) expressed thus, because the ordinates and the abscissas have no regulated ratio. In the geometry of infinitely small quantities, the nature of all curves, whether geometrical or mechanical, can be expressed uniformly by the ratio of the infinitely small portions of the axis to the corresponding differences. The whole difference between geometrical and mechanical curves is that the mechanical ones cannot be expressed by this ratio, whereas the geometrical ones can also be expressed by the ratio of the ordinates and the abscissas'. (*Commentaires sur la* Géométrie *de M. Descartes*, by Father Claude Rabuel, Lyon, 1730, p. 99.)

[77] *La Géométrie*, A.T., VI, pp. 392–3; transl. P. J. Olscamp, pp. 192–3.

the ordinate, but their differentials; and they will give them a new name: transcendental curves.[78]

The definition of the geometrical curve by its algebraic equation marks the new departure in the classification of such curves. Indeed, it makes possible the conception of an entire class of curves, of which those of the first two degrees constitute only a small group.[79] Conversely, the new approach makes it possible to circumscribe other groups inside this class: cubics, quartics, quintics, sextics, etc., and thus to proceed to a sub-classification of curves. The landscape established by means of this new classification not only is incomparably richer, but is organized according to a concept that Descartes introduced: that of 'genre'. Finally this new conception leads to a new question, a technical one this time, the answer to which would in turn stimulate research. It is a matter of determining every time the group to which belongs any curve generated by an 'organic construction', that is, by the intersection of two or more straight lines, or of straight lines and other geometrical curves. Recall that Descartes allows that 'in order to trace all the curved lines which I intend to introduce here, we need to assume nothing except that two or more lines can be moved through one another, and that their intersections determine

[78] This concept of '*Transzendent*' appears in a manuscript of 1675; see H. Breger, 'Leibniz Einführung des Transzendenten', *Studia Leibnitiana*, Sonderheft 14, 300 Jahre 'Nova Methodus' von G. W. Leibniz (1684–1984), Wiesbaden, Franz Steiner, 1986, pp. 116–32, esp. pp. 122–3.

Note however that this term '*Transzendent*' was introduced by Leibniz beginning in the fall of 1673, that is, before the invention of his differential calculus, as E. Knobloch and S. Probst have noted in their introduction to vol. VII, 3 of the *Mathematische Schriften*, '*Ab Herbst 1673 (N. 23) verwendet Leibniz den Ausdruck transcendent für Figuren (figura transcendens), Kurven (curva transcendens), Probleme (problema transcendens), denen keine algebraische Gleichung bestimmten Grades zugeordnet werden kann. Dem entspricht die aequatio transcendens (N. 64)*' (p. xxiv, cf. the text of Leibniz no. 23, pp. 266–7. I thank E. Knobloch for having drawn my attention to this fact).

[79] E. Giusti correctly writes: 'Before Descartes, only named curves were known, that is, particular curves for which one could give an explicit construction by means of geometrical or kinematic operations: after Descartes, the concern will be completely arbitrary lines that are described immediately by means of their equation' ('Le problème des tangentes de Descartes à Leibniz', *Studia Leibnitiana*, Sonderheft 14, 1986, pp. 26–37, esp. p. 26.

other curves – which seems to me no more difficult [than the other two postulates]'.[80]

Indeed we must not forget that Descartes allows as a postulate, so to speak, that geometrical curves are 'organic (*organiques*) curves' in the etymological sense of organon (tool), that is, curves drawn by his famous compasses, and reciprocally. The correctness of this hypothesis was only established in 1876 by Alfred Bray Kempe.[81] Note however that the drawing instruments that Kempe considers are formed by links articulated in a plane, with axes of rotation perpendicular to this plane. One fixes enough points of the instrument so that it has only one degree of freedom (Descartes's continuous movement). One can establish that any point of such an instrument describes an algebraic curve; indeed the coordinates of such a point are tied to the sole parameter by algebraic relations obtained by a sequence of resolutions of triangles, and one therefore obtains an algebraic relation among them by eliminating the parameter.

Kempe's theorem states, reciprocally, that for every plane algebraic curve, one can construct an instrument of the foregoing sort that traces at least part of it. Thus, given a curve of equation $f(x, y) = 0$, assume $x = a \cos \varphi + b \cos \psi$, $y = a \sin \varphi + b \sin \psi$ where a and b are given and φ, ψ are variables; point $M = (x, y)$ describes a certain area of the plane and one will trace the part of the curve that is in this area. One has

$$(1) \qquad f(x,y) = \sum_{i=0}^{n} c_i \cos(m_i\varphi + n_i\psi + \varepsilon_i) = 0$$

with $c_i \in \Re_+^*$, $m_i, n_i \in \mathbf{Z}$ and $\varepsilon_i \in \left\{0, \pm\frac{\pi}{2}, \pi\right\}$.

Now consider the vectors $\overrightarrow{Q_{i-1}Q_i}$ with respective lengths c_i and polar angles $m_i\varphi + n_i\psi + \varepsilon_i$; equation (1) means that point Q_n belongs to the y-axis, since its first member is the abscissa of Q_n.

Kempe devised two types of instruments: the translator and the reversor, which make it possible to bring about, respectively, a translation and a sym-

[80] *La Géométrie*, A.T., VI, p. 389; Descartes, *Discourse on Method, Optics, Geometry, and Meteorology*, transl. P. J. Olscamp, pp. 190–1.

[81] Alfred Bray Kempe, 'On General Methods of Describing Plane Curves of the nth Degree by Linkwork', *Proceedings of the London Mathematical Society*, 7, 1876, pp. 213–16.

metry with respect to a straight line. By connecting a certain number of these instruments, one can bring about the transformation of point M to Q_n or reciprocally. By making Q_n describe the y-axis, the result is that M describes the given algebraic curve.

Descartes's instruments are of a different kind since they involve rulers sliding along certain straight lines.[82] One can nevertheless reproduce the motion of these instruments by means of those of Kempe since the latter can describe a straight line.

But to understand both the concept of 'genre' and the determination of the group to which the curve belongs, let us return to the example that Descartes gives.

Given a curve *EC* drawn by means of two rulers – two straight lines – and of an instrument – a figure. The first ruler is the vertical ruler *AK*; the second is the ruler *GL* which pivots at *G* and turns around *G* without leaving it. Given the figure *LKN* such that *KN* is extended indefinitely without changing direction, that is, remains parallel to itself when *LKN* moves along *AK*; point *C* slides on the rule between *G* and *L*. The intersection of *KN* and of *GL* describes the curve *EC*.

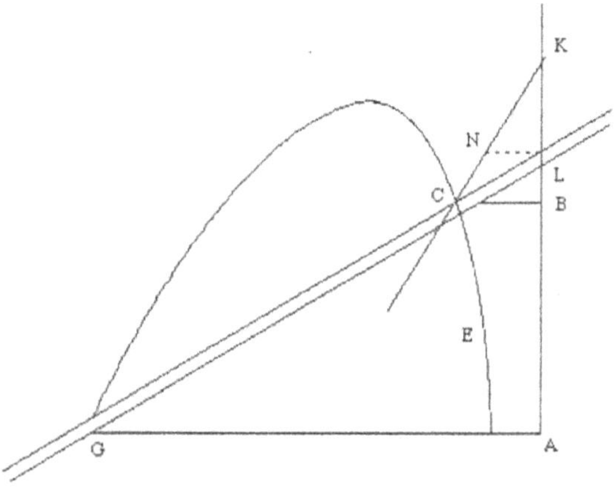

Fig. 19

One thus has a bijection between the sheaf with vertex *G* and the sheaf of straight lines parallel to *KN*. Note that this is a particular case of Poncelet's

[82] *La Géométrie*, A.T. VI, Book II, pp. 391–3; transl. P. J. Olscamp, pp. 191–4.

closure theorem, according to which the intersection point of the two sheaves of straight lines in homographic correspondence describes a conic.

The curve EC thus traced is indeed a branch of a hyperbola. Indeed, note that $AB = x$, $CB = y$ ($CB \parallel AG$), $GA = a$, $KL = b$ and $NL = c$; for the equation of EC, one has

$$y^2 = cy - \frac{c}{b}xy + ay - ac\,,$$

a curve of the second degree, and thus of the 'first genre'.[83]

The concept of the 'genre' of a curve is unquestionably the main tool for classifying algebraic curves. Second-degree curves belong to the first class, curves of the third and fourth degree belong to the second class, and so on. In general, a plane algebraic curve is said to be of 'genre' n when, referred to two coordinate axes, it represents an equation of degree $2n$ or $2n - 1$. This difference goes back to the fact that straight lines were not considered to be curves.

For this concept of 'genre' to be an effective instrument for the classification of curves, the 'genre' must be invariable, whatever its reference system may be. It must also be possible to order the classes. These two requirements seem perfectly natural and necessary when the task at hand is to erect this new classification, clearly *a priori*, on a solid foundation.

About the first requirement, the invariability of the 'genre', Descartes writes:

> [...] although there are many choices which shorten and simplify the equation, nevertheless, no matter in what way we choose the straight lines, the curve will always belong to the same class, as is easily demonstrated.[84]

Descartes thus states that the 'genre' of a curve is independent of its reference frame. He will forcefully rework this idea a year after publishing the

[83] For the time being, Descartes says nothing more about this. In his commentary, Van Schooten gives both its center and its asymptotes, but says nothing about the other branch of the hyperbola. For more details, he cross-references his *De organica conicarum sectionum in plano descriptione*, Leiden, 1646.

[84] *La Géométrie*, A.T. VI, pp. 393–4; transl. P. J. Olscamp, p. 193. This is an important text. The degree of the equation $P(x, y) = 0$ is independent of its reference frame. In 1667 or 1668, Newton affirms the permanence of the 'species' of the curve relative to the angle of the coordinates (see *Mathematical Papers of Isaac Newton*, II, p. 10). Newton gives a magisterial statement of the problem of changing the reference frame (*ibid.*, IV, pp. 383–4). See also Euler, *Introduction à l'analyse infinitésimale*, vol. II, pp. 12–17.

Géométrie, in a letter about the *folium* addressed to Mersenne on August 23, 1638.

Descartes first gives for the equation of this curve $x^3 + y^3 = nxy$; then, after changing the reference frame, the equation $\frac{y^2}{x^2} = \frac{a - x}{a + 3x}$, he writes '[...] that the new line [...] is the same as the other one'.[85] The reason for this is that every affine transformation of coordinate axes does not change the degree of the curve's equation, nor therefore its 'genre'; in other words, the 'genre' is for Descartes affine-invariant. This problem of the permanence of 'genre' will later be tackled and solved by Newton in the years 1667–1678, and discussed in detail by Euler, Cramer, and other successors of Descartes.[86]

Let us turn briefly to the second requirement – ordering the 'genres' – and return to the example of the preceding curve, the branch of the hyperbola. Descartes continues his exposition:

> If in the instrument used to describe the curve we replace the straight line *CNK* by this hyperbola, or some other curved line of the first class terminating the plane *CNKL*, the intersection of this line and the ruler *GL* will describe, instead of the hyperbola *EC*, another curved line, which will belong to the second class. Thus if *CNK* is a circle whose center is *L*, we shall describe the first conchoid of the ancients; and if it is a parabola whose diameter is *KB*, we shall describe the curved line which, as I have already said, is the first and simplest curve required for the problem of Pappus, when there are but five straight lines given in position.[87]

If then one replaces the straight line *KN* by a circle of center *L* and of radius *LK*, one obtains a curve of the superior class ('genre'). Indeed this is the conchoid of Nicomedes, of class 2. And if one replaces *KN* by a parabola, one will obtain a cubic – Descartes's parabola – with the equation

$$(*) \qquad y^3 - 2ay^2 - a^2y + 2a^3 = axy \qquad \text{(trident).}$$

[85] *Correspondance du P. Marin Mersenne*, commencée par Mme. P. Tannery, publiée et annotée par Cornélis de Waard, Paris, Éd. du CNRS, 1963, vol. VIII: *Août 1636–Déc. 1639*, p. 63, see also pp. 40–4.

[86] The implied transformation is the following: $x = \frac{u+v}{\sqrt{2}}$, $x = \frac{u-v}{\sqrt{2}}$, whence

$$u^3 + v^3 = a\sqrt{2}\, u \cdot v .$$

[87] *La Géométrie*, A.T. VI, pp. 394–5; transl. P. J. Olscamp, p. 194.

Descartes then gives the principle of his classification by means of the concept of 'genre'. He writes:

> Now I place curved lines which raise this equation to the square of the square in the same class (*genre*) as those which raise it only to the cube; and those whose equation goes as high as the square of the cube, in the same class as those in which it goes only as high as the supersolid, and similarly for the others.[88]

He thus groups in one and the same class cubics and quartics, quintics, and sextics, etc. The reason for such a grouping is that 'there is a general rule for reducing to a cube all the difficulties that pertain to the square of the square, and to the supersolid all those that pertain to the square of the cube, in such a way that we need not consider the latter more complex than the lower ones'.[89] Whereas this rule is valid in the first case – that of the cube – Jacques Bernoulli will criticize its generalization.[90]

Descartes's main idea is the following: the curves generated by the intersection of one straight line and another straight line, or another curve, always belong to a higher class than that of the straight line or the curve from which they derive. It is this idea that Descartes's contemporaries and successors will criticize. For example, Fermat writes in his *Dissertatio tripartita*:

> Let us imagine, for example, instead of the straight line *CNK* [...] a cubic parabola with vertex *K* and an indefinite axis *KLBA*; if one completes the construction in the spirit of Descartes, it is clear that the constitutive equation of this parabola will be
>
> $$a^3 = b^2e.$$
>
> One can immediately see that the curve *EC* deriving from this supposition only has a biquadratic equation; thus the biquadratic curve is of a higher degree or class than the cubic curve, according to the rule that Descartes himself stated,

[88] Transl. P. J. Olscamp, p. 195, adapted here to follow the original terminology of *La Géométrie*, A.T. VI, p. 395.

[89] *La Géométrie*, A.T. VI, pp. 395–6. Translation note: The above rendering translates more literally the material that the Olscamp translation renders as: 'there is a general rule for reducing to a cube all the difficulties of the fourth degree, and to an equation of the fifth degree all equations of the sixth degree, in such a way that we need not consider the higher ones more complex than the lower' (p. 195).

[90] *Opera*, ed. G. Cramer, Geneva, 1744, vol. II, pp. 676–7.

whereas he expressly states on the contrary that the biquadratic curve and the cubic are of the same degree or class.[91]

Indeed if, in the preceding figure, one replaces the mobile straight line *KNC* by a cubic parabola *CB*,

$$CB^3 = b^2 \cdot KB,$$

one has

$$\frac{CB}{LB} = \frac{GA}{LA};$$

noting that $CB = x$, $AB = y$, $GA = c$, $KL = d$, one thus has

$$LB = KB - d = \frac{x^3}{b^2} - d; \; LA = KB + BA - d = \frac{x^3}{b^2} + y - d,$$

whence the equation

$$x^4 - cx^3 + b^2xy - b^2dx + b^2cd = 0,$$

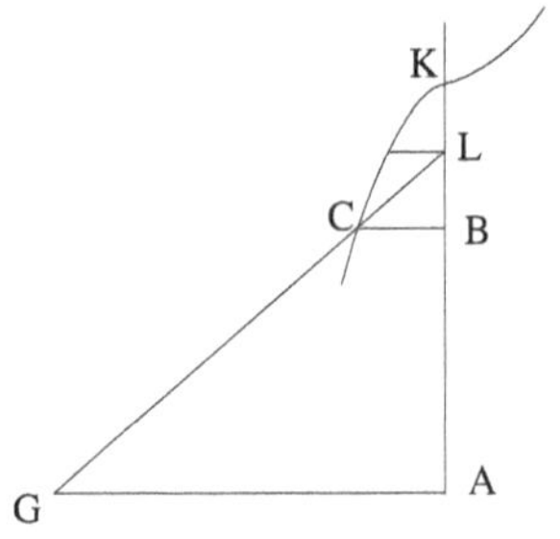

Fig. 20

which defines a curve of degree 4 and 'genre' 2, which Descartes places in the same 'genre' as the cubic that generated it. This point justifies Fermat's critique, which amounts to denouncing a contradiction between the classification that Descartes proposed for algebraic curves, and the principles that he himself defends on the generation of these curves with a motion. This classification of

[91] *Œuvres de Fermat*, III, pp. 112–13.

Descartes's was the butt of yet other criticisms, by Fermat again, Newton, Jacques Bernoulli, and others.

The cubic parabola of equation (*), the trident discovered by Descartes, is, as he proves, a solution to the problem of Pappus for five straight lines.[92] It is precisely to this problem of Pappus, which he studied in the first book of the *Géométrie*, that Descartes returns in the second, now to focus on curved solutions.

This is not the place to cover the problem of Pappus itself;[93] let us consider only the case of five straight lines.

Let AB, IH, ED, GF be four parallel and equidistant straight lines, to which GA is perpendicular. One must find a point C such that CF, CD, CH, CB, CM (the distances perpendicular to the straight lines) are such that

$$CF \cdot CD \cdot CH = CB \cdot CM \cdot a,$$

where a is the distance between AB and IH.

One divides the odd number $2n - 1$ into two groups n and $n - 1$ (for 5, one has 3 and 2), such that the product of the distances to the segments of the first group is in a given ratio to the product of the distances to those of the second, a product completed, as needed, by a given constant factor; this is the problem of Pappus.

In the case of five straight lines, Descartes proves that point C is found on the curve defined by equation (*) generated, as we have seen, by a ruler that

[92] Note that in January 1632, Descartes apparently has not yet discovered the trident. This is what he writes to Golius: 'As I should at least have included an example of five or six straight lines given in position to which I would have applied the required curved line, but I understood the difficulty involved in making the calculation' (A.T. I, p. 234).

[93] Given the very rich secondary literature on the history of this problem, I mention here only a few titles: the sixth chapter of G. Milhaud's classic, *Descartes savant*, Paris, F. Alcan, 1921, pp. 124–84; J. Vuillemin, *Mathématiques et métaphysique chez Descartes*, esp. pp. 99–112; J. M. Bos, 'On the Representation of Curves in Descartes *Géométrie*', *Archive for History of Exact Sciences*, 24, 1981, pp. 295–338; see also Chapter 23 in his book: *Redefining Geometrical Exactness: Descartes' Transformation of the Early Modern Concept of Construction*, New York, Springer, 2001, pp. 313–4. Ch. Sasaki, *Descartes' Mathematical Thought*, Dordrecht/Boston, Kluwer Academic Publishers, 2003, pp. 207 ff, which also includes a rich bibliography.

pivots about a fixed point and a parabola that moves parallel to itself in the same plane.

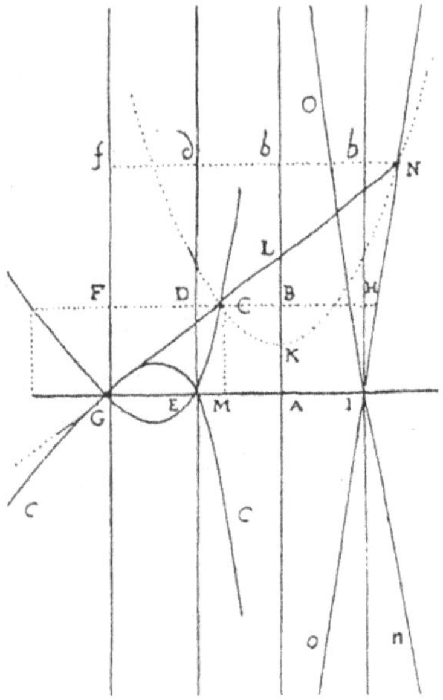

Fig. 21

Given $CB = y$, $CM = x$, $AI = AE = GE = a$, and substituting in the preceding equation, one obtains equation (*).

It is here that Descartes considers 'the intersection of the parabola *CKN* (which is made to move so that its diameter *KL* is always on the straight line *AB*) and the ruler *GL* (which meanwhile turns about the point *G* in such a way that it always passes through the plane of this parabola at the point *L*). And I make $KL = a$, and make the principal right side (that is, the one corresponding to the axis of this parabola) also equal to a, and $GA = 2a$, [...]'.[94] One then has

$$\frac{2a-y}{x} = \frac{GM}{MC} = \frac{CB}{BL} = \frac{y}{BL}$$

(the triangles *GMC* and *CBL* are similar), whence

[94] *La Géométrie*, A.T. VI, pp. 408–9; transl. P. J. Olscamp, pp. 204–5.

$$BL = \frac{xy}{2a - y}$$

and

$$BK = KL - BL = \frac{2a^2 - ay - xy}{2a - y};$$

but BK is a segment of the diameter of the parabola with *latus rectum* a, one therefore has

$$\frac{BK}{BC} = \frac{BC}{a} = \frac{y}{a},$$

whence

$$\frac{y}{a} = \frac{2a^2 - ay - xy}{y(2a - y)},$$

whence equation (*). The point that was sought is therefore C.

Descartes goes on to discuss the 'adjunct' $cEGc$, that is, the parabola symmetrical of the first one in relation to GA; but this is no longer the same algebraic curve.

We will say nothing more about Descartes's discussion in the case of five straight lines. We merely note that, for the problem of Pappus for $2n$ straight lines distributed into two groups of n straight lines with the equations of, respectively, $\varphi_i(x, y) = 0$ $(1 \leq i \leq n)$ for the first group and $\psi_i(x, y) = 0$ $(1 \leq i \leq n)$ for the second group, where the φ_i and ψ_i are polynomials of the first degree, one writes the equation of the problem in the form

$$(1) \qquad \varphi_1(x, y)\,\varphi_2(x, y) \ldots \varphi_n(x, y) = \lambda\psi_1(x, y)\,\psi_2(x, y) \ldots \psi_n(x, y)$$

where λ is a non-zero factor of normalization. Indeed, the distance from one point with coordinates x, y to a straight line with the equation $\varphi(x, y) = 0$ is proportional to $\varphi(x, y)$, which is the fundamental idea behind the Cartesian procedure. One can see that the locus defined by equation (1) is an algebraic curve of degree n. In the case of a locus with $2n - 1$ straight lines, the equation of the problem is

$$(2) \qquad \varphi_1(x, y)\,\varphi_2(x, y) \ldots \varphi_n(x, y) = \lambda a\psi_1(x, y)\,\psi_2(x, y) \ldots \psi_{n-1}(x, y)$$

in which a is a constant segment. The locus is again an algebraic curve of degree n; it can be interpreted as if the last straight line was displaced to infinity, since the ratio of the infinite distances from two points to the straight line at infinity remains constant.

Descartes defines 'the adjunct' by changing the sign of λ, which also yields an algebraic curve of degree n.

Thus, to obtain the trident as a locus of 5 straight lines, one must identify its equation with an equation of the form

$$\varphi_1\varphi_2\varphi_3 - \lambda a \psi_1\psi_2 = 0.$$

The terms of degree 3, which derive from $\varphi_1\varphi_2\varphi_3$, must be reduced to y^3, therefore the straight lines of the first group must be parallel to the x-axis of equation $y = 0$. By writing that the point at infinity in the direction of the x-axis is double, one discovers that $\psi_1 = 0$ or $\psi_2 = 0$ is parallel to this axis, for example $\psi_1 = 0$.

The equation is thus written in the form:

$$(y + \alpha)(y + \beta)(y + \gamma) = \lambda a (y + \delta)(x + vy + w)$$

defining a cubic with a double point at infinity in the direction of the x-axis, with a parabolic branch and a branch tangent to the straight line with the equation $y + \delta = 0$. This seems to explain Descartes's hypotheses.

More generally, consider any cubic C of equation $f(x, y) = 0$, f being a polynomial of degree 3.

Let a_1, a_2, b_1, b_2 be four points of C, any three of which are not aligned, and let $\varphi_1 = 0$, $\varphi_2 = 0$ be the equations of the straight lines joining, respectively, a_1, b_1 and a_2, b_2. These straight lines again cut C, respectively, at points c_1 and c_2. Now let $\psi_1 = 0$, $\psi_2 = 0$, $\psi_3 = 0$ be respectively the equations of the straight lines joining a_1, a_2; b_1, b_2; and c_1, c_2. These straight lines again cut C at the three points a_3, b_3, c_3. The cubics passing through the eight points $a_1, b_1, c_1, a_2, b_2, c_2$, and a_3, b_3 form a linear sheaf, that is, their equations depend linearly on one parameter. They can therefore be written in the form

$$f + \lambda\, \psi_1\psi_2\psi_3 = 0,$$

since $f = 0$ and $\psi_1\psi_2\psi_3 = 0$ are two cubics of the sheaf.

If $\varphi_3 = 0$ is the equation of the straight line joining a_3, b_3, the equation $\varphi_1\varphi_2\varphi_3 = 0$ defines a degenerate cubic of this sheaf. There is therefore a value of λ such that $\varphi_1\varphi_2\varphi_3 = f + \lambda\psi_1\psi_2\psi_3$;[95] one therefore has

$$f = \varphi_1\varphi_2\varphi_3 - \lambda\psi_1\psi_2\psi_3$$

and sees that C is a locus of the problem of Pappus with 6 straight lines.

Descartes's statement is therefore true for the most general cubics.

Conversely, Descartes's statement is not generally true: counterexamples appear already in the fourth degree.

According to the problem of Pappus for $2n$ or $2n - 1$ straight lines, the defined locus is of degree n; for degree 4, one must therefore take 8 or 7 straight lines, and the equation of the locus is of the form

$$\varphi_1\varphi_2\varphi_3\varphi_4 - \lambda\psi_1\psi_2\psi_3\psi_4 = 0$$

or

$$\varphi_1\varphi_2\varphi_3\varphi_4 - \lambda a\psi_1\psi_2\psi_3 = 0$$

where the φ_i and the ψ_i are first-degree polynomials.

We see that each of the straight lines of equation $\psi_i = 0$ meets the curve at the four points where this straight line meets respectively the straight lines of equation $\varphi_1 = 0$, $\varphi_2 = 0$, $\varphi_3 = 0$, $\varphi_4 = 0$. Thus a quartic locus of the problem of Pappus with eight or seven straight lines is met at four points by some straight lines in the plane.

Now, there exist convex quartics that each straight line of the plane cuts in 0 or 2 points; such quartics cannot be defined by the problem of Pappus. This is the case, for example, for the curve of equation $y = x^4$; nevertheless, one considers this curve to be a limiting case of the locus with eight straight lines, of which four are superimposed on the y-axis and three on the straight line to infinity; but this falls outside of the framework that Descartes considers. As another example, one can consider Cassini's ovals, of equation:

$$(x^2 + y^2)^2 + 2a^2(y^2 - x^2) + a^4 - b^4 = 0$$

[95] The result is that $\varphi_1\varphi_2\varphi_3 = 0$ passes through c_3; therefore the straight line $\varphi_3 = 0$ passes through c_3.

with $a, b > 0$; these curves are convex, therefore every straight line cuts them in 0 or 2 points.

By examining all the limiting cases in which some straight lines in the problem are superimposed, one can moreover see that the curve of the solution must have bitangents or tangents of inflection, except in the case of a curve of equation $\varphi^4 - \lambda\psi = 0$, which, by a change of coordinates, reduces to $y = x^4$. Now a convex curve has neither a bitangent nor a point of inflection; thus the oval of Cassini is not a limiting case and constitutes a counterexample even for the limiting cases.

Finally, note that the equation of the locus of the problem of Pappus for $2n$ straight lines (or $2n - 1$, as the case may be) depends on $4n + 1$ coefficients (or $4n - 1$, as the case may be). Now the equation of a curve of degree n depends on $\frac{n(n+3)}{2}$ coefficients; one for $\frac{n(n+3)}{2} > 4n - 1$ for $n \geq 5$ and $\frac{n(n+3)}{2} > 4n + 1$ for $n \geq 6$. It is therefore impossible to reduce the general equation of a curve of degree ≥ 5 (or 6, as the case may be) to that of a locus of the problem of Pappus for $2n - 1$ (or $2n$, as the case may be) straight lines.

All of this indicates that Descartes treated explicitly only the cases that we examined before, namely the loci for 3, 4, or 5 straight lines. He then generalized the result without verifying it.[96]

[96] One referee of this chapter offered the following objection: 'To write that one of the straight lines D present in the equation of Pappus necessarily meets the curve in four points is incorrect: that it meets 4 straight lines is correct, but can two of them, for example, cut each other at the point where they both meet D? This simple case already reduces the number from 4 to 3 and one obviously could go down to 2. There are therefore more numerous parasitic particular cases to examine than the single case in which at least 2 straight lines are superimposed. In short, I am not completely certain that the Cassini ovals are indeed a counterexample'.

I thank the author of this interesting objection, which I answer thus:

Let $\varphi_1\varphi_2\varphi_3\varphi_4 = \lambda\psi_1\psi_2\psi_3\psi_4$ Let that be the equation of a quartic C defined by the problem of Pappus. The φ_i and the ψ_i are affine linear functions of the coordinates x, y. Let us suppose that the straight line D of equation $\psi_i = 0$ meets the 2 straight lines Δ_i defined by the equations $\varphi_i = 0$ $(i = 1, 2)$ at the same point M: the intersection of D and of C in M is of multiplicity 2. That is that D is tangent to C at M, or else M is a double point of C.

Cassini ovals do indeed have two double points at infinity, but they are imaginary (they are the circular points). One can therefore exclude the case of a double point. If D is tangent to C at M, it in addition meets C at the 2 points at which it meets the straight lines

(*Cont. on next page*)

The preceding example illustrates the procedure that Descartes wanted to be systematic. He thus hoped to determine in every case the sought curve. Once he had considered the case of 3 or 4 straight lines that gave him a curve locus of the points of one of the three conics, he went on to the five straight lines and obtained the cubic parabola. Descartes presumed that, by increasing the number of straight lines, he would obtain curves of superior 'genres' ; he even went so far as to believe that the problem of Pappus would allow him to reach all geometrical curves, as he implies in a statement following his completion of the problem with five straight lines:

> As for the lines used in the other cases, I shall not stop to classify them by types, because I did not undertake to cover everything and having explained the method of finding an infinite number of points through which these curves pass, I think I have given sufficient means of describing them.[97]

In Descartes's eyes, the problem of Pappus would therefore allow one, on the one hand, to determine all geometrical curves (an error that Newton will denounce) and, on the other hand, to establish an order that is, as it were, 'genetic' in the sense that each curve solution would serve to find others. But the sentences cited above apparently show that Descartes himself had abandoned this path.[98]

(*cont.*) Δ_j ($j = 3, 4$) but this is impossible for a convex curve. If D was parallel to these straight lines Δ_j, it would be an asymptote of C, which is also impossible.

The objection does not pertain to the quartic of equation $y = x^4$, even though it is convex; indeed, it has a triple point at infinity. If, however, it is defined by the problem of Pappus, the same reasoning implies that D has an equation of the form $x = c$ (it passes through the multiple point) and the same goes for the straight lines D_i defined by the equations $\psi_{i\_} = 0$. Thus the equation is written $\varphi_1\varphi_2\varphi_3\varphi_4 = \lambda\,(x - c_1)(x - c_2)(x - c_3)(x - c_4)$. Each of the straight lines D_i meets C at a single point a finite distance away and at a triple point at infinity; thus 3 of the Δ_j must still have equations of the form $x = d_j$ ($j = 1, 2, 3$) and Δ_4 meets C at the abscissa points c_1, c_2, c_3, and c_4: this is possible only if these 4 values are superimposed. Thus the curve considered is not defined by the problem of Pappus either.

97 *La Géométrie*, A.T. VI, p. 411; transl. P. J. Olscamp, p. 206.

98 All the properties of these curves must therefore derive from their equations or, in the words of Descartes, 'to find all the properties of curved lines one need only know the relation that all of their points have to those of straight lines and the way of drawing other lines that intersect them at right angles at all of these points' (*La Géométrie*, A.T. VI, pp. 412–13). But if Descartes up to now is interested in affine properties, it is the metric that he from now on places in relief, namely those of the normal and of the tangents. But we cannot pause now to discuss this important research. See for example, G. Milhaud,

(*Cont. on next page*)

The Cartesian classification of plane curves[99] nevertheless constituted both a completion of the first classifications and also a new point of departure; it did so not *despite* the criticisms that several of his contemporaries (Fermat) and his successors (Newton and Jacques Bernoulli) levelled at it, but indeed *because* of these critiques and corrections. The central ideas on which this classification is founded became a permanent addition to the mathematical patrimony that Descartes's successors will continue to perfect and to deepen, by actually bringing the Cartesian program to fruition, while adapting it as needed. Such is the case for the idea of defining by its equation the algebraic curve of any 'genre', and not only the first; for the idea of the invariability of the 'genre' and its independence from the reference frame – ideas that are essential to the classification; for the idea of extracting the properties of the curve from its equation, etc. To conclude, let us listen to Cramer and to Le Cozic, the translator of MacLaurin's *Algebra*. Cramer writes:

> The most important thing to notice in this distribution of algebraic lines into orders, is that each line is so well fixed in its order that it never leaves it, no matter what equation represents it. By this I mean that, whereas one can express the nature of an algebraic line by an infinity of different equations, depending on one's choice of origin and of the position given to the axes, yet all of these equations belong to the same order, to which consequently the proposed line must be related.[100]

Paraphrasing MacLaurin, his translator Le Cozic writes:

> To proceed with some order in the search for the main properties of algebraic lines, the latter are distinguished into several orders, according to the degree of the equation, freed from fractions or radicals, that expresses its nature. To prove that this type of distinction is sufficient, we will show that the equation of the same line does not change its degree regardless of any change to the axis, to the origin of the abscissas, and to the angle of the coordinates; consequently, one will be assured that the same line will never be related to different orders.[101]

(*Cont.*) *Descartes savant*, pp. 128 ff. and J. F. Scott, *The Scientific Work of René Descartes (1596–1650)*, London, Taylor & Francis, 1952, pp. 115 ff.

[99] At the end of Book 2 of his *Géométrie*, Descartes briefly brings up skew curves (p. 440). Only at the end of the century, with Parent, and in the next century with Clairaut in his *Recherches sur les courbes à double courbure* (1731) and then Euler in his *Introductio in analysin infinitorum*, 1748, will these curves be studied.

[100] Cramer, *Introduction à l'analyse des lignes courbes algébriques*, pp. 53–4.

[101] *Traité d'algèbre et de la manière de l'appliquer*, Paris, 1753, p. 280.

One could multiply the testimonials, beginning with the writings of Newton, Leibniz, the Bernoulli brothers, Euler, and Bézout: henceforth, when curves are classified, the point of departure is none other than Descartes's classification, reworked and reformed in light of the development of both algebra and the infinitesimal calculus. By completing the program that Descartes put in place, these mathematicians will set out on a new stage in the research on the classification of curves. That, however, is the topic of another study.

7. DEVELOPMENTS OF THE CARTESIAN CLASSIFICATION OF ALGEBRAIC CURVES

To go farther along on the road of the classification of curves is to be able to isolate sub-classes within the whole, in order to make a detailed description and to characterize more precisely the behavior of each: its length, its infinite branches, its singular points, etc. These are precisely the tasks of Descartes's successors, such as Newton, MacLaurin, Euler, or Bernoulli, for example. But Descartes himself, and his junior Fermat even more so, had already taken several steps down this road. The former conceived the seminal idea of giving to the curve an operational role, as it were, leading to the derivation of other curves, which in the final analysis made it possible to speak of a family of curves associated with an initial curve. As to Fermat, he gave a procedure for deriving an infinite family of curves from a given curve, a family that could be ordered thanks to the concept of the arc-length of the curve. Whereas Descartes did not need any new concept, Fermat had to have at his disposal other mathematical means that were altogether absent from Descartes's *Géométrie*.

Indeed, we have seen that Descartes substituted a circle for a straight line in order to obtain a conchoid, then substituted for this straight line a parabola in order to obtain a cubic parabola. To be sure, he neither followed up on this substitution nor did he develop the method. Nevertheless, already present here was the idea of considering a curve in itself as the means of obtaining others, and therefore an entire family.

Twenty-two years after the publication of Descartes's *Géométrie*, Fermat does go farther. Here and elsewhere, as we have shown,[102] he operates along the lines of Descartes, but also against him. He notably devotes to this theme two works published in 1660, but drafted a year earlier: *Propositions à Lalouvère* and *De la comparaison des lignes courbes avec les lignes droites*. In the latter work, Fermat sets himself the task of constructing curves that are 'purely geometrical' – that is, algebraic – with a length equal to a 'given straight line'. This will be the case in particular for a convex curve for which one knows how to construct a running point by means of a straightedge and compass starting from given line segments.[103]

But a rigorous theory for the lengths of the arcs of curves, which is what Fermat proposes to construct, requires one way or the other the introduction of concepts from the differential calculus. In the absence of such a tool, Fermat draws on the idea that underlies the method of Archimedes: the length of a convex curve is 'framed' by the length of polygonal lines or, as Fermat puts it, 'our method by double circumscription'.[104]

My goal here is to examine not the infinitesimal geometry of Fermat, but only his procedure for the derivation of new curves and their classification. To keep the exposition concise and to highlight the stakes clearly, I will draw on straightforwardly anachronistic language.

[102] See below, 'Fermat and algebraic geometry'.

[103] This is how he presents his investigation: 'Never before, as far as I know, have geometers ever made a purely geometrical curved line equal to a given straight line'. He goes on to say: 'Indeed, I will demonstrate the equality to a straight line of a truly geometrical curve, for the construction of which no similar equality has been assumed between another curve and a straight line […]' (III, p. 181).

[104] *Œuvres de Fermat*, III, p. 185. This is how Fermat explains it: 'we will thus have two figures circumscribing [and circumscribed by] the curve, the one greater, and the other smaller, than this curve, and such that the difference between these figures is smaller than any given interval; *a fortiori*, the amount by which the greater [circumscribing] figure exceeds the curve, and that by which the curve exceeds the smaller figure will each be smaller yet'.

Fermat begins by inventing a new geometrical transformation in order to obtain other curves, the length of which is constructed geometrically as an initial curve Γ. Starting from a curve Γ_1 with axis AB and a point C of this curve, the ordinate DN on point N of Γ_2 with abscissa x is equal to the length of the arc CM of Γ_1 included between point C and point M with abscissa x. Thus one can iterate the procedure and build a series of (algebraic) curves Γ_n.

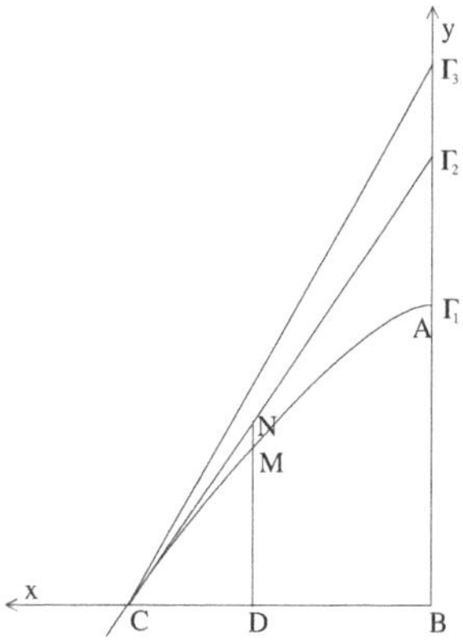

Fig. 22

Note that this construction resembles the construction of the evolute of curve Γ. But Fermat refers the length of the arc CM to DM, the parallel to the axis AB, whereas one must refer the evolute to the tangent to Γ at M.

The curve one obtains thus depends not only on the point of origin C, but also on the axis AB.

Consider with Fermat the curve $\Gamma_1 = \Gamma$ with the equation

$$y_1 = f_1(x) = \frac{1}{\lambda} x^{\frac{3}{2}},$$

and the point C with abscissa a on it. If one marks ds_1 as the element of length of curve Γ_1, the curve Γ_2 will have as its equation

$$y_2 = f_2(x) = \int_a^x ds_1 = \frac{2}{3\mu}\left((\mu+x)\sqrt{\mu(\mu+x)} - (\mu+a)\sqrt{\mu(\mu+a)}\right),$$

with $\mu = \dfrac{4\lambda^2}{9}$.

The element of length of the curve Γ_2 is therefore

$$ds_2 = \sqrt{dx^2 + dy_2^2} = \sqrt{dx^2 + ds_1^2} = dx\sqrt{2 + \frac{x}{\mu}}.$$

More generally, the element of length of the curve Γ_n is

$$ds_n = dx\sqrt{n + \frac{x}{\mu}}.$$

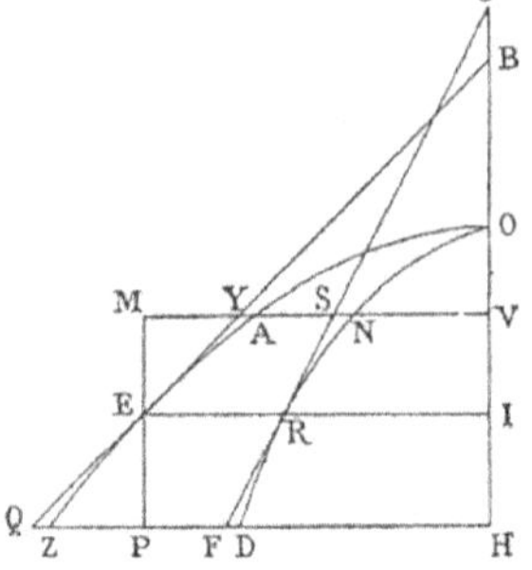

Fig. 23

The calculation of the length of the arcs of curves Γ_n is analogous to that done for curve Γ: Fermat uses the change of variable, $t^2 = \mu(n\mu + x)$, which reduces to calculating the area circumscribed by a translated parabola of parabola $\mathscr{P}$. Note that the axis of curve Γ is indifferently vertical (Fig. 23) or horizontal (Fig. 24).[105]

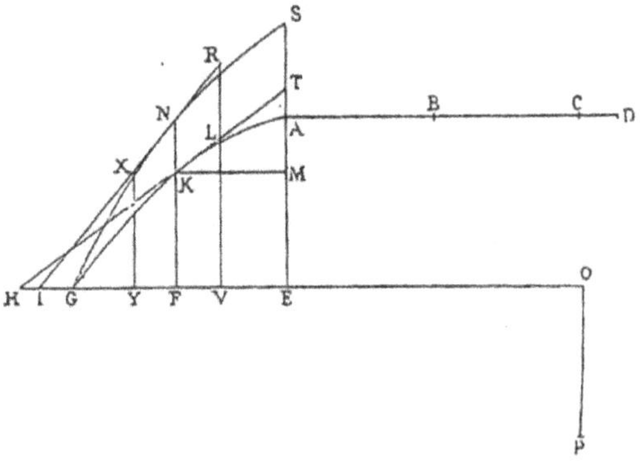

Fig. 24

Fermat faces a genuine difficulty in constructing the tangent to the curve Γ_2. To do this (Fig. 23), he shows that if EB is this tangent, then $MY = RS$. If one uses the notation of the differential calculus, this amounts to the relation

[105] *Ibid.*, reproduction of Figures 129 and 130, pp. 196–7.

$$\frac{MY}{ME} = f_2'(x) = \frac{ds}{dx} = \frac{\sqrt{dx^2 + dy^2}}{dx} = \frac{\sqrt{VI^2 + (RI - VS)^2}}{VI} = \frac{RS}{ME}.$$

But this relation does not bring into play the particular form of the curve Γ. Fermat's demonstration is very elegant and purely geometrical: recall that, for him, the tangent at E to the curve Γ is the unique straight line that passes through E and that remains on the same side as the curve Γ (assumed to be convex).

Finally, the area of the parabola being known, one finds the length ℓ_n of the arc the curve Γ_n is expressed algebraically:

$$\ell_n = \frac{2}{3\mu}\left((n\mu + a)\sqrt{\mu(n\mu + a)} - \mu^2 n\sqrt{n}\right).$$

To this exposition, Fermat drafts an appendix in which he applies the procedure for constructing new curves Γ'_n starting with the curve Γ'_1 of equation $\lambda x^2 = y_1^3$. The curves Γ and Γ'_1 are identical: one passes from one to the other by exchanging the axes of the coordinates. One obtains the curve Γ'_2 by applying Fermat's construction to the curve Γ'_1; the ordinate of point I of Γ'_2 with abscissa x is equal to the length of the arc AO of Γ'_1 included between the vertex and the point with abscissa x, and so on for all curves Γ'_n.

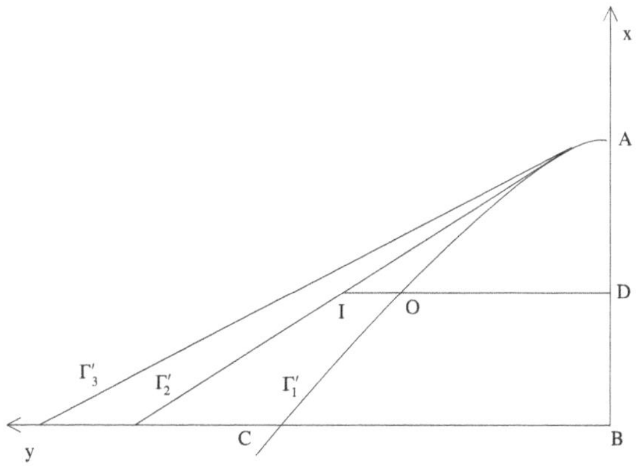

Fig. 25

Thus with this procedure, starting from a semi-cubic parabola, Fermat derives a whole series of curves that are rectified by the iteration of one and the same geometric construction and are comparable.[106]

The classification of algebraic curves is becoming finer and more precise because there is already a way of comparing them. But this important gain was possible precisely by drawing on infinitesimal concepts in the study of curves. Some of these concepts are already present in Archimedes and such successors of his as Ibn al-Haytham; others are new, sometimes still in limbo, but whatever one might say, they are already there. But this use of infinitesimals will soon grow rapidly and massively beginning in the second half of the 17th century, and will contribute effectively to modifying the very nature of the classifications. Here we will restrict ourselves to Fermat's use of such concepts.

We have already pointed out that, in his *Dissertatio*, Fermat demonstrates that the semi-cubic parabola $ky^2 = x^3$ is exactly rectifiable, because its rectification reduces to the quadrature of the ordinary parabola. Fermat's approach is based, on the one hand, on the construction of the tangent, according to his famous method and, on the other, on the examination of the ratio $\frac{ds^2}{dx^2}$. Let us then consider[107] an arc of parabola defined over the interval $[0, a]$. Since $dy = \frac{3}{2}\sqrt{\frac{x}{k}}dx$, one has $ds^2 = dx^2\left[1 + \frac{9}{4k}x\right]$. Let us posit that $t^2 = \frac{4}{9}k\left(\frac{4}{9}k + x\right)$ and $\frac{4}{9}k = \lambda$; then the length ℓ of the arc of the semi-cubic parabola is obtained by

$$\lambda\ell = \lambda\int_0^a ds = \int_0^a \lambda\sqrt{1 + \frac{x}{\lambda}}dx = \int_0^a t\,dt.$$

Since $2t\,dt = \lambda\,dx$, one has

$$\lambda\ell = \frac{2}{\lambda}\int_\lambda^{\sqrt{\lambda(\lambda+a)}} t^2\,dt, \text{ and } \ell = \frac{2}{3\lambda}\left((\lambda + a)\sqrt{\lambda(\lambda + a)} - \lambda^2\right).$$

[106] Note that Fermat, like Descartes before him, iterates a procedure of demonstration by applying this procedure to the result of the previous stage of application of the same procedure. Al-Sijzī had already used this procedure in his treatise on the numerical right triangle (see R. Rashed, *Œuvre mathématique d'al-Sijzī*, vol. I, pp. 171–2).

[107] The point here is to rewrite in other words the steps of Fermat's calculation.

Note that, in Fermat's method, the famous differential triangle appears, and that Lagrange uses this method of rectification.[108]

In the *Appendice* to the *Dissertatio*, the importance of which everyone recognizes, Fermat appeals explicitly to the concept of slope of the tangent when he investigates the length of curves. Thus, Proposition 1 states that two curves having tangents with the same slope at points with the same abscissa have the same length, and moreover are 'similar'. In Fermat's words, 'if one imagines that one superimposes them, they coincide completely and will have, along with their axes, equal or rather identical ordinates'.[109] Rewriting this proposition in a language different from Fermat's, it reads: two functions that have the same derivative over an interval and the same value at the extremity of this interval are identical; in particular they define the same curve. This is obviously nothing less than the fundamental theorem of the integral calculus.

No less important than this theorem, however, are the concepts that Fermat mobilizes in order to demonstrate it. By drawing on infinitesimal considerations, he rigorously demonstrates that the two curves have the same length. But he does not have the tools to demonstrate with the same rigor that the curves are superimposable. His demonstration relies on the remark that, at homologous points, the two curves make the same angle with the ordinate for, he writes, 'the inclination of the curves is measured by that of the tangents'.[110] Fermat did not have available two concepts that would soon be defined: that of curvature and that of the evolute. Conversely, he seems to proceed on the intuition of the fact that a curve is well defined by its intrinsic equation, give or take one displacement.

The second proposition of the *Appendice* states that two curves that are homothetic in a given ratio have lengths in the same ratio. The third proposition states that two curves, deduced the one from the other by orthogonal affinity, have tangents that intersect each other on the axis of affinity. In other words, the image $\phi(T)$, by an affinity ϕ, of the tangent T to a curve Γ at a point M is the tangent to the curve $\phi(\Gamma)$ at the point $\phi(M)$. It is from these propositions that Fermat establishes the 'general theorem' that consists in deriving curves from other curves.

[108] *Théorie des fonctions analytiques*, Paris, Imprimerie de la République, 1797, pp. 222–3.

[109] *Œuvres de Fermat*, III, p. 206.

[110] *Ibid.*, III, p. 205.

From these studies of Fermat, one sees that the first classifications are in the process of completion, and that beginning in the second half of the 17th century mathematicians are engaging in other types of classification that, in order to be effective, will have to wait for the intervention of the differential and integral calculus as such. All told, Fermat's contribution is located at the hinge between the first, ancient classifications and the modern ones, those that will be conceived after the creation of the differential calculus. This pivotal position derives from the studies of asymptotic properties, not only of geometrical curves, but also of the curves that Descartes had excluded from geometry, notably the cycloid and the logarithmic spiral. To these studies, Descartes himself made fundamental contributions, along with Roberval, Fermat, de Beaune, and Pascal. By pursuing that work, their contemporaries and successors would modify both the nature and the range of the classifications.

To refresh our memories, recall that the Cartesian distinction between geometrical curves and mechanical curves became the point of departure for other classifications with more branches. As we have already seen in the case of Fermat, the tendency was very naturally to isolate families from among geometrical curves. Barely three decades after Descartes, sub-classifications were already proposed that presented themselves in terms of a systematic study of each Cartesian '*genre*'. Whereas in the case of the first '*genre*' the classification itself was technical and essentially known even before Descartes, one nevertheless witnesses the appearance, in the course of the systematic investigation, of books in which the properties of curves are examined by starting from their equations. This was absolutely not the case for the 'second *genre*' (the cubics) nor for the superior '*genres*'.

With respect to the 'first *genre*', recall the example of Ozanam and of the three books that he published in 1687. He begins by defining curves of the first class as follows:

> [...] curved lines, for which, having drawn mutually parallel lines, their squares correspond in a given ratio to certain rectangles, such as are the conic sections [...].[111]

[111] Ozanam, *Traité des lignes du premier genre expliquées par une méthode nouvelle et facile*, Paris, 1687, p. 5.

He considers this property 'which is general for these three conic sections, in order to be able to prove all the other properties that are specific to them'.[112] Ozanam then goes on to examine the properties of these curves, which he studies methodically. With respect to classification, he goes no farther than his predecessors. Indeed he considers only the domain of the real numbers and does not consider the degenerate cases. Thus this classification does not include the cases of $y^2 = x^2$, $y^2 = 1$, or $y^2 = 0$. But by drawing on curves to study Diophantine equations of the second degree, he obtains an important result in this systematic research. One can perhaps see here one of the first encounters – admittedly timid – between algebraic geometry and Diophantine analysis. But the true innovation in these sub-classifications occurs in Newton's study of the second 'genre', the cubics, which we will not examine here.[113]

The heirs of the Cartesian classification, such as Leibniz,[114] who were reticent toward the idea of excluding transcendental curves when constructing geometrical problems, took advantage of the active research devoted to them,

[112] *Ibid.*, p. 6.

[113] Newton worried about this problem since the years 1667–68, that is, fewer than 30 years after Descartes, in his *Enumeratio curvarum trium dimensionum* (ed. D. T. Whiteside, in *The Mathematical Papers of Isaac Newton*, vol. II, pp. 10 ff.). In 1704 Newton would publish his famous *Enumeratio linearum tertii ordinis* (*ibid.*, vol. III, pp. 579 ff.), in which he exhibits seventy-two species.

[114] 'I confess that, if one must present the definition of geometrical ones that Descartes gave us, ours will not be such. But just as he rightly criticized the ancients for having excluded from geometrical curves conic sections or the loci that they called linear, likewise must he in turn be criticized for depriving science of a necessary aid by having reduced the name of the geometrical to the analytic' (*De Serierum summis et de quadraturis pars decima*, in Gottfried Wilhelm Leibniz, *Mathematische Schriften*, printed by the Leibniz-Archiv der Niedersächsischen Landesbibliothek Hannover, Dritter Band 1672–1676: *Differenzen, Folgen, Reihen*, Berlin, Academie Verlag, 2003, p. 485, 5–8: 'Fateor si ferenda est definitio geometricarum quam dedit nobis Cartesius, nostra talis non erit: sed quemadmodum ille veteres iure culpat, quod a geometricarum numero conicas, aut certe quos vocabant lineares locos, exclusissent; ita; ille rursus culpandus est, quod geometricarum nomine ad analyticas coarctato; scientiam auxilio necessario privat'). In the introduction by Leibniz, the terms '*algébrique*, *transcendant*' take the place of '*géométrique*, *mécanique*' of Descartes, see H. Breger, 'Leibniz Einführung des Transszendenten' (cited in n. 78).

notably the logarithmic spiral and the cycloid.[115] Since they felt constrained by the geometrical/mechanical opposition, which they nevertheless continued to accept as the basis of the classification, they simply began, as Cramer writes,[116] by intercalating between the algebraic curves and the transcendent ones

> [...] the class (*genre*) of *exponential curves*. This is the name given to curves the nature of which is expressed by equations, in which truthfully no infinitely large or infinitely small quantity enters, but that nevertheless cannot be related to ordinary algebraic equations, because they contain terms that have variable exponents.
>
> One of the simplest curves of this type is the *logarithmic*, represented by the equation $y = ba^x$.

Cramer continues:

> To this genre, or rather to a class intermediate between exponential and algebraic curves, one can relate those that Mr. Leibniz calls *interscendent*. These are the ones in the equation of which one finds several terms with irrational exponents such as the equation $y^{\sqrt{2}} + y = x$.

These are effectively the transcendental curves, which one can nevertheless express without relying on a differential equation, if one admits as granted the definition of a^b for an irrational exponent b, which mathematicians were doing at the time. But with Newton, Leibniz, and their contemporaries, we are already in the domain of the second classification of curves, the topic of a different study. Let us now take stock of the first.

8. CONCLUSION

At first sight, it might seem that for two millenia the classification of curves is a theme imported from philosophy into mathematics, an 'encroachment' of the former on the latter; or also the result of a speculative approach to organizing diverse and scattered strands into a coherent system. It may also

[115] See especially Leibniz, *Quadrature arithmétique du cercle, de l'ellipse et de l'hyperbole*, Latin text edited by E. Knobloch, intro., transl., and notes by M. Parmentier, 'Mathesis', Paris, Vrin, 2004, pp. 124 ff.

[116] Cramer, *Introduction à l'analyse des lignes courbes algébriques*, p. 8.

seem at first sight that this theme originated in the desire of enumerating *a posteriori* the forms in which the curves presented themselves to the mathematician's experience. We can now see that this is not the case. Indeed the preceding investigation shows that the proposed classifications were intimately tied to the mathematical knowledge of the period and to its renewal. At stake were, on the one hand, the knowledge of the technical procedures for generating curves and, on the other, the knowledge of the theoretical means of defining them. Only when the knowledge of the technical procedures conceived for drawing curves was combined with the theoretical means of characterizing them did it become possible to consider an *a priori* classification. That event first took place in the 10th century, and would be constantly reiterated. As one might expect, these combinations modified the nature of the two elements involved: the instruments invented to draw the curves became the incarnations of mathematical formulas that defined the latter, whereas these same formulas took the form of the equations of the curves. In short, precisely this interaction between the technical and the theoretical will constitute the foundation of classifications and will endow the latter with their explanatory power.

But what is the role of philosophical elucidation revealed in such pairs as 'simple-mixed' and 'geometrical-mechanical', for example? Without the slightest doubt, the doctrines on which these oppositions rest are aimed at justifying the proposed classification, in the absence of an authentic mathematical justification that other means alone could supply. Not until the 19th century did this take place. In this case as in many others, the philosophical elucidation finds its origin – better yet, its necessity – in the mismatch between actual mathematical knowledge and that required by the problem at hand.

But let us not delude ourselves: the classifications were hammered out under the pressure of problems imposed by the development of mathematical techniques: constructions that were admissible or not in geometry; proofs of existence and the need for them; relations among geometry and algebra; relations between these two, and infinitesimal geometry (and later the differential calculus).

It is true that the history of the first classifications of curves presents a certain continuity into which one could read a semblance of autonomy. As we have seen, these first classifications agree to restrict themselves to the geometrical framework, that of proportion theory. Even Fermat was still working along these lines. All of these classifications choose a doctrine of motion in order to explain the generation of curves, and all of them go back to the same more-or-less explicit concept of the curve in general, as a 'limit' of polygonal lines. But these common elements, which allow us to distinguish these *first* classifications, must not obscure the deep cleavages that separate them. If indeed mathematicians persist in invoking the exclusive pairs of 'simple-mixed', 'geometrical-mechanical', these oppositions in every case take on a new meaning by embracing the evolution of the mathematical knowledge of curves – before disappearing in the end. It is in just this sense that the classifications echo this knowledge.

Note finally that these first classifications play two crucial roles, which attest to the extent to which they are tied to mathematical activity: the first is heuristic; the second, unifying.

We have pointed out that the classifications often raise new mathematical problems: how to explain the fact that a 'mixed' curve, such as the cylindrical helix, is nevertheless homoeomeric? Is this helix a 'measurable' curve in the sense of al-Qūhī? Or might it be 'nonmeasurable' in the same respect as the quadratrix or the spiral of Archimedes? To which group does such a curve belong? (Descartes more than once raised this question). Starting with an initial curve, can one derive others from it? (a question raised by Descartes, and especially by Fermat). Starting from the general equation of a class of algebraic curves, how can one obtain all the curves that compose it, as for the cubics?

As to the unificatory function, it takes place in the sense that the classification both isolates and integrates. From the 10th to the 12th century, mathematicians isolated 'measurable' curves from those that are not, in order to distinguish from among the former the group of the first two degrees. Separating this last group, however, made it possible to integrate the straight line, the circle, and the conic curves into one and the same theory. By excluding mechanical curves, Descartes for his part was able to gather the algebraic curves into one theory. Even if, for the purposes of justification, mathematicians appeal to a philosophical doctrine – that of motion – the work that they do is in every case genuinely mathematical.

APPENDIX

SIMPLICIUS
On the Euclidean definition of the straight line and of curved lines

Simplicius has said: By his statement, Euclid means '(the) equal to what is between any two given points'. If indeed we assume the two points that are the extremities of the line (given that here he only defined the finite line) and if we consider that the distance between them 'is not a line', then this distance will be equal to the straight line of which these two points are the extremities. If by the line we measure the distances that are between certain points and others, we measure them only by the shortest of lines, which is the shortest (*aqrab*) of the intervals between the separated things and we do not measure them by a line that includes circularity. This is why Archimedes defined it by saying: the straight line is the shortest of lines whose extremities are its extremities; he means that it is the shortest line that joins two points. The measure is only by means of the straight line because it is the only one to be limited. Indeed, among all other lines, absolutely none is limited, given that we can join a point to another point by means of lines that are curved and circular or composite, some of which are longer than others, and we do this all the time. Furthermore, when Euclid defines the genus of the line and when he says that it was 'a length without breadth', he came to speak about species. The species of lines are many – namely, straight lines, circular lines, intermediate ones between circular and straight lines, and that are composed of them. Among the intermediaries, some lines have neither order nor arrangement – that is why geometers do not use them – these are the lines that resemble the shapes of the outlines of animals and which are of curved-in forms and other innumerable ones. Among these, there are lines that geometers use, such as the conic sections, which are the parabola, the hyperbolas, the ellipse, and the helicoidal lines [...] and numerous other lines like these, which include admirable things. But Euclid, to preserve the style and the scope of a prologue, defined only the straight line and the circular one, which are the seeds of all lines.[117]

[117] Text cited by al-Nayrīzī in his *Tafsīr muṣādarāt Uqlīdis*, ms. Qum, Bibliothèque Ayatallah al-Uzma al-Marashi al-Najafi, no. 6526, fol. 2^r, edited and translated by R. Rashed.

DESCARTES'S *GÉOMÉTRIE* AND THE DISTINCTION BETWEEN GEOMETRICAL AND MECHANICAL CURVES

Descartes sometimes complains that his *Géométrie* fell victim to a certain incomprehension. Was this simply part of the man's character? A sign of the author's susceptibility? A way of emphasizing his distance from Viète? Or was it yet another argument in a lively polemic in which he relishes denouncing the Parisians' limitations? There is no doubt a little truth in each of these points, but the essential point lies elsewhere. The reason for this incomprehension, I believe, is much more profound. Descartes himself grasped it as in a chiaroscuro, which takes shape every time conflicts erupt about interpretations of the *Géométrie*. For some, Descartes gives us an algebrization of geometry; for others, he did precisely the contrary. Yet others, in the wake of a judicious analysis, rightly refuse to become prisoners of this dilemma; in this regard, at least two names deserve mention: J. Itard[1] and J. Vuillemin.[2]

One year after the publication of the *Géométrie*, Descartes foresaw that his books would not fail to generate opposite and mutually exclusive reactions from both geometers and algebraists. 'I understand' he writes,[3] 'that it [the *Géométrie*] will be understood with difficulty by those who have not learned analysis before, and I see that those who have, see no merit in it and do their best to despise it as much as they can.' Five months earlier, he had sent to Mersenne the famous line: 'Your analysts understand nothing about my geometry'.[4]

The contempt that Descartes denounced and against which he would often react is in no way the product of his character or the figment of his imagination. Some of his contemporaries also noticed it. Desargues, who

[1] 'La *Géométrie* de Descartes', *Les conférences du Palais de la Découverte*, série D, no. 39, 7 janvier 1956; repr. in J. Itard, *Essais d'histoire des mathématiques*, collected and introduced by R. Rashed, Paris, A. Blanchard, 1984, pp. 269–79.

[2] *Mathématiques et métaphysique chez Descartes*, Paris, P.U.F., 1960.

[3] Letter to Mersenne, 27 July 1638, A.T. II, pp. 275–6.

[4] Letter to Mersenne, 1 March 1638, A.T. II, p. 30.

was not the least of them, has this to say: 'Whatever Monsieur de Beaugrand and others say, I have grounds for suspecting that they do not understand [the *Géométrie*] thoroughly, by which I mean that they do not fully grasp Monsieur Descartes's intentions about his *Géométrie*'.[5] Once set loose, the question will continue to be raised following Van Schooten's translation and Rabuel's commentary:[6] what are 'Monsieur Descartes's intentions concerning his *Géométrie*'? This difficult question will lead us to ponder the reasons that might have prevented the 'analysts', the disciples of Viète, from understanding in depth a book with a genesis and a filiation that are as deliberately algebraic as the theory of equations that it contains. At the same time, this question will lead us to wonder what kinds of obstacles geometers might have encountered when reading a book emphatically entitled *Géométrie*.

Superimposed upon the *Géométrie*'s rather paradoxical place in the history of mathematics is another one that is no less disconcerting. Historians unanimously grant the modernity of Descartes's *Géométrie*; none is so flippant as to reduce it to any of its predecessors. More importantly, no other book of mathematics disputes the role of the *Géométrie* as a symbol of mathematical modernity. Indeed, its date, its impact on contemporaries, the extension that successors such as Newton gave it all make the *Géométrie the* book of the first half of the 17th century. Published in 1637, it was the fruit of research carried out between 1619 and 1636, as attested by Descartes's own *Cogitationes privatae*, a few echoes in his *Regulae*, and several testimonials that Beeckman confided to his journal. Historians are thus correct to treat this book and its author as emblems of the new era. But when it comes to giving some content to this modernity and to define its relations to tradition, historians diverge and the consensus shatters. They find themselves confronting the original question about 'Monsieur Descartes's intentions concerning his *Géométrie*'.

I would like to try to answer this question beginning with the writings of Descartes's predecessors. I restrict myself to medieval mathematicians writing in Arabic, notably ʿUmar al-Khayyām (1048–1131) and the newcomer, Sharaf al-Dīn al-Ṭūsī, who followed him a generation later. Al-Ṭūsī's treatise, entitled *The Equations*, is of considerable importance and has just been restored, analyzed, and integrated into the history of mathe-

[5] Letter to Mersenne, 4 April 1638, in *Correspondance du P. Marin Mersenne*, commencée par Mme Paul Tannery, publiée et annotée par Cornélis de Waard, Paris, éd. du CNRS, 1963, vol. VII, p. 157.

[6] *Commentaires sur la* Géométrie *de M. Descartes*, by R. P. Claude Rabuel, Lyon, 1730.

matics.[7] Recall that al-Khayyām was the first in history to elaborate a geometrical theory of equations with degrees ≤ 3; as to al-Ṭūsī, he wrote the first systematic study of the existence of roots of these equations.

I must emphasize that my intention is in no way to find Descartes's *Géométrie* in the works of his predecessors. By referring to the latter, however, I do wish to *localize* more precisely the novelty of the *Géométrie* in order to understand in what ways it is truly modern, and to establish the ties that bind it to earlier tradition(s). I hope to show on the one hand that, setting aside questions of influence, Descartes's *Géométrie* represents the completion of the tradition that al-Khayyām and al-Ṭūsī inaugurated; and, on the other hand that this accomplishment engages another tradition that certainly finds its source in Descartes's book, but the true foundation of which will be laid in the work of his successors. I am not interested here in whether Descartes indirectly knew the tradition of al-Khayyām. I will analyze only the mathematical projects and their implementation. Also, as the reader will surely appreciate, it is impossible in this study to describe every facet of Descartes's *Géométrie,* and even more so to plumb all the strata imposed upon it. I will quickly restrict myself to two main axes that traverse this magisterial work and around which it is organized. The first consists in reducing a given geometrical problem to an algebraic equation with one unknown; the second reduces the *solution* of the equation to its construction by means of the intersection of two 'geometrical' curves, one of which, as far as possible, will be a circle. The difficulty of grasping the meaning of Descartes's *Géométrie* originates in these two procedures, which are somewhat contradictory even though they are inseparable. Here lies the source of the interpretative controversy that swirls around the book. It is therefore to these two axes that we turn.

1. THE GEOMETRICAL THEORY OF ALGEBRAIC EQUATIONS: THE COMPLETION OF AL-KHAYYĀM'S PROGRAM

One of Descartes's first mathematical acts focuses on the geometrical theory of algebraic equations, that is, on the solution of several third-degree equations by means of geometry. In 1619, he provides a classification of these equations, and solves some of them thanks to his

[7] About the algebra of al-Khayyām and al-Ṭūsī, see R. Rashed and B. Vahabzadeh, *Al-Khayyām mathématicien*, Paris, Librairie Blanchard, 1999; R. Rashed, *Sharaf al-Dīn al-Ṭūsī, Œuvres mathématiques: Algèbre et géométrie au XIIe siècle*, 2 vols, Paris, Les Belles Lettres, 1986.

'compasses', and studies a problem that reduces to a cubic equation. Thanks to Beeckman, we also know that around this time he is concerned with the invention of his 'compasses', an assemblage of articulated stems intended precisely to solve solid problems. The discoveries that he will make in this field during his famous six days of intense creative concentration are in effect four demonstrations, of which he is obviously proud: 'remarkable and completely new, and such on account of my compasses'.[8] The first demonstration focuses on the multisection of the angle into equal parts; the three others are solutions of four cubic equations drawn from among the thirteen equations obtained by combining number, root, square, and cube. Nothing new here: even the language is still that of the German cossists and of Clavius's *Algebra* (1608). It is, moreover, from the latter that he borrows, with minor variations, the modest symbolism he uses. In short, his classification is that of al-Khayyām, his results do not yet reach the generality that characterizes the latter, and his language is that of his masters. The novelty that he claims belongs rather to the order of the imaginary, or perhaps is the result of his still modest mathematical culture. But his genuine novelty lies elsewhere. It consists of one seminal idea that inspires his entire enterprise: all the successive clarifications and formulations until his *Géométrie*, are so many indicators of the genesis and evolution of his algebraic geometry. It is precisely this idea that allowed Descartes, starting from rather modest mathematical knowledge, to aim high and to see far, by opening for him the road he had to travel. Let us follow his steps.

Also in 1619, Descartes writes the following famous sentences, the importance of which is difficult to exaggerate:

> also I hope to demonstrate for continuous quantities that some problems can be solved with straight circular lines alone; that others cannot be solved except by curved lines [other than circles], which, however, result from a single motion and therefore can be drawn with new types of compasses, which I deem to be as exact and geometrical as the common ones used to draw circles; and finally others can be solved only by curved lines generated by motions that differ from one another and are not subordinated to one another, which [lines] are certainly only imaginary: one such is the rather well-known quadratrix. And it is my judgment that one could not imagine anything that could not be solved by such lines. But I hope to establish by demonstration which questions can be solved in this way or that, and not in

[8] Letter to Beeckman, 26 March 1619, in *Descartes. Œuvres philosophiques (1618–1637)*, ed. F. Alquié, Paris, 1963, vol. I, p. 36.

some other way, so that almost nothing will remain to be discovered in geometry.[9]

Descartes himself goes on to comment: 'The task, it is true, is infinite, and cannot be accomplished by one individual, an incredibly ambitious project!' We see that despite a rather humble mathematical knowledge, Descartes launches a program – 'ambitious' by his own admission – the central ideas of which are: a classification of problems organized by the curves used to solve them; a classification of curves organized by the motions with which they are drawn; and finally an unshakable faith, with no clear justification, in the heuristic value of these classifications for exhausting all questions of geometry by means of demonstration.

I will return to these classifications, which only later will be named 'geometrical' and 'mechanical'. For now, I pause on a comparison between the program that Descartes announced and the one that al-Khayyām implemented. One will immediately conclude that the project conceived by Descartes, as he formulates it in 1619, both falls short of and goes beyond that of al-Khayyām.

Indeed Descartes, just like al-Khayyām, concedes that in the last analysis a plane problem reduces to one (or more) equations of the second degree, whose roots are constructible using the properties of the circle and the straight line. But they differ in important respects. Al-Khayyām distinguishes solid problems from supersolid problems, and claims that it is the properties of the conic sections that make it possible to determine the root of cubic equations corresponding to the former and, as Ibn al-Haytham had established in the 10th century, it is the properties of a cubic curve and a conic that make it possible to solve a fifth degree equation. For his part, Descartes does not yet draw any such distinction and alludes in the aggregate to these curves, which will later be called 'geometrical'. Al-Khayyām, on the contrary, alludes to no curve other than a conic (and implicitly a cubic). Descartes speaks about one set of curves being opposed to another and differing according to the type of motion used to trace it. Al-Khayyām, for his part, does not refer to any 'mechanical' curve; in geometry, he 'abhors motion', as it were. In a word, then, in 1619 Descartes stands in the same terrain as al-Khayyām, with mathematical results that are less general, but he is motivated by a more general project in which the dominant idea concerns a class of curves that includes all those with which algebraic

[9] Letter to Beeckman, 26 March 1619, ed. Alquié, vol. I, pp. 38–9; A.T. X, p. 157; translation modified from C. Sasaki, *Descartes's Mathematical Thought*, Boston Studies in the Philosophy of Science, vol. 237, Dordrecht/Boston/London, Kluwer Academic Publishers, 2003, p. 102.

geometry deals. Although still embryonic in this project, these new interests will lead him increasingly to privilege both the concept and the study of the algebraic curve. Other results, however, were required for this seminal idea to keep its promises, become a fertile one, and reach a new stage. At the time, matters were not yet there.

Six years later, in 1625–1626, Descartes composes his *Algèbre*, communicated to Beeckman in 1628, the year in which he also shares his construction of all solid problems. These pieces of information, among others, teach us that Descartes is still involved in this same research already accomplished by al-Khayyām: to reduce solid problems to algebraic equations of the third degree, that can be solved by the intersection of two conics. This task obviously mixes algebraic interests with geometrical interests, which are so intertwined that one would vainly hope to disentangle them. It cannot help but run up against the problem of elaborating and justifying a genuine 'geometrical calculus'. On this point, we do not know precisely what Descartes's idea was in 1628. At the very least, one notices that he is still hesitant.[10] In his *Géométrie*, just like al-Khayyām in his *Algebra*, he proceeds by means of a geometrical calculation based on the choice of a unit length. Whereas the 11th century mathematicians adapted the unity to the dimension in order to respect homogeneity – this constant worry will still haunt Viète, for example – Descartes uses only rectilinear lengths. The conception of the unity that one finds at the beginning of the *Géométrie* was in fact a later acquisition, later than 1628, as far as the later documents reveal it. Until 1628, he mixes the two conceptions, that of al-Khayyām and that of his *Géométrie*. To grasp the difference between these two conceptions, note that, having chosen a unit, the calculation on straight line segments is first of all research aimed at finding the geometrical construction by means of which one can carry out, on these segments, the same operations that one performs on numbers in arithmetic ($\pm$, $\times/\div$, $\sqrt{}$). This idea, which al-Khayyām was the first to express, turns up in al-Ṭūsī, Bombelli, Viète, among others, as well as in Descartes. But, although they all change the unit as a function of dimension, only Descartes, in his *Géométrie*, draws on rectilinear lengths, or as he writes:

> It is to be noted that by a^2 or b^3 or the like, I ordinarily mean only simple lines, although, in order to make use of the names used in algebra, I call them squares, cubes, etc.[11]

[10] G. Milhaud, *Descartes savant*, Paris, 1921, pp. 70 ff.

[11] *La Géométrie*, A.T. VI, p. 371; English translation with introduction by P. J. Olscamp in *Discourse on Method, Optics, Geometry, and Meteorology*, Indianapolis, Bobbs-Merrill Co., 1965, p. 178.

First of all, one observes a clear break, the importance of which several historians have obviously emphasized. And in fact, when one looks more closely, one notes that Descartes's predecessors did not always respect homogeneity and that he himself never completely broke with it, not even in the *Géométrie*. To be sure, it is by omission or negligence that such predecessors as al-Ṭūsī or Bombelli sometimes violated this principle. But a mathematician of al-Ṭūsī's caliber does not neglect a principle if he deems it important for demonstration. As to Descartes, after having enunciated this beautiful opinion, he wastes no time making his first concession to homogeneity on the same page, uttering in addition pertinent remarks on the manner of making homogeneous a formula that is not. His writing then remains homogeneous – for example $z^3 = az^2 + b^2z - c^3$; and it is perhaps this desire for homogeneity that prevents him from setting a polynomial equal to zero (for that, it would be necessary to wait for the middle of the second book of *Géométrie*), and from daring to represent a ratio by a single letter. Descartes was not the sort of fellow to concede to tradition a principle important for demonstration.

Three years later, upon returning from the Middle East with a crop of mathematical manuscripts including an additional copy of al-Khayyām's *Algebra*, the mathematician and Arabist Golius puts to Descartes a problem that will deeply influence his mathematical thinking: the problem of Pappus. It can be rewritten thus: Given a group of $2n$ or $2n - 1$ line segments of known position and not all parallel; divide them into two subgroups, one of which consists of n straight lines, if the number is even; or, of n and $n - 1$ groups if the number is uneven. Find the locus of points such that the product of the distances to the segments of the first subgroup stands in a given ratio to the product of the distances to those of the second, a product to be completed, if necessary, by a given constant factor. The locus of these points is a line of given position. In January 1632, Descartes sends to Golius the sketch of his proposed solution to this problem, which constitutes in broad strokes the one that appears five years later in the *Géométrie*.[12]

I stress briefly only two impacts that this solution to the problem of Pappus had on Descartes's program. It consists first in the obligation of finding equations of curves that answer Pappus's question. This research

[12] See above 'The first classifications of curves', pp. 217–22. Cf. J. Vuillemin, *Mathématiques et métaphysique chez Descartes*, pp. 99–112; as well as H. J. M. Bos, 'On the Representation of Curves in Descartes' *Géométrie*', *Archive for History of Exact Sciences*, 24, 4, 1981, pp. 295–338, esp. pp. 298–302 and 332–8.

had two effects: on the one hand, it extended the theory of algebraic equations much farther than before; on the other, it expanded the domain of algebraic curves. The same research, beginning precisely with line segments, finally led Descartes to show that these 'geometric' curves are the loci of points of straight lines and curves. To clarify, let us consider Pappus's question in the simplest case, that of 3, 4, or 5 straight lines. As Descartes writes, 'We can always discover the required points through simple geometry – that is through the use of ruler and compasses alone'.[13] In other words, for 5 straight lines, one has an equation in x, y such that

$$f(x,y)=xy(x-ay-b)-k\alpha(x-cy-d)(x-ey-f)=0\,;$$

if x (respectively y) is given then y (respectively x) is found by means of the second degree equation. Now if the five straight lines are all parallel

$$f(x)=\alpha x(x-a)-(x-b)(x-c)(x-d)=0,$$

a third-degree equation that generally cannot be reduced to zero.

If Pappus's question concerns 6, 7, 8, or 9 straight lines, analogously one has an equation of the fourth degree, except if the 9 straight lines are all parallel, in which case the equation is of the fifth degree. Next Descartes gives two other groups of straight lines: he writes that 'it is essential to use a curved line of degree still higher than the preceding, and so on to infinity'.[14] In the second book of *Géométrie*, Descartes takes up again the problem of Pappus in order to determine the curve sought in each case. In the first case (with 3 or 4 straight lines), this curve will be the locus of the points of one of the three conics, of the circumference of a circle, or of a straight line;[15] the locus will be a cubic or a quartic in the case of 5, 6, or 7 straight lines; it will be a curve of the fifth or sixth degree in the case of 9, 10, or 11 straight lines. Descartes goes farther; he even believes that the problem of Pappus would thus give him all the geometrical curves, an error that Newton will later denounce.[16] But this error should not obscure

[13] *La Géométrie*, A.T. VI, p. 380; English transl. P. J. Olscamp, p. 185.

[14] *Ibid*., p. 381; English transl. P. J. Olscamp, p. 185.

[15] In fact, without insisting much, Descartes shows that he knows very well that it could only be a matter of not a single curve, but two.

[16] Cf. *The Mathematical Papers of Isaac Newton*, vol. IV: 1674–1684, ed. D. T. Whiteside, Cambridge, Cambridge University Press, 1971, p. 340, where Newton writes: 'Descartes erred further in that he asserted that all curves that he calls Geometrical are useful for the problem of Pappus. // Erravit præterea Cartesius in eo

(*Cont. on next page*)

the essential point: to solve this problem, Descartes proceeds by means of algebraic methods, appealing to the best notation of his era, and without bringing into play the methods of traditional geometry. Better yet, this problem gives him the occasion to generalize his algebraic method. Before 1631, he knew how to solve particular questions, but he probably had not yet conceived of a general procedure that would always work. Now, he has a premonition that the essential point is to obtain the equation of a 'geometrical' curve and the curve itself as a locus of points. Solving the problem of Pappus thus led him to try to isolate the 'geometrical' curves and to express them as the locus of the points that are sought by means of algebraic relations between the coordinates of each of these points, $P(x, y) = 0$, where P is a polynomial. It is therefore thanks to the solution of this problem that Descartes was able to return, but at a much higher level and with a much higher precision, to the questions about the classification of problems and of curves that he had raised in 1619, at the beginning of his mathematical career. Without ambiguity, he now assimilates 'geometrical' curves to 'organic' curves, that is, those that one can draw by means of his 'compasses'. But Descartes does not give the slightest proof of this. Indeed, none will be forthcoming until that of Alfred Bray Kempe (1849–1922) in 1876.

We can now compare more rigorously the net gains of Descartes with those of the mathematician al-Khayyām in the 11th century, from the perspective that we have called the first axis: to reduce a geometrical problem to an algebraic equation with a single unknown. Note first of all that, before confronting the problem of Pappus, Descartes had, like al-Khayyām, solved all equations of the third degree by the intersection of conics. In his *Géométrie*, he proceeds, still like al-Khayyām, to solve all equations of the third and fourth degree by the intersection of two conics, but, for his part, he restricts himself to a given parabola and to a variable circle, depending on the type of equation. But no more than al-Khayyām did, he does not at this time deal with the existence of roots. For the equations of the fifth and sixth degree in which al-Khayyām deliberately takes no interest (indeed he encountered the first case and knew its solution, $x^5 = k$), Descartes conceived a cubic parabola, or a parabolic conchoid, a curve with the equation

$$y^3 - 2ay^2 - a^2y + 2a^3 = xy.$$

(*Cont.*) quod asseruit omnes curvas quas Geometricas vocat utiles esse in Problemate Pappi'. Newton's, it is true, is not very clearly expressed, but it does exist. D. T. Whiteside notes that Newton has left 'a long gap at this point in his manuscript, evidently intending to develop the criticism further' (*ibid.*, p. 340, n. 22).

But, faced with a cubic equation, for example, Descartes like al-Khayyām was reduced to stating that at most it was solid; he could thus not specify the nature of the irrationals that entered into the solution. Neither mathematician evidently suspected that this problem required one to know the decomposition of the associated polynomial into its prime factors over the field of its coefficients. Thus far, then, Descartes is reiterating al-Khayyām's procedure; to be sure, he fine-tunes it, generalizes it, takes it to the very limits of logical possibility; in short, he completes it, but without truly plumbing its substance nor recasting its meaning. Is this also the case when we turn to the second axis of research that runs through the *Géométrie*? This is the question we will now try to answer.

2. FROM GEOMETRY TO ALGEBRA: THE CURVES AND THE EQUATIONS

Without the slightest doubt, the theoretical instrument of the new program is the classification of curves into 'geometrical' and 'mechanical'. Whoever is interested in Descartes's *Géométrie* must first know the origin of this classification and ask what it covers. The most common hypothesis proposed to explain this classification refers to the unprecedented increase in the number of new curves in the 17th century.[17] On this account, it was a new need – the necessity of giving an account of these new objects – that allegedly stimulated Descartes to hammer out his famous distinction. Indeed, one of the most significant points of mathematical research in the first half of the 17th century was this invention of new curves. Consider, for example, all the efforts and debates that surrounded the cycloid. This hypothesis thus appears seductive, even natural, if only it could stand up to the chronology, to the examination of Descartes's mathematical knowledge when he conceived the classification, and to the reception with which his contemporaries greeted it. But this is not the case.

We have just seen that, despite the missing terminology, the first clues about this classification are relatively old, since they go back to the years 1619–1621. For example, Descartes writes the following at the time of the *Cogitationes Privatae*: 'The line of proportions must be conjugated with the quadratrix; [the quadratrix] is indeed born of two nonsubordinated motions, the circular and the straight'.[18] Ten years later, in 1629, Descartes

[17] See for example H. J. M. Bos, 'On the Representation of Curves in Descartes' *Géométrie*', pp. 295–7.

[18] A.T. X, pp. 222–3: 'Linea proportionum cum quadratice conjungenda: oritur enim [quadratrix] ex duobus motibus sibi non subordinatis, circulari & recto'.

affirms about the quadratrix and the cylindrical helix that neither is 'admitted in geometry', 'because [...] one can only draw them in their entirety by joining two motions that are independent of one another'.[19] These few citations confirm what we already knew: the classification is ancient, even though it will later become both clearer and more elaborate. But which new curves did Descartes know at this time? Only one, and in a still rather imprecise manner: the *linea proportionum*; all the other curves he mentions come from the ancients. About this *line of proportions*, he knew at the time only that it was generated, like the quadratrix, by two separate motions, and therefore was a 'mechanical' curve. Only much later will he know the second new 'mechanical' curve, the rolling circle or cycloid: he discusses it only one year after the publication of the *Géométrie*.[20] And even if one assumes that he already had had some echo of it in 1635, his knowledge could only have been about a still poorly studied curve, as Roberval personally attests.[21] Descartes's interest in curves and in their classification could therefore not follow from an increase in the number of new curves. Even more surprising: his interest manifests itself against the background of a traditional knowledge of curves, at least until he wrote the *Géométrie*. Moreover, this classification is credited to Descartes alone. In this particular respect, he indeed owes nothing to his predecessors, whether immediate or distant. Clavius, for example, was far from excluding the quadratrix from geometry,[22] and Viète refuses to consider any curve,

[19] Letter to Mersenne, 13 November 1629, A.T. I, p. 71. Note indeed that the quadratrix is the locus of points M, the intersection of a straight line AM parallel to Ox that is displaced by a uniform motion of translation, and of a straight line OM, rotating about O with a uniform rotary motion.

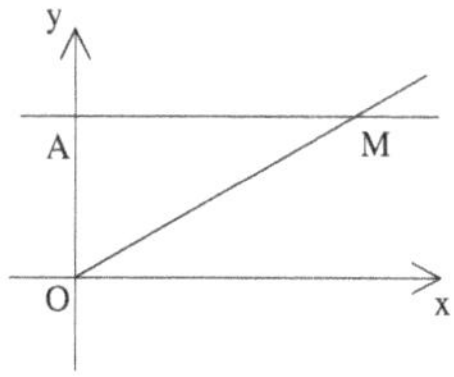

Fig. 26

[20] Letter to Mersenne, 23 August 1638, A.T. II, p. 313: 'One should also note that the curves described by rolling circles (*roulettes*) are completely mechanical lines, and among those that I rejected in my *Géométrie*'.

[21] Letter from Roberval to Torricelli 1647, from which it transpires that the former did not concern himself with this curve before 1634. Cf. *Opere di E. Torricelli*, ed. G. Loria and G. Sassura, Faenza, 1919.

[22] Clavius, *La Géométrie pratique*, p. 918 (end of Book VI of Euclid's *Elements*).

including the conics, as admissible in geometry.[23] As to Descartes's contemporaries themselves, they attached little importance to his classification. Roberval, for example, after having defined mechanical curves differently, that is, as 'curved lines that have a relation only with other curves or partly to straight lines and partly to curves', believes that Descartes excludes them from geometry for no good reason.[24] Even when he adopts Cartesian notions, Roberval persists in rejecting Descartes's classification.[25] Likewise for Mersenne, who knows about the classification since 1629: he does not even bother to refer to it in 1637 in his *Harmonie Universelle*, in a context where it seemed to be particularly relevant.[26] Finally, unless I am mistaken, Fermat nowhere in his correspondence alludes to 'mechanical' curves; they are mentioned only in his *De linearum curvarum* of 1660, where he seems to accept the distinction, but only tacitly, when he writes: 'never, so far as I know, have geometers equated a given straight line to a purely geometrical curved line'.[27]

In short, this classification of curves is therefore a surprising invention. Descartes's predecessors with whom, whether new or distant, he had never broken, had seen no need for it; nor did his contemporaries; and he himself had proposed it under no compulsion whatever, since he did not need to explain new objects, such as new curves. Is this classification an arbitrary one? Absolutely not. He himself sets it at the center of his program, and his successors, who are now under pressure from a growing number of new curves, adopt it and refine it (*e.g.*, Newton and Leibniz). When the young mathematician of genius with his still modest mathematical culture proposes such a classification what then is his motivation? And what content does he perceive in the classification that he has just hammered out?

Descartes himself seems to lift the veil a little about his motives by drawing on a speculative history and by offering us a kind of historical tale. He begins by accusing the ancients of having failed to 'distinguish various degrees among composite lines' beyond the conics. Then he is surprised by

[23] 'Apollonius Gallus', in *Opera mathematica*, recognita Francisci à Schooten. Vorwort und Register von Joseph E. Hofmann, Leiden, 1646; facsimile G. Olms, Hildesheim, New York, 1970, p. 325.

[24] Letter to Fermat, 4 August 1640, in *Œuvres de Fermat*, published by Paul Tannery and Charles Henry, Paris, 1891, vol. II, p. 200.

[25] *Divers ouvrages de mathématique et de physique par Messieurs de l'Académie Royale des Sciences*, Paris, 1693, p. 209.

[26] Vol. II, Book VI, Proposition 45, pp. 408–9; *Correspondance du P. Marin Mersenne*, vol. III, pp. 258–9.

[27] *Œuvres de Fermat*, vol. I, p. 211: 'Nondum, quod sciam, lineam curvam pure geometricam rectæ datæ geometræ adæquarunt'.

a confusion between 'geometrical' and 'mechanical' that he attributes to them. Next, Descartes interprets the omission and makes a conjecture about the confusion:

> [the first curves] they considered happened by chance to be the spiral, the quadratrix, and similar curves, which truly belong to mechanics and are not among the number that I think should be included here; for we can conceive them as being described by two separate movements which have no precisely measurable relation to each other. Yet afterwards they examined the conchoid, the cissoid, and a certain few others which we do accept; nevertheless, perhaps because they did not pay sufficient attention to their properties, they took no more notice of these than of the first. Or else it may have been that – since, as yet they knew only a few thing about conic sections, and there was even much that they did not know about what could be done with the ruler and compass – they believed they should not approach more difficult material.[28]

As customarily happens when telling a historical tale, Descartes here is not being a historian, but trying to present to his readers a few exemplary characteristics/traits of the progress of the mind. It matters little whether the events he evokes are doubtful or even false.[29] By contrast, in the fine texture of his tale, he reveals his own intentions: to go beyond the conics and to distinguish clearly between this class of curves traced and privileged by the algebraists, and all the other curves, under the strict condition of relying on no concept unknown to the ancients. It will therefore be necessary to correct the classification of the ancients, who were deprived of algebra and therefore could not see this cleavage between the classes of curves. The concept to which Descartes has access – at least the one that first presented itself to him – is none other than the ancient concept of motion such as we encounter it in the Aristotelian tradition. In the *Géométrie*, this notion of continuous motion is treated, to be sure, without any apparent kinematic consideration, but also without being clothed in the slightest algebraic dimensions. On the contrary, one will observe that, thus far, the notion of equation of a curve does not enter the picture in order to establish the classification; it is only later that he will call upon it to describe the elements of this class of curves privileged by the algebraists. For the moment, let me insist once again, that a curve such as the quadratrix or the spiral is 'mechanical' because it is generated by two separate

[28] *La Géométrie*, A.T. VI, p. 390; English transl. P. J. Olscamp, p. 191.

[29] A. G. Molland gives a different commentary of this 'historical' text in his article 'Shifting the Foundations: Descartes's Transformation of Ancient Geometry', *Historia Mathematica*, 3, 1976, pp. 21–49, esp. pp. 34–6.

continuous motions, 'which have no precisely measurable relation', and consequently cannot be studied in terms of the theory of proportions. The two motions are rotation and translation. Once he has established the distinction between 'geometrical' and 'mechanical', Descartes in his *Géométrie* can devote himself to geometrical problems and curves. The second book has precisely the goal of studying such curves, and Descartes states in the famous passage in the first few pages that, in order to draw and conceive these curves, he himself knows

> of nothing better than to say that all points of those curves which can be called 'geometrical' – that is, which fall under some precise and exact measure – necessarily have a certain relation to all the points of a straight line; and this relation can be expressed by a single equation for all the points. And when [no term of] this equation is higher than the rectangle of two indeterminate quantities, or else of the square of a single unknown quantity, the curved line is of the first and simplest class, which comprises only the circle, parabola, hyperbola, and ellipse. But when one or both of the equation's two unknown quantities (for there must be two, in order to explain the relation between two points) reaches the third or fourth degree, the curve is of the second class; and when the equation reaches the fifth or sixth degree, the curve is of the third class; and so on for the others, to infinity.[30]

As far as I know, this text gives the most precise formulation Descartes ever wrote about the concept of curve and its relations with the associated equation. It is also valuable for its ambiguities and what it says between the lines; elucidating it will allow us better to grasp the author's thought. We notice immediately that his formulation patently pertains only to 'geometrical' curves, whereas his declared intention was undoubtedly to discuss all curves; in his own words, 'all those that exist in nature'. Everything thus unfolds as if Descartes was *ipso facto* excluding mechanical curves when he was trying to define more precisely the concept of curve – as if 'mechanical' curves were, for him, not real curves. In this case, the only curve that genuinely deserves the name is the one that can be conceived as the locus of points or the one that can be drawn with the help of his 'most general point'. In this passage, Descartes thus seems to imply that there is an essential difference between the two types of curve, which the concept of motion alone does not suffice to bring out clearly. What difference then distinguishes the two classes of curves? This difference seems to refer back to two problems that are intermingled here: that of the construction of points on the curve; and that of the existence of intersection points when the curves cut each other.

[30] *La Géométrie*, A. T. VI, p. 392; English transl. P. J. Olscamp, p. 193.

Indeed Descartes knows that if all the points of a 'geometrical' curve are constructible, it is not the case for a 'mechanical' curve. Better yet, he knows that if the 'geometrical' curve is a conic, all of its points are constructible by means of straightedge and compass: if it is cubic or quartic, its points are all constructible by the intersection of two conics; if it is of the fifth or sixth degree, its points are all constructible by the famous 'second-degree parabola' that he conceived and by a semicircle, and so on. More generally, he knows that for a 'geometrical' curve of order n, one can construct all of its points if one can construct all the points of the 'geometrical' curves of a lower order. This cascade of procedures thus allows one to construct all the points of the 'geometrical' curve. Descartes also knows, but without in any way proving it, that he cannot apply this cascade procedure to the case of 'mechanical' curves: he knew that he could not construct all the points of a 'mechanical' curve, far from it. As will be demonstrated later, the reason for this is that all the constructible points of a small arc of this curve, the arc of a quadratrix for example, 'form a set that is everywhere dense, but not closed, however small the considered arc may be'.[31] Only the points of the ordinate $m/2^n$, where m, n are integers, are obtained by a construction using only the straightedge and compass. Descartes has seen all of this in his own way (and obviously without topological concepts) when he writes:

> there is a great difference between this method of finding several points in order to trace a curved line passing through them, and the method used for the spiral and similar curves. For in the latter we do not find all the points of the required line indiscriminately, but only those which can be determined by some process which is simpler than that which is required for composing the curve. And so, strictly speaking, we do not find any one of its points, that is to say, not any one of those which are so properly points of this curve, that they cannot be found except through it; on the other hand, there is no point on the lines that can be used for the proposed problem which cannot be found among those which can be determined through the method just explained. And because this method of drawing a curved line by finding several of its points indiscriminately only extends to those that can also be described by a regular and continuous movement, we need not entirely reject it from geometry.[32]

It is clear that Descartes *sees* more than he demonstrates the reasons for excluding 'mechanical' curves. His strong intuition rests, at the beginning

[31] Henri Lebesgue, *Leçons sur les constructions géométriques*, preface by M. Paul Montel, Paris, Gauthier-Villars, 1950, p. 15.

[32] *La Géométrie*, A.T. VI, pp. 411–12; English transl. P. J. Olscamp, p. 206.

at least, on two ideas that lie at the center of algebraic geometry since its beginnings in al-Khayyām. The first is a kind of assumed isomorphism between calculation on segments, which represent the real numbers, and Euclidean geometry's constructions by means of the straightedge and compass. From the very beginning of the *Géométrie*, Descartes shows no ambiguity about this point: there he sees perfectly, albeit without proving it, that every 'plane' problem reduces in the last analysis to the solution of a second-degree equation. Second idea: the 'solids' can, in the final analysis, be reduced to a polynomial equation of the third degree. In every case, all the points are constructible. Implicitly then, one will keep only the polynomial equations that can be obtained by means of the intersection of curves of a lower order, but all of whose points are constructible. This time, however, it is the question of existence that will become more pressing, a question that is certainly important in its own right, but also on account of its tight relation to the definition of the entities with which it deals.

This question was raised very early in the history of algebraic geometry: since the solutions to algebraic equations were constructed using the intersection of curves, one had to be sure that the intersection points duly existed. It was therefore necessary to have a proof of the existence of an intersection point in order to deduce the existence of the corresponding root of the equation. Barely broached by al-Khayyām for a single equation, this question of existence has the leading role in al-Ṭūsī's treatise on *The Equations*. What is more, raised as a genuine requirement by the latter, the demonstration of existence leads him to add to al-Khayyām's global analysis of the local behavior of curves, that is, near the point of intersection. In Fermat, there is an attempt similar to that of al-Ṭūsī, but one operating at a higher level. What matters here is that, in all of these contributions, the main argument for affirming the existence of the point of intersection of two curves is the following: in the final analysis, one of the curves has points on one side and the other of the other curve, both being assumed to be continuous, that is, drawn by a continuous motion. When these curves are conics, one deduces from the *symptoma* that the point of intersection belongs to each of the curves.

For his part, Descartes faces this question in a context that is, as it were, more dramatic, not despite the earlier progress, but precisely because of it. There are two reasons for this. Like his predecessors, Descartes privileges the drawing by continuous motion of these same curves. Understandably, in this context, it is the continuous motion alone that assures the continuity of the curves. Unlike his predecessors, however, he recognizes the role of the equation in representing the curve. Now the equation most often only allows him the construction by means of points; and this construction by points, as we have seen, is sufficient only for 'geometrical' curves. This is a highly paradoxical situation from two points of view, even if they are of unequal importance. The first paradox is by far the most profound: Descartes simply did not have the means to resolve this apparent contradiction: for that, he would have needed the Bolzano's demonstration of the theorem about intermediary values in his treatise of 1817. Conversely, Descartes encounters a great difficulty when dealing with 'mechanical' curves insofar as he cannot construct all the points of the curve. What is more, whereas Descartes knows how to solve 'geometrical' problems by the intersection of 'geometrical' curves, he is not up to proceeding analogously for 'mechanical' problems. To solve them, indeed, requires the intersection of curves that are themselves 'mechanical'. Now the latter curves have no algebraic equation; they nevertheless admit a differential algebraic equation that links not the abscissa and the ordinate, but their differentials. In this case, not until Leibniz, and especially successors such as Jacques Bernoulli, did one witness the elaboration of these notions and the transfer of this problem to the new domain of analysis. Until then, no concept of equation can represent a 'mechanical' curve. The situation is thus frankly asymmetrical: whereas one can speak the language of equations for the class of 'geometrical' curves, this is impossible for 'mechanical' curves. About the former, Descartes writes: 'That, in order to discover all the properties of curved lines, it is sufficient to know the relation of all their points to those of straight lines and the way of tracing other lines that cut them at every point orthogonally'.[33] No doubt seems possible: Descartes knew that all the properties of 'geometrical' curves must be drawn from their equation. This was, as we know, a program to which he never devoted himself; it had to await the young Newton. Thus, there is only a short step between what Descartes has just said and the definition of the curve by its equation, *i.e.*, its *characterization* by the latter. Descartes did not take that step, probably prevented from doing so by the previously mentioned asymmetry with 'mechanical' curves. Excluded from geometry, they cannot for Descartes be represented by equations, which renders the task of unification impossible. Powerless to unify the characterization of all curves, Descartes the great proponent of clarity was thus backed into a chiaroscuro corner. This is one of the characteristics of his *Géométrie*, and one of the profound reasons for the conflict among the interpretations of it.

We thus see that what separates the two classes of problems and curves pertains not only to the single concept of movement, but also to the ques-

[33] *Ibid.*, pp. 412–13 in the margin.

tions raised by the construction and by the existence of their points as well as to the power of the equation to define them. Even if Descartes did not make it completely explicit, this classification of problems and curves was nevertheless elaborated over the years, as an instrument geared to exploring the study of certain curves by means of their equations. It is in this sense that the *Géométrie* truly deserves its name. Can one not now find the properties of the curve, notably its tangents and its normals, from the equation? In any case, this is how Descartes's successors read his *Géométrie*. Cramer, for example, writes in his *Introduction à l'analyse des lignes courbes algébriques* of 1750:

> It is especially in the theory of curves that one very clearly senses the utility of a method as general as that of Algebra. No sooner did Descartes – whose inventive spirit shines no less in Geometry than in Philosophy – introduce the way of expressing the nature of curves by algebraic equations than the face of this theory changed.[34]

This is, of course, a regressive reading of the *Géométrie*; it nevertheless truly expresses the potentiality of this second axis of research for Descartes. And in fact, in the wake of al-Khayyām – namely in al-Ṭūsī's research on *maxima*,[35] or more generally on the existence of points of intersection – one can find the study of certain curves by means of their equations. Apart from the conics, however, al-Ṭūsī does not clearly distinguish between polynomial equation and curve. The concept of equation of a curve thus remained limited, and was insufficiently transparent to constitute a program of research. This is, however, precisely the function that it acquires with Descartes, thanks to his extension of the study of curves beyond the conics, and to his distinction of this class of curves that can be studied by means of algebra. The restriction of Descartes's *Géométrie* to 'geometrical' curves is the consequence not of a denial of the existence of 'mechanical' curves, but of the generalization of the concept of 'geometrical' curve; his exclusion of 'mechanical' curves is thus an eminently positive act.

As a powerful tool destined to mark the boundary of algebraic geometry, this distinction between the two classes of curves also offers the means of establishing another opposition, which is important for Descartes's philosophy. He considers geometrical curves and mechanical curves from a

[34] Gabriel Cramer, *Introduction à l'analyse des courbes algébriques*, Geneva, 1750, pp. VII–VIII.

[35] R. Rashed, *Sharaf al-Dīn al-Ṭūsī, Œuvres mathématiques*, vol. I, pp. XVIII–XXXI.

double point of view. As mathematical ideas, they are answerable to the criterion of truth defined as clear and distinct representation. Like the geometrical curve, the mechanical curve is an intellectual object accessible to thought insofar as it represents. The one as well as the other is indubitably true, dependent on self-evidence, which is defined as intellectual clarity and distinction. On the other hand, our knowledge of each is not confused with our knowledge of the other: indeed, they even oppose one another. This opposition is hierarchized: there is, in our knowledge of the geometrical curve, more perfection than in that of the mechanical curve, taking into account the construction of its points, the simplicity of the movement that generates it, and the rigor of the equation that characterizes it. Add to this the cascade procedure, this apodictic chain, regulated by an order of reasons that allows one to know the geometrical curve, altogether and completely, whatever its degree may be. All of these procedures define clear *and* distinct knowledge mathematically, a knowledge opposed to that of mechanical curves, which are, from the second point of view, the object of clear knowledge only. It is this cleavage between 'clear and distinct' on the one hand, and 'clear' on the other, which in the end separated the two classes of curves by consecrating the exclusion of mechanical curves from geometry. Better yet, it is this opposition that isolated geometrical curves.

Briefly sketched, it is these two movements that seem to govern the evolution of Descartes's *Géométrie*. The first is oriented to the completion of a scientific project conceived six centuries earlier in another climate; the second gathers up the beginning of a study of curves in order to create a new program, the realization of which betokens the future, the fountainhead of two new currents: that of algebraic geometry, with Cramer and Bézout; and that of differential geometry, with the brothers Bernoulli.

The *Géométrie* thus illustrates the complexity of the relations between tradition and modernity in the 17th century, and bears witness to the difficulty of establishing a dialectic between these notions. The modernity that Descartes's *Géométrie* represents thus presents itself as the actualization of several potentialities inherited from tradition, even as it was a generator of new potentialities for the future. But could it have been otherwise? If one were to think only in terms of ready-made concepts, one might say that continuities and ruptures are mutually inscribed inside each other. Every discourse on Descartes's *Géométrie* will therefore necessarily be oblique if it ignores or neglects the intimate links that root it in tradition, or also the new possibilities that inhabit it and effectively are realized only once modernity itself has become tradition.

To read Descartes's *Géométrie* is to look upstream towards al-Khayyām and al-Ṭūsī; and downstream towards Newton, Leibniz, Cramer,

Bézout and the Bernoulli brothers. It is then that the *Géométrie* reclaims the place that was always its own: no more than other innovative works does it embody a radical beginning; on the same grounds as the others, it is a way of reworking and adapting, but also correcting, the traditions that it inherited.

– 5 –

DESCARTES'S OVALS

Beginning in the second half of the 18th century, notably with d'Alembert and such successors as Kant, the conditions of possibility of scientific knowledge were often reduced to the application of mathematics to the phenomenon under discussion. As Kant famously proclaimed in the 'Prolegomena' to the *Metaphysical Foundations of Natural Science*:

> I maintain that in every particular theory of nature, one can find only as much proper *science* as there is *mathematics* in it.[1]

Such a conception could only be elaborated and formulated after the development of mechanics by Newton and his successors. In any case, it was alien to the scholars of Antiquity, and notably to the dominant philosophy – and the 'physics' – of Aristotle. For him, mathematics and physics are separate: the former is demonstrative knowledge, while the latter is knowledge of becoming. But this opposition of principle is in no way the radical break that some commentators, starting with Alexander of Aphrodisias, saw in it. Although the question of the application of mathematics to physics was essentially not raised at the time, mathematics nevertheless played two roles: the first, instrumental, was in the '*poietic*' sciences, namely those that focus on the production of useful objects; the second occurred in determining the contours of a phenomenon. Mathematics was thus applied to the rainbow, to the theory of the balance, to the configuration of the universe considered as an *organon*, to mirrors, including burning mirrors, etc., that is, to everything that could be considered a machine. This application had at least the advantage of allowing one to speak mathematically about a localized phenomenon, *e.g.*, about the propagation of rays parallel to the axis of a parabolic mirror, or about the contour of the apparent motion of the moon.

[1] 'Ich behaupte aber, dass in jeder besonderen Naturelehre nur so viel *eigentliche* Wissenschaft angetroffen werden könne, als darin *Mathematik* anzutreffen ist' (I. Kant, *Metaphysische Anfangsgrunde der Naturwissenschaft*, Frankfurt/Leipzig, 1794, p. viii).

This remained the case until the first reform of optics, and more generally of physics, that Ibn al-Haytham (d. after 1040) undertook. The new watchword now becomes 'to combine mathematics and physics' whenever one tries to study any natural phenomenon. It is in the mathematics of material things that resides the possibility of truly knowing them. Ibn al-Haytham is the first scientist who refuses to consider that determinate natural things can be known by concepts alone: for him, true physics is necessarily mathematical. To carry out this program in optics, Ibn al-Haytham was led to break with the ancient tradition of Euclid and Ptolemy, for whom the act of seeing was the same as that of illuminating. It was necessary for him first to distinguish clearly between a physics of light and a psycho-physiology of light. In the vast extent of his work, the application of mathematics to physics takes on several distinct meanings, according to not only the conceptual maturity of the discipline, but also the amenability of the phenomenon to experimentation. The first meaning pertains to an isomorphism of structures, as in the case of geometrical optics, already reformed thanks to Ibn al-Haytham. The second meaning involves the application of mathematics by means of a third discipline that is itself considered mathematized: this is what occurs, for example, when in optics one relies on a mechanical schema. This application takes place when the discipline is not well elaborated conceptually, as is the case in the physical optics of the time. In such an instance, mathematics guarantees the analogies between the third discipline and the phenomena under study: in this case, the mechanical schema of violent motion and the physical phenomenon of propagation. But the application of mathematics can also occur by constructing local models when one is dealing with phenomena inaccessible to direct experimental study, such as the rainbow. Finally, the application of mathematics can take advantage of the objectivity of the technical object, such as happens, for example, in research on the phenomenon of the focalization of light by mirrors, lenses, or simply a phial filled with water. However that may be, all these applications – and herein lies the novelty of Ibn al-Haytham's project – must allow the set-up of an experimental situation, thanks to which one can control the ideal occurrence of the phenomenon or, failing that, its local occurrence. It is this experimental situation that guarantees for the phenomenon under consideration its true plane of existence.

I have on earlier occasions studied the problem of the application of mathematics in Ibn al-Haytham's optics and in many other ancient and modern contexts. I now would like to tackle the study of the reciprocal situation.

1. It is one thing to study the applications of mathematics to optics, their modes, their range, their planes of existence, and the types of experimentation that they underwrite. It is quite another – different although related to it – to wonder what optics brought to mathematical research. If scholars have often inquired into the mathematical fertility of scientific disciplines in the cases of astronomy and ancient and classical* statics, they have rarely done so for optics during the same period. And yet optics intervened with effectiveness and in several ways in the development of many a chapter in mathematics. If one restricts oneself to ancient and classical optics, as I will do here, one must consider, among other chapters, the geometry of conics (and more generally the theory of curves), projective geometry, spherical geometry, and trigonometry in Ibn al-Haytham and his studies on the burning sphere.[2]

Anaclastics is surely among the branches of optics that made the most substantial contributions to the development of mathematics. Research carried out during the last three decades has greatly enriched our knowledge of the history of this discipline. I will begin by summarizing briefly the new historical knowledge before pausing at greater length on the example of Descartes's ovals.

The book of Diocles[3] has made it possible to establish that, beginning in the 3rd century BC, an entire tradition of catoptricians succeeded in developing the geometry of conic sections in a direction different from that of Apollonius. These catoptricians carried out research into the optical properties of conic curves and thus provided a kind of 'characterization' of these curves different from that given by the *symptoma* and later by the equation. The school of Conon of Alexandria already attests to the beginnings of this tradition of catoptrics. At the heart of this school, Dositheos, who later will write Archimedes after Conon's death, takes up the study of the optical properties of the parabola and the parabolic mirror. Among the exact or slightly later contemporaries who continued this work were the geometer Pythion of Thasos, a certain astronomer named Hippodamos, and – last but not least – Diocles. Particularly noteworthy, this research began in the same milieu in which the study of conic sections,

* As explained in the Introduction, I use 'classical' to refer to the mathematical sciences from the beginning of the 9th century to the beginning of the 17th century.

2 See the edition, translation, and commentary in R. Rashed, *Geometry and Dioptrics in Classical Islam*, London, al-Furqān, 2005.

3 *Livre de Dioclès sur les miroirs ardents*, in R. Rashed, *Les Catoptriciens grecs*. I: *Les miroirs ardents*, Collection des Universités de France, Paris, Les Belles Lettres, 2000, first part.

the number of the points of their intersections, etc., was the most active, that is, around Conon in Alexandria; it is to this activity that the prologue to Book IV of Apollonius's *Conics* bears witness.[4] To recall what this involves, let us take up the questions that Diocles treats. He studies the parabola starting from the focus-directrix property, with the aim of learning about its optical properties. His study, in which he draws on the properties of the subtangent and the subnormal, is purely geometrical. Next, he examines the focal property, then goes on to consider a chord perpendicular to the axis of the parabola, showing that turning the arc around this chord generates a parabaloid and that the focus generates a circle whose center is the middle of the chord. Knowing its focus and its directrix, Diocles traces the parabola by points, and finally endeavors to deduce the *symptoma* of the parabola from the focus-directrix property.

This type of research does not end with Diocles; it turns up in other writings preserved in Greek, such as the Bobbio fragment, and especially in the Arabic translations of several Greek treatises. There survives, for example, a treatise by a certain Dtrūms, whose study of the parabola and of its focal property cannot be reduced to any other. Later, in the 6th century, Anthemius of Tralles and Didymos each take up in their own way the study of the parabola and of its optical properties.[5] In the 9th century, the philosopher and mathematician al-Kindī takes up the study of Anthemius, with the intention of making it more rigorous.[6]

It is in Anthemius's treatise on *The Mechanical Paradoxes* and in al-Kindī's treatise *On Solar Rays* that one also finds the studies of the ellipse based on its bifocal property. It is precisely from this property that al-Ḥasan, the youngest of the three Banū Mūsā brothers, develops a complete theory of the ellipse as a plane section of the oblique cylinder and examines the elliptical section by using the cylindrical projection.[7] In his study of the ellipse, al-Ḥasan ibn Mūsā indeed relies upon the bifocal definition $MF + MF' = 2a$ (where a is a semi-major axis). Very schematically, these

[4] See our edition, *Apollonius: Les Coniques*, Tome 2.2: *Livre IV*, Berlin/New York, Walter de Gruyter, 2009.

[5] *Ibid.*, pp. 325 ff.

[6] R. Rashed, *Œuvres philosophiques et scientifiques d'al-Kindī*. Vol. I: *L'Optique et la Catoptrique d'al-Kindī*, Leiden, E.J. Brill, 1997.

[7] R. Rashed, *Les Mathématiques infinitésimales du IXe au XIe siècle*. Vol. I: *Fondateurs et commentateurs*: *Banū Mūsā, Thābit ibn Qurra, Ibn Sinān, al-Khāzin, al-Qūhī, Ibn al-Samḥ, Ibn Hūd*, London, al-Furqān, 1996; English translation: *Founding Figures and Commentators in Arabic Mathematics*. A History of Arabic Sciences and Mathematics, vol. 1, Culture and Civilization in the Middle East, London, Centre for Arab Unity Studies, Routledge, 2012.

rough brush strokes summarize the contributions of anaclastics to the theory of conics up to the middle of the 9th century.

In the 10th century, Ibn Sahl, the predecessor of Ibn al-Haytham, who has the law of refraction – Snel's law, so called – adds to the study of burning mirrors that of plano-convex and bi-convex lenses. He then writes the first known study entirely devoted to the optical properties of the three conic curves and invents an instrument for making a continuous drawing of these curves. Throughout this study, he is particularly interested in determining the plane tangent to a point of the surface generated by the rotation of the curve and the unicity of this plane – research that Ibn al-Haytham will later actively pursue. Ibn Sahl's treatise thus includes the study of the three curves beginning from their focus and from the directrix for the parabola, and of the bifocal property for central conics.

Up to this point, however, the problem is to burn a body by means of luminous rays at a given distance. One then chooses the curve in relation to the source – whether proximate or distant – of the reflection or of the refraction.[8]

2. To answer the same question seven centuries after Ibn Sahl, and although he limits himself to refraction, Descartes invents other curves that do not belong to the family of conics: his ovals. Indeed, he wants to know on which curve a bundle of rays originating from a given point, refracts itself in order to arrive at another point, also given.

This question is already present in the *Excerpta mathematica*, which antedates 1629, that is, a period in which Descartes already knew the law of refraction and was interested in lenses. One finds there a series of essays on ovals, which the editor Paul Tannery rightly judged to be 'a first stab, with their ordinary errors and clumsiness, and without having noted any definitive results'.[9] With the exception of the first essay, which is analytical, all the others are synthetic. In the first, therefore, one can hope to grasp Descartes' intention. He begins by writing:

> Given *A*, *B*, *C* on a straight line, to find the curved line that has *A* as summit, *AB* as its axis, and that is curved in such a way that the rays emanating from

[8] See R. Rashed, *Geometry and Dioptrics in Classical Islam*.

[9] P. Tannery, *Mémoires scientifiques*, published by J. L. Heiberg and H.-G. Zeuthen, Toulouse/Paris, Eds. Privat and Gauthier-Villars, 1926, vol. VI: Sciences modernes 1883–1904, edited by Gino Loria, p. 333.

> point B, after they have undergone a refraction at the latter, continue beyond it, as if they had been coming from point C or inversely.[10]

He continues:

> I take point N halfway between B and C; let the following be the case:
>
> $$NA = a, NB = b, CE + BE = 2a - 2y \text{ and } DA = x.$$[11]
>
> And let x and y be two indeterminate quantities, one of which, remaining indeterminate, will designate all the points of the curved line, and the other will be determined in the manner according to which the curved line must be described. And to find this manner, I look first for the point F, beginning from which, taken as a center, I conceive that the circle that touches the curve in E is described; next, I say that the line BE multiplied by FC is to CE multiplied by BF as <HF is to FG, that is, as> the inclination of the refracted ray in a transparent medium is to the inclination of the same <ray> in another <medium>.[12]

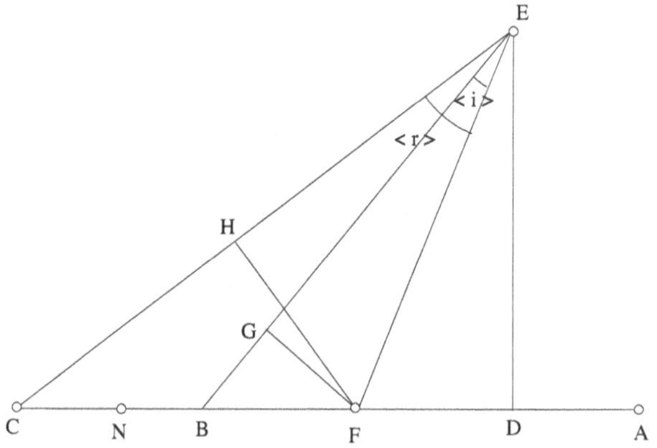

Fig. 27

Descartes thus considers a point E of the curve to be determined and $ED \perp AB$ and he wants the light ray BE to be refracted in E following a ray whose extension passes through C. If the circle (F, FE) is tangent at E to the curve under consideration, then EF is the normal to this curve at E.[13]

[10] *Excerpta Mathematica*, in *Œuvres de Descartes*, ed. by C. Adam and P. Tannery, vol. X, Paris, J. Vrin, 1966, p. 310.

[11] The equation $CE + BE = 2a - 2y$ is arbitrary.

[12] *Excerpta mathematica*, A.T. X, pp. 310–11.

[13] One will find this construction again, but with greater generality, in Book II of the *Géométrie*, in order to determine the tangents of algebraic curves.

He draws $FH \perp EC$ and $FG \perp EB$. The angle of incidence is $B\hat{E}F = i$ and the angle of refraction, r, is equal to $C\hat{E}F$. One thus has

$$\sin i = \frac{FG}{FE} \text{ and } \sin r = \frac{FH}{FE},$$

therefore

$$\frac{FH}{FG} = \frac{\sin r}{\sin i}.$$

But in triangle BFE, one has

$$\frac{\sin i}{BF} = \frac{\sin B\hat{F}E}{EB}$$

and in triangle CFE, one has

$$\frac{\sin r}{CF} = \frac{\sin B\hat{F}E}{EC}.$$

From this, one deduces that

$$\frac{\sin r}{\sin i} = \frac{EB \cdot CF}{EC \cdot BF},$$

whence

$$\frac{FH}{FG} = \frac{EB \cdot CF}{EC \cdot BF}.$$

It follows that this road to determining the curve does not lead to the desired result[14] or, as P. Tannery has remarked:

> If Descartes had had a method inverse of that of his tangents, he would have succeeded in expressing the condition that the normal divide the axis in a given ratio. With the only resources at his disposal, however, he certainly could no more succeed than if, in ordinary coordinates x and y, he had sought the tangent without giving himself the equation.[15]

However that may be, if this road leads nowhere, it remains to find out how Descartes manages to discover his ovals. To answer this question, it is necessary to read what Descartes himself later says about it.

[14] Indeed, if one notes in addition that $NF = c$ and $FE = d$, one has $DF = a - c - x$ and the triangle EFD yields $EF^2 = FD^2 + DE^2$, whence $DE^2 = d^2 - (a - c - x)^2$. Continuing the calculation quickly shows that one gets nowhere.

[15] *Mémoires scientifiques*, vol. VI, p. 334.

Several years later, he indeed returns to the study of ovals, but with a whole new rigor, and he gives the procedure for drawing them. But contrary to what one might expect, instead of setting this study in its natural place, that is, in the *Dioptrique*, and more precisely in the eighth discourse devoted to 'the shapes that transparent bodies must have in order to divert rays through refraction', it is into the *Géométrie* that he integrates it. This move is all the more surprising since this eighth discourse of the *Dioptrique* seeks to examine the conditions such that a ray originating parallel to the focal axis of a central conic passes through one of its foci. There, Descartes studies several cases, those of a lens with an elliptical or hyperbolic surface, as well as those of lenses with two analogous surfaces, or two surfaces, one of which is elliptical and the other hyperbolic. Thus it is clearly in this chapter that Descartes should have placed his examination of ovals, all the more so since he knew that the ellipse and the hyperbola are special cases of the latter. This, however, is not what he does. The only reason to which Descartes appeals to justify the absence of his study of ovals from the *Dioptrique* is that of the 'commodité', that is, the ease or convenience, of his exposition. He writes:

> And, furthermore, we can still imagine an infinity of other lenses, which like the above mentioned, cause all the rays which come from a certain point, or tend toward a point, or are parallel, to be exactly changed from one of these three dispositions to another. But I do not think I need to speak of them here, because I shall be able to explain them afterwards more conveniently in the *Géométrie* [...].[16]

Surely you will grant me that this insufficiently emphasized displacement is not neutral. At the very least, it implies that, from now on, it is the geometric properties of ovals that carry the day, besting the usage that optics makes of them. In other words, the displacement of this discussion is not governed solely by the 'convenience' of the exposition, but by deeper reasons. What are they? Descartes does not place his study of ovals randomly in the *Géométrie*, but precisely at the end of Book 2. He offers no explanation for this choice: in his own words and rather abruptly, he 'adds' this study to the end of the second book. Now the latter is completely devoted to the new theory of curves that are admissible in geometry, that is, algebraic curves. The latter are either generated 'organically' (read: instrumentally), according to Descartes, by his proportional compasses, or are

[16] *La Dioptrique*, in *Œuvres de Descartes*, ed. C. Adam and P. Tannery, vol. VI, Paris, J. Vrin, 1965, p. 185; English transl. in P. J. Olscamp, *Discourse on Method, Optics, Geometry and Meteorology*, Indianapolis, 1965, pp. 141–2.

obtained mathematically as solutions of the famous problem of Pappus. Incidentally, when Descartes found a new curve, he did not fail to verify that it was indeed a solution to the problem of Pappus. Thus he demonstrates that the 'trident', that is, the cubic defined by the equation $y^3 - 2ay^2 - a^2y + 2a^3 = axy$, is a solution of Pappus's problem for five straight lines.[17]

For his ovals, Descartes does not undertake a similar verification, which raises a second difficulty for the historian of Descartes the scientist. He is satisfied with introducing his ovals in the following terms:

> For the rest, so that you may be aware that considering the curve lines here proposed, is not without usefulness, and that they have diverse properties which concede nothing [in value] to those of conic sections, I wish to add here an explanation of certain ovals, that you will see to be very useful for the theory of Catoptrics and Dioptrics.[18]

Thus, to the 'convenience' Descartes had invoked in his *Dioptrique*, he adds here utility without any intrinsic reason justifying the placement of ovals at the end of the second book, and without determining their algebraic nature.

Now we know that, if the most general cubics are indeed solutions of Pappus's problem, the same cannot be said of quartics. Thus, if one limits oneself to ovals, one knows that the ovals of Cassini (1680), for example, those of the equation,

$$\left(x^2 + y^2\right)^2 + 2a^2\left(y^2 - x^2\right) + a^4 - b^4 = 0, \quad \text{with } a, b > 0$$

– convex quartics – cannot be defined by Pappus's problem.[19] The verification alluded to above for the ovals was therefore vital, but Descartes does not even attempt it.

Our question thus becomes precise: apart from 'convenience' and 'utility', what other reasons might have led Descartes to place his ovals at the end of the second book of his *Géométrie*? To answer this question requires that first one know the road that led him to discover them.

[17] See above, 'The first classifications of curves'.

[18] *La Géométrie*, A.T. VI, p. 424; English transl. P. J. Olskamp, p. 215 (slightly edited).

[19] See above 'The first classifications of curves'.

3. About the path he followed, Descartes says almost nothing. But his contemporaries – Fermat and his successors, Huygens,[20] Newton,[21] de L'Hôpital,[22] Reyneau,[23] etc. – had glimpsed it. According to this tradition, Descartes would have discovered his ovals thanks to the inverse method of

[20] Christian Huygens, *Traité de la lumière*, Paris, Gauthier-Villars, 1920, pp. 136–8. Huygens begins his analysis with these terms: 'Pour ce qui est de la manière dont M. Descartes a trouvé ces lignes (les ovales), puisqu'il ne l'a point expliquée, ni personne depuis que je sache, je dirai ici, en passant, quelle il me semble qu'elle doit avoir été' (p. 136). He gives of Descartes's analysis a description in the same spirit, suggested by Fermat's remark.

[21] Isaac Newton, *Philosophiae Naturalis Principia Mathematica*, 3rd edn. (1726) with variant readings, assembled and edited by Alexandre Koyré and I. Bernard Cohen, Cambridge, Mass., Harvard University Press, 1972, vol. I, pp. 344–5. This is what he writes: 'Let *A* be the place from which the corpuscles diverge, *B* the place to which they should converge, *CDE* the curved line that – by revolving about the axis *AB* – describes the required surface, *D* and *E* any two points of that curve, and *EF* and *EG* perpendiculars dropped to the paths *AD* and *DB* of the body. Let point *D* approach point *E*; then the ultimate ratio of the line *DF* (by which *AD* is increased) to the line *DG* (by which *DB* is decreased) will be the same as that of the sine of the angle of incidence to the sine of the angle of emergence. Therefore the ratio of the increase of the line *AD* to the decrease of the line *DB* is given; and as a result, if point *C* is taken anywhere on the axis *AB*, this being a point through which the curve *CDE* should pass, and the increase *CM* of *AC* is taken in that given ratio to the decrease *CN* of *BC*, and if two circles are described with centers *A* and *B* and radii *AM* and *BN* and cut each other at *D*, that point *D* will touch the required curve *CDE*, and by touching it anywhere whatever will determine that curve. Q.E.I.

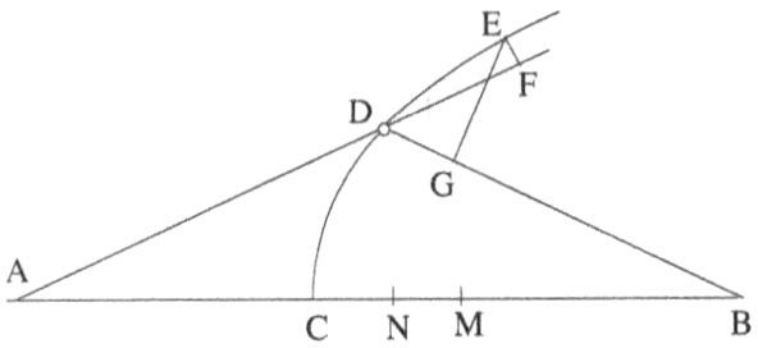

Fig. 28

Corollary I. By making point *A* or *B* in one case go off indefinitely, in another case move to the other side of point *C*, all the curves which Descartes exhibited with respect to refractions in his treatises on optics and geometry will be traced out. Since Descartes concealed the methods of finding these, I have decided to reveal them by this proposition.' (Isaac Newton, *The Principia: Mathematical Principles of Natural Philosophy*, a new translation by I. Bernard Cohen and Anne Whitman, assisted by Julia Budenz, Berkeley, University of California Press, 1999, pp. 626–7.)

[22] Le Marquis de l'Hôpital, *L'Analyse des infiniment petits*, Avignon, 1768, p. 183–4.

[23] C. R. Reyneau, *Analyse démontrée*, 2 vols, Paris, Quillau, 1736–1738, vol. II, pp. xxv and 866.

tangents. If this turns out to be the case, one gets two results at once: one can explain the discovery and one can date the first use of the method. To cite only one representative of this tradition, listen to Fermat:

> Next one could look for the converse of this proposition, and, given the property of the tangent, find the curve to which this property pertains: to which question leads those about the burning lenses proposed by M. Descartes end up.[24]

In June 1638, Fermat thus affirms, in a different language, that it is a problem of integrating a differential equation $f(y, y') = 0$, thus giving priority to Descartes. Later, Newton seems to reach the same conclusion.[25]

If we now return to the *Géométrie*, we note that Descartes distinguishes four types of ovals, whose equations and bipolar coordinates (u, v) are written

$$\left.\begin{aligned} u + kv &= a + kb \\ u - kv &= a - kb \end{aligned}\right\} \quad a + b = d \qquad \text{distance of the poles}$$

$$\text{with } 0 < k < 1$$

$$\left.\begin{aligned} u - kv &= a - kb \\ u + kv &= a + kb \end{aligned}\right\} \quad a - b = d \qquad \text{distance of the poles}$$

These constitute only two ovals, for the first is identical to the fourth, and the second to the third.

Furthermore, he explains how to construct each of them, point by point, by means of a straightedge and compass. Only one example suffices, that of the first oval. Even if the citation is a little long, here are Descartes's own words.

> First, having drawn the straight lines FA and AR, which intersect at point A (at what angle, it does not matter) I take on one of them the point F at random – that is, more or less removed from the point A, according as I want to make these ovals greater or smaller – and from this point F as center, I describe a circle which cuts FA at a point a little beyond A, such as at point 5. Then from this point 5, I draw the straight line 56, which cuts the other line at point 6, so that A6 is less than A5 in any given ratio we may wish (such as that which measures refractions, if we wish to use the oval in optics). After this, I also take point G in the line FA, on the same side as point 5, and at random – that is, by making the lines AF and GA have

[24] *Œuvres de Fermat*, published by Paul Tannery and Charles Henry, Paris, 1891–1922, vol. II, p. 162.

[25] See note 22.

between them any given ratio I might want. Then I make RA equal to GA in the line A6, and from the center G, I describe a circle whose radius is equal to R6. This circle will cut the other at the two points [marked] 1, through which the first of the required ovals must pass. Then again, from the center F, I describe a circle which passes through FA a bit nearer to or farther from point 5 – for example, at point 7 – and having drawn the straight line 78 parallel to 56, from the center G, I describe another circle whose radius is equal to the line R8, and this circle cuts the one that passes through point 7, at the points 1,1, which are again points of the same oval. Thus we can find as many other points as we may wish, by again taking other lines parallel to 78, and other circles with F and G as centers.[26]

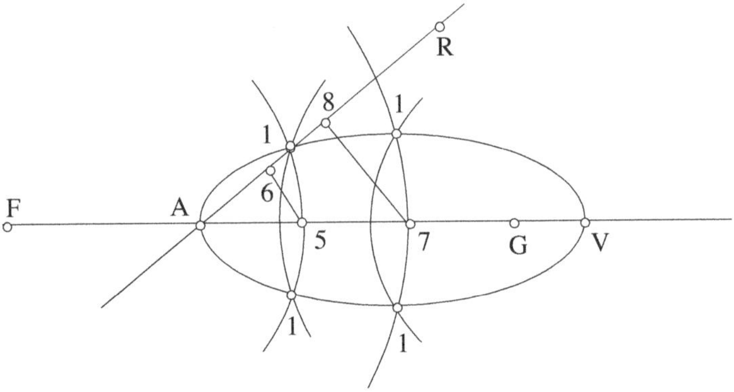

Fig. 29

Thus if one takes F and G and poles and if one posits the coordinates of point 1

$$u = G1 \text{ and } v = F1,$$

one obtains

$$u = G1 = R6 = RA - A6 = RA - k\,A5 = AG - k\,5A$$

and

$$v = F1 = F5 = AF + A5;$$

one has $u + kv = AG + k\,AF$ constant for $AG + AF = d$, the distance of the poles, with $0 < k < 1$.

The preceding exposition from the *Géométrie* leads us to translate our initial question thus: is the path that led Descartes to the discovery of his ovals that of the inverse method of tangents as Fermat, Newton, and others

[26] *La Géométrie*, A.T. VI, pp. 424–5; English transl. P. Olskamp, p. 216.

have suggested? To us the conjecture seems well grounded: let us try to establish it.

One considers two points A and B in a plane, and one seeks a curve separating two media of different transparency in which A and B are respectively located; and such that all of the light rays coming from A are refracted towards B. One assumes that the law of refraction and the index of refraction k are both known.

If Mm is tangent at a point M to the curve that is sought, one projects m at P on AM and at Q on MB, such that the angle of incidence i is equal to $M\hat{m}P$ and the angle of refraction r is equal to $M\hat{m}Q$. Thus one has

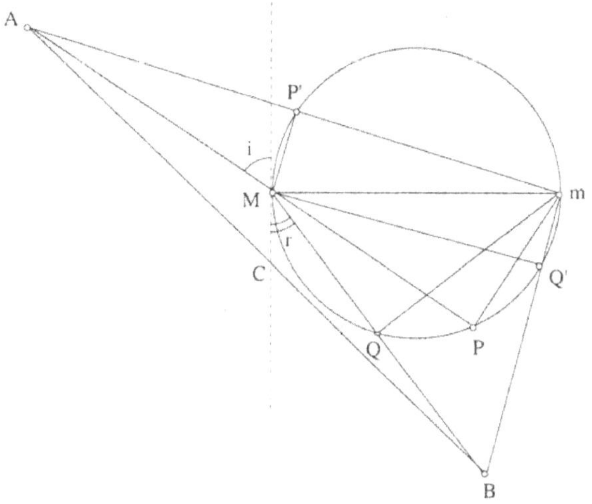

Fig. 30

$$\sin i = \frac{MP}{Mm}, \quad \sin r = \frac{MQ}{Mm},$$

and the law of refraction is written

$$(1) \qquad MP = k \cdot MQ.$$

Also, project M at P' on Am and at Q' on Bm. If Mm is infinitely small, one can assimilate MP to mP' and MQ to mQ'.

Indeed, by *Elements* III.36, one has

$$\frac{PM}{AM} = \frac{P'm}{AP'} \text{ and } \frac{QM}{QB} = \frac{Q'm}{BQ'};$$

and

$$k = \frac{\sin i}{\sin r} = \frac{PM}{QM} = \frac{P'm}{Q'm} \cdot \frac{AM}{AP'} \cdot \frac{BQ'}{QB}.$$

Now $\frac{AM}{AP'} \cong 1$, $\frac{BQ'}{QB} \cong 1$ when Mm tends to 0.

Descartes could perfectly well carry out this type of reasoning, which was commonly used in the tradition of infinitesimal mathematics. Equation (1) then becomes approximately:

$$(2) \qquad mP' = kmQ'.$$

Now mP' is the increase of AM when one passes from M to m, assimilated to a point of the curve, by 'adequality' (*adégalité*), according to Fermat's expression, whereas mQ' is the concomitant diminution of BM. Equation (2) thus signifies that the infinitesimal variation of $z = AM + kBM$ is zero.

It is easy to deduce that the quantity z is a constant. This result is a particular case of the inverse method of tangents, that is, the integration of a differential equation. Here the equation is particularly simple, since it is written $z' = 0$. One will have to await the end of the 18th century to feel the need to demonstrate that this equation implies the constancy of z (theorem of finite increases).

Now the equation

$$(3) \qquad AM + k\,BM = \text{a constant}$$

defines (in bipolar coordinates) a first type of Descartes's ovals. If the curve meets the straight line AB at C between A and B, one has $AB = a + b$ where $a = AC$ and $b = CB$, and the constant of equation (3) is $AC + k\,BC = a + kb$.

If the curve meets the straight line AB beyond A, one has on the contrary $AB = b - a$; and if the curve meets AB beyond B, one has $AB = a - b$. In all of these cases, the equation of the oval is written

$$(4) \qquad AM + k\,BM = a + kb.$$

We note that, in the case of the refraction limit for which the angle r is zero, equation (4) reduces to $AM = a$ and the oval degenerates into a circle of center A.

Another possibility must be considered, in which A and B are on the same side of the normal to the curve at M (this case does not correspond to a physical refraction). mP' and mQ' are then the concomitant increases of AM and of BM, and equation (2) means that the infinitesimal variation of $z = AM - k\,BM$ is zero. From this, one then deduces that

$$(4') \qquad AM - k\,BM = a - kb,$$

using the same reasoning as above.

As we have just reconstituted it, Descartes's procedure shows the first appearance of the inverse method of tangents. It rests completely on intuitive infinitesimal considerations, without making any use of coordinates, as one can easily check in the *Géométrie*.

To demonstrate analytically the equations of Descartes's ovals, one can proceed as follows.

One considers A as the origin and AB as the x-axis of a system of orthogonal coordinates. Let $AB = d$, $AM = \rho$, $BM = \rho'$, whence M is a point of the plane, with Cartesian coordinates x, y; one therefore has

$$\rho = \sqrt{x^2 + y^2}, \quad \rho' = \sqrt{(x-d)^2 + y^2},$$

whence

$$x = \frac{d^2 + \rho^2 - \rho'^2}{2a}, \quad y = \sqrt{\rho^2 - x^2}.$$

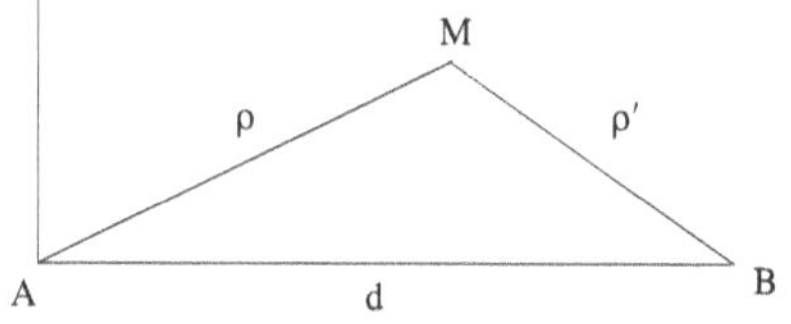

Fig. 31

If $\phi(\rho, \rho') = 0$ is the equation of the sought curve in bipolar coordinates (ρ, ρ'), its equation in Cartesian coordinates is written

$$(5) \qquad f(x,y) = \phi\left(\sqrt{x^2 + y^2}, \sqrt{(x-d)^2 + y^2}\right) = 0.$$

A direction vector of the tangent to this curve has as its components

$$\frac{\delta f}{\delta y} = f'_y \text{ and } -\frac{\delta f}{\delta x} = -f'_x,$$

therefore the sines of the angle of incidence and of the angle of refraction are given, ignoring the sign, by

$$\sin i = \pm \frac{xf'_y - yf'_x}{\sqrt{x^2+y^2}\sqrt{f'^2_x + f'^2_y}}, \quad \sin r = \pm \frac{(x-d)f'_y - yf'_x}{\sqrt{(x-d)^2+y^2}\sqrt{f'^2_x + f'^2_y}}.$$

The sine law is therefore written

$$\rho'(xf'_y - yf'_x) = \pm k\rho((x-d)f'_y - yf'_x). \tag{6}$$

Now one has

$$f'_x = \frac{\delta\phi}{\delta\rho}\cdot\frac{\delta\rho}{\delta x} + \frac{\delta\phi}{\delta\rho'}\cdot\frac{\delta\rho'}{\delta x} = \frac{x}{\rho}\cdot\frac{\delta\phi}{\delta\rho} + \frac{x-d}{\rho'}\cdot\frac{\delta\phi}{\delta\rho'}$$

$$f'_y = \frac{\delta\phi}{\delta\rho}\cdot\frac{\delta\rho}{\delta y} + \frac{\delta\phi}{\delta\rho'}\cdot\frac{\delta\rho'}{\delta y} = \frac{y}{\rho}\cdot\frac{\delta\phi}{\delta\rho} + \frac{y}{\rho'}\cdot\frac{\delta\phi}{\delta\rho'}.$$

After eliminating identical terms and dividing by yd, equation (6) becomes

$$\frac{\delta\rho}{\delta\rho'} = \pm k\frac{\delta\phi}{\delta\rho}. \tag{7}$$

It is this partial derivative equation that one must integrate. One uses the new variables

$$u = \frac{1}{2}(\rho + k\rho'), \qquad v = \frac{1}{2}(\rho - k\rho')$$

such that $\rho = u + v$ and $\rho' = \dfrac{u-v}{k}$.

If $\psi(u,v) = \phi\left(u+v, \dfrac{u-v}{k}\right)$, one has

$$\frac{\delta\psi}{\delta u} = \frac{\delta\phi}{\delta\rho} + \frac{1}{k}\cdot\frac{\delta\phi}{\delta\rho'} \text{ and } \frac{\delta\psi}{\delta v} = \frac{\delta\phi}{\delta\rho} - \frac{1}{k}\cdot\frac{\delta\phi}{\delta\rho'}.$$

Thus, equation (7) with the plus sign means that $\frac{\delta\psi}{\delta v}=0$ and, with the minus sign, that $\frac{\delta\psi}{\delta u}=0$.

In bipolar coordinates, the equation of the sought curve, $\phi(\rho,\rho')=0$, is translated as $\psi(u,v)=0$; whence, in the first case (+ sign), u = constant, and in the second (– sign), v = constant. Finally, one finds the equations

$$\rho + k\rho' = \text{a constant} = a + kb$$

and

$$\rho - k\rho' = \text{a constant} = a - kb$$

which define the two types of Descartes's ovals.

If one returns to Cartesian equations, the preceding equations become

$$\sqrt{x^2+y^2} \pm k\sqrt{(x-d)^2+y^2} = c,$$

that is,

$$\left((1-k^2)\,(x^2+y^2)+2k^2xd+c^2-k^2d^2\right)^2 = 4c^2(x^2+y^2),$$

the equation of a bicircular quartic with two connected components that are convex curves.

4. Descartes gave this equation neither in his *Excerpta Mathematica*, nor in the *Géométrie*, even though he easily could have done so: one need only take as axes *FA* and the perpendicular drawn to point 1 onto *FA*.[27] Descartes gives the curve by a path different from the one he took in his *Géométrie*. The absence of the equation seems to derive less from a weakness than from the very nature of Descartes's procedure. He had obtained his ovals while solving an optical problem; he did not define them by equations, but from their infinitesimal properties. Did he think that the inverse method of tangents was a third way of obtaining curves, some of which are algebraic? In the case studied here, he knew perfectly well that

[27] Cf. *Commentaires sur la* Géométrie *de M. Descartes*, par le R. P. Claude Rabuel, Lyon, 1730, p. 353; *The Geometry of Descartes*, with a facsimile of the first edition, translated from the French and Latin by D. E. Smith and M. L. Latham, New York, Dover Publications, 1954, p. 135; J. F. Scott, *The Scientific Work of René Descartes (1596–1650)*, London, Taylor & Francis, 1952, pp. 125–6.

the curve obtained is algebraic, without needing to make the calculation explicit. The result is that he does not try to demonstrate that these curves can be obtained as solutions to the problem of Pappus.

One argument that supports a positive response is found in the answers that Descartes gave to questions from de Beaune, and notably in the first, the formulation of which is missing from de Beaune.[28] It concerns a geometrical locus defined by a simple relation, if one can trust Beaugrand's testimony from the fall of 1638.

At first, it is a problem of tangent construction, which Beaugrand presents as follows.

Let *A* be the apex of the curve, *S* the middle of *AY*, *E* the projection of any point *M* of the curve, *XM* the tangent at that point; one assumes that *SE*, *EA*, and *ME* are in continuous proportion.

Beaugrand shows that[29]

$$\frac{YE}{SA} = \frac{AE}{XA},$$

which can be rewritten as

$$\frac{SE}{SA} = \frac{XE}{XA};$$

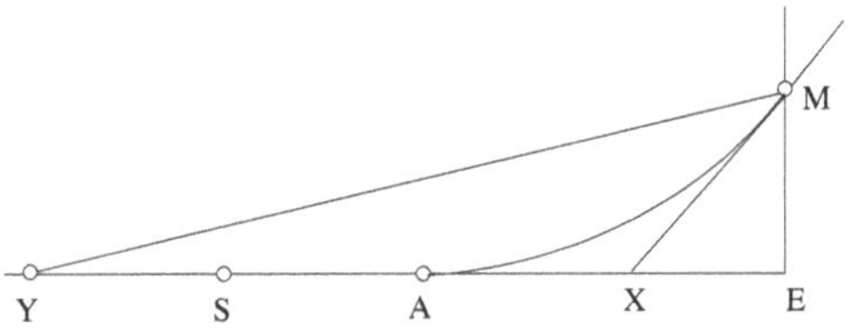

Fig. 32

S and *X* are thus harmonic conjugates with respect to *A* and *E*.

Let $SA = b, AE = x, EM = y, XE = s$, the subtangent; one has

$$(*) \qquad s = \frac{x^2 + bx}{x + 2b}.$$

The problem thus comes down to finding *y* from *s*. It is in this regard that Descartes writes to de Beaune on 20 February 1639:

> About your curved lines, you have sent me a demonstration of a property that seems so beautiful to me that I prefer it to the quadrature of the parabola discovered by Archimedes. For he was examining a given line, whereas you determine the area contained by one that is not yet given.[30]

He continues:

[28] P. Tannery, *Mémoires scientifiques*, VI, pp. 461–5.

[29] *Œuvres de Fermat*, IV, p. 109–10.

[30] *Correspondance*, in *Œuvres de Descartes*, ed. C. Adam and P. Tannery, vol. II, Paris, J. Vrin, 1975, pp. 513–14.

I do not believe it is possible to find in general the converse of my rule for tangents, nor of the one that Mr. Fermat uses, although the usage is in several cases easier than mine. But one can deduce from it *a posteriori* theorems that extend to all curved lines that are expressed by an equation, in which one of the quantities x or y has no more than two dimensions, even though the other were to have a thousand; and I found almost all of them by researching your preceding second curved line; but since I wrote them down only in rough drafts that I did not save, I cannot send them to you.[31]

In order to solve such a problem, Descartes assigns *a priori* a form for the equation of the sought curve, calculates the subtangent, and tries to identify it with the proposed subtangent – s in this case – using the method of indeterminate coefficients; finally, he obtains the equation of the curve, which is a hyperbola. To clarify his procedure, let us apply this algorithm to the case of de Beaune's problem:

Let

$$y = Ax^m + Bx^{m-1} + \ldots + Mx + L$$

and

$$s = \frac{Ax^m + Bx^{m-1} + \ldots + Mx + L}{mAx^{m-1} + (m-1)Bx^{m-2} + \ldots + M}$$

and identify with $\dfrac{x^2+x}{x+2}$ with $b = 1$ in (*) above.

This is impossible, because one will obtain $Ax^{m+1} = mAx^{m+1}$ or $A = 0$; $m = 1$ would yield $s = x + \dfrac{B}{A}$.

Then let

$$y = \frac{Ax^m + Bx^{m-1} + \ldots + Mx + L}{x+p}$$

and

$$s = \frac{\left(Ax^m + Bx^{m-1} + \ldots + Mx + L\right)(x+p)}{\left(mAx^{m-1} + \ldots + M\right)(x+p) - \left(Ax^m + \ldots + L\right)}$$

[31] *Ibid*., p. 514. The curves that Descartes has in mind here have equations of the form $y^2 = P(x)$, where P is a polynomial; they are now called hyperelliptical. Note also that Descartes had sought *a posteriori* certain differential equations that he was trying to integrate: notably some of the form $2yy' = P'(x)$.

and, by identification, one obtains

$$A = mA - A \text{ and } m = 2,$$

therefore

$$y = \frac{Ax^2 + Bx + C}{x + p}$$

and

$$s = \frac{(Ax^2 + Bx + C)(x + p)}{(2Ax + B)(x + p) - (Ax^2 + Bx + C)}.$$

Identify; one obtains

$$Ax^4 + (2A + B + pA)x^3 + (2Ap + 2B + pB + C)x^2 + (2pB + pC + 2C)x + 2pC$$

$$\equiv Ax^4 + (2Ap + A)x^3 + (Bp + 2Ap - C)x^2 + (Bp - C)x.$$

The calculation yields $B = 0$ and $p = 1$ and one has

$$y = A \cdot \frac{x^2}{x + 1},$$

which is the solution of the problem.

Descartes's ovals and de Beaune's first curve are different: the former are quartics, whereas the latter is a hyperbola; the former are algebraic, whereas the hyperbola, which is also algebraic, belongs to a group of problems (those raised by de Beaune) that includes transcendental curves. They nevertheless have several points in common. The ovals as well as this hyperbola were suggested by optics. Beaugrand, who was an eye witness, as it were, claims that the problem of de Beaune was raised during research on the determination of the tangent 'which he [de Beaune] made known that he needed for some purposes related to optics'.[32] Moreover, all of these curves – the ovals as well as those of de Beaune – had been defined by a property characteristic of the tangent and not by their equations. Finally the method that was followed seems to have been the same for all: the inverse method of the tangents.

[32] *Œuvres de Fermat*, IV, p. 110.

In order to obtain these curves, which were assumed to be determinate, Descartes was thinking, already before 1629, of starting from the properties of the tangent drawn to these curves, and not from a property characteristic of their points. At this date, did he already have the method of inverse tangents? Probably not yet, but he may already have glimpsed it 'through a glass darkly', as it were, and intuitively when he was treating ovals. In 1637, the situation is entirely different. To us it seems that he was indeed in possession of this method for obtaining his ovals. But why, then, does he not discuss it in the *Géométrie* when he treats the ovals? Why did he not include it among the methods that he had established in order to obtain curves?

When doing history, everyone is aware of how thorny, if not dangerous, these negative questions are. But if we are granted the right of reasoning about them, we might propose the following explanation: Descartes knew that the method of inverse tangents makes it possible to obtain both algebraic and transcendental curves. Indeed he applies it himself to de Beaune's second problem, which leads to a logarithmic curve, and it was only when he failed that he chose a direct method. In any case, this is what transpires from his famous letter of 20 February 1639. This fact would suffice to exclude this method from the *Géométrie*, which admits only 'geometrical curves'.

But there may be another reasons for not including this method. For Descartes, one had to know how to decide *a priori* if the curve obtained by his method was algebraic or not. This would require that one have methods for integrating differential equations, and therefore to know the connections between differentiation and integration. But nothing indicates that Descartes knew of these methods, which others would invent only later. Moreover, the fact that he was attached to the subtangent was scarcely conducive to such knowledge. To consider only technique, it is much clumsier to manipulate the subtangent than the slope of the tangent. Not until Leibniz would this methodological upheaval take place.

In short, with the inverse method of tangents, Descartes had at his disposal means different from those offered by his compasses and the problem of Pappus in order to obtain other curves; but he did not have the technical tools to decide *a priori* if the curve obtained was algebraic or not. He could use this method here and there, but not in 'general', as he himself states. The philosopher of the *Discourse on Method* certainly could not welcome into his *Géométrie* such a method for obtaining curves. Since he nevertheless knew that his ovals are algebraic curves, he therefore introduced them at the end of Book 2, but 'silently' and with no other explanation. We can now understand why.

– 6 –

DESCARTES AND THE INFINITELY SMALL

Prominent among the recurrent questions in the ancient and classical history of the philosophy of mathematics are the following: how much certainty can be attained by a knowledge of infinity or by a branch of knowledge that makes use of infinity or is concerned with infinite processes? Can the infinite be rationally known? Is it susceptible of rigorous proof?

These problems surfaced early in the development of both of philosophy and the philosophy of mathematics under a variety of guises and names. One variously finds references to conception and proof, imagination and proof, comprehension and proof. Underlying each of these designations, however, is an opposition between demonstration and an act of understanding, not merely that of some sensory faculty. Philosophers and mathematicians had no difficulty agreeing that anything conceivable or understandable could be subjected to proof. But can one subject to proof, and thus endow with certainty, what can neither be conceived nor understood? Or must it remain forever beyond the reach of all demonstration?

Such were the questions that the mathematicians raised, particularly in regard to infinite processes. Geminus (d. *c.* 70 BC) had done so in connection with the asymptote to a hyperbola.[1] It could be demonstrated rigorously that the asymptote and the curve constantly draw ever closer to each other without meeting at infinity. Yet it proved impossible either to form any conception of behaviour that was asymptotic at infinity or to

[1] Cf. R. Rashed, 'Al-Sijzī et Maïmonide: Commentaire mathématique et philosophique de la proposition II-14 des *Coniques* d'Apollonius', *Archives internationales d'histoire des sciences,* no. 119, vol. 37, 1987, pp. 263–96; English transl. 'Al-Sijzī and Maimonides: A Mathematical and Philosophical Commentary on Proposition II-14 in Apollonius' *Conic Sections*', in R. S. Cohen and H. Levine (eds), *Maimonides and the Sciences*, Dordrecht/Boston/London, Kuwer Academic Publishers, 2000, p. 159–72; *Œuvre mathématique d'al-Sijzī*. Volume I: *Géométrie des coniques et théorie des nombres au Xe siècle*, Les Cahiers du Mideo, 3, Louvain/Paris, Éditions Peeters, 2004; 'L'asymptote: Apollonius et ses lecteurs', *Bollettino di storia delle scienze matematiche*, vol. XXX, fasc. 2, 2010, pp. 223–54.

understand it. Following Geminus, Proclus discussed the question at length in his *Commentary on the First Book of the Elements*. Since then, mathematicians and philosophers have continued to pursue the matter in Arabic, in Latin, and subsequently in various vernaculars (*e.g.*, Jacques Peletier, Montaigne, and Francesco Barozzi, to name only three).

In mathematics, the problem arises when one attempts to deal with curves and their asymptotic behaviour without having the means of drawing firm conclusions, that is, before the invention of the differential calculus. For that, one had to handle operational concepts of continuity, limit, and the infinitesimal, at least as something infinitely small, if not as the infinitely small part of a variable quantity, that is, as a differential. As the Marquis de l'Hôpital, following Leibniz, defined it, 'the infinitely small portion the variable quantity of which continually increases or diminishes'.[2] Not until the 19th century did topology and real number arithmetic provide the means of solving the problem definitively.

As to Descartes, he could scarcely avoid the problem of the relationship between proof and conception since he himself had undertaken a fresh study of curves and proposed a new classification of them. The distinction that he established between geometrical curves and mechanical curves served as a powerful tool in demarcating the boundaries of algebraic geometry, and consequently in situating his own *Géométrie* precisely within those boundaries. And in philosophy, this same distinction also allowed him to point out the contrast between ideas that are both clear and distinct and those that are merely clear.[3]

As is well known, Descartes introduced an additional item into the debate: imagination. In both the *Regulae* and the *Meditations*, he embarked on a long explanation of the exact role of the imagination and the circumstances in which it has a contribution to make. But in the interval between these two works, Descartes produced the *Géométrie* and the *Dioptrique*, after which he carried out his research on the cycloid and the logarithmic spiral, that is, on the two new mechanical curves. This chronology is important for the history of our problem. Let us first take a moment to consider the way in which imagination affects the relationship between conception and proof.

Throughout his works, Descartes constantly emphasizes that knowledge is the preserve of the understanding. After all, is not the latter

[2] *Analyse des infiniment petits*, Avignon, 1768, Définition II, p. 2.

[3] Cf. R. Rashed, 'La *Géométrie* de Descartes et la distinction entre courbes géométriques et courbes mécaniques', in J. Biard and R. Rashed (eds), *Descartes et le Moyen Âge*, Études de philosophie médiévale LXXV, Paris, Vrin, 1997, pp. 1–26; translated above.

the faculty to which clear and distinct ideas belong? In his faculty psychology, no other faculty, not even the imagination, is equipped on its own to attain knowledge that can be recognized as true. For that, it must be validated by the understanding. If sensory imagination has a part to play in knowing something, it is as an accessory to the understanding depending on what the object of knowledge is, that is, according to the degree to which the object partakes of extension; hence, 'in imagination, the mind contemplates some corporeal form', whereas 'in intellection, it employs nothing but itself'[4] (Replies to the Fifth Objections). Here is what he wrote in the twelfth of his *Rules*:

> we come to the sure conclusion that, if the understanding deals with matters in which there is nothing corporeal or similar to the corporeal, it cannot be helped by those faculties, but that, on the contrary, to prevent their hampering it, the senses must be banished and the imagination as far as possible divested of every distinct impression. But if the understanding proposes to examine something that can be referred to the body, we must form the idea of that thing as distinctly as possible in the imagination: and in order to effect this with greater ease, the thing itself which this idea is to represent must be exhibited to the external senses'.[5]

Moreover, again in 1643, he was writing to Elisabeth that

> body – that is to say extension, shapes and motions – can be known by the understanding alone, but, much better, by the understanding aided by the imagination.[6]

The fact remains, however, that the role of the imagination is reduced and fizzles altogether when the 'objective reality' of the idea, a reality that results in the understanding's treating it as an object of knowledge, includes the infinite or the infinitely small. Descartes observed that it required a particular effort of mind to imagine even something as simple as a pentagon, to say nothing of a chiliagon..., and that once an object has gone beyond a certain degree of complexity, the imagination is unable to cope with it without becoming confused. In exactly the same way, when it comes to the process of division, it does not take long to run out of steam. In regard to infinity, the issue is quickly settled, since it plays no part in Descartes's *Géométrie*, only in his metaphysics. The infinitely small in mathematics is a different matter. When curves are under discussion, it is

[4] *Descartes, Œuvres philosophiques*, ed. F. Alquié, vol. II, p. 832.

[5] *Descartes, Œuvres philosophiques*, ed. F. Alquié, vol. I, pp. 141–2; *The Philosophical Works of Descartes*, transl. E. S. Haldane and G. R. T. Ross, Cambridge, Cambridge University Press, 1931, vol. 1, pp. 39–40.

[6] *Œuvres de Descartes*, ed. C. Adam and P. Tannery (A.T.), vol. III, p. 691.

indeed there, if only between the lines. As for any assistance that the imagination might have provided in this investigation, Descartes apparently has no explanation to offer, even though he had encountered the infinitely small as early as 1619, and in particular in the research he carried out after 1637. What Descartes does not say, however, is as important as what he says. Let us consider the places where Descartes encountered the infinitely small.

1. The first instance had already occurred in 1619, when Descartes was investigating the law of falling bodies, proceeding by means of indivisibles and an intuitive notion of limit. Much more important, however, was his encounter with the infinitely small during his research on anaclastics, sometime before 1629. The problem at hand involves igniting a body using light rays at a given distance. In the case of refraction, this task entailed determining the curve of lenses – as a function of the source. Descartes had examined the problem first (and rather clumsily) in the *Excerpta mathematica*, before taking it up again in the *Dioptrique* and then in the *Géométrie*. To solve the problem he had to invent new curves: ovals. Indeed, he wanted to know on what kind of curve rays originating from a given point must refract in order to reach a given point.

Descartes did not breathe a word about the path that led him to invent his ovals. However, such contemporaries and successors as Fermat, Huygens, Newton, the Marquis de l'Hôpital, and Father Reynaud, among others, guessed completely and precisely what he did. Here is what Fermat has to say on the subject in 1638:

> The next step was to look for the converse of the proposition and, the property of the tangent being given, to look for the curve that this property must fit: this was the question to which Descartes's questions on burning glasses led.[7]

In short, Fermat means that Descartes raised the problem of integrating a differential equation $f(y, y')=0$, thus acknowledging his elder's priority in discovering the method of inverse tangents. Newton and the other mathematicians reached the same conclusion about Descartes's procedure.

Between 1619 and 1629, then, Descartes had more than once come across the infinitely small without, however, pausing to discuss explicitly his conception of it. What is more, before 1629, he was already thinking of starting from the properties of the tangent drawn to curves assumed to be determinate and not from a characteristic property of their points, that is,

[7] *Œuvres de Fermat*, published by Paul Tannery and Charles Henry, Paris, 1891–1922, vol. II, p. 162.

from a property obtained by equation, or by Apollonius's *symptoma* for conics. Even if at this date he did not yet formally have the method of inverse tangents, he nonetheless had an intuitive premonition of it when he was working on ovals. Later, in 1637, the situation was altogether different: it seems that Descartes was using this method as a means of discovering his ovals, which in the *Géométrie*, however, he was keen to express only in algebraic terms. Although once again, he 'comes forward masked (*larvatus*)' in the famous letter to de Beaune dated February 20, 1639, he does let slip a certain amount of information, as we shall see below.[8]

After 1629, we would expect Descartes to be developing and clarifying his conception of the infinitely small, while he was investigating curves and in particular tangents to curves and the space that curves mark off. On the contrary, so far from making his discovery public, during the following years (1629–1637), he keeps what he knows to himself and throws an even more impenetrable veil over the infinitely small. He had two reasons for doing so, one mathematical, the other philosophical. It is to the former that we turn first.

In the years 1628–1629 Descartes sets up the scaffolding that will lead to his *Géométrie*. It was in fact a two-fold project: to work out a geometrical theory of algebraic equations, thereby completing a programme launched six centuries earlier; and to forge the tools required to study algebraic curves by means of their equations. Indeed, he writes: 'to find all the properties of curved lines, it is sufficient to know the relationship between all their points and those of the straight lines, and the way of drawing other lines that intersect all these points at right angles'.[9] This means that all the properties of geometric curves must be deducible from their equations, including the properties, like those of the tangent, that involve infinitely small quantities. The equation itself is formed from exact ratios.

All signs point to the fact that, to realize this project, Descartes tried to proceed *more algebrico* and to bracket the procedures that resort to infinitesimals. It was therefore necessary to find methods that made it possible to get around the latter where necessary, in order to draw only on finite algebraic procedures, the only ones that will make it possible to reach a knowledge that is sure and exact, *i.e.*, clear and distinct. The most telling example of this approach appears in the method of tangents that he presents in his *Géométrie*. As J. Vuillemin correctly notes:

[8] Cf. R. Rashed, 'Les ovales de Descartes', *Physis*, XLII.2, 2005, pp. 333–54; translated above.

[9] A.T. VI, pp. 412–13.

Descartes tolerates no inexactness in the equations of the *Géométrie*; he therefore lets in no procedure, even though it might be empirically fruitful, if it is not based on a clear and distinct intuition of the entire equation. Such is the case for his method of tangents, which is apparently free of every infinitesimal concept – even though it implicitly contains a theory of the tangent as a limit of the secant – and rests only on the algebraic equality of roots.[10]

Descartes is very direct about the importance that he attaches to this method:

> And I venture to say that, in this, we have the most useful and the most general problem not only that I know, but that I have ever wanted to know in geometry.[11]

His statement has nothing rhetorical about it; rather, it clearly expresses the foundation of his project: henceforth, geometrical curves are defined by their equations. It is the latter that make it possible to know their properties (namely, that of the tangents at each of their points) and that also determine their shape. The remaining task is to invent a method to determine tangents and normals. One can easily understand why this declaration and method of Descartes unleashed so many commentaries. I shall try to be as concise as possible.

In the *Géométrie*, where this method appears for the first time, Descartes locates tangents by constructing normals. He had invented this method in order deliberately to avoid reliance on infinitesimal magnitudes. He does not follow the example of De Beaune, who worked directly on tangents, but he draws on the method of indeterminate coefficients. Let us examine Descartes's path by starting with the example of the parabola, which he had not studied in detail, but by retaining the notation he had used for the ellipse, using the very terms he himself employed.[12]

(E) is a branch of the parabola of vertex A and axis AG. Let C be any point on the curve whose orthogonal projection is point M. We want to find the normal to the curve at point C, that is, the straight line PC.

[10] J. Vuillemin, *Mathématiques et métaphysique chez Descartes*, Paris, PUF, 1960, p. 62.

[11] A.T. VI, p. 413.

[12] *Ibid.*

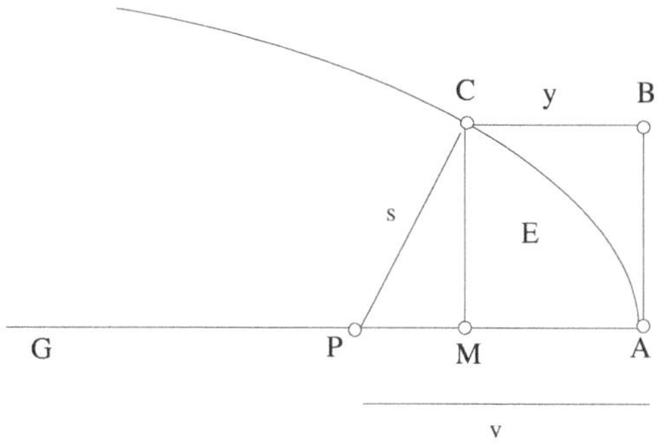

Fig. 33

Let us suppose $PC = s$, $CM = AB = x$, $AM = BC = y$ and $PA = v$. We have $y = rx^2$ by I.11 of Apollonius's *Conics*.

Since Descartes supposes that $y \geq 0$ and $| v - y | \leq s$, the Pythagorean theorem yields

$$x^2 = s^2 - (v-y)^2 \text{ and } y = v - \sqrt{s^2 - x^2}\ ,$$

whence

$$y = r\left(s^2 - (v-y)^2\right) = r\left(s^2 - v^2\right) + 2rvy - ry^2$$

and

$$(*)\quad F(y) = ry^2 - (2rv-1)y - r\left(s^2 - v^2\right) = 0\ .$$

The roots of (*) are the ordinates of the points of the curve and they are the zero values of the second degree polynomial F. Two possible cases arise:

1) Point P is indeed the foot of the normal and the equation $F(y) = 0$ therefore admits a double root y_0 and the polynomial F is factorisable by $(y - y_0)^2$; therefore $F(y) = r(y - y_0)^2$, whence $ry^2 - (2ry - 1)y - r(s^2 - v^2) = r(y - y_0)^2$. By expanding and identifying the terms, we determine the value of v that yields P. This is where Descartes applies what is now called the method of indeterminate coefficients.

2) Point P is not the foot of the normal to the curve at point C; then the equation $F(y) = 0$ has two distinct solutions, which are the abscissas of the two points C and E, intersections of the curve (E) and of the circle of centre P and radius PC; or, as Descartes puts it, 'if point P be ever so much nearer

to or farther from *A* than it should be, this circle must cut the curve not only at *C* but also necessarily at another point.'[13]

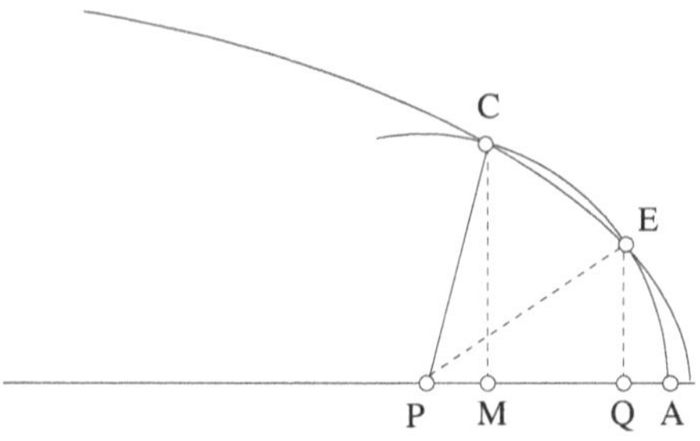

Fig. 34

Here Descartes writes:

> but the closer these two points *C* and *E* are to each other, the smaller the differences between the two roots, and finally the roots are completely equal, if they (the points) are both joined as one, that is, if the circle passing through *C* touches the curve *CE* there without cutting it.[14]

In this last case, the circle 'touches the curve without cutting it'.

To construct the tangent at a given point on a geometric curve involves, first of all, constructing the normal to the curve as the radius of a circle whose centre is on the axis of the abscissa and which intersects the curve at a double point. This method for positioning the tangent would be better described as a 'method for determining the normal'.

This method is valid for curves defined by algebraic equations and is itself formulated in algebraic terms. At first sight, it does not rely on infinitesimal notions such as that of *extrema*. Descartes's reticence about Fermat's method of tangents is understandable, given his conviction that it contaminates the purity of algebra in two respects: firstly, because it rests on the method of *maxima* and *minima* and embodies the notion of 'adequality', *i.e.*, equality at the limit; or, as Albert Girard expressed it in 1625, 'it is not strictly equality as such, but an approximation to

[13] A.T. VI, p. 417.

[14] A.T. VI, p. 418 (my underlining). In his book which has acquired the status of a classic, Y. Belaval seems to be attaching little weight to this expression when he writes: '[...] the discriminant of the equation in x^2 is not an auxiliary quantity, and Descartes does not reduce it to zero by taking it to the limit, he simply posits it as zero' (Y. Belaval, *Leibniz critique de Descartes*, Paris 1960, p. 304).

something'.[15] Second, Fermat seems not to have maintained the distinction between geometrical and mechanical curves, which is so essential to the foundation of algebraic geometry. Just like Descartes, he puts forward his method of tangents for algebraic curves, which is, however, easy to generalize to include mechanical curves, as Fermat would be quick to do. Indeed, as early as August 1638, he writes to Mersenne that his method for determining tangents was, with modest changes, equally applicable to mechanical curves. In 1640, Fermat substitutes the ordinates of the tangents for those of the curves, and the lengths of the tangents discovered, for the arcs of the curves, which will later allow him to determine the tangents to mechanical curves.

But, if we take a closer look at Descartes's method, we notice a certain internal limitation, as well as the more or less clandestine presence of the infinitely small.

1) The method applies within an orthonormal reference frame.
2) Reliance on the Pythagorean theorem involves squaring in order to eliminate the radicals. In this case the degree of the polynomial F broadly attains that of the curve.
3) In factorising by $(y - y_0)^2$, it is necessary to proceed to a certain number of identifications; that number exceeds by one the degree of polynomial F.
4) In addition to these internal operational limitations, a theoretical difficulty arises in the expression 'to touch without cutting', which is not characterized by either its clarity or its precision. Did Descartes mean that the circle would be in contact with the curve? If so he offered no explanation for this concept of contact, which mathematicians and philosophers had constantly discussed since Antiquity. Did he mean that the circle merges with the curve at point C, and that the normal to the circle at this point is also the normal to the curve? Be that as it may, Descartes quickly disposed of the question. This is why the expression 'to touch without cutting' would provoke so many fruitful polemics among such successors as Leibniz and Jean Bernoulli. In the form in which it presents itself, this concept of contact presupposes several others: proximity, neighbourhood, approximation..., and therefore the concept of an infinitely small difference. Now the lapidary phrase, 'to touch without cutting' seemingly manages to avoid bringing up the infinitely small and related concepts, which are nevertheless present, but masked by Descartes's purely algebraic exposition.

[15] Albert Girard, *L'Arithmétique de Simon Stevin de Bruges*, Leiden, 1625, p. 626.

To recapitulate briefly, we can state that Descartes's method of tangents rests mainly on the property of the double point, and on a theorem of Apollonius that he rediscovered and generalised. The double point is obtained by solving a quadratic equation. The theorem in question is Proposition V.31 of Apollonius's *Conics*:

> If a straight line is drawn at a right angle to the extremity of a minimal straight line drawn to any of the conic sections, the aforementioned extremity being the one that is on the conic section, then the straight line drawn is tangent to the section.[16]

This proposition can easily be generalised to convex curves.

But how did Descartes arrive at this theorem? He certainly did not know the fifth book of the *Conics*: the 17th-century reception of the book confirms beyond a doubt that he had heard not so much as a whisper about it. Recall that to establish Proposition V.31, Apollonius had availed himself of a group of propositions in his first book, I.31 to I.37, which easily led to it. Descartes, who was perfectly conversant with the first book of the *Conics*, probably retraced Apollonius's path and then generalised the result. To Descartes, this would have presented no problem at all.

One can readily concede that Descartes was in a better position than anyone else to understand the limitations of his own method as well as the superiority of Fermat's. Moreover, he had admitted to de Beaune that Fermat's method is easier to apply and to use, but leaves something to be desired when it comes to mathematical rigor. The problem then is to find out why Descartes was so keen on his method of tangents.

It would be tempting, if frivolous, to put this preference down to authorial pride, or to some quirk in the scientist's character. It is much more a matter of his firm will to exclude – and rightly so – every object that was not strictly algebraic and every operation that involves an infinitesimal procedure when studying geometrical curves. This bias once again obeys two kinds of reasons: mathematical and philosophical. Indeed, these curves are distinguished from all others by the simplicity of the motions that generate them, by the rigour of the algebraic equations that define them and that characterise all of their points without any exception, and by the precision of the means that allow one to plot them. There is no approximation and no subset of inaccessible points. There is therefore no place for either infinitesimal elements or infinitesimal procedures. One must take care not to introduce any such objects and procedures in order to

[16] *Apollonius*: *Les Coniques*, Tome 3: *Livre V*, historical and mathematical commentary, ed. and transl. from Arabic text by R. Rashed, Berlin/New York, Walter de Gruyter, 2008, p. 318.

preserve algebraic purity, as it were, a purity that Descartes sees as a necessary condition of generality. In his view, geometrical algebra treats only algebraic magnitudes that are necessarily finite. In this field, the understanding makes progress securely and treats no magnitude that it cannot grasp. The understanding cannot conceive of a quantity that is at once different from zero and yet null; nor can it conceive that two quantities whose difference is infinitely small can be equal. These conditions are necessary for our understanding of the properties of these curves to be clear and distinct. This is what philosophy demands.

Thus in the famous controversy that opposed him to Fermat, Descartes tries to rid the latter's method of the infinitesimal and to interpret it along the lines of of two algebraic equations. After having reviewed Fermat's method, he writes to Mersenne on May 3rd 1638, as follows:

> instead of saying simply: they are equalised (*adaequantur*), one should have said: they are equalised (*adaequantur*) in such a way that the quantity to be found by this equation is certainly unique, when one refers it to either the maximal or the minimal quantity, but a unique quantity that comes from the two that might be found by means of the same equation, and which would be unequal, if they were referred to a <line> smaller than the greatest or greater than the smallest.[17]

He comes back to this topic once again in June of the same year in a letter to Hardy in which he writes:

> This, then, is the foundation of the rule in which there are two equations virtually, although it is necessary to mention only the one, because the other serves only to erase these homogeneous elements.[18]

Consequently it was necessary to banish the infinitely small at the risk, as we have seen, of its intruding surreptitiously.

Such a position was fraught with consequences for Cartesian mathematics. Thus, from this point of view, no contribution to the emerging research on the rectification of curves would be forthcoming, which Descartes justifies when he writes:

> for, though [in geometry] no lines are acceptable that are like strings, that is, that become sometimes straight and sometimes curved, because, given that the ratio between straight lines and curves is unknown and even, I believe, beyond human grasp, one could not conclude from this anything exact and certain.[19]

[17] A.T. II, p. 127.
[18] *Ibid.*, p. 173.
[19] A.T. VI, p. 412.

It is perhaps for this reason that Descartes did not try to determine the length of curves. Such research would indeed have required him to proceed by rectification, that is, according to a method that Descartes believed to be beyond human understanding. And besides, even in cases in which the understanding is satisfied with knowledge that is merely clear, like that of mechanical curves, he does not proceed by rectification. Descartes in fact notes that the length of the arc of the logarithmic spiral, from its origin to its extremity, is proportional to the length of its radius vector from its last point. Yet he does not show that this curve is rectifiable (cf. his letter to Mersenne dated 12th September, 1638). Make no mistake, however. Descartes's attitude has nothing to do with that of the ancients, who drew a distinction between curves and straight lines. It is rather a consequence of his own conception of the knowledge of geometric curves and of his position with respect to infinitesimals. This conception and position are eminently positive if the task at hand is to constitute algebraic geometry, but they could only throw up obstacles in the area of infinitesimal analysis.

But, if the infinitely small has no place in clear and distinct knowledge, like that of geometric curves, does it have any role to play in knowledge that is merely clear, like that of mechanical curves? We have shown elsewhere that, if the knowledge is only clear, it is because of the motions that generate these curves, the nature of the equations that define them, and the fact that these equations do not characterise all of their points. These curves are defined by algebraic differential equations.[20] To obtain pieces of an answer, we must turn to Descartes's mathematical writings after 1637. Having completed his grand project, Descartes was free from the powerful algebraic constraints that he had just felt, and was drawn to mathematical research projects in progress, notably those of Roberval and Fermat. He was now available. After 1637 he was ready to return, but at a completely different level, to the questions of his youth, as well as to others pertaining to both mechanical curves and the method of inverse tangents. Tannery[21] and Milhaud have studied his contributions in this domain. J. Vuillemin,[22] and, more recently, Ch. Houzel, have analysed them in detail.[23] They have all emphasized that, in this work, Descartes manipulates the infinitesimal.

[20] See above, 'Descartes's *Géométrie* and the distinction between geometrical and mechanical curves'.

[21] P. Tannery, 'Pour l'histoire du problème inverse des tangentes', 1904, in *Mémoires scientifiques*, published by J. L. Heiberg and H. G. Zeuthen, 1926, vol. 6, pp. 457–77.

[22] G. Milhaud, *Descartes savant*, Paris, 1921, pp. 162 ff.; J. Vuillemin, *Mathématiques et métaphysique chez Descartes*.

[23] C. Houzel, 'Descartes et les courbes transcendantes', in J. Biard and R. Rashed (eds), *Descartes et le Moyen Âge*, pp. 27–35.

Here, I focus on only one question, that of the tangent to the most famous mechanical curve in the 17th century, to wit, the cycloid.

Let us begin by refreshing our memories. The cycloid is the curve described by the trajectory of a point M on the circumference of a circle of radius R that one imagines to roll (without sliding) on an axis Ox, when the circle makes a complete revolution on a straight line segment. Since the movement is uniform, then every displacement of the centre of the circle along a straight line parallel to the base is equal to the angular variation of the circle.[24]

On May 27th 1638, Descartes sends Mersenne his quadrature of a cycloid curve, but his proof is incomplete.[25] Two months later, on 27th July, he sends him with the complete proof. This is the idea on which the latter rests:

> [...] when two figures have the same base and the same height, and when all the straight lines inscribed parallel to the base in the one are equal to those inscribed parallel to the base in the other at similar distances, the one contains as much space as the other.[26]

He goes on:

> Since, if all the parts of one quantity are equal to all the parts of another, the whole is necessarily equal to the whole; and this is a notion so obvious that, I believe, only those possessed of the power to call all things by names that are the opposite of their true ones, are capable of denying it, and claiming that the conclusion can be only approximate.[27]

The tone is firm and bears no resemblance to an approximation; yet here, Descartes does not hesitate to appeal to indivisibles, as his contemporaries since Kepler and Galileo had done. For such a curve, the appeal to the infinitely small is not restricted to the quadrature, but also comes into play in the study of the tangent.

And in fact, scarcely a month later, on August 23rd, Descartes sends Mersenne his method for, as he says, 'finding the tangents to curves described by the motion of a roulette',[28] that is, different kinds of cycloid.

[24] We have arc NM of the circle equal to segment ON. If t is the measurement of angle MCN, the coordinates of point M(x, y) will be $x = R(t - \sin t)$ and $y = R(1 - \cos t) = 2R\sin^2 \frac{t}{2}$; and the equation of the cycloid will be $\sqrt{y(2R - y)}dx = ydy$.

[25] A.T. II, pp. 135–53.

[26] A.T. II, p. 261.

[27] A.T. II, p. 262.

[28] A.T. II, p. 308.

This text displays Descartes's virtuosity in adapting the method applied in his *Géométrie* to mechanical curves.

The method proposed to construct the tangent to the cycloid consists first in determining the normal, as in the case of geometrical curves. In this instance, one obtains the normal thanks to the idea of an 'instantaneous centre of rotation'. According to the idea that Descartes forged, the motion of the generating circle is infinitesimally likened to a rotation at a given instance having as centre the point of contact between the circle and the base. This is a beautiful kinematic method, which Descartes would never have accepted in his *Géométrie*. For mechanical curves, one takes the liberty of using other methods and rights, including the right of letting the imagination roam. Let us pause briefly to examine Descartes' approach.

To construct a tangent at point B of the cycloid, one draws BN parallel to the base DA, intersecting the generating circle CND at N, which is then joined to D, the point of contact between the generating circle and the base. From B, BO is drawn parallel to ND; then BO is the normal sought and the perpendicular BL is the tangent to the curve at B. We may note that BL is parallel to NC, for angle DNC is a right angle.

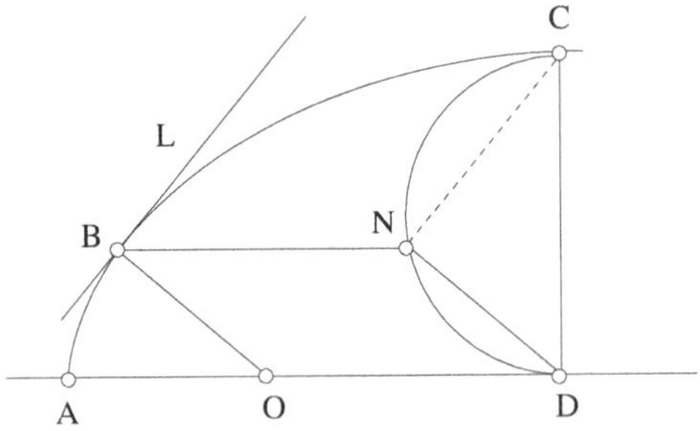

Fig. 35

Later in the letter, Descartes constructs in a similar manner the tangent to the prolate cycloid and the tangent to the curtate cycloid. He goes on to give a demonstration of his method, in which he draws on the concept of an instantaneous centre of rotation. He writes:

> If a rectilinear polygon of any kind is rolled along a straight line, the curve described by any one if its points will be composed of several portions of circles, and the tangents to all the points of each one of these portions of circles will cut at right angles the lines drawn from these points to the point at which the polygon has touched the base in describing the part concerned.[29]

[29] A.T. II, pp. 308–9.

In the case of the hexagon that Descartes considers, the following figure can represent this procedure:

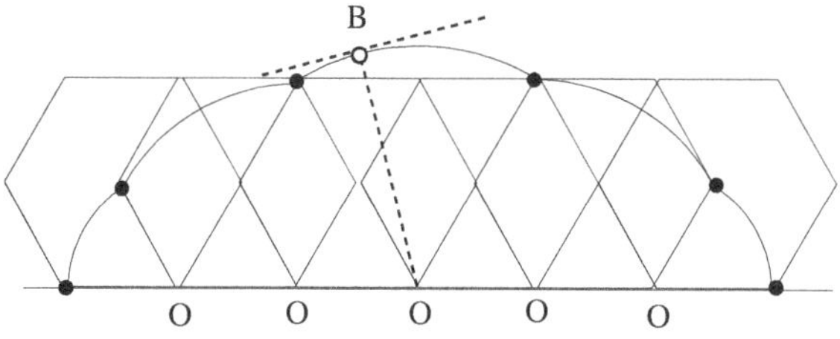

Fig. 36

The curve described by one of the apexes of the polygon as it rolls along is therefore formed by a succession of arcs of circles, the centre of each of which is the point of contact *O* between the polygon and the base, the point about which the polygon pivots. Consequently the normal of a point *B* on the curve is the straight line *BO*, the radius of the arc of the corresponding circle; it passes through the centre of rotation. All the normals to the curve pass respectively through the successive centres of rotation.

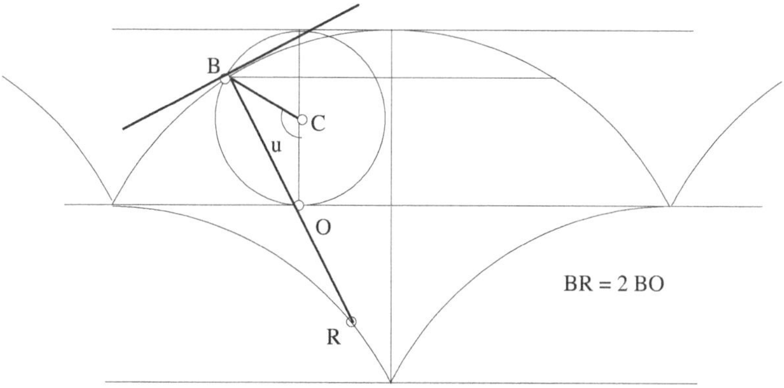

Fig. 37

Descartes continues:

> From which it follows, considering the circular roulette as a polygon with an infinity of sides, that one can see clearly that it must have the same property, that is, that the tangents at each of the points that are on the curve that it describes must cut at right angles the lines drawn from these points to the

points on the base that are touched by it at the same time as it describes them.[30]

Descartes justifies his approach thus: 'Now', he says, 'the same thing happens to a polygon with a hundred thousand million sides, and consequently also to a circle'.[31] If indeed the number of sides of a polygon is increased indefinitely, what is obtained at the limit is a circle, since the length of each side tends to zero. In other words, the circle is a regular polygon with an infinite number of sides, each of a length that is infinitely short. These infinitely short sides determine the curvature of the curve by the angles they form with each other.

Later in the same letter, Descartes points out that his method also applies to the general curves generated by rolling a convex curve with a centre of symmetry, like an ellipse, and therefore to certain geometrical curves.[32] He reminds Mersenne that 'one must also notice that the curves described by roulettes are completely mechanical lines that are among those I rejected from my *Géométrie*; this is why it is no wonder that their tangents are not to be found by the rules I set out there'.[33]

Note therefore that Descartes very naturally combines a strong geometrical intuition and a method of infinitesimal reasoning. This type of reasoning and the procedures that it uses are of the same kind as those employed by Archimedes and the Arab infinitesimalists. Indeed all these geometers made use of the infinitely small, because all of them knew that the tangent is the limit of the secant; but Descartes did not explicitly bring differential notions into consideration any more than his predecessors had. He had himself criticised the use of terms like 'to adequal' and, like the ancients, his infinitely small always had a determinate magnitude that never equalled zero.

If one restricts oneself to such remarks, it will be tempting to assimilate Descartes to his predecessors. Indeed, this is what Y. Belaval does when he writes: 'If he is to be fully understood, Descartes should be attached to the Greek tradition of mathematical demonstration, which avoided all consideration of the infinite as something beyond rational evidence.'[34] And yet, we have seen that he uses the infinitely small in several forms. And, moreover, if Descartes is to be treated as one of the Greek geometers, we

[30] A.T. II, p. 309.
[31] *Ibid.*
[32] A.T. II, p. 312.
[33] A.T. II, p. 313.
[34] Y. Belaval, *Leibniz critique de Descartes*, p. 302.

have no way to explain how he came to use the method of inverse tangents, which had attracted him in his youth and to which he would later return.

Indeed, we have seen that, before 1629, in the *Excerpta mathematica*, in order to obtain a curve, he was thinking of starting from the properties of a tangent drawn to this curve, which was assumed to be determinate, and not at all from a property that characterised the points of the curve. This turning point is of prime importance. Although it is true that Descartes did not yet have the method of inverse tangents, he was already glimpsing it and grasping it intuitively. By 1637, he had got hold of this method employing infinitesimals, thanks to which he discovered his ovals, as I believe I have shown.[35]

Freed from his algebraic constraints, as we have just seen, after 1638 he returned to this method in the well-known correspondence about De Beaune's problems:

> I do not believe that it is possible generally to find the converse of either my rule for tangents, or that which Monsieur de Fermat uses either, though in several cases his is more convenient than mine. But theorems can be deduced from it *a posteriori* that extend to all curves that are expressed by an equation, in which one of the quantities x or y has no more than two dimensions, even though the other might have a thousand; and I found nearly all of them in my earlier search for your second curve; [...].[36]

Recall that this second curve is logarithmic and therefore belongs to the class of mechanical curves.

Descartes is thus stating that, in order to obtain the aforementioned curve, he used the method of inverse tangents, which moreover confirms his solution. He states also that, for the time being, he can use this method in some cases, but not yet generally. He recalls that he has also used it for hyperelliptic curves (*i.e.*, those defined by the equation $y^2 = P(x)$, where P is a polynomial, and that he had sought *a posteriori* to solve certain differential equations of the form $2yy' = P'(x)$. Finally, then, he knows that this method enables him to study geometrical as well as mechanical curves, but he cannot state *a priori* the nature of the curve obtained; still less can he know *a priori* whether or not this curve is the solution to Pappus's problem, like some ovals.

I will not linger here on his solutions to de Beaune's problems, which J. Vuillemin in particular studied. Recall, however, that these solutions led Y. Belaval, who denied that Descartes knew this method of infinitesimals, to write about his solution to the second problem: 'No one will dispute that

[35] See above, 'Descartes's ovals'.

[36] A.T. II, p. 514.

in so doing Descartes invented a technique for solving a differential-equation problem'.[37] He goes on: 'But what conforms to his inventive genius is immediately revealed as contrary to his philosophical genius: his philosophy stands in the way of his consideration of the infinitesimal.' All things considered, it does not seem to me that the problem grows out of a clash between two geniuses; it lies elsewhere. We have just observed that one can use the method of inverse tangents to determine both geometrical curves (*e.g.*, ovals) and mechanical curves (*e.g.*, the logarithmic curve) without being able to tell *a priori* the nature of the curve. This last qualification was enough for Descartes to exclude the method from his *Géométrie*, in which he dreamed of obtaining *all* curves by using his compasses and by solving Pappus's problem. The method of inverse tangents is indeed not a general one, nor does it allow one *a priori* to detect the nature of the curve obtained. The philosopher who wrote the *Discourse on Method* could assuredly not let this method enter his *Géométrie*, but he did not hesitate to apply it outside those bounds, above all to mechanical curves. Thus there is not the least contradiction in Descartes's approach. For the study of mechanical curves, it was simply necessary to invent other methods that applied to all curves, even if such methods were not purely algebraic and resorted to such kinematic notions as the instantaneous center of rotation.

By way of conclusion, we can say that if the infinitely small enters into knowledge that is clear and distinct – the kind of knowledge that human understanding can have of the objects of algebraic geometry – it is only stealthily, without an open invitation. In the course of exposition, terms that might draw attention to its unwelcome presence (such as those reminiscent of inverse tangents) are avoided, or at least, only touched on hastily, in such short-hand expressions as 'to touch without cutting'. But in the case of knowledge that is merely clear, like our knowledge of mechanical curves, which is susceptible to gradations and imperfections, the infinitely small is entitled to play its part. In both cases alike, Descartes avoids any positive discussion of the infinitesimal. Even so, we know from his rejection of differential notions, such as Fermat's 'adequality', that he refused to consider the infinitesimal as a magnitude that tends to zero. We also know that, as much from the ontological as from the logical point of view, the understanding can say nothing about either the infinitely large or the infinitely small; it can only talk about the 'indefinite'.

The position that Descartes took is in no way hostile to the infinitesimal. Neither is it contradictory. Contrary to what one sometimes reads, Descartes did not rule out the infinitely small. For mathematical as

[37] Y. Belaval, *Leibniz critique de Descartes*, p. 310.

well as philosophical reasons, however, he wished to restrict both the domain and the modalities of its use. Indeed, clear and distinct knowledge can only be exact and certain. The understanding includes herein both extended substance and its modes and, in its rigour, thinks through the concept of limit. It is consequently unable to understand the infinitesimal as the result of an infinity of repeated divisions of a finite quantity. It can only think of the infinitesimal as a limit, for, although it may be capable of abstracting such a limit, it is incapable of attributing to it the status of an objective and exact quantity – in Descartes's terms, an algebraic quantity. Even so, although the infinitesimal cannot rightfully be considered a concept in algebraic geometry, it can still have a place elsewhere in mathematics: in the examination not only of mechanical curves, but also curves that do not solve Pappus's problem and about which one cannot say *a priori* whether they are algebraic or mechanical. It is certainly present as the limit of the secant, as an indivisible, and implicitly as the infinitely small value of a variable magnitude. Moreover, it does not seem to me exact to say that it is the presence of the infinitely small that makes the knowledge of mechanical curves imperfect: if that knowledge is imperfect, it is because of their very constitution, for the reasons already advanced – the motions needed to generate them, their differential algebraic equations, the inaccessible points. Such are the reasons that excluded these curves from algebraic geometry (not from geometry as a whole), but that nevertheless legitimized the intervention of the infinitely small.

Finally, if we are to circumscribe fully and rigorously the presence of the infinitely small in Descartes's work and the use to which he puts it, we should refrain from judging his methods with the yardstick of Leibniz's infinitesimal method in the form it took after 1675, the date in which he invented the differential calculus, and from then blaming Descartes for not carrying through a project that was not his. Neither the differential calculus nor, *a fortiori*, its algorithms are to be found in Descartes's work. Conversely, as we have seen, what does appear in his work is the infinitely small in a variety of forms and at various theoretical levels. We are on the eve of a well-disciplined study of the infinitely small. Last, but not least, one should never forget that Descartes was a contemporary of Fermat and Wallis and a predecessor of Leibniz and Newton.

– 7 –

FERMAT AND ALGEBRAIC GEOMETRY

Algebraic geometry came to light as a proper chapter of mathematics when al-Khayyām (1048–1131) gave himself the theoretical means (measure, dimension, geometric calculation, etc.) of a two-fold translation: to reduce geometrical problems – notably solids and supersolids – to algebraic equations, and to solve the latter by the intersection of conic curves. If one were to express this double movement in one catch phrase, one might say that the birth of algebraic geometry takes place at the conjunction of al-Khwārizmī's *Algebra*, Apollonius's *Conics*, and Euclid's *Data*. It is not a coincidence that al-Khayyām cites only these three books. The other research efforts he evokes are those of his predecessors who had already trod the same road that he had taken: al-Khāzin, Abū al-Jūd, Ibn ʿIrāq, al-Qūhī.[1] Nevertheless, one can understand nothing about al-Khayyām's action without following the development of algebra during the century and a half that separates al-Khwārizmī from al-Khayyām in order to grasp the role of the seemingly insurmountable obstacle that the algebraists encountered: to solve by radicals cubic and irreducible biquadratic equations. We also need to understand the impact on the theory of conics itself of the numerous new applications of geometrical constructions,[2] as well as on the

[1] R. Rashed and B. Vahabzadeh, *Al-Khayyām mathématicien*, Paris, Librairie Blanchard, 1999; English translation: *Al-Khayyām mathematician*, New York, Eisenbrauns, 2000. R. Rashed, *Les Mathématiques infinitésimales du IXe au XIe siècle*, vol. III: *Ibn al-Haytham. Théorie des coniques, constructions géométriques et géométrie pratique*, London, al-Furqān, 2000, Notes complémentaires, 4: 'Al-Qūhī et le lemme à la division de la droite par Archimède'; English translation: *Ibn al-Haytham's Theory of Conics, Geometrical Constructions and Practical Geometry*. A History of Arabic Sciences and Mathematics, vol. 3, Culture and Civilization in the Middle East, London, Centre for Arab Unity Studies, Routledge, 2013.

[2] *Les Mathématiques infinitésimales du IXe au XIe siècle*, vol. III.

new chapter the continuous drawing of conic curves that al-Qūhī inaugurated.[3]

Scarcely one half-century after al-Khayyām, Sharaf al-Dīn al-Ṭūsī[4] gives the new discipline a more analytical stamp. Concerned about demonstrating the existence of the points of intersection of curves, he elaborates the concept of the maximum of an algebraic expression over an interval. Henceforth, he is interested in some of the algebraic properties of conic curves.

Some mathematicians – notably Kamāl al-Dīn ibn Yūnus, Athīr al-Dīn al-Abharī, etc., and also members of the school of Isfahan in the 19th century – followed this tradition of al-Khayyām and al-Ṭūsī without going beyond it, however. Only in 1637 and in different climes did new foundational acts occur in algebraic geometry. I am, of course, referring to Descartes's *Géométrie*. In that very same year of 1637, a manuscript treatise by Fermat also begins to circulate. Entitled *Ad locos planos et solidos isagoge*, it also focuses on questions of algebraic geometry. Fermat will write a sequence of other treatises in this field until he produces his famous *Dissertatio tripartita* (before 1650).

Faced with this new beginning of algebraic geometry of the 17th century, one cannot avoid raising two connected questions: what were the relations between the contributions of Descartes and Fermat? Did they replicate, at least in part, the duality between al-Khayyām and al-Ṭūsī, and in what sense? Let us begin by briefly recalling the case of Descartes, treated more fully elsewhere.[5]

In the *Géométrie* we can indeed recognize two movements that guided the evolution of both algebraic geometry and Descartes's own book. The first movement focuses precisely on the completion of al-Khayyām's project: to translate geometrical problems into algebraic equations, then to translate the latter into the language of geometry thanks to a genuine geometrical calculus, in order to solve them by the intersection of curves. Both here and in al-Khayyām's book, algebraic geometry presents itself essentially as a geometrical theory of algebraic equations. By contrast, the second movement collects the beginnings of a study of curves by means of

[3] See R. Rashed, 'Al-Qūhī et al-Sijzī: sur le compas parfait et le tracé continu des sections coniques', *Arabic Sciences and Philosophy*, 13.1, 2003, pp. 9–44; and below, 'The continuous drawing of conic curves and the classification of curves'.

[4] R. Rashed, *Sharaf al-Dīn al-Ṭūsī, Œuvres mathématiques. Algèbre et géométrie au XIIe siècle*, Sciences et philosophie arabes – textes et études, Paris, Les Belles Lettres, 1986, see especially the introduction.

[5] See above, 'Descartes's *Géométrie* and the distinction between geometrical and mechanical curves'.

their equations in order to turn it into a new program, the full realization of which would take place in the future.[6]

Along the lines of this first movement, Descartes's *Géométrie* is deployed, as a theory of algebraic equations, as one also encounters it in al-Khayyām. Moreover, thanks to the *Regulae*, to the testimony of Beeckman, and to the *Cogitationes Privatae,* we know that in 1619 Descartes was already involved with this theory.

To pose questions about the algebraic geometry of Fermat, is first of all to know that he belonged to the tradition outlined by al-Khayyām and Descartes. This question is all the more important because it is not rare to see his contribution interpreted as the product of a meeting of Viète and Apollonius.[7] A study of Fermat's writings, however, shows that he only treats this theory of equations much later. This insufficiently emphasized fact already brings out a significant difference with al-Khayyām and Descartes. And it is indeed true, Fermat's first work in this area, that of 1637, distinguishes itself clearly from this tradition of the geometrical theory of algebraic equations: he is essentially concerned with the equations of the straight line, of the circle, and of the conic curves. His starting point is not the theory of algebraic equations, but research on geometrical loci. To understand the contribution of Fermat to this field is first of all to grasp the precise significance of this difference and of its impact which requires us not to restrict ourselves only to his writings in algebraic geometry. We must analyze his first research in geometrical loci. This is his genuine starting point.

1. THE GEOMETRICAL LOCI AND THE POINTWISE TRANSFORMATIONS

Fermat began his mathematical career by writing *The Restitution of Two Books on Plane Loci by Apollonius of Perga* (*Apollonii Pergaei libri duo de locis planis restituti = De locis planis*). In the *Mathematical Collection*, Pappus reproduces certain propositions from this lost book of Apollonius. Fermat thus tries to *restitute* this book. In this case and in other similar ones,[8] one must not confuse the act of 'restituting' and with that of

[6] *Ibid.*

[7] See for example Michael S. Mahoney, *The Mathematical Career of Pierre de Fermat 1601–65*, Princeton, Princeton University Press, 1973.

[8] For example, compare the reconstitution by Ibn al-Haytham in Book VIII of the *Conics* by Apollonius, in *Les Mathématiques infinitésimales*, vol. III, Chap. I.

'restoring'. 'To restore' is the act of an archaeologist or a historian, which Fermat was not, whereas 'to restitute' is the act of a mathematician who operates according to a criterion of apodicticity. His goal is clear: to invent and rigorously demonstrate the propositions that consolidate the ancient contribution by surpassing it. And in fact, an examination of *The Restitution* shows that we are not in the same domain as Apollonius, but in that of his successors. Let us explain.

About this *Restitution*, Fermat writes to Mersenne (on 26 April 1636):

> [...] I have completely restored Apollonius's treatise *On Plane Loci*. Six years ago I gave to M. Prades, whom you perhaps know, the only copy I had of it, written in my own hand. It is true that the prettiest and most difficult problem, which I had not yet solved, was missing. Now the treatise is complete in every point, and I can assure you that in all of geometry *there is nothing comparable to these propositions*.[9]

When citing this letter, scholars often forget to raise two closely related questions: what ought one understand by 'geometry' and in what sense is there 'nothing comparable to these propositions'? These questions impress themselves on us all the more because these propositions seem at first sight to be very simple. In a word, where ought one locate the radical novelty that Fermat so forcefully claims? Fermat will again take stock of his *De locis planis* scarcely a year and a half later in his first work on algebraic geometry. About it, he writes the following in his *Ad locos planos et solidos isagoge*:

> If this discovery had preceded our already long past restoration of the two books *On Plane Loci*, the construction of the theorems of loci would have been much more elegant; we nevertheless have no regrets about this work, even though it was premature and underdeveloped. Indeed, for scholarship there is some advantage in not robbing posterity of works that are still intellectually inchoate; a work that is initially simplistic and crude is strengthened and grows thanks to new discoveries. It is even important for scholarship to contemplate fully the hidden progress of the mind and the spontaneous development of the craft.[10]

In this beautiful passage, the author warns the reader that between the *De locis planis* and *Ad locos planos et solidos isagoge*, there are intimate connections. Starting from the latter, one could indeed find presented in a

[9] M. Mahoney, *The Mathematical Career of Pierre de Fermat*, p. 112. Emphasis is mine. Cf. *Œuvres de Fermat*, ed. Paul Tannery and Charles Henry, 5 vols, Paris, 1896, vol. II, p. 5.

[10] We are referring to *Œuvres de Fermat*, ed. P. Tannery and C. Henry, vol. II, p. 96.

more elegant manner the theories included in the former. According to Fermat, then, there is a certain continuity between the two treatises; but there is also a break caused by 'a new discovery'. Whatever continuity he alleges is in fact broken somewhere, because these two books are not written on the same plane and do not treat the same objects. To genuinely understand Fermat's procedure, we must examine more *De locis planis* and situate it first in the history of geometry. Before beginning this inquiry, let us recall that Fermat insisted that his book was novel on account of both its procedures and its results. Yet there is a significant difference between Apollonius and his successors from the 9th century on. The latter, unlike the Hellenistic mathematicians, have begun to take as their object the study of the transformations of geometrical loci by homothety, translation, similarity, and inversion. Tied to this practice are the names of Ibn Qurra, Ibn Sinān,[11] al-Qūhī, and Ibn al-Haytham,[12] to keep the list short. This research proved to be of great import for algebraic geometry, as in the case of Sharaf al-Dīn al-Ṭūsī.[13] Now, this is precisely the research that Fermat takes up again in his *De locis planis*. Moreover, one cannot avoid being struck by a similarity between the structure of this book and that of Ibn al-Haytham's (d. after 1040) *The Knowns*. In short and to be clear, what he elaborates here is not a geometry in the manner of Apollonius, but another, different one. Let us now turn to the *De locis planis* itself, which is composed of two books, as the title announces. As attested by the letter by Mersenne cited above, the whole had definitely been completed before April 1636, let us say the beginning of 1636[14] or shortly before. According to this same letter, however, the book was almost complete in 1630 because only one proposition was missing.

[11] See *Les Mathématiques infinitésimales du IXe au XIe siècle*, vol. I: *Fondateurs et commentateurs: Banū Mūsā, Thābit ibn Qurra, Ibn Sinān, al-Khāzin, al-Qūhī, Ibn al-Samḥ, Ibn Hūd*, London, al-Furqān, 1996; English translation: *Founding Figures and Commentators in Arabic Mathematics*. A History of Arabic Sciences and Mathematics, vol. 1, Culture and Civilization in the Middle East, London, Centre for Arab Unity Studies, Routledge, 2012. R. Rashed and H. Bellosta, *Ibrāhīm ibn Sinān. Logique et géométrie au Xe siècle*, Leiden, E. J. Brill, 2000.

[12] *Les Mathématiques infinitésimales du IXe au XIe siècle*, vol. II: *Ibn al-Haytham*, London, al-Furqān, 1993; English translation: *Ibn al-Haytham and Analytical Mathematics*. A History of Arabic Sciences and Mathematics, vol. 2, Culture and Civilization in the Middle East, London, Centre for Arab Unity Studies, Routledge, 2012.

[13] *Sharaf al-Dīn al-Ṭūsī, Œuvres mathématiques* (n. 4).

[14] This is confirmed in another letter to Mersenne dated 15 July 1636.

In another letter, sent to Roberval on Monday, 20 April 1637, Fermat writes: 'The second book has been written for eight years already, and during this time I gave two copies, one to Mr. Despagnet, a counselor to the *Parlement* of Bordeaux, and the other to Mr. de ...' (vol. II, p. 105). If one can trust these remarks – and there is every reason to do so – the second book had already been completed in 1629; and if any book was incomplete, it must have been book one. In other words, Fermat began this work in Bordeaux before 1629 and completed it in 1636.

But what was Fermat's goal in this book at this stage of his life as a mathematician? At once the simplest and also the most certain way of answering is to listen to him. In the preamble to book one, he writes: 'It is this theory (of plane loci), the most beautiful of all geometry, it seems, that we have wrenched from oblivion' (vol. III, p. 3). It is as a geometer, not an algebraist, that he intends to develop this geometrical theory of loci. The key question is to discover how and with which tools.

The plan of the book is simple: to take up, one after the other, Pappus's propositions in the translation of Commandino in order to establish and extend it, and to provide a critical commentary. Thus the first book begins with proposition drawn literally from Pappus:

> If one draws two lines, whether from a given point or from two, whether in a straight line, whether parallel, or making a given angle, and finally having a given ratio or comprising a given rectangle: if the end of the one is on a plane locus given in position, the end of the other will also be on a plane locus given in position, in some cases of the same type, in other cases of a different one, in yet other cases similarly located in relation to the straight line, in yet others contrariwise (vol. III, pp. 3–4).

Fermat begins by dividing this proposition into eight others and notes that 'each of the latter (can be divided) into many cases' (vol. III, p. 4).

The examination of these propositions as well as Fermat's entire approach reveals important facts that have not been emphasized. The propositions treat pointwise transformations, an effective tool in the research on loci; conversely, one sees a structural similarity between this research and that carried out by the mathematicians of the 10th to the 11th century, in particular Ibn al-Haytham in his *The Knowns*. But this research on pointwise transformations is new in relation to Hellenistic geometry. To appreciate this point, one need only rephrase in another language the statements of certain propositions.

In Proposition 1.1, Fermat shows that the homothetic figure of a straight line or a circle is a straight line or a circle. In the case of the straight line, his demonstration is fast and sharp. In the case of the circle, he offers a construction that satisfies the conditions. In each of these cases, he effectively examines only a negative homothety. But his remark, cited above, shows that he deliberately ignores the other cases. Now Ibn al-Haytham in *The Knowns* begins in the same way with one difference: he treats the circle before the straight line.

Fermat proceeds in the same manner for the other propositions. Proposition 1.2 concerns the transform of a straight line or a circle by inversion, a proposition comparable to that of the 10th-century mathematician al-Sijzī.[15] In Proposition 1.3, he shows that the figure similar to a straight line or a circle is a straight line or a circle, which corresponds to Propositions 4 and 5 of *The Knowns*. In 1.4, he considers the transformation of a straight line or a circle by the composite of an inversion and a rotation with the same center. In 1.5, he draws on the composite of a homothety and a translation; in 1.6, on the composite of an inversion and a translation; in 1.7, on the composite of a similarity and a translation; and in 1.8 on the composite of an inversion and a rotation.

This quick glance at the first propositions leaves no doubt about Fermat's intention: to study plane loci – straight lines and circles – by means of pointwise transformations. The other propositions of book one of *De locis planis* confirm this intention. Thus Proposition 2, which is the first proposition of Ibn al-Haytham's *The Knowns*, focuses on the circle as the locus of points at a given distance (the radius) from a given point. Proposition 3, which corresponds to Proposition 6 of *The Knowns*, establishes the property of the potential arc; the following one corresponds to the 10th proposition of Ibn al-Haytham's book. Proposition 5, which corresponds to Propositions 11 and 12 of *The Knowns*, treats the transformation of a straight line or circle by a translation. In the first part of Proposition 6, which is analogous but not identical to Proposition 20 of *The Knowns*, he treats the determination of the locus of points such that the ratio of their distances to two given straight lines is given. The second amounts to determining the locus of points such that a certain linear combination of their distances to two given straight lines is constant. Proposition 7 is devoted to research on the locus of points such that their distances to any given number of given straight lines is connected by a linear relation whose coefficients are known. This locus is evidently a straight line. Fermat calls this proposition 'very beautiful' (vol. III, p. 25).

This quick summary of the content of the first book of *De locis planis* shows indubitably that here Fermat is no longer on the same terrain as

[15] *Les Mathématiques infinitésimales du IXe au XIe siècle*, vol. IV: *Méthodes géométriques, transformations ponctuelles et philosophie des mathématiques*, London, al-Furqān, 2002, Appendix I.

Apollonius, but on that of Ibn al-Haytham and the post-Apollonian mathematicians. Is it not clear that here he is systematically developing certain geometrical transformations – homothety, similarity, inversion, translation – and that he determines the transforms of straight lines and of circles by the latter? Moreover the structure of this book curiously reminds one of that of *The Knowns*, which apparently had the same project. The second book of *De locis planis* is completely devoted to research on geometric loci; always straight lines and circles. This book in effect treats three types of problems.

The first type: one requires that a linear combination of the distances of a variable point to given points be a given area.

The first proposition of the second book illustrates this. Fermat states the proposition as follows:

> If some given points are joined by straight lines to one and the same point, and the difference of the squares of these straight lines be a given area, the point of conjunction will be on a straight line given in position (vol. III, p. 27).

He establishes this proposition by synthesis for two points A and B. Next he considers two other cases, the last of which is stated thus:

> If one joins one point to three others given in a straight line, and if the sum of the squares of the two straight lines thus drawn exceeds the square of the third by a given area, the point will be on a circumference given in position (vol. III, p. 28).

By using a more analytical language, that is, one different from Fermat's, let us consider n points $A_1, A_2, \ldots, A_n$ on a straight line D and the coefficients $\lambda_1, \ldots, \lambda_n$. We are interested in the locus of points N of the plane such that

$$(1) \qquad \sum_{i=1}^{n} \lambda_i \, NA_i^2 = c \qquad \text{with } c \text{ being a given area.}$$

Let a_i be the abscissa of A_i for $i = 1, \ldots, n$ and (x, y) the coordinates of N; $NA_i^2 = (x - a_i)^2 + y^2$, therefore;

$$\sum_{i=1}^{n} \lambda_i NA_i^2 = \sum_{i=1}^{n} \lambda_i \left((x - a_i)^2 + y^2 \right) = \left(x^2 + y^2 \right) \sum_{i=1}^{n} \lambda_i - 2x \sum_{i=1}^{n} \lambda_i a_i + \sum_{i=1}^{n} \lambda_i a_i^2 .$$

Condition (1) is therefore written

$$(2) \qquad \left(x^2+y^2\right)\sum_{i=1}^{n}\lambda_i - 2x\sum_{i=1}^{n}\lambda_i a_i = c - \sum_{i=1}^{n}\lambda_i a_i^2 .$$

If $\sum_{i=1}^{n}\lambda_i \neq 0$, then (2) is the equation of a circle whose center is on the straight line D, with the abscissa

$$\bar{a} = \frac{\sum_{i=1}^{n}\lambda_i a_i}{\sum_{i=1}^{n}\lambda_i}$$

and whose radius r is given by

$$r^2 = \frac{c - \sum_{i=1}^{n}\lambda_i\left(a_i - \bar{a}\right)^2}{\sum_{i=1}^{n}\lambda_i}.$$

This circle is real if the second member is positive, and it is reduced to its center, the barycenter of points A_i, when $\sum_{i=1}^{n}\lambda_i\left(a_i-\bar{a}\right)^2 = c$.

If $\sum_{i=1}^{n}\lambda_i = 0$, equation (2) is reduced to $2x\sum_{i=1}^{n}\lambda_i a_i = \sum_{i=1}^{n}\lambda_i a_i^2 - c$, the equation of a straight line perpendicular to D at the abscissa point with the ratio

$$\frac{\sum_{i=1}^{n}\lambda_i a_i^2 - c}{2\sum_{i=1}^{n}\lambda_i a_i}.$$

As we have seen, Fermat begins by considering the cases with: $n = 2$, $\lambda_1 = 1$, $\lambda_2 = -1$, which yields a straight line perpendicular to D at the abscissa point

$$\frac{a_1^2 - a_2^2 - c}{2\left(a_1 - a_2\right)} = \frac{1}{2}\left(a_1 + a_2 - \frac{c}{a_1 - a_2}\right).$$

Next he considers the case with: $n = 3$, $\lambda_1 = \lambda_2 = 1$ and $\lambda_3 = -1$, which yields a circle whose center has the abscissa $a_1 + a_2 - a_3 = \bar{a}$ and the square of the radius is

$$c-(a_1-\bar{a})^2-(a_2-\bar{a})^2+(a_3-\bar{a})^2=c+2a_3^2+2a_1a_2-2a_1a_3-2a_2a_3.$$

Fermat's demonstration is equivalent to the preceding calculation, but without using algebra.

In Proposition 5 of the second book, he treats a problem of the same type in which n is any number and all the λ_i are equal to 1.

In the second proposition – which is the 9th of *The Knowns* – Fermat establishes that the locus of the points equidistant to two given points is the mediatrix of these two points (this is Proposition 8 of *The Knowns*); and next that the locus of points, the ratio of whose distances to two given points is a circle centered on the straight line joining these two points (the circle of Apollonius; this is Proposition 9 of *The Knowns*).

It is, moreover, plausible to think that these last propositions inspired the foregoing ones, which are extensions of them.

The second type: One requires that a linear combination of the squares of the distances from the variable point to given points be a given area equal to the product of the abscissa of the variable point by the given segment.

Finally, in the third type, the power of the variable point in relation to a given circle is equal to the square of the distance of this point to a given point G. If one considers, instead of point G, a second circle, the problem becomes the quest for the locus of points E, whose powers in relation to two given circles are equal. We know that this locus is the radial axis of the two circles; in the problem that Fermat treats, one of the two circles is reduced to a point G, so that the locus is the radial axis of the bundle of circles containing the circle of diameter AB and of which G is one of the limit points (that is, a point-circle).[16]

Recall that, according to Fermat himself, the second part of the *De locis planis* was completed in 1629. Algebra does not enter into it, even though the presentation could at the very least have been lightened by drawing on algebraic notation. This entire section is in the style of geometrical calculation, with this difference, that the equalities and the operations on geometrical quantities are written with the signs of arithmetic (=, +, −, ...). In the whole of his book, Fermat is doing the work of a geometer informed by algebra, along the lines of his distant predecessor

[16] *Ibid.*

Ibn al-Haytham. Both men intended to put effort into this field of research on geometrical loci, notably by means of pointwise transformations. Thus inflected, this research will lend itself to algebraic treatment.

2. THE EQUATIONS OF GEOMETRICAL LOCI

In the obituary of Fermat that appeared in *Journal des Savants* in February 1665, that is, less than a month after his death, one reads:

> An introduction to loci, both plane and solid, which is an analytical treatise concerning the solution of problems seen before M. Descartes published anything on the subject.

This short passage situates clearly and precisely the *Ad locos planos et solidos isagoge*. The author of the obituary tells us that it is an analytical work (that is, algebraic) written independently of Descartes's *Géométrie*. These two assertions seem to be incontestable. It remains to find out precisely what Fermat wanted to accomplish in this work, which he himself placed both in continuity and discontinuity with *De locis planis*. Fermat himself answers this question in the preamble of the treatise:

> We therefore subject this theory (of plane and solid loci) to an analysis that is proper and peculiar to it, and which opens a general path to research on it (vol. III, p. 85).

Here, there is no ambiguity about the project: to find the equations of curves – the polynomial equations and only them – between two unknown quantities. Fermat himself phrases it as follows:

> Whenever in a final equation one finds two unknown quantities, one has a locus, with the extremity of one of them describing a line that is either straight or curved. The straight line is simple and unique in its genus; the types of curves are indefinite in number, circle, parabola, hyperbola, ellipse, etc. (vol. III, p. 85).

The treatise is therefore explicitly aimed at finding the polynomial equations of curves, which exist in an infinite number of types. But the title of the treatise effectively points only to plane and solid loci. Nevertheless, for Fermat there is no gap between these two statements, since everything reduces to these last two cases. He defines loci thus:

> Whenever the extremity of an unknown quantity that describes the locus follows a straight or circular line, the locus is called *plane*; if it describes a parabola, a hyperbola or an ellipse, the locus is called *solid*; for the other curves, one calls it the locus *of line*. We will add nothing about this last case

for knowledge of the locus *of line* is very easily deduced by means of reductions from the study of *plane* and *solid* loci (vol. III, p. 85).

In this treatise at least, Fermat does not attempt to elaborate a general theory of curves, unlike Descartes in his *Géométrie*. He offers neither any means of isolating algebraic curves *a priori*, nor any method of tracing or drawing them. Despite his allusions to other loci, the only curves that concern him here are those encountered in *De locis planis*.

In the *Ad locos planos et solidos isagoge*, he introduces first the concepts of coordinates and of degree:

> To set up the equations, it is convenient to take two unknown quantities under a given angle that we will ordinarily assume to be right, and to take as given the position and one extremity of one of them; if neither of the two unknown quantities exceeds the square, the locus will be plane or solid (vol. III, p. 86).

Fermat begins with the equations of the first degree in x and y, and shows that the first degree relation with respect to these coordinates defines a straight line. He considers, successively, the case in which the straight line passes through the origin of the coordinates, $ax = by$, then the general case in which the straight line passes through a given point on the axis of the abscissas, $c - ax = by$. Note, however, that throughout the demonstration, he uses the Euclidean language of the *Data*, and the similarity of triangles. Moreover, he scrupulously respects homogeneity. He shares this practice with such predecessors as al-Khayyām and al-Ṭūsī.[17]

With the equation of the straight line in hand, Fermat draws from it, he writes, 'the very beautiful proposition that we have discovered by this means'; and he proposes the statement that follows without the slightest demonstration:

> Given any number of straight lines given in position to which one draws from one and the same point, straight lines under given angles; if the sum of the products of the straight lines thus drawn by the given ones is equal to a given area, the point from which one draws them will be on one of the straight lines given in position (vol. III, p. 87).

Indeed, this is the locus of points M such that

$$\sum_{i=1}^{n} a_i \cdot MH_i = \text{constant},$$

[17] He proceeds similarly when he takes *Elements* VI.26 as an example of searching for the maximum. To reduce the research of these algebraists to the *Data* and the *Elements* on account of this practice is both naive and wrong.

with a_i the given segments, MH_i the segments drawn under the given angles to the given lines d_i, where $i = 1, 2, \ldots, n$; in a different notation, this is rewritten

$$\sum_{i=1}^{n} \lambda_i \left(x\cos\theta_i + y\sin\theta_i - \rho_i\right) = \text{constant},$$

with ρ_i being the distance of point M with coordinates (x, y) to the straight line d_i; $\rho_i = x\cos\theta_i + y\sin\theta_i$.

Fermat next moves on to the second degree and begins with the equation $xy = k^2$, the equation of a hyperbola referred to its asymptotes. During this demonstration, he relies on Book II, Proposition 12 of Apollonius's *Conics*. To this equation, he reduces by translation of the asymptotes the equation of the form

$$k^2 + xy = ax + by.$$

Next he considers the case of a second degree equation homogeneous in x and y and finds as the locus of points a straight line. He should, however, have found either two or none.

The next equation that he considers is $x^2 = ay$. He recognizes it as a parabola whose *latus rectum* is a. He is obviously drawing on Proposition I.11 of the *Conics*. By a suitable translation, he reduces this equation to the equation $k^2 - x^2 = ay$. Likewise the equation $k^2 + x^2 = ay$ yields a parabola.

Fermat then takes up the case in which x^2 and y^2 are both present in the equation. For example, $r^2 - x^2 = y^2$ yields a circle if the ordinates are perpendicular to the axis of the abscissas. To this equation he reduces all those that contain x^2 and y^2 with the same coefficient, and in addition, terms in x and y as well as a constant term; to do so, he uses a suitable translation on the abscissas and on the ordinates.

Next, he gives the general equation of the circle starting from the preceding equation with a change of axes:

$$(a + x)^2 + (b + y)^2 = R^2 \text{ with } R^2 = r^2 + a^2 + b^2;$$

he then points out:

> Using a similar reasoning, one will reduce to it all similar equations. By means of this procedure, we have established all the propositions of the second book of Apollonius *On plane loci*, and we have demonstrated that the

first six take hold for any points whatsoever, which is rather remarkable and was perhaps unknown to Apollonius (vol. III, p. 92).

Next Fermat gives the equation of the ellipse under the form $\frac{r^2 - x^2}{y^2} = k$, k being a given ratio. Here too, is a direct application of Proposition I.12 of the *Conics*. He explains that one can reduce to this equation all cases in which the coefficients of x^2 and of y^2 are not equal – while sharing the same sign if they are in the same member – or else the case in which the angle of the ordinates is not right. In the contrary case, one has a circle.

He goes on to consider the equation of the hyperbola in the form $\frac{r^2 + x^2}{y^2} = k$, k being a given ratio. Here, too, he applies Proposition I.13 of the *Conics*.

Fermat reduces to this equation all those where x^2 and y^2 appear with coefficients having contrary signs eventually associated with terms of lower degrees.

The last case that he treats is the most difficult: terms in x^2, y^2 and in xy appear in the equation of the second degree along with terms of lower degrees.

He treats

$$r^2 - 2x^2 = 2xy + y^2;$$

by adding x^2 to each side, he makes $(x + y)^2$ appear; one obtains

$$r^2 - x^2 = (x + y)^2 = Y^2;$$

one thus has a new ordinate Y and a new axis of abscissa and the equation represents an ellipse.

In other words, perhaps under the influence of Diophantine analysis, Fermat intends to bring out the square of a linear form, in order to reduce it to the preceding cases.

The general equation is written

$$(1) \qquad \alpha x^2 + 2\beta xy + \gamma y^2 + \delta x + \varepsilon y + \theta = 0$$

yielding

$$(2) \qquad (\alpha\gamma - \beta^2)\, x^2 + (\beta x + \gamma y)^2 + \gamma\delta x + \gamma\varepsilon y + \gamma\theta = 0.$$

The quantity $\beta x + \gamma y = Y$ is posited as a new ordinate relative to an axis $\beta x + \gamma y = 0$; depending on the sign of $\alpha\gamma - \beta^2$, the equation

$$(\alpha\gamma - \beta^2)\, x^2 + Y^2 + (\gamma\delta - \beta\varepsilon)x + \varepsilon Y + \gamma\theta = 0$$

represents an ellipse or a hyperbola in the new coordinate system (x, Y).

We have just witnessed an example of a change of coordinate axes, and Fermat tells us, this reduces 'to a triangle of a given species' (vol. III, p. 95).

'As the crowning point of this Treatise', Fermat states a problem analogous to that of Pappus in which one seeks the locus of points such that a linear combination of squares of their distances to straight lines be a constant. This locus is obviously a conic.

The reading of the *Isagoge* brilliantly shows us its intimate connections with the *De locis planis*. Despite his allusion to an indefinite number of geometrical loci, Fermat in fact treats only the loci encountered in the latter book: straight line, circle, conic section. One might add that his systematic use of the method of translation of coordinate axes, of multiplication of coordinates of numerical factors, is nothing but an algebraic translation of the method of pointwise transformations that he puts to work in the *De locis planis*. It is in this sense that the *Isagoge* continues the *De locis planis*. But, whereas in the latter, Fermat is concerned with the geometrical properties of these loci as well as their transformation, in the former his main effort consists in defining these same loci by their equations. Better yet, he presents this translation in a manner deliberately ordered by the degrees of these equations. Finally, his exposition gradually evolves in order to characterize plane and solid geometrical loci. Herein, it seems to me, resides the novelty of the *Isagoge*.

To understand Fermat's procedure and the extent of his project in the *Isagoge* is to grasp his intention and therefore the exact meaning of 'this discovery' to which he refers and that allowed him to move from the *De locis planis* to the *Isagoge*. One might think that his intention here would have been to translate Apollonius's *Conics* into *Logistica speciosa* of Viète. This seems an oversized conjecture in which to dress the *Isagoge* project. If this had indeed been Fermat's aim, one should find in the *Isagoge* the propositions of the *Conics* translated one after the other in algebraic terms. Such an algebraic translation of these geometrical loci – a partial one at least – had been carried out six centuries earlier with the precise goal of studying the intersection of conic sections, with the goal of elaborating a geometrical theory of algebraic equations of the first three degrees. To elaborate such a theory, one begins, following the example of al-Khayyām, with an *a priori* determination of all the forms of these equations, a determination aimed at designating for each one the two curves that suit it. In the *Isagoge,* however, a completely different project appears: if one translates each locus into a polynomial equation with two unknowns, it is essentially to characterize the curve. Everything indicates, moreover, that the role of the polynomial equation is restricted to this characterization insofar as Fermat does not yet use it to deduce other properties of the curve.

Finally, one might think that Fermat's inspiration in the *Isagoge* originates in an intuitive grasp of the relations between Diophantine analysis and polynomial equations with two unknowns. We know that he was interested in Viète's algebra in relation to Diophantine analysis; it is clearly the latter, both for the concepts and for the notations, that plays the role of intermediary between Viète's algebra and the research in which Fermat was engaged, starting with the *Ad locos planos*. The *Isagoge* is thus the product of this combination.

Under these conditions, how ought one situate the *Isagoge* in the history of algebraic geometry? Considered from upstream, it is at once ahead of and behind the corresponding part of al-Ṭūsī's work. Both men translate the locus by means of a polynomial equation; but whereas al-Ṭūsī undertakes this translation to determine the positive roots of algebraic equations, Fermat does so for its own sake in an algebraic study of geometrical loci. Nevertheless, al-Ṭūsī already used the equation to deduce certain properties of the curve: the existence of a maximum for example – which Fermat does not do. Seen from downstream, so to speak, the *Isagoge* remains well on this side of Descartes's *Géométrie*. Whereas the latter attempts to elaborate a theory of algebraic curves and to equip himself with the means to obtain it (the distinction between mechanical and geometrical, for example), Fermat seems to be dealing with a problem that is still specific.

3. SOLUTION OF EQUATIONS BY THE INTERSECTION OF TWO CURVES

A little later, Fermat adds to the *Isagoge* an *Appendix* completely devoted to the question: 'How the solution of solid problems can be deduced from what we have said, and in an elegant manner' (vol. III, p. 96). The date of this *Appendix* is important for two reasons; it marks Fermat's encounter with the geometrical theory of algebraic equations, which is missing from the *Isagoge*, and it raises the question of his acquaintance with Cartesian research in this domain, and particularly with the *Géométrie*. It is important to remember that in 1619 Descartes had already begun to focus on the solution of some cubic equations by using the intersection of conic curves. Also, in 1628 he had divulged his solution of all solid problems by this means, and he had taken up in the *Géométrie* both this solution and the theory that undergirds it. It would be surprising if Fermat had heard not the slightest echo of this Cartesian research before he received a copy of the *Géométrie* in December of 1637. Moreover, Fermat did not discuss this *Appendix* in his correspondence until his letter of 16 February 1638 to Mersenne.[18] One final remark of considerable interest: like Descartes in the *Géométrie*, Fermat claims that, for cubic and biquadratic equations, one can restrict oneself to a parabola and a circle. Incidentally, this restriction is a novelty that goes beyond al-Khayyām, who first elaborated this theory of equations.

Since it was drafted after the *Isagoge* but before the beginning of 1638 (and therefore probably before the end of 1637), the *Appendix* was apparently able to benefit, whether directly or indirectly, from Descartes's research on the theory of equations. The idea expressed in the *Appendix* is the following:

> The most convenient is to determine the question by means of two loci equations, because two line-loci given in position mutually intersect, and the point of intersection, which is given in position reduces the matter of the indefinite to proper terms (vol. III, pp. 96–7).

Fermat immediately gives a first example:

$$x^3 + bx^2 = bc^2,$$

which he reduces to $x^2 + bx = by$ (equation of a parabola) and to $c^2 = xy$ (equation of an equilateral hyperbola). The solutions of the equation are therefore given by the abscissas of the points of intersection of the two curves. Fermat writes 'the method will be the same for all cubic equations' (vol. III, p. 97). For the biquadratic equations, he considers the example: $x^4 + b^3x + c^2x^2 = d^4$ which, by setting x^4 equal to $d^4 - b^3x - c^2x^2$, he reduces to c^2y^2, whence $x^2 = cy$ (equation of a parabola) and $y^2 = \frac{d^4 - b^3x}{c^2} - x^2$ (equation of a circle). The roots are therefore given by the abscissas of the points of intersection of a parabola and a circle.

[18] 'I would be very pleased to know what Mr. de Roberval and Pascal think of my *Topical Isagoge* and of the *Appendix*, if they have seen them both' (vol. II, p. 134).

He notes that 'this same procedure can be used to solve all biquadratic equations; for, by the method of Viète (Cap. I: *De emend.*), one can make the affected term of the cube disappear, and by putting the unknown biquadratic term on one side, and all the remaining terms on the other, one will solve the question by means of a parabola and a circle or a hyperbola' (vol. III, pp. 97–8).

Next he returns to the classical problem of inserting two mean proportionals and he finds anew the construction that Eutocius attributed to Menaechmus. He gives a first solution using a parabola and a hyperbola, and a second solution using two parabolas. And he reminds the reader that these two solutions appear in Eutocius's commentary on Archimedes' *Sphere and Cylinder*.

Fermat then comes to one of the central points of the *Appendix*: '*the construction of all cubic and biquadratic problems by means of a parabola and a circle*' (vol. III, p. 99). He explains that, in order to solve a fourth-degree equation, one must first eliminate the cubic term.

Fermat explains his method with an example that in no way diminishes its generality. He consider the biquadratic equation

$$x^4 - a^3x = b^4$$

which is rewritten

$$x^4 = a^3x + b^4 = f(x).$$

Assume

$$g(x) = (x^2 - c^2)^2 = x^4 - 2c^2x^2 + c^4 = f(x) - 2c^2x^2 + c^4.$$

Let us also assume

$$2c^2 = n^2 \text{ and } g(x) = n^2y^2,$$

whence

$$x^2 - c^2 = ny \qquad \text{(the equation of a parabola).}$$

Moreover

$$g(x) = a^3x + b^4 - 2c^2x^2 + c^4,$$

whence

$$\frac{c^4}{n^2} - x^2 + \frac{a^3x + b^4}{n^2} = y^2 \quad \text{(the equation of a circle).}$$

Fermat reminds the reader that his method is general and works for the biquadratic equation, provided that one eliminate the third-degree term and, for the cubic equation, provided one eliminate the second-degree term. This technique of affine transformation was known and practiced since al-Ṭūsī.

Like Descartes, Fermat reduces the solution of cubic and biquadratic equations to the intersections of parabolas and circles. But in neither the *Isagoge* nor the *Appendix* does he say anything about the cubic curve itself. Moreover, like Descartes but unlike al-Ṭūsī, he does not go on to discuss the existence of the points of intersection.

4. THE SOLUTION OF ALGEBRAIC EQUATIONS AND THE STUDY OF ALGEBRAIC CURVES

Much later, Fermat turns to the study of equations of algebraic curves. He enters through the wide gate, by way of a critique of the classification of equations and of curves proposed by Descartes in his *Géométrie*. To attack the Cartesian classification, however, is to reconsider several ideas in the discipline. The issue is to know in which direction such an approach could modify the project itself.

Fermat presents his critique in two texts with a polemical tone. In the first and by far the more substantial, he challenges the Cartesians. From his tone and word choice, one can infer that Descartes was no longer alive. It was therefore after 1650 that Fermat writes his fundamental treatise, the *Dissertatio tripartita*. The second text is a letter that Fermat addresses to Digby, picking up the same ideas as those of the *Dissertatio*, sometimes in equivalent phrasings.[19] In the first lines of the *Dissertatio*, he addresses the Cartesians as follows:

> It seems paradoxical to say that, even in Geometry, Descartes was only a man; but to recognize this, let the most subtle Cartesians check to see if there is not an imperfection in their master's distribution of curved lines into certain classes or degrees, and whether one does not need to adopt a classification that is more satisfactory and conforms more to the true laws of geometrical Analysis (vol. III, p. 109).

At first sight, then, the *Dissertatio* presents itself as a book against Descartes. Nevertheless, a deeper examination brings us face to face with

[19] Rediscovered by J. E. Hofmann, this letter is reproduced and translated in *Pierre Fermat: La Théorie des nombres*, text translated by P. Tannery, introduction and commentary by R. Rashed, Ch. Houzel and G. Christol, Paris, Blanchard, 1999, pp. 491–7.

the treatise in which one can feel the impact of Descartes's *Géométrie* most strongly and most massively. Is this a paradoxical situation? Not at all, since this treatise is a critical commentary of several of Descartes's ideas. In short, this is a text written in relation to Descartes and against him. Now in similar cases, it sometimes happens that one attributes more in order to criticize more and, by thus skewing the criticized propositions, the critique goes off target. It seems that Fermat did indeed fall into this trap. But for our purposes, it matters much more to learn the fate of the Cartesian project, and of algebraic geometry more generally, once they came into the hands of someone highly trained up to that point in the study of the equations of loci, namely, the hands of Fermat.

We alluded earlier to the two projects that together constituted research in algebraic geometry from al-Khayyām to Descartes: to solve equations by means of curves, and to study the curves by means of their equations. The execution of these projects had already led al-Khayyām, and especially al-Ṭūsī to distinguish two classes of equations and to raise questions about both the possibility and the means of reducing these equations the one to the other. It was at this point that mathematicians distinguished between the equations of plane problems and those of solid problems, and proceeded to reduce biquadratic equations to cubic equations. It was thus known that plane problems are translatable into equations of the first two degrees, and that a second-degree equation reduces to one of the first degree by means of an auxiliary construction using the straightedge and compass.[20]

It was also known that solid problems can be translated by cubic and biquadratic equations in order to reduce the biquadratic equation to the cubic equation, and that one solves cubic equations with the intersection of two second-degree curves. In Chapter IV of his *De emendatione æquationum*, Viète proposed a method called 'the climactic paraplerosis' and Descartes presented an equivalent procedure, of which Fermat claimed that 'Descartes plagiarized it (from Viète)'.[21]

[20] Indeed, the equation $x^2+bx=c$ reduces to $\left(x+\frac{b}{2}\right)^2=c-\frac{b^2}{4}$; the point is to construct the square root of $c-\frac{b^2}{4}$, that is, the mean proportional between the unit length and $c-\frac{b^2}{4}$. These equations of the first two degrees therefore do not present the slightest problem.

[21] *Pierre Fermat: La Théorie des nombres*, p. 495. The methods of Viète and Descartes amount to the following derivation: Given

$$x^4+a_1x^3+a_2x^2+a_3x+a_4=0 .$$

(*Cont. on next page*)

In all of this Fermat concedes that he is following Descartes and his predecessors when they distinguish between planes, solids, and supersolids, when they translate them by means of equations, and finally when they seek a solution involving the straightedge and compass for planes, use conic sections for solids, or proceed by means of curves of a higher degree for supersolid. Fermat believes he is still following Descartes when he writes:

(*Cont.*) Granting that $x = t - \frac{a_1}{4}$, it follows that

$$t^4 + \left(a_2 - \frac{3a_1^2}{8}\right)t^2 + \left(a_3 - \frac{a_1a_2}{2} + \frac{a_1^3}{8}\right)t + a_4 - \frac{a_1a_3}{4} + \frac{a_1^2a_2}{16} - \frac{3a_1^4}{256} = 0\,;$$

given

$$t^4 + at^2 + bt + c = 0\,,$$

where

$$a = a_2 - \frac{3a_1^2}{8}, \quad b = a_3 - \frac{a_1a_2}{2} + \frac{a_1^3}{8}, \quad c = a_4 - \frac{a_1a_3}{4} + \frac{a_1^2a_2}{16} - \frac{3a_1^4}{256}.$$

One then finds the coefficients u, v, v' such that

$$t^4 + at^2 + bt + c = \left(t^2 + ut + v\right)\left(t^2 - ut + v'\right) = t^4 + \left(v + v' - u^2\right)t^2 + u\left(v' - v\right)t + vv'.$$

One must then have $v + v' - u^2 = a$, $u(v' - v) = b$, $vv' = c$; assuming that $u \neq 0$, one has

$$v + v' = a + u^2, \quad v' - v = \frac{b}{u},$$

therefore

$$v' = \frac{1}{2}\left(a + u^2 + \frac{b}{u}\right), \; v = \frac{1}{2}\left(a + u^2 - \frac{b}{u}\right) \text{ and } vv' = \frac{1}{4}\left(\left(a + u^2\right)^2 - \frac{b^2}{u^2}\right) = c;$$

assuming that $u^2 = w$, one obtains the following equation in w:

$$w^3 + 2aw^2 + \left(a^2 - 4c\right)w - b^2 = 0\,.$$

The solution of this cubic equation yields the value $u^2 = w$, whence u, then v and v'. In order to obtain t one must solve two second degree equations.

For the case in which $u = 0$, one has $b = 0$ and the equation is rewritten $(t^2)^2 + at^2 + c = 0$, which is quadratic in t^2.

> Likewise the analyst will in the manner of Viète or of Descartes, albeit with a bit more difficulty, be able to reduce the bicubic equation to the quadratocubic or, in other words, the sixth-degree equation to the fifth-degree equation (vol. III, p. 110).

It is precisely at this stage that, even as he follows Descartes, Fermat separates himself from the latter. He writes:

> But, concerning the fact that, in the aforecited cases in which there is only one unknown quantity, the equations of even degree are reduced to equations of the immediately inferior odd degree, Descartes stated with confidence (page 323 of the *Géométrie* that he published in French) that it was absolutely the same for equations with two unknown quantities. For such are the equations that constitute curved lines; now in these equations, not only does the reduction or lowering in question not work, as Descartes claimed, but also analysts will recognize it to be impossible (vol. III, p. 110–11).

Here we are in the thick of a polemic and therefore at the heart of Fermat's research. He defends two theses against Descartes. First of all, he claims that one can always reduce an equation of one variable of degree $2n + 2$ to an equation of degree $2n + 1$. He concedes that Descartes thought this indeed to be so, but with this crucial difference that the latter wrongly believed that the rule is also valid for the equations of algebraic curves. To put the point in terms that were unknown to Fermat, Descartes would have confused a principal ring and a factorial ring. A reading of the *Géométrie*, however, does not seem to vindicate Fermat.

> [...] there is a general rule, writes Descartes, for reducing to a cube all the difficulties of the square of the square, and to the sursolid (x^5) all those of the square of the cube, in such a way that we need not consider [the higher ones] more composite than the lower.[22]

The text of Descartes evidently develops shy of Fermat's commentary on it. In contrast, one finds a later critique, justified this time, from the pen of Jacques Bernoulli, concerning the lowering of the sixth power to the fifth.[23] Perhaps Fermat drew inspiration from Descartes's statement in order to go much farther and to offer such a general formulation, which from Galois's theory we now know to be incorrect: the symmetrical group of degree $2n + 2$ does not generally contain a subgroup with the index $2n + 1$.

[22] Ed. Adam-Tannery, vol. VI, pp. 395–6; transl. Olscamp modified, p. 195.

[23] Jacques Bernoulli, *Opera*, 1744, vol. II, pp. 676–7.

Did the author of the *Géométrie* believe that the reduction was possible for algebraic curves, as Fermat maintained? Here again, it seems that the latter forces somewhat Descartes's thought. This is what Descartes himself writes:

> [...] I have constructed all plane problems by the intersection of a circle with a straight line, and solid problems also by the intersection of a circle with a parabola, and finally in the same way all problems which are only one degree more complex by making a circle intersect with a line which is only one degree more complex than the parabola; we have only to follow the same method in order to construct all problems that are more complex, to infinity. For in the matter of mathematical progression, once we have the first two or three terms, it is not difficult to find the others.[24]

Although they were not justified, the critiques that Fermat formulated made it possible for him to distinguish two classes of algebraic equations: those that are 'constitutive of curved lines', and those with one unknown. Long concerned, as we have seen, with the equations of geometrical loci, Fermat mobilizes this distinction in order to define curves by their equations, and much more sharply than his predecessors had done. By so doing, he contributes to the realization of Descartes's project.

Next, Fermat states that every equation of degree $2n + 1$ or of degree $2n + 2$ is solved by the intersection of curves of degree $n + 1$. Here, we are on the road that leads to the reciprocal of Bézout's theorem. Later, in 1688, Jacques Bernoulli will state that, for an equation of degree less than or equal to n^2, two curves of degree n are sufficient;[25] and he notes in particular that cubics would allow for the solution of equations up to the 9th degree.

The method that Fermat proposes to solve equations of degree $2n + 1$ or of degree $2n + 2$ draws singularly upon Diophantine analysis. To my knowledge, no one before him had used the techniques of this analysis in algebraic geometry. Here are the rules of his method.

1° One reduces the equation to an even degree, that is, one reduces the equation of degree $2n + 1$ to the degree $2n + 2$, by multiplying by a linear factor, if necessary.

2° One eliminates the first-degree term in order to put the equation in the form

[24] Olscamp translation, p. 269; ed. Adam-Tannery, vol. VI, p. 485.

[25] Jacques Bernoulli, *Animadversio in geometriam cartesianam et constructio quorundam problematum hypersolidorum*, Acta Eruditorum, June1688, pp. 323–30; and see *Opera*, vol. I, pp. 343 ff.

$$(1) \qquad x^{2n+2} + a_1 x^{2n+1} + \ldots + a_{2n} x^2 = p, \text{ with } p \neq 0 .$$

3° One looks for a squared expression such that, when set equal to the left side of (1), it brings about the maximum simplification on the side with the highest powers of x; thus

$$x^{2n+2} + a_1 x^{2n+1} + \ldots + a_{2n} x^2 = \left(x^{n+1} + b_1 x^n + \ldots + b_{n-2} x^3 + a_{n-1} xy\right).$$

We can determine the coefficients $(b_1, \ldots, b_{n-2})$ so as to eliminate all the terms up to x^{n+2} inclusive. The term x^{2n+2} disappears on its own. Next, one takes $b_1 = \frac{a_1}{2}, 2b_2 + b_1^2 = a_2, b_3 + b_1 b_2 = \frac{a_3}{2}, \ldots$

Having carried out the simplification, the remaining term with the highest degree is $x^{n+2}y$. But since x^2 is a coefficient, the result can be written in the form

$$a_{n-1}^2 y^2 + y P_n(x) + Q_n(x) = 0 ,$$

in which P and Q are two polynomials of, at most, degree n. This is the equation of the first curve. The second curve of degree $n + 1$ has as the equation

$$y = R_n(x) + \frac{p}{x} ,$$

in which R_n is a polynomial of degree n.

At least theoretically, therefore, the two curves are easy to construct. But it remains to be noted that, in the equation of the first curve, the ordinate of y appears only in degree 2, and in degree 1 in the equation of the second curve. Moreover, according to Bézout's theorem, these two curves have $(n + 1)^2$ points in common; there are therefore $n^2 - 1$ excess solutions that do not correspond to roots of the proposed equation. These solutions coincide with the point at infinity that is common to the two curves in the direction of the y-axis.

Without a doubt, the origins of Fermat's procedure here lies in Diophantine analysis. It is a technique to which he himself appeals in order to eliminate the higher powers, as one can read in the *Inventum novum*. Diophantus had already appealed to this procedure in his *Arithmetic*: it is the method of chords.[26] Note, however, that in contrast to Diophantus, Fermat begins with an equation in one unknown. At the end of the 10th century, the mathematician al-Karajī proceeded in this fashion to extract the square root of a polynomial assumed to be a perfect square. Whereas for him, the indeterminate in it represented the sequel of the development he was seeking, for Fermat it is the ordinate of a curve's moving point.

To illustrate his method, Fermat provides two examples. He begins by treating a 6th-degree equation and defines two cubic curves to construct its solutions. Next, he considers an 8th-degree equation and defines curves of the 4th degree. He concludes this inquiry by recalling the goal that he was pursuing all along:

> We thus have the solution and the exact and simplest possible construction of the problems of Geometry by loci originating, as the case may be, from curves of various species and appropriate to these problems. The analyst will, however, be free to vary these curves provided that he always stays in the class that is natural for the problems, by solving those of the 8th and 7th degree by curves of the 4th; those of 10th and 9th by curves of the 5th; those of the 12th and 11th by curves of the 6th and so on indefinitely with a uniform method (vol. III, p. 116).

This important text, which suggests that Fermat saw farther than he said he did, concludes with a critique of Descartes. In short, research on the solutions of algebraic equations takes place by means of the intersection of curves, provided that these are of the smallest possible degree. Better yet, the degree of these curves increases uniformly with the degree of the equation. The general rule is that, for an equation of degree $2n+2$, the degree of each of the two curves is $n+1$. Moreover, says Fermat, 'the analyst will be free to vary these curves'. Even if this freedom is limited by the preceding conditions, he believes – this is the important point – that for 'an infinity of special cases', as he states,[27] one can find curves of considerably lower degree. As an example, he gives a problem intimately tied to the very beginnings of algebraic geometry: the insertion of mean proportionals between two given magnitudes. The challenge now is to insert any number whatever of mean proportionals. The first example he brings up is that of six mean proportionals between two segments a and b.

Fermat thus had to solve the equation

$$x^7 = a^6 b;$$

[26] *Diophante: Les Arithmétiques*, Tome III: *Livre IV*, text established and translated by R. Rashed, Collection des Universités de France, Paris, Les Belles Lettres, 1984.

[27] Vol. III, p. 117.

using the same method, he makes the two members equal to x^4y^2b, and obtains the two curves $x^3 = by^2$ and $x^2y = a^3$, from which two parabolas are easy to construct.

He goes on to consider the insertion of 12 means, and solves the equation

$$x^{13} = a^{12}b,$$

by making the two members equal to x^8y^4b, whence $x^5 = y^4b$ and $x^2y = a^3$; he starts over and makes the two members equal to x^9y^3b, whence

$$x^4 = y^3b \text{ and } x^3y - a^4;$$

that is, two curves of only the fourth degree.

He continues by discussing the insertion of 30, 72, ... means. The method is the same, and he knows it is general. This is how it works. To insert p mean proportionals, it is necessary to solve

$$x^p = k, \qquad \text{where } k \neq 0.$$

We can assume that p is a prime number; otherwise, as Fermat indicates, it is reducible to an equation of lower degree. Let us apply the same method and make the two members of the equation equal to

$$x^{qr}y^r;$$

the exponents are chosen from the lowest possible degree. One thus has two curves defined by

$$x^{p-qr} = y^r \text{ and } x^q y = k^{\frac{1}{r}},$$

respectively of degrees sup $(p - qr, r)$ and $q + 1$.

In all the examples that Fermat treats, $p - 1 = 6, 12, 30, 72$, is therefore the product of two successive integers q and $q + 1$, which allows one to choose $r = q$; one then has sup $(p - qr, r) = q + 1$, the degree common to the two curves

$$x^{q+1} = y^q, \quad x^q y = k^{\frac{1}{q}}.$$

If $p - 1 = n^2$, one can then take $q = n - 1$ and $r = n$, which leads to the curves with the equations of, respectively,

$$x^{n+1} = y^2 \text{ and } x^{n-1}y = k^{\frac{1}{n}},$$

of degree $n + 1$ and n respectively.

Fermat then returns to his ever-present preoccupation: to determine the ratio between the degree of the equation and that of the curves. He thinks he can show that this ratio can be as large as one wishes. He writes:

> There is no difficulty in finding a procedure allowing one *to construct a problem the degree of which may be in a ratio larger than any given ratio with the degree of the curves that serve to solve it* (vol. III, p. 120).

Now, according to the preceding, if p is a prime of the form $n^2 + 1$, this ratio is $\frac{n^2+1}{n+1}$. But it is not yet known if there exists an infinity of prime numbers of this form. Fermat justifies this infinity by means of his conjecture; 'the numbers obtained by adding unity to the successive squares that one forms by starting from 2 are always prime', that is, that the numbers of the form $2^{2^\nu}+1$ will be prime. Since Euler, however, we know that this conjecture is wrong, because $\left(2^{2^5}+1\right)$ is divisible by 641.

As one can clearly see, the goal of the *Dissertatio* is to find constructions for the solutions of algebraic equations by means of curves of the smallest possible degree. Fermat hopes that the degree of auxiliary curves increases more slowly than the degree of the equation to be solved. Here we are still far from Bézout's theorem about the intersection of algebraic curves; the systematic study of algebraic curves by means of their equation is not yet the order of the day.

We have just followed succinctly Fermat's road in algebraic geometry. We have yet to raise the question of the exact significance of the path that leads from Fermat's *De locis planis* to the *Dissertatio*. Perhaps we will then be in a position to place Fermat's contribution in his historical perspective. The question is all the more important because some will interpret this pilgrimage in terms of an encounter between Apollonius and Viète, between the *Conics* and the *Logistica speciosa*. It is true that, at first glance, Fermat's first research efforts on the equations of loci corroborate such an interpretation, but the latter soon reveals itself to be insufficient for whomever wishes to understand Fermat's acts and aims. After all, this encounter between the *Conics* and algebra – the founding act of algebraic geometry – had already occurred six centuries earlier when the 10th-century mathematicians al-Māhānī, al-Qūhī, and Abū al-Jūd set to work on the edification of this new chapter in mathematics, which al-Khayyām systematized in the 11th century. Now Fermat's contribution does not reduce to that of al-Khayyām. We believe that we have demonstrated that, if there was an encounter, it was between the writings of Descartes, notably his *Géométrie,* and a geometry in renewal, a geometry concerned with the transformations of the loci of points (notably straight lines and conics) of Diophantine analysis cultivated by both Fermat's predecessors and by himself. Viète was no doubt present but mainly through his notation and his research in Diophantine analysis. It is therefore the confluence of these three traditions that makes it possible to shed light on the unfolding of Fermat's research and to appreciate the direction that it took. Indeed, to forget only one of these traditions is a recipe for the complete miscomprehension of Fermat's contribution.

As we have seen, research on the pointwise transformation of loci began well before Fermat, in the writings of Ibn al-Haytham in particular, as his book *The Knowns* attests. In contrast to his predecessor, however, Fermat was not planning to develop merely one chapter in a geometry undergoing renewal; as he gradually advanced, thanks to his research he wanted to uncover all the equations of loci, notably of conics. It is precisely this orientation that prepared his work in algebraic geometry. That is, when under the influence of Descartes, he devoted himself to the theory of equations and algebraic curves. This encounter with Descartes made it possible for Fermat to innovate in the field. He was then in a position to build a bridge between two domains that heretofore had been separate, in order to make progress on bringing to fruition what had been Descartes's project. Let us explain.

Before Fermat, no one had related Diophantine analysis to algebraic geometry. Certainly Fermat himself does not give a geometrical interpretation of Diophantine analysis. Thus for him, for example, a method like that of the chord, on which he often relies and that he borrows from Diophantus's *Arithmetic,* has no geometrical content. By contrast, he uses Diophantine methods to pursue his research in algebraic geometry, as we have already noted. Now this innovation distinguishes him from his predecessors as well as his contemporaries. In other words, as long as the curves that one studied algebraically were essentially reducible to conics alone, nothing demanded an explicit rapprochement between algebraic geometry and Diophantine analysis. This is precisely the situation of al-Khayyām and even of al-Ṭūsī. By contrast, the mathematicians who were interested in Diophantine analysis outside of this tradition of algebraic geometry, either on account of their context or out of personal interest, obviously could not imagine such a rapprochement. In effect, they were interested in the development either of abstract algebraic calculation (Abū Kāmil, al-Karajī, Bombelli, Viète...) or in number theory (al-Khunjandī, al-Khāzin, Fibonacci in his *Liber quadratorum*, al-Yazdī...). It is Descartes who offers the conditions of possibility for this development: effectively the possibility of treating generally equations of any degree, and of treating more clearly and more algebraically an entire class of curves. It is therefore thanks to Descartes and to his theory of algebraic curves that Fermat was able to insert the Diophantine methods in algebraic geometry. It is precisely this investment that allowed him to go farther than Descartes himself in bringing the latter's own project to fruition.

Finally we point to another road that we have not discussed here, and that leads from al-Ṭūsī to Fermat: analytical research in algebraic geometry – that is, the study of curves by means of their equations. This research touches on maxima, minima, and the contacts of tangents.

Part I

Arithmetic

– 1 –

EUCLIDEAN, NEO-PYTHAGOREAN AND DIOPHANTINE ARITHMETICS: NEW METHODS IN NUMBER THEORY

1. CLASSICAL NUMBER THEORY

Beginning in the 9th century, there existed two very distinct kinds of arithmetic in Arabic. The first was called *ḥisāb* (calculation). In the books of *ḥisāb*, the authors examine successively the decimal system, the sexagesimal system, elementary operations, the extraction of square, cubic, and even higher roots, some elements of algebra, the computation of areas and volumes, etc. The second kind of arithmetic constituted a domain that was inherited from Hellenistic mathematicians and was subsequently extended and transformed, notably under the effect of algebra. This arithmetic often presents itself either under its Greek name – *al-arithmāṭīqī* – when it concerns neo-Pythagorean arithmetic; or as 'the arithmetic books' (*al-maqālāt al-ʿadadiyya*), when it concerns Books VII, VIII and IX of the *Elements*; or also with the name of *Ṣināʿat al-ʿadad* – the art of number or the arithmetic art – when it is concerned with Diophantus's *Arithmetic*. Moreover, we know that before the end of the 9th century, one could read excellent translations of not only Euclid's *Elements*, but also the *Introduction to Arithmetic* of Nicomachus of Gerasa and the first seven books of Diophantus's *Arithmetic*. Note, however, that as time passed the boundaries between these two types of arithmetic became more and more porous. In other words, in treatises of *ḥisāb* one finds some elements of the other arithmetic, of Greek origin, which treats primarily the properties of integers. The latter is our focus here.

Arithmetic research in Arabic thus belongs to two traditions: that of Euclidean and neo-Pythagorean arithmetic, and that of Diophantine analysis. Beginning in the 10th century, number theorists constantly mingled these two traditions. But it will be impossible to understand the extension and transformation of this arithmetic research if one forgets to gauge at every moment the impact of algebra. The introduction of algebraic concepts and methods in every case led to the transformation of one or the other chapter of number theory.

1.1. *Euclidean and neo-Pythagorean arithmetic*

Before the middle of the 9th century, mathematicians had access to two translations of Euclid's *Elements*, and to one translation from the Syriac of Nicomachus of Gerasa. A little later, they would have at hand Thābit ibn Qurra's (d. 901) famous translation of the latter book, as well as his revision of the third translation of the *Elements* by Ḥunayn ibn Isḥāq. The three books of the *Elements* constituted not only a source, but also a model for research, as we will see shortly in the first new contribution, namely that of Thābit ibn Qurra. The *Introduction to Arithmetic*, for its part, yielded topics for research. It is true that, as far as the objects of arithmetic were concerned, mathematicians saw no difference between these books, and for very good reasons. For them, the differences lay in the expositions, on the one hand, and in the methods, on the other. Indeed, it is in Euclid's books that one finds a theory of parity and a theory of the multiplicative properties of integers: divisibility, prime numbers, etc. For Euclid, an integer is nevertheless represented by a line segment, which representation is essential to the demonstration of propositions. Whereas the neo-Pythagoreans shared this concept of integer and focused mainly on the study of these very same properties or others derived from them, yet in their methods and their goals, they distinguished themselves from Euclid. Whereas the latter proceeded by means of demonstrations, the former used induction as their only tool. For Euclid, moreover, arithmetic had no purpose outside of itself whereas, for Nicomachus, it served philosophical and even psychological ends. Mathematicians and philosophers such as Ibn al-Haytham and Avicenna clearly saw this difference of method. The first wrote:

> The properties of numbers exhibit themselves in two ways: the first is by induction, for, if one follows the properties of numbers one by one, and if one distinguishes them, in distinguishing and considering them, one discovers all of their properties. This is shown in the work *al-Arithmāṭīqī* <The *Introduction to Arithmetic* of Nicomachus>. The other way that exhibits the properties of numbers proceeds by means of demonstrations and deductions. All the properties of number grasped by demonstration are contained in these three books of <Euclid> or in what is based on them.[1]

The philosopher Avicenna makes the same point when he writes in the volume of his *al-Shifāʾ* devoted to arithmetic:

[1] *Sharḥ muṣādarāt Uqlīdis*, ms. Istanbul, Feyzullah, no. 1359, fol. 213v.

the treatise of the *Elements* has delivered numerous foundations for the science of numbers, and this art (neo-Pythagorean arithmetic) depends on these foundations when one learns it.[2]

For the mathematicians and philosophers of this era, the point at issue was indeed a difference between methods of demonstration and between the two presentations (the one appears axiomatic and demonstrative, whereas Nicomachus is assertoric and proceeds by example); it was not a difference between the objects of arithmetic. One can therefore understand why some mathematicians, and even those of Ibn al-Haytham's stature, proceeded sometimes by induction, as a function of the problem at hand; so it happens, as we shall see below, that Ibn al-Haytham discusses the 'Chinese theorem' and Wilson's theorem. Conversely, whereas mathematicians of the first rank and some philosophers (notably Avicenna) paid no attention to the philosophical and psychological goals that Nicomachus had assigned to arithmetic, others – second-tier mathematicians, philosophers, physicians, encyclopedists, etc. – took an interest in this arithmetic. Its history thus dissolves into the history of high literate culture in Islamic society, and extensively overflows the confines of this study. We therefore deliberately restrict ourselves to the role that arithmetic played in the development of number theory as a discipline.

Research on number theory in the Euclidean and Pythagorean sense began early, before the end of the 9th century. It was contemporaneous with Thābit ibn Qurra's translation of Nicomachus's book and with his revision of the translation of Euclid's *Elements*. Indeed, Thābit ibn Qurra himself initiated this research in number theory by elaborating the first theory of amicable numbers.[3] This fact has been familiar to historians thanks to F. Woepcke's work in the 19th century,[4] but its true significance only became clear very recently, when we established the existence of an entire tradition that Thābit ibn Qurra inaugurated in the purest Euclidean style and that culminated several centuries later in al-Fārisī (d. 1319), thanks to his application of algebra to the study of the first elementary arithmetic functions. This tradition is studded with many names: al-Karābīsī, al-Anṭākī, al-Qubayṣī, Abū al-Wafā' al-Būzjānī, al-Baghdādī, Ibn al-Haytham, Ibn Hūd, al-Karajī ..., to list only a few of them. Since we obviously cannot provide a detailed description in the few pages devoted to

[2] Ibn Sīnā, *Al-Shifā'*, *al-Ḥisāb* 2, ed. Ibrāhīm Madkūr, Cairo, 1985, p. 17.

[3] See below 'Thābit ibn Qurra and amicable numbers'.

[4] See notably his 1852 article, in which he summarizes Ibn Qurra's *Opuscule*: 'Notice sur une théorie ajoutée par Thābit Ben Korrah à l'arithmétique spéculative des Grecs', *Journal Asiatique*, 4e série, vol. 20, 1852, pp. 420–9; reprinted in *Études sur les mathématiques arabo-islamiques*, Frankfurt-am-Main, 1986, vol. I, pp. 257–66.

this theory, we will therefore attempt merely to sketch the movement to which we have just alluded.

1.2. *Amicable numbers and the discovery of elementary arithmetic functions*

At the end of Book IX of the *Elements*, Euclid presents a theory of perfect numbers in which he demonstrates that the number $n = 2^p (2^{p+1} - 1)$ is perfect – that is, equal to the sum of its proper divisors – if $(2^{p+1} - 1)$ is prime. But neither Euclid nor Nicomachus nor any other Greek author had tried to elaborate an analogous theory of amicable numbers. Thābit ibn Qurra therefore decides to construct that theory.

Indeed, he writes in his treatise *Fī istikhrāj al-aʿdād al-mutaḥābba* (*On the Determination of Amicable Numbers*):

> For perfect numbers of the two types we have mentioned, Nicomachus described a method for determining them, but he did not prove it. As to Euclid, he described the method of determining them and demonstrated it with care in the arithmetic books of his *Elements*; and he placed them at the end of what he had reached in the latter, leading some to think that this was the end-point at which he aimed and the ultimate goal of these books.
>
> As to amicable numbers, I found no one who mentioned them or took the trouble to devote himself to them. Now that the subject has come to my mind and that I have determined a demonstration about them, I should not like – in light of what has been said – to lose it by failing to establish it. I will do so once I have introduced the lemmas necessary for the task.[5]

He goes on to state and to demonstrate in pure Euclidean style what, to date, is the most important theorem for these numbers, the theorem that today bears his name.

Call $\sigma_0(n)$ the sum of the aliquot parts of the integer n (the proper divisors of n), and $\sigma(n) = \sigma_0(n) + n$ the sum of the divisors of n; and recall that two integers a and b are said to be amicable if $\sigma_0(a) = b$ and $\sigma_0(b) = a$.

Ibn Qurra's theorem:

For $n > 1$, *assume* $p_n = 3 \cdot 2^n - 1$, $q_n = 9 \cdot 2^{2n-1} - 1$; *if* p_{n-1}, p_n, *and* q_n *are prime, then* $a = 2^n p_{n-1} p_n$ *and* $b = 2^n q_n$ *are amicable.*

[5] See R. Rashed and C. Houzel, 'Théorie des nombres amiables', in R. Rashed (ed.), *Thābit ibn Qurra. Science and Philosophy in Ninth-Century Baghdad*, Scientia Graeco-Arabica, vol. 4, Berlin/New York, Walter de Gruyter, 2009, pp. 77–151, at pp. 90–2. Cf. Woepcke, 'Notice', 1852, pp. 423–4.

We must emphasize that Ibn Qurra's demonstration rests on a proposition equivalent to *Elements* IX.14, which reads: 'If a number is the least that is measured by prime numbers, it will not be measured by any other prime number except those originally measuring it'.[6] In other words, the smallest common multiple of prime numbers has no prime divisors other than these numbers. He next exploits the properties of the geometric progression of reason 2.

Now the history of the arithmetic theory of amicable numbers from Ibn Qurra to the end of the 18th century at least is limited to an invocation of this theorem, and to its transmission by later mathematicians, and the computation of pairs of these numbers. From a very long list of mathematicians writing in Arabic, we highlight the names of al-Anṭākī, al-Baghdādī, al-Karajī, Ibn Hūd, al-Tanūkhī, and al-Umawī.[7] By their chronological and geographical diversity, these few names sufficiently prove the wide diffusion of Ibn Qurra's theorem, which in 1638 would reappear in Descartes's work. Both the latter and his Arabic predecessors, however, seem to take for granted that Ibn Qurra's method was exhaustive.

As to calculating pairs of amicable numbers, Ibn Qurra appears not to take the trouble to compute any pair beyond (220, 284), not because he was incapable of finding any, but because this Euclidean mathematician had little interest in such calculations. An analysis of his treatise, however, shows that he had also calculated the pair (17296, 18416). Three-quarters of a century later, al-Anṭākī (d. 987) likewise seems not to have calculated other pairs. It is the algebraists who notably undertake this calculation. Thus one finds the pair (17296, 18416), nowadays named for Fermat, in al-Fārisī in the East, in the milieu of Ibn al-Bannā' in the West, and in al-Tanūkhī and many other mathematicians of the 13th century. Al-Yazdī later calculates the so-called 'pair of Descartes', (9363584, 9437056).[8]

[6] *The Thirteen Books of Euclid's Elements*, translated with introduction and commentary by T. L. Heath, New York, Dover Publications, 1956, vol. 2, p. 402 (with subjunctive changed to indicative).

[7] See R. Rashed, 'Matériaux pour l'histoire des nombres amiables et de l'analyse combinatoire', *Journal for the History of Arabic Science*, 6, 1982, pp. 209–78; 'Nombres amiables, parties aliquotes et nombres figurés aux XIIIe et XIVe siècles', *Archive for History of Exact Sciences*, 28, 1983, pp. 107–47; and 'Ibn al-Haytham et les nombres parfaits', *Historia Mathematica*, 16, 1989, pp. 343–52; repr. in *id.*, *Optique et mathématiques. Recherches sur l'histoire de la pensée scientifique en arabe*, Variorum Reprints, Aldershot, 1992, XI.

[8] See below, 'Al-Yazdī and the equation $\sum_{i=1}^{n} x_i^2 = x^2$'.

Although this historical summary is the most complete to date, it nevertheless remains both truncated and blind: indeed, it omits both the role that research on amicable numbers played for the whole of number theory and the intervention of algebra in the latter. Without lingering on the works to which we have alluded above, let us present this intervention of algebra. The famous natural philosopher and mathematician, Kamāl al-Dīn al-Fārisī, composed a *Tadhkirat al-aḥbāb fī bayān al-taḥāb* (*Memoir to Friends to Demonstrate Amicability*), in which he plans deliberately to demonstrate Ibn Qurra's theorem algebraically. This act drove him to conceive the first arithmetic functions and to give himself the training that led him to state for the first time the fundamental theorem of arithmetic. Al-Fārisī also developed the combinatory means necessary for this inquiry, and thus an entire body of research on figurate numbers. In short, his concern is now the elementary theory of numbers such as one encounters it again in the 17th century.

Indeed, throughout his report, al-Fārisī accumulates the propositions necessary to characterize the first two arithmetic functions: the sum of an integer's divisors and the number of these divisors. The report begins with three propositions, the first of which states: 'Every composite number is necessarily decomposable into a finite number of prime factors, of which it is the product'.[9] In the other propositions, he tries to demonstrate – admittedly rather clumsily – the unicity of the decomposition.

In contrast to Ibn Qurra's book, al-Fārisī's exposition does not begin with a proposition equivalent to Euclid IX.14, to say nothing of IX.14 itself. Instead, the author enunciates in turn the existence of a finite decomposition into prime factors and the unicity of this decomposition. Thanks to this theorem and combinatorial methods, one can determine completely the aliquot parts of a number, that is, in al-Fārisī's very own words, in addition to the prime factors, 'every number composed of two of these factors, of three of these factors, and so on, to every number composed of all the factors minus one'.[10]

Following these propositions, al-Fārisī examines the procedures of factorization and the calculation of aliquot parts as a function of the number of prime factors. The most important result in this area is surely the identification of combinations and figurate numbers. Henceforth, everything is in place for the study of arithmetic functions. A first group of propositions pertains to $\sigma(n)$. Even though al-Fārisī in fact treats only $\sigma_0(n)$,

[9] R. Rashed, 'Matériaux pour l'histoire des nombres amiables', p. 264.
[10] *Ibid.*, p. 261.

one sees that he recognizes σ as a multiplicative function. Among the propositions in this group, the following appear:

(1) If $n = p_1 p_2$, with $(p_1, p_2) = 1$, then

$$\sigma_0(n) = p_1\sigma_0(p_2) + p_2\sigma_0(p_1) + \sigma_0(p_1)\sigma_0(p_2),$$

which shows that he knew the expression

$$\sigma(n) = \sigma(p_1)\,\sigma(p_2).$$

(2) If $n = p_1 p_2$, with p_2 prime and $(p_1, p_2) = 1$, then

$$\sigma_0(n) = p_2\sigma_0(p_1) + \sigma_0(p_1) + p_1.$$

(3) If $n = p^r$, where p is prime, then

$$\sigma_0(n) = \sum_{k=0}^{r-1} p^k = \frac{p^r - 1}{p - 1}.$$

Until recently, these three propositions were attributed to Descartes.

(4) Finally, al-Fārisī tries unsuccessfully, as one can readily understand, to establish an effective formula for the case in which $n = p_1 p_2$, with $(p_1, p_2) \neq 1$.

A second group includes several propositions about the function $\tau(n)$, the number of divisors of n.

(5) If $n = p_1 p_2 \ldots p_r$, with $p_1, \ldots, p_r$ distinct prime factors, then the number of aliquot parts of n, written $\tau_0(n)$, is equal to

$$1 + \binom{r}{1} + \ldots + \binom{r}{r-1},$$

a proposition attributed to the Abbé Deidier (1739).

(6) If $n = p_1^{e_1} p_2^{e_2} \ldots p_r^{e_r}$, then

$$\tau(n) = \prod_{i=1}^{r} (e_i + 1)$$

and $\tau_0(n) = \tau(n) - 1$, a proposition credited to John Kersey and to Montmort.

Finally, al-Fārisī demonstrates the theorem of Thābit ibn Qurra. Indeed, he simply needs to show that

$$\sigma(2^n p_{n-1} p_n) = \sigma(2^n q_n) = 2^n[p_{n-1} p_n + q_n] = 9 \cdot 2^{2n-1}(2^{n+1} - 1).$$

This brief analysis of al-Fārisī's report displays the flowering of a new style planted in ancient soil, namely number theory. Indeed, without leaving Euclidean terrain, the mathematicians of the 13th century did not hesitate to draw on the contributions of algebra, most notably combinatorial analysis. And this tendency again came to the fore when mathematicians such as al-Fārisī and Ibn al-Bannā' studied figurate numbers, as we have seen above.[11]

1.3. *Perfect numbers*

If, by their work on amicable numbers, mathematicians were trying also to characterize an entire class of integers, they were pursuing the same goal by studying perfect numbers. From the mathematician al-Khāzin, we know that his 10th-century colleagues were raising questions about the existence of odd perfect numbers, still an unsolved problem at the time. Al-Khāzin writes: 'Among those who ponder <abundant, deficient and perfect numbers>, the following question emerged: whether or not there exists a perfect number among the odd numbers'.[12] At the end of this century and at the beginning of the next, al-Baghdādī obtained some results about these very numbers.[13] Thus he stated: 'if $\sigma_0(2^n) = 2^n - 1$ is prime, then $1 + 2 + ... + (2^n - 1)$ is a perfect number', a rule attributed to the 18th-century mathematician J. Broscius. Ibn al-Haytham,[14] a contemporary of al-Baghdādī's, made the first attempt to characterize this class of perfect numbers by trying to demonstrate the following theorem:

[11] See pp. 153–4 and below, pp. 346 ff.

[12] A. Anbouba, 'Un traité d'Abū Ja'far al-Khāzin sur les triangles rectangles numériques', *Journal for History of Arabic Science*, 3.1, 1979, pp. 134–78, esp. p. 157.

[13] R. Rashed, *Entre arithmétique et algèbre. Recherches sur l'histoire des mathématiques arabes*, Paris, Les Belles Lettres, 1984, p. 267; English transl. *The Development of Arabic Mathematics: Between Arithmetic and Algebra*, Boston Studies in the Philosophy of Science 146, Dordrecht/Boston/London, 1994.

[14] R. Rashed, 'Ibn al-Haytham et les nombres parfaits'; and *Les Mathématiques infinitésimales du IXe au XIe siècle*, vol. IV: *Méthodes géométriques, transformations ponctuelles et philosophie des mathématiques*, London, al-Furqān, 2002.

Given an even number n, the following conditions are equivalent:
1° if $n = 2^p(2^{p+1} - 1)$, with $(2^{p+1} - 1)$ prime, then $\sigma_0(n) = n$;
2° if $\sigma_0(n) = n$, then $n = 2^p(2^{p+1} - 1)$, with $(2^{p+1} - 1)$ prime.

We know that 1° is nothing but Proposition IX.36 of Euclid's *Elements*. Ibn al-Haytham therefore attempts to demonstrate in addition that every even perfect number has the Euclidean form, a theorem that Euler will prove definitively. Note that, just as Thābit ibn Qurra had done for amicable numbers, Ibn al-Haytham does not try to calculate any perfect numbers beyond those that the tradition already knew and had transmitted. This would become the computational job of a class of lesser mathematicians closer to the tradition of Nicomachus of Gerasa, such as Ibn Fallūs (d. 1240) and Ibn Malik al-Dimashqī, among many others.[15] From their writings, we learn that the mathematicians of the time knew the first seven perfect numbers.

1.4. *Equivalent numbers*

In their research into distinguishing various classes of numbers, the mathematicians took an interest in what they called 'equivalent numbers' (*muta'ādila*). Indeed, in the book of Abū Manṣūr ʿAbd al-Qāhir ibn al-Ṭāhir al-Baghdādī (d. 1037), one encounters the following problem about these numbers: 'Given a number, find two numbers such that the sum of each of their parts is equal to that given number.' The point is to find the reciprocal image by σ_0 of the given number a. Here, in his own words, is al-Baghdādī's solution:

> Subtract one from the given number and divide the remainder into <a sum> of two prime numbers, and then into <a sum> of two other prime numbers, and so on as far as one can divide it into two prime numbers. Next, we multiply the two parts in the first division, the one by the other; we multiply the two parts in the second division, the one by the other; and we proceed likewise for the third division, the fourth, and what follows; <the sum> of the parts of these products is equal to the given number.[16]

Two integers a and b are said to be equivalent if $\sigma_0(a) = \sigma_0(b)$. The problem that al-Baghdādī set himself can be translated thus:

Given an integer a, to find all the equivalent numbers linked to a, that is, to find the class of integers defined by $\sigma_0^{-1}(a)$. Al-Baghdādī proceeds as

[15] See R. Rashed, 'Ibn al-Haytham et les nombres parfaits'.
[16] *Kitāb al-Takmila*, ms. Istanbul, Laleli 2708, fol. 79^{r}.

follows: to find p_i, q_i prime ($i = 1, 2, \ldots$) such that $a = 1 + p_i + q_i$ ($i = 1, 2, \ldots$); one obtains

$$\sigma_0^{-1}(a) = \{p_i\ q_i\} = \{b_i\}$$

and the b_i are equivalent numbers.

It is obvious that

$$\sigma_0(b_i) = \sigma_0(p_i q_i) = a \qquad (i = 1, 2, \ldots).$$

Al-Baghdādī gives the example of $a = 57$; ($p_1 = 3$, $q_1 = 53$), ($p_2 = 13$, $q_2 = 43$), whence $b_1 = 159$, $b_2 = 559$; and thus gives only two elements of the reciprocal image.

Many other mathematicians will take up this line of research. Thus al-Zanjānī in his book *ʿUmdat al-ḥisāb*[17] repeats the same definitions, takes up the same example, and finally gives

$$\sigma_0^{-1}(57) = \{159,\ 559,\ 703\}.$$

The treatises of arithmetic (*ḥisāb*) will take up this discussion after that of perfect and amicable numbers. Thus, Muḥammad Bāqir al-Yazdī (d. *c.* 1637) in his *ʿUyūn al-ḥisāb* (*The Fountains of Arithmetic*) follows his famous calculation of the pair of amicable numbers attributed to Descartes with his study of equivalent numbers. He writes:

> We separate any even number into a <sum of> two prime numbers once, then into a <sum of> two other prime numbers, and we take their products. For example, we divide 16 into three and thirteen and take their product; and once into five and eleven, and we take their product; we obtain the two numbers 39 and 55, which are equivalent; the sum of the parts of each is seventeen.[18]

1.5. *Polygonal numbers and figurate numbers*

In his *Introduction to Arithmetic*, Nicomachus of Gerasa treats polygonal numbers and provides a table to generate them. Here is his table, such as it appears in Ibn Qurra's translation, but without the column of units:

triangular number	1	3	6	10	15	21	28	36	45
square number	1	4	9	16	25	36	49	64	81
pentagonal number	1	5	12	22	35	51	70	92	117
hexagonal number	1	6	15	28	45	66	91	120	153
heptagonal number	1	7	18	34	55	81	112	148	189

[17] Ms. Istanbul, Topkapi Saray, no. 3145, fols 1^{v}–84^{r}, at fol. 70^{v}.

[18] Ms. Istanbul, Süleymaniye, Hazinesi no. 1993, fols 1^{v}–120^{v}, at fols 68^{v}–69^{r}.

Triangular, square, pentagonal, hexagonal and heptagonal numbers are determined respectively by the following sums:

$$\frac{1}{2}\sum_{k=1}^{n} k(k+1);\ \sum_{k=1}^{n} k^2;\ \frac{1}{2}\sum_{k=1}^{n} k(3k-1);\ \sum_{k=1}^{n} k(2k-1);\ \frac{1}{2}\sum_{k=1}^{n} k(5k-3).$$

Nicomachus probably knew the rule for generating the table above, which can be rewritten

$$p_n^r = p_{n-1}^r + p_1^{r-1},$$

where p_n^r is the element of the nth line and of the rth column.

The authors of treatises on arithmetic integrated these results into their works, notably al-Baghdādī, Avicenna, Ibn al-Bannā', and al-Umawī, to mention only a few. This arithmetic research would soon expand in two directions. On the one hand, mathematicians took a systematic interest in various sums of integers. This is the approach both of the algebraists (al-Karajī and his successors in particular), and of the Archimedeans for the calculation of the areas of curved surfaces and of the volumes of solids with curved surfaces.[19] In contrast to Nicomachus and his successors, these mathematicians now attempted to demonstrate the formulas. On the other hand, using combinatorial methods, the algebraists went farther and began to study figurate numbers.

Thus treatises of algebra (*e.g.*, Abū Kāmil's, al-Samaw'al's) established expressions[20] such as

$$\sum_{k=1}^{n} k^2 = \frac{1}{3}n^3 + \frac{1}{2}n^2 + \frac{1}{6}n,\quad \sum_{k=1}^{n} k^3 = \frac{1}{4}n^4 + \frac{1}{2}n^3 + \frac{1}{4}n^2.$$

For his part, Ibn al-Haytham demonstrated

$$\sum_{k=1}^{n} k^4 = n\left(\frac{n}{5}+\frac{1}{5}\right)\left(n+\frac{1}{2}\right)\left((n+1)n-\frac{1}{3}\right) = \frac{1}{5}n^5 + \frac{1}{2}n^4 + \frac{1}{3}n^3 - \frac{1}{30}n,$$

using a slightly archaic complete induction. Indeed, his method is valid for any integer power without requiring any additional concepts. The general law that he identified for the calculation of sums of integers n raised to any power is rewritten

$$(n+1)\sum_{k=1}^{n} k^i = \sum_{k=1}^{n} k^{i+1} + \sum_{p=1}^{n}\left(\sum_{k=1}^{p} k^i\right),$$

[19] See below 'The Archimedeans and infinitesimal problems'.
[20] R. Rashed, *The Development of Arabic Mathematics*, pp. 69–76.

so that Ibn al-Haytham could calculate the sum of the ith powers of n first integers.[21]

Recall first that the study of figurate numbers requires prior knowledge of the arithmetic triangle and of elements of combinatorial analysis, which the algebraists of the late 10th century already possessed,[22] as well as a genuinely combinatorial interpretation of these elements, which appears in Naṣīr al-Dīn al-Ṭūsī at least.[23] Kamāl al-Dīn al-Fārisī therefore goes on to a systematic study of figurate numbers of any order. To generate these numbers, al-Fārisī states in an equivalent manner the relation

$$F_p^q = \sum_{k=1}^{p} F_k^{q-1},$$

where F_p^q is the pth figurate number of order q, $F_1^q = 1$.

Using this relation, he constructs the following table by way of example:

their numbers / the sums		1st 2	2nd 3	3rd 4	4th 5	5th 6	6th 7	7th 8	8th 9	9th 10	10th 11
1st	1	3	6	10	15	21	28	36	45	55	66
2nd	1	4	10	20	35	56	84	120	165	220	286
3rd	1	5	15	35	70	126	210	330	495	715	1001
4th	1	6	21	56	126	252	462	792	1287	2002	3003
5th	1	7	28	84	210	462	924	1716	3003	5005	8008
6th	1	8	36	120	330	792	1716	3432	6435	11440	19448
7th	1	9	45	165	495	1287	3003	6435	12870	24310	43758
8th	1	10	55	220	715	2002	5005	11440	24310	48620	92378
9th	1	11	66	286	1001	3003	8008	19448	43758	92378	184756
10th	1	12	78	364	1365	4368	12376	31824	75582	167960	352716

Al-Fārisī then presents an expression equivalent to

$$F_p^q = \binom{p+q-1}{q},$$

[21] R. Rashed, *Les Mathématiques infinitésimales du IXe au XIe siècle*. Vol. II: *Ibn al-Haytham*, London, al-Furqān, 1993, p. 182; English translation: *Ibn al-Haytham and Analytical Mathematics*. A History of Arabic Sciences and Mathematics, vol. 2, Culture and Civilization in the Middle East, London, Centre for Arab Unity Studies, Routledge, 2012, p. 148.

[22] See the chapter 'Algebra and linguistics', above

[23] See the chapter 'Philosophy of mathematics', below.

and thus establishes a link between the figurate numbers of any order and the combinations.

To these theoretical domains in Arabic mathematics, one could add a multitude of results that fit the bloodline of Nicomachus's arithmetic, results that were developed by arithmeticians or algebraists, or simply to meet the needs of other practices, such as magic squares or games of arithmetic. This adds up to a considerable body of results that extend or demonstrate what was already known and that is simply too vast to cover here. One would have to read the arithmetic works of such arithmeticians as al-Uqlīdisī, al-Baghdādī, al-Umawī, etc.; such algebraists as Abū Kāmil, al-Būzjānī, al-Karajī, al-Samaw'al; and such philosophers as al-Kindī, Ibn Sīnā, al-Juzjānī, etc., among a hundred others.

1.6. *The characterization of prime numbers*

One of the main axes of research in number theory has been the characterization of numbers: amicable, perfect, equivalent. It is therefore not surprising that mathematicians returned to prime numbers with an analogous task in mind. This is precisely what Ibn al-Haytham does when solving the problem known as the 'Chinese remainder'.[24] Indeed he wants to solve the system of linear congruences

$$x \equiv 1 \pmod{i_i}$$

$$x \equiv 0 \pmod{p},$$

where p is prime and $1 < i_i \leq p - 1$.

In the course of his inquiry, he gives a criterion for determining prime numbers, now the so-called theorem of Wilson:

If $n > 1$, the following two conditions are equivalent:

1° n is prime

2° $(n - 1)! \equiv -1 \pmod{n}$,

where, in Ibn al-Haytham's words:

> [...] this property is necessary for every prime number, that is, for every prime number, which is a number that is a multiple only of one, if one

[24] R. Rashed, 'Ibn al-Haytham et le théorème de Wilson', *Archives for History of Exact Sciences*, 22.4, 1980, pp. 305–21; repr. in *The Development of Arabic Mathematics*, p. 247.

multiplies the numbers that precede it by the others in the way we have specified, and if one adds one to the product, then if one divides the sum by each of the numbers that precede the prime number, only one remains, and if one divides by the prime number, nothing remains.[25]

Studies of this system of congruences reappear in part among the successors of Ibn al-Haytham in the 12th century, for example, in al-Khilāṭī in Arabic and Fibonacci in Latin.[26]

2. INDETERMINATE ANALYSIS

The first Arabic research on indeterminate analysis – or Diophantine analysis, as it is called nowadays – was very probably undertaken by Abū Kāmil (*c.* 830–900) in the middle of the 9th century, that is, approximately a generation after al-Khwārizmī. It was Abū Kāmil who first conceived of indeterminate analysis as a chapter in the new algebra that his predecessor had invented.

We will therefore begin with the *Algebra* of Abū Kāmil, first to trace rational indeterminate analysis, in order to show next how it became a chapter of algebra, before coming back to the description of what was recently recognized as a fact: the constitution of integer Diophantine analysis in a certain sense against the algebraists, as an integral part of number theory.

2.1. *Rational Diophantine analysis*

Abū Kāmil's project is clear; as he writes:

> We now explain many indeterminate problems that some arithmeticians call *sayyāla* ('fluid'); by this, I mean that one can determine many true solutions with convincing deductions and a clear method; some of these problems circulate among the arithmeticians according to certain types (*bi-al-abwāb*), without their having established the cause from which they proceed. I solved some of these by means of a true principle and an easy and very useful procedure.

Abū Kāmil continues:

> We also explain a large part of what the arithmeticians have defined in their books and that they treated by chapters*, by algebra and by deduction, so

[25] *On the Solution of a Numerical Problem*, in R. Rashed, *The Development of Arabic Mathematics*, p. 250.

[26] *Ibid.*, pp. 244–5.

* Implied: categories of problem.

that he who reads and examines it truly understands it and is not satisfied with simply reciting by heart and imitating its author.[27]

This text is of capital importance, both historically and logically. It attests to the existence of research in Diophantine analysis at the time of Abū Kāmil. The arithmeticians who undertook this inquiry devoted the word *sayyāla* to designating indeterminate equations, which thereby were separated out from the set of algebraic equations. Still according to Abū Kāmil's text, we know that these arithmeticians were satisfied by giving statements about certain types of these equations and the algorithms to solve them, but without worrying about either the reasons for them or the methods needed to establish them. But who are these arithmeticians? We cannot yet answer this question on account of the loss of the writings of many still active algebraists, such as Sind (Sanad) ibn ʿAlī, Abū Ḥanīfa al-Daynūrī, Abū al-ʿAbbās al-Sarakhsī…

In his *Algebra*, therefore, Abū Kāmil plans not to be satisfied with a scattered exposition but to provide a systematic one, in which methods appear in addition to both problem and solution algorithms. It is true that in the last part of *Algebra*, Abū Kāmil treats 38 Diophantine problems of the second degree, 4 systems of indeterminate linear equations, other systems of determinate linear equations, a set of problems that are reducible to arithmetic progressions, and a study of the latter. This set corresponds to Abū Kāmil's double goal: to solve indeterminate problems, and also to solve by means of algebra problems that the arithmeticians were then treating. Note that it is in Abū Kāmil's *Algebra* that one encounters – to my knowledge – for the first time an explicit distinction between determinate and indeterminate problems. Now the examination of these 38 Diophantine problems does not only reflect this distinction; he shows in addition that these problems do not follow each other haphazardly, but according to an order that Abū Kāmil indicated implicitly. Thus the first 25 pertain to one and the same group, for which Abū Kāmil gives a necessary and sufficient condition to determine positive rational solutions. Let us take only two examples. The first problem in this group[28] is rewritten

$$x^2 + 5 = y^2.$$

[27] *Kitāb fī al-jabr wa-al-muqābala*, ms. Istanbul, Beyazit Library, Kara Mustafa, 379, fol. 79ʳ. See *Abū Kāmil, Algèbre et analyse diophantienne*, édition, traduction et commentaire par R. Rashed, Berlin/New York, Walter de Gruyter, 2012, p. 579.

[28] *Abū Kāmil, Algèbre et analyse diophantienne*, p. 581.

Abū Kāmil proposes to give two solutions from among an infinity of rational solutions, according to his own statements. He then posits

$$y = x + u \qquad \text{with } u^2 < 5$$

and takes successively $u = 1$, $u = 2$.

Another example of the same group is Problem 19,[29] which is rewritten

$$8x - x^2 + 109 = y^2.$$

Abū Kāmil then considers the general form

$$ax - x^2 + b = y^2 \tag{1}$$

and writes:

> If you come across problems analogous to this problem, multiply half the number of roots by itself and add this product to the *dirham* (*i.e.* units); if the sum can be divided into two parts each of which has a square root, then the problem is rational and has innumerable solutions; but if the sum cannot be divided into two parts each of which has a square root, then the problem is irrational and without solutions.[30]

This text, which is particularly important in the history of Diophantine analysis, gives the sufficient condition for determining positive rational solutions of the preceding equation. The latter is rewritten

$$y^2 + \left(\frac{a}{2} - x\right)^2 = b + \left(\frac{a}{2}\right)^2;$$

now set $x = \frac{a-t}{2}$, one has

$$y^2 + \left(\frac{t}{2}\right)^2 = b + \left(\frac{a}{2}\right)^2, \tag{2}$$

and the problem is thus reduced to dividing a number into two squares, that is Problem 12 of the same group, which Abū Kāmil has already solved.

[29] *Ibid.*, pp. 616–21.
[30] *Ibid.*, p. 617.

Indeed, suppose that

$$b+\left(\frac{a}{2}\right)^2 = u^2+v^2,$$

with u and v rational. Abū Kāmil sets

$$y = u + \tau$$
$$t = 2\,(k\tau - v);$$

he substitutes these in (2) and finds the values of y, t, and then x. Thus he knows that, if one of the variables can be expressed as a rational function of the other, or in other words, if one can have a rational parametrization, one has *all* the solutions; whereas if the sum leads us to an expression the root of which cannot be evaluated, there is no solution. Put another way, which was unknown to Abū Kāmil, a second-degree curve has no rational point, or is birationally equivalent to a straight line.

The second group is composed of thirteen problems – 26 to 38 – which are impossible to parametrize rationally. Or in language unknown to Abū Kāmil, they define all the curves of type 1. For example, Problem 31[31] is thus rewritten

$$x^2 + x = y^2,$$
$$x^2 + 1 = z^2,$$

which defines a skew quartic, a curve of type 1 in the coordinate space, (x, y, z).

The third group of indeterminate problems is composed of systems of linear equations, such as Problem 39,[32] for example, which is rewritten

$$\begin{aligned} x + ay + az + at &= u,\\ bx + y + bz + bt &= u,\\ cx + cy + z + ct &= u,\\ dx + dy + dz + t &= u. \end{aligned}$$

Focusing this interest on indeterminate analysis, which led to Abū Kāmil's contribution, stimulated another event: the translation of Diophantus's *Arithmetic*. Thus, during the same decade Abū Kāmil was writing his *Algebra* in the Egyptian capital, Qusṭā ibn Lūqā in Baghdad

[31] *Abū Kāmil, Algèbre et analyse diophantienne*, p. 643.
[32] *Ibid*., pp. 654–63.

was translating seven books of Diophantus's *Arithmetic*. The event was crucial, both for the development of indeterminate analysis and for the techniques of algebraic calculations. We have shown that the Arabic version of Diophantus's *Arithmetic* is composed of three books that are also found in the Greek text that has come down to us, and of four books lost in Greek, and that the translation used terminology invented by al-Khwārizmī.[33] The translator not only gave to the *Arithmetic* an underlying algebraic interpretation, but also even gave to Diophantus's book the title *Ṣināʿat al-jabr* (*The Art of Algebra*). Now the Arabic version of the *Arithmetic* stimulated studies and commentaries. We now know that four commentaries on it existed, three of which have not yet been found. From the ancient biobibliographers, we know that Qusṭā ibn Lūqā himself commented on three books of the *Arithmetic*,[34] that Abū al-Wafāʾ al-Būzjānī wanted to demonstrate the propositions, probably the algorithms of Diophantus. In his *Fakhrī*, al-Karajī commented on four books of the *Arithmetic*;[35] his successor al-Samawʾal also commented on Diophantus. Of these four commentaries, only that of al-Karajī came down to us; but we believe that these are not the only commentaries on Diophantus. Beyond these commentaries, however, the algebraists in their various writings discussed indeterminate analysis, the status of which would change with al-Karajī.

Al-Karajī himself treated Diophantine analysis in three works, only two of which have come down to us. He studied indeterminate analysis in *al-Fakhrī*, prior to commenting on Diophantus in that book. He then returned to the subject in *al-Badīʿ*, recalling in his introduction to this book his initial work in *al-Fakhrī*. He composed a third treatise along with these two, but it has not yet been found. As he noted in *al-Fakhrī*, it was a book 'on *al-Istiqrāʾ* [indeterminate analysis]' which he had written in the Persian province of Rayy.[36]

[33] R. Rashed, 'Les travaux perdus de Diophante. I', *Revue d'histoire des sciences*, 27, 1974, pp. 97–122; 'Les travaux perdus de Diophante. II', *ibid.*, 28.1, 1975, pp. 3–30; Diophante, *Ṣināʿat al-jabr* (*L'Art de l'algèbre*), ed. Roshdi Rashed, Cairo, 1975, pp. 13 ff; and Diophante, *Les Arithmétiques*, text established and translated by Roshdi Rashed, 2 vols, Paris, Les Belles Lettres, 1984, vol. III.

[34] *Les Arithmétiques*, ed. R. Rashed, pp. 10–11.

[35] See F. Woepcke, *Extrait du Fakhrī, traité d'algèbre*, Paris, 1853; reprinted in *Études sur les mathématiques arabo-islamiques*, Frankfurt-am-Main, 1986, vol. I, pp. 267–426; and Diophante, *Les Arithmétiques*, ed. R. Rashed, Notes complémentaires. See also R. Rashed and C. Houzel, *Les* Arithmétiques *de Diophante: Lecture historique et mathématique*, Berlin, New York, Walter de Gruyter, 2013.

[36] F. Woepcke, *Extrait du Fakhrī*, p. 74; Woepcke's reading of *bi-al-tattary* must be corrected to *bi-al-Rayy*.

To understand al-Karajī's contribution to indeterminate analysis, one must keep in mind his renewal of algebra, a point that we have emphasized earlier. Indeed, al-Karajī developed indeterminate analysis not only as one of the chapters of algebra, but also as one of the means for the latter to extend algebraic calculation. Al-Karajī writes that Diophantine analysis 'is the pivot of most calculations, and that it is indispensable to all of the chapters'.[37] Thus, after having studied polynomials with square roots and the way of extracting these roots, one moves on to algebraic expressions that have square roots only in potentiality. This is the main object of rational Diophantine analysis, according to al-Karajī, and this is the sense in which it is constituted as a chapter of algebra. The method, or rather the methods, are those required to reduce the problem to an equality between two terms, the powers of which allow us to obtain positive rational solutions. Henceforth, Diophantine analysis is christened with a proper name, *al-istiqrā'*, a term that also includes the emphasized duality, for it designates a chapter as well as a method or set of methods. The term *al-istiqrā'* is derived from the verb *istaqrā*, which meant 'to consider or to examine successively the various cases', before it took on the technical sense of indeterminate analysis. In *al-Fakhrī*, al-Karajī gives the following definition:

> *al-istiqrā'* in calculation is when you encounter an expression of one type or of two types, or of three successive types (that is, algebraic powers) that is not a square literally but is such according to the meaning, and you want to know its square root.[38]

In *al-Badī'*, al-Karajī repeats the same definition and adds: 'I say that *al-istiqrā'* is the relentless pursuit of expressions until you find what you were looking for'.[39]

A simple reading of al-Karajī's explanations and of the chapters he devoted to indeterminate analysis in his two books shows a certain break with his predecessors; al-Karajī's style is different, not only from that of Diophantus, but also from that of Abū Kāmil. In contrast to Diophantus, al-Karajī does not give ordered lists of problems and of their solutions, but he organizes his exposition in *al-Badī'* around the number of terms compo-

[37] *L'Algèbre. Al-Badī' d'al-Karajī*, Manuscript of the Vatican Library *Barberini Orientale 36,1*, edition, with introduction and notes by Adel Anbouba, Publications of the Lebanese University, Section of Mathematical Studies II, Beirut, 1964, p. 8.

[38] *Al-Fakhrī*, ms. Istanbul, Köprulu, no. 950, fols 1^v–151^r; cf. F. Woepcke, *Extrait du Fakhrī*, p. 72.

[39] Ed. Anbouba, p. 62.

sing the algebraic expression and the difference between their powers. In successive paragraphs, for example, he considers:

$$ax^{2n} \pm bx^{2n-1} = y^2,\ ax^{2n} + bx^{2n-2} = y^2,\ ax^2 + bx + c = y^2.$$

His successors will, moreover, borrow this principle of organization. It is therefore clear that al-Karajī's goal was to give a systematic exposition of the subject. On the other hand, he takes farther the task begun by Abū Kāmil, which consists in extracting as much as possible the methods for each class of problems. In *al-Fakhrī*, al-Karajī does not want to give a deep exposition of Diophantine analysis as he understands it, since, as noted, he had already devoted a book to it, and he would later return to it in *al-Badī'*. In *al-Fakhrī*, he merely recalls the principles of this analysis, indicating that they notably pertain to the equation

$$(1) \qquad ax^2 + bx + c = y^2, \qquad\qquad a, b, c \in \mathbf{Z},$$

in which the trinomial in x is not a square, before turning finally to other classes of problems, most of which are indeterminate. These different classes are presented as classes of problems organized from the simplest to the most difficult, in order to 'satisfy him who wants to do exercises (*al-murtāḍ*)'.[40] These are in fact classes of exercises aimed at familiarizing the reader with the procedures 'that direct the problem according to the expression of the inquirer to one of the six canonical forms in order to determine the unknowns from the knowns, which is what calculation substantially is'.[41] In these five classes of problems, al-Karajī makes no pretense to any originality, and borrows most of the problems from Books II, III, IV of Diophantus's *Arithmetic*, a few problems from Book I – as noted in some detail earlier[42] – and more than half of the problems studied by Abū Kāmil. One also encounters other problems that al-Karajī himself may have raised, since they do not appear in these two authors.

It is in *al-Badī'*, which, by his own admission Abū Kāmil writes for an audience more informed and better practiced than the one for which he wrote *al-Fakhrī*, that al-Karajī gives a systematic exposition of the chapter on Diophantine analysis. Thus, after having discussed the types mentioned above, al-Karajī returns to equation (1). He then considers the case in which a (respectively c) is a square, and proposes a change of variable

[40] *Al-Fakhrī*, ms. Köprulu 950, fol. 54r.

[41] *Ibid*.

[42] Diophantus, *Ṣinā'at al-jabr* (*L'Art de l'algèbre*), ed. R. Rashed, pp. 14–19.

$y = \sqrt{ax \pm u}$ (respectively $y = \sqrt{c \pm ux}$). Note that he begins by giving the general formulation before moving on to examples. He then brings up the form

$$ax^{2n} + bx^{2n-1} + c = y^2$$

and proposes to reduce it to (1).

Next, Al-Karajī treats the expression in which the exponents do not follow each other in sequence, such as

$$ax^2 - c = y^2,$$

where a and c are nonsquare, but $\frac{c}{a}$ is square. He proposes the change of variable

$$y = ux - \sqrt{\frac{c}{a}}.$$

Here too, he recalls that by division one can reduce the form

$$ax^{2n} - cx^{2n-2} = y^2$$

to the previous form.

Al-Karajī then studies the equations of the form

$$ax^2 + c = y^2,$$

and gives two examples, the first with $a = 3$, $c = 13$, and the second with $a = 2$, $c = 2$; he points out that, in the two examples, one has $a + c = k^2$. He nevertheless proposes respectively the parametrizations $y = u$ and $y = ux$, and obtains

$$x^2 = \frac{u^2 - c}{a} \text{ and } x^2 = \frac{c}{u^2 - a},$$

which scarcely moves the solution of the problem forward. In commenting on this fact in the French introduction to his critical edition of *al-Badīʿ*, A. Anbouba correctly states:

> But it is clear that al-Karajī does not know about Book VI of Diophantus, which gave the solution to the question, (1) in the case in which $a + c$ is equal to a square (the Lemmas 1 and 2 of the *Arithmetic* pertaining to VI:12 and 13); (2) in the case in which one knows a particular root (lemma

pertinent to VI:15). We have very nearly convinced ourselves that al-Karajī knew neither Books V and VI of the *Arithmetic*, nor the end of Book IV.[43]

Al-Karajī studies many other problems, notably the double equality. Let us simply note the problem

$$x^2 + a = y^2$$

$$x^2 - b = z^2,$$

which defines a curve of type 1 in the coordinate space (x, y, z).

The successors of al-Karajī not only commented on his work, but tried to go beyond it on the path that he had cleared: to extend *al-istiqrā'* to certain cubic equations, and to bring out the methods. Thus, in his *al-Bāhir*, al-Samaw'al comments on *al-Badī'*, and includes in his definition of *al-istiqrā'* equations of the form

$$y^3 = ax + b.$$

Al-Samaw'al then claims that, if one of the values of the right side is cubic in form (that is, that it can have a cubic root), the equation will always have solutions. Note that al-Samaw'al considers the case in which $a = 6, b = 10$; now for this value of a, whatever the value of b, the equation has a solution, since one has $y^3 \equiv y \pmod 6$; but, if $a = 7$, then the equation $y^3 = 7x + 2$ has no solution.

He then considers the equation

$$y^3 = ax^2 + bx,$$

that is, the case in which none of the terms in the right-hand member is in the position of form $3k$. Al-Samaw'al then proposes to find a cubic number m^3 such that one of the following two conditions is verified:

$$am^3 + \left(\frac{b}{2}\right)^2 = z^2 \text{ or } bm^3 + \left(\frac{a}{2}\right)^2 = z^2.$$

This does not advance the solution of the problem, but rather transforms it into another problem that is no simpler.

Now is not the time to follow up on the works of al-Karajī's successors in works of rational Diophantine analysis; suffice it to note that henceforth

[43] *L'Algèbre. Al-Badī' d'al-Karajī*, ed. Anbouba, p. 44.

the latter will be a part of every important treatise on algebra. Thus, from the first half of the 12th century, al-Zanjānī borrows most of the problems of al-Karajī and of the first four books of the Arabic version of Diophantus's *Arithmetic*: Ibn al-Khawwām collects some thirty problems, most of which are indeterminate, as a challenge to mathematicians. At least one third of these problems are impossible if restricted to rational numbers. Among these problems one finds Fermat's equation for $n = 3$ ($x^3 + y^3 = z^3$); some of them are reproduced in the big commentary on the latter's algebra by Kamāl al-Dīn al-Fārisī. This interest in indeterminate analysis will have a long life, and works will relentlessly be devoted to it until the 17th century (with al-Yazdī); contrary to what historians of this episode affirm, they will not die out with al-Karajī.

2.2. *Integer Diophantine analysis*

The translation of Diophantus's *Arithmetic* not only was essential for the development of rational Diophantine analysis as a chapter of algebra, but also contributed to the development of integer Diophantine analysis as a chapter, not of algebra, but of number theory. Indeed, in the 10th century, one witnesses for the first time the constitution of this chapter, not only thanks to algebra, but also against it. The study of Diophantine problems was indeed approached by requiring on the one hand that one obtain integer solutions, and on the other that one proceed by demonstrations of the same as Euclid's in the arithmetic book of the *Elements*. What made possible the beginning of the new Diophantine analysis was this explicit combination – for the first time in history – of the numerical domain restricted to positive integers interpreted as line segments, of algebraic techniques, and of the insistence on demonstration in the pure Euclidian style. As one can readily appreciate, the translation of Diophantus's *Arithmetic* offered these mathematicians not so much methods as certain problems in number theory that were formulated therein, which they did not hesitate to systematize and to examine for their own sake, contrary to what one observes in Diophantus. Such are, for example, the problems involved in representing a number as the sum of squares, congruent numbers, etc. In short, one encounters here the beginning of the new Diophantine analysis in the sense in which one finds it developed later in Bachet de Méziriac and Fermat.[44] It is rather surprising that such an event has escaped the notice of historians,

[44] See R. Rashed, 'L'analyse diophantienne au Xe siècle: l'exemple d'al-Khāzin', *Revue d'histoire des sciences*, 32.3, 1979, pp. 193–222; repr. in *Entre arithmétique et algèbre*, pp. 195–225; English transl. in *The Development of Arabic Mathematics*, pp. 205–37.

even of those who had some knowledge of these mathematicians' works.[45] Faced with this lacuna, other historians of mathematics could only relegate number theory in Arabic mathematics to the doldrums. Perhaps the main reason for ignorance of this chapter lies in the absence of a historical perspective which would have shown that this research in integer Diophantine analysis was the work not of one mathematician, but of an entire tradition that – besides al-Khujandī and al-Khāzin – included al-Sijzī, Abū al-Jūd ibn al-Layth, Ibn al-Haytham, as well as later mathematicians such as al-Samawʾal, Kamāl al-Dīn ibn Yūnus, al-Khilāṭī, al-Yazdī...

The 10th-century authors themselves emphasized this innovation. Thus one of them after having given the principle of generation of rectangular triangles in numbers, writes:

> This is the foundation of knowledge of the hypotenuses of primitive right triangles. I have not found this mentioned in any of the books of the ancients, and none of those among the moderns who wrote books of arithmetic gave an exposition of it, and I know that this was revealed to none of my predecessors.[46]

According to this anonymous report, as to others from al-Khāzin, who was one of the founders of this tradition, mathematicians have introduced the fundamental concepts of this new analysis: that of the primitive right triangle (*aṣl al-ajnās*), that of the generator, and especially that of representing the solution in relation to a certain module. It is true that the new domain is organized around the study of numerical rectangular triangles and of congruent numbers, as well as a variety of problems in number theory linked to these two topics.

After having introduced the foundational concepts for the study of Pythagorean triangles, the author of the anonymous text cited above wonders about the integers that can be the hypotenuses of these triangles; that is, the integers that one can represent as the sum of two squares. He states in particular that every element of the series of primitive Pythagorean triples is such that the hypotenuse belongs to one of these two forms: 5 (mod 12) or 1 (mod 12). Like al-Khāzin after him, he notes, however, that some numbers in this series (49 and 77, for example) are not the hypotenuses of such triangles. The same author knew likewise that certain numbers of the form 1 (mod 4) cannot be the hypotenuses of primitive rectangular triangles.

[45] R. Rashed, 'Nombres amiables, parties aliquotes et nombres figurés aux XIIIe–XIVe siècles'; reprinted in *Entre arithmétique et algèbre*, pp. 259–99; English transl. in *The Development of Arabic Mathematics*, pp. 275–319.

[46] R. Rashed, 'Diophantine Analysis in the Tenth Century: al-Khāzin', pp. 209–10.

As to al-Khāzin, he next provides the analysis of the proposition that was demonstrated only by synthesis in the *Elements*, Lemma 1 to Proposition X.29, namely:

Given (x, y, z) a triple of integers such that $(x, y) = 1$, and x is even. The following conditions are equivalent

1° $x^2 + y^2 = z^2$,

2° there exists a pair of integers $p > q > 0$; $(p, q) = 1$ with p and q of opposite parities, such that $x = 2pq$, $y = p^2 - q^2$, $z = p^2 + q^2$.

Al-Khāzin then solves the equation[47]

$$x^2 = x_1^2 + x_2^2 + \dots + x_n^2.$$

His reasoning is general, even though he stops with the case of $n = 3$. He then considers two fourth-degree equations:

$$x^2 + y^2 = z^4 \text{ and } x^4 + y^2 = z^2.$$

Without pausing longer on these studies of numerical triangles by al-Khāzin and later by Abū al-Jūd ibn al-Layth, let us turn to the problem of congruent numbers, that is, the solutions of the system

$$(1) \qquad \begin{aligned} x^2 + a &= y_1^2, \\ x^2 - a &= y_2^2. \end{aligned}$$

The author of the anonymous text had provided the identities

$$(2) \qquad \left(u^2 + v^2\right)^2 \pm 4uv\left(u^2 - v^2\right) = \left(u^2 - v^2 \pm 2uv\right)^2$$

that make it possible to solve (1) if $a = 4uv\,(u^2 - v^2)$. These identities can be deduced directly from the following equation:

$$z^2 \pm 2xy = (x \pm y)^2;$$

[47] R. Rashed, 'Diophantine Analysis in the Tenth Century: al-Khāzin', pp. 213–16.

indeed, by substituting

$$x = u^2 - v^2,\ y = 2uv,\ z = u^2 + v^2,$$

one obtains (2).

Al-Khāzin then demonstrates the following theorem:

Given a natural integer a, *the following conditions are equivalent*:

1° system (1) admits a solution;

2° there is a pair of integers (m, n) such that

$$m^2 + n^2 = x^2,$$
$$2\,mn = a;$$

under these conditions, a is of the form $4\,uv(u^2 - v^2)$.

It was also in this tradition that the study of the representation of an integer as the sum of two squares was undertaken. Thus, al-Khāzin devotes several propositions of his report to this study. During this important research, he displays a direct knowledge of, on the one hand, Proposition III.19 of Diophantus's *Arithmetic* – and thus of the Arabic version of this book – and, on the other, of the identity already encountered in ancient mathematics

$$(p^2 + q^2)\,(r^2 + s^2) = (pr \pm qs)^2 + (ps \mp qr)^2.$$

Al-Khāzin also tries to find integer solutions of the system of Diophantine equations, such as 'to find four different numbers, such that their sum is a square, and that every sum of two of them is a square',[48] that is

$$x_1 + x_2 + x_3 + x_4 = y^2,$$
$$x_i + xj = z_{ij}^2 \qquad (i < j)\left[\binom{4}{2} \text{ equations}\right].$$

These mathematicians are also the first to raise the question of impossible problems, such as the first case of Fermat's theorem. Indeed, it has long been known that al-Khujandī tried to demonstrate that 'the sum of two cubic numbers is not a cube'. Now, according to al-Khāzin, al-Khujandī's

[48] *Ibid.*, pp. 227–9.

demonstration is flawed.[49] A certain Abū Jaʿfar also tries to demonstrate the following proposition:

> It is impossible that the sum of two cubic numbers be a cubic number, whereas it as possible that the sum of two square numbers be a square number; and it is impossible that a cubic number be divisible into two cubic numbers, whereas it was possible for a square number to be divisible into two square numbers.[50]

Abū Jaʿfar's demonstration is also flawed. Although this demonstration was not established until Euler, the problem nevertheless constantly preoccupied Arab mathematicians who later stated the impossibility of the case $x^4 + y^4 = z^4$.

Research on integer Diophantine analysis and notably on numerical rectangular triangles did not end with its originators in the first half of the 10th century. On the contrary, their successors took it up in the same spirit during the second half of the same century and at the beginning of the next, as is attested by the examples of Abū al-Jūd ibn al-Layth, al-Sijzī, and Ibn al-Haytham. Later, others pursued this research in one way or the other, such as Kamāl al-Dīn ibn Yūnus. Let us pause briefly on the writings of Abū al-Jūd and al-Sijzī.

In a treatise on numerical rectangular triangles, Abū al-Jūd ibn al-Layth takes up the problem of forming the latter, of the conditions necessary for the formation of primitive triangles, and especially establishes tables to inscribe, starting from pairs of integers $(p, p + k)$, with $k = 1, 2, 3...$, the sides of the triangles obtained, their areas, and the ratio of these areas to the perimeters. At the end of his treatise, he also returns to the problem of congruent numbers.

In a treatise entitled *Solution by means of a universal method of the numerical problem that is: how to find two square numbers the sum of which is a square number*[51] his junior al-Sijzī elaborates the geometrical foundations of this theory of numerical rectangular triangles. These foundations indeed had to be consolidated so that one could establish by means of the geometry of integers and without exception, the propositions and the algorithms. This ideal is accessible only if the domain of investigation is restricted to quadratic problems. It is precisely at this field that al-Sijzī

[49] *Ibid.*, p. 231.

[50] *Ibid.*, p. 233.

[51] See R. Rashed, *Œuvre mathématique d'al-Sijzī*. Vol. I: *Géométrie des coniques et théorie des nombres au Xe siècle*, Les Cahiers du MIDEO, 3, Louvain/Paris, Peeters, 2004, Chap. II.

aims his treatise, firmly intending to find, as he says, a ‘universal method’. Thus he devotes all of the first part of the treatise to establishing the general case, namely, the Diophantine equation

$$(*) \qquad v^2 = x_1^2 + \ldots + x_n^2 .$$

His procedure consists in searching for the smallest integer t such that

$$2vt = z^2$$

from which he draws

$$(v+t)^2 = x_1^2 + \ldots x_n^2 + t^2 + z^2 ,$$

and thus finds a number that is the sum of $(n + 2)$ squares. He shows that, if one can solve the cases $n = 2$ and $n = 3$, one can solve the general case.

In effect, al-Sijzī proves the following proposition by means of a slightly archaic finite complete induction:

(P_n): *for every* n, *there exists a square that is the sum of* n *squares.*
Thus, he demonstrates first the case p_2, that is

$$x^2 + y^2 = z^2$$

by analysis and synthesis. His analysis in effect amounts to showing geometrically that

$$y^2 = (z - x)\,(z + x);$$

in his synthesis, he takes the even term, that is y^2,

$$y^2 = 2^k b(2a),$$

then $z + x$ is even and one has

$$z - x = 2^k b \text{ and } z + x = 2a,$$

and one finds

$$z = a + 2^{k-1} b \text{ and } x = a - 2^{k-1} b;$$

thus, one finds a solution for every k such that $k > 0$, $2^{k-1}b < a$, that is, $y^2 > 2^{2k}b^2$, $y > 2^k b$, $y^2 = 2^{k+1}ab$ in particular, if $b = 1$, then

$$y^2 = 2^{k+1}\,a, \quad 2 \leq 2^k < y,$$

hence one has one solution if y is divisible by 2 and $y > 2$; three solutions if y is divisible by 4 and $y > 8$, and, more generally, $2h - 1$ solutions if y is divisible by 2^h and $y > 2^{2h+1}$.

Thus, for this case, al-Sijzī demonstrates in several ways that, for $n = 2$, there exists a square that is the sum of two squares.

For the case p_3, that is,

$$x^2 + y^2 + z^2 = t^2,$$

al-Sijzī introduces a condition that makes the construction less general, namely, $t = x + y$. He then shows that, if one has p_n, then one has p_{n+2}; whence a recurrence for n even and a recurrence for n odd.

Al-Sijzī gives a table, up to $n = 9$, that we reproduce here.

Table: example of squares successively composed of squares

line of roots	line of the square originating in the sum of squares	line of the separate squares	line: composition of a square starting from squares successively
10	100	36 64	square from two squares
11	121	4 81 36	square from three squares
30	900	400 400 64 36	square from four squares
9	81	36 36 1 4 4	square from five squares
45	2025	225 400 400 36 64 900	square from six squares
11	121	4 36 36 36 1 4 4	square from seven squares
55	3025	100 900 225 400 400 36 64 900	square from eight squares
33	1089	484 484 4 36 36 36 1 4 4	square from nine squares

One sees that this table is constructed by means of al-Sijzī's rule of recurrence.

Fragment from the *Anthology of Problems* in number theory[52]

To find a number such that, if one adds to it a known number, one will have a square, and if one subtracts the same number from it, it becomes a square.

Assume that the known number is the number *AK*. Divide it into two halves at *B*. Multiply *AB* by itself, namely *BC*. Add to the latter always 1, namely *BG*. If to the numbers *BC* and *BG*, we add the number *AK*, one obtains the square *CG*, for *BD* and *BE*, which are the complements, are equal to *AK*. If from them [the numbers *BC* and *BG*] we subtract *BD* and *BE*, that is, *IH*, *HB*, *HA*, *BG*, since *HB* is twice, the remainder is *CH*, which is a square.

Example: *AK* is 10. To find a number such that if one adds 10 to it, and if one subtracts 10 from it, one gets a square.

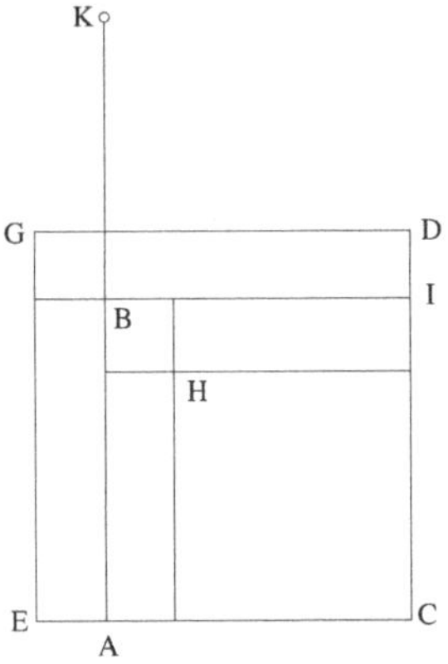

Divide 10 into two halves. One has *AB*, 5; multiply it by itself, one has *BC*, 25. Add to this *BG* which is 1. If we add to it *BE*, *BD* – five, five –, one gets 36, which has a root. If we subtract from it *IH*, which is four, and *HA*, which is four and the double of *HB*, which is two, one has sixteen, which has a root. Q.E.D.

In his *Anthology of Problems*, al-Sijzī tries to solve the following Diophantine problem (double equation):

$$\begin{cases} x + a = y_1^2 \\ x - a = y_2^2 \end{cases}$$

[52] *Œuvre mathématique d'al-Sijzī*. Vol. I: *Géométrie des coniques et théorie des nombres au Xe siècle*, pp. 456–7.

The solution, established geometrically, corresponds to the following algorithm: one calculates $\frac{a}{2}$, squares it, and adds a unit square. To this sum, one adds the number $a = 2 \times \frac{a}{2}$, or else one subtracts the same number from the sum. One obtains the perfect squares $\left(\frac{a}{2}+1\right)^2$ and $\left(\frac{a}{2}-1\right)^2$.

Note that during the demonstration, al-Sijzī makes a segment equal to a surface. Indeed he introduces the unit of length, 1, such that $a = a \times 1$, in order to interpret the segment a as a rectangle. The nonhomogeneous character of the problem makes this detour necessary. Al-Sijzī's solution is not the most general. One finds the general solution in algorithmic form in Diophantus 2.11; recall that it consists in decomposing $\frac{a}{2}$ into two factors u, v and in taking $x = u^2 + v^2$, $y_1 = u + v$, $y_2 = u - v$.

As we have just seen, al-Sijzī's solution is in the spirit of Book II of Euclid's *Elements*.

2.3. *Arithmetic methods in number theory*

As one can ascertain, the works of Abū al-Jūd ibn al-Layth and al-Sijzī in integer Diophantine analysis are indeed in the tradition of al-Khāzin: they borrow the main problems from him and reinforce after a fashion the geometrical methods of proof, all of which firmly establishes its divergence from algebra and from rational Diophantine analysis. Moreover, in the tradition of al-Khāzin and his predecessors, beyond the deliberate use of the Euclidean language of segments to make demonstrations in this field, they occasionally drew on arithmetic arguments like that intended to show that every element in the sequences of primitive Pythagorean triples is such that the hypotenuse is of one of the two forms 5 (mod 12) or 1 (mod 12). Now it is precisely in this direction that Diophantine analysis seems to have moved already in Arabic mathematics, before fully taking this route with Fermat. Instead of using the language of geometry, the goal was to proceed by purely arithmetic means. We do not yet know precisely when this important change of direction took place, but we see it in the works of the later mathematicians. So it is that al-Yazdī devotes a short report to the solution of the same Diophantine equation (*) by purely arithmetic means; in it he studies the different cases as a function of the parity of the x_i and he systematically uses a calculation equivalent to the congruences modulo 4 and modulo 8.[53]

Several results from the work of these mathematicians were transmitted and reappear in Fibonacci's *Liber quadratorum* and sometimes in his *Liber abaci*. It is, however, Fermat's invention of the method of infinite descent in 1640 that will breathe new life into this chapter.[54]

[53] See below, 'Al-Yazdī and the equation $\sum_{i=1}^{n} x_i^2 = x^2$'.

[54] See below.

– 2 –

ALGORITHMIC METHODS

Like their Egyptian and Babylonian predecessors, the Greek mathematicians who carried out arithmetic research invented procedures that one can nowadays call algorithmic. One can make the same point about Chinese and Indian mathematics, as well as many others. Needless to say, these are not algorithms in the strong technical sense in which logicians and programmers use the term; rather, they are procedures that, without being formalized, consist of rules that make possible efficient calculation. In addition, this procedure is iterative in the case of approximations. Consider Pythagorean arithmetic,[1] the so-called arithmetic books of Euclid's *Elements*,[2] and Diophantus's *Arithmetic*.[3] With the exception of the *anthyphairesis*, these procedures were neither formulated nor named. Since they were embedded in the calculations themselves, these procedures came to light only thanks to the historian's research. One good example is the algorithmic method that Diophantus applied to the solution of many indeterminate problems: the 'method of chords' in modern translation. Clearly, if historians want to locate and to unveil most of the algorithms of ancient mathematics, they cannot avoid some form of interpretation.

Outside of arithmetic, however, the appeal to algorithmic methods was very rare. To be sure, Babylonian astronomers had already used linear interpolation, which also appears in Ptolemy. But the fact remains that, in Greek mathematics, algorithmic activity was concentrated in the domain of arithmetic. Schematically, this was the situation before al-Khwārizmī entered the scene.[4] Unprecedented development in the invention and

[1] J. Vuillemin, *Mathématiques pythagoriciennes et platoniciennes*, Paris, Librairie A. Blanchard, 2001.

[2] J. Itard, *Essais d'histoire des mathématiques*, collected and introduced by R. Rashed, Paris, Librairie A. Blanchard, 1984.

[3] Diophantus, *Les Arithmétiques*, transl. and commentary by R. Rashed, 2 vols, Paris, Les Belles Lettres, 1984.

[4] R. Rashed, *Al-Khwārizmī: Le commencement de l'algèbre*, Paris, Librairie A. Blanchard, 2007; English transl. *Al-Khwārizmī: The Beginnings of Algebra*, London, Saqi Books, 2009.

application of algorithmic procedures followed in his wake. Indeed, by comparison with Hellenistic mathematics, Arabic mathematics after al-Khwārizmī reveals a profusion of algorithmic methods, which show up just about everywhere: in arithmetic in the ancient understanding of the term (the 'science of calculation' or *ḥisāb*), but also in algebra, astronomy, and even optics.[5] The number of algorithmic methods and the frequency of their usage are such distinctive characteristics of Arabic mathematics that they already impressed the 19th-century historians. Indeed, precisely this characteristic led some historians who were too subservient to Greek 'geometrism' to label Arabic mathematics computational.

We must now explore the reasons behind this development, the forms that it took, and the new research that it stimulated. One of the main reasons, as we have recently noted, is the birth of algebra. It was algebra that integrated algorithmic approaches into the heart of mathematics, that is, as a demonstrative approach and not merely as a solution procedure. This is a genuine epistemic mutation, one that henceforth creates a cleavage in the history of algebraic methods. Let us pause to reflect on this development.

It is one thing to propose a calculation procedure, a finite set of rules that are applied to one problem after the other; it is quite another to state this finite set of rules independently of any particular instance in order to apply it to a class of problems that are determined *a priori*. In both cases, the goal is, of course, to obtain either a true solution or an effective calculation. Only in the second case, however, can one speak of an algorithm that is not yet formalized, as was already the case with al-Khwārizmī. Indeed, after having defined the primitive terms of the theory of equations of the first two degrees, he determines by combinations of these terms all the equations that must be studied. It is then that he states the operational rules, the application of which makes possible the solution of all these equations in a finite number of steps. These rules allow him in particular to fill in the spaces that were deliberately left open by the givens. The algorithmic method or the algorithm is not presented as a simple solution procedure, but it is such in fact, insofar as it constitutes an integral part of the theory of algebraic equations of the first two degrees. Now this new concept requires another: as one part of a theory, the algorithm must be demonstrable, or at least justifiable. It is precisely this requirement that al-Khwārizmī took up when he wanted to demonstrate the algorithm geometrically, that is, by means of a theory completely devoid of specifics associated with the algorithm.

[5] Cf. below.

After al-Khwārizmī, algebra continuously renewed and extended itself,[6] simultaneously bringing in its wake the development of algorithmic methods. Thus the development of the algebra of polynomials, in the tradition that we have called 'the arithmetization of algebra,' provided theoretical means that proved indispensable for algorithmic research.

With al-Karajī and his successors, one witnesses the generalization of the Euclidean division algorithm to polynomials.[7] An algorithm was invented to extract the square root of a polynomial. The use of tables (to write down the coefficients of polynomials and to carry out arithmetic operations on them by displacing the table's rows and columns) made it possible to manipulate these algorithms easily and effectively.[8] Mathematicians in this tradition invented many other algorithms as they were trying to develop algebraic calculation. And in the other, geometrical tradition of algebra, additional algorithms were also developed to meet the needs of the theory of equations, notably the solution of numerical equations.

Thus algebra evidently not only contributed a new status to algorithmic methods within mathematics, but also fertilized research on algebraic and numerical algorithms. The main concern now was to justify the algorithm in terms of either a general theory, or an *ad hoc* one, as Sharaf al-Dīn al-Ṭūsī would do.

Astronomy would also participate in this movement, most notably from the middle of the 9th century. Note, however, that this development was itself the consequence of a transformation within the discipline, namely the combination of observational astronomy with a mathematical approach to astronomical theory. Such an orientation forces one to take up the problem of interpolation: mathematical astronomers thus invented algorithmic methods to solve the problems involved in interpolating trigonometric functions. The number of these algorithms and the mathematical means put in place to invent them raised new questions, such as the problem of comparing various algorithms to each other in order to find the one with the optimal performance. Al-Bīrūnī and al-Samaw'al, for example, would put much effort into precisely this problem. The theoretical justification of algorithms and the comparison among them to find the best are not only two new themes in mathematical research, but also two specific differences that mark a distinction between finite calculatory procedures and

[6] See above.

[7] Al-Samaw'al, *Al-Bāhir en Algèbre d'al-Samaw'al*, edition, notes and introduction by S. Ahmad and R. Rashed, Damascus, 1972, pp. 22–8 of the French introduction.

[8] *Ibid.*, pp. 28–34.

non-formalized algorithms. These domains were the birthplace of the algorithms that interest us here. After all, every finite calculatory procedure is only one interpretation away from assimilation to an algorithm. Thus, without the aforementioned differences, the history of algorithms would overlap with that of mathematics itself.

1. NUMERICAL EQUATIONS

In the chapter devoted to algebra, we already alluded to the use, by al-Khwārizmī and his successors, of algorithms to solve algebraic equations. Elsewhere, one sees other algorithms put to use for algebraic calculation. Our focus here is exclusively on algorithms invented to solve numerical equations – the 'pure' and the 'affected', to use the terminology of the 17th century, that is, those respectively of the form

(1) $x^n = Q$, where $n \geq 2$ and Q is a natural integer,

and

(2) $\sum_{i=0}^{n} a_i x^{n-i} = 0$, where $a_i \in \mathbf{Z}, a_0 = 1, a_n \neq 0$.

1.1. *The extraction of roots*

The first issue at hand is the very ancient problem of extracting the roots of an integer or rational number, notably the square and cube roots, before the means of extracting the nth root had been invented. In Greek and Arabic mathematics, this invention goes back to al-Karajī, who was the first to establish the binomial theorem. It is therefore after al-Karajī, at the end of the 10th century, that mathematicians will elaborate algorithmic methods to solve equation (1). Without going into many details, let us retrace the stages in the history of these solutions.

As far back as one can go in the history of Arabic mathematics, one finds algorithmic methods for extracting square and cubic roots. One such is of Hellenistic origin (Hero of Alexandria, Theon of Alexandria), whereas others are presumably Indian, and yet others originated with the Arabic mathematicians themselves. Whatever the remote or proximate origins of these algorithmic methods, they were integrated into another mathematics that gave them a new reach by modifying their meaning. So it is that, from the 9th century to the 17th at least, every book of decimal arithmetic (*ḥisāb*) or of algebra included a discussion of the extraction of square and cubic roots, and sometimes more generally of the nth root of an integer. If we emphasize these facts, it is to avoid privileging works such as those of Kūshyār, al-Nasawī or Ibn al-Ḥaṣṣār. The advantage these authors usually enjoy is purely circumstantial: their names appear in the writings of historians simply because their works have been translated into a European language. Our first task will therefore be to retrace at least the most salient points of the tradition to which these works belong, for these are neither the most advanced nor the most profound. Some manuscript texts we have discovered will be of particular value in this undertaking. For obvious reasons, the exposition below cannot touch on all of these 'algorithms'; we therefore discuss only the most important.

Let us begin with al-Khwārizmī. In a book of arithmetic that remains lost and is presently known only from the effects of its Latin translation, al-Khwārizmī proposed, as the mathematician al-Baghdādī (d. 1037) tells us, a formula to approximate the square root of an integer N. If we let $N = a^2 + r$, where a is an integer, this formula is written

$$(1) \qquad \sqrt{N} = a + \frac{r}{2a}.$$

Al-Baghdādī cannot avoid mentioning that this is an approximation by excess,[9] which is far from being satisfactory, as one can readily confirm by trying to find $\sqrt{2}$ and $\sqrt{3}$.

Contemporaneously with al-Khwārizmī, the Banū Mūsā's book *On the Measurement of Plane and Spherical Figures*[10] gave another expression that would later be called 'the rule of zeros', and easily generalized to extract the nth root. The expression is

$$(2) \qquad \sqrt[n]{N} = \frac{1}{m^k}\sqrt[n]{N\, m^{nk}}$$

where m and k are any two integers.

Letting $m = 60$ and $n = 3$ yields the expression of the Banū Mūsā. This rule is found in most books of arithmetic. To give only three examples, it appears in the *Fuṣūl* written by al-Uqlīdisī in 952 to extract square and

[9] *Al-Takmila*, ed. A. S. Saidan, Kuwait, 1985, p. 76.

[10] R. Rashed, *Les Mathématiques infinitésimales du IXe au XIe siècle*. Vol. I: *Fondateurs et commentateurs: Banū Mūsā, Thābit ibn Qurra, Ibn Sinān, al-Khāzin, al-Qūhī, Ibn al-Samḥ, Ibn Hūd*, London, al-Furqān, 1996, p. 56; English translation: *Founding Figures and Commentators in Arabic Mathematics*. A History of Arabic Sciences and Mathematics, vol. 1, Culture and Civilization in the Middle East, London, Centre for Arab Unity Studies, Routledge, 2012, p. 69; cf. also the Latin translation in M. Clagett, *Archimedes in the Middle Ages*, Madison-Philadelphia, 1964, vol. I, p. 350 and his commentary on p. 367.

cubic roots,[11] in the *Takmila* of al-Baghdādī for the cubic root,[12] and in the *Treatise of Indian Arithmetic* by al-Samaw'al (1172/3) for the nth root.

Everything suggests that mathematicians then wanted to find better approximation formulas. Thus al-Uqlīdisī in the aforementioned treatise gives, among other expressions,

$$\sqrt{N} = a + \frac{r}{2a+1}, \tag{3}$$

which will later be called the 'conventional approximation', with $2a + 1$ as 'the conventional denominator', according to the expressions of Naṣīr al-Dīn al-Ṭūsī and later al-Kāshī.

Al-Baghdādī gives the 'conventional approximation' for the cubic root of N, if we let $N = a^3 + r$, where a is an integer

$$\sqrt[3]{N} = a + \frac{r}{3a^2 + 3a + 1}. \tag{4}$$

To avoid getting lost in the details, we leave aside the plethora of formulas that the various mathematicians give both to extract the square and cubic roots and to approximate these roots. We will, however, pause on two contributions from the late 10th century that are linked despite their very unequal significance, namely two that eventually lead to the algorithm of Ruffini-Horner. In his *Arithmetic*, Kūshyār ibn Labbān applies this algorithm, which is very probably of Indian origin. We now know that Ibn al-Haytham not only knew this algorithm, but also tried to give it a mathematical justification. Here we present an exposition of his general approach, but in language that differs from his.

Given the polynomial $f(x)$ with integer coefficients and the equation

$$f(x) = N. \tag{5}$$

Let s be a positive root of this equation, and assume (s_i), $i \geq 0$, a sequence of positive integers such that the partial sums

$$\sum_{i=0}^{k} s_i \leq s;$$

one states that the s_i are parts of s.

[11] Al-Uqlīdisī, *Al-Fuṣūl fī al-Ḥisāb al-Hindī*, ed. A. S. Saidan, 1st ed., Amman, 1973, p. 218 and 313–14; 2nd ed., Alep, I.H.A.S., 1986. English transl., *The Arithmetic of al-Uqlīdisī*, translated and annotated by A. S. Saidan, Dordrecht/Boston, D. Reidel, 1978.

[12] Al-Baghdādī, *al-Takmila*, ed. A. S. Saidan, pp. 76–80 and pp. 84–94.

It is evident that the equation

$$(6) \qquad f_0(x) = f(x + s_0) - f(s_0) = N - f(s_0) = N_0$$

has as roots those of equation (5) minus s_0.

For $i > 0$, let us form by recurrence the equation

$$(7)\, f_i(x) = f(x + s_0 + \ldots + s_i) - f(s_0 + \ldots + s_i)$$

$$= [N - f(s_0 + \ldots + s_{i-1})] - [f(s_0 + \ldots + s_i) - f(s_0 + \ldots + s_{i-1})] = N_i;$$

thus, for $i = 1$, for example, one has

$$f_1(x) = f(x + s_0 + s_1) - f(s_0 + s_1) = [N - f(s_0)] - [f(s_0 + s_1) - f(s_0)]$$

$$= N_0 - [f(s_0 + s_1) - f(s_0)] = N_1.$$

The method, which is named 'Ruffini-Horner' even though Ibn al-Haytham applies and justifies it and Kūshyār uses it, offers an algorithm that allows one to obtain the coefficients of the ith equation from the coefficients of the $(i-1)^{\text{th}}$ equation. Herein lies the main idea of this method.[13]

Let us begin with the extraction of the nth root, which one finds already in the 12th century, if not earlier. One has

$$f(x) = x^n;$$

if one knows the arithmetical triangle and the binomial formula, which, as noted above, al-Karajī[14] produced in the 10th century, there is no need to know Horner's table. The coefficients of the ith equation will then be

$$\binom{n}{k}\left(s_0 + \ldots + s_{i-1}\right)^{n-k} \qquad \text{for } k = 1, \ldots n$$

[13] See our study of Ibn al-Haytham's extraction of the square and cubic roots in *Les Mathématiques infinitésimales du IXᵉ au XIᵉ siècle*. Vol. II: *Ibn al-Haytham*, London, al-Furqān, 1993, Appendix; English transl. *Ibn al-Haytham and Analytical Mathematics*. A History of Arabic Sciences and Mathematics, vol. 2, Culture and Civilization in the Middle East, London, Centre for Arab Unity Studies, Routledge, 2012.

[14] See 'Algebra and its unifying role', above.

(8) and

$$N_i = N_{i-1} - \sum_{k=1}^{n} \binom{n}{k} (s_0 + \ldots + s_{i-1})^{n-k} s_i^k .$$

After these preliminary remarks, let us return to Ibn al-Haytham and Kūshyār for the square and cubic roots. Let

$$f(x) = x^2 = N;$$

one then has two cases:

First case: N is the square of an integer. Let us assume that the root has the form

$$s = s_0 + \ldots + s_h, \qquad \text{with } s_i = \sigma_i \, 10^{h-i} \ (0 \leq i \leq h).$$

The task of the 11th century mathematicians is first to determine h and the numbers σ_i. The formulas at (8) are rewritten

$$2\,(s_0 + \ldots + s_{i-1}), \quad 1, \qquad N_i = N_{i-1} - [\, 2(s_0 + \ldots + s_{i-1})\, s_i + s_i^2].$$

One then determines σ_0 by inequalities

$$\sigma_0^2 \, 10^{2h} \leq N < (\sigma_0 + 1)^2 \cdot 10^{2h}$$

and $\sigma_1, \ldots, \sigma_h$ by

$$\sigma_i = \frac{N_i}{2(s_0 + \ldots + s_{i-1}) \cdot 10^{h-i}} .$$

In these expressions, one calculates the N_i for $(1 \leq i \leq h)$, beginning from N_{i-1}, by subtracting from it $[2(s_0 + \ldots + s_{i-1})s_i + s_i^2]$. For $i = h$, one finds $N_h = 0$.

Second case: N is not the square of an integer. Ibn al-Haytham uses the same method to determine the integer portion of the root and goes on to give as the formula of approximation that of al-Khwārizmī and that of the 'conventional approximation', which are written, respectively, in this notation

$$(s_0 + \dots + s_h) + \frac{N_h}{2(s_0 + \dots + s_h)}$$

and

$$(s_0 + \dots + s_h) + \frac{N_h}{2(s_0 + \dots + s_h) + 1}.$$

He thus not only describes the algorithm, like Kūshyār, but also attempts to give the mathematical reasons for it, by justifying the fact that these two approximations bracket the root.

To extract the cubic root of an integer, the procedure is analogous. Let

$$f(x) = x^3 = N;$$

here also two cases transpire.

First case: N is the cube of an integer. In this case, s_0 is determined such that $s_0^3 < N$. Just like his contemporaries, Ibn al-Haytham then sets $s_1 = s_2 = \dots = s_h = 1$.

The coefficients of the ith equation are rewritten

$$3(s_0 + i)^2, 3(s_0 + i), 1, N_i = N_{i-1} - [3\,(s_0 + (i-1))^2 + 3(s_0 + (i-1)) + 1].$$

If N_i is the cube of an integer, there is some value of i such that $N_k = 0$; that is, such that $(s_0 + k)$ is the root one seeks. Like his contemporaries, Ibn al-Haytham describes in every detail the different steps of the algorithm.

Second case: N is not the cube of an integer. Ibn al-Haytham also gives two formulas that are symmetrical to those already mentioned for the extraction of the square root, and that are rewritten

$$(s_0 + \dots + s_h) + \frac{N_h}{3(s_0 + \dots + s_h)^2}$$

and

$$(s_0 + \dots + s_h) + \frac{N_h}{3(s_0 + \dots + s_h)^2 + 3(s_0 + \dots + s_h) + 1};$$

in the latter, one recognizes the 'conventional approximation'.

Acquired at the beginning of the 11th century, the whole of the preceding methods and results recur not only among the contemporaries of these mathematicians, but in the majority of the many subsequent treatises of arithmetic. These include – among many others – the treatises of al-Nasawī (the successor of Kūshyār), of Naṣīr al-Dīn al-Ṭūsī, of Ibn al-Khawwām al-Baghdādī, of Kamāl al-Dīn al-Fārisī, etc.[15]

Since they knew the arithmetic triangle and the binomial formula, as we have often emphasized, mathematicians encountered no major difficulties in generalizing the preceding methods or in formulating the algorithm for the case of the nth root. In fact, although they are now lost, we know that attempts such as these had already taken place in the 11th century, with al-Bīrūnī and al-Khayyām. Testifying to their activities, the ancient bibliographies contain the titles of the works they devoted to this research, but these remnants say nothing about the mathematicians' methods. It is in his contribution of 1172/3 that al-Samaw'al not only applied the so-called Ruffini-Horner method to extract the nth root of a sexagesimal integer, but also formulated a clear concept of approximation.[16] For this 12th-century mathematician, 'to approximate' is to know a real number by means of a series of known numbers such that the latter differs from the number by a quantity as small as the mathematician wishes. The point is therefore to measure the interval between the irrational n^{th} root and a sequence of rational numbers. After having defined the concept of approximation, al-Samaw'al begins by applying the so-called Ruffini-Horner method to the example

$$f(x) = x^5 - Q = 0,$$

with $Q = 0\,;0,0,2,33,43,3,43,36,48,8,16,52,30.$

This method survived into the 12th century and appeared in many other treatises of 'Indian arithmetic', as they were then called. Even later, it

[15] See H. Suter, 'Über das Rechenbuch des Alī ben Ahmed el-Nasawī', *Bibliotheca Mathematica*, III, 7, 1906/7, pp. 113–19; al-Nasawī, *Nasawī Nāmih*, ed. Abū al-Qāsim Qurbānī, Tehran, 1973, pp. 65 ff. of the Persian introduction to the edition, and 8 ff. of the reproduction of the published Arabic text; Naṣīr al-Dīn al-Ṭūsī, 'Jawāmiʿ al-ḥisāb bi-al-takht wa-al-turāb (Arithmetic Complete, by Board and Dust)', ed. A. S. Saidan, *al-Abhath*, XX, 2, June 1967, pp. 91–164 and 3, Oct. 1967, pp. 213–29, at pp. 141 ff. and 266 ff.; Ibn al-Khawwām, *Al-Fawā'id al-bahā'iyya fī al-qawāʿid al-ḥisābiyya*, ms. British Library, Or. 5615, fols 7^v and 8^r).

[16] See R. Rashed, 'Nombres amiables, parties aliquotes et nombres figurés aux XIIIe-XIVe siècles', *Archives for History of Exact Sciences*, 28, 2, 1983, p. 107–47; English translation, *The Development of Arabic Mathematics: Between Arithmetic and Algebra*, Boston Studies in the Philosophy of Science 146, Dordrecht, 1994, pp. 275–319.

would surface among the predecessors of al-Kāshī, in al-Kāshī himself, and among his successors as well as. To cite only his example, in his *Key of Arithmetic*, he solves

$$f(x) = x^5 - N = 0,$$

with $N = 44\ 240\ 899\ 506\ 197$.

The point is that this is a well-known and diffused method since the 12th century, at least among Arabic mathematicians. Far from being unique, there are many others, all based on knowledge of the binomial theorem, without requiring any appeal to Horner's algorithm. We also want to emphasize the multiplicity and the diffusion of these methods, which circulated not only in the fundamental treatises of arithmetic, but also in those of second-rate commentators and mathematicians. Suffice it to cite one random example taken from among many authors who have never before been studied. The case is that of al-Aḥdab of Kairouan (Kirwan), who lived before 1241 and commented on the text of Abū al-Majd ibn ʿAṭiyya, a second-tier mathematician also from Kairouan. Ibn ʿAṭiyya[17] establishes a method for extracting the nth root, proves it, and gives numerical examples. He thus gives the example of the 5th root of $N = 4\ 678\ 757\ 435\ 232$. Ibn ʿAṭiyya assumes that the root has the form $(a + b + c)$, with $a = \alpha \cdot 10^2$, and $b = \beta \cdot 10$. Here are the main steps of his algorithm:

He first writes $N - a^5 = N_1$, then calculates

$$\sum_{k=1}^{5} \binom{5}{k} a^{5-k}.$$

Next, he multiplies the terms of this expression respectively by b, b^2, b^3, b^4 and b^5 to obtain

$$\sum_{k=1}^{5} \binom{5}{k} a^{5-k} b^k$$

and calculates

$$N_2 = N_1 - \sum_{k=1}^{5} \binom{5}{k} a^{5-k} b^k .$$

Next, he calculates

[17] Ms. London, British Library, 7473, especially beginning from fols 367^r–374^r.

$$\sum_{k=1}^{5}\binom{5}{k}(a+b)^{5-k};$$

he multiplies these terms respectively by c, c^2, c^3, c^4 and c^5 to obtain

$$\sum_{k=1}^{5}\binom{5}{k}(a+b)^{5-k}c^k,$$

and to reach

$$N_3 = N_2 - \sum_{k=1}^{5}\binom{5}{k}(a+b)^{5-k}c^k = 0.$$

When we turn to the extraction of the nth irrational root of an integer, we find an analogous situation. In his *Treatise of Arithmetic*, al-Samaw'al gives a rule for approximating by fractions the non-integer portion of the irrational root of an integer. His procedure amounts to solving the numerical equation

$$x^n = N;$$

he begins by seeking the greatest integer x_0 such that $x_0^n \leq N$. Two cases emerge:

1° $x_0^n = N \Leftrightarrow x_0$ is the exact root that was sought. As we have seen, al-Samaw'al has access to a sure method of obtaining this result when it is possible.

2° $x_0^n < N \Leftrightarrow N^{1/n}$ is irrational. In this case, he states as a first approximation

$$x' = x_0 + \frac{N - x_0^n}{\left[\sum_{k=1}^{n-1}\binom{n}{k}x_0^{n-k}\right]+1},$$

that is,

$$x' = x_0 + \frac{N - x_0^n}{(x_0+1)^n - x_0^n}.$$

This is therefore the generalization of what mathematicians have called the 'conventional approximation'.

This approximation by defect shares the same nature as that of al-Samaw'al's Arabic predecessors, but it is much more general. Indeed, whereas the arithmeticians who had assimilated al-Karajī's results limited the application of this method to powers ≤ 3, the rule here is extended to any power, such as one encounters it among so many later mathematicians, such as Naṣīr al-Dīn al-Ṭūsī and al-Kāshī. Not least, it was to improve these approximations that decimal fractions were explicitly conceived, as the example of al-Samawʾal shows.

1.2. *The extraction of roots and the invention of decimal fractions*

In the middle of the 10th century, al-Uqlīdisī reached an intuitive idea of decimal fractions as he was studying the division of odd integers by 2. He writes:

> [...] the half of one in any place is 5 before it. Accordingly, if we halve an odd number we set the half as 5 before it, the units place being marked by a sign ′ above it, to denote the place.[18]

Although commendable and accompanied by a very convenient principle of notation, this result nevertheless constitutes neither a genuine theory of decimal fractions, nor an explicit recognition of the latter. It merely provides us with an empirical rule for the case of division by 2. Not until the algebraists from the school of al-Karajī does one encounter a fully general theoretical discussion. These mathematicians very naturally felt the necessity of these fractions while pursuing as far as they wished the approximation of the nth irrational root of an integer. To invent these fractions, they put to good use the algebra of polynomials, its rules, and its means of representation. The first known exposition of these fractions, which al-Samawʾal[19] gave in 1172/3, leaves no doubt about either the algebraic means or the anticipated goal and applications. Indeed, in al-Samawʾal's book *al-Qiwāmī fī al-ḥisāb al-hindī*, this exposition immediately follows the chapter devoted to the approximation of the nth root of an integer. The very title of the chapter on decimal fractions is laden with significance:

> Concerning the positing of the unique principle by which one can determine all the operations of 'partition' (*al-tafrīq*), which are division, extraction of the square root, extraction of one side for all powers, and the correction, indefinitely, of all the fractions that appear in these operations.[20]

[18] Al-Uqlīdisī, *Al-Fuṣūl fī al-Ḥisāb al-Hindī*, ed. A. S. Saidan, 1st ed., p. 145; English transl., p. 110.

[19] See R. Rashed, 'L'extraction de la racine $n^{\text{ième}}$ et l'invention des fractions décimales', *Archive for History of Exact Sciences*, 18.3, 1978, pp. 191–243; English transl. in *The Development of Arabic Mathematics*, pp. 85–146.

[20] *Al-Qiwāmī fī al-Ḥisāb al-hindī*, fol. 111v, ed. R. Rashed in *The Development of Arabic Mathematics*, p. 137.

The 'unique principle' to which al-Samaw'al alludes is no other than that already recognized in algebra and that he has already explained in his book *al-Bāhir*, namely that, on either side of x^0, one has an identical structure. It thus suffices to substitute 10^0 for x^0, and for the other algebraic powers, the power of 10, in order to obtain the integers and the decimal fractions. As al-Samaw'al writes:

> Given that the proportional positions beginning with the position of the units [10^0] succeed one another indefinitely according to the proportion of the tenth, we assume, on the other side [of 10^0], that the positions of the parts <of ten succeed each other> according to the same proportion, and the position of units [10^0] is intermediary between the positions of the integers, the units of which likewise are displaced indefinitely, and the positions of the parts that are indefinitely divisible.[21]

Continuing his explanation, al-Samaw'al produces a table that we transcribe below with the substitution of 10^n for verbal expressions and without recording all the positions:

10^{13}	10^{12}...	10^9...	10^6 ...	10^3...	10	10^0	10^{-1} ...	10^{-3} ...	10^{-6} ...	10^{-9} ...	10^{-12}	10^{-13}
	4	3	2	1		0		1	2	3	4	

To write the fractions, al-Samaw'al separates the integer from the fractional part, by noting either the numbers of the different positions, or the denominator

10^0	10^{-1}	10^{-2}	10^{-3}	10^{-4}	10^{-5}	10^{-6}
3	1	6	2	2	7	7

or

3
162277
1000000

In the same algebraic tradition as al-Samaw'al, al-Kāshī (d. 1436/7) much later takes up the theory of decimal fractions and gives an exposition that reveals a great mastery of theory and calculation. He insists on the analogy between the sexagesimal and the decimal systems, and uses fractions to approximate not only real algebraic numbers, but also the number π out to 10^{-16}. What is more, he is, as far as we know, the first to name these fractions *al-kusūr al-aʿshāriyya*, that is, 'decimal fractions'.[22]

[21] *Ibid.*

[22] See al-Kāshī, *Miftāḥ al-Ḥisāb*, ed. A. S. al-Dimirdash and M. H. al-Hifnī, Cairo, 1967, pp. 79 and 121; P. Luckey, *Die Rechenkunst bei Jamshīd B. Masʿūd al-Kāshī*, Wiesbaden, 1951, p. 103; R. Rashed, *The Development of Arabic Mathematics*, pp. 127 ff.

Decimal fractions outlived al-Kāshī in the writings of the 16th-century astronomer and mathematician Taqī al-Dīn ibn Maʿrūf[23] and of al-Yazdī. In al-Yazdī's treatise *ʿUyūn al-ḥisāb*, one cannot avoid noticing a certain familiarity with decimal fractions, even though he prefers to calculate with sexagesimal and ordinary fractions.[24] Several clues suggest that decimal fractions were transmitted to the West before the middle of the 17th century. In a Byzantine manuscript brought to Vienna in 1562, they are called the fractions of 'the Turks'. Al-Kāshī introduces a vertical line that separates the integer from the fractional part – a representation that appears in such Europeans as Rudolff, Apianus, and Cardano. The mathematician Mizraḥi (born in Constantinople in 1455) was using the same sign before Rudolff did. As to the Byzantine manuscript, it states, among other things that 'the Turks carried out multiplications and divisions on fractions using a special calculatory procedure. They had introduced their fractions when they governed our land here'. The example that the mathematician gives leaves no doubt that he was referring to decimal fractions.[25]

1.3. *Numerical polynomial equations*

Not until al-Khayyām and the elaboration of the geometrical theory of cubic equations does one encounter the systematic study of numerical equations in Arabic mathematics. Thereafter, mathematicians occasionally treated one or the other of these equations but, to my knowledge, no one had thought of inventing an effective algorithm to solve them. Thus, according to al-Khayyām's report,[26] some mathematicians – and not the least, since they included al-Būzjānī, al-Qūhī, and al-Ṣāghānī – debated the solution of the equation at the royal court of ʿAḍud al-Dawla.

$$20x^2 + 2000 = x^3 + 200x,$$

which Abū al-Jūd ibn al-Layth would solve. Al-Khayyām himself solved the equation

$$x^3 + 2x^2 + 10x = 20,$$

[23] *Bughyat al-ṭullāb*, fol. 131^r ff.

[24] Cf. Ms. Istanbul, Hazine 1993, for example fols 9^v, 49^{r-v}.

[25] Cf. H. Hunger and K. Vogel, *Ein byzantinisches Rechenbuch des 15. Jahrhunderts*, Vienna, 1963, p. 32, Problem 36.

[26] R. Rashed and B. Vahabzadeh, *Al-Khayyām mathématicien*, Paris, Librairie Blanchard, 1999, pp. 254–6; English transl. *Omar Khayyam. The Mathematician*, Persian Heritage Series no. 40, New York, Bibliotheca Persica Press, 2000 (without the Arabic texts), pp. 173–4.

for which Fibonacci[27] would later give a long solution. Not until al-Khayyām's successor, Sharaf al-Dīn al-Ṭūsī, did anyone invent not only such an algorithm, but also one that was generalizable, as well as provide a theory to justify it. Everything happened as if the absence of a solution by means of radicals, the presence of the solution by means of the intersection of conic curves, and finally the invention of 'analytical' means of ascertaining the existence of positive roots opened wide the gates of the treatment of numerical equations. Henceforth, a treatise called *al-Muʿādalāt* (*On Equations*), like that of al-Ṭūsī, composed *c*. 1180, includes a study of the numerical solution of these very equations. In this treatise, one encounters an algorithm to determine the positive root of quadratic and cubic equations. The use of this algorithm to solve quadratic equations, which were previously solved by means of radicals, evidently shows that al-Ṭūsī had set himself apart from the two types of study, the algebraic and the numerical, and that he intended to treat the numerical solution as a topic in its own right.

To solve equation (2) (p. 368), Sharaf al-Dīn al-Ṭūsī thus invented an optimal algorithm, in the modern meaning of the term, in order to yield an efficient and 'economical' calculation in his search for a positive root.[28] We will describe as briefly as possible al-Ṭūsī's procedure.

After rigorously demonstrating the existence of this root by using algebraico-analytical concepts and methods,[29] he turns to the determination of this root for numerical equations. Let s be this positive root of (2), and let its decimal expansion be

$$s = \sigma_0 10^r + \sigma_1 10^{r-1} + \ldots + \sigma_{r-1} 10 + \sigma_r ,$$

with $s_i = \sigma_i 10^{r-i}$, r the decimal rank of s and $r + 1$ numbers σ_i the digits that constitute root s.

To determine s, one need only calculate successively the digits σ_i.

Two stages make up al-Ṭūsī's method. The first is devoted to the determination of σ_0 and of r. It is to solve this problem that al-Ṭūsī invents an *ad hoc* theory about the dominant polynomials in (2). In the second stage, he forms from (2) an equation that admits $(s - s_0)$ as a root. From

[27] *Omar Khayyam. The Mathematician*, p. 88 and R. Rashed, *The Development of Arabic Mathematics*.

[28] See R. Rashed, *The Development of Arabic Mathematics* and *Sharaf al-Dīn al-Ṭūsī, Œuvres mathématiques. Algèbre et géométrie au XIIe siècle*, 2 vols, Collection Sciences et philosophie arabes – textes et études, Paris, Les Belles Lettres, 1986.

[29] See 'Algebra and its unifying role,' above.

this new equation, he seeks to determine σ_1, in order to form next an equation that admits $((s - s_0) - s_1)$ as a root. He repeats the procedure as often as necessary.

After having calculated σ_0, al-Ṭūsī then defines by recurrence a sequence (E_k) of polynomials by

$$\begin{aligned} &(E_0) \quad f_0(x) = f(x) \\ &(E_k) \quad f_k(x) = f_{k-1}(x + s_{k-1}) \end{aligned} \qquad 1 \leq k \leq r.$$

One can easily confirm that the roots of (f_k), $1 \leq k \leq r$ are those of f_{k-1}, minus s_{k-1}; they are therefore those of (E_0) minus $(s_0 + s_1 + \ldots + s_{k-1})$.

It is moreover evident that s_r is a root of (E_r).

One can therefore formulate al-Ṭūsī's procedure as follows: the equation (E_0) allows one to carry out the calculation in a first time interval s_0, that is, σ_0 and r; and by recurrence, in a second time interval, we form the equations that make possible the calculation, successively, of the σ_k by means of the following equation:

$$(3.1) \qquad s_k = \frac{-f_{k-1}(s_{k-1})}{f_{k-1}^{(1)}(s_{k-1})}.$$

For this procedure to be effective, however, the problem is to find an algorithm to the (E_k) equations by recurrence. Indeed, this algorithm must allow one to calculate the coefficients of equation (E_k) from those of (E_{k-1}). What we have here, as we shall see, is the famous algorithm named for Ruffini-Horner.

Recall briefly that this algorithm is one that makes it possible to calculate systematically, in the simplest and most 'economical' way, the coefficients of an equation whose roots are those of another equation, reduced by a fixed number. One can apply this scheme to our equation in order to reduce one of its roots by its first digit; one applies it again to reduce the corresponding root of the new equation that results from it, also by its first digit, that is, by the second digit of the root under consideration of the initial equation, and so on.

Thus, let the polynomial equation be

$$(3.2) \qquad F(x) = A_0 x^N + A_1 x^{N-1} + \ldots + A_{N-1}x + A_N = 0\,,$$

and let Δ be a fixed number. By change of variable $x \mapsto x + \Delta$, (3, 2) is rewritten

$$(3.3) \qquad (x^N/N!)F^{(N)}(\Delta) + (x^{N-1}/(N-1)!)F^{(N-1)}(\Delta) + \ldots + (x/1!)F^{(1)}(\Delta) + F(\Delta) = 0.$$

One obtains

$$(1/N!)F^{(N)}(\Delta) = A_0 = B_0.$$

Assume that

$$(3.4) \qquad B_i = (1/(N-i)!)F^{(N-i)}(\Delta) \qquad\qquad (0 \le i \le N).$$

It is evident that the roots of (3.3) are those of (3.2), each reduced by Δ.

The next scheme makes possible the systematic formation of all the other elements from those of the first line, which are the coefficients of (3.2).

The only elements remaining to be defined in the following scheme are the $B_{i,k}$. By definition, every element $B_{i,k}$ is the sum of two elements that are directly above it. One can verify that the coefficients of (3.3) just are the diagonal elements in this scheme:

$$\begin{array}{llllll}
A_0 & A_1 & A_2 & \ldots & A_{N-1} & A_N \\[2ex]
 & \dfrac{\Delta A_0}{B_{0,1}} & \dfrac{\Delta B_{0,1}}{B_{0,2}} & & \dfrac{\Delta B_{0,N-2}}{B_{0,N-1}} & \dfrac{\Delta B_{0,N-1}}{B_{0,N} = B_N} \\[3ex]
 & \dfrac{\Delta A_0}{B_{1,1}} & \dfrac{\Delta B_{1,1}}{B_{1,2}} & & \dfrac{\Delta B_{1,N-1}}{B_{1,N-1} = B_{N-1}} & \\[2ex]
 & \ldots & \ldots & & \ldots & \\
 & B_{N-2,1} & & & & \\[3ex]
 & \dfrac{\Delta A_0}{B_{N-1,1} = B_1} & & & & \\[2ex]
A_0 = B_0 & & & & &
\end{array}$$

$$B_0 = (1/N!)F^{(N)}(\Delta) = A_0,\ B_i = B_{N-i,i} = (1/(N-i)!)F^{(N-i)}(\Delta);$$

N, Δ and the coefficients $A_0, \ldots, A_N$ of (3.2) are, by definition, the *inputs* of the scheme and $B_0, B_1, \ldots, B_N$ are the *outputs*.

For this scheme, we will use the notation $SCH(N; \Delta; A_0, \ldots, A_N)$, or simply *SCH*, if there is no risk of confusion. And we will use the notation $SCH(n; \delta; c_0, \ldots, c_n)$ for the result of this scheme for $N = n$, $\Delta = \delta$, $A_i = c_i$.

Al-Ṭūsī's algorithm is the preceding one, making allowances for a few modifications that simplify the preceding scheme.[30] Here, we will go into neither these modifications, nor the tabular translation of the algorithm. It is in these tables that al-Ṭūsī details the different steps of the algorithm.[31] Only one example must suffice to recall the different steps of the algorithm and the successive filling-in of the tables. Given the equation

$$x^3 + ax^2 + bx = N \quad \text{with } a = 12, b = 102, N = 34345395$$

where $r = 2$, $\sigma_0 = 3$, $s_0 = 300$.

Let us begin by writing this equation in the form

$$x^3 + 3a'x^2 + 3b'x = N, \qquad \text{with } a' = 4, b' = 34.$$

Al-Ṭūsī begins with a first table to place $a', b', |-c| = N, \sigma_0$.

I. To place the entries in the table

- write the absolute values of the constant c and determine the decimal positions of the form $3k$, $k \in Z$;
- mark these positions with a 0 above each;
- place $b'10^r$ by placing b' in the position with the decimal rank r;
- place $a'10^{2r}$ by placing a' in the position with the decimal rank $2r$;
- place $\sigma_0 10^{3r}$ by placing σ_0 in the position with the decimal rank $3r$.

II. Calculation to obtain c_1

- calculate $s_0^3 = \sigma_0^3 10^{3r}$, by placing σ_0^3 in the position with the decimal rank $3r$;
- calculate $(a's_0 + b')$, which will be placed in the position with the decimal rank r;

[30] R. Rashed, *Sharaf al-Dīn al-Ṭūsī, Œuvres mathématiques*, vol. I, pp. LXXX–LXXXVI.

[31] *Ibid.*, pp. LXXXIX–XCVII.

- calculate $3(a's_0 + b')\sigma_0$, which will be placed in the position with the decimal rank r;
- deduce $-c_1$, by computing $|c| - 3s_0(a's_0 + b') - s_0^3$.

III. Calculation to obtain b_1

- calculate $a'\sigma_0$, which will be placed in the position with the decimal rank $2r$;
- calculate σ_0^2, which will be placed in the position with the decimal rank $3r$;
- retranscribe $a's_0$ and place it in the decimal rank $2r$;
- deduce $b_1' = (2a's_0 + b') + s_0^2$, by adding $a's_0 + b' + s_0^2$ to $a's_0$, which will be placed in the position with the decimal rank $r - 1$.

IV. Determine $\sigma_1 = \frac{1}{3}E\left(-\frac{c_1}{10^r b_1'}\right)$.

V. Repeat the preceding steps $r - 1$ times to determine $\sigma_2, \ldots, \sigma_r$; here $r = 3$.

If one regroups the different tables, one obtains the following table for the equation. All the operations can easily be recognized. The arrows indicate the position of the part of the decimal rank during the operation.[32]

We have yet to summarize al-Ṭūsī's argument during the determination of $s_0, s_1, \ldots, s_r$. As we have said, al-Ṭūsī proposes for s_0 an *ad hoc* theory that we can call 'the theory of the dominant polynomial'. To explain al-Ṭūsī's idea concisely, still in a language different from his, let us return to (2).

Indeed, the continuity of f shows immediately that f changes sign once over **R**+. More precisely:

$$\text{(4)} \qquad \text{if} \quad \begin{cases} 0 < x_1 < s \\ 0 < x_2 < s \end{cases} \text{or} \begin{cases} x_1 > s \\ x_2 > s \end{cases}, \text{then } f(x_1)f(x_2) > 0.$$

$$\text{(5)} \qquad \text{if} \quad \begin{cases} x_1 < s \\ x_2 > s \end{cases}, \text{ then } f(x_1)f(x_2) < 0.$$

[32] *Sharaf al-Dīn al-Ṭūsī, Œuvres mathématiques*, vol. I, p. XCVI.

(2.0)	$\xleftarrow{r}\sigma_o \xleftarrow{r-1}\sigma_1\,\sigma_2$	3 2 1
(1.0)	$\xleftarrow{3r-2}\sigma_o \xleftarrow{3r-1}\sigma_1$	3 2
(0.0)	$\xleftarrow{3r}\sigma_o$	3
		o o o
(0.1.3)	$\mathrm{N} = -c$	3 4 3 4 5 3 9 5
	$-\xleftarrow{3r}\sigma_o^3 = -s_o^3$	− 2 7
(0.2.3)	$-\xleftarrow{r}(a's_o+b')\sigma_o \times 3$	− 1 1 1 0 6
$(0.3.3)_+$	$\mathrm{N}-(as_o+b)-s_o^3 =$	
(1.1.3)	$-\ c_1$	6 2 3 4 7 9 5
	$-\xleftarrow{3(r-1)}\sigma_1^3 = -s_1^3$	− 8
(1.2.3)	$-\xleftarrow{r-1}(a_1's_1+b_1')\sigma_1 \times 3$	− 5 9 1 0 8 4
$(1.3.3)_+$	$-c_1-(a_1s_1+b_1)s_1-s_1^3 =$	
(2.1.3)	$-\ c_2$	3 1 5 9 5 4
	$-\sigma_2^3 = s_2^3$	1
(2.2.3)	$-(a_2's_2+b_2')\sigma_2 \times 3$	3 1 5 9 5 4
$(2.3.3)_+$	$-c_2-(a_2s_2+b_2)s_2-s_2^3 = 0$	0 0 0 0 0 0
(0.1.2)	$\xleftarrow{r} b'$	3 4
(0.2.2)	$(\xleftarrow{2r} a')\sigma_o = \xleftarrow{r} a's_o$	1 2
(0.3.2)	$\xleftarrow{r}(a's_o+b')$	1 2 3 4
	$\xleftarrow{3r}\sigma_o^2 = \xleftarrow{r} s_o^2$	9
(0.4.2)	$(\xleftarrow{2r} a')\,\sigma_o = \xleftarrow{r} a's_o$	1 2
$(0.5.2)_+$	$\xleftarrow{r}\{(2a's_o+b')+s_o^2\} = \xleftarrow{r} b_1'$	9 2 4 3 4
(1.1.2)	$\xleftarrow{r-1} b_1'$	9 2 4 3 4
(1.2.2)	$(\xleftarrow{2r-2} a')\sigma_1 + (\xleftarrow{3r-2}\sigma_o)\sigma_1$	6 8
(1.3.2)	$\xleftarrow{r-1}(a_1's_1+b_1')$	9 8 5 1 4
	$\xleftarrow{3(r-1)}\sigma_1^2 = \xleftarrow{r-1} s_1^2$	4
(1.4.2)	$(\xleftarrow{2(r-2)} a')\sigma_1 + (\xleftarrow{3r-2}\sigma_o)\sigma_1$	6 8
$(1.5.2)_+$	$\xleftarrow{r-1}(2a_1's_1+b_1')+s_1^2\} = \xleftarrow{r-1} b_2'$	1 0 4 9 9 4
(2.1.2)	b_2'	1 0 4 9 9 4
(2.2.2)	$a'\sigma_2 + (\xleftarrow{r}\sigma_o)\,\sigma_2 + (\xleftarrow{r-1}\sigma_1)\sigma_2$	3 2 4
(2.3.2)	$a_2's_2+b_2'$	1 0 5 3 1 8
(0.1.1)	$\xleftarrow{2r} a'$	4
(1.1.1)	$\xleftarrow{2r-2} a'$	4
(2.1.1)	a'	4

But since $0 < \sigma_0 10^r \le s < (\sigma_0 + 1)10^r$, the inequality

(6) $$f(\sigma 10^r)\cdot f((\sigma+1)10^r) \le 0$$

is verified for $\sigma = \sigma_0$. The equality corresponds to the case $s = \sigma_0 10^r$, that we will set aside in what follows. More precisely, one can state the following result: the only digit that verifies the preceding inequality is σ_0. And one obtains

(7) $$f(\sigma_0 10^r) < 0 \text{ and } f((\sigma_0+1)10^r) > 0.$$

Theoretically at least, it is therefore possible, starting from $f(x)$, to determine σ_0 and r with the help of relation (6). In all these relations, however, we have considered all the terms of f.

By contrast, al-Ṭūsī's major idea is to stop drawing on all the terms and to use only a smaller number of them. And in fact, there exists in general a polynomial f_1, formed from the terms of f, that depends on s and such that the relation

$$(8) \qquad f_1(\sigma_0 10^r) < 0 \text{ and } f_1((\sigma_0 + 1)10^r) > 0$$

is equivalent to relation (7). This polynomial f_1 is a dominant polynomial.

Now, to find the dominant polynomials, one must return to equation (2). A knowledge of the decimal ranks of the numbers $|a_i s^{n-i}|$, $a_i \neq 0$, is indeed essential. But s is unknown; and the decimal ranks and the numbers $|a_i s^{n-i}|$ are as well.

To solve this problem al-Ṭūsī proceeds by comparing the ranks of the parameters,[33] that is, the absolute values of the coefficients a_i contained in polynomial f. This method of the dominant polynomial, however, does not always lead to a result. It is not always easy to obtain this polynomial, and one sometimes gets an incorrect result when al-Ṭūsī's conditions are met.

To understand better al-Ṭūsī's theory of the dominant polynomial, one can consider it, at least at the algorithmic level, as similar to Newton's polygon method.[34]

Now that al-Ṭūsī has obtained the first digit of the root and justified the method applied to determine it, he takes up the other digits. He could have applied the same method of the dominant polynomial. But this time, the method yields a dominant polynomial reduced to the constant term and the first-degree term. Now the coefficient of this last term is nothing but the derivative, which moreover enters into the determination of the maximum of a third-degree polynomial in the theory of algebraic equations al-Ṭūsī has developed. As in the latter, he proceeds first to a change of affine variable and substitutes $s_0 + y$ for x in (2). He then expands $f(s_0 + y)$ and brings out, for the coefficient of y, $f'(s_0)$, which is equivalent to the derivative of s_0. Al-Ṭūsī obtains the highest digit of y, that is, the second digit of the root

[33] *Ibid.*, pp. LXIV–LXVII.

[34] C. Houzel, 'Sharaf al-Dīn al-Ṭūsī et le polygone de Newton', *Arabic Sciences and Philosophy*, 5.2, 1995, pp. 239–62.

that was sought, by taking the integer portion of $\left(\frac{-f(s_0)}{f'(s_0)}\right)$. For the case of the cubic equation that we have considered, this is $\frac{1}{3}E\left(\frac{-c_1}{10^r b_1'}\right)$. The procedure is iterative, and al-Ṭūsī himself uses it for the third digit. Al-Ṭūsī also justifies this procedure, or this part of the algorithm, by his own research in his *Treatise* and by relying on an expression equivalent to the derivative during his research on the *maxima* of algebraic expressions.

Al-Ṭūsī elaborated his algorithm, as well as the justification for it, deliberately for numerical polynomial equations. He even omitted applying it to the case of $x^3 = a$, whereas he would take care to apply it to second-degree trinomial equations as well as to all other cubic equations. He undertook his algorithmic research in natural language and by means of tables. It is true that the latter made possible an effective and parsimonious calculation on polynomials; at the time, this was the way to compensate for a symbolism that had not yet been invented. Like his contemporary al-Samaw'al as well as the successors of al-Karajī, al-Ṭūsī wrote down the coefficients in the cells of a table and undertook his calculation by using displacements in the rows and the columns. Whereas this technique made such a calculation possible, it was obviously unsuited for the expression of the foundational concepts of the algorithm. In sum, armed only with natural language and tables, it would be difficult to imagine going much farther than al-Ṭūsī did. At best, one could deduce what the algorithm already contains, namely its extension to roots with a fractional component. And this is precisely what one encounters in al-Iṣfahānī.[35] For the concept of a truly efficacious language, one would have to wait until late in the 17th century.

Alongside this contribution of al-Ṭūsī's, whose explicit goal, as we have noted, is to obtain an algorithm to solve numerical polynomial equations, mathematicians proposed other algorithmic procedures to solve one equation or the other. Mathematical astronomers in particular had the idea of translating geometrical or trigonometrical problems into algebraic language. This is what al-Bīrūnī did when he wanted to determine the sign of the regular enneagon (the so-called 'nonagon'), which led him to one or the other equation[36]

[35] R. Rashed, *Sharaf al-Dīn al-Ṭūsī, Œuvres mathématiques*, vol. I, pp. 118–25.

[36] See M.-T. Debarnot, 'Trigonometry', in R. Rashed (ed.), *Encyclopedia of the History of Arabic Science*, 3 vols, London/New York, Routledge, 1996, vol. II, pp. 495–538, at pp. 528–30.

$$x^3 = 3x + 1 \quad (x = \cos 20°), \qquad x^3 + 1 = 3x \quad (x = 2\sin 10°).$$

Another famous example is that of al-Kāshī for the calculation of sin 1°.[37] Al-Kāshī's method rests on two relations:

- $\sin(3\vartheta) = 3\sin\vartheta - 4\sin^3\vartheta$
- for $\vartheta = 1°$, $\sin 3° = 3; 8, 24, 33, 59, 34, 28, 15, = q$.

Assuming sin 1° = x, one obtains

$$x = \frac{\sin 3° + 4x^3}{3} = \frac{q + 4x^3}{3} = f(x).$$

The idea of the iterative method he follows is this: when x is sufficiently small, x^3 is negligible, and one has $x_1 = \frac{q}{3}$. The second step of the method consists in writing $x_2 = \frac{q + 4x^3}{3} = f(x_1)$. One then repeats the procedure. The point, therefore, is to form the sequence $x_n = f(x_{n-1})$ for $x_1 > 1$; x_1 is a chosen approximated value. We will not go into all these contributions or follow their filiations here. This is, moreover, a promissory note for forthcoming research. Let us conclude with a final example, that of the treatise of al-Iṣfahānī[38] who, unbeknownst to history as it were, was still working in the wake of al-Ṭūsī. We remarked that he relied on an algorithm founded on the property of the fixed point in order to solve the equation $x^3 + 210 = 121x$.

Did he borrow this method from one of his predecessors, or did he invent it by inspiring himself from works such as those of al-Kāshī or al-Yazdī? We do not yet know. As he presents it, this method relies on the following ideas. One rewrites the preceding equation thus:

$$x = (121x - 210)^{\frac{1}{3}} = f(x).$$

Al-Iṣfahānī then takes $x_1 = 11$, whence

$$y_1 = f(x_1) = (1121)^{\frac{1}{3}} < 11.$$

[37] *Ibid.*, pp. 530–1.

[38] See the chapter on algebra.

He then takes an approximate value of x_1 by defect, namely 10.3; he finds

$$f(10.3) = (1036.3)^{\frac{1}{3}} < 10.3\,.$$

He then takes $x_2 = 10.3$ and $y_2 = f(x_2) = (1036.3)^{\frac{1}{3}}$.

He then takes an approximation of x_2 by defect, namely 10.1. One finds that $f(10.1) = (1012.1)^{\frac{1}{3}} < 10.1$. One then takes $x_3 = 10.1$, and so on; the first terms of this sequence are

$$x_1 = 11 > x_2 = 10.3 > x_3 = 10.1 > \ldots$$

Note that al-Iṣfahānī chooses the value 11 in a slightly different way. Instead of the function f, he considers a function greater than it, namely

$$g(x) = (121x)^{\frac{1}{3}};$$

and he seeks a root x_1 of the new equation $x = g(x)$, which guarantees that $x_1 = 11 > x_0$ if x_0 is the root one seeks.

This algorithm in effect assumes that the function $\ell : x \rightarrow x^{\frac{1}{3}}$, a cubic root, is increasing concave contracting throughout the entire interval $K = [a, b]$, with $a > 1$, and that f is increasing concave. One can then show that $K \subseteq [2,11]$ so that $f[a,b] \subset [a,b]$.

2. INTERPOLATION METHODS

Numerical equations, algebra, and arithmetic research were not the sole domains associated with the invention and study of algorithmic methods. The fields of astronomy, trigonometry, and – to a lesser degree – optics were also open to fruitful research into iterative methods and algorithms. The considerable number of *zījs* – astronomical tables – required for astronomical calculation and composed by the astronomers and mathematicians of classical Islam bear ample witness to this phenomenon. For the most part, these tables contain the values of one or another of the functions beginning with a few initial values, that is, ones determined independently, either by observation or by a clever calculation. These tables can thus pertain to the sine function, the longitude of the planets, the refraction of light rays, ... It is therefore clear that the composition of such tables requires the invention of an interpolation schema. In the history of this invention, however, we must distinguish several different epistemic states. This interpolation schema is most often implicit in the table without any explicit elaboration by the author. At this stage the composition of the table is 'experimental'. At a later stage, the mathematician may himself deliberately make explicit the polynomial formed from these few values and try to improve the approximation by studying some properties of the polynomial. This marks the explicit beginning of a study of finite differences. Arabic mathematics does not reach this stage of research until the end of the 9th century, thanks to attempts to improve linear interpolation. Finally, some mathematicians attempt to demonstrate the existence and unicity of the polynomial. As a proof of existence, the mathematicians of the 10th century gave the explicit construction. Not until several centuries later will the problem of unicity even be raised.

Let us return to the origin, that is, the *Almagest*, which al-Ḥajjāj first translated into Arabic at the beginning of the 9th century. Earlier yet, it seems that Theon's commentary on the *Almagest*, Book 1, was known from an ancient translation.[39] However that may be, it is the second chapter of Book 1 that interests us here, since it contains the table of chords for a circle of radius 60° calculated to three sexagesimal places, with table entries given in half-degrees; that is $x = \left(\frac{1^\circ}{2}\right)$, 1°, $\left(\frac{3^\circ}{2}\right)$, ..., 180°. Moreover, it is often the case that, to use this table, one doubles the arc, which leads to the result

$$\frac{1}{2}\mathrm{Crd}2x = \sin x .$$

It has long been shown that this table was obtained by linear interpolation. Indeed, it was this very method of calculation that mathematicians called (in Arabic) 'the method of the astronomers', a clue that leaves no doubt about its origins. In some Babylonian texts about the risings and settings of Mercury, astronomers proceeded by linear interpolation in the second century BC, as O. Neugebauer has shown. Let us translate 'this method of the astronomers' into our own language: assume that $x_{-1} < x < x_0$ and $d = x_0 - x_{-1} = x_i - x_{i-1}$ for $i = -2, -1, \ldots, n$; the linear interpolation is then rewritten

[39] R. Rashed, 'Greek into Arabic: Transmission and Translation', in J. E. Montgomery (ed.), *Arabic Theology, Arabic Philosophy. From the Many to the One: Essays in Celebration of Richard M. Frank*, Orientalia Lovaniensia Analecta 152, Leuven/Paris, Peeters, 2006, pp. 157–96.

(α) $$y = y_{-1} + \left(\frac{x - x_{-1}}{d}\right)\Delta y_{-1},$$

Δ being the first first-order difference.

The astronomers and mathematicians of the 9th century, such as Ḥabash al-Ḥāsib, improved this method in one respect or the other. From what we now know, however, it was only in the 10th century that one notices a new step forward: the invention of quadratic interpolation and the attempt to establish algorithms on explicit mathematical foundations. Two examples illustrate this new stage: one from Iran, with al-Khāzin; the other from Cairo, with Ibn Yūnus. Let us begin with the latter. In the second half of the 10th century Ibn Yūnus wrote his famous *al-Zīj al-Ḥākimī*, named for the Fatimid caliph al-Ḥākim.[40] This *zīj* is a table of sines to four sexagesimal places, at intervals of 10'. Ibn Yūnus explicitly gives the steps of his algorithm,[41] and not merely the numerical values. This method can be rewritten:

$$y = y_{-1} + \left(\frac{x - x_{-1}}{d}\right)\left[\frac{1}{2}\left(\Delta y_{-1} + \Delta y_0\right) + \frac{1}{2}\left(\frac{x - x_1}{d}\right)\Delta^2 y_{-1}\right].$$

It is obvious that the parabola defined by this equation goes through the points (x_{-1}, y_{-1}).

Al-Bīrūnī gives three methods;[42] the first is that of the linear interpolation discussed above; the others are:

(β) $$y = y_{-1} + \left(\frac{x - x_{-1}}{d}\right)\left[\Delta y_{-2} + \left(\frac{x - x_{-1}}{d}\right)\Delta^2 y_{-2}\right];$$

note that, for the calculation of Δy_{-2} and $\Delta^2 y_{-2,}$ the application of this formula requires that

$$x_{-2} = \left(x_{-1} - d\right) \in \left]0, \frac{\pi}{2}\right[,$$

that is, that $x_{-1} > d$.

[40] C. Schoy, 'Beiträge zur arabischen Trigonometrie', *Isis*, 5, 1923, pp. 364–99 and D. King, *The Astronomical Works of Ibn Yūnus*, Ph.D. dissertation, Yale University, 1972.

[41] C. Schoy, 'Beiträge zur arabischen Trigonometrie', pp. 390–1.

[42] See R. Rashed, 'Al-Samaw'al, al-Bīrūnī et Brahmagupta: les méthodes d'interpolation', *Arabic Sciences and Philosophy*, 1.1, 1991, p. 101–60; repr. in *Optique et mathématiques. Recherches sur l'histoire de la pensée scientifique en arabe*, Variorum Reprints, Aldershot, 1992, XII.

The third method is that of Brahmagupta. Yet al-Bīrūnī's presentation of this method differs slightly from that in the text of Brahmagupta, insofar as one can judge from the English translation of the latter.[43] Indeed, al-Bīrūnī's explanation allows one to rewrite Brahmagupta's formula as follows:

$$(\gamma) \qquad y = y_0 + \left(\frac{x - x_0}{d}\right)\left[\frac{\Delta y_{-1} + \Delta y_0}{2} + \frac{1}{2}\left(\frac{x - x_0}{d}\right)\Delta^2 y_{-1}\right].$$

According to the text of al-Bīrūnī, this method assumes that $x < x_0$, and leads to the formula

$$y = y_0 + \left(\frac{x_0 - x}{d}\right)\left[\frac{\Delta y_{-1} + \Delta y_0}{2} + \frac{1}{2}\left(\frac{x_0 - x}{d}\right)\Delta^2 y_{-1}\right];$$

but this expression yields (γ) if one keeps in mind that $(x_0 - x) > 0$, $\Delta y_{-1} < 0$, $\Delta y_0 < 0$ and $\Delta^2 y_{-1} > 0$. In other words, for al-Bīrūnī, the method of Brahmagupta assumes that $x < x_0$ and that the correction is additive. These conditions do not, however, appear to originate in the text of Brahmagupta such as we now have it.

As al-Bīrūnī himself concedes, the fourth method is also Indian in origin and called *sankalt*, or method of monomials; it is rewritten:

$$(\delta) \qquad y = y_0 - \frac{(x_0 - x)(x_0 - x + 1)}{d(d+1)}\Delta y_{-1};$$

this method proceeds by calculating the increases from x_i to x_{i-1}.

In his various works, al-Bīrūnī does not stop with a simple exposition of one or the other of the preceding methods. The reasons for this are several, notably the work carried out before him not only by al-Khāzin and Ibn Yūnus on quadratic interpolations, but also by Ḥabash and Abū al-Wafā' al-Būzjānī, among others, in this domain. The latter, for example, uses a

[43] Brahmagupta, *The Khaṇḍakhādyaka, an Astronomical Treatise of Brahmagupta*, translated into English with an Introduction, Notes, Illustrations and Appendices by P. C. Sengupta, Calcutta, University of Calcutta, 1934, p. 141.

method more sophisticated than that of Ptolemy for the determination of $\sin\frac{1^\circ}{2}$.[44]

This accumulation of work, of which history has yet to give a properly rigorous account, seems to have stimulated al-Bīrūnī to launch at least three research themes:

• To show that the aforementioned quadratic methods all improve upon linear interpolation.

• To find the inverse of each.

• To justify the algorithm geometrically.

To compare these methods with respect to their performance is another theme that will soon come to the fore and, along with the others, stimulate relatively autonomous research on interpolation algorithms.

About the algorithm that he gives in his *al-Qānūn al-Mas'ūdī*, al-Bīrūnī knows that, for Δy_{-1} in (α), he needs only to substitute $\Delta = \Delta y_{-2} + \frac{x - x_{-1}}{d}(\Delta y_{-1} - \Delta y_{-2})$ in order to obtain (β).

Likewise, in his presentation, al-Bīrūnī suggests that the method of Brahmagupta and, less explicitly, that of the monomials improve upon linear interpolation. Indeed, after having mentioned 'the method of the astronomers', that is, linear interpolation, he introduces the method of Brahmagupta as an improvement upon it. He applies this method to the interval [2°, 5°] and shows its superiority to linear interpolation. Next, he introduces the method of monomials, still applied to the same interval, as 'closer to reason and to exactness'.[45] To take but one case that does not restrict the scope of the discussion, the tables of cotangents show, on the one hand, that linear interpolation yields values that are too large and, on the other, that the first-order difference does not change uniformly. To improve the corrections, the idea of al-Bīrūnī, of Brahmagupta before him, and of the inventor of the method of monomials, had been to replace Δy_0 in (α) by Δ, which depends on x. Starting from Δy_{-1} one thus proceeds by linear interpolation; for $x = x_0$, one has $\Delta = \Delta y_{-1}$ and for $x = x_1$, $\Delta = \Delta y_0$. This linear interpolation over $[x_0, x_1]$ yields

$$\Delta = \Delta y_{-1} + \left(\frac{x - x_0}{d}\right)(\Delta y_0 - \Delta y_{-1}),$$

[44] F. Woepcke, 'Recherches sur l'histoire des sciences mathématiques chez les Orientaux', *Journal Asiatique*, 1860, pp. 281–320.

[45] R. Rashed, 'Al-Samaw'al, al-Bīrūnī et Brahmagupta', p. 139.

whence

$$y = y_0 + \left(\frac{x - x_0}{d}\right)\Delta = y_0 + \left(\frac{x - x_0}{d}\right)\left[\Delta y_{-1} + \left(\frac{x - x_0}{d}\right)\right]\Delta^2 y_{-1};$$

and one thus obtains al-Bīrūnī's formula (β) if one carries out this interpolation over $[x_{-1}, x_0]$.

Consider now the linear interpolation over $[x_{-1}, x_1]$; one now has

$$\Delta = \Delta y_{-1} + \frac{1}{2}\left(\frac{x - x_{-1}}{d}\right)\Delta^2 y_{-1} = \Delta y_{-1} + \frac{1}{2}\left(\frac{x - x_0}{d}\right)\Delta^2 y_{-1} + \frac{1}{2}(\Delta y_0 - \Delta y_{-1}),$$

whence

$$\Delta = \left(\frac{\Delta y_{-1} + \Delta y_0}{2}\right) + \frac{1}{2}\left(\frac{x - x_0}{d}\right)\Delta^2 y_{-1};$$

and one has

$$y = y_0 + \left(\frac{x - x_0}{d}\right)\Delta = y_0 + \left(\frac{x - x_0}{d}\right)\left[\left(\frac{\Delta y_{-1} + \Delta y_0}{2}\right) + \frac{1}{2}\left(\frac{x - x_0}{d}\right)\Delta^2 y_{-1}\right],$$

which is Brahmagupta's formula.

Let us now turn to the method of monomials over $[x_0, x_1]$. Let us divide the interval $[x_0, x_1]$ into d equal parts. To take into account the decrease in the function, which is faster near x_0 than near x_1, as one can see by checking the tables, one considers the cumulative increases of x_1 toward x_0:

$$\varepsilon + 2\varepsilon + \ldots + d\varepsilon = \frac{d(d+1)}{2}\varepsilon = |\Delta y_0|,$$

whence

$$\varepsilon = \frac{2|\Delta y_0|}{d(d+1)}.$$

The correction to be made on y_1 is additive, and corresponds to a cumulative increase over $(x_1 - x)$, where $(x_1 - x)$ is an integer, whence

$$y = y_1 + c$$

with

$$c = \frac{(x_1 - x)(x_1 - x + 1)}{2}\varepsilon = \frac{(x_1 - x)(x_1 - x + 1)}{d(d+1)}|\Delta y_0|$$

and

$$y = y_1 - \frac{(x_1 - x)(x_1 - x + 1)}{d(d+1)} |\Delta y_0| .$$

If one sets $x_1 - x = x_1 - x_0 + x_0 - x = d + x_0 - x$, one has

$$y = y_1 + \left(\frac{x - x_0}{d} - 1\right)\left(1 - \frac{x - x_0}{d+1}\right)\Delta y_0 ,$$

whence

$$y = y_0 + \frac{x - x_0}{d}\left[\frac{2d+1}{d+1}\Delta y_0 - \left(\frac{x - x_0}{d}\right)\frac{d}{d+1}\Delta y_0\right],$$

a relation of the form

$$y = y_0 + \frac{x - x_0}{d}\Delta ,$$

with Δ of the first degree in

$$\frac{x - x_0}{d} .$$

One obtains formula (δ) if one considers the interpolation over $[x_{-1}, x_0]$.

The three methods thus present themselves as three different procedures for improving linear interpolation. In this respect, al-Bīrūnī's method does not differ from Brahmagupta's or from the method of monomials.

Conversely, al-Bīrūnī seeks to determine the inverse interpolations, whether linear or of the second degree. Thus he gives

$$x = x_{-1} + \frac{d(y - y_{-1})}{\Delta y_{-1}}$$

for the linear interpolation, and

$$x = x_{-1} + \frac{d(y - y_{-1})\Delta y_{-1}}{\Delta y_{-1}\Delta y_{-2} + (y - y_{-1})\Delta^2 y_{-2}}$$

for the interpolation (β).

Finally, al-Bīrūnī tries to justify geometrically both the linear interpolation and his own quadratic interpolation. Thus, for the linear interpolation of the sine function,[46] he considers x to be between x_{-1} and x_0, two values known from the table, and he considers on the circle the three angles $ASE = x_{-1}$, $ASO = x$ and $ASH = x_0$, whence $d = x_0 - x_{-1} = \widehat{AH} - \widehat{AE} = \widehat{HE}$, and $x - x_{-1} = \widehat{EO}$. By approximation

$$(*) \qquad \frac{OK}{HL} \approx \frac{\widehat{EO}}{\widehat{EH}} \text{ and } \frac{OK}{HL} \approx \frac{x - x_{-1}}{d}.$$

One has

$$EG = \sin x_{-1} = y_{-1} \text{ and } HI = \sin x_0 = y_0,$$

therefore

$$HL = \sin x_0 - \sin x_{-1} = y_0 - y_{-1} = \Delta\, y_{-1}$$

and

$$OK = \sin x - \sin x_{-1} = y - y_{-1},$$

whence

$$\frac{OK}{HL} = \frac{y - y_{-1}}{\Delta y_{-1}};$$

therefore (*) is rewritten

$$\frac{y - y_{-1}}{\Delta y_{-1}} \approx \frac{x - x_{-1}}{d},$$

whence the relation (α).

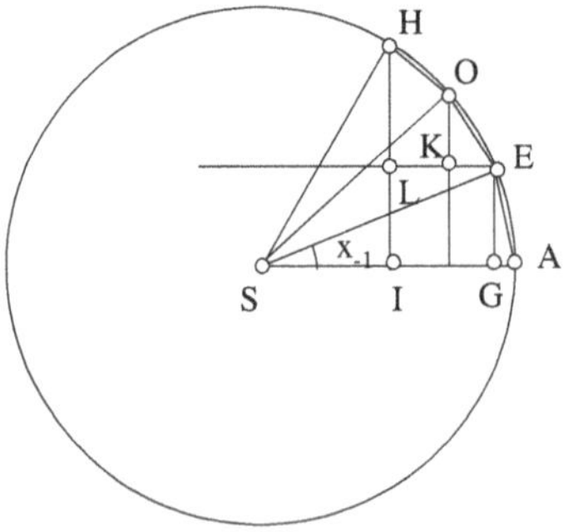

Fig. 38

[46] *Al-Qānūn al-Mas'ūdī (Canun Masudicus)*, 3 vols, Hyderabad, 1954–56; *Al-Maqāla al-Thālitha min al-Qānūn al-Mas'ūdī* (Book 3), ed. Imām Ibrāhīm Ahmad, Cairo, 1985, vol. I, p. 329.

Al-Bīrūnī draws upon analogous notions to justify (β).

The domains studied here are not the only ones in which mathematicians develop algorithms that they both apply and try to demonstrate. Thus, among many others, the rule of the two errors (*ḥisāb al-khaṭa'ayn*), or the method of the double false position, established geometrically by Qusṭā ibn Lūqā and algebraically by al-Samaw'al,[47] are presented in most of the treatises of *ḥisāb*, notably those Muḥammad ibn al-Khawwām al-Baghdādī, Kamāl al-Dīn al-Fārisī, and al-Kāshī, to cite only a few examples. As we now know, moreover, this rule leads to an approximate solution for quadratic problems.

[47] *Al-Bāhir en Algèbre d'as-Samaw'al*, ed. S. Ahmad and R. Rashed, pp. 66–70 (French), pp. 151–63 (Arabic).

– 3 –

THĀBIT IBN QURRA AND AMICABLE NUMBERS*

In his *On the Determination of Amicable Numbers* (*Fī istikhrāj al-aʿdād al-mutaḥābba*),[1] Thābit ibn Qurra demonstrates a theorem that, according to him, allows one to form as many pairs of amicable numbers as one wishes. He considers two series of numbers: $p_n = 3 \cdot 2^n - 1$ and $q_n = 9 \cdot 2^{2n-1} - 1$; his theorem claims that, if for an integer n, p_{n-1}, p_n and q_n are prime numbers, then $2^n p_{n-1} p_n$ and $2^n q_n$ are amicable numbers, that is, that each of them is equal to the sum of the aliquot parts of the other, the aliquot parts of a number being the divisors of this number distinct from the number itself. This demonstration is preceded by nine lemmas or propositions and is written in the purest Euclidean style of Books VII to IX of the *Elements*.

The first three propositions determine the divisors of the product bc of two numbers starting from the divisors of b and the divisors of c; Thābit ibn Qurra follows ancient usage in considering the proper divisors (other than the number itself and 1) rather than the divisors.

In Proposition 1, b and c are assumed to be prime; the proper divisors of bc are then b and c.

Indeed, if d is a proper divisor of bc, there exists a number e such that $bc = de$ and, b being prime, it divides d or e, for example d. Thābit ibn Qurra uses here Proposition 30 of Book VII of the *Elements*, according to which a prime number that divides a product divides one of the factors of this product. Since $b:d = e:c$, e divides c, which is also prime; since d is not equal to bc, e should be equal to c, but then $d = b$.

* In collaboration with Christian Houzel.

[1] *Fī istikhrāj al-aʿdād al-mutaḥābba*, ed. and French transl. R. Rashed, *Thābit ibn Qurra. Science and Philosophy in Ninth-Century Baghdad*, Scientia Graeco-Arabica, vol. 4, Berlin/New York, Walter de Gruyter, 2009, pp. 89–151.

Proposition 14 of Book IX of the *Elements* is related to the latter: it states that if a number a is the smallest number divisible by the prime numbers $p_1, p_2, \ldots, p_n$, the only prime divisors of a are $p_1, p_2, \ldots, p_n$.[2]

In Proposition 2, b is assumed to be prime, but not c. The proper divisors of bc are then classified into three types: 1) b and c; 2) the proper divisors of c; 3) the products of b by the proper divisors of c. This is a generalization of Euclid's reasoning in Proposition 36, the last proposition of Book IX of the *Elements*, in which he had determined the divisors of a number of the form $2^n E$, where E is a prime number.

It is clear that b, c and the divisors of c divide bc: if d is one of them, one has $d{:}c = bd{:}bc$ and, since d divides c, bd divides bc. Reciprocally, let ℓ be any proper divisor of bc, such that $\ell n = bc$. Since b is prime, it divides ℓ or n. If b divides n, the proportion $n{:}b = c{:}\ell$ shows that ℓ divides c and ℓ is therefore equal to c or to one of its proper divisors. If b divides ℓ, n divides c for $b{:}\ell = n{:}c$; thus $b = \ell$ and $n = c$, or else $\ell = bd$ and $c = nd$, where d is a proper divisor of c.

Proposition 3 treats the case in which b and c are both composite. The proper divisors of bc are of six types: 1) b and c; 2) the proper divisors of b; 3) those of c; 4) the products of b by the proper divisors of c; 5) the products of the proper divisors of b by c; 6) the products of a proper divisor of b by a proper divisor of c.

It is clear that b, c and the divisors of b or of c divide bc. If g divides c, the proportion $g{:}c = bg{:}bc$ shows that bg divides bc; one likewise establishes that, if d divides b, dc divides bc. One also has $dg{:}bg = d{:}b$, therefore dg divides bg, therefore it also divides bc.

Reciprocally, let u' be any proper divisor of bc, such that $u'o' = bc$, therefore $u'{:}b = c{:}o'$. If u' and b are mutually prime, u' divides c and b divides o', according to Propositions 20 and 21 of Book VII of the *Elements*, therefore u' is equal to c or to one of its proper divisors.

[2] In R. Rashed, *Entre arithmétique et algèbre. Recherches sur l'histoire des mathématiques arabes*, Sciences et philosophie arabes — Études et reprises, Paris, Les Belles Lettres, 1984. One finds a history of the theory of amicable numbers in Arabic mathematics that we will not repeat here; nevertheless, on p. 264 one reads that Proposition 1 is a particular case of Proposition IX.14 of the *Elements*, but it is more correct to say that the two propositions are related, for that of Euclid treats only prime divisors and not all divisors as does that of Thābit ibn Qurra (English transl. *The Development of Arabic Mathematics: Between Arithmetic and Algebra*, Boston Studies in the Philosophy of Science 146, Dordrecht, 1994). On the diffusion of Thābit's theorem in medieval Hebrew literature, see T. Lévy, 'L'histoire des nombres amiables: le témoignage des textes hébreux médiévaux', *Arabic Sciences and Philosophy*, 6, 1996, pp. 63–87.

Otherwise, u' divides b, or b divides u', or else the greatest common divisor j of u' and b is distinct from u' and from b. In the first case, u' is equal to b or to one of its proper divisors. In the second, o' divides c; let $o'i = c$ and $u' = bi$, where i is a proper divisor of c. Finally consider the third case, in which $b = j\overline{oa}$ and $u' = j\overline{ob}$, where the numbers $\overline{oa}$ and $\overline{ob}$ are mutually prime. One has $\overline{oa} : \overline{ob} = b : u' = o' : c$, therefore $\overline{oa}$ divides o' and $\overline{ob}$ divides c; if $\overline{oa} = o'$, $\overline{ob} = c$ and $u' = jc$ where j is a proper divisor of b and otherwise $\overline{ob} = g$ is a proper divisor of c, and j is a proper divisor of b, and $u' = jg$.

One can summarize by saying that, if ℓ divides a product bc, there exists n such that $\ell n = bc$, therefore $\ell : b = c : n = \ell' : b'$ where $\ell = j\ell'$ and $b = jb'$, j being the greatest common divisor of b and of ℓ. Then ℓ' and b' are mutually prime, therefore ℓ' divides c, and b' divides n, according to Propositions 20 and 21 of Book VII of the *Elements*, and one finds that ℓ is a product of the divisor j of b by a divisor ℓ' of c.

Proposition 4 gives the sum of a geometric progression of reason 2: if $a_0, a_1, \ldots, a_n$ are such that $a_{j+1} = 2a_j$ for $0 \leq j \leq n-1$, then

$$a_n - (a_0 + a_1 + \ldots + a_{n-1}) = a_0;$$

Thābit ibn Qurra specifies that the statement remains valid when $a_0 = 1$, because Euclid never treats the unit as a number. This signifies that $1 + 2 + 2^2 + \ldots + 2^{n-1} = 2^n - 1$.

The proposition is deduced from Proposition 35 of Book IX of the *Elements*, according to which for every geometric progression $a_0, a_1, \ldots, a_n$, one has

$$(a_1 - a_0) : a_0 = (a_n - a_0) : (a_0 + a_1 + \ldots + a_{n-1});$$

here $a_1 - a_0 = a_0$, therefore this ratio is equal to 1:1 and one has

$$a_n - a_0 = a_0 + a_1 + \ldots + a_{n-1}.$$

Proposition 5 takes up by generalizing it, Proposition 36, which is the last proposition of Book IX of the *Elements*. One considers a geometric progression $a_0, a_1, \ldots, a_{n-1}$ of reason 2, where $a_0 = 1$ ($a_j = 2^j$), and by f one designates its sum, namely, $2^n - 1$ according to the preceding proposition. Then if g is an odd prime number

(1) if $g = f$, then $a_{n-1}g$ is a perfect number;

(2) if $g < f$, then $a_{n-1}g$ is an abundant number with an excess of $f - g$;

(3) if $g > f$, then $a_{n-1}g$ is a deficient number with a defect of $g - f$.

Euclid considers only case (1), which gives him perfect even numbers;[3] they correspond to Mersenne prime numbers, of the form, $2^n - 1$, where n must necessarily be prime because $2^{pq} - 1$ is divisible by $2^p - 1$. Recall that numbers are called perfect that are equal to the sum of their aliquot parts (including 1), abundant when the numbers are smaller than the sum of their aliquot parts (the excess being the difference), and deficient when the numbers are greater than the sum of their aliquot parts (the defect being the difference). Like Euclid, Thābit ibn Qurra develops his reasoning on the example $n = 5$ which gives the Mersenne prime number 31 and, in case (1), the perfect number $16 \times 31 = 496$; but he does not make this value explicit, because he treats the example as generic.

According to Proposition 2, the proper divisors of $a_{n-1}g$ are a_{n-1}, g, the proper divisors of a_{n-1}, and the products of g by the proper divisors of a_{n-1}. According to Proposition 13 of Book IX of the *Elements*, the aliquot parts of a_{n-1} are $a_0, a_1, \ldots, a_{n-2}$, and they are pairwise distinct; likewise the numbers $ga_0 = g, ga_1, \ldots, ga_{n-2}$ are pairwise distinct, for they also form a geometric progression of reason 2. It remains to be shown that none of the numbers ga_j is equal to an a_i; if this were the case, using Proposition 11 of Book IX of the *Elements*, one would have for $j \geq i$ (respectively $j \leq i$), $ga_j{:}ga_{j-i} = a_i{:}a_0$ (respectively $ga_j{:}g = a_i{:}a_{i-j}$), therefore $ga_{j-i} = a_0 = 1$ (respectively $g = a_{i-j}$), which is absurd. Note that Euclid does not demonstrate that the aliquot parts he found are pairwise distinct. The sequence $g, ga_1, \ldots, ga_{n-2}, ga_{n-1}$ is a geometric progression of reason 2, therefore, according to Proposition 4, the sum $g + ga_1 + \ldots ga_{n-2}$ is equal to $ga_{n-1} - g$ and therefore the sum of the aliquot parts of ga_{n-1} is equal to

$$a_0 + a_1 + \ldots + a_{n-1} + g + ga_1 + \ldots + ga_{n-2} = f + ga_{n-1} - g = ga_{n-1}$$

if $g = f$; greater than ga_{n-1} if $g < f$ (the excess being $f - g$), and smaller than ga_{n-1} if $g > f$ (the defect being $g - f$). Q.E.D.

Thābit ibn Qurra therefore establishes that the aliquot parts of $a = 2^{n-1}g$ (where g is prime) are $1, 2, \ldots, 2^{n-1}, g, 2g, \ldots, 2^{n-2}g$ and that they are distinct, so that their sum is

$$2^n - 1 + (2^{n-1} - 1)g = a + f - g$$

if $f = 2^n - 1$.

[3] Euler established the reciprocal, that is that every even perfect number has the Euclidean form, that is the form $2^{n-1}(2^n - 1)$ where $2^n - 1$ is prime (see below). We still do not know if there is an infinity of Mersenne prime numbers or if there exist odd perfect numbers.

In Proposition 6, the givens are the same, except that g is no longer a prime number but the product of two distinct odd prime numbers h and i. Then $a_{n-1}g$ is abundant or deficient according to whether $g < f + f(h + i)$ or $g > f + f(h + i)$, equality being excluded; the excess (respectively the defect) of $a_{n-1}g$ is equal to $f + f(h + i) - g$ (or, respectively $g - f - f(h + i)$). Note that in cases (2) and (3) of Proposition 5 along with this Proposition 6 yield partial reciprocals of Euclid's theorem: $2^{n-1}g$ is not a perfect number if g is a prime number different from $2^n - 1$ or if it is the product of two distinct odd prime numbers. Ibn al-Haytham was apparently the first to have tried to demonstrate the reciprocal of Euclid's theorem in complete generality, but he succeeded in establishing only that if $2^{n-1}(2^m - 1)$ is a perfect number, then $m = n$ and $2^m - 1$ is prime, that is, that the number under consideration has the Euclidean form.[4]

According to Proposition 3, Proposition 13 of Book IX of the *Elements*, and Proposition 1, the aliquot parts of $a_{n-1}g$ are $a_0 = 1, a_1, \ldots, a_{n-1}, g, h, i, ga_1, \ldots, ga_{n-2}, ha_1, \ldots, ha_{n-1}, ia_1, \ldots, ia_{n-1}$. Since $(a_0, a_1, \ldots, a_{n-1})$, $(g, ga_1, \ldots, ga_{n-2})$, $(h, ha_1, \ldots, ha_{n-1})$ and $(i, ia_1, \ldots, ia_{n-1})$ are four geometric progressions of reason 2, their terms are pairwise distinct. None of the numbers ga_j $(0 \leq j \leq n - 2)$ is equal to a number ha_k $(0 \leq k \leq n - 1)$ for otherwise one would have, for $k \leq j$ (resp. $k \geq j$), $ga_j{:}ga_{j-k} = ha_k{:}h$ (resp. $ga_j{:}g = ha_k{:}ha_{k-j}$), whence $ga_{j-k} = h < g$ (resp. $hi = g = ha_{k-j}$, whence $i = a_{k-j} = 2^{k-j}$), which is absurd; one demonstrates likewise that none of the numbers ga_j is equal to a number ia_k. None of the numbers ga_j is equal to a number a_k, otherwise one would deduce by the same reasoning that $ga_{j-k} = 1$ if $j \geq k$ or that $g = a_{k-j}$ if $k \geq j$, which is absurd. Likewise also none of the numbers a_j $(1 \leq j \leq n - 1)$ is equal to an ha_k or an ia_k $(0 \leq k \leq n - 1)$, otherwise one would have $a_{j-k} = h$ or i for $j \geq k$ or $1 = ha_{k-j}$ or ia_{k-j} for $k \geq j$, which is absurd. Finally, none of the numbers ha_j is equal to an ia_k, otherwise one would have $ha_{j-k} = i$ or $h = ia_{k-j}$, which is absurd since h and i are distinct prime numbers. According to Proposition 4,

$$g + ga_1 + \ldots + ga_{n-2} = ga_{n-1} - g;$$

the sum of the other aliquot parts of ga_{n-1} is

$$h + ha_1 + \ldots + ha_{n-1} + i + ia_1 + \ldots + ia_{n-1} + 1 + a_1 + \ldots + a_{n-1}$$
$$= f(h + i) + f,$$

[4] See R. Rashed, *Les Mathématiques infinitésimales du* IX*e au* XI*e siècle*, vol. IV: *Méthodes géométriques, transformations ponctuelles et philosophie des mathématiques*, London, al-Furqān, 2002, pp. 192–5 and pp. 320–8.

therefore the sum of all of the aliquot parts of ga_{n-1} is

$$ga_{n-1} + f + f(h + i) - g.$$

Finally one observes that g cannot be equal to $f(h + i) + f = f(h + i + 1)$, otherwise it would be divisible by f and by $h + i + 1$; now, according to Proposition 1, the only proper divisors of g are h and i, and $h + i + 1 > h$ and i.

Thābit ibn Qurra establishes that the distinct aliquot parts of $k = 2^{n-1}hi$ are $1, 2, \ldots, 2^{n-1}, h, 2h, \ldots, 2^{n-1}h, i, 2i, \ldots, 2^{n-1}i, hi, 2hi, \ldots, 2^{n-2}hi$, such that their sum is

$$(2^n - 1)(1 + h + i) + (2^{n-1} - 1)hi = k + f(h + i + 1) - g$$

with $f = 2^n - 1$ and $g = hi$.

More generally, one could establish by the same methods that, if b and c are mutually prime, then the divisors $j\ell$ of bc found in Proposition 3, where j divides b and ℓ divides c, are pairwise distinct. Indeed, if $j\ell = j_1\ell_1$, one has $j{:}j_1 = \ell_1{:}\ell$. Let k be the greatest common divisor of j and j_1; one has $j = ku$ and $j_1 = ku_1$ with u, u_1 mutually prime and $u{:}u_1 = j{:}j_1 = \ell_1{:}\ell$, therefore u divides ℓ_1, and u_1 divides ℓ according to Propositions VII.20 and 21 of the *Elements*. Then u divides j, and therefore b, and ℓ_1 divides c, and one has $u = 1$; likewise $u_1 = 1$ and $j = k = j_1$, therefore also $\ell_1 = \ell$. Consequently, the sum $\sigma(bc)$ of all the divisors of bc is equal to

$$\sum_{j|b,\ell|c} j\ell = \sum_{j|b} j \cdot \sum_{\ell|c} \ell = \sigma(b)\sigma(c).$$

In substance, this is just what Kamāl al-Dīn al-Fārisī would demonstrate at the beginning of the 14th century.[5]

The case in which $b = 2^{n-1}$ and c is odd yields $\sigma(2^{n-1}c) = (2^n - 1)\sigma(c)$ or, if one writes $\sigma_0(a) = \sigma(a) - a$ (the sum of the aliquot parts of a number a),

$$\sigma_0(2^{n-1}c) = (2^n - 1)(\sigma_0(c) + c) - 2^{n-1}c = 2^{n-1}c + (2^n - 1)\sigma_0(c) - c;$$

[5] See R. Rashed, *The Development of Arabic Mathematics*, pp. 287–94 and 'Matériaux pour l'histoire des nombres amiables et de l'analyse combinatoire', *Journal for the History of Arabic Science*, 6.1–2, 1982, pp. 209–78, at pp. 229–66.

$2^{n-1}c$ is therefore perfect if $c = (2^n - 1)\sigma_0(c)$, an abundant number of excess $(2^n - 1)\sigma_0(c) - c$ if $c < (2^n - 1)\sigma_0(c)$ and deficient with a defect of $c - (2^n - 1)\sigma_0(c)$ if $c > (2^n - 1)\sigma_0(c)$. In the first case, if $n \geq 2$, $\sigma_0(c)$ is an aliquot part of c, for $2^n - 1 \geq 3$ and one therefore has $\sigma_0(c) = 1$, which means that c is a prime number and that $c = 2^n - 1$. One has thus demonstrated the reciprocal of Euclid's theorem (later established by Euler).

Propositions 7 and 8 are easy lemmas, in which one considers four numbers a, b, c, and d in geometric progression of reason 2 and one establishes successively that

$$c(d + c)(b + c) = cd(d + a)$$

and that

$$c(b + d + 2c) = d(d + a).$$

Since $b = 2a$, $c = 4a$ and $d = 8a$, this means that $12 \times 6 = 8 \times 9 = 4 \times 18$, but Thābit ibn Qurra demonstrates this by means of proportion theory: $a{:}b = b{:}c = (a + b){:}(b + c) = c{:}d$, whence, if one compounds $(c + d){:}d = (a + c + 2b){:}(b + c) = (a + 2c){:}(b + c) = (a + d){:}(b + c)$ and, consequently $(c + d)(b + c) = d(a + d)$, which one need only multiply by c in order to obtain Proposition 7. For Proposition 8, he observes that $d + 2c = 2d$ and $b = 2a$, whence $d + 2c + b = 2(d + a)$, so that $c{:}d = (d + a){:}(d + b + 2c)$, whence $c(b + d + 2c) = d(d + a)$.

In Proposition 9, the givens are the same and one demonstrates that

$$d(a + d - 1) = c(d(a + d) - 1 - (d + c - 1)(b + c - 1)).$$

Indeed, according to Propositions 7 and 8, one has

$$c((d + c)(b + c) - (b + d + 2c)) = (c - 1)d(a + d)$$

where the first member is equal to $c((d + c - 1)(b + c - 1) - 1)$; thus

$$d(a + d) + c((d + c - 1)(b + c - 1) - 1) = cd(a + d)$$

and

$$d(a + d) - c = c(d(a + d) - (d + c - 1)(b + c - 1)).$$

In addition, subtract c from both members:

$$d(a + d - 1) = d(a + d) - 2c = c(d(a + d) - 1 - (d + c - 1)(b + c - 1)).$$

This identity is moreover a consequence of

$$\begin{aligned} &d(a + d) - 1 - (d + c - 1)(b + c - 1) \\ &= 8a \times 9a - 1 - (12a - 1)(6a - 1) = 18a - 2. \end{aligned}$$

Proposition 10 is Thābit ibn Qurra's theorem; it gives a procedure for constructing pairs of amicable numbers 'at will', which is probably to be understood as infinite in number. One considers a geometric progression of reason 2, $a_0 = 1$, $a_1 = 2$, … , a_n, the sum of which is $g = 2^{n+1} - 1$. One assumes that numbers $h = g + a_n = 3 \times 2^n - 1$ and $i = g - a_{n-1} = 3 \times 2^{n-1} - 1$ are prime; one also assumes that $s = a_{n+1}(a_{n+1} + a_{n-2}) - 1 = 9 \times 2^{2n-1} - 1$, where $a_{n+1} = 2a_n$, is also prime. Therefore the numbers $\ell = hia_n$ and $o = sa_n$ are amicable.

Since $s > g$, the number o is deficient according to case (3) of Proposition 5, and its defect is equal to $p = s - g$. One has

$$p + g + 1 = a_{n+1}(a_{n+1} + a_{n-2}),$$

whence, since $g + 1 = a_{n+1}$,

$$p = a_{n+1}(a_{n+1} + a_{n-2} - 1) = a_{n+1}(g + a_{n-2}).$$

The number hi is smaller than $g(h + i)$ because the difference is $gi + h(g - i)$; according to case (2) of Proposition 5, ℓ is abundant and its excess is

$$\begin{aligned} u = gi + h(g - i) + g &= g(g - a_{n-1}) + (g + a_n)a_{n-1} + g = g^2 + a_{n-1}a_n + g = \\ &= g(g + 1) + a_{n+1}a_{n-2} = a_{n+1}(g + a_{n-2}). \end{aligned}$$

One has

$$o - \ell = a_n(s - hi) = a_n(a_{n+1}(a_{n+1} + a_{n-2}) - 1 - (g + a_n)(g - a_{n-1})),$$

where

$$g + a_n = a_{n+1} + a_n - 1 \text{ and } g - a_{n-1} = a_{n+1} - 1 - a_{n-1} = a_{n-1} + a_n - 1.$$

According to Proposition 9, one therefore has

$$\begin{aligned} o - \ell &= a_n(a_{n+1}(a_{n+1} + a_{n-2}) - 1 - (a_{n+1} + a_n - 1)(a_{n-1} + a_n - 1)) \\ &= a_{n+1}(a_{n+1} + a_{n-2} - 1) = a_{n+1}(g + a_{n-2}). \end{aligned}$$

Thus $o - \ell = p = u$, and $\ell = o - p$ is the sum of the aliquot parts of o whereas $o = \ell + u$ is the sum of the aliquot parts of ℓ. One notes that, in such a pair of amicable numbers, the smaller is abundant, the greater is deficient, and the excess of the abundant number, the defect of the deficient number, and the difference between the two numbers are all equal.

Thābit ibn Qurra shows his reasoning in the example with $n = 4$, which gives $h = 3 \times 16 - 1 = 47$, $i = 3 \times 8 - 1 = 23$ and $s = 9 \times 128 - 1 = 1151$, all prime. The corresponding pair of amicable numbers is

$$\ell = 16 \times 47 \times 23 = 17296, \quad o = 16 \times 1151 = 18416;$$

Kamāl al-Dīn al-Fārisī states it explicitly at the beginning of the 14th century,[6] but it is often attributed to Fermat, who rediscovered it in 1636.[7] Thābit ibn Qurra does not make it explicit, because his example serves as a generic case; that he insists on this generic character is shown by his use of letters to represent the numbers.[8] For $n = 2$, one would have found $h = 11$, $i = 5$, and $s = 71$, which yields the pair (220, 284) known since antiquity.

One verifies that h and i cannot be divisible by the following prime numbers: 3, 7, 17, 31, 41, 43, 73, 79, 89, 103, 109, 113, 127, 137, 151, 157, 199, ... and that s can not be divisible by 3, 5, 11, 13, 19, 29, 31, 37, 43, 53, 59, 61, 67, 73, 83, 89, 97, 101, 107, 109, 113, 127, 131, 139, 149, 151, 157, 163, 173, 179, 181, 193, 197, ...

The following table contains on the first row a list of modules m associated with prime numbers p of the form $2^t m + 1$ in smaller type; the following rows, marked h, i, s indicate remainders r such that, if $n \equiv r \pmod{m}$, the corresponding number in h, i, s is divisible by p. For example, the module $m = 4$, is accompanied by two prime numbers, 5 and 17, because $n \equiv 1 \pmod 4$ implies that h is divisible by 5 and that s is divisible by 17, and $n \equiv 2 \pmod 4$ implies that i is divisible by 5.

[6] R. Rashed, 'Matériaux pour l'histoire des nombres amiables et de l'analyse combinatoire' and 'Nombres amiables, parties aliquotes et nombres figurés aux XIIIe et XIVe siècles', *Archive for History of Exact Sciences*, 28.2, 1983, pp. 107–47; repr. in *The Development of Arabic Mathematics*, pp. 275–319.

[7] M. Mersenne, *Harmonie universelle*, t. I, Paris, 1636.

[8] For failing to grasp this generic character, Thābit is credited with an explicit computation of this pair. He could perfectly well have calculated this pair, as well as the one for $n = 7$, but such was not his intention.

m	$3,_{7}$	$4,_{5,17}$	$10,_{11,41}$	$11,_{23}$	$12,_{13}$	$18,_{19}$	$23,_{47}$	$28,_{29}$	$34,_{137}$	$35,_{71}$	$36,_{37}$	$39,_{79}$
h		1	2	3	8	5	4	23		19	10	
i		2	3	4	9	6	5	24		20	11	
s	0	1	3	9			16		14		2	10

The values 2, 4, 7 of n yield the first three pairs of amicable numbers in the form of Thābit ibn Qurra, the two that we have already mentioned and the third

$$(9363584, 9437056)$$

obtained for $n = 7$ and independently discovered in the 17th century by al-Yazdī and by Descartes. The table allows one to eliminate all subsequent values of n up to 34 inclusive, because for these values, one or the other of the numbers h, i, s is not prime.

Restricting oneself to the first table (12 columns), one also eliminates the numbers of each of the forms: $43q + 30$, $43q + 31$, $43q + 9$, $49q + 29$, $49q + 30$, $51q + 17$, $52q + 35$, $52q + 36$, $53q + 36$, $53q + 37$, $58q + 8$, $58q + 9$, $60q + 54$, $60q + 55$, $66q + 27$, $66q + 28$, etc.

Besides $n = 2$, 4, and 7 (known in the 17th century), the only numbers smaller than 400 that are not excluded by these congruences are $n = 148$, 187, 340, and 391.

Table of the values of p_n for small values of n:

n	1	2	3	4	5	6	7	8	9	10	11	12
p_n	5	11	23	47	5.19	191	383	13.59	5.307	37.83	6143	11.1117

n	13	14	15	16	17	18	19
p_n	$5^2.983$	23.2137	197.499	421.467	5.78643	786431	71.22153

n	20	21	22
p_n	13.241979	5.1258291	$11^2.103991$

Table of the values of q_n:

n	1	2	3	4	5	6	7	8	9	10
q_n	17	71	7.41	1151	17.271	7.2633	73727	294911	7.17.23.431	79.59729

In the notations introduced above, a pair (ℓ, o) is constituted of amicable numbers if one has $\sigma(\ell) = \sigma(o) = \ell + o$. In the case in which $\ell = 2^n k$ and $o = 2^n s$ with k and s odd, this is written

$$(2^{n+1} - 1)\sigma(k) = (2^{n+1} - 1)\sigma(s) = 2^n(k + s).$$

Assume that s is prime and $k = hi$ where h and i are distinct prime numbers; the foregoing condition becomes

$$(2^{n+1} - 1)(h + 1)(i + 1) = (2^{n+1} - 1)(s + 1) = 2^n(hi + s),$$

whence

$$s = hi + h + i \text{ and } hi = (2^n - 1)(h + i) + 2^{n+1} - 1.$$

Set $h = 2^n - 1 + h_1$ and $i = 2^n - 1 + i_1$; the preceding equation yields

$$h_1 i_1 = (2^n - 1)^2 + 2^{n+1} - 1 = 2^{2n},$$

the solutions of which are $h_1 = \varepsilon\, 2^\alpha$ and $i_1 = \varepsilon\, 2^\beta$ with $\varepsilon = \pm 1$, where α and β are natural integers such that $\alpha + \beta = 2n$. Since $h \neq i$, $\alpha \neq \beta$, therefore one of them, for example α, is greater than n and the other, β, smaller than n. Thus $\alpha = n + \gamma$, $\beta = n - \gamma$ with $1 \leq \gamma \leq n - 1$, which requires that $n \geq 2$; thus $2^\alpha = 2^{n+\gamma} > 2^n$ and the fact that h is positive requires that $\varepsilon = 1$. One therefore has

$$h = 2^n - 1 + 2^{n+\gamma} = 2^n(2^\gamma + 1) - 1 \text{ and } i = 2^n - 1 + 2^{n-\gamma} = 2^{n-\gamma}(2^\gamma + 1) - 1,$$

and finally $s = 2^{2n-\gamma}(2^\gamma + 1)^2 - 1$. When $\gamma = 1$, $2^\gamma + 1 = 3$ and one returns to the form given by Thābit ibn Qurra; this more general form, with $1 \leq \gamma \leq n - 1$, was discovered by Euler.

Assume now that $k = h^2$ where h is a prime number, s being always prime; the condition for $(2^n k, 2^n s)$ to be a pair of amicable numbers is written

$$(2^{n+1} - 1)(h^2 + h + 1) = (2^{n+1} - 1)(s + 1) = 2^n(h^2 + s),$$

whence $s = h^2 + h$ is divisible by h, which is impossible. From this, it follows that there is no pair of amicable numbers of the form $(2^n h^2, 2^n s)$ with h and s odd primes.

The pairs of amicable numbers of the form $(2^n hi, 2^n s)$ where h, i, s are odd prime numbers are therefore those for which $n \geq 2$ and there exists an integer γ between 1 and $n - 1$ such that $h = 2^n(2^\gamma + 1) - 1$, $i = 2^{n-\gamma}(2^\gamma + 1) - 1$ and $s = 2^{2n-\gamma}(2^\gamma + 1)^2 - 1$.

Note that the even values of γ must be excluded because they yield values of s that are multiples of 3; when γ is odd, $2^\gamma + 1$ is a multiple of 3, therefore h, i, s cannot be multiples of 3. One also verifies that s is never a multiple of 5, 11, 13 or 19. In its first column, the following table contains the remainders of $\gamma \pmod{24}$ and, facing it, the possible remainders of

$n\,(\text{mod } 24)$; for the other remainders, one of the numbers h, i, s is divisible by $5, 7$, 13 or 17:

± 1	$4, 7, 11, 16, 19, 23$
± 3	$0, 3, 15$
± 5	$7, 8, 11, 16, 19, 23$
± 7	$4, 7, 8, 11, 16, 19, 20, 23$
± 9	$0, 3, 12, 15$
± 11	$7, 11, 16, 19, 23$

Euler's formulas can generate pairs of amicable numbers that were not obtained by Thābit ibn Qurra's formulas only for relatively high values of n ($n \geq 8$ for $\gamma = 5$ or 7, $n \geq 12$ for $\gamma = 9$, $n \geq 15$ for $\gamma = 3$, $n \geq 11$ for $\gamma = 11$, etc.). Finally the amicable numbers are not all given by these formulas, as the example (1184, 1210) discovered by N. Paganini in the 19th century shows. It is still not known whether or not there exists an infinity of pairs of amicable numbers.

– 4 –

FIBONACCI AND ARABIC MATHEMATICS

The 1226 meeting between Emperor Frederick II and the mathematician Leonard of Pisa (a.k.a. Fibonacci) has endlessly fascinated historians. It is rare indeed for an emperor to take the time and the leisure to discuss mathematics, and even rarer when his projects are facing setbacks. Yet this is precisely what Frederick II did when he stopped in Pisa while returning to Sicily after the failure of his undertakings in Lombardy.[1] This event, about which we in fact know very little, could not help but intrigue and tease the imagination. Beyond its legendary aura, however, this meeting made possible another, more recent one, between historians and historians of science. Henceforth, Frederick II is present in histories of mathematics, and Fibonacci appears in every biography of Frederick II.[2] Their conversation has given biographers of the emperor a chance to appreciate the diversity of his intellectual interests and the depth of his concerns; for their part, historians of mathematics have been able to emphasize the incontestable position reserved for the mathematician Fibonacci in the first quarter of the 13th century. From the presence of John of Palermo and (at least temporarily) Theodore of Antioch at the emperor's court,[3] as well as from the latter's correspondence with Arabic scholars,[4] we know that the ruler was

[1] E. Kantorowicz, *Emperor Frederick the Second, 1194–1250*, New York, Frederick Ungar, 1957, authorized English version by E. O. Lorimer 1987, pp. 154–8

[2] Cf. E. Kantorowicz, *Emperor Frederick the Second*; Thomas Curtis von Cleve, *The Emperor Frederick II of Hohenstaufen*, Oxford, Clarendon Press, 1972, pp. 310–12. See also Hans Niese, 'Zur Geschichte des geistigen Lebens am Hofe Kaiser Friedrichs II.', *Historische Zeitschrift*, 108, 1912, pp. 473–540.

[3] Cf. C. H. Haskins, *Studies in the History of Medieval Science*, Cambridge, Mass., Harvard University Press, 1924, notably the chapter on 'Science at the court of the Emperor Frederick II', pp. 242–71. See also M.-T. d'Alverny, 'Translations and Translators', in Robert L. Benson, Giles Constable, and Carol D. Lanham (eds), *Renaissance and Renewal in the Twelth Century*, Cambridge, Mass., Harvard University Press, 1983.

[4] I refer to the philosophical and scientific correspondence. Everyone knows his correspondence with Ibn Sab'īn, an Andalusian mystic. See M. Amari, 'Questions

(*Cont. on next page*)

interested not only in philosophy, astrology, and falconry, but also in such sciences as optics and mathematics.[5] But the conversation with Fibonacci reveals much more: Frederick knew enough mathematics to carry on a discussion with a mathematician. As to Fibonacci, the position that he secured and the games of projection about this conversation would confer on him royal standing among medieval mathematicians. Thus, Kantorowicz saw in him 'the greatest mathematician of his time and of the Middle Ages in general';[6] Haskins, whose horizon differed from that of Kantorowicz, called Fibonacci 'the outstanding scientific genius of the thirteenth century';[7] finally, Kurt Vogel, from yet a third perspective, referred to 'the first great mathematician of the Christian West'.[8] This enthusiastic judgment, which historians of mathematics have unanimously endorsed, is nevertheless not phrased to fit the occasion. But, what precisely does it mean when one alludes to the first great mathematician of the Latin West at the beginning of the 13th century? This question is evidently a central one both for the history of mathematics and for the history of culture, all the more so since, from the 16th century at least, generations of mathematicians would constantly return to the wellspring of Fibonacci's writings.

Before studying this question by analyzing Fibonacci's most important writings, however, let us pause to notice another remarkable feature of this famous conversation, indeed one that suggests a path to follow forward. Without risk of error, one can read into the conversation between the Emperor and the mathematician, an agreement between two great figures belonging to the same cultural world, and whose projects have several points in common. Frederick II and Fibonacci spoke the same language, manipulated the same concepts, and shared several similar values. Indeed both men had from infancy absorbed a certain Mediterranean culture with a predominant Arabic tone, the first in Palermo, the other in Bejaïa (Bougie)

(*Cont.*) philosophiques adressées aux savants musulmans par l'Empereur Frédéric II', *Journal Asiatique*, 5e série, 1, 1853, pp. 240–74. The Arabic text of his correspondence was published by Şerefettin Yaltkaya, with a foreword by H. Corbin, Paris, 1943.

[5] E. Wiedemann, 'Fragen aus dem Gebiet der Naturwissenschaften, gestellt von Frederich II., dem Hohenstaufen', *Archiv für Kulturgeschichte*, II/4, 1914, pp. 483–5. H. Suter, 'Beiträge zu den Beziehungen Kaiser Friedrichs II zu zeitgenössischen Gelehrten des Ostens und Westens, insbesondere zu dem arabischen Enzyklopädisten Kemâl ed-din ibn Jûnis', in *Abhandlungen zur Geschichte der Naturwissenschaften und der Medizin*, 4, 1922, pp. 1–8.

[6] Kantorowicz, *The Emperor Frederick the Second*, p. 153.

[7] Haskins, *Studies in the History of Medieval Science*, p. 249.

[8] Cf. the article by K. Vogel on Fibonacci in C. C. Gillispie (ed.), *Dictionary of Scientific Biography*, vol. 4, p. 604.

where he stayed before the travels that took him to Syria, Egypt, and Sicily. Both men had Arabic teachers; one of them spoke the language, the other seems to have known at least its elements. Moreover, the Emperor and the mathematician each sought to do fundamental and organizational work in the domain of his primary activity. For Fibonacci, that work extended mainly to arithmetic, algebra, and number theory.[9] Our question thus becomes more precise: to reflect on the significance of 'the first great mathematician' of the Latin West is to ask oneself what such work means in the Latin world of the 13th century. For obvious reasons, I cannot discuss all three of the domains I just mentioned; I will therefore consider here only algebra and number theory, focusing on Fibonacci's two most important writings, the *Liber abaci* and the *Liber quadratorum*.

Composed in 1202 and revised in 1228, the *Liber abaci* is a summa of Indian calculation, and of problems in arithmetic and algebra. Significantly, the edition of 1228 is dedicated to Michel Scot,[10] the translator from the Arabic who worked at Frederick II's court. The fifteen chapters of his book include Fibonacci's exposition, which is articulated as follows: he begins by defining the primitive terms of algebra, before moving on to the study of algebraic equations of the first two degrees. He then turns to the study of arithmetic operations on binomials and trinomials associated with these equations, before studying slightly more than ninety problems, of which one part reduces to these equations with rational coefficients, and the second part to these same equations, but with real coefficients.

One already notices that this order is that of the exposition in al-Khwārizmī's *Algebra* (composed *c.* 830) as well as that of his immediate successors, such as Abū Kāmil in his own *Algebra* (composed *c.* 870).[11] Just like al-Khwārizmī, Fibonacci defines only three primitive terms of algebra: the simple number (*numerus simplex*), the square root (*radix*), and the square (*census*). These are the only powers necessary for the study of equations of the first and second degree. Fibonacci introduces these equations in the very terms of al-Khwārizmī: 'six ways, of which three are

[9] Baldassarre Boncompagni, *Intorno ad alcune opere di Leonardo Pisano matematico del secolo decimoterzo*, Rome, 1854.

[10] *Scritti di Leonardo Pisano, matematico del secolo decimoterzo, pubblicati da Baldassarre Boncompagni*, Rome, 1857, vol. I, p. 1.

[11] R. Rashed, *Al-Khwārizmī: Le commencement de l'algèbre*, Paris, A. Blanchard, 2007, Introduction. See our edition of the *Algebra* in *Abū Kāmil: Algèbre et analyse diophantienne*, Scientia Graeco-Arabica 9, Berlin, De Gruyter, 2012.

simple and three compound [...]'.[12] He then enumerates verbally the following six equations.

$$ax^2 = bx, ax^2 = c, bx = c, ax^2 + bx = c, bx + c = ax^2, ax^2 + c = bx.$$

He goes on to discuss the algorithm of solution as well as its geometrical justification. Next comes the study of different problems that are reducible to one or the other of these six equations. Here, contrary to al-Khwārizmī, Fibonacci does not stop with those that have rational coefficients; rather, like Abū Kāmil, he brings in irrational coefficients. These problems are of the following type: to divide 10 into two parts such that the product of the one by the other is equal to the quarter of the product of one of two parts by itself; that is, $x(10 - x) = (1/4)x^2$; or again, to divide 12 into two parts such that the product of the one by 27 is equal to the product of the other by itself, that is, $27(12 - x) = x^2$. Fibonacci treats more than 90 problems in this way. Now an examination of the latter identifies 22 of them as borrowed from al-Khwārizmī's *Algebra* and 53 from Abū Kāmil's *Algebra*. These are the same problems, sometimes with no more than a superficial change of numerical coefficients. This massive borrowing most frequently follows the order of its model. Indeed Fibonacci generally reproduces the order of the problems al-Khwārizmī conceived, but he is even more faithful to the order of Abū Kāmil. The remainder of the approximately 25 problems that he discusses and whose origin we cannot identify are in any case conceived according to the model of problems borrowed from al-Khwārizmī and Abū Kāmil. As to the path that Fibonacci took to gain access to the works of these two algebraists, a simple comparison points to Gerard of Cremona's Latin translation of al-Khwārizmī's[13] works, which Fibonacci surely consulted. Moreover, 'Maumeht', al-Khwārizmī's first name, literally appears in this chapter on algebra in the *Liber abaci*.[14] The case of Abū Kāmil is not so simple, since we know neither the exact date of the Latin translation of his *Algebra*, nor the identity of the translator (probably Gerard of Cremona).

[12] 'sex modis ex quibus tres sunt simplices et tres compositi...' *Scritti di Leonardo Pisano*, vol. I, p. 406.

[13] Cf. *Gerard of Cremona's Translation of al-Khwārizmī's al-Jabr*, a critical edition by B. Hughes in Medieval Studies 48, 1986, pp. 211–63. See also A. Allard, 'The Influence of Arabic Mathematics in the Medieval West', in R. Rashed (ed.), *Encyclopedia of the History of Arabic Science*, 3 vols, London/New York, Routledge, 1997, vol. II, pp. 539–80. See also N. Miura, 'The Algebra in the *Liber Abaci* of Leonardo Pisano', *Historia Scientarum*, 21, 1981, pp. 57–65.

[14] Al-Khwārizmī's first name can be read in the margin of the third part of chapter fifteen.

Fibonacci's *Algebra* thus appears to be a kind of commentary on those of al-Khwārizmī and Abū Kāmil. In fact, however, these two mathematicians represent the initial period of algebra, the first as the founder, the second as the figure who most advanced the work of the first, and in the same spirit. By the time of Fibonacci, algebra had undergone two radical transformations since its creation. The first, which I have recently named the arithmetization of algebra,[15] came to light with al-Karajī, at the end of the 10th century. This transformation led to the elaboration of the algebra of polynomials and to abstract algebraic calculation. There is little doubt that Abū Kāmil, with his research on irrational coefficients and the calculation of quadratic irrational numbers, prepared the ground for the transformation that al-Karajī subsequently effected. Whereas it is not impossible that Fibonacci somehow got wind of al-Karajī's algebra, he surely did not come under the latter's influence, either in his project, or in its realization, contrary to what the eminent historian F. Woepcke[16] believed in the mid-19th century. Fibonacci remains not only tributary to al-Khwārizmī and to Abū Kāmil, but also in a certain sense their contemporary. The second transformation of algebra led to the constitution of algebraic geometry with al-Khayyām (1048–1131) and Sharaf al-Dīn al-Ṭūsī in the second half of the 12th century. No trace of this algebraic geometry appears in Fibonacci, however. Even when one gives him al-Khayyām's equation $x^3 + 2x^2 + 10x = 20$, in his *Flos* he treats it arithmetically to give it an approximate solution.[17] Everything therefore indicates that Fibonacci the algebraist was evolving uniquely and strictly within an earlier stage of Arabic algebra.

But algebra is not the only domain of the *Liber abaci* for which this is the case. The same is true of arithmetic, as Fibonacci himself admits. It remains for us to choose two other examples from the *Liber abaci*, in two other domains, in reference to Arabic works not translated into Latin, or for

[15] R. Rashed, *The Development of Arabic Mathematics*, pp. 22 ff.

[16] F. Woepcke, *Extrait du Fakhrī*, Paris, 1853, pp. 24 ff.

[17] See R. Rashed and B. Vahabzadeh, *Al-Khayyām mathématicien*, Paris, Librairie Blanchard, 1999, pp. 88, 224; English transl. *Omar Khayyam. The Mathematician*, Persian Heritage Series no. 40, New York, Bibliotheca Persica Press, 2000; F. Woepcke, 'Sur un essai de déterminer la nature de la racine d'une équation du troisième degré…', extrait du *Journal de mathématiques pures and appliquées*, XIX, 1854; Boncompagni, *Intorno ad alcune opere di Leonardo Pisano*, p. 6; H. G. Zeuthen, 'Sur la résolution numérique d'une équation du 3e degré par Léonard de Pise', *Bulletin de l'Académie Royale des Sciences et des Lettres du Danemark*, no. 3, janvier-mars 1893, pp. 6–17; See finally H. Henkel, *Zur Geschichte der Mathematik in Altertum und Mittelalter*, Leipzig, 1874; repr. G. Olms, 1965, pp. 292–3.

which no translation has so far reached us. The first is borrowed from numerical computation, the other from classical number theory.

In *Liber abaci*, Fibonacci proposes a method to approximate the cubic root of an integer. For $N = a^3 + r$, he gives[18] an expression equivalent to

$$N^{\frac{1}{3}} = a + \frac{r}{3a^2 + 3a + 1}.$$

Then he writes: 'inueni hunc modum reperiendi radices secundum quod inferius explicabo (I discovered this way of finding roots according to what I will explain below)'.[19] But this method was so widespread in Arabic mathematics ever since the beginning of the 9th century at least, that slightly later mathematicians such as Naṣīr al-Dīn al-Ṭūsī called it 'the conventional method'. Indeed, it already appears in Ibn al-Haytham (d. after 1040) and Abū Manṣūr al-Baghdādī (d. 1037).

The second example is even more striking and is borrowed from a different domain: number theory, from which we want to discuss a problem of linear congruences that Ibn al-Haytham treats and solves. Let us begin by listening to his own words:

> To find a number such that, if one divides it by two, one remains; if one divides it by three, one remains; if one divides it by four, one remains; if one divides it by five, one remains; if one divides it by six, one remains; and if one divides it by seven, nothing remains.[20]

Listen now to Fibonacci:

> There is a number such that, when one divides it by 2, or by 3, or by 4, or by 5, or by 6, the remainder is 1, which is not divisible [by the above numbers], whereas the same number is entirely divisible by 7. Find this number.[21]

This is evidently the same statement of a problem of linear congruences that is known as 'the Chinese remainder theorem' and can be rewritten: *to find an integer* n *such that*

$$\begin{cases} n \equiv 1 (\text{mod}\, m_i) \\ \\ n \equiv 0 (\text{mod}\, p) \end{cases} \qquad \textit{with}\; 1 < m_i \le p - 1$$

where $m_i = 2, 3, ..., 6$.

[18] *Scritti di Leonardo Pisano*, p. 378.

[19] *Ibid.*

[20] *On the Solution of a Numerical Problem*, in R. Rashed, *The Development of Arabic Mathematics*, p. 247.

[21] *Scritti di Leonardo Pisano*, pp. 281–2.

In his own words here is Fibonacci's solution.

> [...] because it is proposed that the remainder is always 1 when it is divided by 2, or 3, or 4, or 5, or 6, when 1 is subtracted from the number the difference is integrally divisible by each of the abovewritten numbers; therefore you find the least common denominator of $\frac{1}{2}$ $\frac{1}{3}$ $\frac{1}{4}$ $\frac{1}{5}$ $\frac{1}{6}$; this number will be 60 which you divide by the 7; the remainder is 4 which should be 6 because the entire number is divisible by the 7; therefore the number which is one less than it when divided by 7 must of necessity have remainder 6 that is 1 less than seven; therefore 60 is doubled, or tripled, or any multiple is taken up to when the number divided by 7 has remainder 6; the multiple will be 5 by which the 60 is multiplied; the result is 300 to which is added 1; there will be 301, and this is the number. Similarly if 420 that is integrally divisible by all of the aforesaid numbers, you will add to the 301 once, or however many times you will wish, then the sought number will always result, namely a number which is integrally divisible by 7, and the remainders are always 1 when it is divided by the others.[22]

Fibonacci repeats the numerical application of the method to find another number with $m_i = 2, ..., 10, p = 11$, and finds 25201. This is what he writes:

> By this method we indeed find another number which when divided by any number from two up to ten always has remainder 1, and is integrally divisible by 11; the number is 25201. Also if 698377681 is divided by any number from 2 up to 23, you will always find that the remainder is 1, and it is truly integrally divisible by 23; this number is found similarly by the abovewritten method.
>
> *number*
> 25201
> *number*
> 698377681

On the Same [Topic]

Again there is a number which when divided by 2 has a remainder 1, and when divided by 3 has a remainder 2, when divided by 4 has a remainder 3, when divided by 5 has a remainder 4, when divided by 6 has a remainder 5, and is truly integrally divisible by 7; therefore the least common denominator of $\frac{1}{6}$ $\frac{1}{5}$ $\frac{1}{4}$ $\frac{1}{3}$ $\frac{1}{2}$ is found, and it will be 60 from which you take 1; there

[22] *Fibonacci's* Liber Abaci*: a Translation into Modern English of Leonardo Pisano's Book of calculation*, translated by L. E. Sigler, Sources and Studies in the History of Mathematics and Physical Sciences, New York, Springer-Verlag, 2002, p. 402.

remains 59. As this is not integrally divisible by 7, you will double the 60, or you will triple it, or you will take another multiple of it until the product is a number which has a remainder 1 when divided by 7; indeed the double of 60, namely 120, when divided by 7 has a remainder 1; when the 1 is subtracted from the 120, there remains 119 for the sought number.

number
119
[to which, if you add 420
once, twice, or as many times as you wish,
you will have the sought number.]

On the Same [Topic]

Also there is a number which when divided by 2 has a remainder 1, when divided by 3 has a remainder 2, when divided by 4 has a remainder 3, and thus so on up to 10; when the number is divided by 10 it has a remainder 9; truly the number is integrally divisible by 11. First indeed you find the least common denominator of $\frac{1}{10}$ $\frac{1}{9}$ $\frac{1}{8}$ $\frac{1}{7}$ $\frac{1}{6}$ $\frac{1}{5}$ $\frac{1}{4}$ $\frac{1}{3}$ $\frac{1}{2}$ which we thus demonstrate to you how to find. First you take 60 which is the least common denominator of the aforesaid fractions $\frac{1}{10}$ $\frac{1}{6}$ $\frac{1}{5}$ $\frac{1}{4}$ $\frac{1}{3}$ $\frac{1}{2}$, and when you multiply it by 7; there will be 420 that you must multiply by 8 and 9; however you leave off multiplying by the 4 that is in the rule for 8, and the 3 which is in the rule for 9 because the least common denominator of $\frac{1}{4}$ $\frac{1}{3}$ is found in the abovewritten 60; therefore you will multiply the 420 by the 2 remaining in the rule for 8; there will be 840 that you will multiply by the 3 remaining in the rule for 9; there will be 2520 which is the least number in which are found all the abovewritten factors, and in geometry it is called the least common multiple of all the numbers which are less than or equal to 10; next you subtract 1 from the 2520; there remains 2519 that is integrally divisible by 11; we have our number without labor; that is, 2519 is the sought number. And when 4655851199 is divided by any number which is less than 23 there will always be remainder 1 less than the number by which it was divided, and it is integrally divisible by 23. And when 698377681 is divided by all the abovewritten numbers up to 22 it always has remainder 1; it is truly integrally divisible by 23.

number
2519
number
4655851199.[23]

[23] *Fibonacci's* Liber Abaci, transl. L. E. Sigler, pp. 402–3.

Fibonacci does not state the main property of these numbers, in contrast to Ibn al-Haytham, for whom it was of primary interest; in his own words: 'this property is necessary for every prime number', that is,

$$n \text{ is prime} \Leftrightarrow (n-1)! \equiv -1 \pmod{n}.$$

Moreover, Fibonacci is not alone in having let slip this fundamental property for characterizing prime numbers, which would later be known as 'Wilson's theorem'. Other algebraists who were contemporaries of Fibonacci also cited this same problem without mentioning this property. But a comparison between the text of Ibn al-Haytham and that of Fibonacci clearly suggests that the latter knew at least a commentary on the former, if not the text itself, but that he could not grasp Ibn al-Haytham's intention: to define a criterion that distinguishes prime numbers.

Thus algebra, numerical calculation, and linear congruences combine with arithmetic to situate the *Liber abaci* in the universe of Arabic mathematics, as much by its borrowings and its developments as by its style and its language. More precisely, the book of Fibonacci belongs not to the most advanced mathematics, such as that of al-Karajī, al-Khayyām and Sharaf al-Dīn al-Ṭūsī, but rather to that of an earlier period. Note further that when he treats a new and promising field of research – linear congruences – Fibonacci lifts out only the most immediate aspect, as do so many Arabic commentators who are not among the most creative.

Does such a characterization apply also to the *Liber quadratorum*, which historians of mathematics rightly consider the most important Latin contribution to number theory before those of Bachet de Méziriac and Fermat? The situation is in fact analogous, for his book belongs to a different Arabic tradition in number theory, one born around the middle of the 10th century and still extending into the period of Fibonacci. This is attested by a report from Kamāl al-Dīn ibn Yūnus,[24] who was also a correspondent of Frederick II and a teacher of Theodore of Antioch,[25] who himself later joined the Emperor's court and exchanged letters with Fibonacci.[26]

[24] Kamāl al-Dīn ibn Yūnus, *Risāla fī bayān annahu lā yumkin an yūjad ʿadadān murabbaʿān fardān majmūʿhuma murabbaʿ*, ms. Paris, BN no. 2467, fols 196v–197v.

[25] H. Suter, 'Beiträge zu den Beziehungen Kaiser Friedrichs II.'.

[26] Fibonacci's answer to a letter from Theodore of Antioch, who was at the time at the court of Frederick II, is thus entitled: 'Epistola suprascripti Leonardi ad Magistrum Theodorum phylosophum domini Imperatoris', in *Scritti di Leonardo Pisano*, vol. II, pp. 247 ff.

The *Liber quadratorum* itself opens with discussions heavily laden with meaning. Fibonacci begins by addressing the Emperor in these terms:

> After being brought to Pisa by Master Dominick to the feet of your celestial majesty, most glorious prince, Lord F[rederick.], I met Master John of Palermo; he proposed to me a question that had occurred to him, pertaining not less to geometry than to arithmetic: find a square number from which, when five is added or subtracted, always arises a square number.[27]

By Fibonacci's own admission, we therefore know that the goal of his book is to solve the system of second-degree Diophantine analysis that is proposed by John of Palermo and rewritten thus:

$$\text{(I)} \qquad \begin{aligned} x^2 + 5 &= y_1^2, \\ x^2 - 5 &= y_2^2, \end{aligned}$$

and that this problem is not algebraic, but geometrical and numerical. After this short paragraph, we observe a combination of facts that can in no way be considered mere coincidences. First of all, John of Palermo, who was at the court of Frederick II, knew both mathematics and Arabic. It is indeed from this language that he translated a treatise on the asymptote to an equilateral hyperbola under the title *De duabus lineis*.[28] He is moreover a colleague of Theodore of Antioch who, as we were just reminded, had studied with Kamāl al-Dīn ibn Yūnus, who was himself practiced in this type of research. In addition, by characterizing the problem as geometrical and numerical, Fibonacci classifies it not in algebra, but in that branch of number theory that treats Pythagorean triples or numerical right triangles, an interpretation that at every point is confirmed by studies of the *Liber quadratorum* itself. Finally, at issue here is not just any question of Diophantine analysis, but a problem that appears, as it were, 'in person' several times in the works of Arabic number theorists and algebraists. So it is that al-Karajī writes: 'If one says a square and if one adds to it five units, one has a square, and if one subtracts five units, one has a square'.[29] Al-

[27] Leonardo Pisano/Fibonacci, *The Book of Squares*, An annotated translation into modern English by L. E. Sigler, Orlando, Florida, Academic Press, Inc., 1987, p. 3. Cf. the French translation by P. Ver Eecke, *Léonard de Pise, Le livre des nombres carrés*, Paris, Blanchard, 1952, p. 1.

[28] This text was established and translated by Marshall Clagett, *Archimedes in the Middle Ages*, vol. IV: *A Supplement on the Medieval Latin Tradition of Conic Sections* (1150–1566), Philadelphia, 1980, pp. 33–61, pp. 335–57.

[29] Al-Karajī, *Al-Badī'*, ed. A. Anbouba, Beirut, 1964, p. 77.

Karajī finds as a solution $x^2 = 1681/144$, which is the solution that appears in Fibonacci.[30] But neither Fibonacci's methods nor his aims are those of al-Karajī. Moreover, this same system appears in a treatise on numerical right triangles composed at least a half-century before al-Karajī; and Fibonacci's method, as well as the problems he discusses in the *Liber quadratorum* and its mathematical style, are very close to what one finds in this treatise, and in similar 10th-century ones devoted to numerical right triangles. In short, the problem that John of Palermo proposed to Fibonacci was borrowed directly or *via* al-Karajī from one of the writings on Diophantine analysis. Fibonacci's research in the *Liber quadratorum* to solve it allows one to glimpse a certain knowledge of the works of number theorists from the middle of the 10th century who were the first to elaborate a new branch of Diophantine analysis: namely, integer Diophantine analysis. We can thus sharpen our question about the *Liber quadratorum*: to what extent is the book integrated into this tradition?

Indeed, one encounters this problem, called *of congruent numbers*, for the first time in an anonymous treatise on integer Diophantine analysis, or more precisely on the theory of Pythagorean triples.[31] The author was fully and justifiably aware of the novelty of his project, which he emphasizes with vigor. But this same problem also reappears in al-Khāzin, a mathematician from the same period, and it is later reproduced by mathematicians such as Abū al-Jūd ibn al-Layth[32] in the latter third of the 10th century. All this is to say that the problem was diffused and transmitted throughout the 10th century. Now, in al-Khāzin's treatise, this problem of congruent numbers is presented in the same form that later appears in Fibonacci as the goal of his treatise. This is how al-Khāzin himself presents it:

> After having introduced the preceding, we reach the goal that we have pursued: how to find a square number such that, if one adds to it a given number and if one subtracts the given number, the sum and the difference are two squares.[33]

Without going through al-Khāzin's procedure, we need only recall that he tries to determine not only such a square, but especially the conditions

[30] *Scritti di Leonardo Pisano*, vol. II, p. 271.

[31] This treatise is translated into French by F. Woepcke, *Recherches sur plusieurs ouvrages de Leonardo de Pise…; Extraits et traduction des ouvrages inédits*, Rome, 1861.

[32] *Fī al-muthallathati al qā'imati al-zawāya*, ms. Leiden, Or. 168/14, fols 132r ff.

[33] R. Rashed, *The Development of Arabic Mathematics*, pp. 221–2.

necessary for solving this Diophantine system; he demonstrates the following theorem:

Given an integer a the following conditions are equivalent:

1) the system

$$\text{(II)}\quad \begin{cases} x^2 + a = y_1^2 \\ x^2 - a = y_2^2 \end{cases} \quad (y_2 < x < y_1)$$

admits a solution;

2) there exists a pair of integers (u, v) called conjugate, that is, such that

$$u^2 + v^2 = x^2$$
$$2uv = a;$$

under these conditions, a is of the form $4k$, where k is not a power of 2; or again, a is of the form $4p\,(2q + 1)$. Al-Khāzin shows clearly that the smaller integer that verifies these conditions is 24: the other integers are multiples of 24.

Note also that al-Khāzin's demonstration rests on a lemma and a proposition established in the treatise. The lemma states that there is no pair of integers, square and odd, whose sum is a square. As to the proposition, it pertains to Pythagorean triples established by synthesis in Euclid's *Elements*, for which al-Khāzin provides the analysis; it states:

Given (x, y, z), a triple of integers such that $(x, y) = 1$, x is even, the following conditions are equivalent

(III) 1) $x^2 + y^2 = z^2$

2) there exists a pair of integers (p, q) such that $p > q > 0$, $(p, q) = 1$ and p and q are of opposite parity such that

$$x = 2pq \qquad y = p^2 - q^2 \qquad z = p^2 + q^2.$$

Finally, after having considered several problems of Pythagorean triples and solved the system of congruent numbers, al-Khāzin treats the representation of a number as the sum of squares, and in this connection establishes for the first time the double identity known since the Babylonians:

$$\text{(IV)}\qquad (p^2 + q^2)(r^2 + s^2) = (pr \pm qs)^2 + (ps + qr)^2.$$

We now come to the *Liber quadratorum*, and quickly examine Fibonacci's procedure. He begins by establishing

$$\sum_{k=1}^{n}(2k - 1) = n^2$$

in order to show that $[1 + 3 + ... + (2n - 1)^2 - 2]$ and $(2n - 1)^2$ are the two sides of a numerical right triangle. He establishes an analogous property for even numbers. In the third proposition, he comes back to an application of III. In Proposition 6, he gives IV, which is one of the first known decompositions of quadratic forms. In Proposition 12, he gives a form of II, by establishing that for $(u, v) = 1$, one has

$$uv\,(u + v)\,(u - v) = 24\,k, \quad k = 1, 2, ...$$

In Proposition 16, he returns to system I, and chooses (u, v) such that $4\,uv\,(u + v)\,(u - v)$ is a square multiple of 5. He takes $u = 5$, $v = 4$, where

$$4\,uv\,(u + v)\,(u - v) = 720 = 12^2 \cdot 5 = 25 \cdot 6 \cdot 5.$$

It is not necessary to linger further in order to see that these results are very close to those of the 10th century mathematicians, and much more importantly, that they fit into an identical mathematical context: the theory of Pythagorean triples. This conclusion is in no way novel; Gino Loria, an eminent historian with undoubted admiration for Fibonacci, has already proposed it. Loria, who wrongly believed that Fibonacci was the first to have found the double identity (IV),[34] wrote after studying the *Liber quadratorum*:

> If it seems difficult to deny that the example of Muḥammad ibn Ḥusayn led Leonard of Pisa to the research that we have summarized above, the latter's dependence on the former appears even less doubtful when one turns to the next section of the *Liber quadratorum*, which treats 'congruent numbers'.[35]

No one knew, however, that this Muḥammad ibn al-Ḥusayn was none other al-Khāzin.

The *Liber quadratorum* thus belongs to this tradition of 10th century mathematicians who gave birth to integer Diophantine analysis. At the

[34] G. Loria, *Storia delle Matematiche*, Milan, Ulrico Hoepli, 1950, p. 233: 'Va ancora rilevato che il nostro matematico stabilisce la doppia identità

$$(a^2 + b^2)\,(c^2 + d^2) = (ac + bd)^2 + (bc - ad)^2 = (ad + bc)^2 + (bd - ac)^2;$$

ritrovata da Bachet de Méziriac, fu aplicata da Viète, Cauchy e da molti altri, ma, in memoria di chi per primo la scoperse, meriterebbe di recare il nome di *Teorema di Fibonacci*'.

'One should also note that our mathematician established the double identity

$$(a^2 + b^2)\,(c^2 + d^2) = (ac + bd)^2 + (bc - ad)^2 = (ad + bc)^2 + (bd - ac)^2;$$

rediscovered by Bachet de Méziriac, it was applied by Viète, Cauchy, and many others. In memory of its first discoverer, however, it would deserve to bear the name *Theorem of Fibonacci*'.

[35] *Ibid*., p. 234.

moment, however, we know of no Latin translation of these writings, no more than we do for the case of Ibn al-Haytham's treatise.

In concluding this exposition, we therefore see 'that the first great mathematician' of the Latin West presents himself, not only in arithmetic, but also in algebra and in number theory, as carried by the current of the first period of Arabic mathematics, that of the 9th–10th centuries. This conclusion finds confirmation in the presence of other Arabic mathematicians from this period in Fibonacci's work, for example, Aḥmad ibn Yūsuf.[36] In this regard, however, Fibonacci's work does not differ from that of other Latin mathematicians of his day, such as Jordanus de Nemore's *De numeris datis*. Yet in contrast to these others, Fibonacci evidently had direct access to the various traditions of Arabic mathematical writing. It was therefore inevitable that his contribution appeared much superior to the Latin writings of his own day, whether in the fields that it embraced or in the results that it presented. But this man who, when viewed from upstream, is tied to the Arabic mathematics of the 9th–10th centuries, is, when seen from downstream, a scholar of Latin mathematics from the 15th–16th centuries. In any event, his work proved to be a source of inspiration and renewal for Latin mathematics.

[36] Indeed, Fibonacci borrows certain problems from the work on proportions by this mathematician from the late 9th to the beginning of the 10th century; cf. *Scritti*, pp. 118–19. This is how Fibonacci concludes the borrowed problem: '[...] et Ametus filius ponat decem et octo combinationes ex ea in libro, quem de proportionibus composuit' (p. 119). On Aḥmad ibn Yūsuf, see notably D. Schrader's article in the *Dictionary of Scientific Biography*, vol. I, pp. 82–3.

– 5 –

FIBONACCI AND THE LATIN EXTENSION OF ARABIC MATHEMATICS

In the middle of the 19th century, thanks to recent works by Cossali,[1] Libri,[2] and especially Boncompagni,[3] F. Woepcke[4] became the first to study what Fibonacci's *Flos* and *Liber quadratorum* owed to Arabic mathematics. His explanations were accepted and adopted by many historians who wrote on Fibonacci: Gino Loria, A. Youshkevitch, E. Picutti, for example. In an earlier study that examined not only the two preceding books, but also the *Liber abaci*, I tried to show that 'the first great mathematician of the Christian West', as K. Vogel[5] called him, turns out to be carried by the current of Arabic mathematics, not in general (as people like to repeat), but only from the first period, that is, the mathematics of the 9th and the first half of the 10th century.[6] It is to this tradition that Fibonacci seems to have had access, particularly the writings of al-Khwārizmī and Abū Kāmil in Latin translation. To these names, one should also add those of a few other mathematicians whose writings have also translated into Latin, such as Aḥmad ibn Yūsuf, the Banū Mūsā, al-Nayrīzī... One reaches

[1] Pietro Cossali, *Origine, trasporto in Italia, primi progressi in essa dell' algebra*, Parma, 1797.

[2] Guglielmo Libri, *Histoire des sciences mathématiques en Italie*, vol. II, Hildesheim, Georg Olms, 1967.

[3] Baldassarre Boncompagni, *Tre scritti inediti di Leonardo Pisano*, Firenze, 1854.

[4] See in particular *Extrait du Fakhrī, Traité d'algèbre par Aboū Bekr Mohammed Ben Alhaçan Alkarkhī*, précédé d'un mémoire sur l'algèbre indéterminée chez les Arabes, Paris, Imprimerie Nationale, 1853; repr. Hildesheim, Georg Olms, 1982.

[5] K. Vogel, 'Fibonacci', *Dictionary of Scientific Biography*, vol. IV, 1971, pp. 604–13.

[6] See above, 'Fibonacci and Arabic mathematics'. On the *Liber Abaci*, see also R. Franci, 'Il *Liber Abaci* di Leonardo Fibonacci, 1202–2002', *La Matematica nella Società e nella Cultura*, Bollettino della Unione Matematica Italiana, Serie VIII, vol. V–A, Agosto 2002, pp. 293–328. See also O. Terquem, 'Sur Léonard Bonacci de Pise et sur trois écrits de cet auteur publiés par Balthasar Boncompagni', *Annali di scienze matematiche*, vol. 7, 1856.

this conclusion very naturally, by examining several chapters of the *Liber abaci*. Calculations on roots in the fourteenth chapter can be understood perfectly in light of the works cited above. Better yet, of the ninety problems that Fibonacci studied in the fifteenth chapter, seventy-nine are borrowed from the books of al-Khwārizmī and Abū Kāmil. By 'borrowing', I mean a repetition of the problem that is identical to the original or contains a few insignificant variants, such as a change of numerical coefficients. Note also that the problems that, according to Woepcke, Fibonacci allegedly borrowed from al-Karajī, the mathematicians from the end of the 10th century, or from Diophantus *via* the latter, are all found in Abū Kāmil's book. Contrary to Woepcke's belief, there is no evidence that Fibonacci knew either al-Karajī's *al-Fakhrī* or Diophantus's *Arithmetic*.

These borrowings (and many others, no doubt) are certainly important for situating Fibonacci's contributions to the history of mathematics. To stop here, however, would be to obscure another facet of his contribution and thus to miss its true significance. Not only did Fibonacci borrow entire chapters from these mathematicians, but also his work presents itself in some sense as an *extension* into Latin of the Arabic mathematics of the first period. By this, I mean the invention of new results, but within the framework of the inherited *mathesis*, and without any rupture from it. The key question is: In which sense and according to which style did this extension take place? It goes without saying that a full and definitive answer to such a question lies in the future, for it will require a better knowledge of Latin translations from the Arabic and especially of the Arabophone communities of Italy as well as the completion of a genuine critical edition of the *Liber abaci*. In the meantime, we should toe to incontestable facts and avoid arbitrary resemblances, to say nothing of abusive ones. In what follows, I propose nothing more than the first sketch of an answer to this question.

To me, the most propitious terrain for such an inquiry seems to be the debate between Fibonacci and the mathematicians at the Hohenstaufen court. The likes of John of Palermo and Theodore of Antioch not only knew Arabic, but were obviously in touch with mathematical research written in this language. By way of illustration, John of Palermo translated into Latin an anonymous Arabic treatise on the asymptote to an equilateral hyperbola. Now, we know that mathematicians such as al-Sijzī at the end of the 10th century had already formulated this problem, which would later become an object of research.[7] In all probability, then, the translated trea-

[7] R. Rashed, 'Al-Sijzī et Maïmonide: Commentaire mathématique et philosophique de la proposition II.14 des *Coniques* d'Apollonius', *Archives internationales d'histoire*

(*Cont. on next page*)

tise was written after the end of the 10th century. As to Theodore of Antioch, he was himself an Arab, a student of the Mosul mathematician Kamāl al-Dīn ibn Yūnus (1156–1241), himself a student of the great algebraist, Sharaf al-Dīn al-Ṭūsī, who pushed and developed in a new direction al-Khayyām's work in algebraic geometry. Recall also that the mathematician Kamāl al-Dīn ibn Yūnus was one of the Arab correspondents of Frederick II. In short, these mathematicians at the emperor's court had access to Arabic mathematical works for which we know of no Latin version, and it was from these writings that they drew the questions they put to Fibonacci. These were truly difficult questions, drawn from works that Fibonacci very probably did not know, or else knew only from their statements and their style, in a debate to which the Emperor himself served as arbiter. All of these elements prompt us to see here a challenge that Fibonacci could not resist accepting, and thus forced him to excel, if not to out-do himself. In other words, here we are speaking no longer of borrowings, but only of the pursuit of inventive research. Precisely in this sense is the terrain most favorable for our inquiry.

I. John of Palermo asks Fibonacci to solve the equation

$$x^3 + 2x^2 + 10x = 20 , \qquad (1)$$

by requiring in addition that the solution be 'ex his que continentur in X° Libro Euclidis [drawn from what is contained in Book X of Euclid]'.[8]

Why this equation and this solution? Clearly the problem raised is neither easy nor innocent: John of Palermo no doubt knew that the solution was not in any way easy, perhaps because he had wrestled with it himself, perhaps because he knew the history of this equation, or finally, perhaps, for both reasons at once.

Indeed, an identical equation with the very same coefficients appears in the *Treatise of Algebra* by al-Khayyām (1048–1131), as Woepcke long ago noted. Al-Khayyām writes in this regard: 'One will determine the side of the cube according to what we have explained by means of conic sections [...] The square (*murabba*ʿ) of that side will then be the square (*māl*, *i.e.*

(*Cont.*) *des sciences*, no. 119, vol. 37, 1987, pp. 263–96; repr. in *Optique et mathématiques: Recherches sur l'histoire de la pensée scientifique en arabe*, Variorum reprints, Aldershot, 1992, XIII.

[8] Boncompagni, *Tre scritti inediti di Leonardo Pisano*, p. 3.

the root) looked for'.[9] In fact, al-Khayyām solves this type of equation by the intersection of a circle and a hyperbola.[10] Moreover, this type of equation was among those that the mathematicians in the tradition of al-Karajī tried to solve by radicals. The 12th-century mathematician al-Sulamī proposed to eliminate the second-degree terms by means of an affine transformation, then by imposing a condition on the coefficients of the first-degree terms, in order to make this new equation amenable to the simple extraction of a cubic root.[11] Thus, for the equation

$$x^3 + ax^2 + bx = c,$$

one sets $x = y - \frac{a}{3}$; the equation is rewritten

$$y^3 + \left(b - \frac{a^2}{3}\right)y + \left(\frac{2a^3}{27} - \frac{ba}{3} - c\right) = 0.$$

One imposes $b = \frac{a^2}{3}$, a method that cannot work here. It was therefore necessary to find another method.

Whatever John of Palermo's source(s), he only compounded the difficulty by requiring Fibonacci to proceed by means of Book X of the *Elements*. In effect, he was demanding that the latter interpret the book algebraically – unless Fibonacci somehow already knew about this interpretation of Arabic mathematics, undertaken by al-Karajī and pursued by such successors as al-Samaw'al, for example. In the present state of our knowledge, however, nothing supports such a supposition. Indeed, Chapter 14 of the *Liber abaci* shows that Fibonacci's knowledge of calculation by radicals does not go beyond that found in Abū Kāmil's *Algebra*; and unfortunately his own work on Book X has not reached us. But John of

[9] R. Rashed and B. Vahabzadeh, *Al-Khayyām mathématicien*, Paris, Librairie A. Blanchard, 1999, p. 225, 4–5; English version without the Arabic text: *Omar Khayyam: The Mathematician*, Persian Heritage Series no. 40, New York, 2000, p. 159.

[10] *Ibid.*, pp. 55–8 and p. 185; English version, pp. 55–8 and 141–2.

[11] R. Rashed, 'Les commencements de l'algèbre', dans *Entre arithmétique et algèbre. Recherches sur l'histoire des mathématiques arabes*, Paris, Les Belles Lettres, 1984, pp. 17–29, at p. 28; English transl.: *The Development of Arabic Mathematics: Between Arithmetic and Algebra*, Boston Studies in the Philosophy of Science, 156, Dordrecht/Boston/London, Kluwer Academic Publishers, 1994, p. 17.

Palermo required even more: that the solution be formed by Euclidean radicals. We now know that this is impossible, but neither the emperor's mathematician nor Fibonacci was aware of this.

From this unusually delicate situation, the Pisan mathematician will extricate himself superbly. In fact, he combines Euclidean arithmetic and algebra, like many number theorists from the 10th century: al-Khāzin, al-Khujandī, Abū al-Jūd, al-Sijzī. Had he come under their influence? Perhaps, but it is also possible that, being familiar with the algebra of al-Khwārizmī and Abū Kāmil, he found the same way to answer John of Palermo.

Let us begin by summarizing his procedure in a different language.

Equation (1) is rewritten

$$\frac{x^3}{10}+\frac{x^2}{5}=2-x.$$

One sees immediately that $1 < x < 2$, and that x is not an integer.

Let $x=\frac{m}{n}$, with m and n mutually prime, whence equation (1) is rewritten

$$m^3+n\left(2m^2+10mn-20n^2\right)=0.$$

Thus n divides m^3; since it is prime with m, the repeated application of Euclid's lemma shows that it divides m, contrary to the hypothesis; therefore, x is not a rational number.

Fibonacci then shows that x can be neither a Euclidean irrational nor a combination of Euclidean irrationals.

Assume that $x=\sqrt{n}$, where n is not square. Equation (1) is rewritten

$$(2) \qquad x=2\frac{10-x^2}{10+x^2}=\sqrt{n};$$

$\sqrt{n}$ is therefore rational, which is absurd.

Assume that $x=\sqrt[4]{n}$: therefore equation (2) is rewritten

$$10\sqrt[4]{n}+\sqrt[4]{n}\cdot\sqrt{n}=2\left(10-\sqrt{n}\right),$$

and thus the first of the two medials is equal to the apotome, two irrational quantities the heterogeneity of which Euclid had demonstrated.

One falls into a similar contradiction if one assumes that $x=\sqrt{m+\sqrt{n}}$ or $x=\sqrt{\sqrt{m}+\sqrt{n}}$.

Thus, Fibonacci demonstrates that no Euclidean radical can satisfy equation (1). He concludes by giving an approximation of x.[12]

Fibonacci answered John of Palermo well, but without thinking of going beyond the question the latter had raised. Let me explain. Historians have noticed how, during this demonstration, the appeal to Euclidean arithmetic to discuss this cubic equation led to most interesting results:

• Fibonacci showed that if the root of equation (1) is non-integer, it cannot be rational either. The result is doubly important – in itself, but also on account of the method used to establish it. It is the same method traces of which already appear in the construction of the irrationality of the diagonal of a square in Greek mathematics. As a treatise by al-Sijzī[13] attests, the same method also appears in the works of the 10th-century mathematicians on integer Diophantine analysis. After becoming generalized in the 19th century, it will make possible the demonstration that every principal ring is completely closed.

• He proves also that an irreducible cubic equation cannot be solved by any combination of the irrationals in Book X, that is, quadratics. Note that Fibonacci did not attempt to find other constructible irrationalities which do not appear in Book X of the *Elements*. In *al-Badīʿ*, however, al-Karajī, like his successor al-Samawʾal in *al-Bāhir*, explained that there is an infinity of kinds of irrationals beyond those of Book X of the *Elements*.

• Fibonacci's demonstration is geometrical, that is in the style and terminology of Book X, even if his means are arithmetical.

II. The second 'challenge' that John of Palermo set for Fibonacci is reported by the latter himself in the prologue to his *Liber quadratorum*. His opening words to the Emperor are worth quoting again:

> After being brought to Pisa by Master Dominick to the feet of your celestial majesty, most glorious prince, Lord F[rederick.], I met Master John of

[12] See Boncompagni, *Tre scritti di Leonardo Pisano*, pp. 1–54; F. Woepcke, 'Sur un essai de déterminer la nature de la racine d'une équation du troisième degré', *Journal de mathématiques pures et appliquées*, XIX, 1854. See also *Il 'Flos' di Leonardo Pisano dal Codice E.75 P. Sup. della Biblioteca Ambrosiana di Milano*, Traduzione e Commenti di Ettore Picutti, *Physis*, Anno XXV, Fasc. 2, 1983, pp. 293–387; H.-G. Zeuthen, 'Notes sur l'histoire des mathématiques', *Oversigt over det Kongelige Danske. Videnskabernes Selskabs. Forhandlinger og dets Medlemmers Arbejder* (*B.A.R.S.L.D.*, no. 3 Janvier-Mars), Copenhagen, 1893, pp. 1–17.

[13] See R. Rashed, *L'Œuvre mathématique d'al-Sijzī*, vol. I: *Géométrie des coniques et théorie des nombres au Xe siècle*, Les Cahiers du MIDEO, 3, Louvain/Paris, Peeters, 2004, pp. 171–2.

Palermo; he proposed to me a question that had occurred to him, pertaining not less to geometry than to arithmetic: find a square number from which, when five is added or subtracted, always arises a square number. Beyond this question, the solution of which I have already found, I saw, upon reflection, that this solution itself and many others have origin in the squares and the numbers which fall between the squares.[14]

Fibonacci could not have been more explicit about the object of his inquiries, nor about their target: the object, he says, belongs both 'as much to geometry as to number'. It starts from the problem of 'congruous' numbers, as he put it himself, and requires that one examine the properties of square numbers. At stake then, is research in integer Diophantine analysis such as it emerged in the 10th century. Being both Euclideans and readers of Diophantus, these 10th-century mathematicians, like al-Khāzin, thought of arithmetic as concerned with integers, and as represented by line segments. Contrary to Diophantus's *Arithmetic*, such a representation made it possible to respect the norms of demonstration, such as the arithmetic books of the *Elements* defined and practiced it. In short, they read Diophantus in light of the *Elements*, while being also well informed about al-Khwārizmī's *Algebra*. At issue in this tradition was not merely the presentation of algorithms alone, nor the solution of Diophantine problems, but rather the *demonstration* of the solutions.

Before situating Fibonacci in relation to this tradition, let us turn to the *Liber quadratorum*. Fibonacci begins by showing that one can obtain square numbers as the sum of odd prime integers beginning with one $[n^2 = \sum_{k=1}^{n}(2k-1)]$. He then moves to the Diophantine problem: 'Find two numbers so that the sum of their squares makes a square formed by the sum of the squares of two other given numbers'.[15]

Assume that the two given numbers are u and v,

$$u^2 + v^2 = r^2.$$

Fibonacci then considers the two segments EZ and ED such that

[14] Leonardo Pisano/Fibonacci, *The Book of Squares*, An annotated translation into modern English by L. E. Sigler, Boston, Academic Press, 1987, p. 3. See also E. Picutti, 'Il *Libro dei Quadrati* di Leonardo Pisano e i problemi di analisi indeterminata nel Codice Palatino 557 della Biblioteca nazionale di Firenze. Introduzione e Commenti', *Physis*, Anno XXI, 1979, pp. 195–339.

[15] Fibonacci, *The Book of Squares*, p. 18.

$$EZ^2 + ED^2 = DZ^2 .$$

If $DZ = r$, the problem is solved; if not, suppose first that $DZ > r$; and let a segment I be equal to r. From ZD, remove the segment $TZ = I$ and drop the perpendicular TK; then ZK and TK are the numbers that were sought.

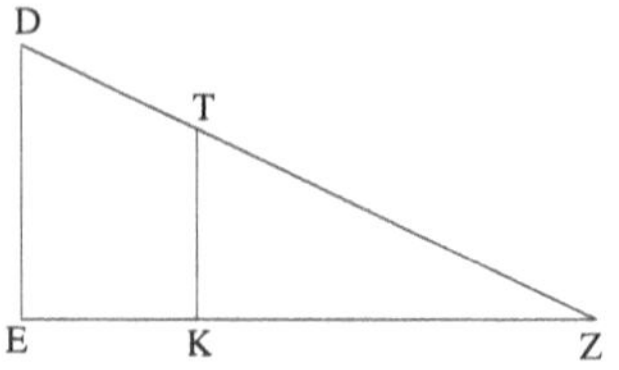

Fig. 39

Indeed, the similarity of the two triangles, DEZ and TKZ, yields

$$\frac{KZ}{EZ} = \frac{ZT}{ZD} \text{ and } \frac{KT}{ED} = \frac{ZT}{ZD},$$

whence

$$KZ = \frac{ZT}{ZD} \cdot EZ \text{ and } KT = \frac{ZT}{ZD} \cdot ED;$$

whence

$$KZ^2 + KT^2 = \left(\frac{ZT}{ZD}\right)^2 \left(EZ^2 + ED^2\right) = ZT^2 = r^2 .$$

One reasons analogously if $DZ < r$, but one extends ZD. We have just translated Fibonacci's procedure.

If Fibonacci 's reasoning holds, it is on account of a supplemental proposition missing from his book. If one can find x and y such that $x^2 + y^2 = a^2$, where a is the given number, then $x_1 = \frac{r}{a} x$ and $y_1 = \frac{r}{a} y$ yield a solution to the problem.

Fibonacci's method is neither that of Diophantus in his *Arithmetic*, nor the algebraic one of an Abū Kāmil, for example. Not only does he fail to establish the supplemental proposition, but his method is geometrico-arithmetical. Geometry is bodily present insofar as Fibonacci appeals to similar triangles, for example. But neither is this method that of the theoreticians of the new Diophantine analysis, such as al-Khāzin or al-Sijzī.

The latter did, to be sure, represent integers by line segments, but only to apply purely arithmetic or arithmetico-algebraic methods. Even when some of them, such as al-Sijzī, happened to draw upon geometry, it was always in the manner of the algebraists. Finally note that, in his *Zététiques*, Viète borrows this method from Fibonacci. He writes:

> To find in numbers two squares, the sum of which is equal to a given square. Given a number, F quadratum [F^2]. One must find two squares the sum of which is F quadratum.
>
> Assume a numerical right triangle of hypotenuse Z, base B, perpendicular side D. Given a right triangle similar to it with hypotenuse F. One has Z to F equals B to the other base, which is therefore equivalent to $\frac{BF}{Z}$ and likewise Z is to F as D is to the other perpendicular, which is therefore equivalent to $\frac{DF}{Z}$. This is why the squares of $\frac{BF}{Z}$ and of $\frac{DF}{Z}$ will have a sum equal to the given F quadratum, which was to be done.[16]

In his *Ad logisticam speciosam notae priores*, however, Viète returns to the method of the algebraists and shows that the two methods are equivalent once the missing proposition is restored.

Our question about the very status of Fibonacci's contribution in the *Liber quadratorum* thus becomes more precise: is the geometrical inflection of his work circumstantial or essential?

To answer this question, let us begin by remembering that Fibonacci uses a very long and laborious demonstration to establish an important proposition, which can be rewritten in other terms:

Given (p, q, r, s) four integers such that $\frac{p}{q} \neq \frac{r}{s}$ and that $p^2 + q^2 = m$ and $r^2 + s^2 = n$, where m and n are nonsquare integers. One has

$$(1)\ mn = (p^2 + q^2)(r^2 + s^2) = (pr + qs)^2 + (ps - qr)^2 = (pr - qs)^2 + (ps + qr)^2.$$

$$(2)\ mn^2 = (pr + qs)^2 + (ps - qr)^2 = (ps + qr)^2 + (pr - qs)^2 = p^2(r^2 + s^2) + q^2(r^2 + s^2)$$

$$(3)\ m^2n^2 = (pr + qs)^2 + (ps - qr)^2 = (ps + qr)^2 + (pr - qs)^2$$
$$= p^2(r^2 + s^2) + q^2(r^2 + s^2) = r^2(p^2 + q^2) + s^2(p^2 + q^2).$$

[16] *Opera mathematica*, Leiden, 1646; repr. Hildesheim, Georg Olms, 1970, p. 62.

This proposition is a consequence of one of the first known decompositions of quadratic forms. Already known by the Babylonians, one finds it thinly veiled in Problem 3.19 of Diophantus's *Arithmetic*, and it is stated and established by 10th-century mathematicians such al-Khāzin.

Fibonacci's demonstration is in the pure style of the arithmetic books of the *Elements*.

Let us turn now to the next proposition, which states: 'Find two numbers which have the sum of their squares equal to a nonsquare number which is itself the sum of the squares of two given numbers';[17] which is rewritten

$$x^2 + y^2 = c = a^2 + b^2,$$

where c is not square and a and b are rational.

Let us formulate Fibonacci's demonstration in a different language.

Given u and v such that

$$u^2 + v^2 = r^2 \text{ and } \frac{u}{v} \neq \frac{x}{y};$$

and given $k = r^2 \cdot c$. According to the preceding proposition, k is rewritten

$$k = \left(x^2 + y^2\right)\left(u^2 + v^2\right) = \left(xu - yv\right)^2 + \left(xv + yu\right)^2 = \left(xv - yu\right)^2 + \left(xu + yv\right)^2.$$

Next Fibonacci considers the triangle ABC right at C and such that $AB = \sqrt{k}$, $AC = m$, $BC = n$, where $m = xu - yv$ and $n = xv + yu$.

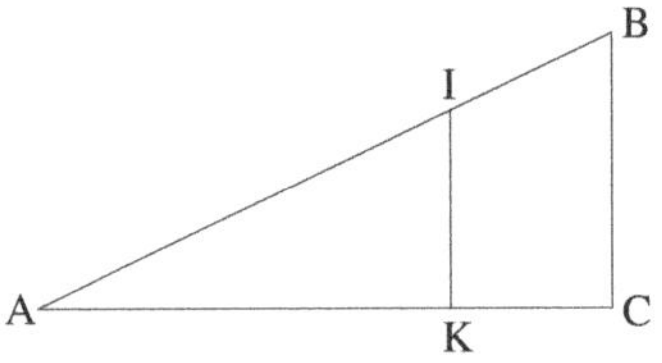

Fig. 40

On AB, take the segment $AI = \sqrt{c}$ and drop the perpendicular IK onto AC. The two segments AK and IK give the solution. Indeed

[17] Fibonacci, *The Book of Squares*, p. 36.

$$\frac{AK}{AC}=\frac{AI}{AB}=\frac{\sqrt{c}}{\sqrt{k}}=\frac{1}{r},$$

whence

$$AK=\frac{m}{r}=\frac{xu-yv}{r} \text{ and } IK=\frac{n}{r}=\frac{xv+yu}{r};$$

whence the result.

As one can see, the demonstration is geometrico-arithmetical. Here, too, Fibonacci draws upon similar triangles. To grasp the difference, let us compare Fibonacci's solution to some others.

In effect, this problem is none other than 2.9 of Diophantus, who posits

$$x = a + t \text{ and } y = st - b,$$

whence

$$t=\frac{2(bs-a)}{s^2+1},$$

whence

$$x=\frac{as^2+2bs-a}{1+s^2} \text{ and } y=\frac{bs^2-2as-b}{1+s^2}.$$

Regardless of whether one gives it an arithmetico- or geometrico-algebraic interpretation, this method indeed differs from Fibonacci's.

Abū Kāmil applies a method that resembles that of Diophantus. Thus he posits

$$x = a + t \text{ and } y = b - st,$$

and calculates t, then x and y.

This is the method that one finds again among algebraists such as al-Karajī.[18] Finally, in his *Zététiques* [4.2], Viète gives Fibonacci's method before giving that of Diophantus.

Next, Fibonacci establishes the following propositions before returning to the problem of congruous numbers:[19]

[18] Woepcke, *Extrait du Fakhrī*, 3.37, p. 100.

[19] Integer n is congruous in Fibonacci's sense if $n = ab(a + b)(a - b)$ with $(a + b)$ is even; or if $n = 4\,ab(a + b)(a - b)$ with $(a + b)$ odd.

- $\sum_{i=1}^{n} i^2 = \frac{n(n+1)(2n+1)}{6}$.

- $\sum_{i=1}^{n} (2i-1)^2 = \frac{4n(2n-1)(2n+1)}{12}$.

• If a and b are two integers such that $a > b$, $(a, b) = 1$ and $(a + b) = 2k$, then $ab(a + b)(a - b)$ is a multiple of 24 (24 is the smallest congruous number). One shows that this will also be the case for $4ab(a + b)(a - b)$ if $(a + b)$ is odd, a congruous number.

Everything is now ready for Fibonacci to take up John of Palermo's challenge and to solve the system thus.

$$(1) \qquad \begin{aligned} x^2 + a &= y_1^2 \\ x^2 - a &= y_2^2 \end{aligned} \qquad \text{with } y_1 > x > y_2.$$

Let us begin by examining Fibonacci's method, before returning to the history of the problem in order to compare this method with that of his 10th-century predecessors.

Fibonacci's main idea is to reduce the problem to finding three squares in arithmetic progression. This is an interesting idea, which in a sense spontaneously springs to mind. Indeed, from equation (1), one obtains

$$(2) \qquad y_1^2 - x^2 = x^2 - y_2^2.$$

Let us therefore posit that

$$y_2^2 = \sum_{i=1}^{n}(2i-1), \quad x^2 = \sum_{i=1}^{n+m}(2i-1), \quad y_1^2 = \sum_{i=1}^{n+m+k}(2i-1), \text{ with } k < m.$$

Equality (2) is rewritten

$$(3) \qquad (n+m+k)^2 - (n+m)^2 = (n+m)^2 - n^2,$$

because $\sum_{p=1}^{n}(2p-1) = n^2$.

From (3) one has

$$y_2 = \left|k^2 + 2mk - m^2\right|, \quad x = m^2 + k^2, \quad y_1 = m^2 + 2mk - k^2,$$

and one has

$$a = 4mk(m+k)(m-k). \tag{4}$$

This last step is one that Fibonacci did not truly establish.

From the preceding proposition, one knows that $a = 24$ is the smallest congruous number.

The solution of equation (4) is not immediate. Stated in a different language, the issue is to find the rational points on an elliptical curve.

To situate Fibonacci's contribution, let us remember some history. This problem has very deep roots, since it goes back to research on numerical right triangles in Babylonian mathematics. In various forms, one also finds it in Diophantus's *Arithmetic*. Earlier we alluded to Problem 3.19 of the *Arithmetic*, which is rewritten:

$$\begin{aligned} x^2 + a_i &= y_i^2, \\ x^2 - a_i &= y_{i+1}^2 \end{aligned} \qquad (y_{i+1}^2 < x^2 < y_i^2)$$

and $i = 1, 2, 3, 4$; with the condition

$$x = a_1 + a_2 + a_3 + a_4.$$

This problem amounts to finding four congruous numbers corresponding to x^2 and such that their sum is equal to x.

Diophantus certainly does not raise the problem of congruous numbers for itself, but encounters it during his research on square numbers. His solution draws primarily on a relation that he established between the Pythagorean triples (a, b, c) and a congruous number; in his own words: 'Since the square of the hypotenuse of every right triangle, augmented or diminished by the double product of the sides constituting the right angle, forms a square [...]':[20]

$$(*) \qquad c^2 \pm 2ab = (a \pm b)^2 .$$

Next, Diophantus looks for four right triangles having the same hypotenuse, a problem equivalent to representing a square as the sum of two squares in four different ways. Neither the problem such as Fibonacci states it, nor the method that he follows to solve it owes anything at all to Diophantus's *Arithmetic*, about which Fibonacci knew nothing. Had it been otherwise, Fibonacci would have rephrased the problem into that of finding a numerical right triangle for which $2ab$ is equal to the number a_i. The path that Diophantus took is well known, and would consist in positing

$$u^2 + v^2 = c, \quad u^2 - v^2 = a, \quad 2uv = b,$$

whence

$$4uv(u+v)(u-v) = a_i .$$

Not until the 10th century, when integer Diophantine analysis diverged from rational Diophantine analysis did the problem of congruous numbers arise for its own sake. It was then presented as a problem in number theory, to which several mathematicians would devote sustained attention. One of the first was al-Khāzin, who devoted several works to integer Diophantine analysis, all of which treat numerical right triangles and the problems deduced from them, such as the first case of Fermat's theorem. Yet neither al-Khāzin nor the mathematicians of this era used a specific word to distinguish these numbers. Fibonacci was the first to call them 'congruous'.

One of the theorems that al-Khāzin demonstrates concerns congruous numbers and can be stated as follows. The following conditions are equivalent:

1° system (1) admits a solution;

2° there exists a pair of integers (u, v) such that

$$u^2 + v^2 = x^2, \; 2\,uv = a;$$

under these conditions, a has the form $4k$, and k is not of the form 2^n.

[20] Diophante d'Alexandrie, *Les Arithmétiques*, *Les six livres arithmétiques et le Livre des nombres polygones*, Œuvres traduites pour la première fois du grec en français avec une introduction et des notes par Paul Ver Eecke, Paris, Librairie A. Blanchard, 1959, p. 108.

Since I have examined al-Khāzin's demonstration elsewhere,[21] suffice it here to mention only the ideas that he puts into play.

System (1) gives us by addition and subtraction

$$x^2 = \frac{y_1^2 + y_2^2}{2}, \quad a = \frac{y_1^2 - y_2^2}{2}.$$

Given $y_1 = u + v$ and $y_2 = u - v$, one has

$$x^2 = u^2 + v^2 \text{ and } a = 2uv.$$

Earlier, however, al-Khāzin had demonstrated by analysis and synthesis the lemma to Proposition X.29 of the *Elements*, which Euclid had established by synthesis alone. One knows that there exists a pair of integers (p, q) with opposite parities such that $p > q$ and $(p, q) = 1$ such that

$$u = p^2 - q^2, \ v = 2pq \text{ and } x = p^2 + q^2,$$

and one has

$$\left(p^2 + q^2\right)^2 + 4pq(p-q)(p+q) = \left(p^2 - q^2 + 2pq\right)^2,$$

$$\left(p^2 + q^2\right)^2 - 4pq(p-q)(p+q) = \left(p^2 - q^2 - 2pq\right)^2.$$

System (1) is thus identically verified if

$$a = 4pq(p - q)(p + q).$$

Al-Khāzin notes that the smallest integer a that verifies this equality is 24. He does not stop here but for the use of algebraists, he proposes a second method to solve (1), that is, to obtain rational numbers. This method requires the solution of the equation

$$x^2 + y^4 = z^2.$$

Other mathematicians in this tradition – notably Abū al-Jūd ibn al-Layth – also deal with the problem of congruous numbers.

[21] R. Rashed, 'L'analyse diophantienne au Xe siècle: l'exemple d'al-Khāzin', *Revue d'histoire des sciences*, 32, 1979, pp. 193–222; English transl. in *The Development of Arabic Mathematics*, pp. 205–37; and *Histoire de l'analyse diophantienne classique: D'Abū Kāmil à Fermat*, Berlin/New York, Walter de Gruyter, 2013.

For their part, the algebraists will soon take up this same problem. At the end of the 10th century, al-Karajī in his *al-Badī‘* formulates precisely the problem that John of Palermo would later put to Fibonacci, namely

$$(5)\qquad \begin{aligned} x^2 + 5 &= y_1^2, \\ x^2 - 5 &= y_2^2. \end{aligned}$$

Here are al-Karajī's steps. From (5) one pulls

$$y_1^2 - y_2^2 = 10,\ y_1 + y_2 = u \text{ and } y_1 - y_2 = v,$$

whence

$$y_1 = \frac{u+v}{2},\quad y_2 = \frac{u-v}{2} \text{ and } y_1^2 - y_2^2 = uv;$$

whence $v = \frac{10}{u}$.

Let us make the substitution in (5), and set $u = \frac{3}{2}$.

$$x^2 = \frac{1}{4}\left(u^2 + \frac{10^2}{u^2}\right) = \frac{1681}{144} = \left(\frac{41}{12}\right)^2,$$

$$y_1^2 = \left(\frac{1}{2}\left(u + \frac{10}{u}\right)\right)^2 = \frac{2401}{144} = \left(\frac{49}{12}\right)^2,$$

$$y_2^2 = \left(\frac{u^2 - 10}{2u}\right)^2 = \frac{961}{144} = \left(\frac{31}{12}\right)^2.$$

These numbers are identical to the ones Fibonacci obtained in the *Liber quadratorum*. His method is different and draws its inspiration from the same idea he had applied earlier: to find three squares, x^2, y_1^2, y_2^2 in arithmetic progression with a difference of 5; that is, to solve

$$y_1^2 - x^2 = x^2 - y_2^2 = 5,$$

for rational numbers (5 is not divisible by 24); or, what amounts to the same thing, to solve

$$Y_1^2 - X^2 = X^2 - Y_2^2 = 5k^2$$

for integers.

Fibonacci takes $k^2 = 144$ and thus $5k^2 = 720$, a multiple of 24, and therefore divisible by 24. He obtains for X, Y_1, Y_2: 41, 49, 31, and for x, y_1, y_2 the values that al-Karajī had found, that is

$$x = \frac{41}{12}, \quad y_1 = \frac{49}{12}, \quad y_2 = \frac{31}{12}.$$

By Fibonacci's own admission, the *Liber quadratorum* was conceived and composed with the goal of solving the problem of congruous numbers, and in fact, two-thirds of its propositions are directly tied to this research. The rest consists of problems that are merely variations on the central question. Thus, in the following proposition, Fibonacci shows that, for two integers p, q such that $p > q$, if $(p + q)$ is even, then $\frac{p}{q} \neq \frac{p+q}{p-q}$.

At the end of this proposition, he writes a sentence that has often excited the imagination of some historians:[22] 'nullus quadratus numerus potest esse congruum' – 'no square number can be a congruous number'.[23]

Without falling into the trap of anachronism, let us concede that Fibonacci's statement reflects a difficulty that he felt intuitively. It is perhaps this same difficulty, moreover, that earlier had incited al-Khāzin to a proliferation of methods. Let us stay within the latter's perspective. As we noted he was dealing with Pythagorean triples. In this context, the problem of congruous number is explicitly posed as follows: how to decide if a non-square integer is the area of a numerical right triangle? We must emphasize again that this context is neither that of al-Karajī, nor strictly that of Fibonacci. With his theorem, al-Khāzin gives a necessary and sufficient condition. Nevertheless, this theorem does not allow one always to solve the problem. If, for example, one wants to show that 1 is not a congruous number, one is reduced to proving the impossibility of $x^4 - y^4 = z^4$ for positive integers, which was not demonstrated until Fermat. Let us return to Fibonacci's statement.

From the preceding statement, one immediately obtains, as Fibonacci noted,

$$\frac{p}{q} = \frac{p+q}{p-q} \Rightarrow pq(p+q)(p-q) \text{ and } 4pq(p+q)(p-q)$$

are squares.

[22] K. Vogel for example (cited n. 5).

[23] Boncompagni, *Tre scritti di Leonardo Pisano*, p. 98; Fibonacci, *The Book of Squares*, transl. L. Sigler, p. 83.

The reciprocal should then pass through equation $x^4 - y^4 = z^2$. Indeed, $pq(p + q)(p - q)$ is the area of a Pythagorean triangle with factors that are pairwise prime. If the product is a square, all the factors must be as well. Let us therefore posit $p = x^2$, $q = y^2$, $(p + q) = u^2$, $(p - q) = v^2$, with u, v both odd, and $(u, v) = 1$. Then $x, y, z = uv$ solve $x^4 - y^4 = z^2$.

Did the mathematician of Pisa notice such a difficulty ? We think not. By examining some values, however, he may have seen the property stated above.

Between Fibonacci's text and this analysis starting from Pythagorean triples, therefore, lies an abyss that some have tried to leap with one bound.

Thus the group of remaining problems (less than a third) includes ones that also appear in al-Karajī's *al-Badī*ʿ, but that Fibonacci solved by means of congruous numbers:

$$\begin{array}{ll} x^2 + nx = y_1^2, & x^2 + a = y_1^2 \\ x^2 - nx = y_2^2, & x^2 - na = y_2^2 \end{array} \quad \text{with } n = 1, 2, \ldots;$$

it concludes with another system of Diophantine equations that Theodore of Antioch proposed to Fibonacci.

One can certainly compare this procedure to that of al-Khāzin in one of his treatises. Indeed, he too recalls that the goal (*al-gharad*) of the first part of this treatise is to solve the problem of congruous numbers. But the analogy stops there.

As we have noted, for al-Khāzin, the problem of congruous numbers arises during research on Pythagorean triples that he intended as systematic. By forgetting this context, one is condemned to understanding nothing about the choice of representing numbers by line segments, and especially al-Khāzin's deliberate preference for methods that are purely arithmetic or arithmetico-algebraic – in other words, the two foundations on which integral Diophantine analysis builds. Yet it is indeed to the latter that the problem of congruous numbers belongs, as does the method that al-Khāzin used to solve it.

The *Liber quadratorum*, by contrast, is in no sense a treatise on Pythagorean triples. In it, Fibonacci is concerned with congruous numbers. His method consists, on the one hand, in obtaining the squares x^2, y_1^2, y_2^2 as sums of arithmetic progressions and, on the other, in returning to squares in arithmetic progression. To distinguish the two contributions concisely, one might characterize them as follows. Al-Khāzin treats Pythagorean triples by arithmetico-agebraic means, or even by Diophantine methods reinvigorated by integer arithmetic, in order to found integer Diophantine analysis. By contrast, Fibonacci focuses on the problem of congruous numbers and on several related problems, by drawing on the methods of Euclidean and neo-Pythagorean arithmetic without recourse to algebraic means. This is an original procedure, but it also makes his occasionally defective demonstrations longer and heavier.

Nevertheless, unlike al-Karajī, both Fibonacci and al-Khāzin are already squarely in the domain of number theory. Indeed al-Karajī treats rational solutions and, as we have seen, chooses a Diophantine method that he interprets algebraically.

Fibonacci's procedure evokes in particular the path that he himself took in his study of al-Khayyām's equation. His first move was therefore to draw upon Euclid's arithmetic books in order to establish that the root cannot be a rational number, and this independently of John of Palermo's requirement that he proceed via Book X of the *Elements*.

How ought one understand the causes and consequences of this Euclidean inclination? I believe that two terms capture the essentials of Fibonacci's situation: isolation and originality, where the first is in some sense the cause of the second. Let me explain.

Most, if not all, of the propositions of the *Liber quadratorum* were known by Fibonacci's Arabic predecessors, who had widely diffused them. But Fibonacci's demonstrations were different. Was there a Latin translation that dealt with these questions in number theory? No evidence so far allows one to say so. But the case is not unique. I cite in particular Ibn al-Haytham's study of a problem of linear congruence that is now known as the 'Chinese remainder theorem' and that appears in the *Liber abaci*.[24] There is, however, no known trace of a Latin translation of Ibn al-Haytham's text or of any of his commentators. There can be no doubt that Fibonacci trained himself to put his energy into the Arabic mathematics translated into Latin, or that he had some sort of direct access to the latter, whether oral (as was the case for John of Palermo and Theodore of Antioch) or written. Equally incontestable, however, this gifted mathematician was doubly isolated: not only with respect to the active mathematical production in this domain since the second half of the 10th century; but also with respect to the advanced contemporary research that was at the time thriving in the cities of the Muslim East. In this isolation, one can nevertheless glimpse the ferment of an unmistakable originality. Since he stood outside the tradition of integer Diophantine analysis that was emerging in the middle of the 10th century and would prosper among such

[24] See above, 'Fibonacci and Arabic mathematics', pp. 416–8:

$$n \equiv 1 (\mathrm{mod}\, m_i), \quad \text{with } 1 < m_i \leq p-1, \quad \text{where } m_i = 2, 3, \ldots, 6$$
$$n \equiv 0 (\mathrm{mod}\, p).$$

mathematicians as Kamāl al-Dīn ibn Yūnus, Fibonacci drew upon the means at his disposal, namely the *Elements* and Euclidean and neo-Pythagorean arithmetic. Isolated also from the tradition of algebraic geometry, he once again turned to these same books. This is how he came to develop certain lines of research within Arabic mathematics, but in different directions and by other means. This is precisely what we mean by characterizing Fibonacci's work as 'the Latin extension of Arabic mathematics'.

AL-YAZDĪ AND THE EQUATION $\sum_{i=1}^{n} x_i^2 = x^2$

The 10th century witnessed the birth of two relatively distinct traditions in Diophantine analysis. The first is that of the algebraists who, following the works of Abū Kāmil and of the translation of seven books of Diophantus's *Arithmetic*, developed Diophantine rational analysis as an integral chapter of algebra. From al-Karajī and his successors until the 16th century, this chapter is not simply a component of every substantial treatise in algebra: it also gets its own proper name, *al-Istiqrāʾ*. With al-Karajī, this term, which literally means 'induction', acquired the technical sense of indeterminate rational analysis. The second tradition is that of mathematicians such as al-Khujandī, al-Khāzin, Abū al-Jūd, al-Sijzī, among others, who deliberately broke from the preceding tradition and chose a style decidedly different from that of Diophantus's *Arithmetic*. They allowed only integer solutions and they required demonstrations. Understandably, the latter were carried out by means of line segments and proportion theory. This is how the new, or integer, Diophantine analysis was first conceived in the 10th century. As we have noted, however, numerical right triangles from the beginning became a favorite domain of research in this tradition. Indeed they often stand at the origin of most of the mathematical questions that the aforementioned mathematicians raise in all their works. Already in the 10th century, they tried to formulate a theory of these triangles, and to express the results in relation to certain modules.[1] Among the problems that are often taken up after al-Khāzin, one encounters that of the general solution, with integers, of the Diophantine equation

$$* \qquad x_1^2 + x_2^2 + \ldots + x_n^2 = x^2 .$$

[1] R. Rashed, 'L'analyse diophantienne au Xe siècle: l'exemple d'al-Khāzin', *Revue d'histoire des sciences*, XXXII/3, 1979, pp. 193–222; English translation in *The Development of Arabic Mathematics: Between Arithmetic and Algebra*, Boston Studies in Philosophy of Science 156, Dordrecht/Boston/London, Kluwer Academic Publishers, 1994, pp. 205–37.

Throughout this research, one finds here and there the thinly veiled desire of reaching the solution arithmetically. It is here that the historian of Diophantine analysis confronts one of the most important questions, an unavoidable one: when and how in this episode did one begin to proceed by purely arithmetic demonstrations? It is thanks to this requirement, and with the help of infinite descent, that Fermat brought about the transformation that made his name in number theory.

We want to show that, before this transformation, the history of the new Diophantine analysis evinces partial progress on the path towards purely arithmetic demonstration. With textual evidence in hand, we will see that the 16th-century mathematician al-Yazdī[2] made an advance of this type in relation to the solution of the equation (*). The arithmetic character of his procedure pertains to his reliance on a calculation equivalent to the congruences (mod 8) and (mod 4). We cannot yet assert, however, that this contribution originates with al-Yazdī himself, or whether it reflects a more ancient phase of Diophantine analysis in Arabic. Too many texts still remain unstudied for us to offer a clear answer. The example of al-Yazdī, nevertheless, does allow us to conclude that at the end of the 16th century, the flame of Arabic mathematical research, although weakened, had not gone out.

Muḥammad Bāqir al-Yazdī wrote a voluminous treatise called *The Fountains of Arithmetic*,[3] which is still unedited. In this volume, al-Yazdī contributes to Euclidean number theory as well as to rational Diophantine

[2] Muḥammad Bāqir al-Yazdī was an Iranian mathematician who died around 1637. We know very little about his life except that he was an important mathematician, as were his son and grandson. Written in Arabic, his mathematical work includes an important treatise called *ʿUyūn al-ḥisāb* (*The Fountains of Arithmetic*), in the tradition of al-Karajī and of his successors until al-Kāshī; it therefore contains a substantial chapter on indeterminate analysis. It is also in this book that al-Yazdī studies amicable numbers, and lists the pair that is now called 'of Descartes' (R. Rashed, *The Development of Arabic Mathematics*, p. 286; as well as 'Matériaux pour l'histoire des nombres amiables et de l'analyse combinatoire', *Journal for the History of Arabic Science*, vol. 6, nos. 1–2, 1982, pp. 209–78). He also wrote a commentary on Book X of the *Elements*, as well as other mathematical books (see Agha Buzrug Tihrānī, *Ṭabaqāt aʿlām al-shīʿa: al-Rawda al-naḍra fī ʿulamāʾ al-mīʾa al-ḥādiya ʿashra*, Beirut, 1990, pp. 75–6; see also A. Qurbānī, *Biography of Mathematicians from the Islamic Period* (in Persian), Teheran, University Press of Teheran, 1365 H., pp. 436–41.

[3] The large number of manuscripts of this text suggests that it was a teaching textbook. We worked with manuscript no. 1993 Hazinesi, Süleymaniye, Istanbul.

analysis. In addition, he composes a report, heretofore unknown,[4] that is devoted precisely to the solution of equation (*).

Among other things, al-Yazdī proves the following lemmas (we follow the order of the text):

LEMMA 1: *If* n *is odd, then* $n^2 \equiv 1 \pmod 8$.

Indeed

$$(2n+1)^2 = 4n(n+1) + 1 \equiv 1 \pmod 8.$$

This lemma is fundamental for the congruences that al-Yazdī uses next.

LEMMA 4: *Given an odd number* n *that is neither prime nor the square of a prime, then for every pair* (d_1, d_2) *of associated divisors* $(d_1 > d_2)$, *one has*

$$n = d_1 \cdot d_2 = \left(\frac{d_1 + d_2}{2}\right)^2 - \left(\frac{d_1 - d_2}{2}\right)^2.$$

LEMMA 5: *For an even number* n *to be the difference of the squares of two integers, it is necessary and sufficient that it be greater than 4 and have the form*

$$n = (2m+1)\,2^k \qquad \text{with } k \geq 2.$$

From 4 and 5, al-Yazdī shows how to represent a number in several ways as the difference of two squares – of different parities if the number is odd; of the same parity if it is even.

LEMMA 6: *Let* n *be an odd number such that* $n \not\equiv 1 \pmod 8$, *then* $a_1^2 + \ldots + a_n^2$ *cannot be a square if* $a_1, \ldots, a_n$ *are odd numbers.*

[4] The manuscript of this text is found in Teheran, Majlis Shūrā Library, no. 171, catalogued under the title (in Persian) 'Gloss of Muḥammad Bāqir Yazdī on the Commentaries of the *Spherics*, Text of Naṣīr al-Dīn al-Ṭūsī'. This title is suggested by a gloss at the end of the manuscript. See our critical edition of this text in *Historia Scientiarum*, 4–2, 1994, critical apparatus, p. 101.

Al-Yazdī's text evidently does not have any connection to the *Commentary on the Spherics*. The text consists of four folios; the last page contains only the copyist's gloss, according to which the gloss pertains to the aforementioned commentary. The writing is *nastaʿlīq*, in the same hand; there is nothing in any marginal note to suggest that the copyist revised his text based on his model (*ibid*., pp. 92–101).

Indeed, for every $1 \leq i \leq n$, $a_i^2 \equiv 1 (\mathrm{mod}\, 8)$, therefore $a_1^2 + \ldots + a_n^2 \equiv n \not\equiv 1 (\mathrm{mod}\ 8)$. Moreover $a_1^2 + \ldots + a_n^2$ is odd since n is. It is therefore not a square.

Al-Yazdī calculates

$$\frac{1}{4}\left(a_1^2 + \ldots + a_n^2\right) = \sum_{i=1}^{n} \alpha_i (\alpha_i + 1) + \frac{n}{4}, \text{ if } \alpha_i = 2\alpha_i + 1;$$

in this expression, the summation that appears in the second term is even, and $\frac{n}{4}$ is not of the form $2k + \frac{1}{4}$ with k integer. The second term can therefore not be ¼ of a square.

LEMMA 7: *Let* $n \equiv 1(\mathrm{mod}\ 8)$*; if* $a_1, \ldots, a_{n-1}$ *are given odd numbers, then there exists an odd* a_n *such that* $a_1^2 + \ldots a_{n-1}^2 + a_n^2$ *is a square.*

Indeed the sum $a_1^2 + \ldots + a_{n-1}^2 \equiv n - 1 (\mathrm{mod}\, 8)$, therefore it is divisible by 8; one posits

$$a_n = \frac{1}{4}\left(a_1^2 + \ldots + a_{n-1}^2\right) - 1.$$

Thus

$$a_1^2 + \ldots + a_{n-1}^2 + a_n^2 = \left[\frac{1}{4}\left(a_1^2 + \ldots + a_{n-1}^2\right) + 1\right]^2.$$

LEMMA 8: *Let* n *be even such that* $n \not\equiv 0(\mathrm{mod}\ 4)$ $[\Leftrightarrow n \equiv 2(\mathrm{mod}\ 4)]$*; then* $a_1^2 + \ldots + a_n^2$ *cannot be a square if* $a_1, \ldots, a_n$ *are odd numbers.*

Indeed, $a_1^2 + \ldots + a_n^2 \equiv n \equiv 2 (\mathrm{mod}\, 4)$; but a square is congruent to 0 or 1 (mod 4).

LEMMA 9: *Let* $n \equiv 0(\mathrm{mod}\ 4)$; *if* $a_1, \ldots, a_{n-1}$ *are given odd numbers, there exists an odd* a_n *such that* $a_1^2 + \ldots + a_{n-1}^2 + a_n^2$ *is a square.*

Indeed, $a_1^2 + \ldots + a_{n-1}^2 \equiv (n-1) \equiv -1 (\mathrm{mod}\, 4)$. One posits

$$a_n = \frac{1}{2}\left[\left(a_1^2 + \ldots + a_{n-1}^2\right) - 1\right].$$

This is an odd number, and one has

$$a_1^2 + \ldots + a_{n-1}^2 + a_n^2 = 2a_n + 1 + a_n^2 = \left(a_n + 1\right)^2.$$

Note: if $a_1^2 + \ldots + a_n^2 = b^2$, with $a_1, \ldots, a_n$ odd, and if c is any number, one has

$$(a_1c)^2 + \ldots + (a_nc)^2 = (bc)^2.$$

One applies this note to the specific case in which $c = 2^p$.

LEMMA 10: *If* n *is even such that* n $\not\equiv 0 (\text{mod } 4)$ [$\Leftrightarrow n \equiv 2 (\text{mod } 4)$], *and if* $a_i^2 = \alpha_i^2 2^{2p}$, *where* α_i *are odd, then it is impossible that* $a_1^2 + \ldots + a_n^2$ *be a square*.

Indeed $s_n = \sum_{i=1}^{n} a_i^2 = 2^{2p} \sum_{i=1}^{n} \alpha_i^2 = 2^{2p} s'_n$, with s'_n the sum of n odd squares. For s_n to be a square, it is necessary and sufficient that s'_n be one. But according to lemma 8, if $n \not\equiv 0 (\text{mod } 4)$, one has $s'_n \not\equiv 0 (\text{mod } 4)$; s'_n is not a square, s_n is not one either.

LEMMA 12: *If* n *is even such that* n $\not\equiv 0 (\text{mod } 4)$ – *or odd such that* n $\not\equiv 1$ (mod 8) – [$\Leftrightarrow$ *if* n $\not\equiv 0, 1, 4 (\text{mod } 8)$] *and if* $a_1^2 + \ldots + a_n^2$ *is a square, then the highest power of 2 that divides the* a_i (*that is, the dyadic valuations*) *is not the same for all* i; *for example the* a_i *cannot all be odd. Moreover, the number of odd squares is* $\equiv 0$ *or* $\equiv 1 (\text{mod } 4)$.

Indeed, if this number is r, with $a_1, \ldots, a_r$ being odd, one has

$$a_1^2 + \ldots + a_r^2 \equiv r (\text{mod } 4)$$

and

$$a_{r+1}^2 + \ldots + a_n^2 \equiv 0 (\text{mod } 4),$$

since the $a_j, j \geq r + 1$ are even. But a square is always $\equiv 0$ or $\equiv 1 (\text{mod } 4)$.

Using examples, al-Yazdī develops a method for constructing a summation of $r \equiv 0$ or $\equiv 1 (\text{mod } 4)$ odd squares and of any number of even squares, such that this sum is a square.

1st case: an even square and r odd squares.

One takes $a_1, \ldots, a_r$ as odd, and one uses the formula

$$tu + \left(\frac{t-u}{2}\right)^2 = \left(\frac{t+u}{2}\right)^2.$$

One therefore seeks t, u, such that $a_1^2 + \ldots + a_r^2 = tu$ and $\frac{t-u}{2}$ is even. In other words, one decomposes $a_1^2 + \ldots + a_r^2 \equiv r(\text{mod } 4)$ into a product of two factors t, u such that $t \equiv u(\text{mod } 4)$. One must therefore choose t, u such that $r \equiv tu(\text{mod } 4)$ and $t \equiv u(\text{mod } 4)$; also $r \equiv t^2(\text{mod } 4)$, thus $t \equiv r(\text{mod } 2)$ and $t \| (a_1^2 + \ldots + a_r^2)$. If $r \equiv 0(\text{mod } 4)$, one writes $a_1^2 + \ldots + a_r^2 = 4\alpha$ and $t = 2\tau$, $u = 2\tau + 4k$; one must have $\alpha = \tau(\tau + 2k)$, that is, α is the product of two numbers with the same parity. In other words, if τ is even, α is divisible by 4. If $r \equiv 1(\text{mod } 4)$, one writes $a_1^2 + \ldots + a_r^2 = 4\alpha + 1$, and $t = 2\tau + 1$, $u = 2\tau + 4k + 1$; one must have $\alpha = \tau^2 + \tau + 2k\tau + k$, which always admits the solution $\tau = 0, k = \alpha$.

Al-Yazdī takes as his example $r = 5$, $a_1 = 3$, $a_2 = 5$, $a_3 = 7$, $a_4 = 9$ and $a_5 = 11$, which yields

$$a_1^2 + \ldots + a_5^2 = 285 = 3 \times 95 = 5 \times 57 = 15 \times 19;$$

one can therefore choose

$$(t, u) = (3, 95), (5, 57), (15, 19),$$

and one has:

$$a_1^2 + \ldots + a_5^2 + 46^2 = 49^2,$$
$$a_1^2 + \ldots + a_5^2 + 26^2 = 31^2,$$
$$a_1^2 + \ldots + a_5^2 + 2^2 = 17^2.$$

2nd *case*: method for any r and two even squares.

The task is to find any v^2 such that $\left(\frac{t+u}{2}\right)^2 + v^2$ is a square and to repeat the procedure.

Let us take up the preceding example. One starts with one of the squares s_n obtained for any r and a single even square, on the condition that it not be the square of a prime number; given

$$s_n = a_1^2 + \ldots + a_5^2 + 46^2 = 49^2,$$
$$s_n = tu.$$

Whence

$$s_n + \left(\frac{t-u}{2}\right)^2 = \left(\frac{t+u}{2}\right)^2;$$

in the example, $t = 343$ and $u = 7$.

For any number of even squares, one repeats the process as many times as necessary.

We have just seen that al-Yazdī devotes his treatise to the solution of equation (*) by purely arithmetic means. He studies different cases as a function of the parity of the x_i, starting with the case in which all x_i are odd. He systematically uses a calculation equivalent to the congruences mod 4 and mod 8. These apparently new results were evidently al-Yazdī's response to one of his concerns: to characterize the squares with the properties of congruences – or, in another language, to take into account dyadic properties; and thus to establish arithmetic propositions arithmetically. Borrowed from congruences, these arguments did not seem trivial to the mathematicians of the 17th century, at least to those who were effectively working in number theory. Thus Bachet de Méziriac drew precisely on congruences modulo 4 and modulo 8.[5] In his famous letter XII, probably from 1638,[6] Fermat himself takes pride in having shown that no number of the form $8k - 1$ is a sum of less than four squares. He writes 'This proposition leads to remarkable consequences … [which] in any case seemed to have fruitlessly tested Bachet's genius and efforts'.[7]

[5] Bachet de Méziriac notes in his commentary on Problem V.12 (V.9 in P. Tannery's edition) of Diophantus's *Arithmetic*, that prime integers of the form $4n + 1$ are decomposable into two squares. In his commentary on V.14 (V.11 in P. Tannery's edition), he shows that every integer of the form $8n + 7$ cannot be in integers either a square or a sum of two or three squares. Cf. *Diophanti Alexandrini Arithmeticorum libri sex*, Paris, 1621, pp. 311–12.

[6] J. Itard, 'Les méthodes utilisées par Fermat en théorie des nombres', *Revue d'histoire des sciences*, III, 1949; repr. in J. Itard, *Essais d'histoire des mathématiques*, revised and introduced by R. Rashed, Paris, Librairie Blanchard, 1984, pp. 229–34; see also A. Weil, *Number Theory, an Approach through History*, Boston, Basel, Stuttgart, Birkhaüser, 1983, p. 61.

[7] *Œuvres de Fermat*, vol. III, French translation by P. Tannery, Paris, Gauthier-Villars, 1896, p. 288.

For the Diophantine problems and for the sums of squares, one sees here the same phenomenon noted elsewhere in the very same al-Yazdī, but in a different arithmetic, that of the Euclidean tradition: until the 1630s research in Arabic is still alive; what is more, both for the topics broached and the results obtained, one notices a certain parallelism with the research pursued in Latin and in French. Mathematicians who knew nothing of each other's existence – al-Yazdī on the one hand, Bachet and Fermat on the other – treated the same questions and reached results that, although not identical, were nevertheless analogous. Not surprisingly, all of them started from Diophantus's *Arithmetic* and, more or less directly, from the works of the 10th-century mathematicians, a substantial portion of whose results Leonardo of Pisa or Fibonacci had taken up.[8] This is therefore a privileged moment in the history of mathematics, at the hinge between two eras, marking the cleavage between the end of one tradition and the beginning of a new one. Indeed one should not exaggerate the parallelism to which we alluded above: although al-Yazdī is not without originality, he is walking on a path previously laid out in the 10th century in order to answer questions already raised by al-Khāzin. A little later, thanks to the invention of the infinite descent, Fermat answers questions raised by these mathematicians (Fermat's theorem for $n = 3, 4$), but formulates and solves others that his predecessors had never conceived (notably the emergent study of certain quadratic forms), and he fulfills al-Yazdī's fondest wish for a purely arithmetic demonstration. This cleavage, which also characterizes Fermat's own work, seems to mark the moment when, on the Mediterranean, the sun of mathematics sets in the East in order to rise in the West.

[8] See above, 'Fibonacci and Arabic mathematics'.

– 7 –

FERMAT AND THE MODERN BEGINNINGS OF DIOPHANTINE ANALYSIS

Fermat's arithmetic research belongs not to one but to two traditions: that of Euclidean and neo-Pythagorean arithmetic, and that of Diophantus's analysis. It is true, however, that number theorists have constantly mingled these two traditions since the 10th century.[1] This is also the case in Fermat's work, notably before the 1640s. In the accounts that historians give of Fermat's contribution to number theory, it is nevertheless not rare that the second tradition obscures the first. Of the several reasons for this, two matter here. The first pertains to the importance of the results that Fermat established in Diophantine analysis; the second follows from the flowering of this analysis, not only among the heirs of Fermat – Euler and Lagrange in particular – but especially recently in algebraic geometry. As can sometimes happen even in studies with historical pretensions, Fermat's mathematics are nevertheless presented outside of all historical determination. One can therefore empathize with the difficulty that the historian faces who confronts a work as foundational of a new domain and as fruitful for new currents of research as Fermat's, even as the correct perspective requires that it be seen also as the product of two traditions. In such a case, caution demands that one follow these traditions, which both entered into the formation of the work, and that one also trace their entanglement, in order to follow this formation step by step and to situate Fermat's contribution in history. This is all the more necessary because the history of each of these traditions is distinct and unique, for the very reason that each had a different intention and goal. To be concise, the history of the Euclidean and neo-

[1] See R. Rashed, 'Théorie des nombres et analyse combinatoire. L'analyse diophantienne au Xe siècle: l'exemple d'al-Khāzin', in *id.*, *Entre arithmétique et algèbre. Recherches sur l'histoire des mathématiques arabes*, Paris, Les Belles Lettres, 1984, pp. 195–225; English translation in *The Development of Arabic Mathematics: Between Arithmetic and Algebra*, Boston Studies in Philosophy of Science 156, Dordrecht/Boston/London, Kluwer Academic Publishers, 1994, pp. 205–37; and *Histoire de l'analyse diophantienne classique: D'Abū Kāmil à Fermat*, Berlin/New York, Walter de Gruyter, 2013.

Pythagorean tradition – in spite of the difference between their two styles of arithmetic – can be rewritten as research on integers and on the criteria for determining some of their classes. It studies, among other things, questions of parity, of divisibility, criteria for prime numbers, perfect numbers, amicable numbers..., by means of demonstrations analogous to those encountered in the arithmetic books of Euclid's *Elements*, or inductively, as among the neo-Pythagoreans. This tradition will undergo an important transformation directly tied to algebra, when algebraic demonstrations are substituted for demonstrations of the Euclidean type. This transformation occurred around the 12th–13th centuries among the Arabic algebraists. Matters are completely different for Diophantus's analysis. The history of this chapter is much more complex; already well before Fermat, it is characterized by transformations that are even more profound, on account of algebra and such 10th century mathematicians as al-Khāzin. Let me simply note here that this history can be rewritten as that in which algorithmic procedures acquired diverse and precise mathematical significations: algebraic, purely arithmetic, geometric. Indeed, between these algorithmic procedures and the mathematical meanings that they are supposed to capture, there is an internal relation, even if it is not always apparent. In my view, it would be illusory to characterize the algorithmic research of an ancient or even a more recent mathematician as being independent of all mathematical content. This is to say, to speak of purely algorithmic research in Diophantus, Fermat, and Euler, for example, is highly anachronistic and inexact. That said, it nevertheless remains the case that the need to conquer the mathematical meaning of every algorithm became an unavoidable necessity beginning in the 9th century in algebra, that is, when one began to require a mathematical justification of algorithmic methods. This event was absolutely crucial, and it can be dated rather precisely to *c.* 830. Indeed in his *Algebra*, written around this date, al-Khwārizmī (to whom the concept of 'algorism' – subsequently corrupted to 'algorithm' – owes its name) formulates for the first time in history the idea that an algorithm must be mathematically justifiable, if not demonstrable. A genuine mathematical discipline, like the algebra that he had just founded, could not be satisfied to be merely algorithmic: it also had to be demonstrative. Only demonstration could confer on the algorithm the dimension that it lacked: apodicticity. In the last analysis, it is apodicticity that distinguishes an algorithm from a simple empirical procedure or a recipe. In addition, it is also demonstration, or its justification, that gives the algorithm a precise meaning. From the 9th century on, this new requirement suffuses Arabic mathematical research, including the most advanced. Diophantus's analysis not only was a part of this project, it was also precisely this question that stood at the origin of the different interpretations of Diophantus's *Arithmetic*, and of the different styles of Diophantine analysis that we have tried to distinguish.[2]

We will therefore follow the paths traced by the two traditions mentioned above. But to grasp the way in which Fermat uses them and to avoid arbitrarily introducing demarcations that run the risk of separating what he insisted on mixing, let us begin by following him step by step to understand the evolution of his arithmetic research. To a notable extent, we can do this thanks to his correspondence, which allows us to enter into this genealogical research.

For the most part, we do not know how Fermat acquired his mathematical training. But we do know that, before 1636, he studied Viète's *Zététiques*, Diophantus's *Arithmetic*, and probably Bachet's commentaries. To this one must add a deep familiarity with Euclidean arithmetic. One therefore expects a combination in which 'la spécieuse' (the 'new algebra' of Viète), Diophantine analysis, and more or less algebraized Euclidean methods enter. This is precisely what one sees when one examines his mathematical studies up to September 1636. At the time, this combination nevertheless had a dominant emphasis, which had a clear effect on his concept and practice of Diophantine analysis. Indeed, the core of his research belonged to the tradition of Euclidean arithmetic, and his knowledge of Diophantine analysis was still modest. This is what we will show thanks to his correspondence.

Whether it is a matter of its formulation or its solution, the first problem we encounter sends us back to Diophantus and Viète. At issue is to find two numbers, each of which is composed of three squares, as is their sum. It is about this very problem that Fermat will later state:

> That a number, composed of three squares only in integers, can never be divided into two squares nor even in fractions, no one has yet demonstrated; and this is what I am working on, and I believe that I will bring it to an end. This knowledge is of enormous utility and it seems to me that we do not have enough principles to bring it to an end.[3]

[2] R. Rashed, *Diophante: Les Arithmétiques*, Collection des Universités de France, 2 vols, Paris, Les Belles Lettres, 1984.

[3] 'Or, qu'un nombre, composé de trois quarrés seulement en nombres entiers, ne puisse jamais être divisé en deux quarrés, non pas même en fractions, personne ne l'a jamais encore démontré et c'est à quoi je travaille et crois que j'en viendrai à bout. Cette connoissance est de grandissime usage et il semble que nous n'avons pas assez de principes pour en venir à bout.' Letter X – Tuesday 2 September 1636, of Fermat to Mersenne, vol. II, p. 58.

For now, we simply note this requirement of an integer solution. Here we are in palpable contact with Fermat's goal and with the limits of his knowledge. At issue is a problem in integer Diophantine analysis, not the analysis of the algebraists; conversely, the problem at hand is easily soluble by congruences modulo 8. It seems that Fermat did not yet have sufficient knowledge of this congruence and of this type of problem. In this regard he writes: 'we do not have enough principles in order to bring it to an end; M. de Beaugrand agrees with me about this'.[4]

If the preceding problem shows us the nature and the limits of Fermat's knowledge of Diophantine analysis at that time, the second problem reveals the domain in number theory on which he is working the most. Both a letter addressed to Mersenne and the preface that the latter wrote to his *Harmonie universelle*[5] indicate that Fermat in this period is treating primarily the problems of the divisors of an integer, their sum, perfect numbers, amicable numbers, etc. – questions originating directly from the Euclidean and neo-Pythagorean traditions. Thus, in order to form subduplicate numbers – that is, the integers n such that $\sigma_0(n) = 2n$; σ_0 being the summation of the proper divisors, with $\sigma(n) = \sigma_0(n) + n$ the sum of the divisors – Fermat proceeds as follows:

If an integer n is such that $2^n - 1 = p(2^{n-3} + 1)$, with p prime, then $3p \cdot 2^{n-1}$ is a subduplicate integer. Indeed,

$$\sigma(3p \cdot 2^{n-1}) = 4(p+1) \cdot (2^n - 1) = 4p(2^n - 1) + 4p(2^{n-3} + 1)$$
$$= p \cdot 2^{n+2} + p \cdot 2^{n-1} = 9p \cdot 2^{n-1};$$

in other words, the sum of the divisors of the integer must be its triple, and the sum of its proper divisors its duplicate.

[4] *Ibid.*, p. 58.

[5] In the letter dated Tuesday, 24 June 1636 addressed to Mersenne, Fermat writes: 'A long time ago already, I sent to Mr. Beaugrand the proposition about aliquot parts, together with the construction for finding an infinity of numbers of the same nature' (vol. II, p. 20). P. Tannery and C. Henry have linked this letter to two excerpts from Mersenne's general preface to his *Harmonie Universelle* (1636), in which the latter writes: '... and I would add Monsieur Fermat, Counselor to the *Parlement* of Thoulouze, to whom I owe the remark that he made about the two numbers 17296 and 18416, whose aliquot parts mutually remake each other...' (vol. II, p. 21). In the second passage, an excerpt from the *Seconde Partie de l'Harmonie Universelle* (1637), Mersenne gives the procedure, probably from Fermat, for forming subduplicate numbers.

Fermat's procedure is ingenious, but ineffective. He is vulnerable to the criticism that Descartes sent him.[6] If one posits $2^{n-3} = x$, p is then written $p = \frac{8x-1}{x+1}$; for $n = 3, x = 1$. Since p is a monotone increasing function of x, it tends towards 8 when x tends to infinity; p varies from $\frac{7}{2}$ to 8 and therefore cannot have the values 5 and 7. And thus, Fermat's rule can only give two subduplicate numbers: 120 and 672, corresponding respectively to $n = 4$, $p = 5$ and $n = 6, p = 7$. This is why we called it ineffective.

The third problem encountered in the correspondence confirms what we already stated earlier about the domain Fermat was privileging at the time, and about his arithmetic knowledge. It concerns a rule to characterize and find amicable numbers, that is, the integers a and b such that $\sigma_0(a) = b$ and $\sigma_0(b) = a$. The rule can be stated as follows:

Let n be an integer such that $3 \cdot 2^{n-1} - 1 = p_{n-1}$, $3 \cdot 2^n - 1 = p_n$, and $9 \cdot 2^{2n-1} - 1 = q_n$; if p_{n-1}, p_n, q_n all three are prime numbers different from 2, then the numbers $A = 2^n \cdot p_{n-1} \cdot p_n$ and $B = 2^n q_n$ are amicable.

This rule is none other than the one that Thābit ibn Qurra established in the 9th century, and that which Kamāl al-Dīn al-Fārisī re-established algebraically at the end of the 13th century, and that was diffused broadly, as we have shown.[7] Moreover, the pair of amicable numbers (17296, 18416), called 'Fermat's pair', had been discovered earlier on several occasions.[8]

But what was Fermat's method? Did he proceed by means of proportion theory, as in Euclidean arithmetic, to which these problems pertain? Or by means of algebra, like his Arabic predecessors from the end of the 12th century on, and as the case of al-Fārisī in the 13th century illustrates? Barely a few months later, Fermat offers at least an indirect answer in the letter of 16 December 1636, in which he writes:

[6] *Correspondance du P. Marin Mersenne*, commencée par Mme Paul Tannery, publiée et annotée par Cornélis de Waard, vol. VII, Paris, 1962, Letter XVII – 27 May 1638, from Descartes to Mersenne, pp. 237–8, at p. 263 and following.

[7] R. Rashed, 'Nombres amiables, parties aliquotes et nombres figurés aux XIIIe et XIVe siècle', in *id.*, *Entre arithmétique et algèbre. Recherches sur l'histoire des mathématiques arabes*, Paris, Les Belles Lettres, 1984, pp. 259–99; English translation in *The Development of Arabic Mathematics: Between Arithmetic and Algebra*, Boston Studies in Philosophy of Science 156, Dordrecht/Boston/London, Kluwer Academic Publishers, 1994, pp. 275–319, at pp. 278 f.

[8] *Ibid.*

> As far as numbers and their aliquot parts are concerned, I found a general method to solve all questions by means of algebra, about which I plan to write a little Treatise.[9]

At this stage, one can lift out the following characteristics of Fermat's research: as in Euclidean arithmetic, it focuses on integers; but in contrast to Euclid, Fermat proceeds by means of algebra, without, however, neglecting the Diophantine problems that also treat integers. This is to say that he participates in two traditions at once, that developed by mathematicians such as al-Fārisī, by the study of divisions by means of algebra, and that at work since al-Khāzin, and later borrowed by Fibonacci, on integer Diophantine analysis. So far Fermat has produced no new result, and conceived no new method. What is more, his knowledge seems to be inferior to that of his predecessors.

One should therefore not be surprised that Fermat from the outset positions himself on the terrain of integer Diophantine analysis, and that he tackles the problems that such predecessors as al-Khāzin and Fibonacci, among others, had most heavily worked, namely, numerical right triangles. Thus, in the same letter of 16 December cited above, he poses the question of finding three right triangles whose areas constitute respectively the sides of a right triangle.

To understand Fermat's solution, we must return to his *Observation* 29 on Diophantus's *Arithmetic*, which is probably contemporaneous with his letter. In this *Observation*, Fermat tries to find two numerical right triangles, the areas of which are in a given ratio $\frac{a}{b}$. Let us then assume (m, n), (p, q) are the pairs of generators of two triangles respectively. One immediately has

$$bmn(m-n)(m+n) = apq(p-q)(p+q).$$

Fermat identifies two of the factors of the second member with two of the first. Let us summarize in the following table the results that this method can yield.

Areas proportional to	a	b	
	$a+b;\ 2a-b$	$a+b;\ a-2b$	1
	$a-b;\ 2a+b$	$a-b;\ a+2b$	2

[9] In his letter of 22 September 1636 to Roberval, he seems to be referring to this method, when he recalls that he communicated it with Mr. Despagnet (Letter XVIII, vol. II, p. 93).

Pairs of generators	$6a$; $2a + b$	$4a + 2b$; $b - 4a$	3
	$6a$; $2a - b$	$4a - 2b$; $b + 4a$	4
	$2a + 4b$; $a - 4b$	$6b$; $a + 2b$	5
	$2a - 4b$; $a + 4b$	$6b$; $2b - a$	6

Fermat obtains all of them except (3) and (5). Overlooking sign changes or permutations of a and b, there are in fact only two possible cases.

In this same *Observation* 29, Fermat wants to deduce from the preceding 'a method to find three right triangles whose areas are proportional to three given numbers, on condition that the sum of two of these numbers be four times the third';[10] the solution of this problem can be found from the preceding one. Indeed given these last three numbers a, b, c, let us associate the solutions (2) to a and b and (6) to a and c in the table. One will then obtain $a - b = 2a - 4c$ and $2a + b = a + 4c$; the two equations are reducible to $4c = a + b$.

Let us then return to the problem raised in the letter mentioned above and find a right triangle (a, b, c) such that $4c = a + b$. If (m, n) is a generating pair, one obtains $8mn = 2m^2$, and therefore $4n = m$. By setting $n = 1$, one obtains the triangle (8, 15, 17). Finally, by setting $a = 17$, $b = 15$, $c = 8$, the three sought triangles are formed respectively by the pairs (2, 49), (2, 47), (1, 48).

This is Fermat's first research on numerical right triangles. One can note that the questions are traditional – to form right triangles from others – and that the method is algebraic: the identification of factors. Fermat himself presents his *Observation* 29 as a restitution and explanation 'of Diophantus's method, which Bachet did not understand'.[11] But this time, as in the 10th century, the use of algebra by a mathematician concerned with integers inside the framework of Euclidean arithmetic will raise new questions very unlike those of Diophantus in his *Arithmetic*. As we shall see, this is the theoretical complexity that now hovers for the second time over the history of the invention of integer Diophantine analysis. For now, note that in *Observation* 23, which is slightly later than *Observation* 29, Fermat seems to be trying to form from one Pythagorean triple an infinity of others with the same area.

[10] *Œuvres de Pierre Fermat*. I: *La théorie des nombres*, Textes traduits par Paul Tannery, Introduits et commentés par R. Rashed, C. Houzel, G. Christol, Paris, Albert Blanchard, 1999, p. 142.

[11] *Ibid*., p. 141.

Research in number theory from the years 1636–1640 on

Until the end of 1636, neither Fermat's knowledge in number theory, nor the results he obtained, nor yet the methods he follows, surpass in any way those of his predecessors – quite the contrary. From the end of 1636 to the beginning of 1640, Fermat's correspondence displays nothing fundamentally new: the problems he raises are analogous to those of the preceding period or to extensions of them. Fermat still continues to be concerned with Diophantus's analysis and Euclid's arithmetic.

Several clues nevertheless reveal that this is a very important period: Fermat begins to lay the foundations for a new departure, a turning point in his career as a number theorist and in the history of the discipline. The clues are assembled in his famous Letter XII, the one that J. Itard, with the competence that was his hallmark, dated to the beginning of June 1638.[12]

In this letter, Fermat formulates several impossible problems. Let us here quote this one:

> To find a numerical right triangle such that its area is a square.[13]

In this letter, written in the style of a controversy, Fermat seems to recognize that the problems he is posing are impossible. Nevertheless, nothing allows one to infer that he knew the demonstration of this impossibility, such as he established it in *Observation* 45 on Diophantus, that is, by infinite descent. Let us emphasize again that nothing allows us to state that Fermat knew the method of infinite descent in June 1638. In May 1640, the situation seems different. In a challenge thrown to Frenicle around that date, the impossible problems surface again. This time, everything suggests that Fermat has the demonstration. Let us read what he wrote to Mersenne:

> If [Frenicle] answers you that, up to a certain number of numbers, he has tested that these questions have no solution, you can be sure that he is proceeding by tables.[14]

If Fermat speaks in this way, it is probably because he has solid demonstrations, or at least an idea about them, and that consequently he has already conceived the method of infinite descent. In any case, the conjecture is not a bold one.

Beyond this, however, what were Fermat's concerns in number theory during this period? He is interested in Euclidean arithmetic, in prime num-

[12] J. Itard, 'Les méthodes utilisées par Fermat en théorie des nombres', in *Essais d'histoire des mathématiques*, réunis et introduits par R. Rashed, Paris, Librairie Blanchard, 1984, pp. 229–34.

[13] Letter XII, vol. II, p. 65; vol. III, p. 287.

[14] Letter XXXIX, from Fermat to Mersenne, vol. II, p. 195.

bers, and in one case of what will become his 'little theorem'. In May (?) 1640, he writes:

> I have found several shortcuts to find perfect numbers and I tell you in advance that there is none of either 20 or 21 digits, which destroys the opinion of those who had thought that there was one in the confines of every group of ten; such as one from 1 to 10, another from 10 to 100, another from 100 to 1000, etc.[15]

Now this problem is equivalent to that of recognizing that prime numbers have the form $2^p - 1$. But for $p = 31$, one has a prime number, and the perfect number is $2^{30}(2^{31} - 1)$. For $p = 37$, the perfect number is $2^{36}(2^{37} - 1)$; it has a number of digits given approximately by 73log2, which is clearly many more than 22 digits. It is therefore clear that Fermat could answer without knowing if this number is perfect or not.

More important yet is the letter of June (?) of the same year in which Fermat states: 'Here are three propositions that I found, on which I hope to erect a great edifice'.[16] The propositions are the following:

(1) If n is composite, then $2^n - 1$ is as well.
(2) If p is prime, then $2^p - 1 \equiv 1 \bmod 2p$.
(3) If p is prime, the only prime divisors of $2^p - 1$ have the form $2kp + 1$.

Fermat continues:

> Here are three very beautiful propositions that I found and proved with considerable difficulty: I can call them the foundations of the discovery of perfect numbers.[17]

It is moreover reasonable to follow Fermat when he asserts that he demonstrated them. If indeed, one considers the first proposition and one assumes $n = pq$, one has in this case

$$2^{pq} - 1 = (2^p)^q - 1.$$

Now, according to the identity $a^m - b^m = (a - b)(a^{m-1} + \ldots + b^{m-1})$, one immediately concludes that $2^n - 1$ is divisible by $2^p - 1$; whence the result. The identity one obtains was known well before Fermat.

To demonstrate the second proposition, one need only write

[15] *Ibid.*, vol. II, p. 194.
[16] Letter XL, by Fermat to Mersenne, vol. II, p. 198.
[17] *Ibid.*

$$2^n = (1+1)^n = \sum_{k=0}^{n} \binom{n}{k} = 2 + 2n + 2n\frac{n-1}{2} + 2n\frac{(n-1)(n-2)}{6} + \ldots$$

If $n = p$ prime, one has

$$2^p \equiv 2 \bmod 2p,$$

whence

$$2^{p-1} \equiv 1 \bmod p.$$

This algebraic demonstration has all the odds of being similar to Fermat's. In 1636 he already knew the rules for forming figurate numbers by multiplication. Indeed, as we have shown, all of these rules were well known and used before Fermat.

The third proposition is also demonstrable with the tools Fermat had at his disposal, if one can concede that he already had his little theorem, at least for the specific case in which $a = 2$ in the statement:

For p prime, and with a prime with p, $a^{p-1} \equiv 1 \bmod p$.

In this case, the relation $2^n - 1 \equiv 0 \bmod p$ makes n a divisor of $p - 1$ where $p = kn + 1$. Since 1 and p are odd, therefore $p = 2kn + 1$.

Thus, we have enough clues to think that, in the middle of 1640, Fermat had both infinite descent and his little theorem, at least for the special case in which $a = 2$. In August of the same year, as we shall see below, he had also found the theorem on the representation of an integer as the sum of two squares. Everything therefore indicates that the year 1640 marks the beginning of a conceptual and technical transformation, the effects of which reach all the domains of numbers, including that of numerical right triangles.[18] Beginning in 1640, the study of the latter finds itself therefore a *vaster arithmetic*, which is not the case for the other mathematicians working in this area, such as Frenicle. This integration will itself gradually inflect the study of these triangles: instead of taking an interest only in the properties of the numbers that compose them, one focuses more and more on problems associated with the representation of integers by combinations of squares. Since the movement that we have just summarized led to a genuine revolution in number theory, we must pause here.

[18] Our conclusion corroborates but pushes back the date by two years that of J. Itard when he writes: 'in short, one could say: in 1638, Fermat is in possession of his method of infinite descent, but he does not have his own method of factorization. His little theorem gives him such a method in 1640.' ('Les méthodes utilisées par Fermat en théorie des nombres', p. 232.)

Toward the end of this period, and throughout the year 1640, everything unfolds as if the interest directed to the properties of integers, from the points of view of both Euclidean arithmetic and the new Diophantine analysis, was gradually increasing, culminating in a reconfiguration of the field of arithmetic. Let us follow Fermat's correspondence a little farther.

On 25 December 1640[19] – Fermat formulates the following problem:

Given the integer n, find the number of solutions of $x - y = n$ and of $xy = z^2$. He then writes:

> And there is only this difference, that in this question all prime numbers except 2 are useful, which is not the case in the preceding one about hypotenuses.[20]

He continues:

> Now, to find all the triangles and also the stated numbers in this question, the matter is rather easy, about which I will write you (Mersenne) separately, if you wish.[21]

To understand Fermat's assertion, let us rewrite the preceding system

$$n = x - y, \quad y^2 + ny - z^2 = 0;$$

whence

$$n^2 + 4z^2 = t^2.$$

For this numerical right triangle, consider the case in which n is odd; one has $n = p^2 - q^2$ and $z = pq$. One will therefore have as many solutions as there are decompositions of n into two factors; in this case, x and y are each square.

More generally, if one posits

$$n = \lambda\left(p^2 - q^2\right) \text{ and } z = \lambda pq,$$

one will have

$$t^2 = \lambda^2\left(p^2 + q^2\right)^2, y = \lambda q^2, x = \lambda p^2.$$

[19] Letter XLV, from Fermat to Mersenne, vol. II, pp. 212–17.

[20] 'Et n'y a que cette différence, qu'en cette question tous les nombres premiers hormis 2 sont utiles, ce qui n'est pas en la précédente des hypoténuses' (*ibid.*, p. 216).

[21] 'Or, pour trouver tous les triangles et aussi les dits nombres en cette question, la chose est assez aisée, de quoi je vous (= Mersenne) écrirai séparément, si vous voulez' (*ibid.*, pp. 216–17).

One therefore obtains as many solutions as there are decompositions of n into products of three factors, since $n = \lambda (p - q)(p + q)$. More precisely, every product of three distinct factors yields three solutions.

One can see clearly that Fermat is interested no longer in integer solutions as such, but in their properties, that is, in the structure of these solutions. This becomes more pronounced, and six months later – in June 1641 – in a letter to Mersenne on the 15th of the month, he returns to a result that practitioners of integer Diophantine analysis had already obtained in the 10th century.[22] Indeed, he returns to numbers of the form $12k \pm 1$, $12k \pm 5$ and $10k \pm 1$, and he writes:

> In the progression of 3, all the prime numbers that differ by one from a multiple of 12 measure only the powers –1. Such are: 11, 13, 23, 37, etc.
>
> In the same progression, the prime numbers that differ by 5 from a multiple of 12 measure powers +1. Such are: 5, 17, 19, etc.
>
> In the progression of 5, all the prime numbers that end in 1 or 9 measure only powers –1. Such are: 11, 19, etc.
>
> The ones that finish in 3 or 7 measure powers +1. Such are: 7, 13, 17, etc.[23]

Regardless of how one judges the correctness of these claims, note that if Fermat treats numerical right triangles, it is to find the properties of the pertinent families of integers. On this point, however, Fermat is not very far from his 10th century predecessors, a conclusion that the preceding example confirms. But this research into the extension of the new 10th-century Diophantine analysis should not obscure the step that Fermat is taking. The issue for him is to lift out certain quadratic forms encountered during his research on numerical right triangles and to begin the process of studying them. Whereas the preceding example shows his interest, partial and indirect to be sure, in quadratic remainders, other examples from 1641 suggest that Fermat is beginning to turn toward the study of forms. If this turns out to be the case, the distance he has covered is immense.

On June 15, 1641, Fermat writes to Frenicle:

> There are triangles whose lesser sides differ only by one.[24]

In other words, if $c < b < a$ is one of them, another will be

$$2a + 2c + b < 2a + 2c + b + 1 < \text{the hypotenuse}.$$

[22] R. Rashed, *The Development of Arabic Mathematics*, p. 210.
[23] Letter XLVII – Saturday 15 June 1641, from Fermat to Mersenne, vol. II, p. 220.
[24] Letter XLVIII, vol. II, p. 224.

As is often the case, Fermat does not explain. Let us therefore comment on this assertion, in order to display the means to which he appealed. Indeed the equation to be solved is

$$(*) \qquad c^2 + (c+1)^2 = a^2.$$

Positing that $a = \frac{p}{q}c + 1$, one gets

$$c\left(p^2 - 2q^2\right) = 2q(q - p).$$

If $p^2 = 1 + 2q^2$, the solution has been found.

This is the Diophantine algorithm interpreted as 'the chord method'.[25] Through point ($a = 1$, $c = 0$) of the curve with equation (*), one draws a straight line with slope $\frac{p}{q}$, and one finds the second point of intersection, $c = \frac{2q(p-q)}{p^2 - 2q^2}$, to be an integer if $p^2 - 2q^2 = 1$.

Note therefore that one is led to a Fermat equation that, in the simplest case, has the coefficient 2. In the general case of equation $y^2 = \alpha x^2 + \beta x + \gamma$, which Euler will study later, the same method leads one to the general equation of Fermat, $y^2 - \alpha x^2 = \pm 1$.

In other words, this problem boils down to a study of the form $x^2 - 2y^2$. The big question is to discover how Fermat might have proceeded.

Given $a = y$, $2c + 1 = x$; the problem amounts to finding integer solutions to the equation

$$(1) \qquad 1 + x^2 = 2y^2.$$

If u and v satisfy the same equation, then

$$(2) \qquad 1 + u^2 = 2v^2.$$

One can therefore write

$$(x - u)(x + u) = 2(y - v)(y + v)$$

which is solved by assuming

$$y + v = 2\alpha\beta, \; x + u = 4\alpha p,$$
$$y - v = 2pq, \; x - u = 2\beta q,$$

[25] See Diophante, *Les Arithmétiques*, ed. R. Rashed, vol. III; R. Rashed and C. Houzel, *Les* Arithmétiques *de Diophante: Lecture historique et mathématique*, Berlin/New York, Walter de Gruyter, 2013.

whence

$$y = \alpha\beta + pq,\ x = 2\alpha p + \beta q,$$
$$v = \alpha\beta - pq,\ u = 2\alpha p - \beta q.$$

By rewriting equation (1)

$$2y^2 - x^2 = 1$$

or equation (2)

$$2v^2 - u^2 = 1,$$

one obtains

$$(2\alpha^2 - q^2)(\beta^2 - 2p^2) = 1,$$

and therefore

$$(2\alpha^2 - q^2)(2p^2 - \beta^2) = -1.$$

One of the factors is therefore equal to 1 and the other to -1. For example

(3) $$2\alpha^2 = 1 + q^2, \quad \beta^2 = 1 + 2p^2.$$

Reciprocally, if α, β, p and q satisfy equation (3), x and y are solutions of equation (1). Specifically, if $p = 2$, $\beta = 3$, $\alpha = y_0$, $q = x_0$, one obtains

(4) $$x = 4y_0 + 3x_0, \quad y = 3y_0 + 2x_0.$$

This is the general solution, for, from equation (4) one gets

$$x_0 = 3x - 4y, \quad y_0 = 3y - 2x,$$

and x_0, y_0 are solutions, if x and y are. One thus obtains a linear transformation (4) which leaves the curve of equation (*) invariant.

Returning now to the right triangles $y = a, x = b + c = 2c + 1$, one has

$$a = 3a_0 + 4c_0 + 2 = 3a_0 + 2b_0 + 2c_0,$$
$$b = 2a_0 + 3c_0 + 2 = 2a_0 + 2c_0 + b_0 + 1 = 2a_0 + 2b_0 + c_0,$$
$$c = 2a_0 + 3c_0 + 1 = 2a_0 + 2c_0 + b_0$$

A reasoning by descent starting from x, y, by noting that $x_0 < \frac{x}{5}$, as well as $x \geq 3$, shows that, by iterating the operation of descent, one must reach $x_0 = 1$, and one will therefore deduce all the solutions of $x_0 = 1$, $y_0 = 1$, by iterating (4).

The preceding discussion suggests that in 1641, during his research on numerical triangles, Fermat advances farther than his predecessors, since he goes back to the study of equations of the preceding type: what we now call Fermat equations. This is the orientation that henceforth seems to govern his research. In order to show this, one need only read more of the same letter. Indeed, Fermat notes that one could handle in the same way research on right triangles whose lesser sides differ by a given number. In this case, one is likewise led to solve

$$(5) \qquad x^2 + 1 = 2y^2.$$

Fermat goes on to state that 'there are triangles for which the least side always differs by a square from each of the other two'.[26] One must therefore solve

$$m^2 - n^2 = 2mn + p^2,$$
$$m^2 + n^2 = 2mn + q^2$$

where (m, n) is the pair of generators for this triangle.

One deduces from this that

$$2n^2 = q^2 - p^2, \quad q^2 = p^2 + 2n^2.$$

If one gives q the form $u^2 + 2v^2$, its square will also have this form; whence $p = u^2 - 2v^2$, $n = 2\,uv$, and then

$$(m - n)^2 = q^2 \text{ and } m = n + q = u^2 + 2v^2 + 2uv.$$

But Fermat gives as a rule: to form a triangle in which $b_0 - c_0$ is a square; then, starting from this triangle, to apply twice the given rule, that is,

$$c = 2a_0 + 2c_0 + b_0.$$

[26] Letter XLVIII, from Fermat to Frenicle, vol. II, p. 224.

The second triangle will answer the question if the primitive does; otherwise, the first triangle obtained will do. Why then propose a method different from the one we just gave?

Indeed, in a primitive triangle, one knows that the hypotenuse is odd and that the two sides have opposite parities. Or if

$$c_1 = 2a_0 + b_0 + 2c_0,$$
$$b_1 = 2a_0 + 2b_0 + c_0,$$
$$a_1 = 3a_0 + 2b_0 + 2c_0,$$

one has $b_1 - c_1 = b_0 - c_0$ and it will always be square; and $a_1 - c_1 = a_0 + b_0$; whence,

> if c_0 is even, $a_0 - c_0$ is a square, and the triangle (a_0, b_0, c_0) will answer the question;
> if c_0 is odd, b_0 is even, c_1 is even, and the triangle (a_1, b_1, c_1) will answer the question.

Finally, in the same letter Fermat raises the following problem: 'Given a number, find how many times it can be the sum of the two lesser sides of a right triangle',[27] which amounts to solving

$$k(m^2 + 2mn - n^2) = a, \qquad \text{where } a \text{ is a given number.}$$

An examination of this problem shows that we have come back once again to a Fermat equation. Let us divide a by one of its divisors, that is, let $a = \alpha k$, then solve

$$\alpha = 2m^2 - (m - n)^2;$$

then α must be odd.

Here, the forms in question are $x^2 - 2y^2$ and $2x^2 - y^2$, that is, forms to which the descent applies. From one of Legendre's observations, we know that $p = 8k + 1$ (which is prime) divides $x^2 - 2y^2$. If $p = 8k + 7$, one has

$$(6) \qquad x^{8k+6} - 1 = \left(x^{4k+3} - 1\right)\left(x^{4k+3} + 1\right) \equiv 0 (\text{mod}\, p).$$

Now equation (6) has $8k + 6$ roots.

If $x = 2$, one has

[27] *Ibid.*, vol. II, p. 226.

$$(2 \cdot 2^{4k+2} - 1)\,(2 \cdot 2^{4k+2} + 1) \equiv 0(\text{mod}\, p).$$

But the form $2x^2 + y^2$ has no prime divisor of the form $8k + 7$; therefore $2 \cdot 2^{4k+2} - 1 \equiv 0(\text{mod}\, p)$, and p is a divisor of the form $2x^2 - y^2$. Whence, by descent, every prime number of one or the other form $8k + 1$ and $8k + 7$ is of the form $2x^2 - y^2$; whence the solution.

Everything seems to suggest that, around this period, Fermat is circling around these mathematical meanings and this type of demonstration. At least he has ideas about them. Everything also suggests that during the same period, he is formulating problems about numerical right triangles that reduce to the quadratic form $x^2 \pm 2y^2$. One can therefore see that research on numerical right triangles is increasingly being inflected towards a study of the forms that Fermat was beginning to undertake, together with demonstrations by infinite descent. Our conclusion seems to be confirmed by the several years that follow. In this connection, one need only read the letter of 27 January 1643, in which Fermat responds to questions from M. de Saint-Martin. To keep the discussion brief, let us bring up only the problem that he raises for Mersenne in the letter of 16 February 1643:

> To find two right triangles whose areas are in a given ratio, such that the two small sides of the larger triangle differ by one.[28]

This problem is clearly linked to the form $2x^2 - y^2$. One can give as a solution (20, 21, 29), (12, 35, 37). It remains to be discovered whether the problem has other solutions, and how many there are. One can appreciate the difficulty of these questions.

So it was that Fermat's double arithmetic membership – in both the Diophantine and the Euclidean traditions – led him to conceive of integer Diophantine analysis, the second such occurrence in history. Was he also influenced by the tradition of al-Khāzin and by the 10th-century school *via* Fibonacci's *Liber quadratorum*, which belongs to precisely this tradition? Did he know Fibonacci's work directly or *via* Bachet? These open-ended questions should be studied for their own sake. This new Diophantine analysis encompassed a privileged domain: numerical right triangles and the problems associated with them. With all of the strength he had, Fermat tackled this domain in order both to improve the methods he had inherited and to invent new ones. As we have seen, it was while carrying out this research that, beginning in 1640, he was led to the study of certain quadratic forms.

[28] Letter LV, from Fermat to Mersenne, vol. II, pp. 251–3, at p. 252.

Part II

Geometry

– 1 –

THE ARCHIMEDEANS AND PROBLEMS WITH INFINITESIMALS

The history of infinitesimal geometry, notably the part that treats areas and volumes, surfaces and solid curves, has one distinctive characteristic. After having reached a high peak of achievement in the writings of Archimedes, work on the subject stopped for more than a millennium, to start again in the 9th century, this time in Arabic civilization. During the next two centuries, the mathematicians of the time scaled even higher summits, before progress once again halted, no less suddenly than it had done in the third century BCE. A third beginning took place in the 17th century, and since this time, progress has continued unabated. Because it is so often insufficiently recognized, this historical development deserves close study in order to determine why the subject experienced two beginnings and two endings.

An anomaly never occurs alone, however. Even the writings of Archimedes on infinitesimals appear in a somewhat paradoxical light from the 9th century on. Except for complete translations of *The Measurement of the Circle* and *On the Sphere and Cylinder*, very few of his works were known in Arabic. Despite the small number of translated texts, however, the field would undergo a rapid expansion of further research and reach new heights. In fact, for reasons tied to the preservation and transmission of the Archimedean corpus, only a small number of Archimedean texts were known in late Antiquity. Thus Eutocius seems to be aware of only the two works just mentioned. The situation was the same in the world of Classical Islam; all the evidence points to the fact that the translators could not have known the Archimedean corpus as extensively as they knew the works of Euclid and Apollonius. Without fear of contradiction, one can state that the mathematicians at that time did not know the works of Archimedes on infinitesimals, except for what appears in the two works we have cited. An examination of their works in this field establishes this fact, which is further confirmed directly by one of them, the 12th-century mathematician

Ibn al-Sarī.[1] The mathematicians of Classical Islam did not know, either directly or indirectly, the *Quadrature of the Parabola*, or *On Conoids and Spheroids*, or *On Spirals*, or *The Method*. This is an important fact, since Archimedes' method for the sums of integrals, which completes the method of exhaustion, is applied only in *On Conoids and Spheroids* and *On Spirals*, that is, in treatises unknown to Arabic mathematicians. Likewise, the latter did not know about Archimedes' methods either for determining the areas of segments of parabolas or for finding the volumes of segments of paraboloids, which are also given in these two works.

The fundamental issue is to discover how the two works, *The Measurement of the Circle* and *On the Sphere and Cylinder*, were received in the world of Arabic mathematics. Indeed, their reception undoubtedly plays a part in determining the characteristics that distinguish the Arabic Archimedeans. To begin, these two books were received and studied by mathematicians who were engaged in the investigation of conic sections and, soon after these two works by Archimedes became available, also had access to Apollonius's *Conics*. Such was certainly the case after the time of the Banū Mūsā brothers. These three mathematicians were also interested in other mathematical disciplines that are often today called 'applied': astronomy, statics, mechanics, optics. It should be noted that such interests gave birth to the development of research on conics: their optical properties, tracing out their point locus by using geometric transformations, tracing their continuous locus with dedicated instruments, and also research into projective methods. It should therefore not be surprising that the mathematicians of the time introduced methods different from those of Archimedes, and systematized procedures that appear only in scattered places in Archimedes' works. In point of fact, these Arabic mathematicians, who were intent on the study of pointwise transformations and projections, did indeed combine such methods with infinitesimal ones. Finally, consider a fact that would influence research in infinitesimal geometry: these same mathematicians also knew about the works of their algebraist colleagues. Taken as a whole, this knowledge will in some sense mould the conceptions and methods of the Arabic Archimedeans, and determine the course of their investigations into infinitesimal geometry. Under these circumstances, one would expect to see them not only deepen their understanding of asymptotic behaviour and of infinitely small objects, but also expand the field of inquiry into other domains of infinitesimal geometry

[1] R. Rashed, *Les Mathématiques infinitésimales du IXe au XIe siècle*. Vol. II: *Ibn al-Haytham*, London, 1993, pp. 498–510; English translation: *Ibn al-Haytham and Analytical Mathematics*. A History of Arabic Sciences and Mathematics, vol. 2, Culture and Civilization in the Middle East, London, 2012.

that Archimedes had not expressly treated, namely isoperimetric and equal-area figures, solid angles, and lunes. It is also crucial to understand that these Archimedeans were not, like Eutocius, mere commentators on Archimedes, but also emulators of his work. It is also important to note that at that time, very few commentaries and writings on infinitesimal geometry were available in Arabic; for our part we know only of al-Kindī's on *The Measurement of the Circle*, al-Māhānī's on *On the Sphere and Cylinder*, and Naṣīr al-Dīn al-Ṭūsī on both of these treatises. Below, we shall consider briefly but systematically the different fields of infinitesimal geometry in mathematics during the period of Classical Islam.

1. CALCULATING INFINITESIMAL AREAS AND VOLUMES

1.1. *The Pioneers*

The earliest known text of Archimedean mathematics in Arabic is the philosopher and mathematician Abū Isḥāq al-Kindī's commentary on Proposition 3 of *The Measurement of the Circle* (the approximation of π).[2] This commentary, *The Epistle from al-Kindī to Yūḥannā ibn Māsawayh on the Approximation of the Ratio of the Circumference to the Diameter* (*Risālat al-Kindī ilā Yūḥannā ibn Māsawayh fī taqrīb al-dawr min al-watar*), was published by al-Kindī before 856. By this date, the influence of algebra on al-Kindī's work is already evident. He used the vocabulary of algebra for some expressions and he also employed ratios of numbers and segments, that is, ratios that no Greek mathematician would have allowed. Note also that this commentary by al-Kindī was eventually known in a Latin translation, as apparently attested by the famous 'Florence versions'.[3] But the truly innovative work in Archimedean geometry began with the contemporaries of al-Kindī, the three brothers Muḥammad, Aḥmad, and al-Ḥasan, known collectively as the Banū Mūsā. They published a famous work, the *Book on the Measurement of Plane and Spherical Figures* (*Kitāb maʿrifat misāḥat al-ashkāl al-basīṭa wa-al-kuriyya*), the inaugural great work of the Arabic Archimedeans.

The merest glance at the Banū Mūsā's treatise clearly reveals a work in the Archimedean tradition, even though it is not modelled on Archimedes' *On the Sphere and the Cylinder*. What is more, the Banū Mūsā do not adopt Archimedes' approach, even though certain fundamental ideas

[2] R. Rashed, 'Al-Kindī's commentary on Archimedes' *The Measurement of the Circle*', *Arabic Sciences and Philosophy*, 3, 1993, pp. 7–53.

[3] *Ibid*.

remain the same. The treatise contains 18 propositions devoted to four themes: the measurement of the circle, the measurement of the surface and volume of the sphere, Hero's formula for the area of a triangle, the matter of two mean proportionals and the trisection of the angle. At the end of their treatise, the Banū Mūsā declare their own assessment of their work, writing:

> Everything we describe in this book is our own work, with the exception of knowing the circumference of the circle from the diameter, which is the work of Archimedes, and the position of two magnitudes between two others such that [the four] are in continued proportion, which is the work of Menelaus, as stated earlier.[4]

This statement, which has generally been overlooked, deserves our attention. We shall therefore consider briefly the work of the Banū Mūsā to understand better the significance of this statement and also to locate their precise position within the Archimedean tradition. The Banū Mūsā began by proving the following proposition:

Given a circle of circumference p *and a line of length l, two cases arise*:

1) *if $l <$ p, a polygon of perimeter* $\mathrm{p_n}$ *can be inscribed in the circle, such that*

$$l < \mathrm{p_n} < \mathrm{p}.$$

2) *if $l >$ p, the circle can be circumscribed by a polygon of perimeter* q_n, *such that*

$$\mathrm{p} < \mathrm{q_n} < l.$$

The proofs of these two propositions assume the existence of both a circle of given circumference p and a regular polygon. The Banū Mūsā grant the existence of a circle of given circumference. For the regular polygon, they appeal to Proposition XII.16 of Euclid's *Elements*: 'Given two circles about the same center, to inscribe in the greater circle an equilateral polygon with an even number of sides that does not touch the smaller circle'. On the other hand, note that for a regular n-sided polygon to satisfy the conditions of the problem, the perpendicular distance from its center to the midpoint of one of the sides, its apotome a_n, must satisfy the condition:

[4] See Chapter I of *Mathématiques infinitésimales du IXe au XIe siècle*. Vol. I: *Fondateurs et commentateurs*, London, 1996, p. 132; English translation: *Founding Figures and Commentators in Arabic Mathematics*. A History of Arabic Sciences and Mathematics, vol. 1, Culture and Civilization in the Middle East, London, 2012, p. 109.

$$r_1 < a_n < r_2 \Leftrightarrow r_1 < r_2 \cos \pi/n < r_2 \Leftrightarrow p_1/p_2 < \cos \pi/n < 1,$$

where r_1, r_2 and p_1, p_2 are, respectively, the radii and circumferences of the two concentric circles (the existence of the integer n being dependent on the continuity of the cosine function; Fig. 41). It follows that, contrary to what some have asserted, appealing to Euclid XII.16 is insufficent fully to establish the proposition, the proof of which must be completed by means of a homothety (enlargement), a technique known to al-Ḥasan ibn Mūsā.[5]

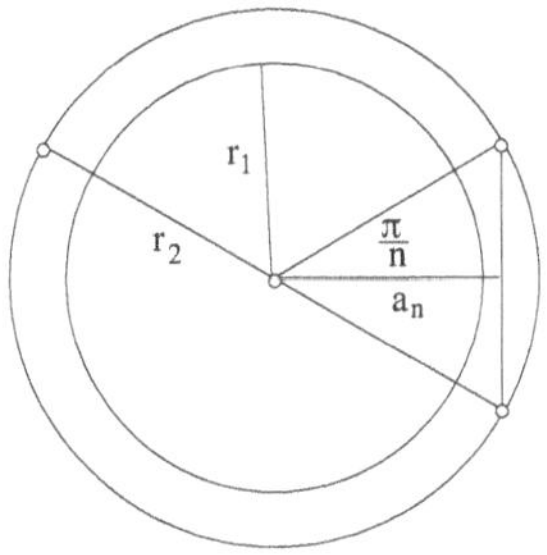

Fig. 41

In the next proposition, the Banū Mūsā use a *reductio ad absurdum* to prove that the area of a circle is 'the product of its semi-diameter and its semi-circumference', that is, $S = r \cdot \frac{p}{2}$. Note that in this proof, they do not compare S to $S' > S$ or S to $S'' < S$ in order to show that both $S' > S$ and $S'' < S$ result in contradictions, as Archimedes had done; instead, they consider only the circumference and examine $p' < p$ and $p'' > p$, that is, they compare lengths.

Having determined the area of the circle, the Banū Mūsā go on to approximate π using the method of Archimedes, which they acknowledge. Yet the approach of the Banū Mūsā, as just outlined, differs from that of Archimedes in several places. The first difference concerns the application of the method of exhaustion itself and its complement, proof by contradiction.

We have already pointed out that the Banū Mūsā avoided the most difficult part of the method of exhaustion (the passage to the limit, to use our language) by using Euclid XII.16 and completing the argument with a homothety – the proof of which does, however, require a limit process argument. As to the proof by contradiction, we have already noted that they compared lengths – not areas, as Archimedes had done. Finally, the Banū

[5] *Ibid.*, p. 132.

Mūsā did not follow Archimedes in determining the area of the circle by comparing it with another figure, such as a right-angle triangle with one side equal to the radius and the other equal to the circumference; rather, they find it directly as the product of two magnitudes. Given these circumstances, it is easy to see why their proof is shorter than that of Archimedes.

The key question is whether the Banū Mūsā took this approach deliberately or as a simple response to the situation at hand. To answer this question, we must look at their treatment of the second topic: determining the surface and volume of a sphere. They began by proving the following propositions:

PROPOSITION 11 – *The surface area of a frustum of a cone with parallel bases is given by*

$$S = \frac{1}{2}\, l(p_1 + p_2),$$

where l *is the length of the sloping side and* p_1, p_2 *are the circumferences of the two bases, respectively.*

PROPOSITION 12 – *If a quadrant of a circle* A_1B *is divided into* n *equal arcs by the points* $A_2, A_3, \ldots, A_n$, *then*

1° $$A_1B_1 + 2\sum_{k=2}^{n} A_kB_k = B_1E$$

2° $$B_1M^2 < \frac{1}{2} BA_n \left(B_1A_1 + 2\sum_{k=2}^{n} B_kA_k \right) < B_1B^2 .$$

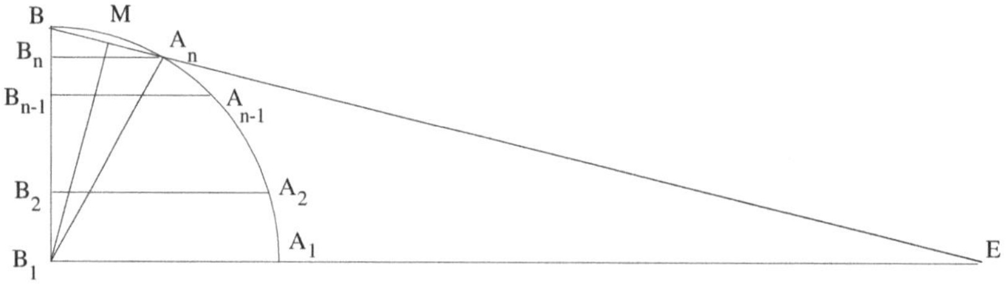

Fig. 42

After some work and rearranging, 1° and 2° can be rewritten

1° $$2\sum_{k=1}^{n} \sin k \cdot \frac{\pi}{2n} = \cot an \frac{\pi}{4n} + 1$$

2° $$\cos^2\frac{\pi}{4n} < \cos\frac{\pi}{4n} < 1 \text{ (valid for all } n\text{).}$$

In Proposition 13, the Banū Mūsā then consider a semicircle *ABD* with center *M* and radius R_2 in which one inscribes a regular polygonal line with an even number of sides. Inside the latter, one inscribes another semicircle. By rotation about an axis through the center and perpendicular to the diameter, one generates a hemisphere, an inscribed solid of revolution of surface area *S* composed of a cone and several frusta of cones, and a second concentric hemisphere inscribed within the solid of revolution. The Banū Mūsā show that

$$2\pi R_1^2 < S < 2\pi R_2^2;$$

where R_1 and R_2 are, respectively, the radii of the inscribed and circumscribed circles. Note that the hypotheses concerning the plane figure lying in the plane *ABD* derive from Proposition 12 and that the proof uses Propositions 11 and 12, which depend on *Elements* XII.16.

The Banū Mūsā can now use an argument by contradiction twice here: first in Proposition 14, to obtain the surface of the hemisphere, 'twice that of its great circle', as they put it, *i.e.* $S = 2\pi R^2$; second, to deduce the volume of the sphere as 'the product of its semi-diameter and one-third the surface area', that is, $\frac{4}{3}\pi R^3$.

Here too, as in the case of the measurement of the circle, we find differences between the approach of the Banū Mūsā and that of Archimedes. The first concerns the method of exhaustion. Here again the Banū Mūsā use *Elements* XII.16 completed by a homothety – not, as once scholars once thought, *Elements* XII.17, which considers a polyhedron enclosed between concentric spheres. The Banū Mūsā considered a solid inscribed in the hemisphere, made up of a cone and frusta of cones, whose surface lies outside the inner hemisphere but inside the outer hemisphere. Such a solid is generated from a regular line polygon inscribed in a great circle of the sphere that does not touch a great circle of the second sphere inscribed within the first, that is, proceeding always from *Elements* XII.16. This method dispenses with the need for a passage to the limit for the series of sines given above. Here again, when it comes to determining the volume of the sphere, the method of contradiction is carried out on surface, not volumes. Finally, the volume of the sphere is given as the product of two magnitudes – not, as in Archimedes, in terms of another volume: 'four times the cone that has its base equal to the greatest circle in the sphere and its height equal to the radius of the sphere'.

The Banū Mūsā's work had a variety of important consequences for mathematics. First, in the Arabic world, this initial stage, which was quickly superseded, had some impact on mathematical research, but most especially on teaching. Beginning with the work of the Banū Mūsā's collaborator Thābit ibn Qurra; for more than a century and a half thereafter, research was pursued on the measure of curved surfaces and curved volumes. This tradition rediscovered the method of integral sums and simultaneously developed geometrical methods for facilitating the application of the method of exhaustion, for example, using affine transformations. Ibn al-Haytham later took up the measure of the sphere, now using integral sums (Darboux sums). The treatise of the Banū Mūsā would survive above all as a manual for teaching, evidence for which can be found in the many surviving Arabic manuscript copies of Naṣīr al-Dīn al-Ṭūsī's commentary of this treatise and which were clearly intended for this purpose.

In the Latin world, the situation was altogether different. Once translated by Gerard of Cremona, Latin copies of the Banū Mūsā's treatise constituted the essential reference work on research by Archimedean mathematicians, along with Archimedes' *Measurement of the Circle*, itself rendered into Latin from an Arabic version. Among others, Fibonacci, Jordanus de Nemore, the anonymous author of *Liber de triangulis*, and Roger Bacon all consulted the treatise.

We have noted that, however excellent in other respects, the most recent discussions on the Banū Mūsā's *Book on the Measurement of Plane and Spherical Figures* err in believing that they relied on *Elements* XII.16 alone, and especially when it is claimed that the brothers used *Elements* XII.17. It is by failing to notice the very explicit use of geometric transformations – homotheties – that these modern commentators have fallen into error: they have examined the contributions of the Banū Mūsā starting from the geometry of the Ancients. One need only consult al-Ḥasan's book on the ellipse to understand the explicit role of enlargement in the study of geometric transformations. Precisely this feature distinguishes Arabic geometry from its Hellenic and Hellenistic origins. This thesis, which is fundamental for us, has several consequences: in particular it obliges us to rewrite the history of geometry in this era. To mention only what interests us primarily here, we can see that, in order to develop an infinitesimal geometry, the Archimedean tradition from the outset had to redirect geometrical research in order to deal with pointwise transformations. Seen in such a light, the history of this tradition becomes clear, and the position of the Banū Mūsā within it is illuminated. Let us then consider their second book, which in our view, must be studied in order to make sense of the first.

Al-Ḥasan ibn Mūsā's purpose is to determine the area of an ellipse and of elliptical sections. Ibn Mūsā, as we know, did not possess yet an intelligible version of Apollonius's *Conics*. I mean that Ibn Mūsā's research was also aimed at the study of conic sections, beginning with a study of plane sections of a cylinder. Not until after al-Ḥasan's death did his brother Aḥmad come across the Eutocius edition of the first four books of the *Conics*, which made it possible to translate a manuscript containing seven of the books into Arabic (the Greek original of the eighth was already lost). In other words, al-Ḥasan ibn Mūsā in his book attempted to reach two goals; the first was, so to speak, Archimedean: to determine an area bounded by a curve; the second was in the tradition of Apollonius (even though al-Ḥasan did not truly know the *Conics*): a study of the geometric properties of curves. It is in this book that al-Ḥasan ibn Mūsā combines concepts of projection and of affine orthogonal transformations for applications of *Elements* XII.2 and of the method *reductio ad absurdum*: here we see the first inflection in the meaning of Archimedean mathematics. Let us consider this development briefly.

According to Ibn al-Samḥ (born in 1035 in Cordoba), whose book summarizes al-Ḥasan ibn Mūsā's lost treatise, the latter proceeded as follows. He began with an 'elongated circular figure' defined by the bifocal property $MF + MF' = 2a$, where $2a$ is the length of the major axis, in order to establish that the plane section of a cylinder of revolution, cut by a plane not parallel to its base – an ellipse – has the very same properties as the former figure. He then went on to determine the axes of the ellipse in order to study the properties of its chords, its chord midpoint distances to the arc, etc. The deduction graph shows that the first set of six propositions pertains the 'elongated circular figure', its vertices, its center, its diameters, its chords, its minor axes, its inscribed circle with the minor axis as diameter, as well as its circumscribed circle with the major axis as diameter. The next set of five propositions concerns the ellipse as a plane section of a cylinder, as well as its identification with the 'elongated circular figure'. Finally, a set of eight propositions treats the area of the ellipse. We shall now summarize these sets of propositions.

The fifth proposition of the first set of propositions can be expressed as:

If to a point M of the 'elongated circular figure' is associated on the inscribed circle a point T with the same ordinate ($MT \perp BD$ at K), then

$$MK^2 = KT^2 + (OA - MF)^2.$$

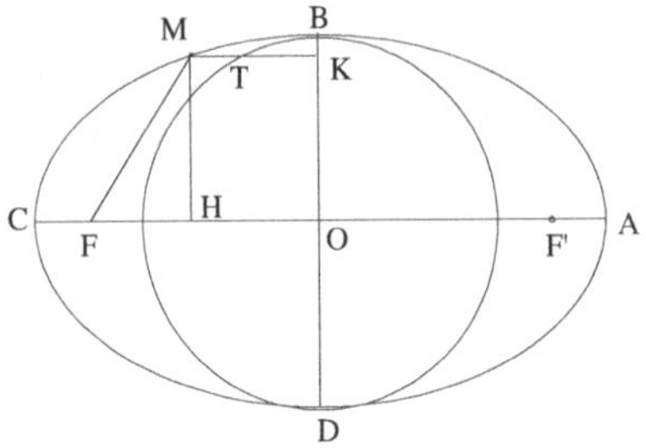

Fig. 43

Then, using $MK = x$ and $MH = y$ as a coordinate system and letting $AC = 2a$, $BD = 2b$, and $2c = FF' = KT = X$; and setting $MF = a - \frac{cx}{a}$, the above relation becomes:

$$(*) \qquad x^2 - \frac{c^2}{a^2}x^2 + y^2 = b^2 .$$

Dividing throughout by b^2 yields once again the equation of the ellipse.

The sixth proposition deals with the orthogonal affine transformation pertaining to the minor axis

$$\frac{MK}{TK} = \frac{OA}{OB} \Leftrightarrow \left[\frac{x}{X} = \frac{a}{b}\right];$$

from (*) one has

$$x^2\left(1 - \frac{c^2}{a^2}\right) = b^2 - y^2 = X^2 \Leftrightarrow b^2x^2 = a^2X^2 ,$$

and so

$$\frac{x}{X} = \frac{a}{b},$$

which is an orthogonal affine transformation – an elongation – with axis BD and scale factor $\frac{a}{b} > 1$, in which the ellipse $ABCD$ is the image of the circle with diameter BD.

The second set of propositions deals with the ellipse as a plane section of a right cylinder. After recalling the property that the section of a right circular cylinder by a plane not parallel to its base, is an ellipse whose center lies on the axis of the cylinder, al-Ḥasan ibn Mūsā, considers a family of curves that deform continuously from the circle to the ellipse. He then considers the orthogonal affine transformation associated with the minor axis (Proposition 7):

Let *ADBG* be an ellipse and *EDZG* its inscribed circle. If a line parallel to *AB* cuts *GD* at *H*, the circle at *K*, and the ellipse at *T*, then

$$\frac{HT}{HK} = \frac{AB}{GD} = \frac{a}{b}.$$

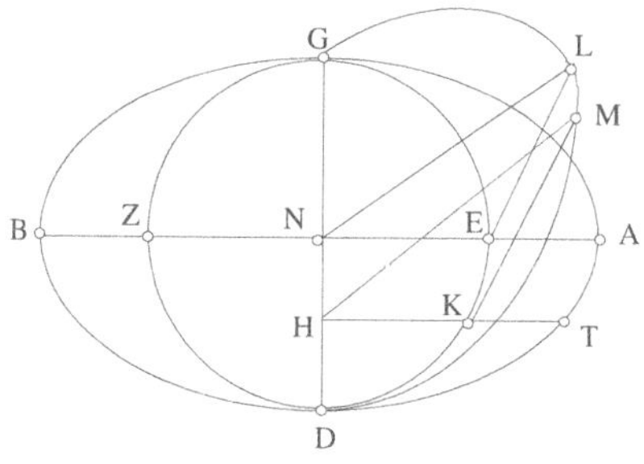

Fig. 44

In the course of his proof, al-Ḥasan ibn Mūsā considers the ellipse *ADBG* as a rabatment of the ellipse *DLG* (see Fig. 44), which is found by making *ADBG* pivot around *DG* onto the plane perpendicular to the axis of the cylinder at *N*. The circle *DEG* is the cylindrical projection of the ellipse *DLG* onto this same plane.

In the next proposition – the eighth – he considers a transformation perpendicular to the major axis. The ellipse is then the image of a circumscribed circle under an orthogonal affine transformation with scale factor $\frac{b}{a}$, that is, a contraction.

In other words, and to use modern algebraic symbolism, if we let the ellipse be

$$\mathscr{E} = \left\{(x, y),\ \frac{x^2}{a^2} + \frac{y^2}{b^2} = 1\right\} \quad \text{where } a > b,$$

then the ellipse can be transformed into two circles

$$X^2 + Y^2 = b^2 \text{ and } X^2 + Y^2 = a^2$$

by means of the mappings

$$\psi:(X,Y)\to(x,y):\begin{cases}x=\dfrac{a}{b}X\\ y=Y\end{cases}$$

$$\phi:(X,Y)\to(x,y):\begin{cases}x=X\\ y=\dfrac{b}{a}Y\end{cases}$$

where ψ, ϕ are a one-way elongation and a one-way contraction, respectively. In this way, al-Ḥasan ibn Mūsā shows that the two figures, an elongated circular figure and the ellipse, are superposable (Proposition 9).

By these conceptual means, al-Ḥasan ibn Mūsā also went on to determine the area of the ellipse. He showed first, by means of an elongation, that the ratio of the area of a polygon inscribed in the ellipse to the area of a polygon inscribed in a circle is equal to the ratio of the axes of the ellipse. He then refined his results to show that the area of an ellipse is πab.

This briefly describes al-Ḥasan ibn Mūsā's approach while simultaneously shedding some light on the first book of the Banū Mūsā and the results that these Archimedean mathematicians obtained amongst their followers. The use of a homothety in the first book now seems entirely natural in the context of this geometry; the use of pointwise transformations by the successors of the Banū Mūsā, like Thābit ibn Qurra, his grandson Ibrāhīm ibn Sinān, and many others, has its origin in the works of the Banū Mūsā. In this way, our image of the Banū Mūsā changes somewhat: they are no longer to be seen as a pale reflection of Archimedes, mere commentators on his work, but as mathematicians in their own right who, in order to begin a new tradition, reshaped the old one. Their successors were eminent mathematicians who made rapid progress. By doing so, however, they paradoxically obscured the importance of the Banū Mūsā's work. Historians of mathematics also contributed to this outlook by attributing to others, especially to Thābit ibn Qurra, some of the achievements of the Banū Mūsā, and thus blending what were two distinct portraits: the one transmitted by Latin mathematics, and the other found in Arabic mathematics.

We have just seen that, at the very moment when the two texts of Archimedes were being translated into Arabic, the mathematicians of the 9th century were seeking alternative acceptable methods of obtaining Archimedes' results and of making new discoveries. The search for new paths of discovery was not confined to the Archimedean scholarship of the Banū Mūsā and their successors; it was also found in their works on mechanics and astronomy. Furthermore, the same tendency is evident in other mathematical disciplines – arithmetic, theory of numbers, algebra, trigonometry, projective methods, etc. – and also in other scientific disciplines, such as optics with al-Kindī, and statics with Ibn Qurra.

This new spirit of scientific inquiry that was developing in the 9th century in the tradition of Hellenic science, but with greater freedom – the freedom to invent, to criticize, and to look to other sources such as Indian ones – explains at least in part the ever-surprising phenomenon of the enormous number of translations from the ancient heritage.

The contemporaries and successors of the Banū Mūsā actively continued research in the areas they had explored. Thus, Thābit ibn Qurra (826–901), the collaborator of the Banū Mūsā made major contributions to the subject. In quick succession, he published three treatises: one on the area of the segment of a parabola, the second on the volume of a paraboloid of revolution, and the third on sections of the cylinder and its surface area. These works by Thābit ibn Qurra would themselves lead to new departures for mathematics, both by reducing the number of preliminary lemmas required, and by improving the method. Thus the measure of the parabola was taken up by al-Māhānī, Ibrāhīm ibn Sinān, and Ibn Sahl; the measure of the paraboloid was treated by al-Qūhī and later by Ibn al-Haytham. This in itself demonstrates the seminal role of Thābit ibn Qurra's works in this field as in many others.

In the first of his treatises, *On the Measure of the Section of a Cone Called a Parabola* (*Fī misāḥat qiṭʿ al-makhrūṭ alladhī yusammā al-mukāfiʾ*),[6] which sought to determine the area of a segment of a parabola, Thābit ibn Qurra, who was unaware of Archimedes' treatment of the subject, began by proving twenty-one lemmas, fifteen of which were arithmetic. These lemmas mainly concerned the sums of numerous arithmetic progressions. For example, he proved the results:

$$2\sum_{k=1}^{n} k^2 = \frac{1}{2}\sum_{k=1}^{n}(2k-1)^2 + n^2 + \frac{n}{2};\ \sum_{k=1}^{n}(2k-1)^2 = \frac{4n^3}{3} - \frac{n}{3};\ \ldots$$

After eleven arithmetic lemmas, Thābit ibn Qurra stated four lemmas on sequences of line segments, sequences whose upper bounds need to be found. Thus, in Lemma 14, he proved:

Let a, b be line segments such that $\frac{a}{b}$ is known; then there exists an $n \in \mathbf{N}^*$ such that the sequence $(u_k)_{1 \leq k \leq n}$ of n successive odd numbers

[6] Rashed, *Mathématiques infinitésimales*, vol. I, Chap. II.

beginning with 1, and the sequence $(v_k)_{1\leq k\leq n}$ of n successive even numbers beginning with 2 satisfies

$$(*) \qquad \frac{n}{v_n \cdot \sum_{k=1}^{n} u_k} < \frac{a}{b}.$$

Here Thābit ibn Qurra introduced an approximation that he went on to use in the next proposition in order to find subdivisions of segments. The proposition stated:

Let AB, H be two line segments and a, b two line segments such that $\frac{a}{b}$ is given. For any given n,

1° there exists a partition $(A_k)_{0\leq k\leq n}$ with $A_0 = A$, $A_n = B$ and such that

$$\frac{A_k A_{k+1}}{A_{k+1} A_{k+2}} = \frac{u_{k+1}}{u_{k+2}} \qquad (0 \leq k \leq n-2),$$

with $(u_k)_{1\leq k\leq n}$ being the sequence of successive odd numbers beginning with 1;

2° there exists a sequence of line segments $(H_j)_{1\leq j\leq n}$ with $H_n = H$ and such that

$$\frac{H_j}{H_{j+1}} = \frac{v_j}{v_{j+1}} \qquad (1 \leq j \leq n-1),$$

with $(v_j)_{1\leq j\leq n}$ being the sequence of successive even numbers beginning with 2.

If n satisfies condition (*), then

$$\frac{n\, A_0A_1 \cdot \frac{H_1}{2}}{AB \cdot H} < \frac{a}{b}.$$

The proof depends on the partition of a given line segment into a sequence of segments proportional to the terms of a given numerical sequence, and thus on the generalization of the preceding proposition, which introduced approximation, to sequences of segments, leading to the generalization of determining the upper bound of a sequence of ratios of segments.

It is after these fifteen lemmas – eleven purely arithmetical ones and four concerning sequences of line segments – that Thābit ibn Qurra began his calculation of the area of a parabola segment. To do this, he proved four propositions. These lemmas and propositions show that Thābit ibn Qurra understood perfectly and rigorously the concept of the upper bound of a sequence of real square numbers, and also the uniqueness of this upper bound. Thābit ibn Qurra identified the upper bound in the following way:

Let BAC be a parabola segment with diameter AD. Let S be the area of the parallelogram with base BC associated with the parabola. Then for all $\varepsilon > 0$ there exists a partition $A, G_1, G_2, \ldots, G_{n-1}, D$, of the diameter AD, such that

$$\text{area } BAC - \text{area of the polygon } BE_{n-1}AF_1F_2 \ldots F_{n-1}C < \varepsilon.$$

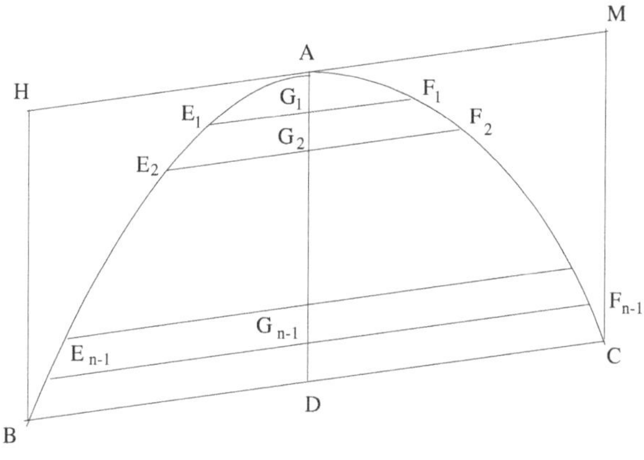

Fig. 43

Thābit ibn Qurra proves rigorously that $\frac{2}{3}$ of the area $BHMC$ is the upper bound of the areas of the derived polygons. He then comes to his theorem:

The area of the parabola is infinite but the area of any one of its segments is equal to two-thirds of the area of the parallelogram associated with this segment.

The proof is argued along the following lines:

Let $\mathscr{S}$ be the area of the segment of the parabola $\mathscr{P}$ and let S be the area of the parallelogram associated with this segment.

If $\frac{2}{3}\, S \neq \mathscr{S}$, there are two cases:

1. $\mathscr{S} > \frac{2}{3}\,S$. Let $\varepsilon > 0$ be such that

$$(1) \qquad \mathscr{S} - \frac{2}{3}\,S = \varepsilon.$$

From Proposition 18, for this ε, there exist an N such that for all $n > N$ the polygon $\mathscr{P}_n$ of area S_n satisfies

$$(2) \qquad \mathscr{S} - S_n < \varepsilon;$$

from (1) and (2), it follows that

$$\left(\frac{2}{3}\,S + \varepsilon\right) - S_n < \varepsilon,$$

whence

$$\frac{2}{3}\,S < S_n.$$

But, from Proposition 17, we have

$$\frac{2}{3}\,S > S_n,$$

which is a contradiction.

2. $\mathscr{S} < \frac{2}{3}\,S$. Let $\varepsilon > 0$ be such that

$$(3) \qquad \frac{2}{3}\,S - \mathscr{S} = \varepsilon.$$

From Proposition 19, for this ε, there exist an N such that for all $n > N$ the polygon $\mathscr{P}_n$ of area S_n satisfies

$$(4) \qquad \frac{2}{3}\,S - S_n < \varepsilon;$$

from (3) and (4), it follows that we have

$$(\mathscr{S} + \varepsilon) - S_n < \varepsilon,$$

whence

$$\mathscr{S} < S_n.$$

But, $\mathscr{P}_n$ is inscribed in $\mathscr{P}$, so $S_n < \mathscr{S}$, which is a contradiction. Hence, from both contradictions, it follows that

$$\frac{2}{3} S = \mathscr{S}.$$

This theorem depends on establishing the uniqueness of the upper bound and the proof essentially uses the properties of the upper bound. In fact, the assignment is to establish that $\frac{2}{3} S = \mathscr{S}$, given that

1) $\mathscr{S} = \sup.\, (S_n)_{n\geq 1}$
2) $\frac{2}{3} S = \sup.\, (S_n)_{n\geq 1}$.

Suppose, by *reductio ad absurdum*, $\mathscr{S} \neq \frac{2}{3} S$. There are two cases:

a) $\mathscr{S} > \frac{2}{3} S$, then there exists $\varepsilon > 0$ such that $\mathscr{S} - \frac{2}{3} S = \varepsilon$. But from 1) $\mathscr{S}$ is the least upper bound of the S_n, so for this ε, there exists an S_n such that

$$S_n > \mathscr{S} - \varepsilon,$$

hence

$$\frac{2}{3} S < S_n;$$

which is absurd since, from 2), $\frac{2}{3} S$ is an upper bound of the S_n.

b) $\mathscr{S} < \frac{2}{3} S$, then there exists $\varepsilon > 0$ such that $\frac{2}{3} S - \mathscr{S} = \varepsilon$. But from 2) $\frac{2}{3} S$ is the least upper bound of the S_n, so for this ε, there exists an S_n such that

$$S_n > \frac{2}{3} S - \varepsilon,$$

hence

$$\mathscr{S} < S_n;$$

which is absurd since from 1), $\mathscr{S}$ is an upper bound of the S_n.

Of course we do not claim that either Thābit ibn Qurra, or his predecessors, or his successors down to the 18th century had defined the concept of an upper bound. On the other hand, it seems to us that he used the properties of upper bounds as a guiding principle when measuring areas bounded by convex borders.

In fact, in Ibn Qurra's approach, we can discern the fundamental idea that underpins Riemann integration. For the particular case in which the diameter is the axis of the parabola, Thābit ibn Qurra's method is to consider a partition $\sigma = AG_1G_2 \ldots G_{n-1}$ of the diameter AD, so as to obtain the sum

$$S_\sigma = \sum_{i=1}^{n} \left(AG_i - AG_{i-1}\right) \frac{G_{i-1}F_{i-1} + G_iF_i}{2}$$

and to show that $\forall\ \varepsilon > 0, \exists\ \sigma$ such that area $ACD - S_\sigma < \varepsilon$, and to prove finally that S_σ converges to that area.

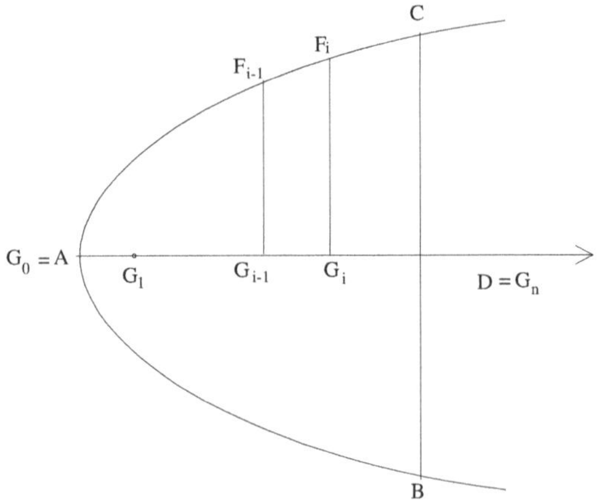

Fig. 46

Let us translate the preceding argument into the language of analysis: let x_i be the abscissa of G_i and let $y = f(x)$ be the equation of the parabola. S_σ can then be written

$$S_\sigma = \sum_{i=1}^{n} \left(x_i - x_{i-1}\right) \frac{f(x_{i-1}) + f(x_i)}{2};$$

but since

$$f(x_{i-1}) \le \frac{f(x_{i-1}) + f(x_i)}{2} \le f(x_i)$$

and f is continuous, we can deduce that

$$\frac{f(x_{i-1}) + f(x_i)}{2}$$

is a value that f takes at point ξ_i of the interval $[x_{i-1}, x_i]$. But S_σ can then be written in the form

$$S_\sigma = \sum_{i=1}^{n} (x_i - x_{i-1}) f(\xi_i)\,; \quad x_{i-1} \le \xi_i \le x_i,$$

which is none other than the sum used in the definition of the Riemann integral of a function f. We note finally that, given the definition of the parabola, the quadrature achieved by Ibn Qurra is equivalent to calculating the value of the integral $\int_0^a \sqrt{px}\, dx$. The historian of mathematics M. A. Youschkevitch had this to say about Ibn Qurra's procedure:

> Thanks to this procedure, Ibn Qurra revived a method for calculating integral sums that had fallen into obscurity. Furthermore, by using this same procedure, Ibn Qurra had, for the first time, effectively evaluated an integral $\int_0^a x^n\, dx$ for a fractional value of the exponent n, namely $\int_0^a x^{1/2}\, dx$. In doing this, he proceeded, also for the first time, by a subdivision of the integral interval into unequal parts. It was by an analogous procedure, consistent with dividing the abscissa axis into segments forming a geometric sequence, that P. Fermat, in the middle of the 17th century, undertook the quadrature of curves of the form $y = x^{m/n}$, with $(m/n) \neq 1$.[7]

After having calculated the area of the parabola, Thābit Ibn Qurra went on to find the volume of a paraboloid of revolution. This entailed a change from the two-dimensional plane to three-dimensional space. First, he needed to establish thirty-six lemmas, arranged in groups. Finally, he proved the following theorem:

> *The volume* v *of a parabolic cupola* ABC *with axis* BD *is half the volume* V *of the cylinder with height* h *whose circular base has diameter* AC,
>
> $$\mathrm{v} = \frac{1}{2}\mathrm{V} = \frac{1}{2}\pi\,\mathrm{h} \cdot \frac{\mathrm{AC}^2}{4}.$$

[7] M. A. Youschkevitch, 'Note sur les déterminations infinitésimales chez Thābit ibn Qurra', *Archives internationales d'histoire des sciences*, vol. 17, no. 66, 1964, p. 43.

Thābit ibn Qurra's approach here was similar to the way he determined the area of a parabola segment. He used a subdivision of the diameter of a part of the parabola arranged in segments proportional to successive odd numbers. The points of the parabola associated with this division were then abscissas proportional to the squares of integers and the ordinates were proportional to successive integers.

These points determine:

in the plane:	in space:
a polygon inscribed in the parabola and decomposed into trapeziums	a solid of revolution inscribed in the paraboloid and decomposed into conical type solids
s area of the parabola	v volume of the paraboloid
S area of the associated parallelogram	V volume of the associated cylinder
s_i area of a trapezium	v_i volume of a conic solid

Thābit shows that, given $\varepsilon > 0$, an N can be found such that, for all $n > N$, one has

$$\frac{2}{3}S - \sum_{i=1}^{n} s_i < \varepsilon \text{ (17 and 19)} \qquad \frac{V}{2} - \sum_{i=1}^{n} v_i < \varepsilon \text{ (32 and 35)}$$

$$s - \sum_{i=1}^{n} s_i < \varepsilon \text{ (18)} \qquad v - \sum_{i=1}^{n} v_i < \varepsilon \text{ (33 and 34).}$$

In other words, he showed that:

$$\frac{2}{3}S = \text{upper bound of } \sum_{i=1}^{n} s_i \qquad \frac{V}{2} = \text{upper bound of } \sum_{i=1}^{n} v_i$$

$$s = \text{upper bound of } \sum_{i=1}^{n} s_i \qquad v = \text{upper bound of } \sum_{i=1}^{n} v_i \,.$$

By *reductio ad absurdum*, he then showed the uniqueness of the upper bound, that is:

$$s = \frac{2}{3}S \text{ (20)} \qquad v = \frac{V}{2} \text{ (36).}$$

The contributions that Ibn Qurra made to this area of mathematics did not end here. He also published a substantial volume, *The Book on the Sections of the Cylinder and its Surface Areas* (*Kitāb fī quṭūʿ al-usṭūwāna wa-basīṭihā*), in which he explored different sections of the right and oblique cylinder, and then determined the area of the ellipse and the area of elliptical segments, discussed the greatest and least sections of the cylinder and their axes, and determined the area of that part of the surface bounded by two plane sections.

This treatise, like the two earlier ones, is not only significant in the history of infinitesimal geometry, but also one of the most important texts in geometry. In fact, since he showed how to use pointwise geometrical transformations, he was responsible for taking geometry in a new direction, and thereby enriching algebra. Evidence for this can be found in the works of Ibrāhīm ibn Sinān, Ibn Sahl, al-Qūhī, Ibn al-Haytham, and Sharaf al-Dīn al-Ṭūsī, among others.[8]

It is not possible here to give a full account of the results and proofs contained in this rich and profound work. We shall present only two of his propositions, numbers 14 and 31, so as to give a flavour of the work and to illustrate his ideas.

In Proposition 14, Ibn Qurra proves the following:

If S *is the area of the ellipse* $\mathscr{E}$ *with axes* 2a *and* 2b, *and* Σ *the area of the circle* E *with radius* $r = \sqrt{ab}$, *then* S = Σ.

We give here a translation of his proof in modern notation, using the following symbols:

S area of the ellipse $\mathscr{E} \rightarrow S_n$ area of polygon P_n inscribed in $\mathscr{E}$
Σ area of the circle equivalent to $E \rightarrow \Sigma_n$ area of polygon Π_n inscribed in E
S' area of the circumscribing circle $\mathscr{C} \rightarrow S'_n$ area of polygon P'_n inscribed in $\mathscr{C}$.

The proof is by contradiction, considering the two cases:

Case I: if $S > \Sigma$, then

(1) $S = \Sigma + \varepsilon$.

Let P_n be the polygon of 2^{n+1} sides inscribed in the ellipse $\mathscr{E}$ obtained from P_{n-1} by doubling the numbers of vertices, the new vertices being located where the diameters through the midpoints of the opposite sides of P_{n-1} meet the ellipse. The first polygon P_1 is a rhombus with vertices at the ends of the axes of the ellipse. If S_n is the area of P_n, then we have, successively

[8] See R. Rashed, *Les Mathématiques infinitésimales*, vol. I, pp. 458 ff.; English transl. pp. 333 ff.

$$S_1 > \frac{1}{2}S \Rightarrow S - S_1 < \frac{1}{2}S$$
$$S_2 - S_1 > \frac{1}{2}(S - S_1) \Rightarrow S - S_2 < \frac{1}{2^2}S$$
$$\dots$$
$$S_n - S_{n-1} > \frac{1}{2}(S - S_{n-1}) \Rightarrow S - S_n < \frac{1}{2^n}S;$$

thus for ε defined by (1), there exists an $n \in \mathbf{N}$ such that $\frac{1}{2^n}S < \varepsilon$, whence

$$S - S_n < \varepsilon$$

$$S_n > \Sigma.$$

Now consider the circle $\mathscr{C}$ and the polygon P_n' obtained from $\mathscr{E}$ through the orthogonal affinity with ratio $\frac{a}{b}$. Let S_n' be the area of P_n' and let S' be the area of $\mathscr{C}$,

$$\frac{S_n}{S_n'} = \frac{b}{a} = \frac{ab}{a^2} = \frac{\Sigma}{S'};$$

but $S_n > \Sigma$, whence $S_n' > S'$, which is impossible.

Case II: if $S < \Sigma$, then

$$\frac{S}{S'} < \frac{\Sigma}{S'};$$

so

(2) $$\frac{\Sigma}{S'} = \frac{S}{S' - \varepsilon'}.$$

For the circle $\mathscr{C}$ and polygons P_n' referred to above, we have, successively:

$$S' - S_1' < \frac{1}{2}S'$$
$$S' - S_2' < \frac{1}{2^2}S'$$
$$\dots$$
$$S' - S_n' < \frac{1}{2^n}S',$$

thus for ε' defined by (2), there exists an $n \in \mathbf{N}$ such that $\frac{1}{2^n} S' < \varepsilon'$, whence

(3) $$S' - S'_n < \varepsilon'.$$

If P_n is the polygon inscribed in $\mathscr{E}$ corresponding to P'_n, by means of an orthogonal affinity with ratio $\frac{b}{a}$, we have

$$\frac{S_n}{S'_n} = \frac{\Sigma}{S'} = \frac{S}{S' - \varepsilon'}.$$

But from (3),

$$S'_n > S' - \varepsilon',$$

which means that $S_n > S$, which is absurd. From a) and b), we therefore deduce that $S = \Sigma$.

Of course, we can see that the ellipse $\mathscr{E}$ can be transformed into the circle $\mathscr{C}$ by an orthogonal dilation f with ratio $k_1 = \frac{a}{b}$ and from the circle $\mathscr{C}$ with radius a into the circle E of radius r, where $r^2 = ab$, by a homothety h of ratio

$$k_2 = \frac{r}{a} = \frac{\sqrt{ab}}{a} = \sqrt{\frac{b}{a}}.$$

Hence $E = h \circ f(\mathscr{E})$ and the transformation $h \circ f$ preserves area since $k_1 \cdot k_2^2 = 1$.

The aim of Proposition 14 is to establish precisely this property in the case of the ellipse, which he proves by contradition.

Using our notation, Ibn Qurra uses $\frac{\Sigma}{S'} = \frac{b}{a} = k_2^2$ and shows that $\frac{S_n}{S'_n} = \frac{b}{a} = \frac{1}{k_1}$ whatever the choice of n, whence

$$S = \Sigma \Leftrightarrow \frac{S}{S'} = \frac{\Sigma}{S'} \Leftrightarrow \frac{S}{S'} = \frac{S_n}{S'_n}.$$

Ibn Qurra's method corresponds to the two following steps:

(a) $\frac{S_n}{S'_n} < \frac{S}{S'}$, therefore $\frac{S_n}{S'_n} = \frac{S - \varepsilon_1}{S'}$ (1).

It can be shown that $\exists\, P_n \subset \mathscr{E}$, such that $S - \varepsilon_1 < S_n < S$. But $f(P_n) = P'_n \subset \mathscr{C}$ satisfies (1) and hence $S'_n > S'$, which is impossible.

(b) $\frac{S_n}{S'_n} > \frac{S}{S'}$, therefore $\frac{S_n}{S'_n} = \frac{S}{S' - \varepsilon_2}$ (2).

It can be shown that $\exists\, P'_n \subset \mathscr{C}$, such that $S' - \varepsilon_2 < S'_n < S'$. But $f^{-1}(P'_n) = P_n \subset \mathscr{E}$ satisfies (2) and hence $S_n > S$, which is impossible. It has therefore been shown that

$$\frac{S}{S'} = \frac{S_n}{S'_n}.$$

Hence, beginning with the property of affine orthogonality, that the ratio of the areas S'_n and S_n of the two corresponding polygons P_n and P'_n whatever the value of n, is equal to the ratio $\frac{a}{b}$ of the affinity, Ibn Qurra is able to deduce that the same holds for the area S of the ellipse $\mathscr{E}$ and the area S' of the circle $\mathscr{C}$. This amounts to saying that the ratio is preserved during the passage to the limit, or symbolically:

$$\frac{S_n}{S'_n} = \frac{b}{a}$$

and

$$\frac{S}{S'} = \frac{\lim S_n}{\lim S'_n} = \lim \frac{S_n}{S'_n} = \frac{b}{a}.$$

We may note that Luca Valerio used this type of assumption as the basis of his method.[9] This method does not involve the sums of integrals.

Finally, we should note that Archimedes obtained the same result earlier in *On Conoids and Spheroids*. This work was, however, not known to any mathematicians of the time, including Thābit ibn Qurra. Comparing these two works has a double advantage for us: first, we are in a better position to appreciate the contribution of the 9th-century mathematicians and, second, we can better understand their awareness of the Archimedean corpus at that time.

Moving on to Proposition 31, Thābit ibn Qurra now proves:

[9] Luca Valerio, *De centro gavitatis solidorum libri tres*, Bologna, 1661, Book II, Propositions I–III, pp. 69–75.

The surface area Σ of a segment of an oblique cylinder taken between two parallel plane sections is given by

$$\Sigma = \mathrm{p} \cdot l$$

where p *is the circumference of the minimal ellipse and l the length of the generator between the two sections.*

The proof likewise proceeds by using a double contradiction.

Let $\mathcal{E}$ be one of the sections, K its center and $2a$ its major axis. We consider two cases:

Case I:

If $\Sigma < p \cdot l$, there is a length g, $g < p$ such that $\Sigma = g \cdot l$.

Let h be such that $g < h < p$, then there exists an area ε such that $\Sigma + \varepsilon = h \cdot l$ and so $\varepsilon = l(h - g)$.

Now construct an ellipse $\mathcal{E}_1 = \varphi(\mathcal{E})$, where φ is an homothety – an enlargement with center K and ratio $\frac{a_1}{a}$ such that $1 > \frac{a_1}{a} > \frac{h}{p}$. The circumference p_1 of the ellipse $\mathcal{E}_1$ is such that $\frac{p_1}{p} = \frac{a_1}{a}$ from Proposition 26, hence $\frac{p_1}{p} > \frac{h}{p}$ and so $p_1 > h$.

Let P_n be a polygon inscribed in $\mathcal{E}$ and not in contact with $\mathcal{E}_1$, and let P'_n be its projection onto the other plane section, and p_n the perimeter of each of them. If Σ_n is the surface area of the prism with end faces P_n and P'_n, then $\Sigma_n = p_n \cdot l$. But $p_n > p_1 > h$ and so $\Sigma_n > hl$.

$$(1) \qquad \Sigma_n > \Sigma + \varepsilon.$$

a) If $\frac{\varepsilon}{2} \geq s$, the areas s and s' of the two bases which are the minimal ellipses being equal, we have $\varepsilon \geq s + s'$, and so $\Sigma_n > \Sigma + s + s'$. The surface area of the prism inscribed in the cylinder will be greater than the total surface of the cylinder, which is absurd.

b) If $\frac{\varepsilon}{2} < s$, we can impose on a_1 the additional condition $\frac{a_1^2}{a^2} > \frac{s - \frac{\varepsilon}{2}}{s}$, but since s_1 is the area of $\mathcal{E}_1$, then $\frac{s_1}{s} = \frac{a_1^2}{a^2}$, and so $s - s_1 < \frac{\varepsilon}{2}$.

If s_n is the area of P_n, and s'_n the area of P'_n, we have

$$s_n = s'_n\,,\, s > s_n > s_1,\, s - s_n < \frac{\varepsilon}{2} \text{ and } \varepsilon > (s - s_n) + (s' - s'_n).$$

From (1) this gives $\Sigma_n > \Sigma + (s - s_n) + (s' - s'_n)$, which is absurd.

From a) and b) we deduce that $\Sigma \geq p \cdot l$.

Case II:

If $\Sigma > p \cdot l$, there exists a length g, $g > p$ such that $\Sigma = g \cdot l$.

Let $h, p < h < g$ and ε be the area such that $\Sigma = h \cdot l + \varepsilon$.

Let $\mathscr{C}_1 = \varphi(\mathscr{C})$, where φ is an homothety with center K and ratio $\frac{a_1}{a}$ such that

$$\frac{a_1}{a} < \frac{h}{p} \text{ and } \frac{a_1^2}{a^2} < \frac{s + \frac{\varepsilon}{2}}{s}.$$

If p_1 is the perimeter of $\mathscr{C}_1$, we have $\frac{p_1}{p} = \frac{a_1}{a}$ and so $p_1 < h$.

Let there be inscribed in $\mathscr{C}_1$ a polygon P_n having no points in common with $\mathscr{C}$. Expressed in the notation of the first part, we have $\Sigma_n = p_n \cdot l$. But $h > p_1 > p_n$ and so $\Sigma_n < h \cdot l$ and it follows that

$$\Sigma > \Sigma_n + \varepsilon. \tag{2}$$

But

$$\frac{s_1}{s} = \frac{a_1^2}{a^2},$$

and so

$$s_1 < s + \frac{\varepsilon}{2};$$

however,

$$s_1 - s > s_n - s,$$

whence

$$s_n - s < \frac{\varepsilon}{2}.$$

We know from the convexity of the cylinder that

$$\Sigma_n + (s_n - s) + (s'_n - s') > \Sigma,$$

and so

$$\Sigma_n + \varepsilon > \Sigma,$$

which contradicts (2). Hence we have $\Sigma \leq p \cdot l$.

From the results of Cases I and II, we deduce that $\Sigma = p \cdot l$.

We should add that, up to this point, the only surfaces that had been evaluated were those of the right cylinder and the sphere (Archimedes, *On the Sphere and the Cylinder*). Ibn Qurra was the first to consider the surface of an oblique cylinder, which today is evaluated by means of an elliptic integral (with an elliptical base of circumference p).

This proposition constitutes a step in the direction of determining the lateral surface of an oblique cylinder with circular bases and also the surface of the whole of an oblique cylinder contained between two plane sections, whether they are parallel or not. It is precisely this result that Ibn Qurra establishes in the propositions that follow.[10]

1.2. *The Heirs*

For the Banū Mūsā and Thābit ibn Qurra, work in infinitesimal geometry was based on solid foundations and had also made considerable progress. These mathematicians accumulated a substantial corpus of knowledge, sufficient to provide a foundation for new departures. The time had come for the heirs of their work to take up the challenge. This corpus went well beyond the awareness of certain results and theorems; it also comprised mathematical methods, both newly discovered and rediscovered. We have witnessed the development of two types of methods. Geometrical methods based on transformations were already evident in the work of al-Ḥasan ibn Mūsā, as well as that of his brothers Muḥammad and Aḥmad, and also Thābit ibn Qurra's final treatise. On the other hand, we have seen how Thābit ibn Qurra had reintroduced the concept of integral sums. To be sure, this idea is already present in Archimedes, but not in any of the works translated into Arabic at the time. It was the intensive study of *Measurement of a Circle* and *On the Sphere and Cylinder*, the two Archimedean works then available in Arabic, that led a mathematician of the stature of Thābit ibn Qurra to the rediscovery of the method of integral sums. In Thābit ibn Qurra, however, integral sums are more general than in Archimedes, to the extent that Thābit used subdivisions of the interval that were not necessarily regular. As for his treatment of the paraboloid, in which he always used integral sums, he did not consider cylinders of equal height, as Archimedes had, but a cone and frusta of cones with heights proportional to successive odd numbers beginning with 1.

[10] See R. Rashed, *Les Mathématiques infinitésimales*, vol. I, pp. 493 ff.

One could therefore anticipate that in the period that followed, when research into mathematics accelerated by growing both more intense and more extended, mathematicians did their best to improve the demonstrations of their predecessors and to develop both the method of integral sums and the method of geometrical transformations.

The first to follow this path was al-Māhānī (*d. c.* 860), who took up again the area of the parabola, and found a proof much shorter than Thābit ibn Qurra's. Unfortunately, since al-Māhānī's text has never been found – if indeed it still exists – we cannot know precisely how he proceeded.

The second heir of this tradition was Thābit ibn Qurra's grandson, Ibrāhīm ibn Sinān (909–946). This mathematician of genius, who lived only thirty-eight years, was, in his own words, not pleased that al-Māhānī had produced 'a study more advanced than that of my grandfather, without there being, amongst us, one whose work surpasses it'.[11] Thus he thus wanted to provide a demonstration that was shorter than the proof not only of his grandfather, which had required twenty lemmas, but also of al-Māhānī.

Indeed, Ibn Sinān published a particularly economic and elegant treatment of the area of the parabola (*Kitāb fī misāḥat qiṭʿ al-mukāfiʾ*). His central idea, which he set out to establish at the beginning, was that the ratios of areas remain invariant under affine transformations. For this, all he needed was three propositions.

PROPOSITION 1 – *Let there be two convex polygons* $A = (A_0, A_1, \dots, A_n)$ *and* $B = (B_0, B_1, \dots, B_n)$. *Let the points* $A_1, A_2, \dots, A_{n-1}$ *be projected to the points* $A'_1, A'_2, \dots, A'_{n-1} = A_n$ *by a projection parallel to* $A_{n-1}A_n$ *onto the line* A_0A_n, *and let the points* $B_1, B_2, \dots, B_{n-1}$ *be projected to the points* $B'_1, B'_2, \dots, B'_{n-1} = B_n$ *by a projection parallel to* $B_{n-1}B_n$ *onto the line* B_0B_n. *Then if*

$$\frac{A_0A'_1}{B_0B'_1} = \dots = \frac{A'_{n-2}A_n}{B'_{n-2}B_n} = \lambda$$

and

$$\frac{A_1A'_1}{B_1B'_1} = \dots = \frac{A'_{n-1}A_n}{B'_{n-1}B_n} = \mu ,$$

then one has the following ratio of proportion between the areas of the triangles and the areas of the polygons

[11] R. Rashed and H. Bellosta, *Ibrāhīm ibn Sinān. Logique et géométrie au Xᵉ siècle*, Leiden, E.J. Brill, 2000, p. 18.

$$\frac{\mathrm{tr}(A_0, A_{n-1}, A_n)}{\mathrm{p.}(A_0A_1, \ldots, A_n)} = \frac{\mathrm{tr}(B_0, B_{n-1}, B_n)}{\mathrm{p.}(B_0B_1, \ldots, B_n)}.$$

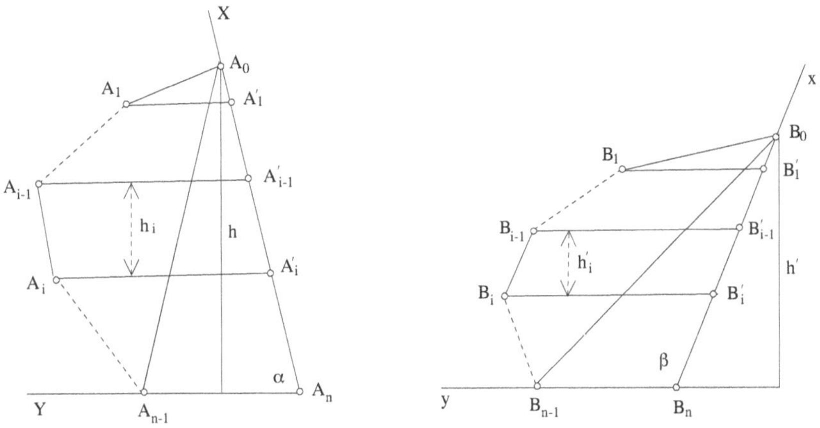

Fig. 47

To establish this proposition, Ibn Sinān uses the transformation T defined in Proposition 1, which is an affine transformation. He shows that T preserves the ratio of areas for the case of triangles and polygons.

The second proposition is:

PROPOSITION 2 – *The ratio of the areas of two segments of a parabola is equal to the ratio of their two corresponding triangles.*

Here, Ibn Sinān shows that under the affine transformation the ratio of the area of a segment of a parabola to that of its corresponding triangle is the same as the corresponding ratio of their images under projection. In other words, it depends on the result that an affine transformation preserves ratios of areas, even when, as here, they are curvilinear. To establish this, Ibn Sinān uses the axiom of Archimedes to show that it is possible to inscribe in the segment of the parabola a polygon whose area differs by as little as one wishes from that of the segment of the parabola.

Having proved these two lemmas, Ibn Sinān is now able to prove his main result concerning the ratio of the area of a segment of a parabola to that of its associated triangle. For this he did not need to use an infinitesimal argument but only the fact that the ratio was independent of the segment being considered, which is what he established in his third proposition.

PROPOSITION 3 – *The area of a segment of a parabola is four-thirds of its associated triangle.*

Thus, to improve his grandfather's proof and to reduce the number of propositions required from twenty to three, Ibn Sinān's strategy was based on a combination of affine transformations and infinitesimal methods.

In the second half of the 10th century, al-Qūhī also achieved a similar economy by reducing to just three propositions the thirty-six propositions of Thābit ibn Qurra for calculating the volume of a paraboloid of revolution. But whereas Ibn Sinān had extended the use of transformation geometry that had appeared in the work of Thābit ibn Qurra, al-Qūhī picked up and extended the other strand of mathematics in that work, which led to the rediscovery of integral sums, present in the work of Archimedes, but unkown at that time to the Islamic mathematicians.

For the paraboloid of revolution, Archimedes had considered cylinders of the same height. In contrast, Thābit ibn Qurra had used frusta of adjacent cones, the bases of which determine a partition of the parabola's diameter (which generates the paraboloid), the steps of which are proportional to the sequence of odd numbers beginning with one, and the heights of which are the steps of this partition. In order to reduce the number of propositions that Thābit ibn Qurra used to just three, as was his boast, al-Qūhī had to rediscover independently the integral sums that had appeared in the work of Archimedes, particularly when he needed to prove that he could make the difference between the inscribed and circumscribed cylinders as small as he wished. Here are his three propositions.

PROPOSITION 1 – *Let there be a paraboloid of revolution with vertex* X *and axis* XF *and any partition of the axis by abscissae points* $(b_i)_{0\le i\le n}$, *with* $b_0 = 0$ *and* $b_n = XF$. *Let* $(I_i)_{2\le i\le n}$ *and* $(C_i)_{1\le i\le n}$ *be respectively the volumes of the inscribed and circumscribed cylinders associated with this partition and let* V *be the volume of the cylinder associated with the paraboloid. Then*

$$\sum_{i=2}^{n} I_i < \frac{1}{2}V < \sum_{i=1}^{n} C_i \quad \textit{for all } n \in \mathbf{N}^*.$$

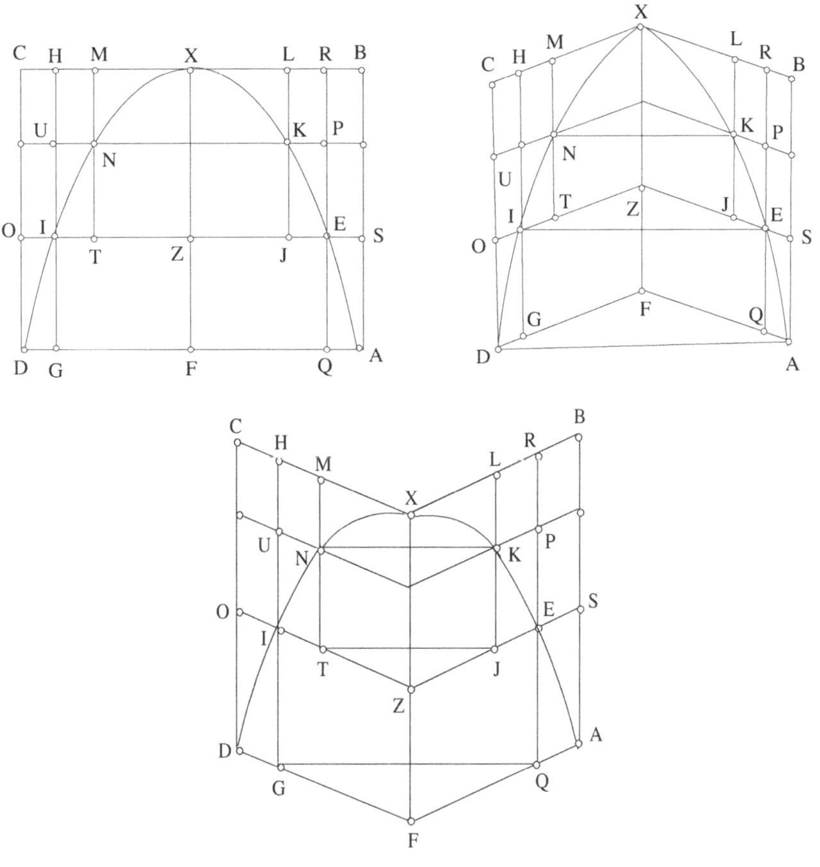

Fig. 48

PROPOSITION 2 – *Consider a segment of a paraboloid between any two sections perpendicular to the axis, and let* I, C *be the volumes of the corresponding inscribed and circumscribed cylinders respectively. If this segment is cut by a third section half-way between the two sections, this will determine two inscribed cylinders with volumes* I_1, I_2, *and two circumscribed homologous cylinders with, respectively, volumes* C_1 *and* C_2*; then we have*

$$(C_1 - I_1) + (C_2 - I_2) = \frac{1}{2}\ (C - I).$$

$C - I = v$ (ring HGEC),

$C_1 - I_1 = v$ (ring NLMC),

$C_2 - I_2 = v$ (ring LKGS).

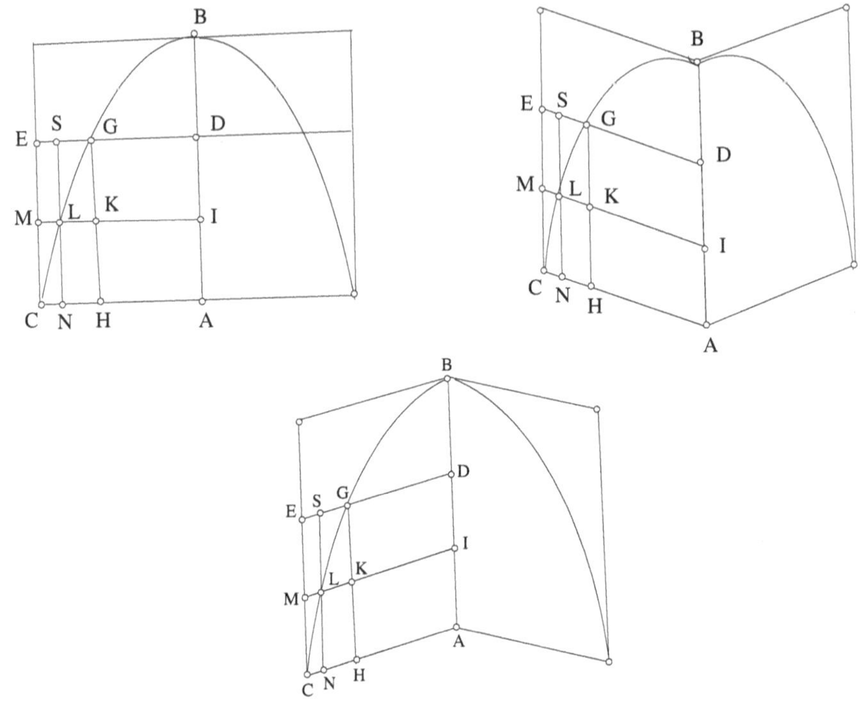

Fig. 49

Al-Qūhī's proof proceeds as follows. Let $(b_i)_{0 \le i \le n}$ be a partition of the axis XF and let the volumes of the inscribed and homologous circumscribed cylinders be I_i and C_i, where $I_1 = 0$. If we then consider the sequence $(c_j)_{0 \le j \le 2n}$, with $b_0 = c_0$, $b_n = c_{2n}$, where $c_{2i+1} = \frac{b_i + b_{i+1}}{2}$, and let the volumes of the associated cylinders be respectively $(I'_j)_{1 \le j \le 2n}$, $(C'_j)_{1 \le j \le 2n}$, then

$$\sum_{j=1}^{2n} (C'_j - I'_j) = \frac{1}{2} \sum_{i=1}^{n} (C_i - I_i).$$

Finally, al-Qūhī proves:

PROPOSITION 3 – *If* P *is the volume of a segment of a paraboloid and* V *the volume of its associated cylinder, then*

$$\mathrm{P} = \frac{1}{2}\mathrm{V}.$$

The proof is as follows. If we let respectively $\left(I_i^q\right)_{1\le i\le n\cdot 2^q}$ and $\left(C_i^q\right)_{1\le i\le n\cdot 2^q}$ be the volumes of the cylinders associated with the partition of the axis, we know from the preceding proposition that

$$\sum_{i=1}^{n.2^q}(C_i^q - I_i^q) = \frac{1}{2}\sum_{i=1}^{n.2^{q-1}}(C_i^{q-1} - I_i^{q-1})$$

for a fixed n and any q in $\mathbf{N}^*$.

This allows al-Qūhī, thanks to a rider to X.1 of the *Elements*, to claim that, after a certain number of operations, one has

$$(*) \qquad \sum_{i=1}^{n.2^q}(C_i^q - I_i^q) < \varepsilon .$$

Stated in a different language, he shows that for all $\varepsilon > 0$, there exists an N such that, for all $q > N$, the inequality (*) holds. But we know that if P is the volume of the paraboloid, then

$$P - \sum_{i=1}^{n.2^q} I_i^q < \sum_{i=1}^{n.2^q}(C_i^q - I_i^q),$$

and so

$$P - \sum_{i=1}^{n.2^q} I_i^q < \varepsilon.$$

Now, if $P = \frac{V}{2} + \varepsilon$, then we have $\frac{V}{2} < \sum_{i=1}^{n.2^q} I_i^q$, which is impossible given Proposition 1. If $P = \frac{V}{2} - \varepsilon$, we can reason similarly, since

$$\sum_{i=1}^{n\cdot 2^q} C_i^q - P < \sum_{i=1}^{n\cdot 2^q}\left(C_i^q - I_i^q\right) < \varepsilon ,$$

and so

$$\sum_{i=1}^{n\cdot 2^q} C_i^q - \left(\frac{V}{2} - \varepsilon\right) < \varepsilon ,$$

whence

$$\sum_{i=1}^{n.2^q} C_i^q < \frac{V}{2},$$

which is also impossible from Proposition 1, and therefore

$$P = \frac{V}{2}.$$

Al-Qūhī's proof is short and direct thanks to Proposition 1, which directly compares the sums of inscribed and circumscribed cylinders to the volume of the large cylinder. This obviated the need for him to evaluate these sums as Archimedes had done, by reducing them to the sum of a geometric progression. The proof here depends on the inequalities $u_i - u_{i-1} < 2C_i$ and $u_i - u_{i-1} > 2I_i$, which come from considering equal cylinders such as *QGHR* and *SBCO* (see Fig. 48), which are neither inscribed nor circumscribed, and thus do not establish themselves *a priori*.

Proposition 2 establishes that when the partition is refined by dividing each interval by two, the excess of the circumscribed cylinders over the inscribed cylinders is itself halved. This strategy plays the same role as Proposition 19 in Archimedes' *On Conoids and Spheroids*.

Al-Qūhī's method of using integral sums appears to be similar to Archimedes' method, but it is applied differently. Everything points to al-Qūhī's having himself rediscovered the method of integral sums.

1.3. *Later developments*

Starting from the work of the Banū Mūsā, and especially of Thābit ibn Qurra, Ibn Sinān developed a line of research that effectively combined geometric transformations and infinitesimal methods. And, as we have seen, al-Qūhī rediscovered the method of integral sums. These mathematicians thus left to their successors not only another perspective of research into infinitesimal geometry, but also other methods. The next generation of Islamic mathematicians will be quick to exploit and to take up the problems solved by their predecessors, and also to add solutions to new problems.

Ibn Sahl seems to have been the first to take up the challenge.[12] He once again tackled the problem of the quadrature of the parabola, but his treatise unfortunately remains undiscovered. However, given his status

[12] R. Rashed, *Géométrie et dioptrique: Ibn Sahl, al-Qūhī et Ibn al-Haytham*, Paris, 1993. English version: *Geometry and Dioptrics in Classical Islam*, London, 2005.

among mathematicians of his period, his knowledge of the works of al-Qūhī, and his many mathematical contributions, which we have succeeded in reconstructing, we believe that he must have proceeded by the use of integral sums. This conjecture, which seems entirely justifiable for the reasons just advanced, gains added weight from the fact that his successor Ibn al-Haytham applied himself to the problem of the measure of the sphere and the paraboloid, without reconsidering that of the parabola, as if that had already been established by the same methods.

To Ibn al-Haytham, known in the west as Alhazen (*d.* after 1040), fell the task of bringing the mathematical tradition of a century and a half to its culmination. With him, as we shall see, the calculation of curvilinear areas and the volumes of curved figures attained a level comparable to that which would be encountered at the beginning of the 17th century in an altogether different climate.

Ibn al-Haytham revisited the proof of the volume of the paraboloid of revolution, but he did not stop there. He also determined the volume of a parabaloid generated by a rotation about an axis perpendicular to the axis of the parabola. In addition, he used infinitesimal methods to determine the volume of the sphere. But before giving a necessarily brief description of his approach, let us note a fundamental trait in the work of this mathematician, physicist, and astronomer. Whereas such predecessors as al-Qūhī had used infinitesimal methods for determining surface areas and volumes of curved figures, and whereas they had also used such methods for finding the centers of gravity of these shapes, Ibn al-Haytham dealt the entire set of problems of this type: infinitesimal calculations of surface areas and volumes, centers of gravity, isoperimeters and equal surface areas, and the solid angle, and also entered into a new chapter loosely connected with these: the calculation of lunes. Even more, both in his studies here and elsewhere, he used true differential methods. Everything suggests that he had traveled through most of the regions that would later constitute the continent of analysis.

The structure of Ibn al-Haytham's treatise on *The Measurement of the Paraboloid* (*Maqāla fī misāḥat al-mujassam al-mukāfi'*) is simple but significant. In his introduction, he recalls the works of his predecessors Thābit ibn Qurra and al-Qūhī. He blames the former for having 'followed a course without any plan, forcing him to weave a path through his explanation that was both long and laboriously difficult'.[13] About al-Qūhī, he simply states that his treatise 'albeit less cluttered and easier to follow,

[13] See R. Rashed, *Les Mathématiques infinitésimales*, vol. II, p. 208; English translation p. 177.

includes a proof for only one of the species of paraboloid'.[14] Ibn al-Haytham follows this introduction with a section devoted only to arithmetic lemmas necessary for his proofs. The next section first treats the paraboloid of revolution before moving on to the second type of paraboloid. In a final part, he discusses the method that he had applied in that chapter.

Thus Ibn al-Haytham begins by proving, with the aid of a finite recurrence, a general formula for calculating the sum of n integers raised to any given power. This law can rewritten as

$$(n+1)\sum_{k=1}^{n} k^i = \sum_{k=1}^{n} k^{i+1} + \sum_{p=1}^{n}\left(\sum_{k=1}^{p} k^i\right), \qquad \text{for } i = 1, 2, \ldots$$

This result enables him to establish an inequality that he needs to apply throughout the work:

$$\frac{8}{15}n(n+1)^4 \leq \sum_{k=1}^{n}\left[(n+1)^2 - k^2\right]^2 \leq \frac{8}{15}(n+1)(n+1)^4 \leq \sum_{k=0}^{n}\left[(n+1)^2 - k^2\right]^2.$$

Ibn al-Haytham's proof of this lemma is very long, but serves to illustrate the high level of his achievement in this branch of arithmetic, as well as the his virtuosity as a mathematician.[15]

Ibn al-Haytham then goes on to determine the volume of the paraboloid of revolution. He considers three cases, depending on whether the angle *ACB* (see Fig. 50) is right-angled, acute, or obtuse. We will examine the first case, closely following his own argument.

Letting V be the volume of the circumscribed cylinder and v the volume of the paraboloid, we shall show that

$$v = \frac{1}{2}V.$$

First, suppose that this is not the case and that $v > \frac{1}{2}V$, so that

$$v - \frac{1}{2}V = \varepsilon.$$

[14] *Ibid.*

[15] See R. Rashed, *Les Mathématiques infinitésimales*, vol. II, pp. 182–5.

Let M be the midpoint of AC and draw $MU \parallel BC$, cutting the parabola at E and BH at U. Draw $SEO' \parallel AC$, cutting BC at O' and AH at S. Let $[EC]$ represent the solid generated by the rotation of $MCO'E$ about AC and let the other solids formed by rotation be similarly represented. Then we have

(1) $$[HE] + [EC] = \frac{1}{2}V \text{ and } [BE] + [AE] = \frac{1}{2}V.$$

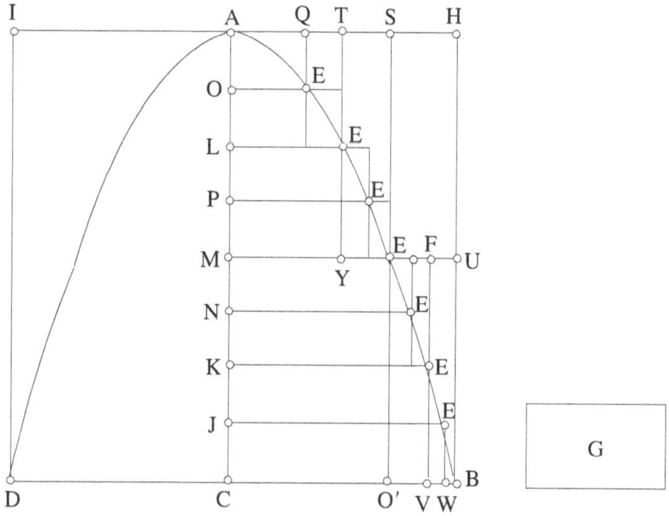

Fig. 50

The construction can now be repeated beginning with point L, the midpoint of AM, and then with K, the midpoint of MC. This gives

$$[SE_l] + [ME_l] = \frac{1}{2}\,[MS] = \frac{1}{2}\,[AE],$$

$$[UE_k] + [E_kO'] = \frac{1}{2}\,[UO'] = \frac{1}{2}\,[BE];$$

Therefore

(2) $$[SE_l] + [ME_l] = [UE_k] + [E_kO'] = \frac{1}{2}\,[AE] = \frac{1}{2}\,[BE] = \frac{1}{4}V.$$

Repeating the construction again, for the points O, P, N, J, the midpoints respectively of AL, LM, MK, KC. Then the sum of the volumes of the eight solids will be half of (2), that is, $\frac{1}{8}V$.

Proceeding in the same way, that is, by removing solids of the type (1) and (2) from the circumscribed cylinder, we shall remove from V, successively

$$\frac{1}{2}V, \frac{1}{2}\left(\frac{1}{2}V\right), \frac{1}{2}\left(\frac{1}{2}\left(\frac{1}{2}V\right)\right),$$

and so on. After a finite number of these operations, we shall necessarily arrive at a remainder smaller than ε, by virtue of Euclid's lemma X.1 (or Ibn al-Haytham's theorem).

Let us assume that we have carried out the partition to the point at which the remainder is less than ε.

Let V_n be the volume of the solids remaining after n steps, then $V_n < \varepsilon$ and let v_n be the volume of the portion of the solids inside the paraboloid, then $v_n < V_n$ and $v_n < \varepsilon$, and so $v - v_n > \frac{1}{2}V$, from our hypothesis. But from the properties of the parabola, we have

$$\frac{AC}{AM} = \frac{CB^2}{EM^2}$$

and so

$$BC^2 = 2\,EM^2.$$

Likewise,

$$\frac{BC^2}{AC} = \frac{JE_j^2}{AJ} = \frac{OE_0^2}{AO} = \frac{JE_j^2 + OE_0^2}{AC}$$

and so

$$JE_j^2 + OE_0^2 = BC^2 = 2EM^2.$$

One likewise proves that

$$KE_k^2 + LE_l^2 = BC^2 = 2EM^2$$

and so on.

If we now relabel our points E with a numerical subscript, so the points on the parabola are given by $E_0 = A, E_1, \ldots, E_n = B$, with $n = 2^m$, and we also label the corresponding points on the axis

$$F_0 = 1, \ldots, F_{\frac{n}{2}} = M, \ldots, F_n = C,$$

we have

$$\overline{E_iF_i}^2 + \overline{E_{n-i}F_{n-i}}^2 = \overline{BC}^2 = 2\overline{EM}^2 \qquad (0 \leq i \leq n)$$

and

$$\overline{E_1F_1}^2 + \ldots + \overline{E_{\frac{n}{2}-1}F_{\frac{n}{2}-1}}^2 + \overline{E_{\frac{n}{2}+1}F_{\frac{n}{2}+1}}^2 + \ldots + \overline{E_{n-1}F_{n-1}}^2 = \frac{1}{2}(n-1)\overline{E_nF_n}^2$$

and hence

$$\sum_{i=1}^{n-1} \overline{E_iF_i}^2 = \frac{1}{2}(n-1)\overline{E_nF_n}^2.$$

Now let $S_i = \pi\overline{E_iF_i}^2$ $(1 \leq i \leq n-1)$ be the areas of the disks of radius E_iF_i and S_n the area of the disk of radius $E_nF_n = BC$, then since these areas are proportional to the square of the radii, we have

$$\sum_{i=1}^{n-1} S_i = \frac{1}{2}(n-1)S_n.$$

Let W_i be the volume of the cylinder with base S_i and height $h = \frac{1}{n} AC$ and W_n the volume of the cylinder with base S_n and height h; then

$$\sum_{i=1}^{n-1} W_i = \frac{1}{2}(n-1)W_n;$$

but

$$\frac{1}{2}(n-1)W_n < \frac{1}{2}V$$

since $V = n\ W_n$; hence

$$\sum_{i=1}^{n-1} W_i < \frac{1}{2}V;$$

but

$$\sum_{i=1}^{n-1} W_i = v - v_n > \frac{1}{2}V,$$

which is impossible, and so

(3) $$v \leq \frac{1}{2}V.$$

Suppose now that $v < \frac{1}{2}V$, that is, $v + \varepsilon = \frac{1}{2}V$. We proceed as before, successively subtracting half the volume of the cylinder, then half the remainder, and continuing until the remaining volume V_n is less than any given ε. Let u_n be the part of V_n lying outside the paraboloid, then we have $u_n < V_n$, therefore $u_n < \varepsilon$, and so

$$v + u_n < \frac{1}{2} V;$$

but

$$v + u_n = \sum_{i=1}^{n} W_i ,$$

therefore

$$\sum_{i=1}^{n} W_i < \frac{1}{2} V .$$

But we have shown that

$$\sum_{i=1}^{n-1} W_i = \frac{1}{2}(n-1)W_n ;$$

but

$$\sum_{i=1}^{n-1} W_i = \sum_{i=1}^{n} W_i - W_n ,$$

therefore

$$\sum_{i=1}^{n-1} W_i - W_n = \frac{1}{2}(n-1)W_n ,$$

and so

$$\sum_{i=1}^{n} W_i - \frac{1}{2} W_n = \frac{n}{2} W_n = \frac{1}{2} V ;$$

hence

$$\sum_{i=1}^{n} W_i > \frac{1}{2} V ,$$

which is impossible, and so

(4) $$v \geq \frac{1}{2} V.$$

From (3) and (4), we therefore have, finally

$$v = \frac{1}{2} V,$$

for the case in which $A\hat{C}B = \frac{\pi}{2}$.

For the other two cases, where $A\hat{C}B$ is acute or obtuse, Ibn al-Haytham reduces them to the first case by means of an affine transformation, changing oblique axes to rectangular axes. More precisely still, he associates, point by point, the figures of each of these two cases to the figure of the first case and uses the fact that the relations within each figure are preserved under the transformation. What we have here, then, is the method of integral sums and an application of the method of exhaustion, with both placed on a solid arithmetical base.

Ibn al-Haytham's style becomes stunningly apparent when he deals with the second type of paraboloid, that obtained by a rotation about a line orthogonal to its axis. We present his method here using modern terminology, the better to understand his approach.

Let the paraboloid be generated by a rotation of the plane ABC of the parabola $x = ky^2$ about the ordinate BC (see Fig. 51). Let $AC = c$ and $BC = b$ and let $\sigma_n = (y_i)$, $0 \leq i \leq n = 2^m$, be a partition of $[0, b]$ with step $h = \frac{b}{2^m} = \frac{b}{n}$.

Let M_i be the points of the parabola with coordinates (x_i, y_i). Then, if $r_i = c - x_i$, we have

$$r_i = k\left(b^2 - y_i^2\right) = kh^2\left(n^2 - i^2\right).$$

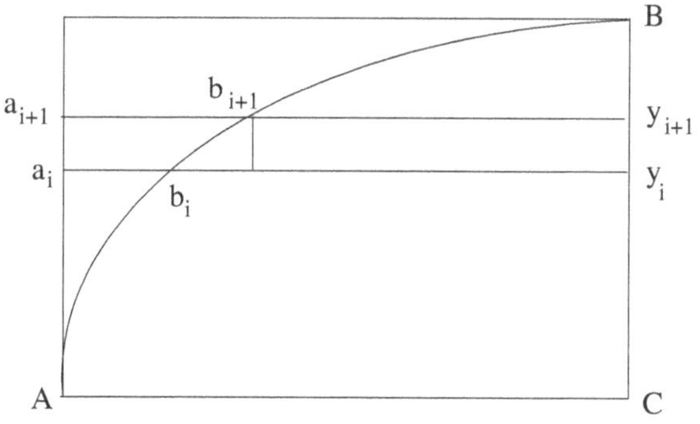

Fig. 51

Now, summing the inscribed and circumscribed cylinders between the ordinates, generated by rotation about AC, we obtain

$$I_n = \sum_{i=1}^{n-1} \pi k^2 h^5 \left(n^2 - i^2\right)^2 \text{ and } C_n = \sum_{i=0}^{n-1} \pi k^2 h^5 \left(n^2 - i^2\right)^2.$$

By the inequality (1) he had established in his arithmetic lemmas, Ibn al-Haytham therefore finds

$$I_n \le \frac{8}{15} V \le C_n$$

where $V = \pi k^2 b^4 \cdot b$ is the volume of the total circumscribing cylinder.

To express it in a modern form, let $g(y) = ky^2$ be continuous over $[0, b]$ and let $v(p)$ be the volume of the paraboloid, then Ibn al-Haytham's calculation amounts to:

$$v(p) = \lim_{n \to \infty} \sum_{i=1}^{n} \pi k^2 h^5 \left(n^2 - i^2\right)^2,$$

so

$$v(p) = \lim_{n \to \infty} \sum_{i=1}^{n} \pi k^2 \left(b^4 - 2b^2 y_i^2 + y_i^4\right) h,$$

so

$$v(p) = \pi \int_0^b k^2 \left(b^4 - 2b^2 y^2 + y^4\right) dy,$$

hence

$$v(p) = \frac{8}{15} \pi k^2 b^5 = \frac{8}{15} V,$$

where V is the volume of the circumscribed cylinder.

Ibn al-Haytham finally investigated the behaviour of the solids enclosing the surface of the paraboloid as the number of points of the partition is increased indefinitely. This presents the problem of the variation of the ratios between these infinitesimal solids, that is those parts that are exterior or interior to the paraboloid. For paraboloids of the first type, these parts will have equal volumes, but this is not the case for the second type of paraboloid.

Letting v_n, u_n be respectively the small internal and external volumes enclosing the paraboloid, he shows that

$$v_n = v - \sum_{k=1}^{n-1} W_i = W$$

and

$$u_n = V_n - W = u.$$

Let $u(m)$, $W(m)$ be the values of u and W corresponding to the mth partition, $n = 2^m$. Ibn al-Haytham shows that

$$\frac{u(m+1)}{W(m+1)} > \frac{u(m)}{W(m)}$$

and so the ratio increases as the points of the partition increase.[16]

After this treatise on paraboloids, Ibn al-Haytham considers the measurement of the sphere in a treatise by that name (*Qawl fī misāḥat al-kura*) in which he applies the same method as he had used for the paraboloids. As before, he begins by establishing the arithmetic lemmas he will need. He recalls that

$$\sum_{k=1}^{n} k^2 = \left(\frac{n}{3}+\frac{1}{3}\right)n\left(n+\frac{1}{2}\right)$$

and so establishes the inequality

$$\frac{n^3}{3}+\frac{n^2}{2} < \sum_{k=1}^{n} k^2 \leq \frac{n^3}{3}+\frac{2}{3}n^2 .$$

Using the same method as before, he shows that the volume of the sphere is equal to two-thirds the circumscribed cylinder. Let us examine this method, using the integral calculus in order to bring out the ideas on which it is based.

In order to find the volumes of revolution about any given axis, Ibn al-Haytham takes slices of inscribed and circumscribed cylinders, whose axes are the same as the axis of revolution. This allows him to find over- and under-approximations of the volume to be determined, by using integral sums – Darboux sums – of the function corresponding to the curve generating the solid of revolution. For the volume of the sphere, for example, he considered

$$I_n = \sum_{i=1}^{n-1} \pi y_i^2 (x_i - x_{i-1}) = D(f, \sigma_n, m_i)$$

$$C_n = \sum_{i=1}^{n} \pi y_{i-1}^2 (x_i - x_{i-1}) = D(f, \sigma_n, M_i).$$

[16] See R. Rashed, *Les Mathématiques infinitésimales*, vol. II, p. 196–200.

We note that the function f is monotonic, so that m_i and M_i are the values of f at the extremities of the ith interval of the partition; f being the function defined by

$$f(x) = \pi (R^2 - x^2) = \pi y^2,\ m_i = \inf_{x_{i-1} \le x \le x_i} f(x) = \pi y_i^2, \text{ and } M_i = \sup_{x_{i-1} \le x \le x_i} f(x) = \pi y_{i-1}^2,$$

m_i and M_i being the points with ordinates y_i and y_{i-1}.

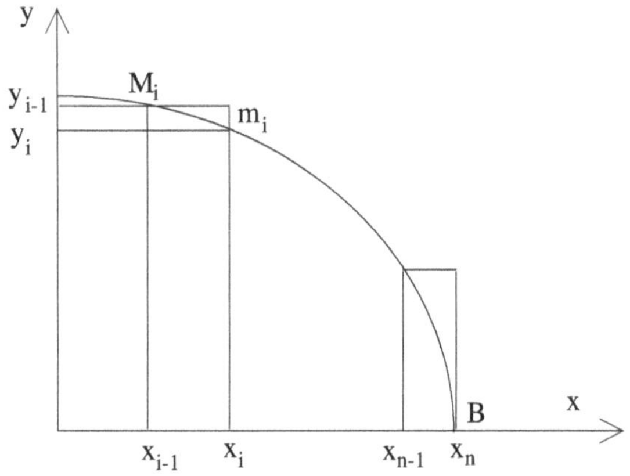

Fig. 52

Moreover, Ibn al-Haytham then uses the inequalities

$$I_n < v < C_n$$

and shows that, for all $\varepsilon > 0$, there exists an N such that, for $n \geq N$, we have

$$v - I_n < \varepsilon, \quad C_n - v < \varepsilon,$$

which proves that I_n tends to v and similarly for C_n. Thus we certainly have

$$v = \int_0^R f(x)\, dx.$$

In other words, Ibn al-Haytham's calculation is equivalent to a simple Cauchy-Riemann integral.

But this mathematical equivalence should not obscure the following question: having once determined these volumes by this integral method, why did Ibn al-Haytham never explicitly set out the general method for determining other volumes and surface areas? Certainly we cannot offer a satisfactory answer based on an examination of Ibn al-Haytham's mathematical requirements – it is not as if he needed just these results, and having obtained them saw no need to go further; in fact, in his works of mathematics, optics, and astronomy, he never had any need to calculate the volume of the paraboloid, nor the volume of a hyperboloid of revolution, for example. It must therefore be the method itself that accounts for an absence of generalization.

We may note that Ibn al-Haytham – like his predecessors who determined surface areas – always had recourse to another solid of known volume to which he could compare the volume of the solid to be determined. This prior knowledge was in no way merely an artefact of the method: it allowed Ibn al-Haytham, like his predecessors, to carry out a direct and exact calculation of the limits of the corresponding upper and lower integral sums (Darboux). Now, in the most general case, these solids of comparison do not necessarily exist, which makes the method adopted by Ibn al-Haytham unsuitable for carrying out effectively the summations of upper and lower integral sums. Ibn al-Haytham's method is thus characterized by an internal limitation. We must, however, be careful not to exaggerate the significance of this limitation, which will recede with the much more massive introduction of arithmetic calculation. If the use of a reference volume for determining unknown volumes is hallmark of the Archimedean tradition, the increasingly arithmetical cast of the investigations in the Arabic tradition shows that we are moving beyond the Archimedean heritage. It was not geometry alone that guided Ibn al-Haytham's approach: he was already using arithmetic, and his lemmas had been conceived by thinking arithmetically about geometrical figures.

In this investigation, we can already see the development of methods and techniques from this domain in Arabic mathematics. As we have seen, in his investigation of the paraboloid, Ibn al-Haytham achieved results that historians would later attribute to Kepler and Cavalieri, for example. However, progress in this domain stopped at this point, probably for lack of an effective symbolism.

2. THE QUADRATURE OF LUNES

Among the problems concerned with finding the areas of curved figures, the exact quadrature of lunes – surfaces bounded by two circular arcs – is one of the most ancient. According to the late evidence supplied by Simplicius, the 6th-century commentator on Aristotle, this problem can be traced back to Hippocrates of Chios in the 5th century BC. In his commentary on Aristotle's *Physics*, Simplicius provides a long quotation from Eudemus's *History of Geometry* that gives an account of Hippocrates' quadratures of certain lunes.[17] This passage, which raises several other philological and historical problems that we cannot address here, is the only known source for the history of the problem in Greek mathematics. It also indicates that the context in which the problem of the quadrature of certain lunes was posed was the problem of the quadrature of the circle.

Approximately five centuries after Simplicius, Ibn al-Haytham returned to this same problem on several occasions, first in connection with the quadrature of the circle, and then for its own sake. In fact, he wrote three memoirs, only one of which has so far been studied: his memoir on the quadrature of the circle, another succinct study of the quadrature of lunes, and later, another treatise in which he obtained results that were later attributed to mathematicians of the 17th and 18th centuries. It is for lack of knowledge about Ibn al-Haytham's works, particularly this third treatise, that historians have in good faith made erroneous judgments about his contribution to this topic.

Everything suggests Ibn al-Haytham's point of departure was the text attributed to Hippocrates of Chios. In his first memoir, *A Treatise on Lunes* (*Qawl fī al-hilāliyyāt*), he begins by writing: 'Upon my examining [...] the shape of the lune, equal to a triangle as mentioned by the Ancients [...]'.[18] Later, in a second memoir, *An Exhaustive Treatise on the Figures of Lunes*, Ibn al-Haytham refers to his first text: 'I wrote a brief treatise on the figures of lunes according to specific methods'.[19] Furthermore, Ibn al-Haytham's works incorporate Hippocrates of Chios's results. Did he, perhaps, know the latter from an Arabic translation of Simplicius's commentary on Aristotle's *Physics*? Unfortunately no known documents allow us to give a definitive answer. Certainly Ibn al-Haytham refers to the 'ancients' in his first treatise, but strictly speaking he did not reproduce any of Hippocrates' figures. His first result does, however, rest on a slight generalization of one of Hippocrates' propositions, which Simplicius cites from a text by Alexander of Aphrodisias, thereby rather complicating the problem. This is Proposition 3 of his first treatise, which also appears in his memoir *The Quadrature of the Circle* (*Qawl fī tarbīʿ al-dāʾira*), and again as Proposi-

[17] T. Heath, *A History of Greek Mathematics*, 2 vols, Oxford, 1921, vol. 1, pp. 191–201 and O. Becker, *Grundlagen der Mathematik*, 2nd ed., Munich, 1964, pp. 29–34.

[18] See R. Rashed, *Les Mathématiques infinitésimales*, vol. II, pp. 69–81; commentary pp. 32–4; English transl. p. 93.

[19] See R. Rashed, *Les Mathématiques infinitésimales*, vol. II, p. 103; English transl. p. 107.

tion 8 of his second treatise. Whatever may be the genesis of his ideas, let us turn to Ibn al-Haytham's two memoirs on lunes.

In both his treatises, Ibn al-Haytham's approach to studying lunes bounded by any arcs whatever, amounts to finding equivalent areas. He usually introduces circles generally equivalent to sectors of the circle given in the problem and expressed as fractions of that circle. He justifies the existence of the circles he introduces, which are to be added to, or subtracted from, polygonal areas, so as to obtain an area equivalent to that of the lune, or the sum of two lunes.

In his first short treatise, in Propositions 1, 2 and 5, he uses a semicircle *ABC* to determine lunes L_1 and L_2 bounded by arcs *AB* and *BC*, and by a semicircle (see Fig. 53). He takes the arc *AB* as being equal to a sixth of the circumference, and he establishes the results:

$$L_1 + \frac{1}{24}\text{circle}(ABC) = \frac{1}{2}\text{tr}(ABC)$$

$$L_2 = \frac{1}{2}\text{tr}(ABC) + \frac{1}{24}\text{circle}(ABC)$$

$$L_2 + \frac{1}{2}\text{tr}(ABC) = L_3 + \frac{1}{8}\text{circle}(ABC),$$

where L_3 is a lune similar to L_1, such that $L_3 = 2L_1$.

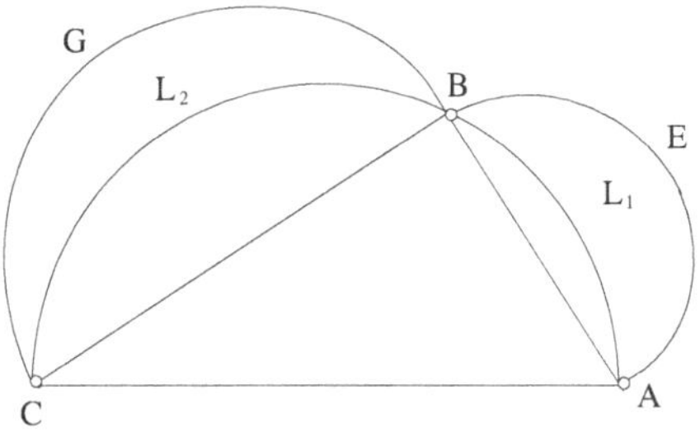

Fig. 53

In Proposition 3 of this treatise, Ibn al-Haytham generalizes the proof of Hippocrates' result by letting point *B* lie anywhere on the semicircle *ABC*:

$$L_1 + L_2 = \text{tr}(ABC),$$

and in Proposition 4, he considers the ratio of two similar lunes.

Note that in these propositions the lunes L_1 and L_2 are associated with three semicircles *ABC*, *AEB*, and *BGC*.

Ibn al-Haytham's first treatise thus reads as following Hippocrates of Chios's line of research. This is also true for the part of his *The Quadrature of the Circle* that concerns lunes.[20] In his proof, Ibn al-Haytham, like Hippocrates, uses the fact that the area of a circle is proportional to the square of its diameter, and also uses Pythagoras's theorem. Both cases consider the lune associated with the right-angled isosceles triangle. Even if Ibn al-Haytham's reasoning is slightly more general than that of Hippocrates, this does not significantly alter the similarity between the two approaches to the problem. As a point of interest, note that the important issue in Ibn al-Haytham's memoir on the quadrature of the circle is not the results he obtained for lunes (unlike in the first memoir) but the clear difference he establishes between the existence of a square equivalent to a circle – for us, the existence of a transcendental ratio – and the constructibility of that square or ratio.[21]

The situation is profoundly different in his second treatise devoted to lunes, *An Exhaustive Treatise on the Figures of Lunes* (*Maqāla mustaqṣāt fī al-askhāl al-hilāliyya*).[22] Not only does Ibn al-Haytham obtain more general results here, but his approach is different: he takes the problem of the quadrature of lunes back to fundamentals, recasts it in terms of trigonometry, and attempts to deduce different cases in terms of the properties of a trigonometric function, just as Euler would do more precisely much later.

From the very beginning of his treatise, Ibn al-Haytham explicitly recognizes that the calculation of the areas of lunes involves the sums and differences of areas of sectors of circles and of triangles, for which he needs to compare the ratios of angles and the ratios of line segments. This is why he begins with four lemmas about a triangle *ABC*, right-angled at *B* in the first lemma and obtuse in the other three, which shows that the essential point henceforth comes down to a study of the function

[20] Cf. H. Suter, 'Die Kreisquadratur des Ibn el-Haiṯam', *Zeitschrift für Mathematik und Physik, Historisch-litterarische Abteilung*, 44, 1899, pp. 33–47.

[21] R. Rashed, 'L'analyse et la synthèse selon Ibn al-Haytham', in *Mathématiques et philosophie de l'antiquité à l'âge classique*, Paris, 1991, pp. 131–62; reprod. in *id.*, *Optique et mathématiques. Recherches sur l'histoire de la pensée scientifique en arabe*, Aldershot, 1992, Variorum Reprints, XIV.

[22] See R. Rashed, *Les Mathématiques infinitésimales*, vol. II, pp. 102–75; commentary pp. 37–68.

(1) $$f(x) = \frac{\sin^2 x}{x} \qquad 0 < x \leq \pi.$$

These lemmas can be expressed as:

1° If $0 < C < \frac{\pi}{4} < A < \frac{\pi}{2}$, then $\frac{\sin^2 C}{C} < \frac{2}{\pi} < \frac{\sin^2 A}{A}$; it is evident that if $C = A = \frac{\pi}{4}$, then $\frac{\sin^2 C}{C} = \frac{\sin^2 A}{A} = \frac{2}{\pi}$.

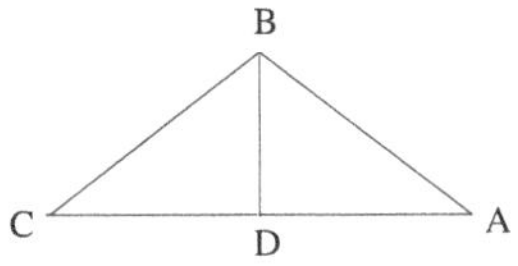

Fig. 54

2° Let $\pi - B = B_1$, if $C < \frac{\pi}{4} < B_1 < \frac{\pi}{2}$, then $\frac{\sin^2 C}{C} < \frac{\sin^2 B_1}{B_1}$.

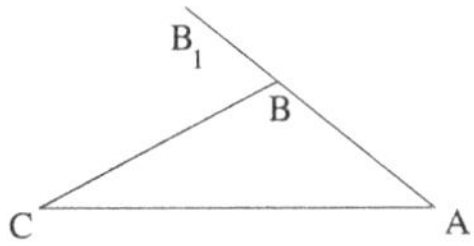

Fig. 55

3° Again, with $\pi - B = B_1$: if $A \leq \frac{\pi}{4}$, then $\frac{\sin^2 A}{A} < \frac{\sin^2 B_1}{B_1}$.

4° Here, Ibn al-Haytham wanted to consider the case $A > \frac{\pi}{4}$; but the study is incomplete. He shows that for a given A, a B_0 can be found, such that

$$B_1 \geq B_0 \Rightarrow \frac{\sin^2 A}{A} > \frac{\sin^2 B_1}{B_1}.$$

This incomplete study seems to have kept Ibn al-Haytham from recognizing the equality

$$\frac{\sin^2 A}{A} = \frac{\sin^2 B_1}{B_1}.$$

Ibn al-Haytham proves these lemmas by comparing arc lengths, areas, and sides of triangles.

Note that these lemmas, because they link the question of the quadrature of lunes to trigonometry, change the perspective of the problem and allow particular cases to be considered together. But incompleteness mentioned above masks the possibility of determining those lunes that are susceptible to quadrature. Let us now consider briefly the main propositions of Ibn al-Haytham's second treatise.

In nine propositions – numbers 8 to 16 – the lemmas are associated pairwise and in all cases, the three arcs *ABC*, *AEB* and *BCG* are similar. Let O, O_1, and O_2 be the centers of the corresponding circles (see Fig. 56) and let $A\hat{O}C = A\hat{O}_1B = B\hat{O}_2C = 2\alpha$, $A\hat{O}B = 2\beta$, and $B\hat{O}C = 2\beta'$; with $\beta \leq \beta'$ and $\beta + \beta' = \alpha$.

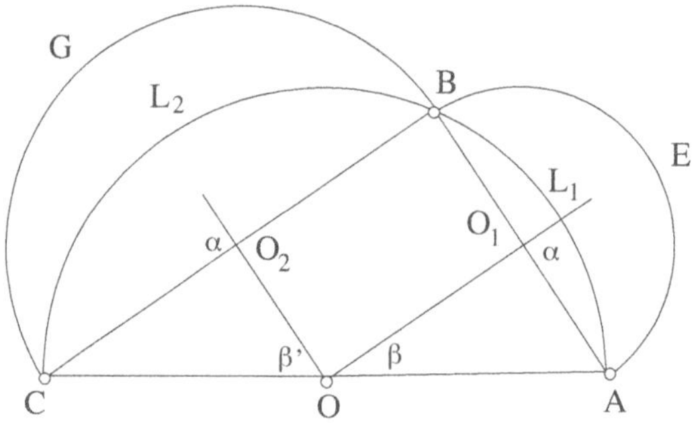

Fig. 56

The lune L_1 is characterized by (α, β) and the lune L_2 by (α, β'). Consider the case in which $\alpha = \frac{\pi}{2}$; we have the following results:

1. Given any (β, β') such that $\beta + \beta' = \frac{\pi}{2}$, then $L_1 + L_2 = \text{tr}\,(ABC)$.

2. If $\beta = \beta' = \frac{\pi}{4}$, then $L_1 = L_2 = \frac{1}{2}\,\text{tr}\,(ABC)$; in this case $\frac{\alpha}{\beta} = \frac{2}{1}$, and this is the only lune susceptible of quadrature studied by Ibn al-Haytham.

For $\beta < \beta'$ we have

$$L_1 = \frac{1}{2}\,\text{tr}\,(ABC) - \text{circle}\,(N),$$

$$L_2 = \frac{1}{2}\,\text{tr}\,(ABC) + \text{circle}\,(N);$$

the circle N depending on the ratio of $\frac{\alpha}{\beta}$.

3. If $\beta = \frac{\pi}{6}$, then $L_1 = \frac{1}{2}$ tr $(ABC) - \frac{1}{24}$ circle (ABC); in this case $\frac{\alpha}{\beta} = \frac{3}{1}$.

If $\beta' = \frac{\pi}{3}$, then $L_2 = \frac{1}{2}$ tr $(ABC) + \frac{1}{24}$ circle (ABC); in this case $\frac{\alpha}{\beta'} = \frac{3}{2}$.

Up to this point, Ibn al-Haytham has only used his Lemma 1 in his proofs; for the next proposition, he needed three additional lemmas. The key idea is to take points M, N on the chord AC such that $A\hat{B}C = B\hat{M}C = A\hat{N}B = \pi - \alpha$, and to define a point P on AB and a point Q on BC such that $NP \,||\, OA$ and $MQ \,||\, OC$ (see Fig. 57). The results cannot be established from the triangle ABC, as in the previous propositions.

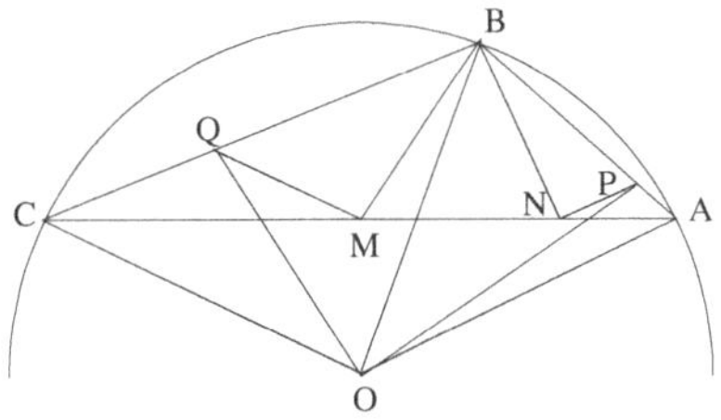

Fig. 57

For any pair (β, β'), such that $\beta + \beta' < \frac{\pi}{2}$, Ibn al-Haytham defines two circles K, Z, such that the areas

$$L_1 + L_2 + (K) = \text{quadraliteral } (OPBQ)$$

$$L_1 + (Z) = \text{tr } (OPB);$$

and he examines the following cases:

If $\beta = \beta'$, then $(Z) = \frac{1}{2}\ (K)$, $L_1 = L_2$, $L_2 + (Z) = \text{tr } (OQB) = \text{tr } (OPB)$.

If $\beta' < \frac{\pi}{4}$, then $(Z) < (K)$, $L_2 + (K) - (Z) = \text{tr } (OQB)$.

If $\beta' > \frac{\pi}{4}$, there are three cases

$(Z) < (K)$, then $L_2 + (K) - (Z) = \text{tr}\,(OQB)$, and $L_2 < \text{tr}\,(OQB)$,

$(Z) = (K)$, then $L_2 = \text{tr}\,(OQB)$,

$(Z) > (K)$, then $L_2 = \text{tr}\,(OQB) + (Z) - (K)$, and $L_2 > \text{tr}\,(OQB)$.

Ibn al-Haytham then gave some examples to illustrate these results, after which he proved these propositions:

4. If $\alpha = \frac{\pi}{3}$, $\beta = \beta' = \frac{\pi}{6}$, then $\frac{\alpha}{\beta} = \frac{2}{1}$ and we have

$$L_1 = L_2 = \frac{2}{3}\ \text{tr}\,(ABC) - \frac{1}{18}\ \text{circle}\,(ABC).$$

5. If $\alpha = \frac{\pi}{3}$, $\beta = \frac{\pi}{12}$, $\beta' = \frac{\pi}{4}$, then $\frac{\alpha}{\beta} = \frac{4}{1}$, $\frac{\alpha}{\beta'} = \frac{4}{3}$; in this case, the required circle is not a fraction of circle (ABC).

6. If $\alpha = \frac{\pi}{3} + \frac{\pi}{8}$, $\beta = \frac{\pi}{8}$, $\beta' = \frac{\pi}{3}$, then $\frac{\alpha}{\beta} = \frac{11}{3}$, $\frac{\alpha}{\beta'} = \frac{11}{8}$; in this case, the required circle is not a fraction of circle (ABC).

Excepting Proposition 21, in the propositions that follow, Ibn al-Haytham considers figures consisting of the sums and differences of lunes, segments of circles, and triangles. In Proposition 21, he indicates one property of a lune whose two arcs belong to two equal circles. This property is derived from the translation that maps one circle onto the other, a property that was studied by Ibn al-Haytham in his treatise *Analysis and Synthesis*.[23]

With this second treatise of Ibn al-Haytham's, the study of the quadrature of lunes takes a new path, which will later lead to Euler: it displaces the problem in the direction of trigonometry, and in a certain sense recognizes its dependence on the function (1) above, $f(x) = \frac{\sin^2 x}{x}$, where $0 < x \leq \pi$.

[23] R. Rashed, 'L'analyse et la synthèse selon Ibn al-Haytham' and *Les Mathématiques infinitésimales du* IXe *au* XIe *siècle*, vol. IV: *Méthodes géométriques, transformations ponctuelles et philosophie des mathématiques*, London, al-Furqān, 2002, Chap. II.

3. EQUAL PERIMETERS AND EQUAL SURFACE AREAS: A PROBLEM OF *EXTREMA*

Infinitesimal geometry, as noted earlier, arose from areas other than just the determination of surface areas and volumes of curved figures. Isoperimeters and equal areas was one such study. Whereas investigations into minimum perimeters and minimum surface areas were closely linked to the science of astronomy, the subject began to attract the attention of mathematicians to such an extent that it become itself a chapter in the history of infinitesimal geometry.

Of all plane shapes of a given perimeter, the circle has the largest area; and of all solids with a given surface area, the one with the greatest volume is the sphere. This is the statement of a proposition about boundary conditions that has been of interest to mathematicians and astronomers since Antiquity. Astronomers needed to establish the sphericity of the heavens, and mathematicians attacked the problem in order to provide the proof that the astronomers lacked.

If one can believe the 5th-century commentator Simplicius, the question of isoperimetry is of considerable antiquity. Simplicius wrote:

> It has been shown, not only before Aristotle, where it appears as a proven [proposition], but also by Archimedes, and in a more detailed (πλατύτερον) way by Zenodorus, that, of the isoperimetric figures, that which is the greatest among plane figures is the circle, and, among solids, it is the sphere.[24]

This important, if late, text shows, as Schmidt noted,[25] that these fundamental propositions were known before Zenodorus. This idea encouraged Mogenet to attribute to Zenodorus only an outline of the theory of isoperimeters in 'a general way' and to take that as an argument for placing Zenodorus in the third century BC.[26] This position remains controversial. All that can be said for the present is that Zenodorus lived after Archimedes and before Pappus, that is, between the second century BC and the third century of our era.

This interval of half a millenium was bound to generate controversy in the absence of any of the supplemental evidence for which historians still hope. But the controversy that has raged since the beginning of the 20th century has not prevented historians from agreeing that Zenodorus had

[24] Simplicius VII, 4/2, lines 12–17.

[25] W. Schmidt, 'Zur Geschichte der Isoperimetrie', *Bibliotheca Mathematica*, 2, 1901, pp. 5–8.

[26] J. Mogenet, 'Les isopérimètres chez les Grecs', *Scrinium Lovaniense*, Mélanges historiques, Louvain, 4e série, 24, 1961, pp. 69–78.

indeed himself treated the mathematical problem and had found a proof. Fortunately for us, Theon of Alexandria summarizes Zenodorus' work in his *Commentary on the First Book of the Almagest* where, after stating the isoperimetric problem, he says that he will 'give the proof of these propositions in a summary taken from the proofs by Zenodorus in his book *On Isoperimetric Figures*'.[27] Theon goes back to Zenodorus to quote the famous formula given by Ptolemy in this first book:

> Since, among figures having an equal perimeter, those with the greatest number of sides are the greatest, among plane figures, the circle that is the greatest, and among solids, the sphere, and the sky is greatest of all bodies.[28]

Commentators on the *Almagest* since Theon did not let this formula stand without feeling the need for further comment, since it incorporates a fundamental doctrine of astronomy, cosmology and philosophy: the nature of the sphericity of the heavens and of the world. As we have remarked, there was a need for a convincing proof. It was also natural that the problem should evoke the interest of other mathematicians, such as Hero of Alexandria and also Pappus, who addressed the problem in Book V of his *Collection*.[29] Both the *Commentary* of Theon and the *Almagest* itself were known to astronomers and mathematicians in the 9th century, who instigated a new tradition of study of isoperimeters and equal areas. The first to tackle the question was al-Kindī. He treated the problem in his 'book on spheres'. In his *The Grand Art* (*Fī ṣināʿat al-uẓmā*), he wrote:

> In the same way that the largest of figures inscribed in a circle is that which has the most angles, and the greatest of solid figures having equal plane surfaces is the sphere, as we have explained in our book *On Spheres*, the sky is therefore the greatest of all other bodies, and it is spherical since it must have the greatest figure.[30]

The 13th-century bibliographer Ibn Abī Uṣaybiʿa cites, among al-Kindī's books, a title dedicated explicitly to this problem: *The Sphere is the Greatest of the Solid Figures*. However that may be, his book *The Grand Art* itself testifies to the strong influence of Theon of Alexandria on al-Kindī.

[27] A. Rome, *Commentaires de Pappus et Théon d'Alexandrie sur l'Almageste*, II, Roma, Cité du Vatican, 1936, p. 33.

[28] *Claudii Ptolemaei opera quae exstant omnia. I. Syntaxis mathematica*, ed. J. L. Heiberg, Leipzig, 1898, p. 13, lines 16–19.

[29] Pappus d'Alexandrie, *La Collection mathématique*, French transl. by Paul Ver Eecke, Paris and Bruges 1933, t. I, pp. 239 *sq*.

[30] Ms. Istanbul, Aya Sofya 4860, fol. 59v.

Many were to contribute to the study of isoperimeters and equal surface areas: we can name, for example, the astronomer Jābir ibn Aflaḥ and the mathematician Ibn Hūd. But the two who most transformed investigation into the problem, each in his own different way, were al-Khāzin and Ibn al-Haytham. These are the two principal figures presently known to us. A reading and analysis of their writings reveals the gulf between them. Whereas al-Khāzin developed his work from the scholarship of the past, Ibn al-Haytham, half a century later, completed this work and touched the shores of the future.

3.1. *Al-Khāzin: the mathematics of the Almagest*

Abū Jaʿfar al-Khāzin (mid-10th century) had himself written a *Commentary on the First Book of the Almagest* (*Sharḥ al-maqāla al-ūlā min al-Majisṭī*). Doubtless it was on account of Ptolemy's famous assertion on isoperimetry that he included in this *Commentary* a kind of treatise aimed at providing a rigorous proof.[31] Al-Khāzin is clear about this; he intends to justify Ptolemy's claim, not by calculation (*ḥisāb*), but by means of geometry. An examination of his text shows that, whether or not he knew Zenodorus's results as given in Theon's summary, his approach to the proof took a different direction, as we shall see. Al-Khāzin's key idea, of which he was perfectly aware, was that of all convex figures of a given type (triangle, rhombus, parallelogram, ...), the most symmetric instantiates an *extremum* for a certain magnitude (area, ratio of areas, perimeter, ...). The procedure is as follows: having set a fixed parameter, the figure is changed so as to make it symmetrical with respect to a certain line segment. Thus, upon fixing the perimeter of a parallelogram, the figure is transformed into a rhombus by making it symmetrical about a diagonal; in doing so, the area is increased. Not only does this idea distinguish the contribution of al-Khāzin from that of his predecessors, but also we cannot understand his contribution without being aware of this idea.

Al-Khāzin's treatise follows a simple structure. The first part deals with isoperimeters, the second with equal areas. These two parts rely on unarticulated notions and unstated axioms. One of the former is the notion of convexity. In fact, all the polygons and polyhedra that al-Khāzin considers in his treatise are convex. Among others, important axioms of convexity, which he does not declare are:

A_1 If a convex polygon is inscribed in a circle, then its perimeter is less than that of the circle.

[31] See R. Rashed, *Les Mathématiques infinitésimales*, vol. I, chap. IV.

A_2 If a convex polygon circumscribes a circle, then its perimeter is greater than that of the circle.

A_3 If a convex polyhedron is inscribed in a sphere, then its area is less than the surface of the sphere.

A_4 If a convex polyhedron circumscribes a sphere, then its area is greater than that of the sphere.

These axioms are needed in order to establish an important lemma (the eighth) and a proposition (19). Let us now look, briefly, at how al-Khāzin's treatise is presented.

3.1.1. *Isoperimeters*

Al-Khāzin first proves eight lemmas before his theorem on isoperimeters. The first four deal with isosceles and equilateral triangles, and he shows that the area of an equilateral triangle is greater than that of any isosceles triangle of the same perimeter. The fifth lemma shows that the area of an equilateral triangle is greater than that of any other triangle of the same perimeter. In the course of his proof, al-Khāzin proves a result already established by Zenodorus and Pappus, namely, that 'among isoperimetric figures having the same number of sides, the greatest is that which is equilateral and equiangular'. The sixth lemma compares a parallelogram to a square with the same perimeter, and in the seventh lemma al-Khāzin takes the example of a regular pentagon, changes it into an irregular pentagon of the same perimeter, and shows that the second has a smaller area than that of the first.

Before continuing, let us compare this approach with that of Zenodorus. The Greek mathematician begins by comparing any triangle to an isosceles triangle with the same base and the same perimeter, in order to arrive at the lemma: 'The sum of two similar isosceles triangles, not having equal bases, is greater than the sum of two non-similar isosceles triangles which are isoperimetric to the two similar triangles.' By 'isoperimetric' here, we are to understand that the sums of the sides are equal, excluding the bases. Now, this lemma of Zenodorus is incorrect, and it is quite surprising that neither Pappus nor Theon remarked on the error.[32] In fact, to use modern symbolism, this lemma comes down to finding the maximum of $ax + by$, given that $\sqrt{a^2+x^2}+\sqrt{b^2+y^2}=1$. A maximum is found when

[32] J. L. Coolidge, *History of Geometrical Methods*, Oxford, 1940; reprod. Dover, 1963, p. 49.

$ax' + by' = 0$, from which $x' = b$ and $y' = -a$; hence, from differentiating the second, we have:

$$\frac{bx}{\sqrt{a^2+x^2}} = \frac{ay}{\sqrt{b^2+y^2}};$$

now, if we put $x = au$ and $y = bv$, this reduces to:

$$\frac{u}{a\sqrt{1+u^2}} = \frac{v}{b\sqrt{1+v^2}},$$

whereas the assertion of the lemma requires that $u = v$.

Might this error have been the reason for al-Khāzin's method of attack on the problem?

In Lemma 8, al-Khāzin went on to consider convex polygons with inscribed and circumscribed circles.

Everything is now in place for establishing the isoperimetric properties of regular polygons before finally considering the circle. First of all, he proves:

> If two regular polygons P_1 and P_2 have, respectively, n_1 and n_2 sides, with $n_1 > n_2$, and have the same perimeter, then the area of P_1 is greater than that of P_2.

He then proves the theorem:

> *Of all plane figures, regular convex polygons, and the circle that have the same perimeter, it is the circle that has the greatest area.*

Al-Khāzin's method for dealing with isoperimeters is therefore to proceed by: a) comparing regular polygons that have different numbers of sides but the same perimeter; b) comparing a regular polygon to a circle of the same perimeter with the aid of a similar polygon circumscribing the circle. This approach is common to both al-Khāzin and Zenodorus. It is an approach that we may characterize as static, in the sense that we have, on the one hand, the given polygon and, on the other, the circle. We shall see that Ibn al-Haytham later uses a) to establish b) by considering the circle as the limit of a sequence of regular polygons, which is a dynamic approach to the problem. To sum up, even if the method chosen by al-Khāzin differs from that of Zenodorus or Pappus, it belongs essentially to the same genre of approach, whereas that taken by al-Haytham is of an altogether different species.

3.1.2. *Equal areas*

The second part of al-Kāzin's treatise deals with the same problem of determining an *extremum*, but this time for the area of solids. This part also begins by establishing a number of lemmas, nine in all, which deal with the surface area and volume of the pyramid, and similarly of the cone and the frustum of a cone. The first lemma deals with the area of a regular pyramid and the second concerns the volume of a regular pyramid containing an inscribed sphere. In the third lemma, al-Khāzin considers the surface area and volume of a cone of revolution. In the fourth lemma, he takes up the following problem: given a circle C, to construct two similar polygons of area S_1 and S_2, respectively, the one circumscribing C and the other inscribed in C, such that the ratio $S_1/S_2 = k$ is equal to some given value k.

In the fifth lemma, al-Khāzin gives an expression for the surface area of a cone and proceeds, in Lemma 6, to do the same for the frustum of a cone. Lemma 7 is deduced from Lemma 6 and reads:

> *If a regular rectilinear polygonal is inscribed in a circle of area* S_1 *and circumscribes a circle of area* S_2, *then the area* S *generated by rotating the polygonal line about one of its axes satisfies* $4S_2 < S < 4S_1$.

Recall that Archimedes had obtained the same results for a solid obtained from a regular polygon inscribed in a sphere, where the number of sides of the polygon is a multiple of 4 (Propositions 27–30 of *On the Sphere and Cylinder*). The Banū Mūsā later treated the same problem for a solid defined by a polygonal line inscribed in a semicircle, where the number of sides is even (Propositions 12 and 13 of their *Book on the Measurement of Plane and Spherical Figures*). This is precisely the case treated by al-Khāzin. Later, Johannes de Tinemue in his Proposition 5 would also treat the same problem,[33] starting from a regular polygon inscribed in a circle, the number of sides being a multiple of 4 or even simply an even number.

Al-Khāzin concludes by establishing two fundamental propositions. In the first, he showed that the surface area S of a sphere is equal to four times the area of its great circle. Archimedes had given the same result in I.33 of *On the Sphere and Cylinder* and the Banū Mūsā did so as Proposition 14 of their *Book on the Measurement of Plane and Spherical Figures*.

[33] M. Clagett, *Archimedes in the Middle Ages*, vol. I: *The Arabo-Latin Tradition*, Madison, 1964, pp. 469 *sqq*.

In the second proposition, al-Khāzin revisits the proof that the volume of a sphere is $V = \frac{1}{3} R \cdot S$, where R is the radius and S the surface area.

To establish this, he applies the method of contradiction and uses Euclid XII.17;[34] this is equivalent to the method formerly used by the Banū Mūsā. Finally, he proves the theorem:

Of all convex solids having the same surface area, the sphere is the one with the greatest volume.

The proof runs as follows: Let there be a sphere, with center O, radius R, surface area S and volume V, and let there be a polyhedron with the same area S and volume V_1 which one supposes circumscribing another sphere of radius R' and surface area S'. Then we have

$$V_1 = \frac{1}{3} S \cdot R'.$$

The surface area S' is less than that of the polyhedron, therefore $S > S'$, and so it follows that

$$R' < R \text{ and } \frac{1}{3} S \cdot R' < \frac{1}{3} S \cdot R,$$

that is, $V_1 < V$.

Note that the nature of the polyhedron is not specified, but it must circumscribe the sphere, which is the case for a regular polyhedron. But the proof given here cannot be applied to any arbitrary polyhedron or solid.

As we have just seen, al-Khāzin's approach was not the same for three-dimensional space as it is for the plane. For solid figures, he compared not polyhedra having the same area but polyhedra having a different numbers of faces, but he got his result by using a formula relating the volume of the sphere to its surface area, a formula that was obtained by approaching closer to the sphere with non-regular polyhedra. Ibn al-Haytham's method of attacking the problem was, as we shall see, quite different.

3.2. *Ibn al-Haytham: a new theory*

Half a century after al-Khāzin, Ibn al-Haytham addressed the same problem of isoperimeters and equal areas, but he recast it to fit his own domain of research, that of asymptotic behaviour. His brief was not a new

[34] See R. Rashed, *Les Mathématiques infinitésimales*, vol. I, pp. 771–5.

introduction to the *Almagest*, serving as justification for revisiting Ptolemy's famous claim in the first book of his work. Instead, he wrote a treatise dedicated entirely to the problem itself and explicitly bringing to it this asymptotic characteristic. Ibn al-Haytham declares that he intends to establish that 'of all the figures with similar <and equal> perimeters, the circle has the greatest <area>; and, of all the polygons, that which is closest to being circular in shape is greater than that which is less circular in shape'.[35] More generally, he seeks to prove that among 'solid and plane figures those having a shape close to being circular are greater than those whose shape is far from being circular'.[36] Let us examine Ibn al-Haytham's approach.

3.2.1. *Isoperimeters*

Ibn al-Haytham lays out the question of isoperimeters in three propositions and one lemma. Here, in order, are the three propositions.

PROPOSITION 1: Let p be the perimeter of a circle of area A, and p' the perimeter of a regular polygon of area A':

if $p = p'$, then $A > A'$.

PROPOSITION 2: Let P_1, P_2 be two regular polygons with n_1, n_2 sides, with perimeters p_1, p_2, and areas A_1 and A_2, respectively:

if $p_1 = p_2$ and $n_1 < n_2$, then $A_1 < A_2$.

PROPOSITION 3: Let P_1, P_2 be two regular polygons inscribed in the same circle, using the previous notation:

if $n_1 < n_2$, then $p_1 < p_2$ and $A_1 < A_2$.

For this third proposition, Ibn al-Haytham first established a lemma, equivalent to

$$\frac{\alpha}{\beta} = \frac{\sin\alpha}{\sin\beta}, \qquad \text{for } \frac{\pi}{2} > \alpha > \beta.$$

[35] See R. Rashed, *Les Mathématiques infinitésimales*, vol. II, p. 394; English translation pp. 309–10.

[36] *Ibid.*, p. 385; English translation p. 305.

We see that in order to establish the isoperimetric property, Ibn al-Haytham shows that the area of the circle is a sort of 'limit' of an increasing sequence of areas of regular polygons. He assumes the limit exists, which is in any case assured by Archimedes' *Measurement of the Circle*. Ibn al-Haytham's approach here is intentionally 'dynamic' and thus evidently different from that of his predecessors. This explains the author's statement at the beginning of his treatise: 'Mathematicians have mentioned this notion [isoperimetry and equal surface area] and have used it. Nonetheless, none of their proofs have come down to us'.[37] It is, however, hardly credible that Ibn al-Haytham was unaware of the results of his predecessors, including those of al-Khāzin. It would seem, therefore, that his intention is to alert the reader to the novelty of his own approach. And it is precisely this new approach that he used to deal with the case of solids.

3.2.2. *Equal surface areas*

After presenting his solution to the isoperimetric problem, Ibn al-Haytham went on to consider the problem of equal areas, using an analogous method. But the passage from the plane to three-dimensional space is not simple. In fact, one obstacle prevents Ibn al-Haytham from reaching his goal, but not from writing one of the most advanced mathematical texts before the mid-17th century.

Analogously to his approach in the first part, Ibn al-Haytham wished to demonstrate the two propositions:

1. Of two regular polyhedrons with similar faces and having the same total area, the one with the larger number of faces has the greatest volume.

2. Of two regular polyhedrons with similar faces inscribed in the same sphere, the one with the larger number of faces has the greatest total area and the greatest volume.

Ibn al-Haytham thought that, just as he had done with isoperimeters, he would be able to establish the greatest-area problem as the limit of a sequence of areas of polyhedra; that is, he could approach the sphere from an infinite series of polyhedra that it circumscribes. The obstacle is that the two propositions given above concern regular polyhedra with a finite number of faces. The passage from this case to the sphere is therefore not rigo-

[37] *Ibid*., p. 386; English translation p. 306.

rously established. The core of Ibn al-Haytham's treatise rests essentially on demonstrating these two propositions.

Ibn al-Haytham begins by establishing five lemmas (numbers 6 to 10 in the treatise). All of these lemmas are based on inequalities between the ratios of solid angles and the ratios of areas. As far as we know, this is the first important and extensive application of solid angles and therefore the first substantial study of some of their properties (to be discussed below). By using these lemmas, Ibn al-Haytham is able to establish the two propositions with complete generality. But the lemmas apply only to polyhedra with triangular faces, that is, to tetrahedra, octahedra, and icosahedra, since the number of faces of a regular polyhedron with square or pentagonal faces is fixed (six or ten, respectively). The first lemma shows that if regular tetrahedra, octahedra, and icosahedra have the same total area, then their volumes increase in the following order: tetrahedron, octahedron, icosahedron. The second proposition shows that if regular tetrahedra, octahedra, and icosahedra are inscribed in the same sphere, their volumes will increase in the same order. The issue is thus clearly not to approach the sphere from an infinite sequence of polyhedra. Such an oversight on the part of a mathematician who knew Euclid's *Elements* better than anyone is disconcerting. How could he not see that his polyhedra reduce to those of Euclid and that their number is finite? This should not, however, prevent us from admiring the mathematical richness of this treatise. For the moment, let consider how he proved his two propositions.

The idea behind his proof of the first proposition can be summarized thus:

Let A, B be respectively the centers of the spheres circumscribing the two polyhedra; let AE, BG, be respectively the distances from the center to the plane of one of the faces (see Fig. 58) and let the areas of the polyhedra be respectively S_A, S_B and their volumes V_A, V_B. Then we have

$$V_A = \frac{1}{3} S_A \cdot AE \quad \text{and} \quad V_B = \frac{1}{3} S_B \cdot BG .$$

Let n_A, n_B be respectively the number of faces of each polyhedron, and let $n_B > n_A$.

Ibn al-Haytham now proves that $BG > AE$ and thus $V_B > V_A$. His proof consists in comparing the lengths of AE and BG.[38] We shall follow his reasoning in order to illustrate his approach and the style of his argument.

[38] See R. Rashed, *Les Mathématiques infinitésimales*, vol. II, pp. 379–81.

On EC, take a point K such that $GH = EK$ and on ED, a point L such that $EL = GI$. If $AE = BG$, the pyramid $AEKL$ will be equal to the pyramid $BGHI$ and thus solid angle (A, EKL) = solid angle (B, GHI).

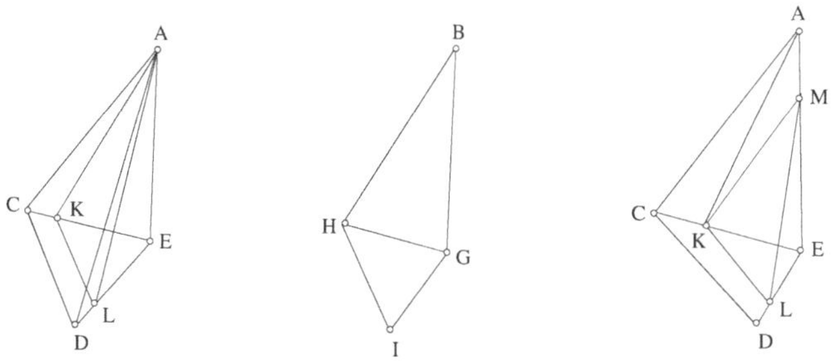

Fig. 58

From Lemma 2 of his Proposition 4, it is known that

$$\frac{\text{solid angle}(A, ECD)}{4\pi} = \frac{\text{pyr.}(AECD)}{V_A} = \frac{\text{area}(ECD)}{S_A}$$

and

$$\frac{\text{solid angle}(B, GHI)}{4\pi} = \frac{\text{pyr.}(BGHI)}{V_B} = \frac{\text{area}(GHI)}{S_B}.$$

Since $S_A = S_B$, we can deduce that

$$(*) \qquad \frac{\text{area}(ECD)}{\text{area}(GHI)} = \frac{\text{solid angle}(A, ECD)}{\text{solid angle}(B, GHI)},$$

therefore

$$\frac{\text{area}(ECD)}{\text{area}(GHI)} = \frac{\text{solid angle}(A, ECD)}{\text{solid angle}(A, EKL)},$$

which is absurd, according to Lemma 8. Therefore, $AE \neq BG$.

Now suppose $BG < AE$, in which case there is a point M on AE such that $EM = BG$, and we have

$$\text{solid angle } (M, EKL) = \text{solid angle } (B, GHI).$$

But we would then have $E\hat{M}K > E\hat{A}K$, $E\hat{M}L > E\hat{A}L$ and (from the apex angles of two isosceles triangles with the same base, respectively, *KE*, *EL*, and *KL*) $K\hat{M}L > K\hat{A}L$. Hence

$$E\hat{M}K + E\hat{M}L + K\hat{M}L > E\hat{A}K + E\hat{A}L + K\hat{A}L,$$

and thus solid angle (*M*, *EKL*) > solid angle (*A*, *EKL*). But, from Lemma 8,

$$\frac{\text{area}(ECD)}{\text{area}(EKL)} > \frac{\text{solid angle}(A, ECD)}{\text{solid angle}(A, EKL)};$$

therefore

$$\frac{\text{area}(ECD)}{\text{area}(EKL)} > \frac{\text{solid angle}(A, ECD)}{\text{solid angle}(M, EKL)},$$

that is,

$$\frac{\text{area}(ECD)}{\text{area}(BHI)} > \frac{\text{solid angle}(A, ECD)}{\text{solid angle}(B, GHI)},$$

which is absurd in light of (*).

Hence, by this double contradiction, $BG > AE$, and it follows that $V_B > V_A$.

To prove his second proposition, Ibn al-Haytham considers two regular polyhedra P_1, P_2 inscribed in the same sphere, with areas S_1, S_2, volumes V_1, V_2 and number of faces n_1, n_2, respectively, and with $n_1 > n_2$. If *A* is the center of the sphere, there will be n_1 regular pyramids, each with the vertex *A*, associated with the faces of P_1 and similarly n_2 pyramids with vertex *A* associated with the faces of P_2.

Let P'_1 be one of the regular pyramids of P_1 with vertex solid angle α_1, base area s_1 and height h_1, and similarly let α_2, s_2, h_2 be these elements of a pyramid P'_2 of P_2.

Now, we have: $n_1\alpha_1 = n_2\alpha_2 = 4\pi$, and since $n_1 > n_2$, then $\alpha_1 < \alpha_2$. Let us suppose that the pyramids P'_1, P'_2 have the same axis *AH*. Since $\alpha_1 < \alpha_2$, the solid angle at the vertex of P'_1 lies inside the solid angle at the vertex of P'_2 and so the edges of P'_1 cut the sphere beyond the base of P'_2. The bases of the two pyramids are parallel, and where they cut the sphere circles on the sphere circumscribe them; from this, we deduce that $s_1 < s_2$ and $h_1 > h_2$.

On the other hand, we have

$$\frac{\alpha_1}{4\pi} = \frac{s_1}{S_1} = \frac{1}{n_1} \quad \text{and} \quad \frac{\alpha_2}{4\pi} = \frac{s_2}{S_2} = \frac{1}{n_2},$$

whence

$$\frac{\alpha_2}{\alpha_1} = \frac{s_2}{S_2} \cdot \frac{S_1}{s_1} = \frac{s_2}{s_1} \cdot \frac{S_1}{S_2}.$$

But, it has already been established (Proposition 9) that $\frac{\alpha_2}{\alpha_1} > \frac{s_2}{s_1}$, therefore $\frac{s_2}{s_1} \cdot \frac{S_1}{S_2} > \frac{s_2}{s_1}$, and hence $S_1 > S_2$.

It is known that $V_1 = \frac{1}{3} S_1 h_1$ and $V_2 = \frac{1}{3} S_2 h_2$, so from $S_1 > S_2$ and $h_1 > h_2$, we deduce the desired result that $V_1 > V_2$.

Finally, it should be emphasized, as we have said, that although Ibn al-Haytham has proved the theorem in a perfectly general manner, the result can apply only to the tetrahedron, octahedron, and icosahedron, since these are the only regular polyhedra with the same faces.

Ibn al-Haytham went on to prove the following corollary:

> Let P_1, P_2 be two regular polyhedra inscribed in the same sphere having respectively n_1, n_2 faces where the faces of the polyhedra are regular polygons with n'_1 and n'_2 sides. If $n_1 > n_2$ and $n'_1 > n'_2$, then $S_1 > S_2$ and $V_1 > V_2$.

What this states is that, since the polyhedra are regular, if a tetrahedron, a cube, and a regular dodecahedron are inscribed in the same sphere, their total surface areas and their volumes increase in that order.

We have just seen that Ibn al-Haytham's intentionally dynamic approach to solving the problem of equal surface areas, conceived by analogy with the successful method of treating isoperimeters, foundered because of the fact that there is only a finite number of regular polyhedra.

This work of Ibn al-Haytham, and that of al-Khāzin before him, make by far the most important contributions to the study of isoperimetry and equal surface areas found in the mathematics of classical Islam. To our knowledge, no other work matches it. The work of later mathematicians, like Ibn Hūd, Jābir ibn Aflaḥ, and Abū al-Qāsim al-Sumaysāṭī, among others, was not at the level attained by al-Khāzin, and certainly did not equal that of Ibn al-Haytham. While Ibn Hūd did study the isoperimetric problem in his book *al-Istikmāl* (*The Completion*), he only reproduced Ibn al-Haytham's proof, with some small variations.[39] As for al-Sumaysāṭī, he added nothing of substance to the results of al-Khāzin.[40] Jābir ibn Aflaḥ,

[39] See R. Rashed, *Les Mathématiques infinitésimales*, vol. I, pp. 1014–27.

[40] *Ibid.*, pp. 777–8 and 830–3.

the astronomer of Andalusia, handled only the equal area problem, and considered only regular polyhedra in his proof.[41] It was quite probably from the Latin translation of this latter book that Bradwardine borrowed a proposition appearing in his *Geometria Speculativa*, Book II, which also turned up later in Cardano's *De Subtilitate*: 'Of all plane isoperimetric figures, having the same number of sides and equal angles, the greatest has its sides equal'. This is precisely al-Khāzin's sixth proposition.

Only further historical research may reveal whether there were other contributions to this mathematical topic at the level of al-Khāzin and Ibn al-Haytham, and in particular whether the latter's work was taken up and developed by other Islamic scholars. Were any elements of this work transmitted to Latin scholars? Without wishing to prejudge future discoveries, it nevertheless seems to us doubtful that much further progress could have been made without suitable analytical tools that had yet to be invented; Ibn al-Haytham's work on the solid angle confirms this.

4. THE THEORY OF THE SOLID ANGLE

In his research on the total surface areas of solids, Ibn al-Haytham, as we have seen, developed a theory of the solid angle. This was in fact the first real theory of the solid angle after the modest work of Euclid. Ibn al-Haytham's contribution to the theory would not be equaled, let alone superseded, for several centuries. Ibn al-Haytham's work in the 11th century greatly enriched the development of Archimedean research into a new branch of geometry. We should, therefore, sketch out the history of the studies into the solid angle prior to Ibn al-Haytham so as better to locate his own contribution.

The first study of the solid angle that has come down to us is that of Euclid in the *Elements*. In Book XI, which treats solid geometry, Euclid defines the solid angle and elaborates a theory in Propositions 20–23 and 26, before applying the theory in Propositions 27, 36 and 37 of the same book, and then in the *Scholia* of Book XIII (the so-called 'Book XIV'). This theory, we should immediately point out, is based less on the solid angle as such, than on the ratios of the plane angles which form it.

Euclid begins by defining the solid angle:

> A solid angle is the inclination constituted by more than two lines which meet one another and are not in the same surface, towards all the lines.

[41] *Iṣlāḥ al-Majisṭī*, ms. Escorial 390, fol. 12^{r-v}.

Otherwise: A solid angle is that which is contained by more than two plane angles which are not in the same plane and are constructed to one point.[42]

This is the eleventh of the 28 definitions that begin Book XI. It is clear in this double definition that the first is based on the definition of a plane angle in Euclid I Def. 8 (a plane angle is the inclination to one another of two lines in a plane which meet one another and do not lie in a straight line). It is also quite evident that, taken as such, this definition raises all sorts of problems. When we consider a solid angle formed by the inclination of three curved lines, the latter, taken in pairs, do not form a plane angle because the lines are not coplanar. The second definition, in its turn, seems only to speak of one sort of solid angle (a polyhedral angle): the solid angle at the vertex of a cone of revolution, for example, would be excluded. Whatever it may be, this double definition has caused a great deal of ink to flow and doubts have been raised as to whether the first definition is actually authentic.[43] The second proposition is the one retained in the Arabic tradition in different copies of Isḥāq-Thābit's translation of the *Elements*. This is the definition that Ibn al-Haytham reproduces in his work (*Book on the Solution of Doubts of Euclid in the Elements*).[44] The first definition is found, however, in a fragment containing the preface to Book XI, the translation of which is explicitly attributed to the father of Isḥāq, that is to Ḥunayn ibn Isḥāq (the fragment was intercalated between the pages of the ms. Malik 3586, Teheran).

Euclid proved, in order, the following five propositions, which provide the basis of his theory of solid angles:

PROPOSITION 20 – *If a solid angle be contained by three plane angles, any two, taken together in any manner, are greater than the remaining one.*

PROPOSITION 21 – *Any solid angle is contained by plane angles less than four right angles.*

On this proposition, Heath justly remarks: 'It will be observed that, although Euclid enunciates this proposition for *any* solid angle, he only proves it for the particular case of a *trihedral* angle'.[45] Euclid's proof depends on I.32 and consequently on the parallel postulate. However, we

[42] *Euclidis Opera Omnia*, Leipzig, 1945, vol. VII: *Euclidis optica, opticorum recensio Theonis, catoptrica, cum scholiis antiquis*, edidit I. L. Heiberg, p. 2; *The Thirteen Books of Euclid's Elements*, vol. I-III, English translation and commentary by T. L. Heath, Cambridge, 1926, vol. III, p. 261.

[43] *The Thirteen Books of Euclid's Elements*, ed. T. L. Heath, vol. III, pp. 267–8.

[44] *Kitāb fī ḥall shukūk kitāb Uqlīdis fī al-Uṣūl*, ms. Istanbul, University 800, fol. 157^r.

[45] *The Thirteen Books of Euclid's Elements*, ed. T. L. Heath, vol. III, p. 210.

know, since Lobachevsky and Bolyai, that spherical geometry does not depend on the parallel postulate.

PROPOSITION 22 – *If there be three plane angles of which two, taken together in any manner, are greater than the remaining one, and they are contained by equal straight lines, it is possible to construct a triangle out of the straight lines joining the extremities of the two equal straight lines.*

This proposition serves as a lemma to prove the most important of this set of propositions, namely:

PROPOSITION 23 – *To construct a solid angle out of three plane angles two of which, taken together in any manner, are greater than the remaining one; thus the three angles must be less than four right angles.*

PROPOSITION 26 – *On a given straight line, and at a given point on it, to construct a solid angle equal to a given solid angle.*

Without having defined the equality of two solid angles in terms of their measurement, Euclid proves this last proposition by assuming the equality of two trihedral angles; as Heath put it: 'This proposition again assumes the equality of two trihedral angles which have the three plane angles of the one respectively equal to the three plane angles of the other taken in the same order.'[46] In other words, Euclid supposes, without proof, that two solid trihedral angles are equal if the plane angles enclosing them are equal to each other in pairs. This implicit hypothesis that two solid angles are equal if the plane angles constituting them are respectively equal to each other is not, however, always valid. To be convinced of this, consider a solid angle made by four plane angles meeting at a vertex; it is easy to see that an infinite number of different solid angles can be formed in this way. Nonetheless, the statement of Proposition 26 carries no restrictions. The difference between the generality of the proposition and his limitation of the proof to solid trihedral angles alone is a clue. It shows that in the *Elements* Euclid considered only solid angles that were trihedral. This was not the only restriction. We have just seen that Euclid attempted to establish the equality of two solid angles without, however, having explicitly defined the concept of equality. It follows that Euclid supposes that the solid angle has a magnitude of the same genus as other magnitudes, like the plane angle, for example. But at no time did he give any rules for treating this magnitude, for example for comparing solid angles to each other with the aid of the solid angle formed by three plane right angles at a vertex.

[46] *Ibid.*, vol. III, p. 329.

This comparison, even if restricted to trihedral angles, is independent of the sum of the plane angles which define the solid angle. Hence, we would look in vain to find in the *Elements* any such study of the solid angle as a magnitude. In Euclid, the theory of the solid angle remains rather weak.

After Euclid, the history of the solid angle is sparse. The only information we have comes from the second part of Book V of Pappus's *Collection*, dedicated to solid figures. Of the thirteen semi-regular polyhedra, he writes:

> In fact, if, for the polyhedra whose solid angles are enclosed by three plane angles, one simply counts the number of plane angles possessed by all the bases [faces] of the polyhedron, it is evident that the number of solid angles is one third of the number obtained; whereas for polyhedra whose solid angles are enclosed by four plane angles, if one counts all the angles of the bases of the polyhedra, the number of the solid angles is one quarter of the number obtained. Finally, in the same way for polyhedra whose solid angles are enclosed by five plane angles, the number that expresses the quantity of solid angles is one fifth of the quantity of plane angles.[47]

What we have here, as Pappus himself reminds us, is a reference to the thirteen polyhedra discovered by Archimedes and given in his lost book.[48] This means that, after Euclid, Archimedes must at least have referred to the solid angle in his work on the semi-regular polyhedra. But our knowledge is limited to this remark by Pappus: we cannot know more without access to this lost work. Such was, it seems, the extent of research into the solid angle until the subject was explored again in the 9th century. Since then, even though no comprehensive history of the solid angle has been written, we can observe two lines of research. The first is based on Euclid XI, with the intention of improving on the proofs given there; the second is directly linked to approximating solids by convex polyhedra. Examples of the first type are found in the works of al-Sijzī, Ibn al-Haytham, and doubtless others; whereas the second direction of research belongs to Ibn al-Haytham.

As far as we know, the first Islamic mathematician who showed an interest in the solid angle was Aḥmad ibn ʿAbd al-Jalīl al-Sijzī in the second half of the 10th century. Al-Sijzī considered the solid angle on at least two occasions. In his *Introduction to the Science of Geometry* (*al-Madkhal ilā ʿilm al-handasa*), he classified different types of solid angle: 'one is comprised of a single surface, the second is comprised of a surface

[47] *La Collection mathématique*, French transl. Paul Ver Eecke, vol. I, p. 274.
[48] *Ibid.*, pp. 272–3.

and a plane, and the third is comprised of planes'.[49] Al-Sijzī provided examples of each type: the angle at the vertex of a cone and that at the summit of an ellipsoid; the vertex of a half-cone; the angle comprised of three plane angles – a trihedral angle – and whose sum is less than four right angles.

In a second memoir, a *Treatise on the Resolution of the Doubt in Proposition Twenty-Three of the Eleventh Book of the Treatise of Euclid* (*Risāla fī ḥall al-shakk alladhī fī al-shakl al-thālith wa-al-ʿashrīn min al-maqāla al-ḥādiyya ʿashara min kitāb Uqlīdis fī al-Uṣūl*),[50] al-Sijzī examines Euclid's proposition and lemma. In the preface to the text, he addresses other mathematicians who have raised doubts about Euclid's proof. In other words, at that time, al-Sijzī was not the only mathematician interested in the solid angle, albeit from a Euclidean perspective.

The second mathematician was Ibn al-Haytham himself. In his *Book on the Resolution of Doubts Concerning the* Elements *of Euclid* (*Kitāb fī ḥall shukūk kitāb Uqlīdis fī al-Uṣūl*), he also returned to Euclid's proof in order to correct it.[51] These manuscripts are as interesting for what they have to tell about the history of Book XI of the *Elements* as they are for information about the solid angle; these studies do not, however, radically alter the nature of the Euclidean theory.

The third study is by far the most novel and the most important. It is in Ibn al-Haytham's book on isoperimetry and equal surface areas that he elaborates a theory of the solid angle. How then was he led to do this?

As we have seen above, one of the aims of his book was to demonstrate the maximal surface area property of the sphere. Following in the footsteps of his predecessors like al-Khāzin, Ibn al-Haytham had opted for a strategy employing infinitesimals that approach the volume of the sphere by a sequence of volumes of convex polyhedra. His unfortunate choice of regular polyhedra changes nothing about the intention that lay behind his strategy, but the approach does require him to compare the volumes of convex polyhedra. Such a comparison is more manageable if it involves the solid angle and if, consequently, one compares solid angles with each other. But comparisons are only possible among magnitudes and by means of magnitudes. Thus it was necessary to conceive of the solid angle explicitly as a magnitude, and therefore as subject to those operations applicable to magnitudes, including the theory of proportions. Ibn al-Haytham established many lemmas on the solid angle in order to prove the extremal

[49] Ms. Dublin, Chester Beatty 3652, fol. 7ᵛ.
[50] *Ibid.*, fol. 33ʳ–34ᵛ.
[51] Ms. Istanbul, University 800, fol. 161ʳ–165ᵛ.

property of the sphere. But these lemmas themselves constitute a new theory of the solid angle. Let us see how he proceeded to establish the theory.

In such a situation, the most natural method, without necessarily being the simplest, would be to proceed by analogy: to begin with the theory of the plane angle in order to generalize it to three dimensions. But we should not be naïve: analogy here has only a heuristic value; it is a means of discovery. The move from the plane to three-dimensional space has too many pitfalls to be simple. In fact many valuable properties of the plane angle turn out to be invalid in three dimensions. For Ibn al-Haytham, the knowledge he had of the relations between chords and the angles at the center of the circle provided him with strong intuitive ideas but not, even so, a theory of the solid angle.

This mathematician found himself in an entirely new situation: he was within the landscape of Archimedean geometry, but using the tools of spherical geometry. This is a far cry from the world of Euclid. To explain what we mean, let us examine the elements of his theory.

Ibn al-Haytham began by recalling a result of Archimedes that he proved again in his treatise *On the Measurement of the Sphere* (*Qawl fī misāḥat al-kura*): The volume of the sphere is equal to $\frac{4}{3}\pi r^3$, where r is the radius. Following this, he wrote:

> Any regular polyhedron inscribed within a sphere is such that, if planes are drawn from the center of the sphere passing[52] through the sides of one of its bases, then these planes divide off a sector from the sphere whose ratio to the entire sphere is equal to the ratio of the spherical surface at the base of this sector to the entire surface of the sphere, and is also equal to the ratio of the solid angle, that is the angle at the center of the sphere which is surrounded by the surfaces of a regular pyramid[53] whose lines are straight and whose base is one of the bases of the polyhedron, to the eight solid right angles which is the sum of all the solid angles at the center of the sphere and which are also at the center of any regular polyhedron, as the sphere and the surface of the sphere are divided by these planes into equal parts.[54]

In this text Ibn al-Haytham introduces the concept of the solid angle with its relationship to spherical surfaces and spherical sectors. More precisely, he proposes that if a sphere has surface area A and volume V, and a

[52] Lit.: planes to the sides.

[53] Lit.: rectangular cone.

[54] *On the Sphere which is the Largest of all the Solid Figures*, in R. Rashed, *Les Mathématiques infinitésimales*, vol. II, p. 401–2; English translation, pp. 313–14.

sector of the sphere with solid angle α has surface area s and volume v, then

$$(1) \qquad \frac{v}{V} = \frac{s}{A} = \frac{\alpha}{8D},$$

where D is a solid right angle, and each of these fractions is equal to $\frac{1}{n}$ if n is the number of faces of the polyhedron.

It is only after this that Ibn al-Haytham defines and constructs a solid right angle:

> As for the angles, if a great circle is drawn within the sphere, and if two diameters are drawn within this circle which cross each other at right angles, and if a perpendicular to this plane is drawn passing through its center, and if this is extended on both sides until it meets the surface of the sphere, and if <perpendicular> straight lines are drawn from these two extremities onto the extremities of the two diameters, then they form eight equal pyramids within the sphere whose vertices are at the center of the sphere and whose angles at the vertices are equal. Each of these angles is called a 'solid right angle', and the sum of these angles is equal to the sum of the angles of any polyhedron inscribed within the sphere.[55]

Ibn al-Haytham concludes then that $v = \frac{1}{3} sr$. We can already see that Ibn al-Haytham has not chosen to use the Euclidean definition of the solid angle, instead using a polyhedron inscribed in a sphere. Each face of the polyhedron is associated with a regular pyramid whose vertex is at the center of the sphere, B say. By (1) he defines the solid angle at the vertex B, a spherical surface, and a sector of the sphere.

Ibn al-Haytham then goes on to attempt to establish the following propositions:

PROPOSITION 1: Let A be the center of a sphere and let $P_1(ABCDE)$ and $P_2(AHFG)$ be two pyramids, where B, C, D, E, F, G, H lie on the surface of the sphere; consider the two pyramids P_1, P_2 with solid angles α_1, α_2 respectively. Let the faces of these pyramids be extended so as to cut the surface of the sphere to determine spherical surfaces s_1, s_2 and spherical sector volumes v_1, v_2 respectively. Then

$$\frac{\alpha_1}{\alpha_2} = \frac{s_1}{s_2} = \frac{v_1}{v_2}.$$

[55] *Ibid.*, p. 403; English translation, p. 314.

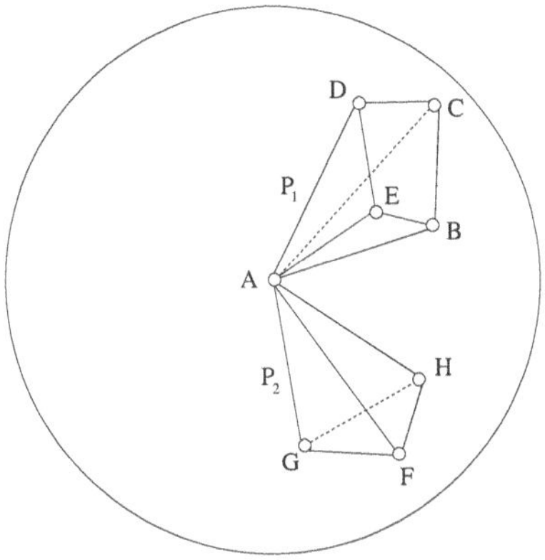

Fig. 59

If we take nP_1 pyramids, the portion of the sphere associated with them will have a spherical surface area ns_1 and spherical sector volume nv_1, with a total solid angle $n\alpha_1$, and the same is true for P_2. Ibn al-Haytham then states:

If $nv_1 > nv_2$, then $n\alpha_1 > n\alpha_2$ and $ns_1 > ns_2$.
If $nv_1 < nv_2$, then $n\alpha_1 < n\alpha_2$ and $ns_1 < ns_2$.
If $nv_1 = nv_2$, then $n\alpha_1 = n\alpha_2$ and $ns_1 = ns_2$.

If $n\alpha_1 > n\alpha_2$, then $ns_1 > ns_2$ and $nv_1 > nv_2$.
If $n\alpha_1 < n\alpha_2$, then $ns_1 < ns_2$ and $nv_1 < nv_2$.
If $n\alpha_1 = n\alpha_2$, then $ns_1 = ns_2$ and $nv_1 = nv_2$.

We note that the explanations given by Ibn al-Haytham do not in any way constitute a proof of his assertion that:

$$\frac{\alpha_1}{\alpha_2} = \frac{s_1}{s_2} = \frac{v_1}{v_2}.$$

PROPOSITION 2: Let $ABCD$ be a pyramid such that $A\hat{B}C \geq \frac{\pi}{2}$ and $A\hat{B}D \geq \frac{\pi}{2}$, E is a point on BD such that $A\hat{E}C \geq \frac{\pi}{2}$ or $A\hat{C}E \geq \frac{\pi}{2}$, then

$$\frac{\text{area}(DBC)}{\text{area}(EBC)} > \frac{\text{solid angle}(A, BDC)}{\text{solid angle}(A, EBC)}.$$

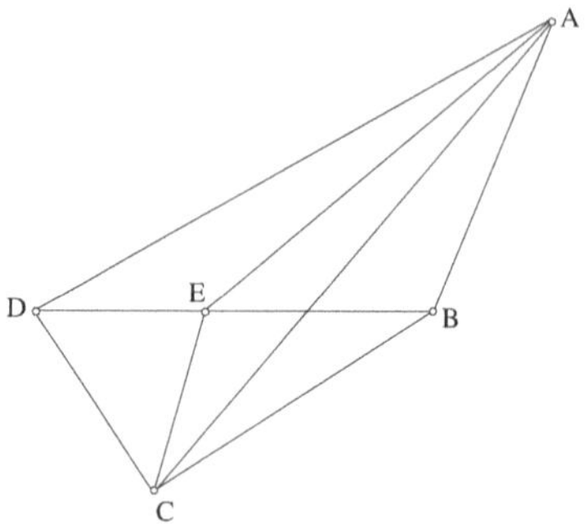

Fig. 60

Consider a sphere Σ of center A and radius AB cutting AC, AD, AE at H, I, L, respectively, such that $AB = AH = AI = AL$. Thus lying in the planes BAC, BAD, ACD, ACE, we have arcs BH, BLI, HI, HGL respectively. The line segment BL lies in the plane BAD and cuts AD at K (since $A\hat{B}L$ is acute and $B\hat{A}D$ is acute). The arc LGH lies on Σ, so K is outside the sphere and $AK > AI$. The conical surface with vertex B defined by the arc LGH cuts the plane ADC following an arc KFH, since any straight line BG cuts this plane at a point F exterior to Σ, the arc KFH lies entirely outside the sphere, except for the point H. Thus the sector of the sphere $AILGH$ lies inside the solid $AKFHGL$, bounded by planes and a part of the conical surface, since the part GF of the generator lies outside Σ; and the sector of the sphere $ALHB$ is greater than the solid $ALHB$, bounded by the planes and another part of the conical surface, since the part BG of the generator lies inside Σ.

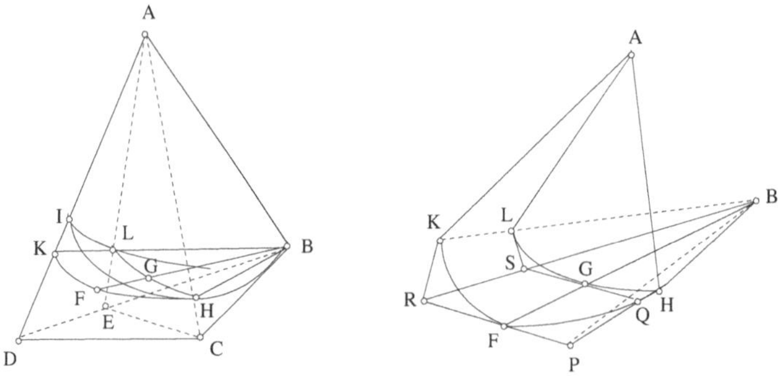

Fig. 61

We now have

$$\left.\begin{array}{l}\text{sect. }(A,ILH) < \text{sol. }(A,KFHGL)\\ \text{sect. }(A,LHB) > \text{sol. }(A,HGLB)\end{array}\right\} \Rightarrow \frac{\text{sect. }(A,ILH)}{\text{sect. }(A,LHB)} < \frac{\text{sol. }(A,KFHGL)}{\text{sol. }(A,LGHB)}.$$

By composition, this gives

$$(*) \qquad \frac{\text{sect. }(A,IHB)}{\text{sect. }(A,LHB)} < \frac{\text{sol. }(B,AKFH)}{\text{sol. }(B,AHGL)}.$$

In the course of his proof of Lemma 6, Ibn al-Haytham introduces another proposition, namely:

$$(**) \qquad \frac{\text{area tr. }(AEC)}{\text{area sect. }(ALGH)} \leq \frac{\text{area tr. }(ADC)}{\text{area sect. }(AKFH)}.$$

In other words, the conical projection from center B of the plane AEC onto the plane ADC increases some ratios of star-shaped areas in relation to A. The proof of this proposition reduces, by the method of contradiction, to the case where we consider areas of triangles with the vertex at A. Let us examine the steps of this proof.

Let us suppose that

$$(1) \qquad \frac{\text{area}(AEC)}{\text{area sect.}(ALGH)} > \frac{\text{area}(ADC)}{\text{area sect.}(AKFH)};$$

then there exists an area L_a (a fourth proportional) such that

$$(2) \qquad \frac{\text{area}(AEC)}{L_a} = \frac{\text{area}(ADC)}{\text{area sect.}(AKFH)}.$$

Hypothesis (1) can therefore be written: $L_a >$ area sect. ($ALGH$) and thus there exists a polygon $LSQH$ circumscribing the circular arc LGH, such that

$$(3) \qquad \text{area }(ALSQH) < L_a.$$

Ibn al-Haytham considers the case where this polygon has three sides, LS, SQ, QH tangent to the arc at points L, G, H, respectively. The polygon $LSQH$, projected onto the plane ADC, is mapped to the polygon $KRPH$, whose sides KR, RP, PH are tangents at K, F, H, respectively, to KFH (the arc of a conic), which is itself the projection of the circular arc LGH (see Figs 61 and 62).

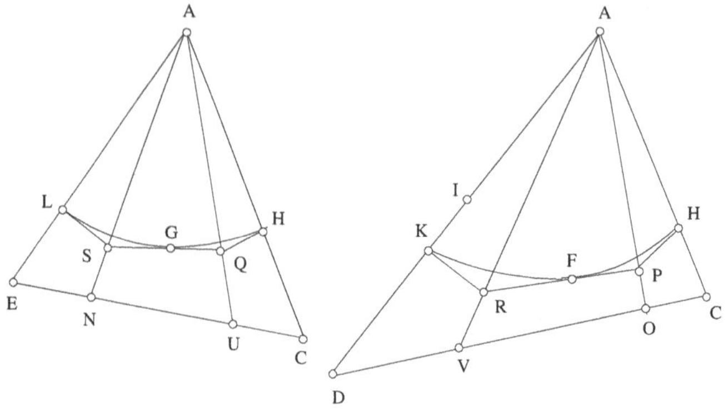

Fig. 62

Note that Ibn al-Haytham is certainly aware of the fact that a conical projection preserves contact, which recalls the properties of the tangent plane of the conic, known to Ibn Sahl.[56]

The inequality (*) thus reduces to

$$(4) \qquad \frac{\text{area}(ADC)}{\text{area}(AKRPH)} > \frac{\text{area}(AEC)}{\text{area sect.}(ALSQH)}.$$

But the ratio on the right hand side of (4) is, according to (3), greater than

$$\frac{\text{area}(AEC)}{L_a} = \frac{\text{area}(ADC)}{\text{area sect.}(AKFH)}$$

from which we deduce that

$$\text{area}(AKRPH) < \text{area sect.}(AKFH),$$

which is absurd, since the polygon circumscribes the curved arc.

Ibn al-Haytham then asserts that the following inequalities result from inequality (4):

$$(5) \qquad \frac{\text{area}(AEN)}{\text{area}(ALS)} < \frac{\text{area}(ADV)}{\text{area}(AKR)}; \quad \frac{\text{area}(ANU)}{\text{area}(ASQ)} < \frac{\text{area}(AVO)}{\text{area}(ARP)};$$

$$\frac{\text{area}(AUC)}{\text{area}(AQH)} < \frac{\text{area}(AOC)}{\text{area}(APH)}.$$

[56] R. Rashed, *Geometry and Dioptrics in Classical Islam*, London, 2006; *Géométrie et dioptrique*, pp. XVIII–XXXIX.

Unfortunately Ibn al-Haytham's last claim is not true in every case.[57] The proposition is, however, true whenever the points C, D are on the same side of the plane perpendicular to ABD through AB. This condition escaped Ibn al-Haytham, but his error did not affect the propositions that followed and that depend on it. Ibn al-Haytham somehow always manages to situate himself in favourable conditions.

The other propositions on the solid angle follow:

PROPOSITION 3: Let $ABCD$ be a pyramid such that AB is perpendicular to the plane BCD, with $B\hat{C}D \geq \frac{\pi}{2}$. If E lies on CD, between C and D, then

$$\frac{\text{area}(DBC)}{\text{area}(EBC)} > \frac{\text{solid angle}(A, BDC)}{\text{solid angle}(A, EBC)}.$$

The argument is as follows. In the plane ABE, construct a perpendicular to AE at E to cut AB at G. Since $B\hat{C}E \geq \frac{\pi}{2}$, then $BE > BC$, and since $AB \perp$ plane BCD, we have $AE > AC$. If BC is produced to E' such that $BE' = BE$, then $A\hat{E}'G = A\hat{E}G = \frac{\pi}{2}$, and so $A\hat{C}G$ is obtuse. Hence the pyramid $AGCD$ fulfils the conditions of Proposition 2. For the case in which $A\hat{E}G$ is a right angle, we have pointed out that the situation is ambiguous; the lemma, however, remains true.[58]

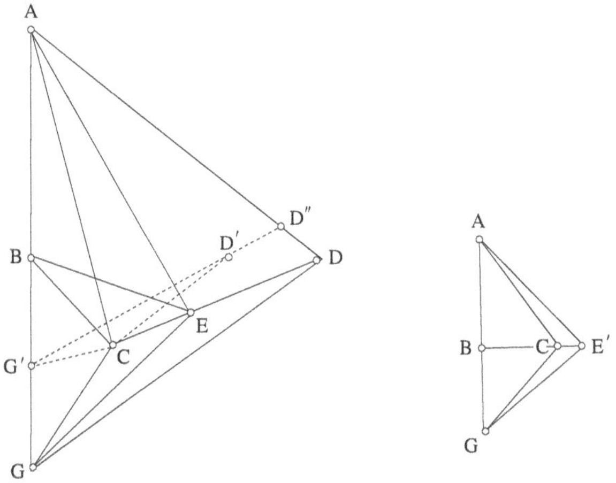

Fig. 63

[57] See R. Rashed, *Les Mathématiques infinitésimales*, vol. II, pp. 346 *sqq*.

[58] *Ibid*., pp. 370 *sqq*.

First we show that $A\hat{C}D \geq \frac{\pi}{2}$:

• If $B\hat{C}D = \frac{\pi}{2}$, then $CB \perp CD$ whereas CD is orthogonal to AB, therefore CD is perpendicular to the plane ABC, and consequently $CD \perp AC$, *i.e.* $A\hat{C}D = \frac{\pi}{2}$.

• If $B\hat{C}D > \frac{\pi}{2}$, we can draw CD' inside $B\hat{C}D$ such that $D'C \perp BC$, and so CD' is perpendicular to the plane ACB and $AC \perp CD'$; thus $A\hat{C}D' = \frac{\pi}{2}$ and $D' \in \,]BD[$.

Now, in the plane ACG, draw $CG' \perp AC$, where $G' \in \,]BG[$. Then the plane $CD'G'$ is perpendicular to AC and meets the plane ABD in the line $G'D'$, which meets AD between A and D, at D'', say. We then have $A\hat{C}D'' = \frac{\pi}{2}$, and hence $A\hat{C}D > \frac{\pi}{2}$.

Ibn al-Haytham then continues the argument thus:

$$\frac{\text{area}(GCD)}{\text{area}(GCE)} > \frac{\text{solid angle}(A, GCD)}{\text{solid angle}(A, GCE)};$$

but

$$\frac{\text{area}(GCD)}{\text{area}(GCE)} = \frac{CD}{CE} = \frac{\text{area}(DBC)}{\text{area}(EBC)}$$

and

$$\text{solid angle } (A, GDC) = \text{solid angle } (A, BCD)$$

$$\text{solid angle } (A, GCE) = \text{solid angle } (A, BCE),$$

from which

$$\frac{\text{area}(DBC)}{\text{area}(EBC)} > \frac{\text{solid angle}(A, BCD)}{\text{solid angle}(A, BCE)}.$$

PROPOSITION 4: Let $ABCD$ be a pyramid such that AB is perpendicular to the plane CBD with $BC = BD$. If $EG \parallel CD$ then

$$\frac{\text{area}(CDB)}{\text{area}(EBG)} > \frac{\text{solid angle}(A, BCD)}{\text{solid angle}(A, BEG)}.$$

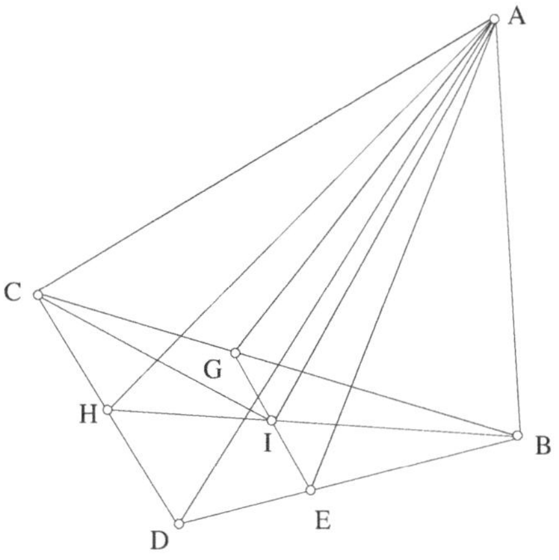

Fig. 64

Since $EG \parallel CD$, the triangle BGE is isosceles. If I is the midpoint of EG, then $BI \perp EG$ and BI cuts DC at H, the midpoint of CD. But $AB \perp CBD$ and so ABC and ABH are both perpendicular to CBD. But, since $BH = ABH \cap BCD$, and GI in BCD is perpendicular to BH, then GI is also perpendicular to the plane ABH; hence $A\hat{I}G = \frac{\pi}{2}$, likewise $A\hat{H}C = \frac{\pi}{2}$.

Since we have $A\hat{I}H$ and $A\hat{I}C$ both obtuse and $B\hat{H}C = \frac{\pi}{2}$, we can apply Proposition 2, with $A\hat{I}C$ obtuse. Again we have the doubtful case, but it is valid here, since the points C, H both lie on the same side of the plane perpendicular to ABH passing through AB (CH is parallel to this plane). Hence

$$\frac{\text{area}(BCH)}{\text{area}(BCI)} > \frac{\text{solid angle}(A, BCH)}{\text{solid angle}(A, BCI)}.$$

The same Proposition 2, with $A\hat{I}G$ a right angle (this is true in all cases, $A\hat{I}G$ playing the same role as $A\hat{C}E$ in Proposition 2), gives

$$\frac{\text{area}(CBI)}{\text{area}(IBG)} > \frac{\text{solid angle}(A, BCI)}{\text{solid angle}(A, BIG)}$$

and multiplying together both sides of these last inequalities gives

$$\frac{\text{area}(HBC)}{\text{area}(IBG)} > \frac{\text{solid angle}(A, BCH)}{\text{solid angle}(A, BIG)}.$$

Finally, doubling the areas and solid angles, since *AHB* is a plane of symmetry, we have the result

$$\frac{\text{area}(DBC)}{\text{area}(BEG)} > \frac{\text{solid angle}(A, BCD)}{\text{solid angle}(A, BEG)}.$$

The integral calculus now makes it possible to establish this result directly, without first having to prove Proposition 2.

PROPOSITION 5: Let P_1, P_2 be two right pyramids with the same vertex *A* whose bases are similar but unequal regular polygons, and let them be inscribed in a sphere of center *A*. Let P_1 have the greater base. Then

$$\frac{\text{solid angle at } A \text{ of } P_1}{\text{solid angle at } A \text{ of } P_2} > \frac{\text{base of } P_1}{\text{base of } P_2}.$$

In al-Haytham's proof, the key idea, here, as elsewhere, is to measure the solid angle in terms of the area of the surface of the sphere it intercepts.[59]

PROPOSITION 6: Let P_1, P_2 be two right pyramids with the same vertex *A* whose bases are regular polygons of sides n_1, n_2 and let them be inscribed in a sphere of center *A*. Let the areas of their bases be s_1, s_2.

If $n_1 > n_2$ and $s_1 < s_2$, then

$$\frac{\text{solid angle at } A \text{ of } P_2}{\text{solid angle at } A \text{ of } P_1} > \frac{s_2}{s_1}.$$

These, then, are the elements that Ibn al-Haytham conceived for a theory of the solid angle. As we have seen the principal unifying ideas of the theory are: the solid angle is a magnitude with the same status as other magnitudes; and the size of a solid angle can be measured by the area of the spherical surface it intercepts, which allows one to define the ratios between polygons inscribed in a sphere and the solid angles at the center of the sphere subtended by the polygons.

[59] R. Rashed, *Les Mathématiques infinitésimales*, vol. II, p. 373–7 and *Les Mathématiques infinitésimales du IXe au XIe siècle*, vol. III: *Ibn al-Haytham. Théorie des coniques, constructions géométriques et géométrie pratique*, London, 2000, pp. 942–3: English transl., *Ibn al-Haytham and Analytical Mathematics*, pp. 288–95.

At this point, any analogy with the theory of the plane angle no longer holds.

An examination of the proofs of the foregoing properties shows that Ibn al-Haytham combines conical projections with infinitesimal processes. This episode thus marks one of the most advanced and sophisticated chapters of mathematics of the time. As far as we know, the theory of the solid angle was not treated again for six centuries, until Descartes's *Progymnasmata de solidorum elementis* and Florimond de Beaune's *La doctrine de l'angle solide*: the latter pursued his own path; the former handled ideas close to those of Ibn al-Haytham. Not until Euler and the Abbé de Gua does one witness further advances in the theory, thanks to the invention of the differential and integral calculus.

– 2 –

THE TRADITIONS OF THE *CONICS* AND THE BEGINNING OF RESEARCH ON PROJECTIONS*

INTRODUCTION

From the middle of the 9th century, mathematicians began to proceed by geometrical transformations much more than they had before. Some of the best evidence of this trend is found in the works of al-Farghānī, the Banū Mūsā brothers (especially the youngest, al-Ḥasan), and Thābit ibn Qurra. One century later, the names of Ibn Sahl, al-Qūhī, and al-Sijzī, for example, are linked to the study of these mathematical concepts, which quickly replace the objects studied. A careful reading of their writings shows indeed that geometers were interested not only in the study of figures, but also in the relations that united them. It is true that transformations had made a modest showing before the 9th century: Archimedes and Apollonius, among others, drew upon them. The former, in his *Conoids and Spheroids*, had brought orthogonal affinity into play, but Arabic mathematicians did not know this work. If we can believe Pappus, Apollonius allegedly had used certain transformations in his *Plane Loci*, a book that the 9th–10th century mathematicians did not have, but from which they certainly knew indirectly at least a few propositions from it (see 'Archimedeans and the problems with infinitesimals').

In the 9th century, however, the use of geometrical transformations grew much more frequent and their field of application much more extensive. The difference between ancients and moderns is significant: among the former, certain transformations arise in the course of demonstrations; among the latter, it is a new point of view that emerges. The first chapter in which this new orientation of geometry is pronounced will soon get the name of 'science of projection (*ʿilm al-tasṭīḥ*)'.[1] This chapter of geometry separated itself from astronomy and was established when it became neces-

* In collaboration with Philippe Abgrall.

[1] See especially R. Rashed, *Geometry and Dioptrics in Classical Islam*, London, al-Furqān, 2005 and *Les Mathématiques infinitésimales du IXe au XIe siècle*.

sary to give a firm foundation to the procedures for representing the sphere exactly in order to construct astrolabes. We must remember two significant historical facts. In the middle of the 9th century, questions of projection were already topics of discussion, not to say controversy, in which mathematicians such as the Banū Mūsā, al-Kindī, al-Marwarūdhī (the astronomer of the caliph al-Ma'mūn), and al-Farghānī, among others, took part. In addition, it has not been sufficiently emphasized that mathematicians who knew about the recent translation of Apollonius's *Conics* raised or debated these questions of projection. This cross-fertilization between research on projections and the geometry of conic sections occurred in, among others, al-Farghānī's *al-Kāmil*, which devotes an entire chapter to the geometry of projections. In it al-Farghānī offers the first truly geometrical study of conic projections. From al-Farghānī to al-Bīrūnī in the 11th century, through al-Qūhī and Ibn Sahl, one witnesses both a deployment and a clear confirmation of this geometrical research. Moreover, in two of al-Bīrūnī's works – *Tasṭīḥ al-ṣuwar wa-tabṭīḥ al-kuwar* (*The Plane Projection of Figures* <= *Constellations*> *and of the Spheres*), *Istī'āb al-wujūh al-mumkina fī ṣan'at al-asṭurlāb* (*On All Possible Methods for Constructing the Astrolabe*) – he retraces in his own way this history of projective methods. He makes an inventory of the different kinds of projection known in his day, and elaborates a few of them himself. He not only takes credit for the invention of the cylindrical projection, but also illustrates the projection that is today called zenithal equidistant projection, reminding the reader that it was the focus of several controversies.

Overall, in their research on projections, mathematicians took several paths. The most frequent of these was that traced by theoretical research on the stereographic projection, a conical projection of the sphere from one of its poles. In it, they included the study and the demonstration of its properties, accompanied by a discussion of its applicability to the astrolabe, that is, the exactitude of its representation of a sphere on a plane. Other paths consisted in generalizing, in the context of this astrolabe, this projection by displacing its pole along the axis or off the axis, or in considering cylindrical and conic projections of the sphere in all of their generality. Finally, yet other paths were explored to elaborate new projections, independently of the studies we just mentioned, and among which one finds the projection nowadays called zenithal equidistant or globular projection. This new chapter on projections, for which Ptolemy's *Planisphere* is at best only a distant ancestor, was also enriched by geometrical research carried out on sundials by many a geometer, including Thābit ibn Qurra[2] and his grandson, Ibn Sinān.[3] Add to this the fact that, beginning with al-Ḥasan ibn Mūsā and especially Thābit ibn Qurra in the 9th century, an entire tradition of research on the cylindrical projection comes into being in the context of works on cylindrical sections, and that this tradition in fact also stands at the origins of the chapter on pointwise transformations.

1. CYLINDRICAL PROJECTIONS

1.1. *Al-Bīrūnī's testimony and his priority claim*

In several of his works, al-Bīrūnī claims credit for the invention of cylindrical projection. In his *Chronology* (*al-Athār al-bāqiyya ʿan al-qurūn al-khāliyya*), after having asserted al-Ṣāghānī's priority in generalizing the conic projection by displacing its pole along the axis of the sphere,[4] he claims priority for cylindrical projection.

> There is a type that I have called cylindrical. I have no evidence that, before me, any of the specialists of this art ever mentioned it. It consists in passing straight lines and planes through the circles and the points of the sphere, parallel to the axis. One thus produces in the plane of the equator [literally: the diurnal plane] only straight lines, circles, and ellipses.[5]

This is the orthographic projection. A passage on the same subject appears in his treatise on the *Plane Projection of Figures* (*Tasṭīḥ al-ṣuwar*), already cited; here, the author admits having gotten the idea of the cylindrical projection from a critical reading of al-Farghānī's *al-Kāmil*, in which the latter asserted the impossibility of such a projection:

> As to the cylindrical projection, it came to my mind on account of the abundance of absurdities in which al-Farghānī engaged at the end of his book, concerning the refutation of the astrolabe in the shape of a melon. I believe that I am the first to have attained this projection which I have called the [cylindrical] projection for a reason I will not go into here. It consists of an intermediate species, neither from the north nor from the south, and by

[2] Thābit ibn Qurra, *Œuvres d'astronomie*, edited and translated by R. Morelon, Paris, 1987.

[3] R. Rashed and H. Bellosta, *Ibrāhīm ibn Sinān, Logique et géométrie au Xe siècle*, Leiden, 2000.

[4] See below Section 2.4.

[5] Al-Bīrūnī, *Al-Athār al-bāqiya ʿan al-qurūn al-khāliya, Chronologie orientalischer Völker*, ed. C. E. Sachau, Leipzig, 1923, p. 357.

which one can project the stars of the celestial sphere in their totality on the plane of the celestial equator or on the plane of any given great circle.[6]

In his great book *On All Possible Methods for Constructing the Astrolabe* (*Istīʿāb al-wujūh al-mumkina fī ṣanʿat al-asṭurlāb*), al-Bīrūnī describes more precisely the principles of this projection, which he now calls 'perfect' (*kāmil*), because it makes possible the representation of the entire celestial sphere:

> This projection is constructed from the intersections of the equatorial plane with the lateral surfaces of cylinders with <a right circular section> and of cylinders with a right elliptical section with parallel sides (generatrices) and parallel to the axis of the sphere. Indeed, whenever one makes the surfaces of cylinders subject to the preceding condition pass through the circumferences of the celestial circles <parallel to the plane of the equator>, these surfaces cut the equatorial plane following circles that are parallel and equal to the sizes of the celestial circles. And when, through the circumferences of circles inclined to the sphere, whether large or small, one passes cylinders with a right elliptical section in the aforementioned position, the intersection forms on the equatorial plane ellipses of different positions and sizes.[7]

Al-Bīrūnī does not abandon the context of orthographic projection, perhaps because his project consists in studying workable methods for the astrolabe. We shall see that, one half-century before him, the mathematicians al-Qūhī and Ibn Sahl considered cylindrical projections more generally, but al-Bīrūnī very likely did not know of his predecessors' contributions.

He does not mention either the scientist who, under the stimulus of the Banū Mūsā (especially of al-Ḥasan, the youngest of them) studied the cylindrical projection in the context of their works on the cylinder and its plane sections. These works go back to antiquity: Serenus of Antinoeia's treatise *On the Cylindrical Section and on the Conic Section* applies Apollonius's method to the case of the cylinder, and thus develops the sketch of a theory of plane sections of the right or oblique cylinder with a circular base, which are ellipses. This work had no known sequel until this area of research was revived in 9th-century Baghdad.

[6] Al-Bīrūnī, *Tasṭīḥ al-ṣuwar wa-tabṭīḥ al-kuwar*, ed. A. Saidan in *Majalla ʿIlmiyya, al-Jāmiʿa al-Urduniyya*, vol. 4, nos. 1 and 2, 1977, pp. 7–22, at p. 14.

[7] Ms. Leiden 1066, fol. 93^{r}.

1.2. *Al-Ḥasan ibn Mūsā's study of the ellipse*

According to the biobibliographers, al-Nadīm and al-Qifṭī, al-Ḥasan ibn Mūsā in the 9th century composed a treatise on the ellipse called *The Elongated Circular Figure (Kitāb fī al-shakl al-mudawwar al-mustaṭīl)*, which is unfortunately lost. We have witnesses that attest to its authenticity, and contribute a few details about its content. The first are his brothers Muḥammad and Aḥmad, who, in their short treatise on *The Lemmas to the Book of Conics* (*Muqaddamāt kitāb al-Makhrūṭāt*), mention that al-Ḥasan composed a work on the generation of elliptical sections as well as on the demonstration of their areas.

> Thanks to his capabilities in geometry and the prominence of his standing in [this field], it became possible for al-Ḥasan ibn Mūsā to examine the science of the section of the cylinder cut by a plane not parallel to its base, and such that the line that surrounds the section does so completely. He then discovered the science, and the science of the fundamental properties associated with it, relative to diameters, axes and chords, and he discovered the science of its area.[8]

As if it were necessary, this testimony is confirmed by the late-10th-century mathematician al-Sijzī, who, in his treatise on *The Description of Conic Sections* (*Risāla fī waṣf al-quṭūʿ al-makhrūṭiyya*), cites the title of the work, attributing it to the Banū Mūsā without distinguishing them, and summarizes the procedure they applied for the continuous drawing of the ellipse by means of the bifocal property.[9] Al-Ḥasan very probably had only approximate knowledge of Apollonius's treatise, insofar as the version he had was not very readable, not to say incomprehensible. It was probably only after his death that his brother Aḥmad brought back from Damascus Eutocius's redaction, which then made it possible to understand *The Conics*.

Nevertheless, we know that these works by al-Ḥasan ibn Mūsā inspired a genuine research tradition on cylindrical sections beginning with the use of projections, among other geometrical transformations. Among these, one must count the magisterial study of Thābit ibn Qurra, a collaborator of the Banū Mūsā, and, a century later, that of Ibn al-Samḥ, the student of Maslama al-Majrīṭī in Cordoba, one of whose treatises takes up al-Ḥasan's

[8] *Apollonius: Les Coniques*, Tome 1.1: *Livre I*, historical and mathematical commentary, ed. and transl. from Arabic text by R. Rashed, Berlin/New York, Walter de Gruyter, 2008, Appendix I, p. 505, 4–8.

[9] R. Rashed, *Œuvre mathématique d'al-Sijzī*, vol. I: *Géométrie des coniques et théorie des nombres au Xe siècle*, Les Cahiers du Mideo, 3, Louvain/Paris, Éditions Peeters, 2004, p. 246.

results on this question. We will therefore return to al-Ḥasan's contribution by studying the works of his two successors.

1.3. *Thābit's treatise on the cylinder*

In his *On the Sections of the Cylinder and on its Lateral Surface* (*Kitāb fī quṭūʿ al-usṭūwāna wa-basīṭihā*), Thābit ibn Qurra goes farther down the path that al-Ḥasan ibn Mūsā had opened, and refers directly to him in his introduction.

> We shall continue by discussing [first] the area of a cylindrical section, which was determined by Abū Muḥammad al-Ḥasan ibn Mūsā – may God be pleased with him – and which is the ellipse that belongs to conic sections, and [second] the area of the kinds of portions of this section.[10]

In contrast to his predecessor, Thābit knew Apollonius's *Conics* very well, having translated Books V to VII and revised the translation of others. He drew inspiration from it for his own project, which is therefore distinct from al-Ḥasan's. Thābit elaborated a theory of the cylinder and of its plane sections analogous to that of Apollonius for the cone, but by using additional means, namely geometrical transformations: projections, homothetic transformations, and affinities. This same characteristic stands out when one compares the works of Thābit and of Serenus. The many analogies between them show that Thābit was immersed in Serenus's work, all the more so because his knowledge of the *Conics* allowed him direct access to results that Serenus had borrowed from Apollonius. One can, however, see the two paths diverge when Thābit ibn Qurra introduces projections. In effect, it was the means that al-Ḥasan ibn Mūsā put to work, utilized in the context of Apollonius's (and therefore Serenus's) method, that made it possible for Thābit to blaze his own trail. Even though this possibility was an option for both of them, thanks to their general definitions of the cone and the cylinder by generatrices, neither Apollonius nor Serenus took the projective path.

It is in Proposition 7 of his own treatise that Thābit introduces for the first time a cylindrical projection p of a plane (P) on a plane (P') parallel to (P), in the direction of (AE).

[10] *Les Mathématiques infinitésimales du IXe au XIe siècle*, vol. I: Fondateurs et commentateurs: *Banū Mūsā, Thābit ibn Qurra, Ibn Sinān, al-Khāzin, al-Qūhī, Ibn al-Samḥ, Ibn Hūd*, London, al-Furqān Islamic Heritage Foundation, 1996, p. 500; English translation: *Founding Figures and Commentators in Arabic Mathematics*. A History of Arabic Sciences and Mathematics, vol. 1, Culture and Civilization in the Middle East, London, Centre for Arab Unity Studies, Routledge, 2012, p. 381.

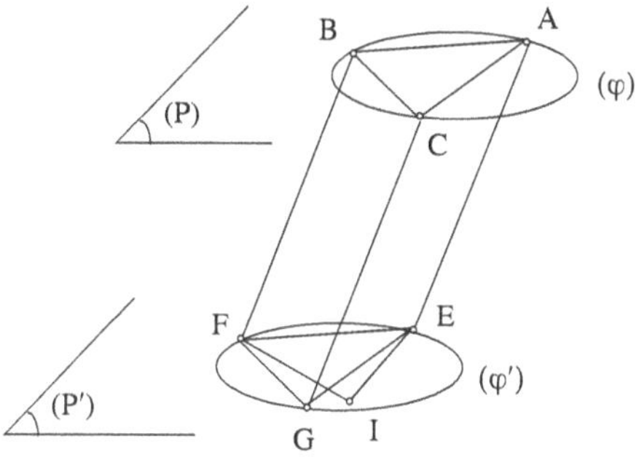

Fig. 65

This projection is indistinguishable from the translation of vector AE, but what follows proves that this is indeed a projection, notably in Proposition 10, where (P') is now no longer parallel to (P). In Proposition 7, Thābit shows that if a figure (φ) is contained in plane (P) and if $(\varphi') = p(\varphi)$, then the two figures (φ) and (φ') are similar. He considers two points A and B of (φ), and their images E and F through p. Then E and F are on figure (φ') and $ABFE$ is a parallelogram, which has the consequence that $AB = EF$ and allows one to superpose (*waḍaʿa ʿalā*) the segments $[AB]$ and $[EF]$. By *reductio ad absurdum* he shows that figure (φ) is superposed (*inṭabaqa ʿalā*) upon figure (φ') by supposing that a point C of (φ) is superposed, in the plane (P'), on a point I that is not on (φ'). By considering point $G = p(C)$, which by hypothesis is on (φ'), he reaches a contradiction because G and I must be the same. Hence the conclusion.

In the next proposition, he applies this result to the case in which (φ) is a circle. He thus uses the cylindrical projection to study the nature of the intersections of a cylinder's lateral surface with a plane parallel to the bases. Indeed, if a plane cuts a cylinder parallel to its bases, then the section one obtains is the image of one of the bases through the cylindrical projection in the direction of the axis of the cylinder, and therefore it is similar to the bases, that is circular, and even equal.

This Proposition 8 corresponds to Proposition I.4 of the *Conics*, and Thābit could have applied Apollonius's method to the case of the cylinder and shown that every point L of the section verifies $ML = \frac{d}{2}$, if M is the point where the secant plane meets the axis, and d is the diameter of the base, using Proposition 1.

In Proposition 9, which corresponds to *Conics* I.5 and concerns subcontrary or antiparallel sections, he obviously does not use a cylindrical projection. In this case, he naturally draws on Apollonius's method, which brings into play 'the equation' of the circle in relation to one of its diameters and to the tangent to one of the endpoints of its diameter, namely, the equation $y^2 = x(d - x)$, if d is the diameter.

Conversely, he reintroduces the cylindrical projection in Proposition 10, to demonstrate that the projection of a circle (*ABC*) of center *D* and contained in plane (*P*) on a plane (*Q*) not parallel to (*P*), is a circle or an ellipse. In this proposition, as in Proposition 7, he does not consider the cylinder, but only the projection *p* of (*P*) onto (*Q*) in the direction (*CG*). This time, the demonstration relies on the characterization of the ellipse that results from the reciprocal of *Conics* I.21. Thābit distinguishes two cases, according to whether or not plane (*Q*) passes through the center *D* of the circle to be projected.

In the first case (Fig. 66), plane (*Q*) passes through the center *D* of circle (*ABC*); it therefore cuts plane (*P*) according to a diameter (*AB*) of this circle. Let *E* be any point of circle (*ABC*) and $F = p(E)$, let *C* be one of the endpoints of the diameter of (*ABC*) perpendicular to (*AB*), and let (*CG*) be the direction of projection, with $G = p(C)$, then $(CG) \parallel (EF)$.

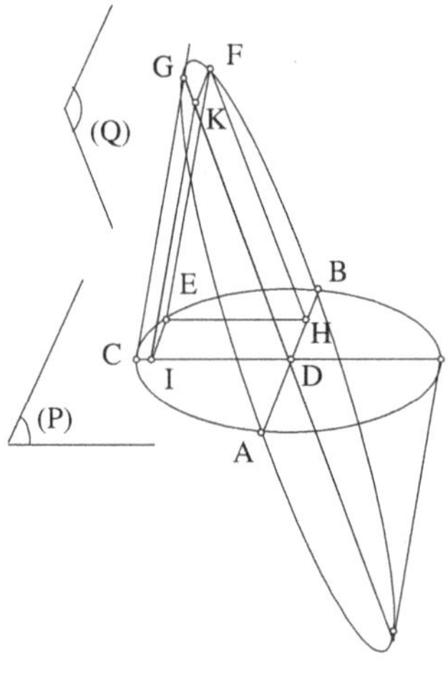

Fig. 66

Thābit ibn Qurra constructs the rectangle *EHDI* that leans on the two perpendicular diameters under consideration, and then draws point $K = p(I)$. He shows that *K* is on (*CG*), and then that $(FK) \parallel (HD) \parallel (EI)$, therefore $FH = KD$. Consequently the triangles *FEH* and *GCD* are similar. Thus $\frac{EH^2}{HF^2} = \frac{CD^2}{DG^2}$, whence $\frac{AH \cdot HB}{HF^2} = \frac{AD \cdot DB}{DG^2}$, which characterizes the ellipse with center D and one of whose axes is AB, according to the reciprocal of *Conics* I.21.

For the case in which plane (Q) does not pass through D, Thābit introduces plane (Q'), which is parallel to (Q) and passes through D. He then decomposes projection p into two projections p' of (Q') onto (Q), and p'' of (P) onto (Q'). According to the first case, the image of (ABC) is determined in (Q') by p''; the last remaining step is to apply projection p' between two parallel planes by using Proposition 7.

In Proposition 12, which opens the second part of his work devoted to measuring the area of the ellipse and portions of it, Thābit will again draw on projections to prove that a plane cuts two cylinders with the same axis and the same base planes according to two homothetic sections.

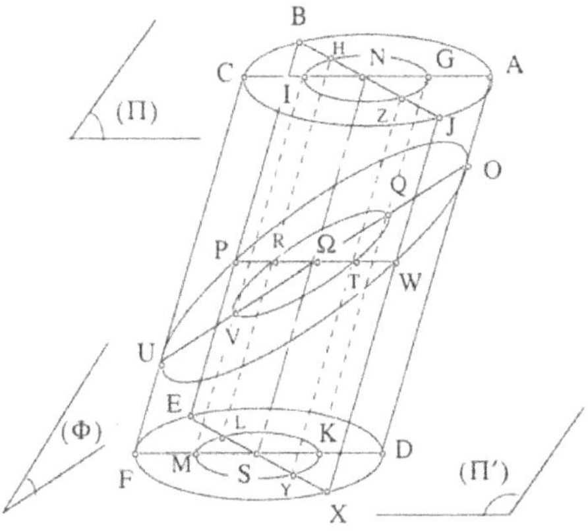

Fig. 67

He considers the cylinder (C_1) with, as bases circles (ABC) and (DEF) of diameter d_1, and the cylinder (C_2) with, as bases the circles (GHI) and (KLM) with diameter d_2, such that circles (ABC) and (GHI), which have the same center N, are contained in the same plane (Π), and that circles (DEF) and (KLM), which have the same center S, are contained in the same plane (Π'). The two cylinders therefore have NS as their axis. If the plane (Φ) cuts the two cylinders without intersecting their bases, then the two sections obtained, (OPU) in (C_1) and (QRV) in (C_2), are similar, that is, that the first is the image of the second by means of the homothetic transformation, whose center Ω is their common center located on (NS), and whose ratio is that of the diameters of the base circles $\frac{d_1}{d_2}$.

The problem arises only if plane (Φ) is neither parallel nor antiparallel to planes (Π) and (Π'), that is, if the sections (OPU) and (QRV) are ellipses with axes, $(2a_1, 2b_1)$ and $(2a_2, 2b_2)$, respectively.

Thābit ibn Qurra begins by establishing the homothetic ratio. He shows that $\frac{OU}{QV} = \frac{AC}{GI} = \frac{DF}{KM}$, by considering the projection p of (Π) onto (Φ), parallel to the axis (NS), and by placing himself in the principal plane of the cylinder – the one containing the axis (NS) and the diameter (AC) of the circle (ABC) – which cuts plane (Φ) along the straight line (OU), the major axis of the ellipse (OPU), and (QV), the major axis of the ellipse (QRV). Thus $\frac{OU}{QV} = \frac{2a_1}{2a_2} = \frac{d_1}{d_2}$.

Next he shows that, for every diameter (PW) of the section (OPU), which is co-linear with diameter (RT) of section (QRV), one finds the same equality of ratios, namely $\frac{PW}{RT} = \frac{d_1}{d_2}$. Indeed, by means of projection p, applied this time in the plane containing the axis of the cylinder and the diameter (BJ) of circle (ABC) such that $p(BJ)$ = (PW), he obtains $\frac{PW}{RT} = \frac{BJ}{HZ}$. But since the two base circles in plane (Π) are concentric, (ABC) is the image of (GHI) by means of the homothetic transformation $h\left(N, \frac{d_1}{d_2}\right)$, which he expresses with the equality $\frac{BJ}{HZ} = \frac{AC}{GI} = \frac{d_1}{d_2}$. In other words, at this point in his reasoning, the author has demonstrated that if δ_1 and δ_2 are two co-linear diameters of the sections (OPU) and (QRV) respectively, then $\frac{\delta_1}{\delta_2} = \frac{2a_1}{2a_2} = \frac{2b_1}{2b_2} = \frac{d_1}{d_2}$. He concludes with a permutation of the ratios in the central equation: $\frac{2a_1}{2b_1} = \frac{2a_2}{2b_2}$ (equation 1), and by referring to Proposition VI.12 of the *Conics*. In this proposition, Apollonius shows that, if two ellipses have as axis $2a_1$ and $2a_2$ respectively, and as associated *latera recta* c_1 and c_2 such that $\frac{2a_1}{c_1} = \frac{2a_2}{c_2}$ (equation 2), then they are similar, and reciprocally. In fact, equation (1), which Thābit ibn Qurra uses to characterize the two similar ellipses, was not established by Apollonius, but one can easily show that it is equivalent to equation (2).

If $2b_1$ and $2b_2$ are the minor axes of the ellipses, one has $4b_1^2 = 2a_1 \cdot c_1$ (according to *Conics*, I, Second Definitions III), whence $\frac{a_1^2}{b_1^2} = \frac{2a_1}{c_1}$ and likewise $\frac{a_2^2}{b_2^2} = \frac{2a_2}{c_2}$.

Thus: $\frac{2a_1}{c_1} = \frac{2a_2}{c_2} \quad \Leftrightarrow \quad \frac{2a_1}{2b_1} = \frac{2a_2}{2b_2}$.

In this Proposition 12, everything takes place as if Thābit were characterizing the similarity of the two ellipses by their correspondence in the homothetic transformation $h\left(\Omega, \frac{2a_1}{2a_2}\right)$, by decomposing the latter into three transformations: the two projections p and p' and the homothetic transformation $h\left(N, \frac{d_1}{d_2}\right)$ between the base circles.

Together, the propositions contained in Thābit ibn Qurra's treatise on *The Sections of the Cylinder and on its Lateral Surface* (*Kitāb fī quṭūʿ al-usṭūwāna wa-basīṭihā*) that we have just presented, constitute a very important contribution to the history of projections. Excepting the lost treatise of al-Ḥasan ibn Mūsā, known only from Ibn al-Samḥ's summary,[11] this is the first time that the concept of cylindrical projection appears and that a theoretical study is devoted to it, applied here to the case of a circle. Thābit studied the cylindrical projection of a circle onto a parallel plane and onto a nonparallel plane, in order to use it as a geometrical transformation, just as one would a homothetic transformation or an affinity (dilation or contraction), in order to solve the various problems that he encountered when studying the nature or the measurement of cylindrical sections.

1.4. *Ibn al-Samḥ's study of plane sections of a cylinder and the determination of their areas*

At the turn of the 10th century, Ibn al-Samḥ (368 H./979; 426 H./1035), a disciple of the famous mathematician Maslama al-Majrīṭī (d. 398 H./1007–1008), composed a 'great book of geometry that treats exhaustively all of its parts relevant to the straight, arched, or curved line', if we can trust Ṣāʿid al-Andalūsī's remarks in his *Categories of Nations* (*Ṭabaqāt al-umam*).[12] Ibn al-Samḥ's book is unfortunately lost, but the text that discusses the cylinder and its plane sections, and has come down to us in a

[11] See below.

[12] Ed. Būʿalwān, Beirut, 1985, p. 170.

Hebrew version, is very likely an excerpt from it.[13] Since the attribution is not in doubt, this text proves Ibn al-Samḥ's interest in a topic that we have already encountered among the preceding authors. In it, moreover, the Andalusian mathematician picks up on al-Ḥasan ibn Mūsā's treatise of on the ellipse. We would like to situate both this text in the tradition of studies on the cylinder, and the cylindrical projection in Ibn al-Samḥ's research.

A comparative study of the texts of Thābit ibn Qurra, Serenus, and Ibn al-Samḥ and the evidence for the lost treatise of al-Ḥasan ibn Mūsā lead us to the following conclusions. Everything suggests that Ibn al-Samḥ did not know Thābit's treatise, but that he started from the study of al-Ḥasan, to whom he remained closer than to Thābit himself. The direct comparison of the studies of Thābit and of Ibn al-Samḥ shows, on the one hand, that their projects, their methods, and even their lexica are different. Whereas Thābit elaborates a theory of the cylinder by adapting Apollonius's model for the case of the cone, and therefore begins from the definition of the cylinder by means of its bases and generatrices, Ibn al-Samḥ starts from the bifocal definition of the ellipse in order to show that the figure obtained in this way has the same properties as the figure produced by the intersection of a cylinder and a plane. To designate this figure defined by its foci, he uses the expression 'elongated circular figure', just as the youngest of the Banū Mūsā does, in contrast to Thābit. Conversely, the latter treats the case of the antiparallel section, which appears nowhere in Ibn al-Samḥ's text. From the testimony of his brothers, we know that, in his treatise on the ellipse, al-Ḥasan ibn Mūsā was concerned with ellipses, diameters, chords, and sagittas which are precisely the topics of Ibn al-Samḥ's last three propositions. Thus al-Ḥasan's study became the starting point for the research of his two successors, but Thābit ibn Qurra diverged from this project by using some results from the *Conics*, whereas Ibn al-Samḥ, even though he worked a century and a half later, carried out his research in the same spirit as al-Ḥasan.

On the other hand, a comparison of the two contributions also reveals commonalities: the concepts of projection, whether tied or not to two orthogonal affinities that transform each of the inscribed and circumscribed circles into the ellipse itself, and the calculation of the area of the ellipse from Proposition XII.2 of the *Elements*, and the apagogic method. These are borrowings from al-Ḥasan's research.

[13] See R. Rashed, *Founding Figures and Commentators in Arabic Mathematics*, Chap. VI.

To conclude this comparison, note that there is no connection between the texts of Ibn al-Samḥ and Serenus. Thābit alone was interested in the latter, on account of his borrowing from Apollonius's theory, as noted out earlier.

In Ibn al-Samḥ, the idea of the cylindrical projection underlies the cylinder. To begin, let us go over the various definitions of the cylinder that one encounters at the beginning of his text. First of all, in the general introduction of the part that survives, the author gives the definitions of the main solids: the sphere, the cylinder, and the cone, repeating Euclid's definitions in Book XI of his *Elements*. According to the definition he gives, the cylinder is 'what one obtains by fixing the side of a rectangle, such that it does not move, and by making the rectangle as a whole pivot about that side until it returns to its initial position'.[14] Ibn al-Samḥ considers here the right cylinder with a circular base. Then, in the next paragraph, he gives a more general definition:

> Given two round figures of any perimeter, located in two parallel planes; let their centers be determined and joined by a straight line. One makes a straight line revolve around the two figures parallel to the axis that joins their centers, until it returns to its initial position. What this straight line parallel to the axis describes is the cylinder.[15]

The author has now defined the oblique as well as the right cylinder, on condition that the two curves are deducible the one from the other by translation. Ibn al-Samḥ returns to this condition a little later by defining 'figures of similar positions'. If the round figures under consideration are circles, one returns to the definitions of Thābit ibn Qurra and of Serenus. In fact, according to the translator of the text, these round figures designate circles or ellipses, and in the next paragraph, Ibn al-Samḥ distinguishes four types of cylinders, associated two by two. First, the right cylinder with a circular base is associated with the oblique cylinder with an elliptical base, that is, one can generate the second from the first; to do so, one need only cut the latter with two parallel planes that are parallel to each other but not parallel to the bases. Likewise, if one cuts an oblique cylinder having a circular base with two planes perpendicular to its axis, the two elliptical sections generate, together with the cylindrical surface which they circumscribe, a right cylinder with an elliptical base.

[14] R. Rashed, *Founding Figures and Commentators in Arabic Mathematics*, p. 668.
[15] *Ibid*.

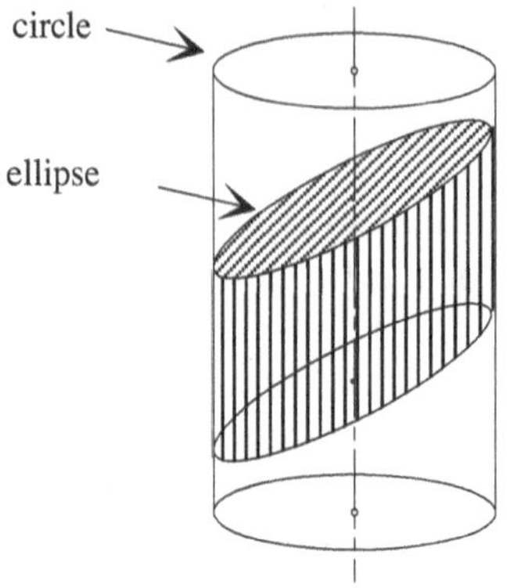

Fig. 68a

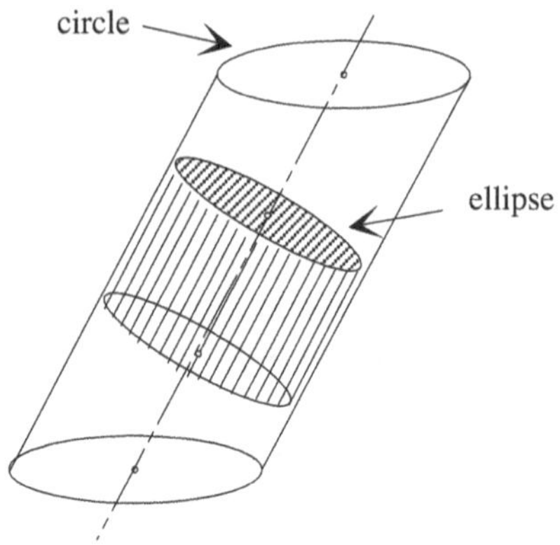

Fig. 68b

Ibn al-Samḥ notes next that, by proceeding inversely, one can also generate the two kinds with the circular base from the two kinds with an elliptical base. This property, which Ibn al-Samḥ highlights, reveals the underlying cylindrical projection that associates circles and ellipses, which are sections of the same cylindrical surface. This projection will appear in the course of Proposition 7 of his exposition, on which we will comment below.

Following this introduction, he moves on to the chapter devoted to the cylinder itself, and further generalizes his definition to cylinders whose base are closed curves that have a center of symmetry. This time, however, he details the conditions that these two curves must verify to generate a cylinder: they must be equal, of the same shape, and of similar positions which in today's terminology means that they are deducible from one another by translation. To do this, he begins by considering two closed curves that each have a center of symmetry C_1 and C_2, respectively, equal and of the same form, and two points M_1 and M_2, located respectively in the plane portion defined by each of these curves. The two points are said to be of similar position, when the straight lines drawn from M_1 to the curve C_1 are equal to their homologous counterparts drawn to C_2 from M_2. Nowadays one would say that, in relation to their respective polar coordinates M_1 and M_2, C_1 and C_2 have the same equation. Here, Ibn al-Samḥ does not count the polar angle in relation to an initial axis, but he compares the angles that two radius vectors make. If the curves C_1 and C_2 are located in parallel planes, and if they are cut by a plane passing through points M_1 and M_2 with similar positions along two equal straight lines, then the two curves are similar in position. Beginning from the two curves C_1 and C_2, of similar positions, he defines a cylinder generated by a straight line that turns by leaning on C_1 and C_2 while staying parallel to M_1M_2. In Proposition 7 of Thābit's treatise,[16] he studies explicitly the image of a plane figure by translation, which turns out to be a projection of a plane on a parallel plane. This proposition constitutes a kind of reciprocal of the preceding definition. The methods that the two mathematicians use are different, but the proximity of their results and the generality of Ibn al-Samḥ's definition, reinforce the underlying idea of projection in the latter's work.

Throughout the rest of his treatise, Ibn al-Samḥ returns to the special case of his first definition, and treats the case of the right cylinder with a circular base to study its plane sections – first the circle, then the ellipse, as a function of whether the cylinder is cut by a plane parallel, or not, to its bases.

Upon considering the structure of his treatise on the cylinder, one can distinguish three levels. At the initial level, one finds three independent parts in which the author demonstrates the properties first of the circle, then of the ellipse as an 'elongated circular figure' obtained from the bifocal definition, and finally of the ellipse as a plane section of the cylinder. At the second level, the author confronts the various properties to identify, among others, the ellipse-section with the elongated-bifocal circular figure. At the third level, he focuses, on the one hand, on the area of the ellipse, then on the other hand, on chords, sagittas, and finally diameters, which he calculates. Note in passing that Proposition 7, which concerns the properties of the ellipse defined as a plane section, plays a central role in this research, and is a prerequisite for seven propositions of the second level. It is precisely in this proposition that the use of the cylindrical projection appears clearly.

Ibn al-Samḥ introduces the ellipse as the plane section of a cylinder by recalling certain definitions and by specifying a property that he considers to be known: the section of a right cylinder with a circular base by a plane (P) that is parallel to its bases and that passes through the center of the ellipse obtained as a section of the same cylinder by a plane (Q) not parallel to (P), is a circle equal both to the base circle and to the circle inscribed in the ellipse, and having as a diameter the smallest diameter of the ellipse (that is, its minor axis GD).

[16] See above.

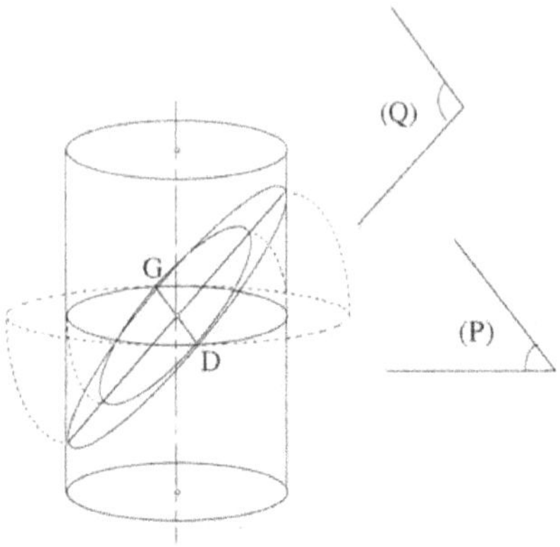

Fig. 69

The author adds that, if one makes plane (Q) rotate around the diameter of GD to bring it to (P), then the circle inscribed in the ellipse is superposed on the circle that is the section of the cylinder by (P). Thus, Ibn al-Samḥ justifies the fact that the circle of plane (P) is both the rabatment of the small circle of the ellipse and the orthogonal projection of the latter.

In Proposition 7, which immediately follows this statement, Ibn al-Samḥ posits the ellipse $ABGD$, with axes $AB = 2a$ and $GD = 2b$, with $AB > GD$, and its inscribed circle DEG of diameter GD. He tries to demonstrate that the ellipse is the image of the circle by orthogonal affinity of axis GD and of ratio $\frac{a}{b}$, that is, he shows that $\frac{HT}{HK} = \frac{AB}{GD} = \frac{a}{b}$. To do this, he will put in place the configuration of the preceding property by considering the right cylinder for which one of the bases is the circle DEG. Then, he can make the ellipse $ABGD$ pivot around GD, until it arrives on the cylindrical surface, which one obtains by making point A describe a circular arc of center N located in a plane perpendicular to DG at N, until point L, which belongs to the perpendicular to the plane $ABGD$ at E.

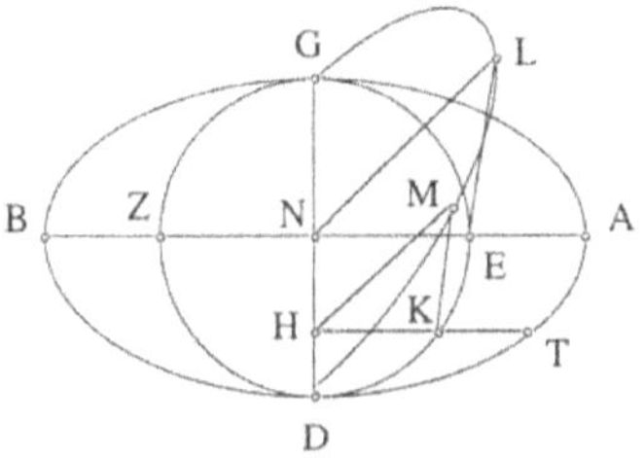

Fig. 70

One then only needs to show that point T also describes a circular arc of center H until point M located on the ellipse, which is a section of the cylinder. The triangles LNE and MHK are right and similar, therefore $\frac{LN}{NE} = \frac{MH}{KH}$. But $LN = AN = a$, $NE = NG = b$ and $MH = HT$. The result follows.

As with the preceding property, one shows that the ellipse $ABGD$ is therefore the rabatment of ellipse DLG, and the circle DEG is the cylindrical projection of the same ellipse in plane $ABGD$. We therefore find a rabatment and a projection underlying the orthogonal affinity. In this reasoning, the cylinder makes only a partial appearance, in a supporting role for the projection, leaving the main stage for geometrical transformations.

Ibn al-Samḥ will use this result when he wants to construct an ellipse defined by its foci and equal to an oblique section of a right cylinder with a circular base, and inversely in order to demonstrate the equivalence of the two figures. He relies on the orthogonal affinity to calculate the area of the ellipse, as Thābit ibn Qurra had done (and undoubtedly al-Ḥasan ibn Mūsā as well); we will encounter this same configuration again in Propositions 19 and 20, which pertain to the sagittas and the chords of the ellipse.

1.5. *The theory of projections: al-Qūhī and Ibn Sahl*

Some years before the study of Ibn al-Samḥ that we have just presented, al-Qūhī wrote a work that is apparently devoted to the astrolabe and offers a much more general exposition of projections. This is the first theory of the method of projections, or alternatively, of a local projective geometry of the sphere, amply completed by the commentary of his contemporary Ibn Sahl. As a matter of fact, in his *Treatise on the Art of the Astrolabe by Demonstration* (*Kitāb ṣanʿat al-asṭurlāb bi-al-burhān*), to be discussed in detail in the second part of this chapter, al-Qūhī is not interested in the practical problems that concerned the artisans who built astrolabes; he considers only the underlying geometrical theory. In the first chapter of the first book, he presents the method of projections, on which more than half of Ibn Sahl's commentary focuses. Since the astrolabe serves to study the rotation of the celestial sphere around one axis by projecting the latter on a movable surface superposed on a fixed surface, the two mathematicians were led to study generally the projection of a sphere of known axis BC onto a surface that may, or may not, be of revolution, and to distinguish, in their geometrical study, two cases for the surface of revolution, according to whether its axis is, or is not, parallel to the axis BC of the sphere. Al-Qūhī and, following him, Ibn Sahl then define the cylindrical projections in a direction parallel, or not, to the axis of the sphere, and the conical projections starting from a vertex belonging, or not, to this axis. To our knowledge this is the first occurrence of the expression 'cylindrical projection' – 'orthogonal' or 'oblique'; recall, however, that the concept is already found in Thābit ibn Qurra.[17] In al-Qūhī's own words:

> [...] the projection of the sphere is divided into two: one cylindrical, the other conical. The cylindrical projection is that which, starting from circles of the sphere, yields the cylinders with parallel axes falling on the surface (onto which the sphere is projected) and, starting from lines and points that are on the sphere, yields surfaces and straight lines parallel to these axes.[18]

Although rooted in problems associated with the construction of the astrolabe, the method that the two contemporary scientists present has freed itself from this context. Thus we find a classification of all the cylindrical and conic projections applied to the sphere, whereas only one of these – the stereographic – is necessary for the astrolabe. This classification draws on the nature of the support of the projection, on the one hand, and the nature of the projection lines, on the other. As we have said, the study of projections nevertheless remains linked to the context of the celestial sphere's motion. This is why Ibn Sahl, in covering al-Qūhī's discussions in greater detail and complementing them fulsomely, studies not only these projections but also the way in which the different cases allow the movable surface of the astrolabe to turn, even as it remains superposed on the fixed surface. From the point of view of their nature, this can happen only if these are surfaces of revolution. The author begins by considering the case in which the surface of the astrolabe is a plane, a case in which every perpendicular to this plane is then an axis for this plane. In this paragraph, we treat only the case of the cylindrical projections; but the two mathematicians whose research we are discussing treated conic projections in parallel. We will come back to these below.

Two situations occur depending on whether the axis BC of the sphere is or is not superposed on an axis of the surface. Ibn Sahl shows that in the case in which the axis BC is superposed on no axis of the plane surface, that is, when BC is not perpendicular to the surface, then the mobile surface does not remain superposed on the fixed surface during its motion. In the case in which the two axes are indistinguishable, he introduces the two conceptions of cylindrical projection:

[17] See above.

[18] R. Rashed, *Géométrie et dioptrique au Xe siècle: Ibn Sahl – al-Qūhī et Ibn al-Haytham*, Paris, Les Belles Lettres, 1993, pp. 191–2 and *Geometry and Dioptrics in Classical Islam*, p. 880.

1. The cylindrical projection of direction D parallel to BC (Fig. 71a)

If the axis BC of the sphere is also the axis of revolution of the mobile surface that it pierces at A, this point is the projection of points B and C. The rotation of any point M of the sphere around BC carries with it that of its projection M' around A, therefore around the axis BC. The mobile surface, which is the set of points M', remains superposed on its initial position, therefore superposed on the fixed surface. Note that, in the case in which the surface of the astrolabe is a plane, the projection thus defined is an orthogonal or an orthographic projection.

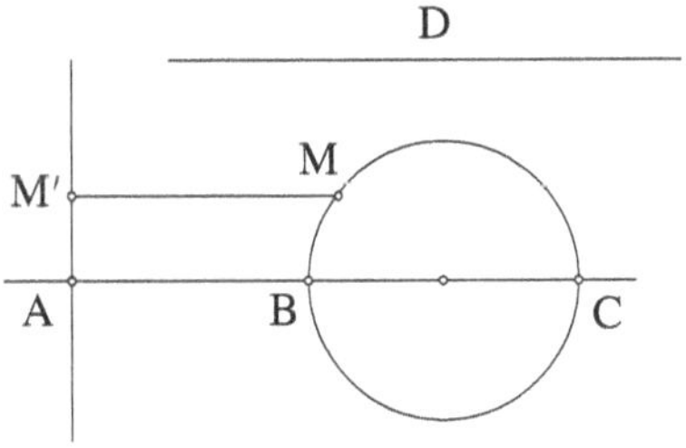

Fig. 71a

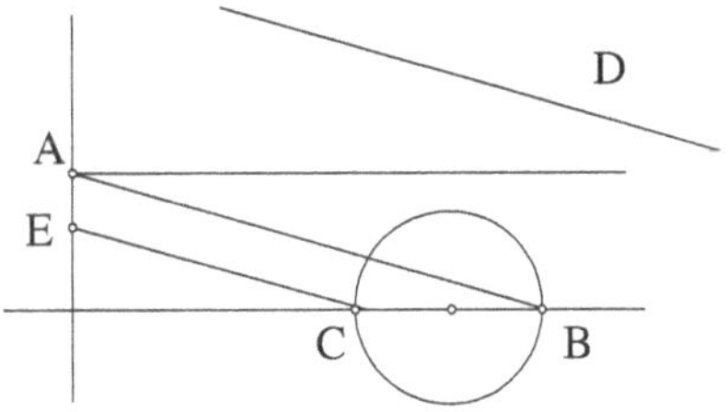

Fig. 71b

2. The cylindrical projection of direction D not parallel to BC (Fig. 71b)

Let A be the projection of pole B, and E be that of pole C on the sphere; B and C are fixed during the motion of the instrument, therefore A and E are as well. If M' is the projection of a point M of the sphere, the rotation of M around BC causes for M' an elliptical, therefore noncircular, trajectory. The surface on which the sphere is projected can therefore not turn around axis BC, because it has two fixed points, A and E.

If the two surfaces of the astrolabe are surfaces of revolution about axis $A\Delta$ but are not planes, then the mobile surface can only remain superposed on the fixed surface in the case in which $A\Delta$ and BC are indistinguishable. The situation is therefore compatible with the cylindrical projection parallel to BC.

After this detailed study of the different classes of projections and of the various conditions that the surfaces must fulfill for the instrument to be conceived, Ibn Sahl goes on to detail several properties of projections. First of all, the projection onto the surface of the astrolabe is obtained by the intersection of two surfaces. In the present case, if D is the direction of the cylindrical projection, then the latter associates a cylindrical surface with every circle of the sphere, the plane of which does not contain D or is not parallel to it. The projection of such a circle will be obtained by the intersection of this cylindrical surface with the surface of the astrolabe, whether cylindrical or conic. These intersections generally are not plane curves. For the case in which the plane of the circle contains the straight line *D*, or is parallel to it, the projection then associates with a circle a plane parallel to *D*, and the circle is projected according to the intersection of this plane with the surface of the astrolabe.

In his commentary on al-Qūhī's text, Ibn Sahl then returns to the concept of projecting line. He explains that, for the projection with which we are now concerned, the projecting line of any point is a straight line parallel to *D*, and the surface that is projecting any line *L* is generated by parallels to *D* drawn through every point of *L*, except if this line is a straight line parallel to *D*, in which case it is its own projecting line.

If the projection has as its direction the axis *BC* of the sphere, then the cylinder that projects a circle Γ of diameter *DE* cuts the sphere according to another circle Γ' of diameter $D'E'$; the two circles therefore have the same projection (Fig. 72a). The projection of any point of the spherical cap Γ is superposed on that of a point of the cap of base Γ'.

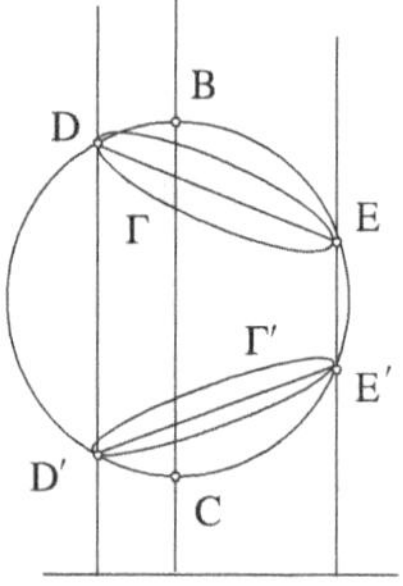

Fig. 72a

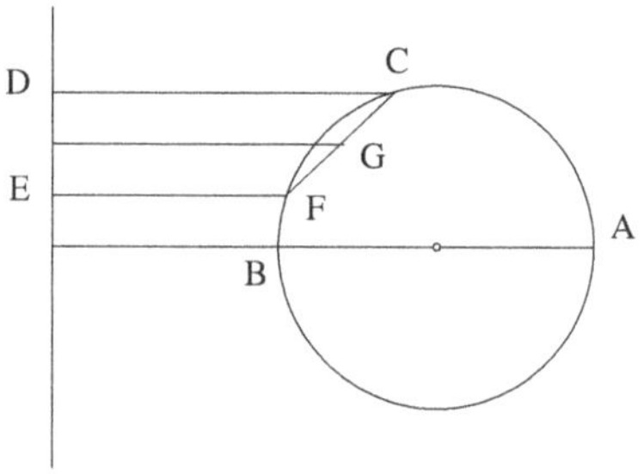

Fig. 72b

To complete his commentary on the cylindrical projection, Ibn Sahl studies the projection of a circle onto the plane surface, and states that it is a conic section, on condition, of course, that the plane of the circle not contain the direction *AB* of the projection, or not be parallel to this direction. He then leans on Thābit ibn Qurra's treatise, *On the Section of Cylinders*, which we introduced above, and in particular on Proposition 10, to show that if one projects the circle of diameter *CF*, from the sphere, onto the plane of the astrolabe, then the image is a conic section, an ellipse with minor axis *DE*. For this he considers the cylinder *CDEF* (Fig. 72b).

In the period that concerns us here, cylindrical projections were as important conceptually as conic projections, insofar as neither the one nor the other is tied directly to specific constructions, and this even though one conic projection, namely the stereographic, is more tightly linked than any other with the construction of the astrolabe. We now turn to conic projections and their development between the 9th and the 11th centuries.

2. CONIC PROJECTIONS

One of the mathematical sciences necessary for constructing maps of the heavens, the Earth, and the seas is 'the science of the projection of the sphere' (*ʿilm tasṭīḥ al-kura*). This science is as indispensible to the construction of astrolabes as it is to cartography. Although the need to construct maps in fact goes back to antiquity, not until the formation of the Islamic world did a science of the subject appear. This is why the field of studies of the astrolabe became such a large domain of inquiry for research on projections.

2.1. *Ptolemy's* Planisphere

The name of Hipparcus of Nicea, an astronomer of the 2nd century BC (*c.* –160 to –125), is associated with the oldest known study that explains the method of representing the celestial sphere on a plane, which would become the stereographic projection.[19] Nevertheless, the most ancient surviving text that treats this method is Ptolemy's *Planisphere*. According to O. Neugebauer, the main goal of this work is to demonstrate how it is possible to solve problems of spherical trigonometry by means of plane trigonometry alone. In his treatise, Ptolemy tries to 'represent' the elements of the sphere (the ecliptic, the circles parallel to the equator, the meridian circles) on a plane, in order to 'obtain a configuration that conforms to what appears on the solid sphere'.[20] As a matter of fact, he does not use the term projection, nor any other term that denotes a geometrical element linked to the transformation. To begin, he traces on the plane a circle that represents the equator and tries to place 'the other circles of the sphere correctly' in relation to it. He concedes that the diameters of this circle represent its meridians, and its center represents the north pole. From this he deduces that the parallel circles located to the north of the equator on the sphere must necessarily be represented inside the circle *ABGD* that is taken as the equator in the plane, and that the parallel circles located to the south must necessarily lie outside *ABGD* (Fig. 73). He then constructs two circles

[19] O. Neugebauer, 'The Early History of the Astrolabe – Studies in Ancient Astronomy IX', *Isis*, no. 40.3, 1949, pp. 240–56.

[20] C. Anagnostakis, *The Arabic Version of Ptolemy's Planisphaerium*, Ph.D. thesis, Yale University, 1984.

concentric to the circle *ABGD*, starting from the equal arcs *GZ* and *GH*, located on one side and the other of point *G*, by drawing the straight lines *DTZ* and *DHK* and by taking *ET* and *EK* as the radii of these two circles. He states that these two circles 'correspond' to two circles of the sphere located on one side and the other of the equator and equidistant from it. He does not demonstrate this, however; rather, it is a part of what he 'posits' in his representation and which must correspond to what appears on the sphere.

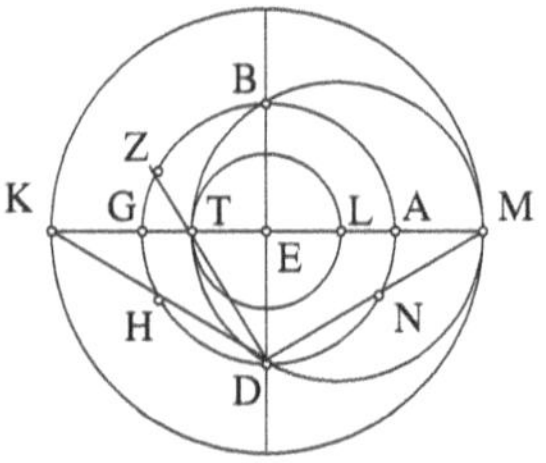

Fig. 73

Conversely, he demonstrates that the circle that is inclined (represented by the circle whose center is the middle of *TM*) and tangent to two circles at *T* and at *M*, cuts the equator *ABGD* in half at points *B* and *D*. To this effect, he constructs *DM*, which cuts *ABGD* at *N*. Then the arcs *AN*, *GH*, and *GZ* are equal, therefore *N* and *Z* are diametrically opposed on the equator; therefore the angle *MDT* is right, and the circle with diameter *TM* passes through *D*.

Thereafter, positing that the center *E* of circle *ABGD* represents the pole of the sphere, he demonstrates that every straight line passing through *E* represents a meridian of the sphere and cuts the ecliptic at two points *Z* and *H*, to correspond to points diametrically opposed on the sphere.

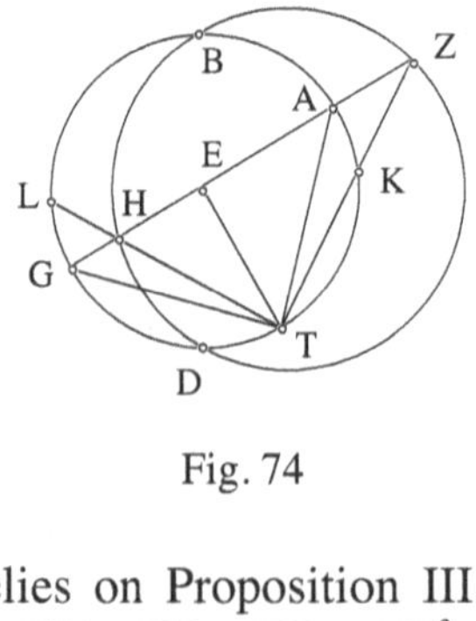

Fig. 74

His demonstration relies on Proposition III.35 of Euclid's *Elements*, according to which $ZE \times EH = ED \times EB = ET^2$, with *ET* perpendicular to

ZH. Therefore according to the reciprocal of Proposition VI.8 of the *Elements*, the angle *ZTH* is right, therefore equal to *ATG*. Thus the arcs *AK* and *LG* are equal, therefore the circles that are parallel to the equator and pass through the points represented by *Z* and *H* are symmetrically arranged in relation to the equator. These points are therefore diametrically opposed on the sphere.

Next he demonstrates that a horizon *GTAH* cuts in two halves the equator *ABGD* as well as the ecliptic *BTDH*; in other words, the points *T*, *E*, and *H* are aligned. This demonstration also relies on Proposition III.35 of Euclid's *Elements*.

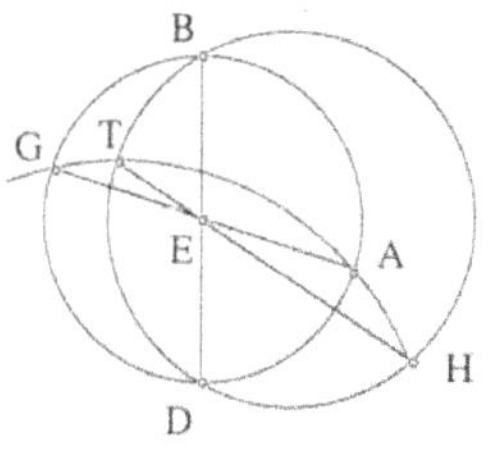

Fig. 75

These are the foundations of the method presented in the *Planisphere*. To name the operation that he uses to flatten the sphere, Ptolemy uses the words 'to trace', 'to represent', 'to correspond'. He tries to show that the principles of his representation are compatible with what happens on the sphere. This is the representation one obtains if one applies a stereographic projection to the sphere, but the *Planisphere* contains no mathematical treatment of this projection.

A reading of Federico Commandino's 16th-century commentary on the *Planisphere*[21] fuels this interpretation. Ptolemy does not explicitly define the plane of the equator as the plane of projection, since he does not define the projection. Commandino emphasizes this fact by saying that 'he supposes that Ptolemy represents the circles in the plane of the equator'. In fact, in his commentary, Commandino verifies mathematically – at least for the first few propositions – that the representation Ptolemy proposed is consistent with the stereographic projection. It is he who introduces the cross-section of the sphere, and he verifies that the elements traced by Ptolemy in the plane are indeed the images that one obtains in the plane of the equator after a conic projection of the elements of the sphere from the south pole.

[21] *Ibid.*, pp. 145–70.

In the version of the *Planisphere*'s text that has come down us, the style changes slightly in the third part. At paragraph 16, one finds an element that pertains to the projection, when Ptolemy takes a point as the hidden pole, that is, the pole of projection. At paragraph 18, when the author considers the intersections of the planes of two circles of the sphere, he implicitly uses a rabatment. Finally, in paragraph 19, he determines the image of a circle, parallel to the ecliptic, and passing through the pole, by considering the intersection of the plane of the circle and the equatorial plane. As one can see in this section, both the language and the geometrical operation change: we are dealing with projections. This fact leads us to suppose that the version we have today, via its Arabic translation, is not authentic.

2.2. *Al-Farghānī*'s *treatise*, al-Kāmil fī ṣanʿat al-asṭurlāb

In the middle of the 9th century, one finds a conjunction of conditions necessary for the science of projections to separate from astronomy: a multiplication of studies on different kinds of projections, as well as the use of Apollonius's *Conics*. The main problem that astronomers as well as geographers encountered was to be able to conceive a projection to represent precisely the sphere, and the lines and circles traced on it; for the mathematicians, however, it was necessary to establish this conception on solid geometrical foundations. The encounter between the geometry of conics and the problem of the projections necessary to elaborate a theory of the astrolabe among others will occur in al-Farghānī's *al-Kāmil*, which marks the founding act of the new geometrical discipline of projections.[22]

In his treatise on the theory of the astrolabe, al-Farghānī devotes an entire chapter to its geometrical foundations: 'Introduction to the geometrical propositions by which one demonstrates the figure of the astrolabe'.[23] His use of the qualifier 'geometrical' is fundamental. As was usual at the time for treatises on the astrolabe, al-Farghānī begins by treating the problem of a conic projection of a sphere, or its exact representation, before turning to a study of the astrolabe proper as an astronomical instrument. But he crosses an essential threshold by carrying out a purely geometrical study of conic projections. The problem can be translated into the following terms:

[22] See R. Rashed, 'Les mathématiques de la terre', in G. Marchetti, O. Rignani and V. Sorge (eds), *Ratio et superstitio*, Essays in Honor of Graziella Federici Vescovini, Textes et études du Moyen Âge, 24, Louvain-la-Neuve, FIDEM, 2003, pp. 285–318.

[23] Mss London, British Library, or. 5479, 5593; Kastamonu, 794.

Let A be the pole of the conic projection, and (Γ) be a circle of center ω in plane (P); the conic projection of circle (Γ) onto plane (Q) is the intersection of the plane and the conic surface defined by the vertex of A and the circle (Γ).

The nature of the conic section obtained – circle, ellipse, parabola or hyperbola – depends on the choice of plane (Q). For the construction of the astrolabe, the question is: how to choose (Q) so that the projection of a circle (Γ) is a circle (Γ')? The first book of Apollonius's *Conics* gives an indirect answer to this question, in that he never treats the issue of projection in his fundamental work. The problem is formulated in other terms that pertain to the nature of the intersection of the conic surface and a plane. In Proposition 4, where $(Q) \parallel (P)$, he shows that the intersection is a circle whose center is aligned with the center of the first circle and the vertex of the cone (Fig. 76a). If one translates this proposition into the later language of projection, the intersection (Γ'), and therefore the image of (Γ) by the conic projection from pole A, is the transformation of (Γ) through the homothetic transformation h of center A, such that $(Q) = h(P)$. Thus $\omega' = h(\omega)$ and A, ω, and ω' are aligned. But such was neither Apollonius's intention, nor his manner of proceeding.

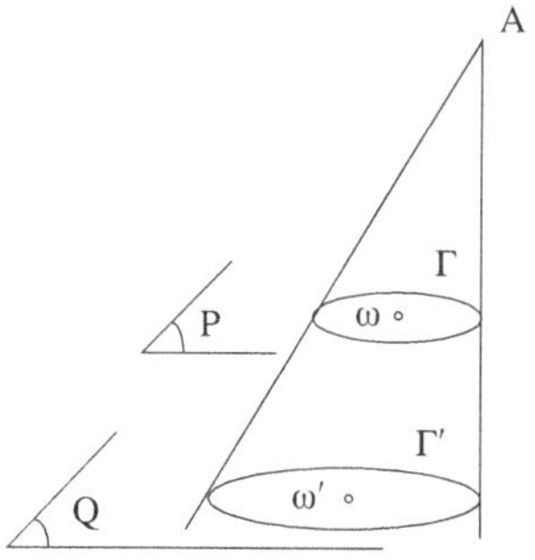

Fig. 76a

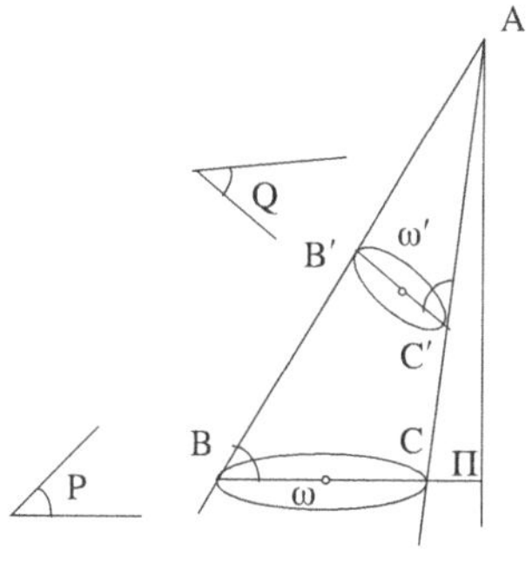

Fig. 76b

In Proposition 5, Apollonius considers the principal plane of the cone, that is plane (Π), which contains axis $A\omega$ of the cone and the perpendicular dropped from A onto the plane of the circle. This plane is a plane of symmetry for the conic surface. Let BC be the diameter of (Γ) in plane (Π). Then Apollonius considers a plane (Q) perpendicular to (Π) and that cuts it along $B'C'$ such that the angle $AC'B'$ is equal to the angle ABC (Fig. 76b). He shows that the intersection of the cone by plane (Q) is a circle of diameter $B'C'$. This proposition is known by the name 'subcontrary' or 'antiparallel sections'.

Al-Farghānī knew Apollonius's work. He drew broad inspiration from it, but by taking the new point of view of projections on account of the context of his study, which concerns the construction of the astrolabe. He begins his geometrical study by demonstrating the following lemma:

Given a circle of diameter AG, the tangent at G to this circle and any chord BC. The projections from pole A of points B and C onto the tangent are respectively I and K. Then the triangles ABC and AIK are similar.

Indeed, the angles AGB and ACB are equal, because they are inscribed and intercept the same arc AB, and the angles AGB and AKG are also equal, because they have the same complement, angle BAG. Thus angles ACB and AKG are equal (Fig. 77). Likewise, one can show that angles CBA and KIA are equal.

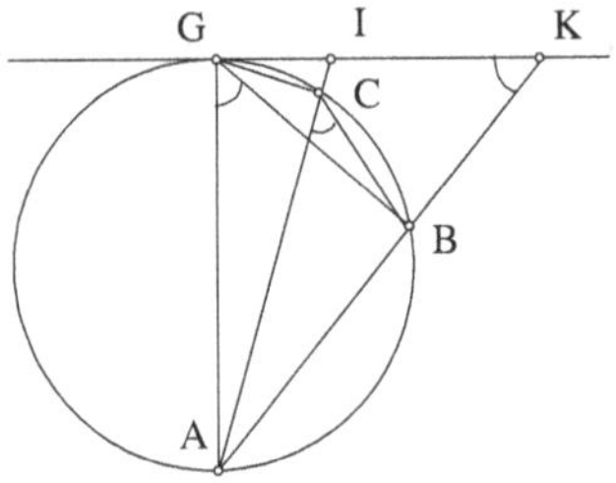

Fig. 77

In another language, one can interpret the result as follows:

Note that in the triangles GAK and GAI, respectively of height GB and GC, one has $AG^2 = AB \times AK = AC \times AI$; thus, through the inversion T of pole A and of the power AG^2, one has

$$I = T(C) \text{ and } K = T(B).$$

By the equalities of the angles of the two triangles, the points B, C, I, K belong to a circle invariant through the inversion T.

This lemma amounts to stating that: the conic projection of a chord, from pole A, onto the tangent at the point diametrically opposed to the pole, is a segment of the tangent such that the endpoints of the chord and of the segment are on a circle invariant through the inversion T with the same pole A, which transforms the given circle into a tangent straight line.

In the second proposition, al-Farghānī treats the problem of the projection of a circle on a sphere onto a plane tangent to that sphere.

Let AG be the diameter of a sphere, and let (Q) be the plane tangent to it at G. One considers the cone of the vertex A which has as its base circle (Γ) with center ω. Let BC be its diameter in the plane (Π) defined by A, G, and ω, a plane of symmetry for the sphere, the cone, and plane (Q). The intersection (Γ') of the conic surface (A, Γ) with plane (Q) also admits plane (Π) as plane of symmetry. The figure in plane (Π) is that of the lemma and GIK is the tangent at G.

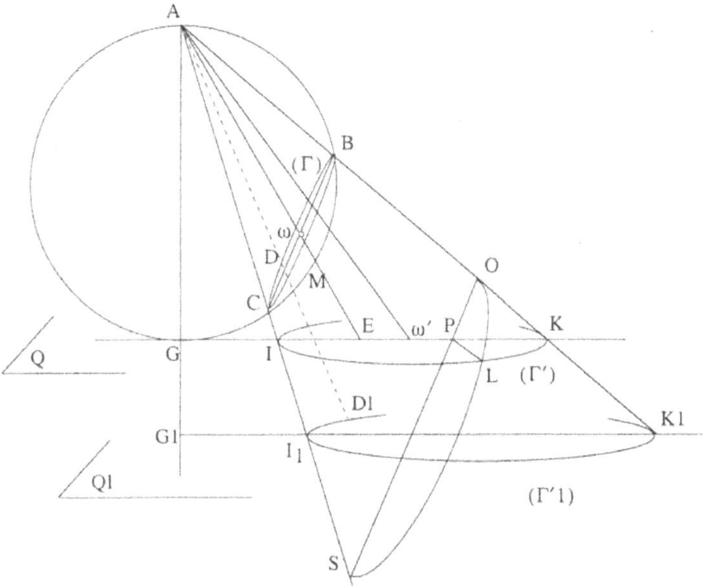

Fig. 78

Al-Farghānī begins by studying (Γ'), the projection of (Γ). Given that L is a point of (Γ'), the plane passing through L and parallel to the plane (Γ) cuts the conic surface along a circle SLO, of diameter OS, a circle that is the homothesis of circle (Γ). The planes IKL and SLO are both perpendicular to plane (Π), therefore their intersection is as well. Also, the diameter OS cuts IK at P, and one therefore has $PL \perp IK$ and $PL \perp OS$.

Now, according to this lemma, the angles ACB and IKO are equal; but $SO \parallel BC$, whence the angles ACB and ISO are equal, therefore the angles IKO and ISO are also equal, and the two triangles KPO and SPI are similar. From this one therefore deduces $\frac{KP}{PS} = \frac{PO}{PI}$, which also has as a consequence $PI \times PK = PS \times PO$.

But OLS is a circle of diameter OS and $PL \perp OS$, therefore $PL^2 = PS \times PO$. Therefore, for every point L of the curve (Γ'), with $PL \perp IK$, one has $PL^2 = PI \times PK$, and the curve (Γ') is a circle of diameter KI.

In a third proposition, al-Farghānī shows that the straight line $A\omega$ that passes through the center of (Γ) cuts IK at a point E that is not the center of (Γ'). Indeed, the straight line $A\omega$ cuts the great circle $ABCG$ at M and one has: arc $CM <$ arc MB, because $AB < AC$ and ω is the middle of BC, therefore angle $CAM <$ angle MAB. Through A, one draws the half-line $A\omega'$ such that the angle $BA\omega'$ is equal to angle $CA\omega$. According to the lemma, the angles $AK\omega'$ and $AC\omega$, are equal, therefore the triangles $AC\omega$ and ABC are similar to triangles $AK\omega'$ and AIK respectively; one therefore has

$$\frac{AC}{AK} = \frac{C\omega}{K\omega'} = \frac{BC}{IK}.$$

Now $C\omega = \frac{1}{2}BC$, therefore $K\omega' = \frac{1}{2}IK$, the point ω' is the center of (Γ'), and one has $\omega' \neq E$.

Al-Farghānī has thus demonstrated that, if one considers a sphere of diameter AG, and the plane (Q) tangent at G, the conic projection from pole A onto the plane (Q) of every circle (G) traced on the sphere is a circle (Γ').

It is clear that every plane parallel to the plane tangent at G will cut the conic surface along a circle. One thus obtains the section that Apollonius had studied in Proposition I.5 of the *Conics* and that he calls 'the subcontrary section'.[24] The method of demonstration is the same: he uses the property of the height of a right triangle that leads to the equation of a circle (the power of a point):

$$\left[PL^2 = PI \times PK\right] \Rightarrow \left[y^2 = x(d-x)\right], \text{ if one posits } \overline{IK} = d, \overline{IP} = x, \overline{PL} = y.$$

Despite this similarity, there are also differences that are equally important. On the one hand, Apollonius's proposition does not pertain to the sphere; and, on the other, al-Farghānī is not concerned with the 'subcontrary section'. In other words, the object and the goal of the research are different in each case, even if the procedures of demonstration are borrowed.

[24] R. Rashed, *Apollonius: Les Coniques*, Tome 1.1: *Livre I*, p. 272. *Apollonius of Perga, Conics, Books I–III*, transl. by R. Catesby Taliaferro, Santa Fe, Green Lion Press, 2000, p. 9.

Indeed, consider the evolution of al-Farghānī's research beginning with the lemma, and note that the mathematicians as well as the later bibliographers will integrate it into 'the science of projections', despite the absence of a terminology of projections, which will soon appear later. If one keeps in mind the foregoing along with the author's goal, one can easily see that we are no longer on the terrain of Apollonius, even though the *Conics* furnished the means to begin this research on conic projections. Conversely, if the idea of inversion seems latent in the lemma, it is such in the proposition as well. Indeed, the plane tangent at G is the image of the sphere under the inversion T of pole A and of power $AG^2 = p$, and every point D of circle (Γ) has as its image a point D_1 on circle (Γ'), such that $\overline{AD} \times \overline{AD_1} = AG^2$.

The two circles (Γ) and (Γ') belong to the same sphere that admits as a great circle in plane (Π), the circle that circumscribes a quadrilateral *BCIK*. This sphere is invariant under the inversion T. In a word, al-Farghānī succeeds in showing that the projection of a sphere that has point A as pole onto the plane tangent to the diametrically opposite point, or onto a plane parallel to this plane, is the stereographic projection.

After al-Farghānī's work in the 9th century, this research was carried out by many other prestigious mathematicians, including Abū al-ʿAlāʾ ibn Karnīb, Abū Yaḥyā al-Māwardī, Ibn Maʿdān, Ibn Sinān. To appreciate all the development of this new discipline during approximately one century, one need only examine the writings of al-Qūhī, Ibn Sahl, and al-Sijzī. The first two were closely linked in their researches insofar as Ibn Sahl wrote a commentary on al-Qūhī's treatise. We now turn to their texts.

2.3. *Al-Qūhī's treatise and Ibn Sahl's commentary on it*

Al-Qūhī's *Treatise on the Art of the Astrolabe by Demonstration*, mentioned above about the cylindrical projection, has not reached us entire. It appears in two books, the second of which is stripped of three of the seven original chapters, namely Chapters 3–5. Also missing from Book 2 is a large part of the demonstration of the sixth and last proposition of Chapter 2, which had to be reconstituted for the edition.[25]

The first book, composed of four chapters, is complete. All the problems presented in it are solved by synthesis. In Chapter 1, after a global presentation of the instrument in an intentionally rigorous style, the author presents, in the most general form available at the time, the first elements of a theoretical study of the projection of the sphere which, although it

[25] R. Rashed, *Geometry and Dioptrics in Classical Islam*.

originates in a context linked to the construction and use of the astrolabe, tries to free itself completely from the latter and to become purely geometrical. The exposition is nevertheless so short and condensed that the development of its content will be the focus of most of Ibn Sahl's commentary. It is from this exposition and the commentary on it that we elicited the remarks on cylindrical projections reconstituted above. The two mathematicians jointly treat conic projections. Recall that, at first, Ibn Sahl takes up the case in which the surface of the astrolabe is a plane. From the two situations that then emerge, according to whether the axis *BC* of the sphere is superposed or not on an axis of the surface, only the case in which the two axes are superposed is compatible with the motion of the instrument. In this case, the author goes into detail on two conceptions of conic projection:

1. The conic projection originates from a point *D* on the axis *BC* (Fig. 79a)

If $D \neq B$ and if $D \neq C$, then *A* is the projection of the two points *B* and *C*. If $D = B$, *A* is the projection of *C*, and if $D = C$, *A* is the projection of *B*. Since *B* and *C* are fixed, *A* is as well, and it is the only fixed point of surface *A*. The latter can therefore spin on the other surface.

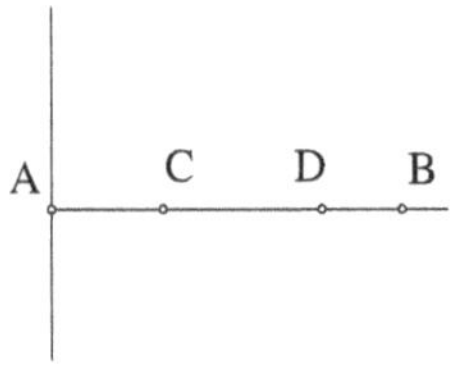

Fig. 79a

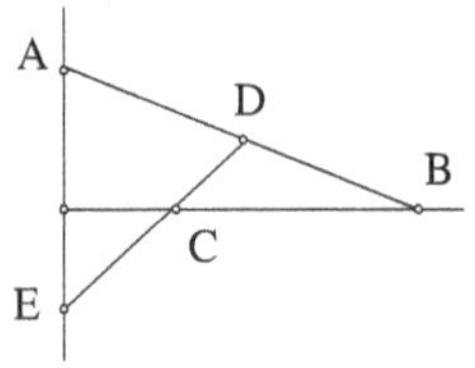

Fig 79b

2. The conic projection originates from a point *D* which is not on the axis *BC* (Fig. 79b)

In this case, the poles *B* and *C* have different projections, given *A* and *E* not on *BC*. The surface therefore has two fixed points *A* and *E*, and consequently cannot turn by remaining superposed on the other surface.

When the two surfaces of the astrolabe are of revolution around axis $A\Delta$, but are not planes, one returns to the case of the conic projection in which the pole is on *BC*.

Ibn Sahl studies the concept of projecting surface in the case of the conic projection, in parallel with the case of the cylindrical projection that we have already presented. In a conic projection originating from a point *B*, the projecting surface of a circle is generally a conic surface with a vertex *B*, except if *B* is in the plane of the circle; in this case the projecting surface is on the plane itself.

In a conic projection, if the vertex *S* of the cone is on the axis *BC*, then the cone projecting a circle *Γ* of diameter *DE* cuts the sphere along another circle *Γ′* of diameter *D′E′*; the two circles therefore have the same projection (Fig. 80). The projection of any point of the spherical cap with base *Γ* is superposed on that of the point of the cap with base *Γ′*.

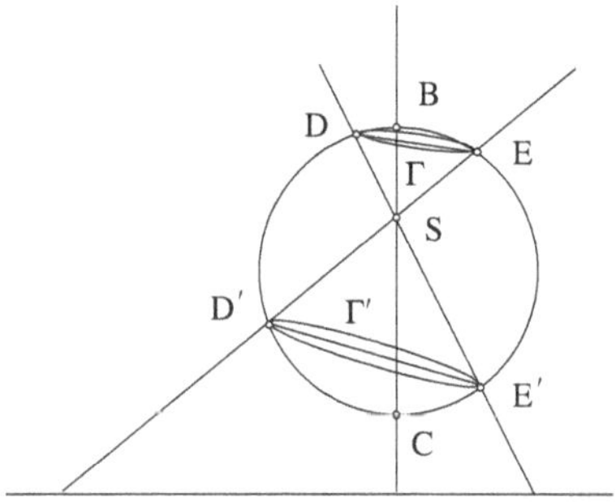

Fig. 80

Taking the same route he had adopted for the cylindrical projections, and after considering – in order to set them aside – the exceptions involving the circles the plane of which contains the vertex of the cone, Ibn Sahl examines the projection of a circle of diameter *CF* onto a plane surface and perpendicular to the axis *AB* of this sphere, where the vertex of the cone is point *G* of axis *AB*. Two cases come up: either $G \in [AB]$, or $G \in [AX)$.

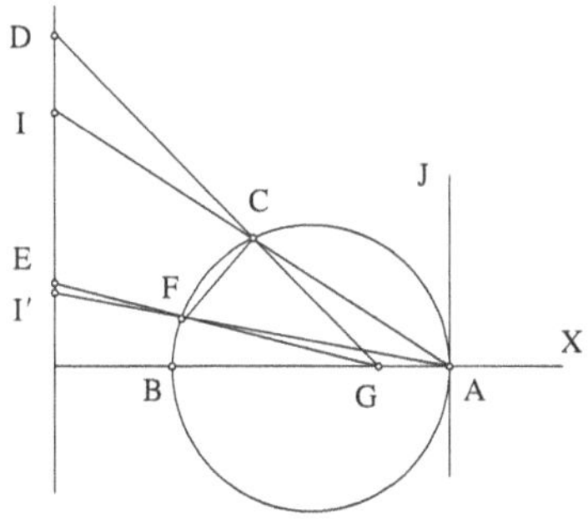

Fig. 81a

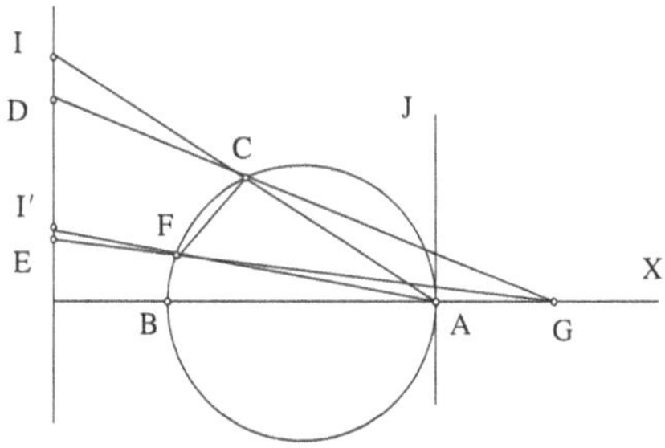

Fig. 81b

In the first case (Fig. 81a), angle *GFC* is greater than angle *AFC*, and angle *AIE* is greater than angle *GDE*; and in the second case (Fig. 81b), it is the contrary, namely, angle *GFC* is smaller than angle *AFC*, and angle *AIE* is smaller than angle *GDE*. In both cases, if *AJ* is the tangent at *A* to the circle of diameter *AB*, $AJ \parallel DE$; therefore angles *AFC*, *IAJ*, and *AIE* are equal. In the first case, therefore, angle *GFC* is greater than angle *GDE*, and in the second, it is smaller. Therefore, according to Apollonius, since *CF* is a circle, its conic projection *DE* is a noncircular conic section.

Ibn Sahl deliberately avoids treating the cases in which *G* is found at *A* or at *B*, which correspond to the stereographic projection. Indeed, by contrast, this projection is the focus of the entire remainder of al-Qūhī's treatise, which studies it with great care.

Thus, still in Chapter 1, al-Qūhī continues his exposition by demonstrating the fundamental property of the stereographic projection, namely that the image of every circle of the sphere is either a circle or a straight line, the latter case occurring when the circle to be projected passes through the pole of projection.

One wishes to project onto plane (*P*), which is perpendicular to axis *AD* of the sphere, a circle of diameter *BC* that does not pass through pole *A* of the projection. If points *B* and *C* are projected onto (*P*) at *E* and *G*, then the projection that was sought is the circle of plane (*P*), with diameter *EG*. Note that the author always seeks the greatest possible generality, taking plane (*P*) at random: it is neither the plane of the equator, nor the plane tangent to the sphere at *D*.

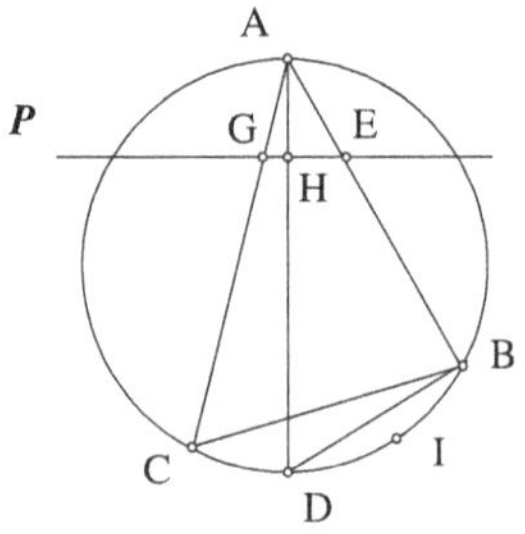

Fig. 82

To prove this property, which al-Farghānī had already demonstrated,[26] al-Qūhī will very explicitly use Proposition I.5 of the *Conics*, to which we alluded above. The fundamental property of the stereographic projection is indeed a direct application of the property of subcontrary or antiparallel sections. The plane (*ABCD*) of the meridian of the circle to be projected contains the triangle *ABC* and is perpendicular both to plane (*P*) and to the plane of the circle. It is these hypotheses that make it possible to find the fundamental configuration of Apollonius, based on the main plane of the cone. Here, it is the plane of the meridian of the circle to be projected that

[26] See above.

plays this role. One can therefore apply Proposition I.5 of Apollonius's *Conics*, once triangles *AGE* and *ABC* have been shown to be similar. This last point is a consequence, on the one hand, of the similarity of the right triangles *ABD* and *AHE* and, on the other hand, of the equality of the angles *ACB* and *ADB*. One can therefore conclude that the intersection of the cone with plane (*P*), which is perpendicular to the axis of the sphere, is a circle.

If the circle of diameter *BC* passes through pole *A*, then the circle is projected according to the intersection of its plane and of plane (*P*), therefore along a straight line.

From here to the end of the treatise, al-Qūhī does not leave the framework circumscribed by this projection, which he voluntarily introduced as a particular case of conic projections of the sphere; that is, the case in which the pole of the projection is identical to one of the poles of the sphere. The author nevertheless insists on the general character of this first chapter and on a certain independence of the projection with respect to the astrolabe. Indeed, beyond his treatment of the projections we have just discussed, he considers the projection of the sphere 'onto a plane to which the axis of the sphere is perpendicular'. Only in the next chapter will he project the sphere 'onto the plane of the astrolabe'. This expression appears for the first time at the beginning of Chapter 2 of Book 1. It nicely reveals the difficulty that al-Qūhī faces, for he cannot separate this line of research from the context of problems raised by the construction of the astrolabe, that is, the 'land of its birth'. He then presents the astronomical terminology specific to the astrolabe (Chapter 2), and reminds the reader of the two procedures for constructing the astrolabe, from the north pole and from the south.

In Chapter 3, he gives an exposition of and demonstrates the constructions of the *muqanṭarāt*, the projections of the circles of equal height for a given horizon. The problem consists in finding, on the plane of the astrolabe, the projection of a circle parallel to the horizon, known from its angular distance α between its pole and the pole of the sphere, that is, α is the complement of the latitude of the place for which the horizon is defined. This circle labeled Γ has as a diameter *IK* on the sphere and is itself defined by the angular distance to its pole, labeled β, which therefore represents the complement of the height of this circle on the horizon.

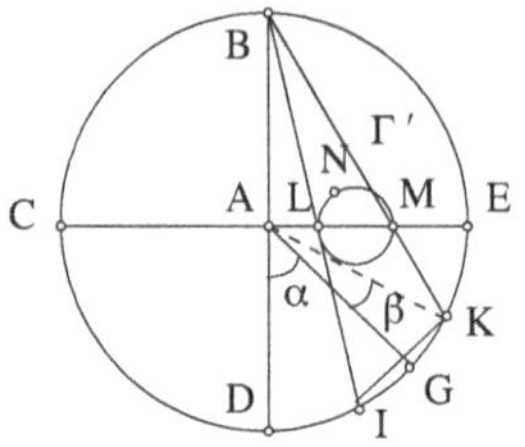

Fig. 83

On the circumference (*BCDE*), the meridian of circle Γ, one places point G, the pole of the known horizon and therefore also of the circle. Then one places point I, such that arc GI is equal to b, and point K is symmetrical to I in relation to AG. Circle Γ therefore has IK as its diameter. If the points L and M are the projections of I and K respectively, then the circle Γ' of diameter LM and passing through N is the projection of Γ onto the plane of the astrolabe.

The operation of projection that the author carries out consists of two stages: 1) the projection onto the equatorial plane; 2) the rabatment onto the plane of the figure as plane of the astrolabe. As is the case for all the problems in Book 1, al-Qūhī presents the construction on the plane of the astrolabe, then demonstrates that the elements projected onto the equatorial plane correspond by rabatment to those constructed on the plane of the astrolabe. This concept of rabatment, which he uses throughout his work, appears here for the first time. The author's systematic usage of this concept in Book 1 of his work marks a change in both the lexicon he uses and the style of his demonstrations.

In the following chapter, the fourth and last of Book 1 of his treatise, al-Qūhī turns to the construction of azimuthal circles (*sumūt*), projections of circles of height for the given horizon.

Assuming the meridian (*BCDE*) in the plane of the astrolabe, and a horizon through its poles G and I, that is, angle α, he proposes to construct the image of a circle of height, labeled Λ. This great circle of the sphere, which passes through poles G and I of the known horizon, is defined by the angle γ that it forms with the meridian of the horizon, that is, the complement of the azimuthal arc. To do so, he considers an auxiliary circle of the sphere, parallel to the horizon, therefore with poles G and I, and of diameter KL in the plane of the meridian. In effect, he considers the rabatment of this circle onto the plane of the astrolabe, about KL. Let us call it Γ.

Thanks to this rabatment, al-Qūhī can represent the angular distance γ by means of a point S of this circle, such that arc LS is equal to γ. He takes pains to consider all the cases pertaining to this auxiliary circle Γ, beginning with the general case of a circle that does not pass through the pole of projection and that is not the horizon. Then, after noting the simplification of the construction when one chooses the horizon, he treats the case in which this parallel circle is the one which passes through pole B of the projection. This discussion emphasizes the author's desire to carry out an exhaustive study and, let us note once again, within a completely geometrical and theoretical framework.

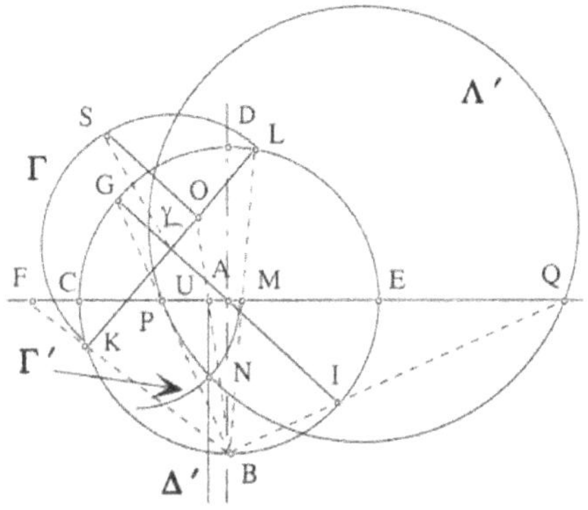

Fig. 84

Al-Qūhī shows that the projection of the circle of height considered is the circle Λ', which passes through points P and Q (projected from the poles of the horizon, G and I), and through point N, which is the intersection of the circle Γ' (the projection resulting from the rabatment of Γ onto the plane of the astrolabe) with the straight line Δ', which is perpendicular to CE at U (the projection of point O itself obtained by the orthogonal projection of S onto the diameter LK of circle Γ).

Note that the way in which al-Qūhī distinguishes the projection on the plane of the astrolabe and the projection on the equatorial plane of the sphere plays an essential role in the elaboration of the geometrical theory of projections. Indeed, the relation between these two consists of a rabatment. He carries out the constructions on the plane on the astrolabe, which he considers to be superposed on the plane of the figure, and his demonstrations throughout the last two chapters of the first book consist in establishing that these constructions conform to the theory of stereographic projection.

With the exception of Chapter 7, which is a sort of appendix in which al-Qūhī demonstrates several lemmas that he has used in the preceding chapters, the problems of the second book are all aimed at constructing the astrolabe and therefore the projected sphere, starting from the data of its plane and of three specific elements that he will consider systematically among the possible data of the representation on the plane of the astrolabe. This is what appears in the three surviving Chapters (1, 2, and 6) and one can suppose that the lost chapters were similar. The structure of Chapters 1 and 2 is completely similar. Both consist of six problems for which al-Qūhī solves by synthesis. As we shall see, thanks to the results of Book 1, he reduces the construction of the astrolabe to the determination of the center and radius of the sphere.

In Chapter 1 of the second book are grouped together problems for which the author grants himself an image point in the plane of the astrolabe, that is in the plane of the equator, that is the rabatment of the equatorial plane on the meridian plane, and the angular distance between its homologue to the pole of the sphere. By the homologue of an element (point, circle or straight line) from the plane of the astrolabe, al-Qūhī means what we call in modern language its antecedent by projection (point or circle), situated on the sphere. The datum of the angular distance of this homologue to the pole of the sphere amounts to giving its latitude, therefore to situating it on a circle parallel to the equator of the sphere. (Book 1, Chapter 2). The third given element is then taken from among the following: 1) the pole of the sphere, 2) the center of the sphere, which is also that of the astrolabe, 3) the radius of the sphere, 4) the distance, on the plane of the astrolabe, from the pole to the point the homologue of which is at a known angular distance from this pole, 5) the distance, on the plane of the astrolabe, from the center of the sphere to a point the homologue of which is a known angular distance from the pole of the sphere, 6) another point image and the angular distance from its homologue to the pole of the sphere.

Chapter 2 of the second book, for its part, regroups the problems for which are given a *muqanṭara* on the plane of the astrolabe and the angular distance from the pole of its homologue to the pole of the sphere. The datum of this angular distance goes back to the datum of the horizon with which this parallel circle is associated, and therefore of the latitude of the place for which the astrolabe is designed, since a horizon has the same poles as the circles that are parallel to it. Thus, for the six problems, one finds exactly the same third elements given in the same order as occurred in the first chapter.

In Chapter 6, we find once again the same variation of the third datum, which is added to the datum of a point in the plane of the astrolabe and of its homologue with respect to a known horizon. This time, the author is content to list the cases, and to develop only the solution to the first problem, to which he proceeds by analysis, and for which Ibn Sahl will give the synthesis in his commentary. One is therefore entitled to think that the problems of the three lost chapters of his book displayed the same style as the first two because Ibn Sahl, who surely had his contemporary's complete treatise, does not bother to comment on them.

Although al-Qūhī treats the astrolabe in this work, in fact throughout the entire treatise he is studying the mathematical structure that underlies the astronomical instrument, that is, the stereographic projection of a sphere, defined by its center and its radius, on the equatorial plane. The characteristic elements of this object (its pole and its support) are given by the center, the radius of the sphere, and the plane of projection. His treatise is thus divided between the direct study of the projection (Book I), and its reciprocal study (Book II). In other words, in Book I, knowing the characteristic elements, how does one construct the image of a point or of a circle? In Book II, knowing certain images, how does one recover the elements that are characteristic of the projection? Even if a reading of certain passages from Book I may induce us to think of one of the many studies carried out during the 10th century and aimed at explaining the functioning of the instrument, and its design, the work is in no sense addressed to artisans. The style is that of a mathematical treatise. The title already makes its intention clear: the author plans to *demonstrate* all the constructions that he will carry out. And, indeed, he gives statements of these constructions in the form of problems, offers solutions by either analysis or synthesis, and then demonstrates them. In addition, this treatise is perfectly organized, and its methodical character leaves no doubt. In it, al-Qūhī systematically solves geometrical problems raised by the instrument he is studying. He gathers together the propositions that rely on the same basic configuration and, whenever possible, reduces some problems to those he has already solved. In effect, his treatise is one of pure geometry, on a completely original subject. This is why this text, accompanied by Ibn Sahl's commentary, constitutes a significant testimony to the geometrical activity of the 10th century.

2.4. *Al-Ṣāghānī's study of the projection of the sphere*

At about the same time, al-Ṣāghānī was composing a treatise that he called *On the Manner of Projecting the Sphere on the Plane of the Astrolabe* (*Kitāb kayfiyyat tasṭīḥ al-kura*), in which he generalized the conic projection of the sphere, by displacing the pole of projection along the axis of the sphere, developing in 12 chapters his new theory as well as its applications to practical construction by craftsmen. He studies the construction, on the plane of the instrument, of the projections of the parallels at the equator (*madārāt*) and of the meridians, of the projections of a horizon, of its parallels, located in the hemisphere, bounded by that horizon and containing the north pole, since this hemisphere represents the visible part of the celestial vault, and of circles of height on this horizon, as well as the construction of projections of the zodiac and of the fixed stars – the rete – by means of the modes of projection derived from the stereographic projection. Indeed, all throughout his treatise, he considers conic projections, but instead of projecting the sphere on the equatorial plane from one of its poles, he displaces the pole of projection along the axis and discusses all the cases that emerge. In his introduction, al-Ṣāghānī claims priority for these projections. In his treatise on *The Plane Projection of the Figures* (*Tasṭīḥ al-ṣuwar*), al-Bīrūnī mentions conic projections for which the pole is on the axis of the sphere, whether at the poles themselves or outside them, but he does not mention al-Ṣāghānī's name. The latter notes specifically that no one had yet studied the drawing of the conic sections on the plane of the astrolabe; perhaps at the time he was writing, he did not know about al-Qūhī's work or about Ibn Sahl's commentary on it, which we have just discussed.

In the first six chapters, al-Ṣāghānī develops the theoretical side of the methods of projection of circles that constitute the network of horizontal coordinates: the circles parallel to the horizon and the circles of height, the projections of which yield the *muqanṭārāt* and the azimuths (*sumūt*), on the tympanum of the astrolabe. In the last three chapters, by contrast, after having presented three ways of tracing the rete, he gives three practical methods, that he calls 'of the artisan', that make it possible to trace this network. We will deal here with the purely geometrical part of this treatise. What constitutes the innovation of this work, of which its author is keenly aware as we emphasized, is the study of conic projections that do not transform the circles of the sphere into other circles on the plane of the astrolabe. Although he does not say so explicitly, there can be no doubt that al-Ṣāghānī knew the main property of the stereographic projection (namely, that the circles of the sphere are projected as circles or straight lines onto the plane of projection) and that he deliberately stays away from the framework of this projection, since, whenever it occurs, he voluntarily omits the case in which the pole of projection is one of the two poles of the sphere. He is interested only in the conics as projections of the circles of the sphere, and we believe that this follows from a thorough knowledge of Apollonius's book, and from an example of the application of the theory of conics, which are both so characteristic of the 10th-century geometers' activity.[27] During the theoretical development, the author carries out reasonings in space, returning as much as possible to planes, by means of rabatment procedures that al-Qūhī had systematized in his treatise. The most frequent rabatment occurs about the meridian and consists in superposing the equatorial plane, the support of the astrolabe, onto the plane of the horizon's meridian.

The construction of *muqanṭārāt*: After having presented the four lemmas, studied the projection of the tropics, and discussed the different cases that presented themselves as a function of the position of the pole of projection, al-Ṣāghānī turns to the exposition of the projection of circles parallel to the horizon. He confronts several situations that he will treat successively in the different chapters.

First situation: Let A be the south pole and C the north pole of the sphere with center E, $[GH]$ is the diameter of the horizon, F the orthogonal projection, onto the axis of the sphere, of the southern endpoint G of this diameter and O the pole of projection. Then all of the *muqanṭārāt* are represented as ellipses when the astrolabe is northern and O is on the open interval of $[FA)$ not including point A or when the astrolabe is southern and when O is external to the sphere.

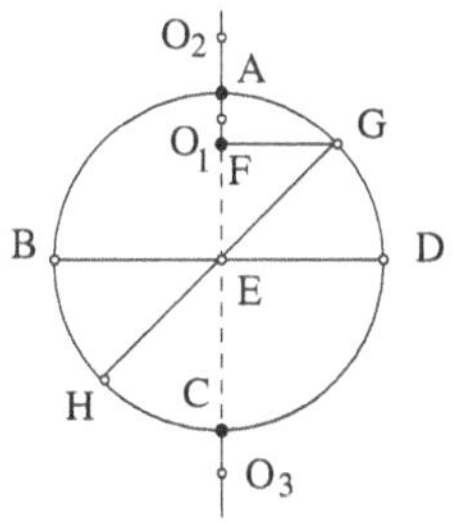

Fig. 85

He gives three cases of figure; but the theorem, which constitutes Chapter 3 of his work, has a single statement, namely, that the image of the circle of diameter GH by means of the conical projection from pole O is the ellipse of axis IK and of parameter p, where I and K are the projections of G and H, respectively, and where p is defined by the relation:

[27] See above 'The Archimedeans and problems with infinitesimals' and *Ibn al-Haytham's Theory of Conics, Geometrical Constructions and Practical Geometry*. A History of Arabic Sciences and Mathematics, vol. 3, Culture and Civilization in the Middle East, London, Centre for Arab Unity Studies, Routledge, 2013.

$\frac{MO^2}{MH \times MG} = \frac{IK}{p}$, when M is the point of intersection of GH with the parallel to BD passing through O. He refers to the constructions that Apollonius solved at the end of Book I of the *Conics*, to ascertain that the ellipse is well defined, and he emphasizes that, in the first case, $p < IK$, which makes IK the major axis of the ellipse; the two other cases yield the inverse: $p > IK$, therefore IK is the minor axis. These follow from Lemmas 3 and 4 of Chapter 1.

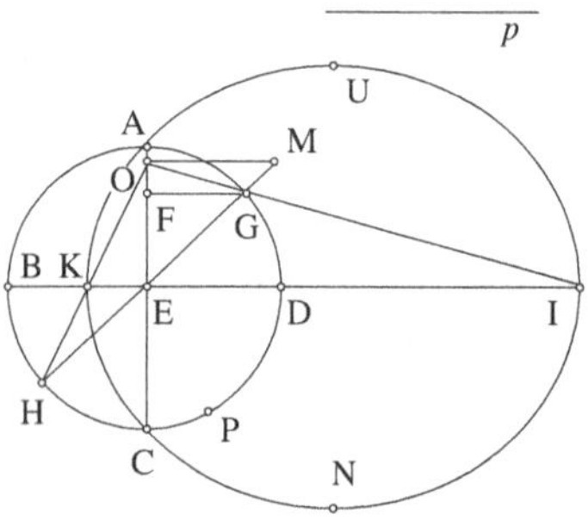

Fig. 86a

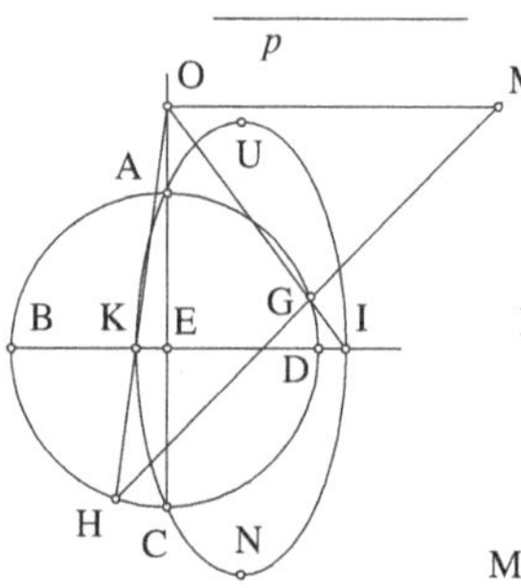

Fig. 86b

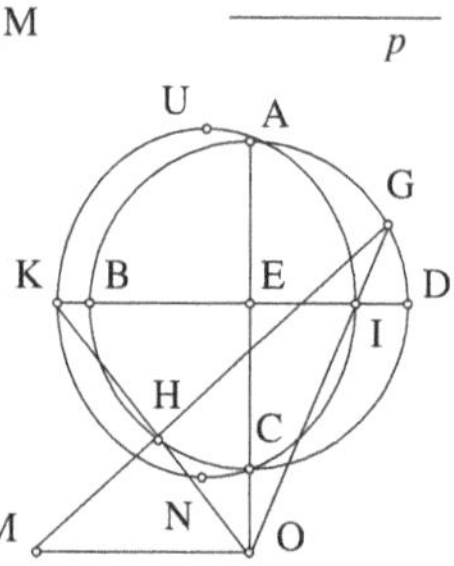

Fig. 86c

To demonstrate this theorem, he considers a cone of vertex O and having as base the circle of diameter GH, and shows that the first two lemmas create the conditions for applying Proposition I.13 of Apollonius's *Conics*, which defines the ellipse.

Second situation: If, given the same data, and if the pole of projection is now at F, then the horizon is projected as a parabola of vertex S, of axis DS, and of parameter p, where S is the projection of the northern endpoint H of the diameter of the horizon, and where p is defined by the relation:

$$\frac{GH^2}{GF \times FH} = \frac{p}{SF}.$$

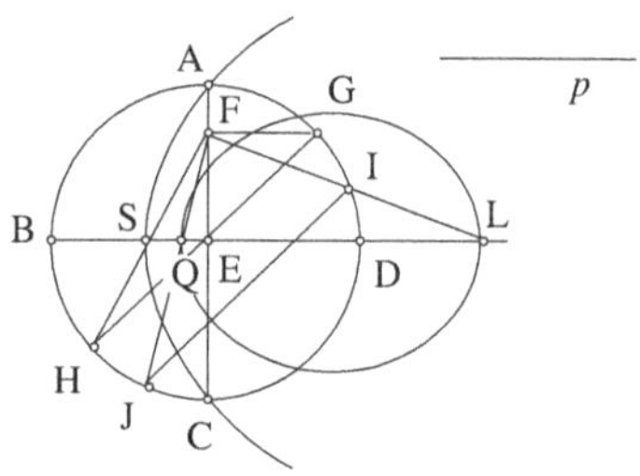

Fig. 87

All the other *muqanṭārāt* are ellipses in the plane of the astrolabe. Once again, the author refers to Apollonius to construct the parabola and to demonstrate the result.

Third situation: If one uses the same data, if the pole of projection O is found between F and E, then the horizon projects as a hyperbola of axis SQ, with vertex Q, diameter SQ, and parameter p, where S and Q are the projections of G and H, respectively, and where p is defined by the relation: $\dfrac{UO^2}{GU \times UH} = \dfrac{QS}{p}$, when U is the point of intersection of GH with the parallel to BD passing through O.

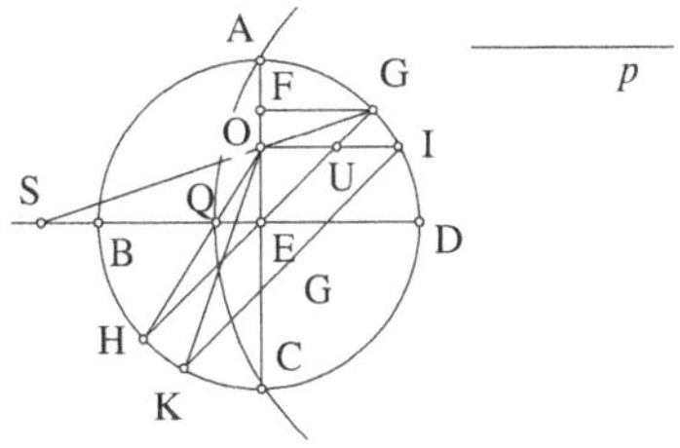

Fig. 88

As to the other *muqanṭārāt*, let IK be the diameter of circle Γ parallel to the horizon and such that OI is parallel to BD. Then, according to the second situation, its image by projection on pole O is a parabola in the plane of the astrolabe, the image of every circle parallel to the horizon located between this horizon and Γ will be a hyperbola, and the image of every parallel circle located beyond Γ will be an ellipse with respect to the horizon.

Fourth situation: now the pole of projection is at F' and all the other data stay the same. The astrolabe then is of a southern type, because the pole of projection is found, like the north pole, on the same half-line that issues from the center of the sphere. Let M be the point where the straight line HF' pierces the sphere. Thus there exists a circle Γ, parallel to the horizon, one diameter of which will have one of its endpoints at M. Moreover, let us call Λ the circle parallel to the horizon, of which the diameter IK passes through F'. Then the images of the horizon and of Γ are parabolas in the plane of the astrolabe; the images of the parallel circles included between these two are hyperbolas, with the exception of Λ, the image of which is a straight line; and the images of the circles parallel to the horizon and located above Γ, that is, between it and the pole of the horizon, are ellipses (Fig. 89a).

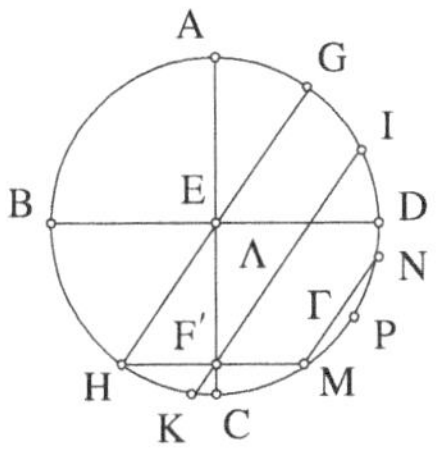

Fig. 89a

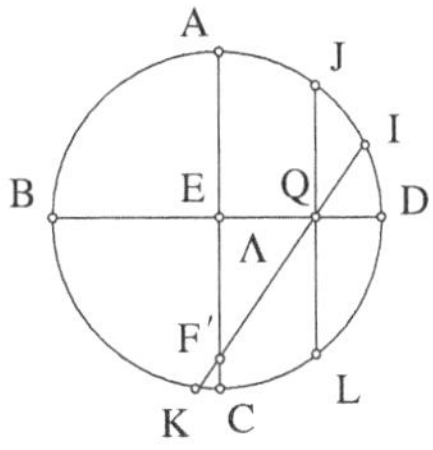

Fig. 89b

In this situation, al-Ṣāghānī explains only the case of the hyperbola and the straight line. For the projection of a circle located between the horizon and Γ and distinct from Λ, he goes back to the definition of the characteristics of the conic section: axis, vertex, *latus rectum*, and diameter, just as he had done in the preceding proposition. He then demonstrates that the image of Λ is the segment JL, since the cone of vertex F' and of base Λ is then reduced to the disk of circumference Λ, for which the intersection with the plane of projection has a rabatment along JL in the plane of the astrolabe (Fig. 89b).

Fifth situation: to conclude this study of *muqanṭārāt*, al-Ṣāghānī considers the case in which the pole of projection is at the center E of the sphere. He therefore considers the cone of center E and with the circular base Γ of diameter IK, parallel to the horizon of diameter GH.

Then the image of Γ is the intersection of the cone and of the plane of projection, whose rabatment in the plane of the astrolabe is formed by two segments ES and ES', where S and S' are the points of intersection of the circle of the meridian ($ABCD$) with the perpendicular to BD passing through L, when the diameter IK cuts BD at L.

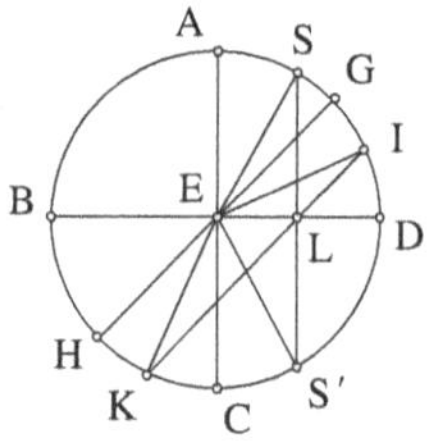

Fig. 90

Al-Ṣāghānī adds that, if the segment IK does not meet the segment BD, then the circle of diameter IK is not projected. By contrast, he says nothing about the case of the horizon itself, nor about the case in which the endpoint I of the diameter of the parallel circle is superposed on point D. This study of the projection, the pole of which is the center of the sphere, sheds light on his intention, if there were any doubt about it. By treating a case with no application to the astrolabe because of its trivial character, he shows clearly that his research is theoretical and that specific cases have significance as 'limiting cases', to establish the generality of a theory of conic projections of the sphere.

2.5. *The construction of the* sumūt

As he had done for the study of the *muqanṭārāt*, al-Ṣāghānī begins his study of the construction of the azimuths (*sumūt*) with the demonstration of four lemmas that yield results in spherical trigonometry. He carries out his reasoning in space, and although he discusses the nature of the projection of a circle of height Γ, he has determined the characteristics of the conic only with respect to the planes of the sphere given in space. In the next chapter, the sixth, which is specifically devoted to the construction of the *sumūt* in the plane of the astrolabe, he will define the conic of this plane, notably by using two rabatments, the same two on which he relied in the first lemma. Now he no longer has the parameter of the conic, as he did for the *muqanṭarāt*; this is why he no longer refers to Book I of Apollonius's *Conics*, but relies on an ordinate drawn from the section onto the axis.[28]

In the first three propositions of chapter six, he projects the first circle of height Γ_0, of diameter *FH*, which is orthogonal to meridian (*ABCD*) of the horizon, by considering the different positions of the pole of projection, but limiting himself to northern projections.

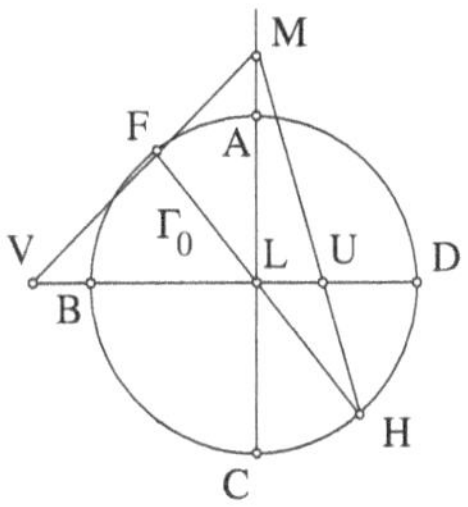

Fig. 91a

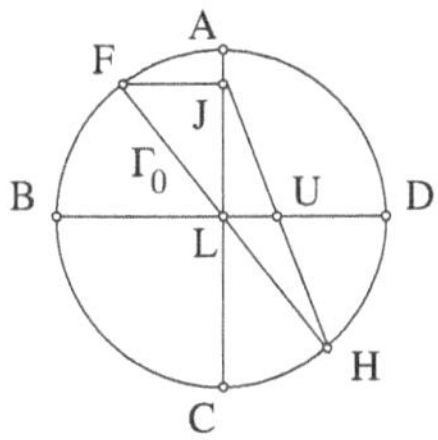

Fig. 91b

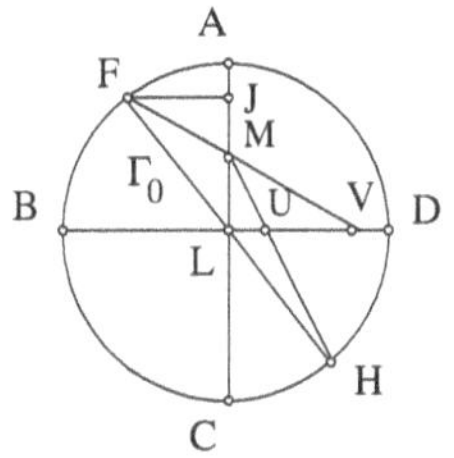

Fig. 91c

Since Γ_0 has (*ABCD*) as its meridian, it can be treated as a horizon. Therefore the axis of the conic obtained by the projection of Γ_0 has *BD* as its support. Al-Ṣāghānī nevertheless does not apply the results of his study of the *muqanṭarāt*, since he does not determine the parameter of the conic. In every case, however, *AL* is an ordinate. If *U* is the projection of *H*, and *V* is that of *F*, when it exists, then in the first case, the pole of projection *M* is external to the sphere, therefore the image of Γ_0 is the ellipse of axis *UV*. In the second case, the pole of projection is at *J* and the image is the parabola of vertex *U* and of axis *UV*. In the third case, the pole of projection *M* is between *J* and *L* and the image is the hyperbola of vertex *U* and of diameter *UV*.

[28] P. Abgrall, 'La géométrie de l'astrolabe au Xe siècle', *Arabic Sciences and Philosophy*, vol. 10.1, 2000, pp. 7–77, at pp. 24–7.

In the fourth proposition, al-Ṣāghānī considers a circle of any given height Γ, and assumes that the pole of projection M is external to the sphere. First of all, he constructs the straight line LG that forms the angle ε with BD, and the points R and E of the circle ($ABCD$), such that the arcs DR and BE are both equal to δ, by referring what was posited and demonstrated in the first lemma about the construction of azimuths. Then the points I and S are taken at the intersections of ME and MR with BD. One thus obtains the points U and V on LZ, which is perpendicular to LG, such that $LU = LS$ and $LV = LI$.

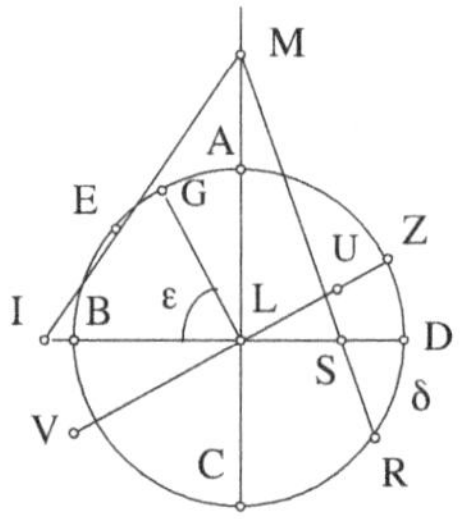

Fig. 92

According to Lemma 3, al-Ṣāghānī knows that the image of Γ is an ellipse. In fact the construction that precedes includes the two rabatments of Lemma 1. On the one hand, the meridian of Γ undergoes a rabatment, about AC, onto the meridian ($ABCD$). Then the straight line PT, the intersection of the meridian of Γ and of the equator, undergoes a rabatment along BD, and by projecting E and R, one obtains the length IS of the axis of the ellipse being sought. One obtains the position of this axis with the second rabatment, that of the plane of the equator about BD, by which PT undergoes a rabatment onto LZ. This is why, by rotating IS about L, one obtains the axis of ellipse UV. In the same rabatment about BD, the point S becomes superposed onto G, since the two arcs BG and BS are equal to ε. Therefore the ellipse has UV as its axis and one of the ordinates is GL. In this text, al-Ṣāghānī does not flesh out these rabatments; he defines the stages of the construction and refers to Lemma 3 for the demonstration.

In the fifth proposition, he treats the case in which the projection is a parabola by following the same method of rabatments, and one encounters this case when the pole of projection is the orthogonal projection, J' of E on the axis. He discusses briefly the other types of conics one obtains, as a function of the pole of projection with respect to J', for a circle of fixed height. He adds that, for a given pole of projection, all the azimuths (*sumūt*) – pass through two fixed points, that is, the projections of poles on the horizon. We encounter the fact that the *sumūt* form a bundle of circles with points of base. Al-Qūhī demonstrates this result indirectly by determining the locus of the centers of the *sumūt*, in Chapter 4 of Book 1 of his treatise.[29] Al-Ṣāghānī concludes his study with two propositions in which he treats the case in which the pole of projection is the center of a sphere.

The studies devoted to the astrolabe originating in the stereographic projection were very numerous from the 9th century on, and the works by al-Bīrūnī that we have already cited offer much evidence for them. We find more evidence in a work by Ibn al-Sarī, known under the name of Ibn al-Ṣalāḥ (d. 548 H./1154), called *The Projection of a Surface of the Sphere* (*Tasṭīḥ basīṭ al-kura*). His work is divided into two parts, theoretical and 'practical', which treat the way of constructing the astrolabe and the way of dividing the circles and the lines on the tympanums of the instrument. The author cites a good number of his predecessors in order to criticize their approaches to the construction of the astrolabe. In a series of unfair criticisms, Ibn al-Sarī denounces the absence of explanations aimed at the practical construction of the astrolabe in Ptolemy's book on the projection of the surface of a sphere (probably the *Planisphere*) and Pappus's commentary on it. He blames al-Farghānī, in the book to he calls 'theoretical principles and construction of the astrolabe' (known under the title *al-Kāmil*), not only for devoting too little space to theory and too much to practical construction, but also for inadequacies. According to Ibn al-Sarī, al-Farghānī gives useless lemmas, leaves out others that are indispensable, but above all berates him for not having read Apollonius, which seems very prejudicial to this type of research. Finally, Ibn al-Sarī criticizes four works – one by Ḥabash al-Ḥāsib on the northern astrolabe, a treatise by al-Bīrūnī (probably *al-Istīʿāb*, or *Tasṭīḥ al-ṣuwar*), one chapter from a book by Kūshyār on the use of astrolabe, and a treatise on the astrolabe by Ibn al-Samḥ. He blames them all for giving no space to theory, and for treating only the construction of the instrument.

As we can see, in this era and into the 12th century, this area of research stimulated debates, and even quarrels. So it happened that, from the 9th century, other projections considered in the domains of cartography and of the representation of the heavens also became the focus of controversies and criticisms. In his *The Plane Projection of Figures* (*Tasṭīḥ al-ṣuwar*), al-Bīrūnī recalls certain debates and quarrels about the advantages of the different projections. In particular, about a projection *al-mubaṭṭakh* (in the

[29] R. Rashed, *Géométrie et dioptrique*, pp. 202–4; *Geometry and Dioptrics*, pp. 896–8.

form of a melon), which is a zenith-equidistant projection referred to one of the poles of the ecliptic, he writes:

> It is possible to transfer what belongs to the sphere onto a plane by another method, which, in several copies of his *al-Kāmil*, Abū al-ʿAbbās al-Farghānī attributes to Yaʿqūb ibn Isḥāq al-Kindī, and in several other copies to Khālid ibn ʿAbd al-Malik al-Marwarūdhī, which is called the melon-shaped astrolabe. A book by Ḥabash about its construction has been found [...].[30]

This projection, one of the main contributions to which came down to us from Ḥabash ibn Ḥāsib,[31] was criticized by the Banū Mūsā, and especially by the eldest, Muḥammad, as a way of constructing the astrolabe. Al-Farghānī, the very author of one of al-Bīrūnī's sources, also objected to this projection.[32] For the construction of the astrolabe, several other projections were elaborated by such first-rank mathematicians as al-Sijzī, al-Bīrūnī himself, Kūshyār, and many others. Given the great extent of these studies, we have paid attention mainly to those that, in our opinion, influenced the history of projective geometry.

[30] Al-Bīrūnī, *Tasṭīḥ al-ṣuwar wa-tabṭīḥ al-kuwar*, ed. A. Saidan, pp. 13–14.

[31] E. S. Kennedy, P. Kunitzsch and R. P. Lorch, *The Melon-Shaped Astrolabe in Arabic Astronomy*, Stuttgart, 1999.

[32] *Ibid.*, pp. 178–81.

THE CONTINUOUS DRAWING OF CONIC CURVES AND THE CLASSIFICATION OF CURVES

1. INTRODUCTION

The continuous drawing of conic curves requires not only the invention of an instrument, or of a mechanical system capable of drawing these curves, but also the elaboration of concepts in the theory of conics capable of explaining and controlling the use of this instrument. It is therefore a type of research in which *techne* and *theoria* are tightly intertwined. In this respect, it has nothing to do with another kind of research, the goal of which is completely practical: to discover only the instrument. An example of the latter is the 'ruler that curves inward' used by Diocles to join the points in the arc of a cissoid.[1] By contrast, the issue that concerns us now involves delving into the conceptual work necessary to explain the structure and operation of the instrument. Not until the second half, if not the last third, of the 10th century does one encounter this *mathematical* research on instruments designed to draw curves. Not until then does one therefore witness the emergence of a new chapter in geometry, the importance and even the existence of which have eluded historians. This fact is all the more remarkable because prior to this period, the history of mathematics reveals nothing comparable to this research. To be sure some engineers had already broached more or less directly the question of the continuous drawing of conics without, as far as we know, ever elaborating the relevant mathematical theory. Indeed they merely invoked a procedure without undertaking the geometrical research capable of explaining its construction and usage. However that may be, our knowledge of the history of continuous drawing before the advent of Arabic mathematics is reduced to one unique fact and one remnant, both of which go back to the engineers of the 6th century. We know that Anthemius of Tralles traced the ellipse using the so-called 'gardener's method'. This procedure obtains a continuous movement

[1] R. Rashed, *Les Catoptriciens grecs*. I: *Les miroirs ardents*, edition, translation, and commentary, Collection des Universités de France, Paris, Les Belles Lettres, 2000, p. 84 and p. 133.

by using a string, the extremities of which are fixed at the foci, and that one tightens with a stylus that traces a curve. The procedure relies on the following property: the sum of the distances of the foci to every point of the ellipse is constant and equal to the length of the major axis.[2] As to the remnant, it comes from a phrase of dubious authenticity, which is attributed to Eutocius and implies that Isidore of Miletus[3] had invented an instrument to trace the parabola. This phrase famously states: 'The parabola is traced by means of the instrument invented by our teacher, Isidore of Miletus, the engineer, and described by him in his commentary on Hero's treatise entitled *On Vaulting* (□ῶ□ Ἥρωνος Καμαρικῶν)'.[4] As its name □□□□□□□□ suggests, this instrument seems to be a compass, although this is not certain. Now this very same phrase appears also in the Arabic translation of Eutocius's *Commentary* on Archimedes' *On the Sphere and Cylinder*, which the mathematician al-Qūhī knew. Although he is the first to write on this type of compass, al-Qūhī does not mention this statement; on the contrary, he claims that nothing of the sort came down from the ancients on this subject. His young contemporary al-Sijzī, who was inclined to erudition, reconstructs a rather legendary history of the 'perfect compass' that evidently starts from the phrase attributed to Eutocius. Without any additional historical or linguistic information, he interprets διαβήτης as this 'perfect compass.' The most we can say is that if the Byzantine engineers Anthemius and Isidore encountered the problem of continuous drawing in their work, they never raised it for all three curves at the same time; nor did they provide a geometrical theory of the instrument.

All the clues thus suggest that, if the Greek mathematicians came across the problem of continuous drawing, they never settled on it as a topic nor gave it a place at the heart of their research. Until the middle of the 10th century, this same situation still obtains among the Arabic mathematicians. To be sure, scientists like Ibn Sinān (296/901–335/946) implemented research on pointwise construction,[5] but not yet on continuous drawing. Several decades later, this very topic of research moves into the

[2] R. Rashed, *Œuvres philosophiques et scientifiques d'al-Kindī*. Vol. I: *L'optique et la catoptrique d'al-Kindī*, Leiden, E. J. Brill, 1997, pp. 678–79; and *Les Catoptriciens grecs*, pp. 250 and 292.

[3] P. Tannery, 'Eutocius et ses contemporains', *Bulletin des Sciences Mathématiques*, 2e série, vol. VIII, 1884, pp. 315–29; repr. in *Mémoires Scientifiques*, vol. II, pp. 118–36.

[4] Eutocius, Commentaire de *La Sphère et le Cylindre d'Archimède*, in *Archimède, Commentaires d'Eutocius et fragments*, Text, French translation and commentary by C. Mugler, Collection des Universités de France, Paris, 1972, IV, p. 62, lines 3–4.

[5] R. Rashed and H. Bellosta, *Ibrāhīm ibn Sinān. Logique et géométrie au Xe siècle*, Leiden, E. J. Brill, 2000, pp. 245 ff.

foreground. In fact, three of Ibn Sinān's most prestigious successors invent new instruments for continuous drawing, and elaborate the theoretical means necessary to underpin it. They are: al-Qūhī in his *Kitāb fī al-birkār al-tāmm* (*The Book on the Perfect Compass*);[6] Ibn Sahl in his *Kitāb al-Ḥarrāqāt* (*The Book on Burning Instruments*)[7] and al-Sijzī in a short treatise that was discovered only recently, the *Risāla fī ʿamal al-birkār al-tāmm* (*Treatise on the Construction of the Perfect Compass*).[8]

The fact that three contemporary mathematicians of the first rank are interested in the same problem is an event that deserves to be described and analyzed. Whereas Ibn Sahl conceives of a mechanical system founded on the properties of the foci and the directrix to trace only the three conic sections, al-Qūhī invents the perfect compass to trace all the lines that he calls 'measurable, *qiyāsiyya*',[9] that is, the three conics, but also the circle and the straight line; al-Sijzī, for his part, seeks to perfect this instrument in order to trace similar sections as well.

These three mathematicians thus managed to form, together if not equally, the nucleus of a new research tradition that will thrive for two more centuries. Neither the importance of this event nor al-Qūhī's foundational role escaped the notice of his immediate successors. This is what al-Bīrūnī (973–after 1050) writes in his *Fī Istīʿāb al-wujūh al-mumkina fī ṣanʿat al-aṣṭurlābī* (*On All the Possible Methods for Constructing the Astrolabe*):

> A group of eminent moderns took pains to trace conic sections according to Apollonius's exposition in the book of *Conics*, such as Ibrāhīm ibn Sinān, Abū Jaʿfar al-Khāzin and many others. Each of them tried to find the successive points of their perimeters. But Abū Sahl Bijan ibn Rustum al-Qūhī wrote a book to trace them by means of the perfect compass. Indeed, he called it perfect because of the possibility of constructing the straight line and the circular line that he assumed and each of the three conic sections, an effective construction that dispenses with connecting and adapting numerous points on their perimeters.[10]

[6] F. Woepcke, 'Trois traités arabes sur le compas parfait', *Notices et extraits des manuscrits de la Bibliothèque Impériale et autres bibliothèques*, vol. 22, 1874, pp. 1–175; R. Rashed, *Geometry and Dioptrics in Classical Islam*, London, al-Furqān, 2005.

[7] R. Rashed, 'A Pioneer in Anaclastics. Ibn Sahl on Burning Mirrors and Lenses', *Isis*, 81, 1990, pp. 464–91 and *Geometry and Dioptrics in Classical Islam*.

[8] R. Rashed, 'Al-Qūhī et al-Sijzī: sur le compas parfait et le tracé continu des sections coniques', *Arabic Sciences and Philosophy*, 3.2, 2003, pp. 9–44.

[9] Cf. below.

[10] Ms. Leiden 1066, fol. 49^r–v^.

These three mathematicians thus truly inaugurate a research tradition that others will in turn join, such as Ibn al-Haytham, al-Bīrūnī, Hibat Allāh al-Baghdādī, Muḥammad ibn al-Ḥusayn, among others.[11]

The key question is thus to know why this problem of the continuous drawing of conic curves surfaced during the second half of the 10th century and not before, and with such vigor that several very prominent mathematicians devoted their efforts to solve it. In fact, the genuine reasons for this development lie in the new directions that Arabic mathematical research took during the 9th–10th centuries. Let us quickly review some of these, as well as some of the mathematical chapters that drove the topic of continuous drawing.

The first chapter concerns geometrical construction by the intersection of conics.[12] The issue is no longer what it was in ancient geometry, *e.g.*, in that of Eutocius, where isolated problems arose sporadically and were solved by the intersection of curves, conic or otherwise. One now has a method to explore the domain of geometrical problems: these are mostly solid, but eventually also (and, in a sense, uselessly) plane and are constructed by means of conic curves alone, to the exclusion of all others. In the context of this new chapter, some mathematicians study the existence of solutions and their number, generally with great care. Carried out by analysis and synthesis, this research rests on the asymptotic and local properties of conics, and in particular their contact. Cultivated by mathematicians from the middle of the 9th century on, this new chapter became a domain of active research with the mathematicians of the second half of the 10th century – specifically al-Qūhī, Ibn Sahl, al-Sijzī, Abū al-Jūd etc. With these mathematicians is born a new criterion of admissibility: henceforth, the construction by means of conic sections is a construction admissible in geometry, on a footing equal to that of construction by straight edge and compass. Moreover, these same geometers, who during their constructions proceeded by means of geometrical transformations – similarity, translation, homothetic transformation, affinity – took steps to introduce motion into geometrical statements and demonstrations.[13] This continuous motion

[11] R. Rashed, *Geometry and Dioptrics in Classical Islam.*

[12] R. Rashed, *Les Mathématiques infinitésimales du IXe au XIe siècle*, vol. III: *Ibn al-Haytham. Théorie des coniques, constructions géométriques et géométrie pratique*, London, al-Furqān, 2000; English translation: *Ibn al-Haytham's Theory of Conics, Geometrical Constructions and Practical Geometry*. A History of Arabic Sciences and Mathematics, vol. 3, Culture and Civilization in the Middle East, London, Centre for Arab Unity Studies, Routledge, 2013.

[13] R. Rashed, *Les Mathématiques infinitésimales*, vol. III and vol. IV: *Méthodes géométriques, transformations ponctuelles et philosophie des mathématiques*, London, al-Furqān, 2002.

– translation, rotation… – is brought about by one instrument or the other, and is always reproducible with precision. It was evidently necessary to broach the theoretical and practical study of procedures for reproducing the motion and thus for drawing conic curves. No sooner said than done.

But this chapter on constructions is not the only one to require such a study. It is indeed in this period that one began to solve certain cubic equations by the intersection of conic curves. Here too it was necessary to generate the curve by motion in order to ensure its continuity, which was a fundamental concept when discussing the problem of the existence of points of intersection of curves.[14]

Beyond these orientations internal to geometric research and the elaboration of these new chapters, however, the additional requirement of new results in applied mathematics enters the stage: Ibn Sahl's systematic study of burning mirrors (parabolic, ellipsoidal) and of plano-convex and biconvex lenses, and also of the anaclastic properties of the three conics,[15] theoretical and practical research on astrolabes,[16] and sundials.[17] In short, it is not a coincidence that Ibn Sahl and al-Qūhī were the first to conceive this new chapter on continuous drawing. Indeed, was it not Ibn Sahl who elaborated the first geometrical theory of burning instruments, that is, mirrors and lenses? Was it not al-Qūhī who wrote the first book devoted to the geometry of the astrolabe? Were not both of them in the avant-garde of their era's research on the theory of conics and its applications? We must therefore pause to examine their respective writings.

2. IBN SAHL: A MECHANICAL DEVICE TO TRACE CONIC SECTIONS

In his *Kitāb al-Ḥarrāqāt*,[18] Ibn Sahl starts from the property of the focus and the directrix to invent a device devoted to continuous drawing. This device is composed of two parts, one of which is not deformable whereas the other is, without, however, varying in length. The first part is constituted of rigid rulers and pulleys, whereas the second is made of a flexible belt, that is, one that can take on one shape or the other while pre-

[14] See above, First part: Algebra.

[15] See R. Rashed, *Geometry and Dioptrics in Classical Islam*.

[16] See above, 'The Traditions of the *Conics* and the Beginning of Research on Projections'.

[17] R. Rashed and H. Bellosta, *Ibrāhīm ibn Sinān. Logique et géométrie au Xᵉ siècle*.

[18] R. Rashed, 'A Pioneer in Anaclastics. Ibn Sahl on Burning Mirrors and Lenses', and *Geometry and Dioptrics in Classical Islam*.

serving a constant length. This belt circumscribes the pulleys, the function of which is to avoid rupturing the belt. To this, one adds a stylus that is placed at the center of the movable pulley and that will trace the studied curve.

Thus, to trace the arc of parabola AB, of focus F, and of directrix D, Ibn Sahl invents a device composed of a ruler that takes the place of the directrix, a T-square of constant length $HH' = l$, where H is the apex of the T-square's right angle, and H' is placed on the ruler. Thus, when the T-square slides on the ruler, H describes a straight line Δ, which will be perpendicular to the axis of the section. At point H, one fixes the end of a belt that can be flexed with no change of length. At a point M of HH', one places a movable pulley and at point F an immobile pulley on which one fixes the other end of the belt. The latter must be kept taut by means of a stylus placed at point M. Ibn Sahl then demonstrates that, if M is such that $MF = MH'$, then M is the current point of the parabola. Indeed, he verifies that

$$MF + MH = l.$$

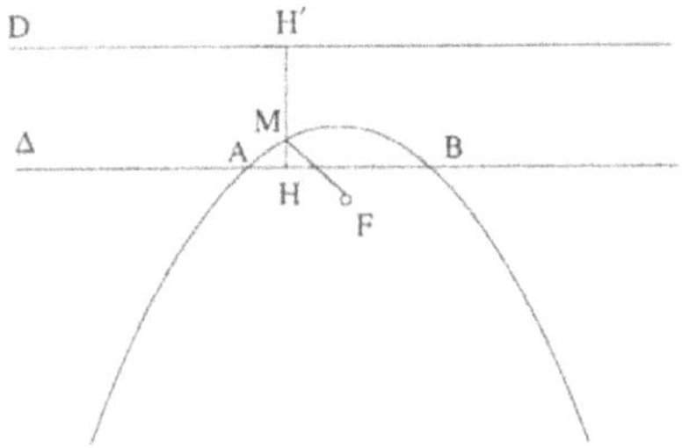

Fig. 93[19]

In the case of the ellipse, the fixed length l is that of the major axis; F and F' are the two foci. This time, the device that Ibn Sahl proposes differs from the famous 'gardener's method' only by the use of three pulleys, two of which are fixed at the two foci respectively, whereas the third, placed at M, is movable. The belt of length l surrounds the pulley M and its extremities are fixed at pulleys F and F'. The point M is the current point of the ellipse; indeed he proves that

$$MF + MF' = l.$$

[19] R. Rashed, *Géométrie et dioptrique*, pp. LXXIX–LXXX and pp. 10–15.

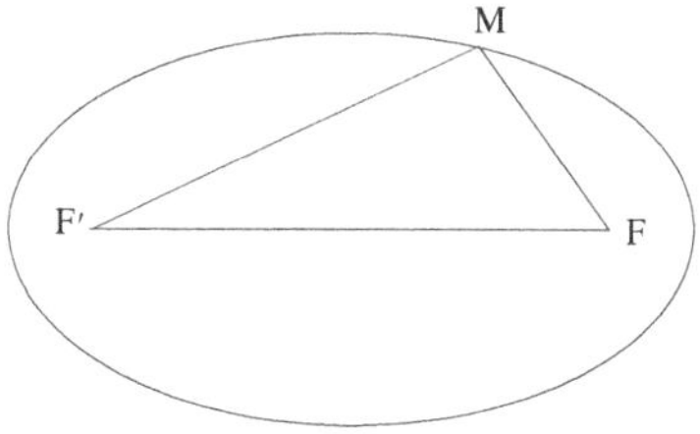

Fig. 94a

Finally, to trace the hyperbola with foci F and F' with a transverse axis of length $2a$, Ibn Sahl conceives a devise composed of two rulers and two pulleys. The first ruler FF' is fixed, the second FS is movable: it pivots around focus F. He then considers a belt. one extremity of which is fixed on an immovable pulley placed at F', and the other is at point S of the movable ruler. Around this last pulley passes the belt, which must be kept taut by means of a stylus that leans at M on the ruler.

If now $FS = l$, a constant length, the length of the belt will be $l' = l + 2a$. In this case the point of the stylus traces hyperbola MB when the ruler pivots around F. Indeed, point M will verify that $MF' - MF = 2a$, whence $(SM + MF') - SF = 2a$.

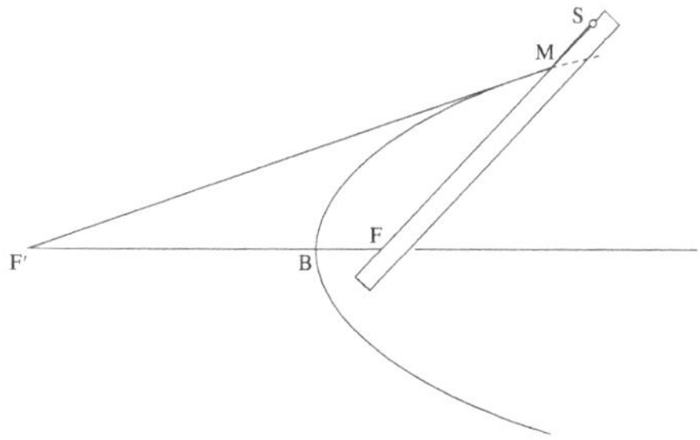

Fig. 94b[20]

3. Al-QŪHĪ: THE PERFECT COMPASS

Ibn Sahl's contemporary, al-Qūhī, invents another instrument that makes possible the continuous drawing of conic curves, as well as the circle and the straight line. Thanks to this instrument, he writes, it is easy 'to construct astrolabes on plane surfaces or surfaces of revolution, as well as

[20] R. Rashed, *Géométrie et dioptrique*, pp. LXXX–LXXXI and pp. 31–9.

to construct sundials on any surface, and likewise to construct all the instruments on which the lines that are the intersections of a conic surface and of any surface are situated'.[21] To this compass, al-Qūhī devotes an entire geometrical treatise in two books. In the first, he elaborates a theory of the instrument, before moving on to its applications in the second. Note that despite his preliminary statements on the utility of the instrument for astrolabes, sundials, etc., all the applications contained in the second book are purely geometrical.

The first book of the treatise begins with a description of a perfect compass and continues with the explanation of its use to trace on a given plane the various following curves: straight line, circle, parabola, ellipse, hyperbola (one- and two-branched).

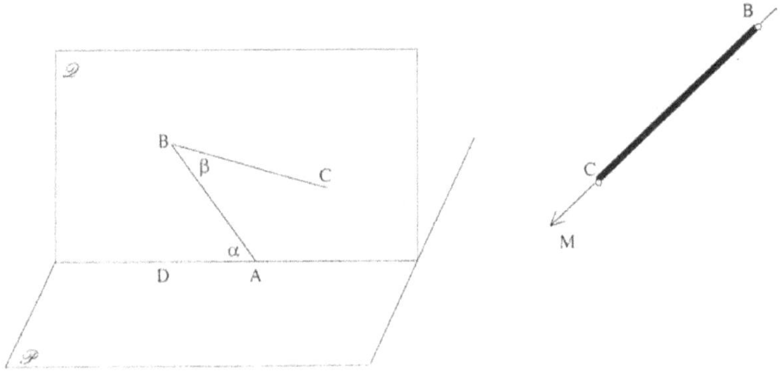

Fig. 95

The perfect compass has four articulated parts:

• A plane part, that is the base and contains 'the straight line of center' *AD*.

• 'The axis of the compass', *AB*, which can turn about a point *A* called the 'center' while remaining in plane $\mathcal{Q}$, a plane perpendicular to plane $\mathcal{P}$ along the straight line *AD*; its position is determined by the angle at the center of the compass $\alpha = D\hat{A}B$.

• 'The straight line of the apex', *BC*, which can turn about point *B*, called the 'apex'. First one considers its initial position in plane $\mathcal{Q}$ at the time one chooses the apex angle $\beta = A\hat{B}C \leq \frac{\pi}{2}$. Then one makes it rotate about the axis to generate a plane surface, if the angle at the apex is right, and a conic surface of revolution, if it is not.

[21] R. Rashed, *Geometry and Dioptrics in Classical Islam*, pp. 796–7.

• The straight line BC is itself the support of a drawing pen that can slide on this straight line so that one of its extremities – for example, M – comes into the plane to trace the sought figure.

The figure $DABC$ is that of the compass; $\mathcal{Q}$ is the plane of the compass.

Everything therefore depends on the angle β formed by the axis of the compass with the branch BC of the latter and on the angle α that the axis makes with the base of the compass. The drawing pen has a variable length, which allows its point to stay in contact with the plane on which one intends to trace the conic sections. The axis itself remains fixed.

Consider the case in which α is obtuse. Let BK be the perpendicular dropped from B onto the secant plane; BK is thus perpendicular to the base of the compass. The nature of the traced curve depends upon angle CBK; one has

$$A\hat{B}K = \alpha - \frac{\pi}{2}, \quad C\hat{B}K = \beta + \alpha - \frac{\pi}{2}.$$

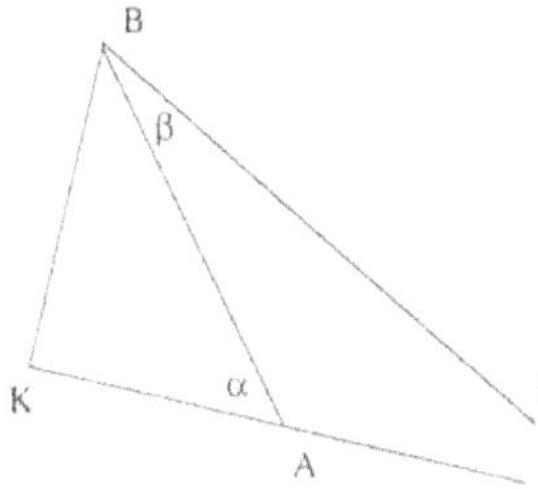

Fig. 96

If therefore

$C\hat{B}K < \frac{\pi}{2} \Leftrightarrow \beta + \alpha < \pi$, the section is an ellipse;

$C\hat{B}K = \frac{\pi}{2} \Leftrightarrow \beta + \alpha = \pi$, the section is a parabola;

$C\hat{B}K > \frac{\pi}{2} \Leftrightarrow \beta + \alpha > \pi$; the section is a hyperbola.

Al-Qūhī's idea is that the nature of the curve that one wishes to trace on the plane depends on the initial position of the compass with respect to this plane. This position is characterized by the length l of the axis of the compass, by the angle β that the axis makes with the movable branch, and by the angle α that the axis makes with the base of the compass. The issue, therefore, is to study the correspondence between the elements (l, β, α) on the one hand, and the elements that characterize the conic section on the other: diameter and *latus rectum* for the ellipse and the hyperbola, and *latus rectum* for the parabola.

Al-Qūhī goes on to establish the following propositions:

- If $\alpha = \frac{\pi}{2}$ and $\beta = \frac{\pi}{2}$, then the drawing pen does not meet plane $\mathscr{P}$ and it traces nothing.
- If $\alpha \neq \frac{\pi}{2}$ and $\beta = \frac{\pi}{2}$, the straight line *BC* generates a plane that cuts plane $\mathscr{P}$ along a straight line that will be traced by the drawing pen.
- If $\alpha = \frac{\pi}{2}$ and $\beta < \frac{\pi}{2}$, the traced curve is a circle with $r = AB \tan \beta$.
- If $\alpha = \beta < \frac{\pi}{2}$, the traced curve is then a parabola with axis *AD* and apex *D*.
- If $\alpha < \beta < \frac{\pi}{2}$, the first end of the drawing pen describes a branch of the hyperbola of apex *D* and the other end describes the other branch of the hyperbola of apex *E*.
- If $\beta < \alpha < \frac{\pi}{2}$, the drawing pen will trace an ellipse.

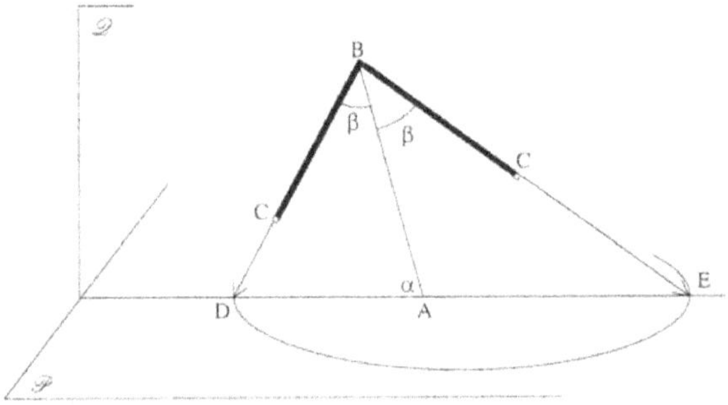

Fig. 97

To illustrate al-Qūhī's procedure, take the last case as an example: consider the nappe of the conic surface with axis *BA* and apex angle β, generated by the half-line *BC*. Plane $\mathscr{P}$ cuts all the generatrices of this nappe; the intersection is an ellipse with axis *DE*. The point *D* is the end of the drawing pen when it is in plane $\mathscr{Q}$, with α and β on the same side of *AB*. Point *E* is the end of the drawing pen, also in plane $\mathscr{Q}$, when α and β are on one side and the other of *AB*. During the rotation of *BC* about *AB*, the drawing pen will trace the entire ellipse.

Thus the perfect compass makes possible the continuous drawing on a given plane of a straight line, a circle, or any conic section, whether in relation to an axis or to a diameter. The reasonings and the constructions that al-Qūhī indicates in no case bring in the foci; they rely instead on the properties Apollonius established in the first book of the *Conics* for the plane sections of a cone with a circular base; these are properties that concern a diameter, the *latus rectum* associated with it, and the angle formed by this diameter with the direction of the ordinate lines.

The second book of the treatise is entirely devoted to the solution of problems concerning the continuous drawing of curves.

For each problem, al-Qūhī provides the preceding elements for the curve that he wants to trace, and then wonders how to determine the figure of the perfect compass by specifying the size of the elements. Let us briefly take the example of the parabola: to trace a parabola with a given apex *B*, the diameter *BC*, *latus rectum D*, and angle *E* of the ordinate lines that correspond to the diameter *BC*.

Two cases present themselves, according to whether angle *E* is right or not.

First, *E* is right; the straight line *BC* is then the axis of the parabola. But al-Qūhī had demonstrated in the first book of his treatise that one must have $\alpha = \beta$. If $\mathscr{H}$ is the length of the axis of the compass, it is necessary that

$$\cos\alpha = \cos\beta = \sqrt{\frac{D^2}{16.\mathscr{H}^2}+1} - \frac{D}{4.\mathscr{H}}.$$

There is no point in reproducing here the calculation that led al-Qūhī to this condition.[22] One need only remember the beginning of this calculation. Al-Qūhī begins by drawing a semicircle with any diameter *GH* and by placing on it the point *K* defined by

$$\frac{GK^2}{GH{\cdot}HK} = \frac{D^2}{4.\mathscr{H}^2}.$$

One has $KI \perp GH$. Angle *KHI* will be the angle sought for the compass. One makes the drawing pen slide until its end meets the straight line of center. The compass is then represented by triangle *BLM* such that *LM* of length $\mathscr{H}$ is the axis; point *M* is the center, the straight line of the apex is *LB* if the angle of the apex and the angle of the center are on the same side of axis *BM*; it is *LN*, parallel to *BM* if these angles alternate. One places the straight line of center *MB* on the given straight line *BC*. If one now makes

22 R. Rashed, *Œuvre mathématique d'al-Sijzī*.

LB rotate about *LM*, the drawing pen carried by *LB* traces out the sought parabola.

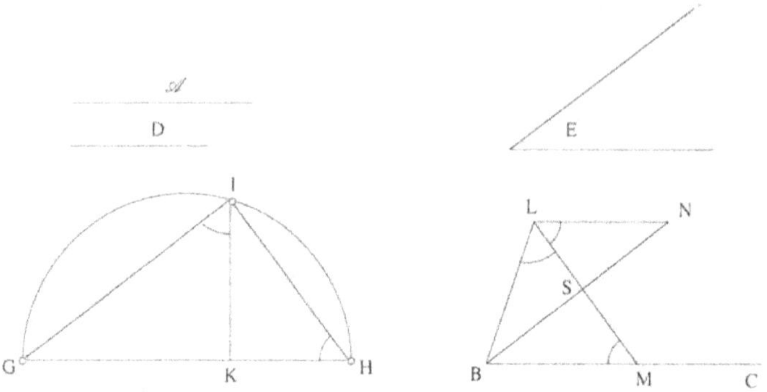

Fig. 98

Now if the angle *E* is not right, the straight line *BC* is in this instance any diameter of the sought parabola. The method al-Qūhī applies in this case consists in defining the axis of this parabola, its apex, the *latus rectum* corresponding to it, to reduce it to the preceding case.

Let us go over al-Qūhī's procedure. Through a point *G* taken on one of the sides of a given angle *E*, one draws *GH* perpendicular to the other side at *H*; and let *I* be the middle of *EH*. One sets

$$\frac{D}{N}=\frac{EG^2}{GH\cdot HI};$$

which defines the length *N*.

Now take a point *K* on the extension *CB*, such that $K\hat{B}L = G\hat{E}H$, $KL \perp BC$ and $KL = N$. The triangles *LKB* and *GHE* are similar, and $LK = N$; the triangle *LKB* is therefore known, the length *KB* is known, and $KB=\frac{N}{\tan\hat{E}}$.

Let *O* be such that $LO \parallel BC$ and $BO \parallel KL$, and *M* is the middle of *LO*. One defines a length *S* by $\frac{OB}{S}=\frac{OM}{OB}$. With the compass whose axis has length $\mathscr{A}$, one can then trace the parabola of diameter *MO*, of apex *M* and of *latus rectum S*, such that the angle that the diameter *LO* makes with the ordinate line is right, that is, *MO* is its axis, as al-Qūhī had established in the first case. This parabola passes through *B*, because $OB^2 = S \cdot OM$ (abscissa *OM*, ordinate *OB*). It is obvious that *BC* parallel to the axis *MO* is a diameter.

It remains to be demonstrated that the *latus rectum* associated with BC is D and that the angle that the ordinate makes with the axis is angle E, that is, that BL is tangent to the parabola at B.

Take (Mx, My) as reference, the parabola of axis Mx and of *latus rectum* S has as an equation $y^2 = Sx$.

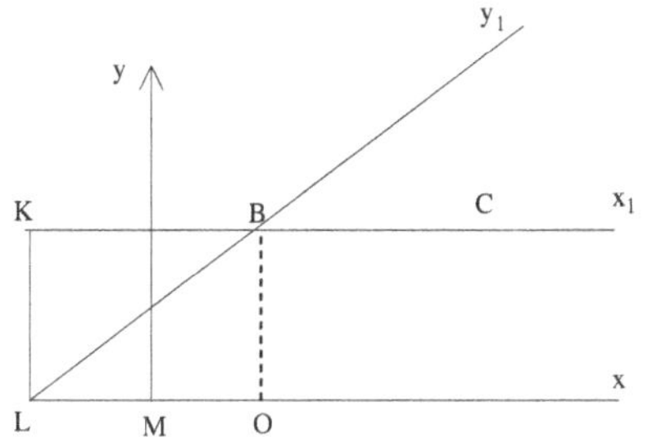

Fig. 99.1

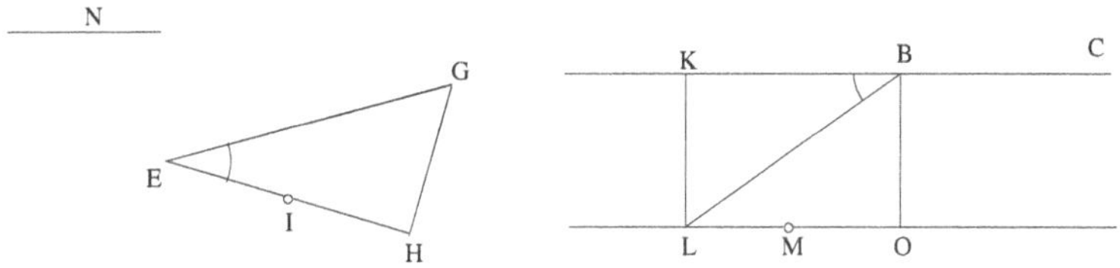

Fig. 99.2

The apex M of the parabola is the middle of OL, therefore the straight line BL is tangent to the parabola at B (*Conics*, I.33).

For the oblique axes reference (Bx_1, By_1), one has the equation $y_1^2 = Tx_1$, if T is its *latus rectum*. The coordinates of M are $x_1 = LM$, $y_1 = LB$, whence $LB^2 = T \cdot LM$. Let us show that $T = D$. One has $KL = N$, whence

$$LB = \frac{N}{\sin E} \quad \text{and} \quad LM = \frac{1}{2} LO = \frac{1}{2} \frac{N}{\tan E};$$

so

$$T = 2N \frac{\tan E}{\sin^2 E} = N \frac{GE^2}{GH \cdot HI} = D.$$

Al-Qūhī then treats the following problem in an analogous manner: to trace a hyperbola, knowing its transverse diameter, the associated *latus rectum*, the angle formed by the transverse diameter and the ordinate line when the length of the axis is given.

He then moves on to the problem of drawing, in a given plane, an ellipse for which one knows the diameter, the associated *latus rectum* and the angle formed by this diameter with its ordinate line.

In all of these problems, he shows that the angles α and β are known, and thus determine the perfect compass.

Note that the three problems al-Qūhī treats correspond to Propositions I.52–56 of Apollonius's *Conics*.[23] Indeed, everything suggests that al-Qūhī took, by way of exercise as it were, the problems from the end of Book I of the *Conics*, in which Apollonius was examining the construction of conic sections, in order to show how one can undertake their continuous drawing by means of a perfect compass. In this way, he perhaps wanted to show that the new instrument and the new methods are both necessary and effective to solve Apollonius's own problems.

4. Al-SIJZĪ: THE IMPROVED PERFECT COMPASS

One young contemporary of al-Qūhī's was al-Sijzī who, as noted earlier, was very familiar with his elder's writings. There can be no doubt that he knew thoroughly the latter's treatise on the perfect compass. To be persuaded of this fact, one need only read carefully his *Risāla fī waṣf al-quṭūʿ al-makhrūṭiyya* (*Short Treatise on the Description of Conic Sections*), as well as his *Risāla fī ʿamal al-birkār al-tāmm* (*Treatise on the Construction of Perfect Compass*); the dependence is obvious even though al-Qūhī's name appears nowhere in them.[24] This is not the only important point, however. The two preceding titles are the best witnesses to the fact that the problem of drawing curves had become sufficiently central for one and the same author to devote two treatises to it. Moreover, in his relatively late book, *Fī waṣf al-quṭūʿ al-makhrūṭiyya*, one sees that al-Sijzī is as concerned with the drawing by points, the continuous drawing by means of patterns constructed for this purpose, as well as the continuous drawing by means of the perfect compass. We know, moreover, that he had devoted an elegant study to drawing the hyperbola by means of points,[25] and written a work on the perfect compass that also treats continuous drawing.

[23] R. Rashed, *Apollonius: Les Coniques*, Tome 1.1: *Livre I*, Berlin/New York, Walter de Gruyter, 2008.

[24] R. Rashed, *Œuvre mathématique d'al-Sijzī*.

[25] R. Rashed, 'Conceivability, Imaginability and Provability in Demonstrative Reasoning: al-Sijzī and Maimonides on II.14 of Apollonius' *Conics* Sections', *Fundamenta Scientiae*, vol. 8, no. 3/4, 1987, pp. 241–56.

In this last work, the author apparently wants to modify al-Qūhī's invention in order to expand what it can do: thus in addition to drawing the curves al-Qūhī had considered, the modified compass will also be in a position to trace similar sections. The proposed modifications focus on the way of articulating both the axis and the branch of a compass on the devices that allow one to adjust its angle, and finally on the means of lengthening or shortening the axis or the branch that carries the drawing pen.

Thus, al-Sijzī begins by presenting a procedure for varying the length of the axis of the compass, not to mention the length of the second branch that carries the drawing pen. He implies, however, that the point of the drawing pen must remain in contact with the plane on which one wishes to trace the conic section.

Indeed he considers two tubes, *AN* and *AS* articulated at point *A*, the apex of the compass. In tube *AN*, there is a stem *AB* that will be the axis of the compass, and in tube *AS*, the stem *AC* will carry the drawing pen. The articulation of the two tubes at point *A* must allow tube *AS* to rotate around *A* so that one can choose the angle *BAC* and allow *AS* to be carried along with the tube *AN* in the rotational motion around axis *AB*.

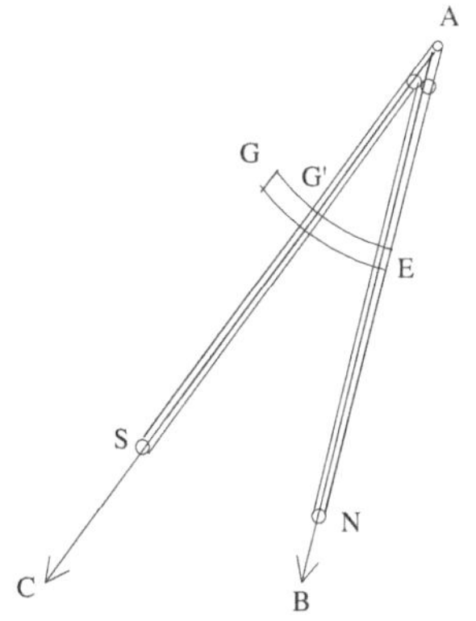

Fig. 100

Next al-Sijzī considers an arc *GE* from the circle of center *A*, that goes through two notches *E* and *G′*, respectively, on the two walls of the tube *AN* and of the tube *AS* and that slides on two grooves on these walls. One must nevertheless have a device that allows one at any moment to stop the sliding and preserve the chosen separation for the two tubes, before making the tube *AS*, which holds the drawing pen, rotate with an angle *BAC* of fixed size. The angle *BAC* is measured by the ratio of arc *EG* to *AE*. Since arc *EG* is fixed, one need only make *E* and *G* slide in their respective notches in order to add the distance *AE* and to block the spread.

Al-Sijzī then indicates three methods to adjust the length of the stem *AB*, which is the axis of the compass. Take the first as an example. Assume that stem *AB* is cut at point *P*, that is, that it is separated into two parts, part *AP* that one assumes fixed, and part *PB* that one can make slide in tube *AN* by means of a peg or handle placed at point *P*. In this case, one can go from position *PB* to position P_1B_1. This assumes that there is a rectilinear slit in the wall of tube *AN* so that the peg *P* slides in this slit. The axis of the compass thus reaches length AB_1 which corresponds to the desired distance in order to bring point B_1 onto the plane on which one wants to trace the conic section.

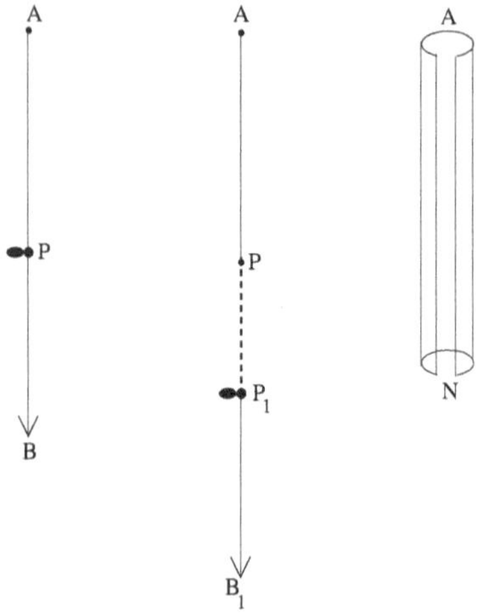

Fig. 101

Al-Sijzī then gives a procedure to make the stem that carries the drawing pen slide in the tube that is the second branch of the compass. Finally he presents a third procedure in which the length of the axis remains constant and the stem of the drawing pen continues to be assumed mobile. One can thus express the idea that emerges from this work of al-Sijzī. Whatever model one chooses, the perfect compass allows one to trace continuously the three conic sections, as well as similar sections in addition to the circle and the straight line.

To understand this last point, recall briefly that Apollonius, in *Conics* VI.11, shows that all parabolas are similar: for two parabolas of *latera recta* c and c', the ratio of similarity is c/c'. In Proposition VI.12, he shows that the necessary and sufficient condition for two ellipses (or two hyperbolas) of diameters d and d' and of *latera recta* c and c' to be similar is that $(d/d') = (c/c')$, which can written $(c/d = c'/d')$.

Consider only the case for the drawing of the parabola.

Assume that plane *ABC* is given, corresponding to the initial position of the compass, and assume that, in this plane, the axis *AB* of the compass is given in both position and size.

The plane Π on which the compass can trace a conic section is perpendicular to this plane along a straight line *XBY* on which the axis of the section is found. Let $AB = l$, $A\hat{B}Y = \beta$, and $B\hat{A}C = \alpha$. The question is therefore the following: given these hypotheses, can one trace two similar sections?

According to Apollonius, as we have just noted, all parabolas are similar. So let us now turn to the given plane Π and to the compass; such a compass allows one to draw only one parabola. Indeed, if Π is given, β is known and so is α, because $\alpha + \beta = \pi$. Let us determine the *latus rectum* c of this parabola with apex I. Let M be the point of the curve such that $MB \perp IY$; so $MB \perp AB$ (because $\Pi \perp ABC$) and $B\hat{A}M = B\hat{A}C = \alpha$; whence

$$MB^2 = c \cdot BI \text{ and } MB^2 = l^2 \cdot \tan^2 \alpha;$$

now, $BI = l/2 \cos \alpha$ (isosceles triangle IAB); from this, one deduces

$$c = 2l \frac{\sin^2 \alpha}{\cos \alpha}.$$

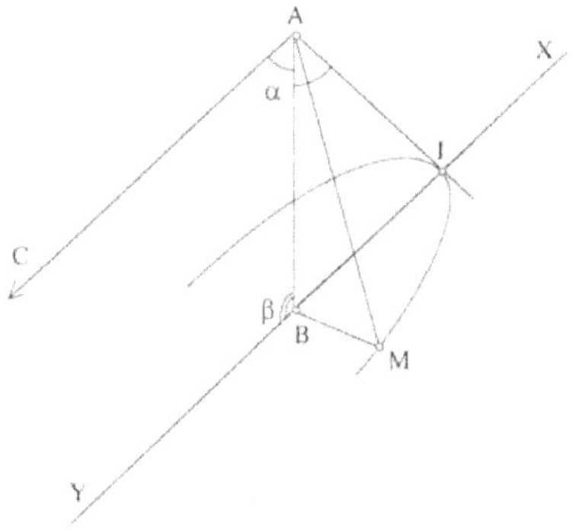

Fig. 102

By taking a length l' ($l' \neq l$ or $l' = l$) as the axis AB of the compass, one will trace in Π' (defined by $A\hat{B}Y' = \beta'$ with $\alpha' = \pi - \beta'$) a parabola with *latus rectum* $c' = 2l' \frac{\sin^2 \alpha'}{\cos \alpha'}$, whence the ratio of similarity $\frac{c'}{c}$. Thus one can trace by means of the compass two similar parabolic sections in the same plane Π. This is the idea that al-Sijzī recalls briefly at the end of his work.

Thus, in the wake of al-Qūhī, al-Sijzī establishes generally that the nature of the drawn curve depends upon the initial position of the compass in relation to the plane on which one wants to draw it.

Research on the perfect compass and on continuous drawing does not stop with al-Sijzī, however. As we have said, other mathematicians will pursue it for more than two centuries.

5. CONTINUOUS DRAWING AND CLASSIFICATION OF CURVES

Research on continuous drawing, as we have noted, was a response, among other things, to the necessity for mathematicians of this era to ensure the continuity of curves. At the time, the only means at their disposal was to introduce motion into geometry. Now it did not take long for these new concerns to steer this research toward the major problem of classifying curves as a function of the kind and the number of motions that participate in drawing them. Since this is seminal research, its considerable importance deserves to be emphasized.

Al-Qūhī groups the curves traced by the perfect compass – the straight line, circle, conic sections – under a generic name: 'the *qiyāsiyya* lines' which we will translate as 'measurable lines'. At the beginning of his second book, he writes:

> This is a treatise on the instrument called the perfect compass, which contains two books. The first one deals with the demonstration that it is possible to draw measurable lines (*qiyāsiyya*) by this compass – that is, straight lines, the circumferences of circles, and the perimeters of conic sections, namely parabolas, hyperbolas, ellipses, and opposite sections.[26]

Now this expression 'measurable lines' will be used throughout the entire tradition: it reappears in al-Sijzī, Hibat Allāh al-Baghdādī, Ibn al-Ḥusayn… In his 19th-century French translation, however, F. Woepcke rendered this same expression by 'regular lines,' thus throwing a veil over this important question.

The plural adjective *qiyāsiyya* is derived from the verbs *qāsa*, *yaqīsu*, or *qāsa*, *yaqūsu*, which both express the idea of measure – whence the literal translation above: 'measurable lines'. This same term *qiyāsī* (in the singular) also has a figurative sense: it is used of a woman who walks in regular fashion. Now it is precisely this figurative sense that F. Woepcke chose when he referred to 'regular lines.' This translation is not very satisfactory, not only because it unnecessarily privileges the figurative sense, but also it is ambiguous and lacks precision. If one wants to speak about a regular curve in the sense in which one understands this since the 19th century

[26] R. Rashed, *Geometry and Dioptrics in Classical Islam*, p. 726.

tury, one would have to exclude the straight line. But al-Qūhī very precisely lists the latter among these *qiyāsiyya* lines.

What then does al-Qūhī mean by 'measurable lines'? According to the geometrical terminology of the time, they are lines (that is, magnitudes) that are subject to proportion theory. Such is precisely al-Qūhī's implication: it is a matter of the lines generated by a single continuous motion – that of the branch of the perfect compass – and to which one can apply the theory of proportions. This is the case for the straight line and the circle, but also for the three conic sections characterized by the *symptomata* and the properties of the focus and the directrix.

Thus al-Qūhī has just established a classification of curves: the measurable ones, and the others. But he has also left behind a distinction with deep roots in the tradition, that between a straight line, on the one hand, and curves (including the circle), on the other.

Not only did al-Sijzī seize upon al-Qūhī's gains, but he also will confirm them. Thus, in a book entitled *Kitāb al-Madkhal ilā ʿilm al-handasa* (*Book of Introduction to Geometry*), he carries out different classifications. When he comes to lines, he distinguishes three kinds: measurable lines (the circle, the straight line, and the conics); nonmeasurable lines that have an order (*niẓam*) and a regularity (*tartīb*); finally, the nonmeasurable lines that have neither order nor regularity. According to al-Sijzī, the first are generated by a unique continuous motion and are 'geometrical', that is, one can use them in geometry. The second are generated by two continuous motions; they are no longer 'geometrical' but 'mechanical.' The third, which are also generated by two continuous motions, are not even mechanical. The example of mechanical curves that he gives is the cylindrical helix. It is indeed a skew curve generated by a motion of uniform rotation about an axis and by a uniform translation parallel to the axis. This is what he writes:

> As to the curve, the cylindrical helix *(al-khaṭṭ al-lawlabī)*, which is used in mechanics (*al-ḥiyal*) and not in geometry, for it is not measurable (*ghayr qiyāsī*) but has an order and a regularity; it is generated by the motion of a point following a straight line and following a circle, in common with the cylinder.[27]

He goes on:

> Given the cylinder *ABCD* whose two bases are *AB* and *CD*. If we imagine that point *A* moves by uniform motion along the straight line *AC* and that the cylinder rotates about the two centers of its bases with uniform motions, the line *AEGHID* is generated, which is a cylindrical helix. As to the line that

[27] Ms. Dublin, Chester Beatty 3652, fol. 4ʳ.

has no order, it therefore has neither limit (*ḥadd*, which is also translated by 'definition'), nor end, and is used in none of the arts; this is why it is neither described nor defined.[28]

Al-Sijzī traces the cylindrical helix, but gives no example of nonmeasurable curves with no order or regularity. Perhaps he had in mind such curves as the quadratrix or the spiral.

About the meaning of this distinction between measurable and nonmeasurable curves, there is not the slightest shadow of a doubt. If need be, one can find an additional proof of their meaning in al-Sijzī's use of these terms when he defines the angles: the nonmeasurable angles are precisely the curvilinear angles and the angle of contingence ('horn angle'), whereas measurable angles are those that one can study by means of proportion theory.[29]

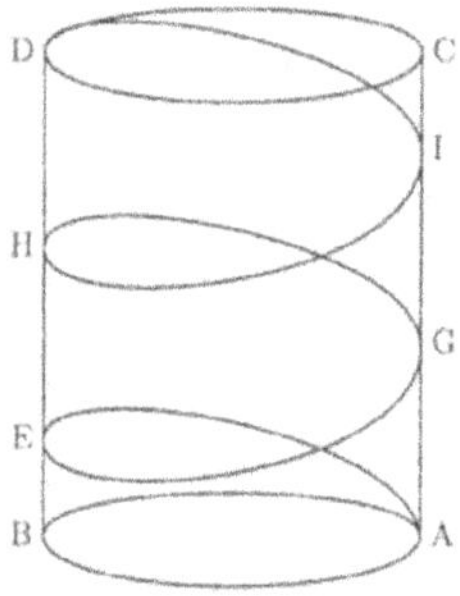

Fig. 103

This seminal research on the concept of curves by means of the concept of motion and the number of motions, as well as the separation between geometrical curves and mechanical curves, according to whether it is possible or not to apply the theory of proportions to them, is of major importance to the history of geometry, and notably much later for algebraic geometry. The crucial question is to find out what became of this chapter, whose history we have just sketched, in the mathematics beyond the 10th–11th centuries.[30]

[28] *Ibid.*

[29] *Ibid.*, fol. 68r.

[30] See above, 'The first classifications of curves'.

– 4 –

THĀBIT IBN QURRA ON EUCLID'S FIFTH POSTULATE*

1. INTRODUCTION

Two works by Thābit ibn Qurra on the theory of parallels have come down to us. Both treatises are well known and have already been the subject of several commentaries and translations.[1] Here we intend to comment on them, the better to situate Thābit ibn Qurra's contribution following the recent publication of his works in astronomy and mathematics.[2]

The first of these treatises is called *If One Draws Two Straight Lines According to Two Angles Less than Two Rights, They Meet*[3] and the second, *On the Demonstration of Euclid's Famous Postulate*.[4]

* In collaboration with Christian Houzel.

1 A. P. Youshkevitch and B. Rosenfeld, *The Theory of Parallels in the Medieval East, 9th–14th Centuries* (in Russian), Nauka, 1983; Kh. Jaouiche, *La Théorie des parallèles en pays d'Islam: contribution à la préhistoire des géométries non euclidiennes*, Paris, 1986; I. Tóth, 'Das Parallelenproblem im Corpus Aristotelicum', *Archive for the History of Exact Sciences*, vol. 3, nos. 4–5, 1967, pp. 249–422.

2 Thābit ibn Qurra, *Œuvres d'astronomie*, edited and translated by R. Morelon, Paris, 1987; R. Rashed, *Les Mathématiques infinitésimales du IXe au XIe siècle*, vol. I: *Fondateurs and commentateurs: Banū Mūsā, Thābit ibn Qurra, Ibn Sinān, al-Khāzin, al-Qūhī, Ibn al-Samḥ, Ibn Hūd*, London, 1996, chap. II, pp. 140–673 (English translation: *Founding Figures and Commentators in Arabic Mathematics*. A History of Arabic Sciences and Mathematics, vol. 1, Culture and Civilization in the Middle East, London, Centre for Arab Unity Studies, Routledge, 2012), and *Les Mathématiques infinitésimales du IXe au XIe siècle*, vol. IV: *Méthodes géométriques, transformations ponctuelles et philosophie des mathématiques*, London, 2002, App. I, pp. 687–765; *id.*, *Geometry and Dioptrics in Classical Islam*, London, 2005.

3 Text established on the basis of ms. Paris, Bibliothèque nationale 2457, fols 156v–160r. See R. Rashed (ed.), *Thābit ibn Qurra. Science and Philosophy in Ninth-Century Baghdad*, Scientia Graeco-Arabica, vol. 4, Berlin, Walter de Gruyter, 2009, pp. 42–63.

4 Text established on the basis of the manuscripts Cairo, Dār al-Kutub, Riyāḍa 40, fols 200v–202r and Istanbul, Aya Sofya 4832, fols 51r–52r. See R. Rashed (ed.), *Thābit ibn Qurra. Science and Philosophy in Ninth-Century Baghdad*, pp. 64–73.

Unfortunately, we have no way of knowing the order in which he composed these two works. We will therefore analyze Thābit ibn Qurra's demonstrations and compare them. Before that, however, we must sketch the other known attempts to prove this postulate before Thābit ibn Qurra.

As is well known, the difficulties presented by the theory of parallels were already noticed well before the era of Euclid, for example in certain texts by Aristotle.[5] These difficulties are linked to the definition of parallelism and to the possibility of drawing parallels. Euclid defined parallel straight lines as 'straight lines which, being in the same plane and being produced indefinitely in both directions, do not meet one another in either direction' (*Elements*, I, Def. 23).[6] This definition obviously raises serious problems. On the one hand, it is negative; as such, logicians will not accept it. On the other hand, it brings in a previously undefined concept, namely that of infinity (indefinite extension). Under these circumstances, it would therefore seem impossible to verify parallelism. Euclid, however, proved capable of demonstrating that a certain angular property implies parallelism (*Elements*, Propositions I.27 and 28):

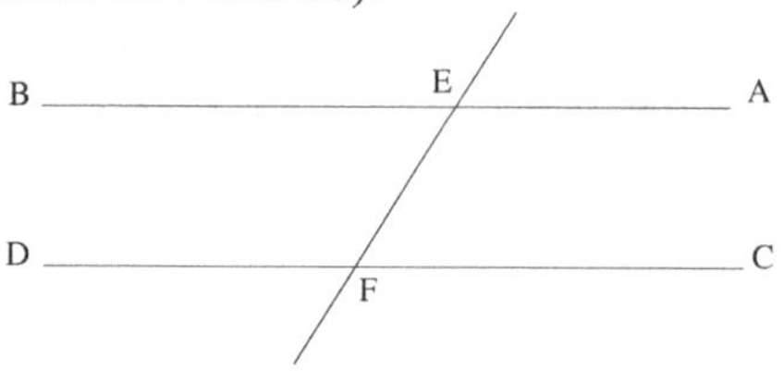

Fig. 104

AB and *CD* are two straight lines in a plane, and *EF* is a transverse line that meets *AB* at *E* and *CD* at *F*. If the alternate interior angles *BEF* and *EFC* are equal, or if the two angles *AEF* and *EFC* form two right angles, then *AB* and *CD* meet neither in one direction nor in the other; hence they are parallel.

Euclid needs the reciprocal property to reduce parallelism to an angular property (positive and entirely in the finite domain). He states this reciprocal property in Proposition I.29, but his demonstration requires a postulate, the famous fifth postulate: in the same situation (Fig. 105), if the angles *AEF* and *EFC* add up to less than two right angles, 'two straight lines, if

[5] *Anal. prior.*, B 16, 65a4 and 17, 66a11; see also I. Tóth, 'Das Parallelenproblem'.

[6] *The Thirteen Books of Euclid's Elements*, translated with introduction and commentary by T. L. Heath, 2nd ed., 3 vols, New York, Dover Publications, 1956, vol. 1, p. 154.

produced indefinitely, meet on that side on which are the angles less than the two right angles' (*Elements*, Book I, Postulate 5).[7]

This postulate is obviously *ad hoc* and its statement has the form of a theorem (the reciprocal of I.28). From the Hellenistic period on, therefore, it drew many objections, which are known from Proclus's commentary on the first book of Euclid from the 5th century of our era. Geometers thus tried to *demonstrate* the fifth postulate. The grounds for such demonstrations were another definition of parallels and implicit or explicit hypotheses that, to the geometers, seemed more natural than Euclid's postulate. It is in his commentary that Proclus summarizes the propositions of Posidonius, Geminus, and Ptolemy, as well as his own.[8]

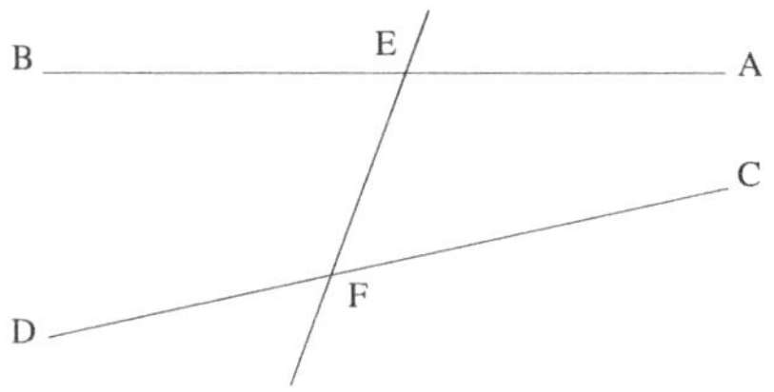

Fig. 105

The most frequent definition proposed to replace Euclid's is a property that implies parallelism: two straight lines in a plane are called parallel if they preserve the *same distance* between them, that is, if the distance from a point *M* of *AB* to *CD* does not depend on the choice of *M* on *AB*. In other words, parallelism is replaced by *equidistance*. This definition has its own problems: one must prove that equidistant straight lines exist and that the property is symmetrical, that is, that if *AB* is equidistant to *CD*, *CD* is equidistant to *AB*. In the *Elements*, once Proposition 29 is demonstrated, one can establish the existence first of parallelograms, then of rectangles; it is then easy to demonstrate that parallels are equidistant.

We know of one detailed attempt to demonstrate Euclid's postulate from a fragment of the 6th-century commentator Simplicius; the Greek text is now lost, but it had been translated into Arabic and was cited by al-Nayrīzī.[9] In this fragment, Simplicius credits his friend Aghānis with a demonstration of the postulate. Unfortunately, we know nothing about this

[7] Heath translation, vol. 1, p. 155.

[8] Proclus, *in Eucl. I*, ed. Friedlein 191.16–193.9; *Proclus: A Commentary on the First Book of Euclid's* Elements, transl. G. R. Morrow, Princeton, Princeton University Press, 1970, pp. 150–1.

[9] See ms. Leiden, Or. 399/1, fols 16r–17r.

mysterious Aghānis, not even his Greek name. He defines 'parallelism' by equidistance, and after allowing the symmetry of this property, he establishes that the segment that gives the distance between two straight lines is perpendicular to each of them. As far as we know, Thābit ibn Qurra was the first to establish this important property. Indeed, as we shall see below, Thābit ibn Qurra demonstrates rigorously the symmetry of equidistance once the existence of equidistant straight lines is admitted. The demonstration of the statement corresponding to I.29 is therefore easy:

One assumes that *AB* and *CD* are equidistant and considers a transverse line *EG*. One draws *EI* perpendicular to *CD*, and *GK* perpendicular to *AB*; by hypothesis, *EI* = *GK*, so that the right triangles *GKE* and *EIG* are equal. Then the alternate interior angles *EGK* and *GEI* are equal.

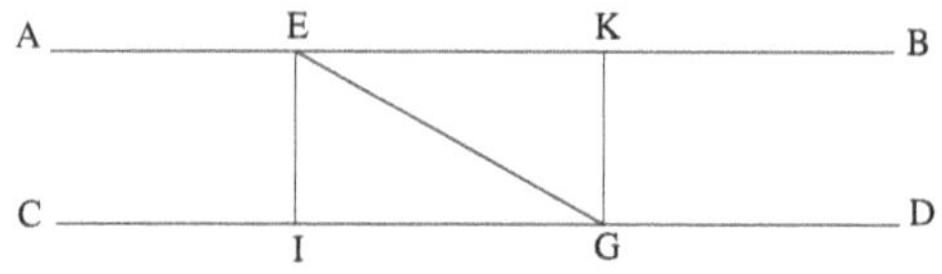

Fig. 106

To demonstrate the fifth postulate directly, it is necessary to determine how far one must extend the straight lines *AB* and *CD* in Fig. 107 in order to be certain that they meet. Aghānis uses Lemma X.1 of Euclid's *Elements*, a form of what is nowadays called 'Archimedes' axiom'; he also tacitly uses the fact that a straight line that meets the side of a triangle and that is parallel to another side necessarily meets the third side (a special case of what will later be called 'Pasch's axiom').

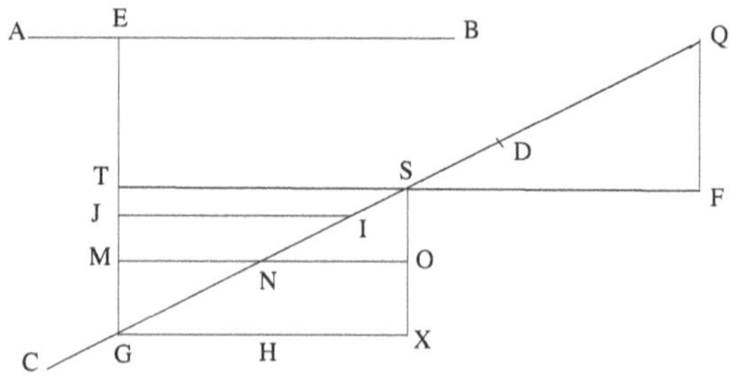

Fig. 107

From a point *I* of *CD*, one draws *IJ* 'parallel' to *AB*. One divides *EG* in two at *T*, then *TG* in two at *M*, and so on. By Lemma X.1, after a large enough number n of steps, one reaches a division point (call it M) such that $MG < JG$. The 'parallel' MN to AB meets CD at N between G and I (Pasch's axiom). Let Q be the extension of CD such that $GQ = 2^n GN$ (here $4GN$, since one has assumed that $n = 2$); let S be such that $GS = 2GN$. 'Parallel' to EG, draw SX, which meets the extension of MN at O. The triangles MNG and ONS are equal (*Elements*, I.29 and I.15), such that $SO = GM = XO$ (parallelogram) and $SX = TG$. Then, TS is 'parallel' to GX, and therefore to AB; proceeding in the same fashion, one demonstrates that EQ is 'parallel' to AB, that is, that Q is on AB (the unicity of the parallel line).

The main components of this demonstration will recur among several other authors: the definition of 'parallelism' by equidistance, the demonstration of the statement corresponding to I.29 for equidistant lines (implicitly) assumed to exist, and the demonstration of the fifth postulate by appeal to the axioms of Archimedes and Pasch.

2. THĀBIT IBN QURRA'S FIRST TREATISE

Let us now return to the treatises that Thābit ibn Qurra devoted to the fifth postulate, beginning with *If One Draws Two Straight Lines According to Two Angles Less than Two Rights, They Meet*. It must be emphasized that contrary to his predecessors, Thābit ibn Qurra in this treatise tries to justify the existence of equidistant straight lines. To this end, he draws upon the motion of rectilinear translation. In Thābit ibn Qurra, as in his master al-Ḥasan ibn Mūsā, motion becomes a necessary foundation of geometry. Recall that in his other mathematical treatises, notably that devoted to cylindrical sections, Thābit ibn Qurra had also introduced the concept of motion.[10] Here, Thābit ibn Qurra uses the following considerations to justify the necessity of motion: both equality and measurement presuppose the possibility of displacing and superposing figures. For him, both the definition of the circle and Euclid's third postulate implicitly appeal to a motion of rotation about a fixed center; by analogy, Thābit ibn Qurra allows the motion of rectilinear translation of a rigid body as a primitive notion in geometry. He states the corresponding postulate: every point of a body subjected to a motion of rectilinear translation describes a straight line in the direction of translation.

In the first proposition, Thābit ibn Qurra proves the existence of equidistant straight lines as follows: one considers two straight lines AB and CD

[10] Rashed, *Founding Figures and Commentators in Arabic Mathematics*, Chap. II, pp. 333–458.

in a plane such that the segments *AC* and *EF* are equal and that the angles *ACD* and *EFD* are equal. Then *AG* and *EH*, the perpendiculars to *CD*, are equal and, for every point *I* of *AB*, the perpendicular *IK* to *CD* is equal to *AG* and *EH*.

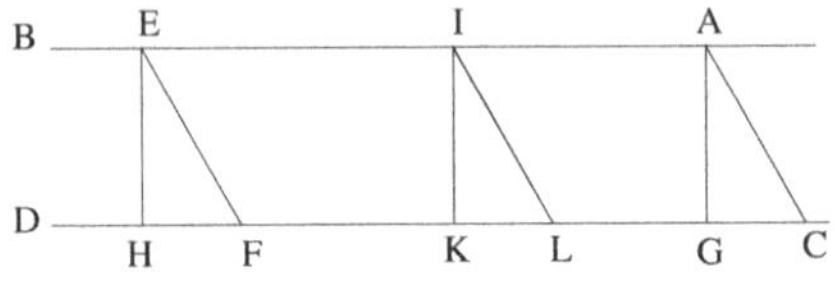

Fig. 108

To demonstrate this proposition, Thābit ibn Qurra observes that the motion of translation of triangle *ACG* along *CG* brings it onto triangle *EFH* or onto triangle *ILK*.

Propositions 2 and 3, which are reciprocals of each other, state the properties of a certain quadrilateral (an isosceles trapezium); these properties will be the foundation of ʿUmar al-Khayyām's work on the fifth postulate.[11] The quadrilateral *ABCD* has equal base angles, $B\hat{A}D$ and $C\hat{D}A$ and its sides *AB* and *DC* are equal; therefore $A\hat{B}C = D\hat{C}B$. Inversely, if $\hat{A} = \hat{D}$ and $\hat{B} = \hat{C}$, then *AB* = *CD*.

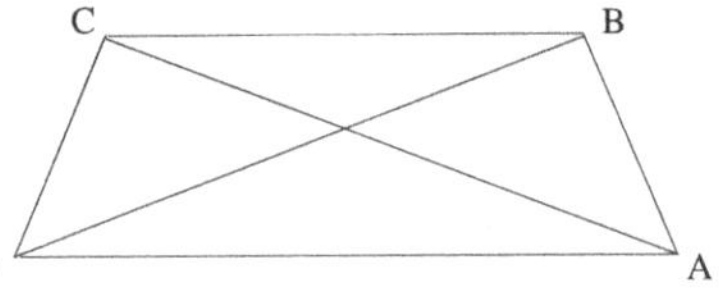

Fig. 109

Indeed, if $\hat{A} = \hat{D}$ and *AB* = *DC*, the triangles *ABD* and *DCA* are equal (*Elements*, I.4), such that *AC* = *DB*; from this, it follows that the triangles *ABC* and *DCB* are equal (*Elements*, I.8), therefore $\hat{B} = \hat{C}$.

To demonstrate the reciprocal, one assumes that $AB \neq DC$, for example, that $AB > DC$ and by deducing a contradiction. Let *AE* = *DC* be transposed onto *AB*; by the direct property, $A\hat{E}C = D\hat{C}E$, which is smaller than $D\hat{C}B$. But, as an angle exterior to triangle *CBE* (*Elements*, I.16), $A\hat{E}C$ must be larger than $A\hat{B}C$ and one has assumed that $A\hat{B}C = D\hat{C}B$.

[11] R. Rashed and B. Vahabzadeh, *Omar Khayyam. The Mathematician*, Persian Heritage Series no. 40, New York, Bibliotheca Persica Press, 2000, p. 225 (Proposition 1), p. 230 (Proposition 4).

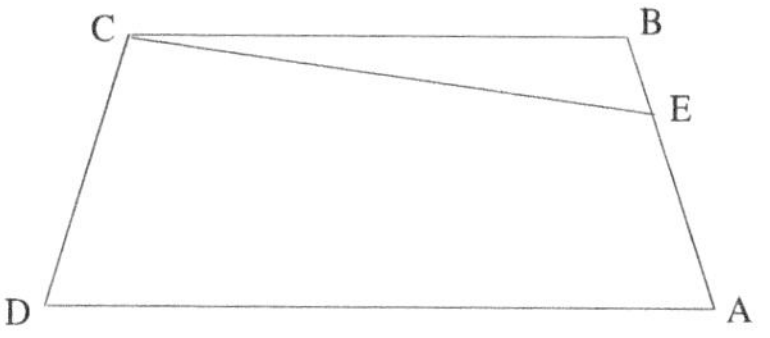

Fig. 110

In the fourth proposition, Thābit ibn Qurra demonstrates the symmetry of the equidistance property for two straight lines. Let the latter be *AB* and *CD*; one assumes that *EG* and *FH*, the perpendiculars to *CD* drawn from the points *E* and *F* of *AB*, are equal. Then *EG* and *FH* are perpendicular to *AB*.

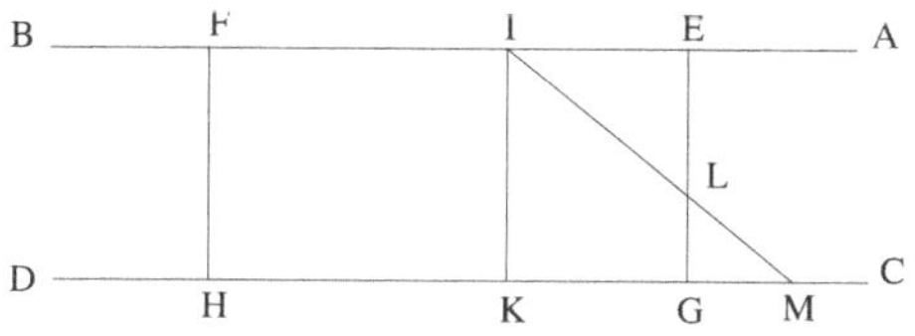

Fig. 111

First of all, a perpendicular *IK* to *CD* drawn from a point *I* on *AB* cannot meet *EG*: if it met *EG* at a point *L*, the triangle *LGM* would have two right angles, at *G* and at *M*, which contradicts *Elements* I.17. By Proposition 1, we know that *IK*, the perpendicular to *CD*, is equal to *EG* and by Proposition 2, that $K\hat{I}E = G\hat{E}I$. Likewise $K\hat{I}F = H\hat{F}I$ and $G\hat{E}F = H\hat{F}E$, so that $K\hat{I}E = K\hat{I}F$ and the angles at *I* are right, then the angles *GEI* and *HFI*, which are equal to them, are also right angles.

Proposition 5 characterizes equidistant straight lines as being those that have a common perpendicular; it is founded on a property that will be the core of Ibn al-Haytham's demonstration of the fifth postulate (existence of the rectangle[12]).

One draws *AC* and *BD* perpendicular to a given *AB* and, from a point *E* on *AC*, one drops the perpendicular *EF* to *BD*. Then angle *FEA* is right and *EF* is equal to *AB*.

[12] Ibn al-Haytham, *Sharḥ Muṣādarāt Kitāb Uqlīdis*, ms. Istanbul, Feyzullah 1359, fols 170^{v}–176^{r}.

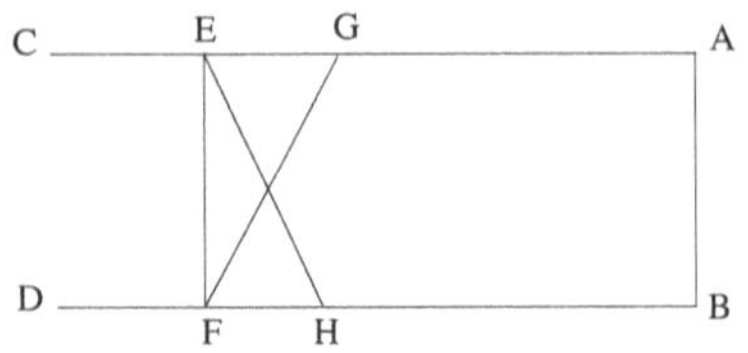

Fig. 112

Let us demonstrate that $AE = BF$; otherwise $AE > BF$ if not $AE < BF$. If $AE > BF$, let $AG = BF$, with G on AE; since GA and FB are perpendicular to AB, one knows by Proposition 4 that they are also perpendicular to FG. Thus $G\hat{F}B$ is a right angle, as is $E\hat{F}B$, which is absurd. If $AE < BF$, let $BH = AE$, with H on BF; by Proposition 4, the angle EHB is right, and therefore equal to EFH, which is absurd, since an exterior angle of a triangle is greater than each of the opposite interior angles (*Elements*, I.16).

Thus $AE = BF$ and by Proposition 2, the angle AEF is equal to the right angle BFE; by Proposition 3, it follows that $EF = AB$.

In Proposition 6, Thābit ibn Qurra proves the equality of the alternate interior angles formed by a line that is transverse to two equidistant straight lines.

One assumes that straight lines AB and CD have in common a perpendicular EF and one considers a transverse line GI; then the alternate interior angles AHI and FIH are equal.

To demonstrate this, let L be the center of HI; drop the perpendicular LM onto AB. The extension of ML meets CD at a point N, for it penetrates into the quadrilateral $EHIF$ and can exit through neither EF nor EH (a form of Pasch's axiom implicitly enters the picture here). By Proposition 5, the angle at N is right, like the angle at M, and one has $M\hat{L}H = N\hat{L}I$, such that the right-angled triangles LNI and LMH are equal and that their angles at H and I are equal.

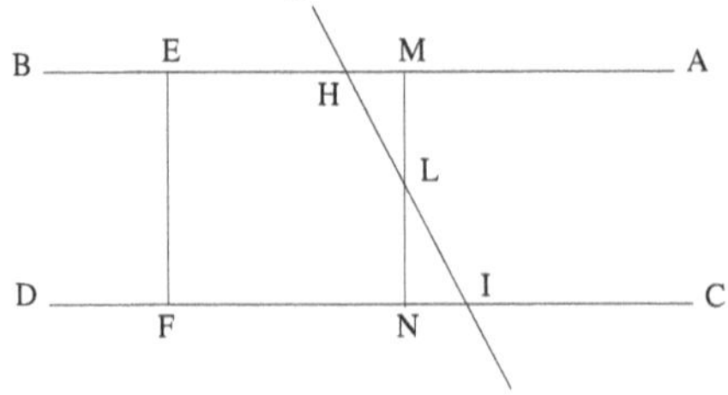

Fig. 113

This proposition is intended to replace Proposition I.29 of the *Elements* in which Euclid proves that 'a straight line falling on parallel straight lines makes the alternate angles equal to one another, the exterior angle equal to the interior and opposite angle, and the interior angles on the same side equal to two right angles'.[13]

Thābit ibn Qurra concludes his treatise with a demonstration of Euclid's postulate along lines analogous to those of Aghānis, that is, by using the axioms of Archimedes and Pasch.

Let AC and BD be two straight lines such that the sum of the angles CAB and DBA is less than two rights; then at least one of the two angles, $D\hat{B}A$ for example, is acute. Let AE be perpendicular to BD; one chooses a point F on AC and one drops the perpendicular FG onto AE. According to the axiom of Archimedes, there exists a multiple AH of AG greater than AE, let's say $4AG = AH > AE$.

Let I, K, H on AE be such that $AG = GI = IK = KH$ and let L, M, N on AC be such that $AF = FL = LM = MN$; let us raise perpendicular IS to AE and drop the perpendicular FS onto IS. One can apply Proposition 5 to the quadrilateral $GISF$ to prove that its angle at F is right and that $FS = GI = AG$. Since the angle AFG is acute, FL remains exterior to rectangle $GISF$. The straight lines FS and AI have a common perpendicular SI, therefore, by Proposition 6, the transverse line AF forms with them equal angles GAF and SFL; the triangles AFG and FLS are therefore equal, and angle FSL is equal to the right angle AGF. Consequently, the points ISL are aligned and LI is perpendicular to AE.

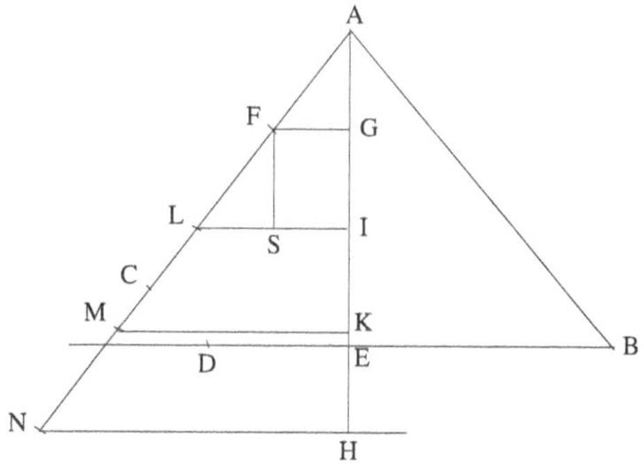

Fig. 114

Likewise one proves that MK and NH are perpendicular to AE; since BE is also perpendicular to AE, it cannot meet HN. Therefore the extension

[13] Heath translation, vol. 1, p. 311.

of *BD*, which penetrates the triangle *ANH*, must meet *AN* ('Pasch's axiom').

The properties established in this proposition are:

1) The straight lines that join the points of division corresponding to each other on the straight lines *AH* and *AN* are equidistant. This property is obviously an immediate consequence of Proposition VI.2 of the *Elements*; at this stage, however, one cannot use the results of this Book VI, since they rely on the theory of parallels.

2) The property expressed by Pasch's axiom: straight line *BD* cuts the plane in two half planes, and every straight line joining a point from one of the half-planes to a point on the other cuts *BD*. As far as we know Thābit ibn Qurra is the first to have deliberately used this property. In contrast, Aghānis makes no mention of such a property, even though it is indispensable.

Finally, note that whereas Aghānis uses the axiom of Archimedes in its multiplicative form (Lemma X.1), Thābit ibn Qurra uses it in its additive form.

As we have just seen, Thābit ibn Qurra in this text draws on the concept of motion as a primitive geometrical concept and also on the axioms of Archimedes and Pasch; the combination of these elements makes Thābit ibn Qurra's exposition stand out from those of his predecessors and will give Ibn al-Haytham and ʿUmar al-Khayyām their new starting points. It is in this respect that Thābit ibn Qurra's contribution lays the foundation for a multi-century tradition of research on the theory of parallels.

3. THĀBIT IBN QURRA'S SECOND TREATISE

Thābit ibn Qurra's second treatise is entitled, *On the Demonstration of Euclid's Famous Postulate*. The fundamental figure of this demonstration is a pair of straight lines 'that get neither closer nor farther from each other' (*i.e.*, equidistant straight lines). In the first proposition of this treatise, Thābit ibn Qurra takes up Proposition I.28 of Euclid's *Elements*: if two straight lines form equal alternate interior angles with a transverse line, they are *parallel*. Thābit ibn Qurra states that such lines 'get neither closer nor farther from each other', that is, that they are equidistant.

The hypothesis is the equality of angles *AEG* and *DGE*; Thābit ibn Qurra superposes *EA* on *GD* and *EG* on *GE* such that *GC* falls onto *EB*, by the symmetry of center (the center of *EG)*. He assumes implicitly that this symmetry preserves these angles. If *EB* and *GD* approach each other near *B* and *D*, then *EA* and *GC* should approach each other near *A* and *C* and the two straight lines get closer to one another in both directions, a property that Thābit ibn Qurra rejects as absurd. Likewise, he shows that if the two straight lines get farther from one another in the same direction, they should also get farther from one another in the other, which he also rejects as absurd.

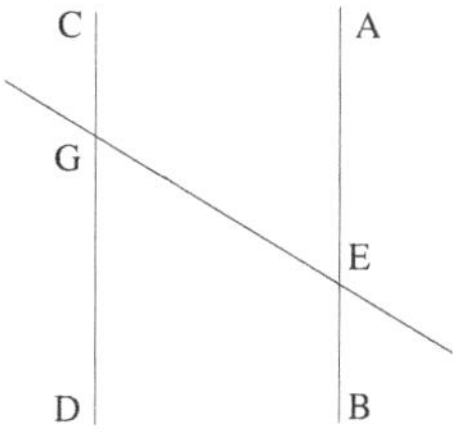

Fig. 115

The postulate on which Thābit's demonstration rests – if two straight lines get closer in one direction, they get farther in the other (false in elliptical geometry, however) – will also enter into al-Khayyām's and Naṣīr al-Dīn al-Ṭūsī's[14] demonstrations of Euclid's fifth postulate.

The second proposition is the reciprocal of the first and it corresponds to Proposition I.29 of Euclid's *Elements*: if two straight lines get neither closer to nor farther from each other, they form with a transverse line equal alternate interior angles, a statement in which 'parallel straight lines' is replaced by 'straight lines that get neither closer to nor farther from one another'.

Indeed, if for example $A\hat{E}G$ is smaller than $E\hat{G}D$, one draws HGI such that $E\hat{G}I = A\hat{E}G$; then GI is between CD and AB. But, by the first proposition, HI and AB get neither closer to nor farther from one another, and this is absurd, for the same applies to CD and AB.

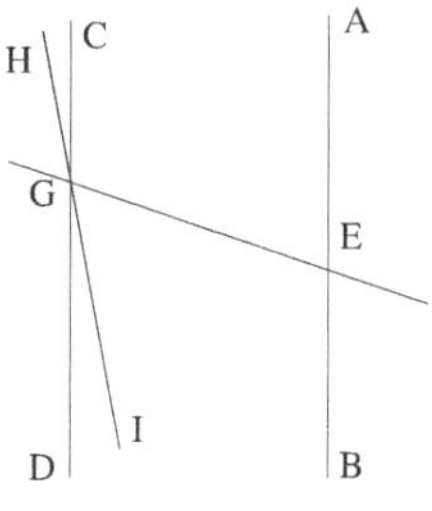

Fig. 116

[14] *Risāla fī Sharḥ mā ashkala min Muṣādarāt Kitāb Uqlīdis*, in Rashed and Vahabzadeh, *Omar Khayyām. The Mathematician*, pp. 223–8 and *al-Risāla al-Shāfiya*, no. 8, vol. II, pp. 4–14 of *Rasā'il al-Ṭūsī*, 2 vols, Hyderabad, 1359 H.

This reasoning recalls that of Proclus in his attempt to demonstrate Euclid's postulate;[15] here, Thābit implicitly grants the unicity of a straight line that passes through a given point and that gets neither closer to nor farther from a given straight line.

Proposition 3 takes up *Elements* I.33, which proves the existence of parallelograms by replacing 'parallel straight lines' by 'straight lines that get neither closer to nor farther from one another'. One considers two straight lines *AB* and *CD* that get neither closer to nor farther from one another and that are equal.

For the angles *BAD* and *CDA* are equal (Proposition 2) such that the triangles *ADB* and *DAC* are equal: $B\hat{D}A = C\hat{A}D$ and $BD = CA$. Then the straight lines *AC* and *BD* get neither closer to nor farther from one another according to Proposition 1.

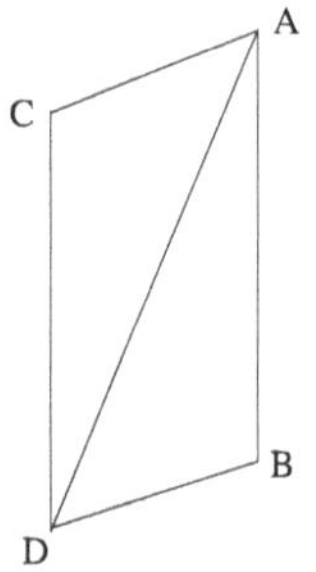

Fig. 117

The fourth proposition gives the property of the straight line that joins the centers of two sides of a triangle: this is a special case of *Elements* VI.2, if one replaces 'parallel straight lines' with 'straight lines that get neither closer to nor farther from one another': let *ABC* be a triangle and let *D*, *E* be the centers of *AB* and *AC* respectively. Then *DE* and *BC* get neither closer to nor farther from one another and *DE* is equal to half of *BC*.

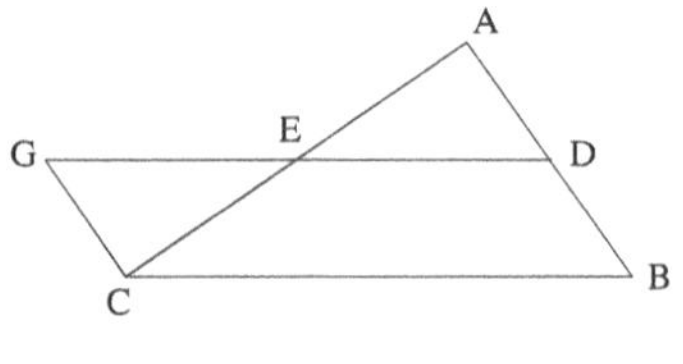

Fig. 118

[15] Proclus, *loc. cit.*, 371.23–373.2; see also Rashed and Vahabzadeh, *Omar Khayyām The Mathematician*, pp. 231–2, Proposition 6.

One extends DE by $EG = DE$ and draws CG. The triangles AED and GEC are equal (*Elements*, I.4), therefore $A\hat{D}E = C\hat{G}E$ and $CG = AD = BD$; thus AB and GC get neither closer to nor farther from one another (Proposition 1) and one can apply Proposition 3 to prove that BC and DG get neither closer to nor farther from one another and that $DG = BC$.

This proposition will be used to demonstrate that the straight lines joining the corresponding points of division on two given straight lines get neither closer to nor farther from one another.

Finally, Thābit ibn Qurra demonstrates Euclid's postulate by using once again the axioms of Archimedes and Pasch.

With the transverse line EG, the two straight lines AB and CD form two angles BEG and DGE less than two rights, and one wishes to demonstrate that they meet in the direction of B and D.

Let GH be the straight line passing through G that gets neither closer to nor farther from AB; one chooses a point I on CD and one draws IK, which gets neither closer to nor farther from EG. One extends GI by $IL = GI$ and GK by $KH = GK$; according to Proposition 4, IK and LH get neither closer to nor farther from one another and IK is equal to half of LH, that is, $LH = 2IK$. By iterating this construction, one thus obtains a segment LH greater than EG by virtue of Archimedes' axiom, which in this instance is used in its multiplicative form. Let M on HL be such that $HM = GE$; according to Proposition 3, EM and GH get neither closer to nor farther from one another and they are equal. This must mean that M is on the extension of EB, on account of the already granted unicity of the straight line passing though E that gets neither closer to nor farther from (GH); the extension of EB then meets CD before LH by the same variant of Pasch's axiom as in the preceding demonstration: M and E are in two different half-planes determined by (CD), therefore (ME) cuts (CD) between M and E.

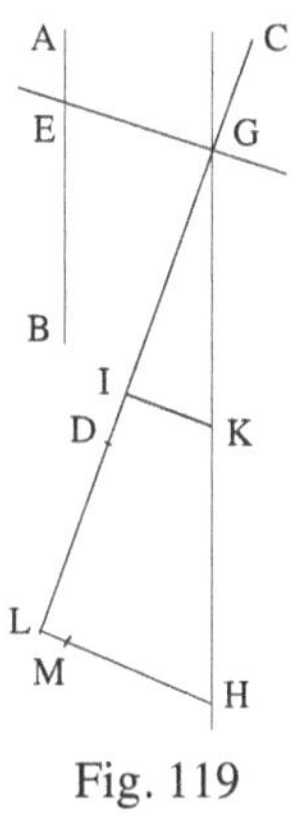

Fig. 119

Thābit ibn Qurra's two demonstrations are founded on the same concept of equidistance, which replaces parallelism in the Euclidean sense. But this should not obscure the differences between them. The first proves the existence of equidistant straight lines by using a motion of rectilinear translation and by appealing to the additive form of Archimedes' axiom and to a variant on Pasch's axiom. The second, by contrast, starts from an implicit principle according to which two straight lines can neither converge simultaneously on both sides of a transverse line, nor diverge simultaneously on both sides, in order to prove the existence of straight lines that neither converge nor diverge. In this second demonstration, Thābit ibn Qurra does not introduce motion; and although he uses again Archimedes' axiom and the same variant of Pasch's axiom used in the first demonstration, it is the multiplicative form of the first axiom that he brings into play. It would be wrong, as some have done, to reduce both of Thābit ibn Qurra's procedures to that of Aghānis, insofar as Thābit in the first treatise demonstrates the existence of equidistant straight lines as well as the fundamental property of the symmetry of equidistance.

Finally, the foundational figure of Thābit ibn Qurra's first demonstration is the so-called quadrilateral of al-Khayyām (or Saccheri), which also appears in Ibn al-Haytham's demonstration. One can thus see the crucial role that this demonstration plays both in the history of the theory of parallels and in the foundation of the tradition that Thābit inaugurates, a tradition that many mathematicians, including the two just cited, will take up.

Part III

Application of mathematics

Astronomy and optics

– 1 –

THE CELESTIAL KINEMATICS OF IBN AL-HAYTHAM

1. INTRODUCTION

1.1. *The astronomical work of Ibn al-Haytham*

Ever since Pierre Duhem at least, historians of astronomy have agreed on the importance of Ibn al-Haytham's contribution to the study of celestial kinematics. Some have paid particular attention to his criticisms of Ptolemy, which later gave rise to the construction of new planetary models. But Ibn al-Haytham is seen merely as restricting himself to criticism, not participating in construction. Other historians have seen his contribution as synthesizing the *Almagest* with an Aristotelian cosmology. But a careful historical reading of Ibn al-Haytham's writings, including new texts previously left out of consideration, shows that both of these pictures of him are inaccurate. Instead, we find that Ibn al-Haytham tried to carry out a reform of astronomy, excluding all cosmology and developing the study of celestial kinematics.[1]

However, such a reading requires us to consider Ibn al-Haytham's astronomical work as a whole, in order to define the limits of his concerns and to exclude the writings incorrectly ascribed to him, which distort any assessment of his contribution.

The early bio-bibliographers – al-Qifṭī, Ibn Abī Uṣaybiʿa and an anonymous predecessor – tell us that Ibn al-Haytham wrote twenty-five astronomical works.[2] This means that one quarter of the eminent

[1] See R. Rashed, *Les mathématiques infinitésimales du IXᵉ au XIᵉ siècle*. Vol. V: *Ibn al-Haytham: Astronomie, géométrie sphérique et trigonométrie*, London, al-Furqān, 2006; English translation by J. V. Field under the title *Astronomy and Spherical Geometry: The Novel Legacy of Ibn al-Haytham*, London, 2014.

[2] The first critical examination of what is known about Ibn al-Haytham and his writings is given in R. Rashed, *Les mathématiques infinitésimales du IXᵉ au XIᵉ siècle*. Vol. II: *Ibn al-Haytham*, London, 1993, together with a summary in the form of a table listing all his works, including those on astronomy (pp. 511–35); English translation: *Ibn al-Haytham and Analytical Mathematics*. A History of Arabic Sciences and Mathematics, vol. 2, Culture and Civilization in the Middle East, London, Centre for Arab Unity Studies, Routledge, 2012, pp. 394–423.

mathematician's works were concerned with astronomy. In other words, he wrote twice as many works on this subject as he did on optics, the field with which his name is forever associated. The number of these writings alone indicates the enormity of Ibn al-Haytham's accomplishment and the sheer importance of astronomy in his life work.

The writings that have come down to us show clearly that, although the author's primary concerns are theoretical and mathematical, he neglected no part of astronomy. Several treatises relate to technical applications of astronomy, others to methods of astronomical calculation, yet others to procedures for making astronomical observations, and so on. One can nevertheless divide his writings into four groups, on the basis of both surviving texts and the lost ones for which the books of early bibliographers mention titles.

The first group consists of about ten treatises in which Ibn al-Haytham is concerned with technical problems: *Hour Lines* (*Fī khuṭūṭ al-sāʿāt*), *Horizontal Sundials* (*Fī al-rukhāmāt al-ufuqiyya*),[3] *The Direction of Mecca* (*Fī samt al-qibla bi-al-ḥisāb*),[4] *The Exact Determination of the Pole* (*Fī istikhrāj irtifāʿ al-quṭb ʿalā ghāyat al-taḥqīq*), *The Exact Determination of the Meridian* (*Fī istikhrāj khaṭṭ niṣf al-nahar ʿalā ghāyat al-taḥqīq*), *The Correction of Astrological Operations* (*Fī taṣḥīḥ al-aʿmāl al-nujūmiyya*),[5] and so on.

The second group consists of two treatises on astronomical observation: conditions for making observations, the errors that may occur in observation, and so on.

The third group of writings is concerned with such various questions and diverse problems as those relating to parallaxes, the Milky Way, and so on.

The fourth group, concerned with astronomical theory, can in turn be divided into three subgroups:

In the writings in the first of these, Ibn al-Haytham discusses the work of Ptolemy in three books that are of great historical and theoretical interest:

1. *Doubts concerning Ptolemy* (*Fī al-shukūk ʿalā Baṭlamiyūs*)[6]

[3] See our edition, translation and commentary of these two treatises in *Les mathématiques infinitésimales*, V, Part II, Chap. I and II.

[4] See A. Dallal, 'Ibn al-Haytham's Universal Solution for Finding the Direction of the *Qibla* by Calculation', *Arabic Sciences and Philosophy*, 5.2, 1995, pp. 145–93.

[5] See R. Rashed, *Les mathématiques infinitésimales*, vol. V, Appendix. II, p. 895.

[6] *Al-Shukūk ʿalā Baṭlamiyūs* (*Dubitationes in Ptolemaeum*), ed. A. I. Sabra and N. Shehaby, Cairo, The National Library Press, 1971.

2. *Corrections to the Almagest* (*Fī tahdhīb al-Majisṭī*)

3. *Resolution of Doubts concerning the Almagest* (*Fī ḥall shukūk fī kitāb al-Majisṭī*)

Of these three books, only the first and the third have come down to us.

In the writings in the second subgroup, Ibn al-Haytham examines individual celestial motions:

1. *The Winding Motion* (*Fī ḥarakat al-iltifāf*)

2. *Resolution of Doubts concerning the Winding Motion* (*Fī ḥall shukūk ḥarakat al-iltifāf*)[7]

3. *The Motion of the Moon* (*Fī ḥarakat al-qamar*)

Only the last two texts of this subgroup survive.

The third subgroup includes four titles:

1. *The Different Altitudes of the Wandering Stars* (*Fī ikhtilāf irtifāʿāt al-kawākib*)

2. *The Ratios of Hourly Arcs to their Altitudes* (*Fī nisab al-qusiyy al-zamāniyya ilā irtifāʿātihā*)

3. *The Configuration*[8] *of the Motions of Each of the Seven Wandering Stars* (*Fī hay'at ḥarakāt kull wāhid min al-kawākib al-sabʿa*)

4. *The Configuration of the Universe* (*Fī hay'at al-ʿālam*)

The first of these texts has come down to us, but the second has been lost. Only a part of the third survives;[9] the fourth is not to be identified with the apocryphal text of the same title.[10]

This simple summary shows very clearly that, apart from *The Configuration of the Universe* (of doubtful authenticity, as noted), the treatises on *The Different Altitudes of the Wandering Stars*, and on *The Configuration of the Motions of Each of the Seven Wandering Stars*, this major body of astronomical work is far from being well known.

Notice also that, in the three books in which Ibn al-Haytham mentions Ptolemy or the *Almagest*, he does so in order to criticize the work, for he

[7] A. I. Sabra, 'Maqālat al-Ḥasan ibn al-Haytham fī ḥall shukūk ḥarakat al-iltifāf', *Journal for the History of Arabic Science*, 3.2, 1979, pp. 183–212, 388–92.

[8] The Arabic *hay'a* could be translated equally by 'configuration' or 'model'.

[9] See our edition, translation and commentary of Treatises 1 and 3 in *Les mathématiques infinitésimales*, V, Part I.

[10] See R. Rashed, '*The Configuration of the Universe*: a Book by al-Ḥasan ibn al-Haytham?', *Revue d'histoire des sciences*, 60.1, 2007, pp. 47–63, and *Les mathématiques infinitésimales*, vol. V, Appendix I.

mentions 'Doubts', 'Corrections', and the 'Resolution of Doubts'. If to that we add the criticism of Ptolemy put forward in *The Resolution of Doubts concerning the Winding Motion*, it is no exaggeration to describe Ibn al-Haytham's researches as explicitly and deliberately designed as criticism and projects for reform. It remains to be seen when this project of reform was actually conceived, and what its outcome was. Here our task is complicated by the fact that some treatises are lost, and by the difficulty of dating the extant writings. We know that *The Doubts concerning Ptolemy* was promised at the end of *The Resolution of Doubts concerning the Winding Motion*. We also know that *The Resolution of Doubts concerning the Almagest* was completed after August 1028, the date when Ibn al-Haytham finished *The Halo and the Rainbow*, which he cites.[11] Lastly, we know that these four books must have been composed at different times. The order of composition therefore is: *The Winding Motion*, *The Resolution of Doubts concerning the Winding Motion* and, finally, *The Doubts concerning Ptolemy*. Like *The Resolution of Doubts concerning the Almagest*, these three treatises were all composed before 1038, as we learn from the list of Ibn al-Haytham's writings up to that date. It would therefore appear that around 1028, and certainly before 1038, Ibn al-Haytham was actively engaged with astronomy.

Although we cannot discuss the content of the lost *Corrections to the Almagest*, the titles of these works make it obvious that Ibn al-Haytham took a critical stance, a characteristic common to all the titles we have mentioned so far. Even in his book *The Motion of the Moon*, also composed before 1038, where he makes a point of explaining the difficulties in Ptolemy as the result of a first reading, Ibn al-Haytham does not refrain completely from making criticisms. In other words, far from being merely incidental, his criticisms express his dissatisfaction with Ptolemy's astronomy. To illustrate the full measure of his radical criticisms of Ptolemy, consider what Ibn al-Haytham replies to an anonymous scholar who had criticized his treatise *The Winding Motion*:

> From the statements made by the noble Shaykh, it is clear that he believes in Ptolemy's words in everything he says, without relying on a demonstration or calling on a proof, but by pure imitation (*taqlīd*); that is how experts in the prophetic tradition have faith in Prophets, may the blessing of God be

[11] In fact, Ibn al-Haytham himself transcribed his book *The Halo and the Rainbow* (*Fī al-Hāla wa-qaws quzaḥ*) in the month of Rajab 419 of the Hegira (August 1028). Ibn al-Haytham refers to this book and to his *Optics* in his *Resolution of Doubts concerning the Almagest* (*Fī ḥall shukūk fī kitāb al-Majisṭī*); see ms. Aligarh, ʿAbd al-Ḥayy no. 21, fol. 12^r and ms. Istanbul, Beyazit, 2304, fol. 8^v.

upon them. But it is not the way that mathematicians have faith in specialists in the demonstrative sciences. And I have noted that it pains him (*i.e.* the noble Shaykh) that I have contradicted Ptolemy, and that he finds it distasteful; his statements suggest that error is alien to Ptolemy. Now there are many errors in Ptolemy, in many passages in his books, among others, what he says in the *Almagest*: if one examines it carefully, one finds many contradictions. He (*i.e.* Ptolemy) has indeed laid down principles for the models he considers, then he proposes models for the motions that are contrary to the principles he has laid down. And this not only in one place but in many passages. If he (*i.e.* noble Shaykh) wishes me to specify them and point them out, I shall do so.

I resolved to write a book to establish the truth in the science of astronomy; in it I show the contradictory passages in the *Almagest*, then the correct passages, and I show how to correct the [faulty] passages. He made many mistakes in the *Book on Optics*, one of which was a mistake in the proof concerning the shape of mirrors, which shows how uncertain his grasp was.

As for his *Book on Hypotheses*, if one examines the notions he propounded in the second chapter and the models he put forward using spheres and parts of spheres, the demonstration [of the models] is immediately seen to be contradicted and discredited. In my reply I have shown his error in regard to the two parts of the sphere, which he postulated for the epicycle, and I have explained it with an irrefutable demonstration; and I have shown that, in whatever cases one postulates for the [two] parts of spheres, one obtains an indefensible impossibility.[12]

This radical critique has led many historians to believe that Ibn al-Haytham's purpose was limited merely to criticism, or 'aporetic', as it is sometimes characterized.[13] This is not so, however. During this same

[12] Ms. St Petersburg, no. B1030/1, fol. 19^v.

[13] Because of this clearly stated intention to criticize, some historians have followed S. Pines in believing that Ibn al-Haytham belongs to an ancient aporetic tradition. Thus we find our mathematician placed in the same category as the eminent physician al-Rāzī, the author of the famous *Doubts concerning Galen*. This taxonomy overlooks an important but unnoticed difference that specifically separates Ibn al-Haytham, al-Rāzī and many others in a very wide range of disciplines, from this so-called aporetic tradition. Indeed, it is one thing to raise difficulties and criticize solutions, quite another to criticize for constructive purposes. In every kind of innovative research, criticism is an integral part of the heuristic procedure. For instance, Ibn al-Haytham's doubts and criticisms were not put forward as arguments for a principle, but as statements the mathematician strove to prove mathematically and with the help of controlled observations. More importantly still, these doubts and criticisms cannot be understood except in the light of what, in a sense, is Ibn al-Haytham's final work: *The Configuration of the Motions of Each of the Seven Wandering Stars*. It is thanks to his endeavours to provide a firmer footing for Ptolemy's astronomy by ridding

(*Cont. on next page*)

period (before 1038), Ibn al-Haytham had worked on a problem that would later prove fundamental: the altitudes of planets in the course of their motion. Moreover, in all of his other critical writings except the *Doubts concerning Ptolemy*, Ibn al-Haytham tries to solve particular problems encountered in the *Almagest*, notably those that are not yet connected with the work's theoretical structure. In other words, even at this stage, the criticism is also a heuristic strategy. This will become still clearer when we examine the consequences. It is in the course of this research, and after carrying out further work to bring it to maturity, that Ibn al-Haytham conceived the idea of writing his monumental book *The Configuration of the Motions of the Seven Wandering Stars*, in which he sets out the details of his new astronomy. This is to say that this last book – in which he again takes up the problem of altitudes – is the culmination of the critical and inventive research he carried out during at least two decades before 1038, and which was very probably not published until shortly after that date.

Now, by an ironic recent coincidence, our mathematician al-Ḥasan ibn al-Haytham has confidently been credited with a commentary on the *Almagest* written in strictly Ptolemaic terms by a man with almost the same name, a philosopher called Muḥammad ibn al-Haytham[14] who was interested in the sciences but not himself a mathematician. Confusion naturally peaks when one cites this text to introduce a deliberately critical book, such as the *Doubts*. Such confusion inevitably creates an error of perspective that makes it impossible to understand al-Ḥasan ibn al-Haytham's astronomy.

As noted earlier, however, Ibn al-Haytham suffers from another misapprehension on the part of historians of astronomy. For centuries, he

(*Cont.*) it of its internal inconsistencies that Ibn al-Haytham discovers that to prepare the way for this reformulation he needs to separate an account of the motions (*i.e.* celestial kinematics) from cosmology. In short, in Ibn al-Haytham's case, it is not possible to separate doubts and criticisms from the conscious aim of making fundamental reforms. See S. Pines, 'Ibn al-Haytham's Critique of Ptolemy', in *Actes du dixième Congrès international d'histoire des sciences*, 1, no. 10, Paris, 1964, pp. 547–50 and *id.*, 'What was Original in Arabic Science', in A. C. Crombie (ed.), *Scientific Change*, Leiden, 1963, pp. 181–205.

[14] In the introduction to the printed edition of *al-Shukūk* (note 6), A. Sabra believes he can shed light on the critical text of this book by drawing upon the *Commentary on the Almagest* of Muḥammad ibn al-Haytham, a book which follows Ptolemy to the letter. This strange endeavor stems from the longstanding confusion between Muḥammad ibn al-Haytham and al-Ḥasan ibn al-Haytham. In this regard, see R. Rashed, *Les mathématiques infinitésimales*, II, pp. 8–19; III, pp. 937–41 and IV, pp. 957–9.

has been supposed to be the author of the book called *On the Configuration of the Universe* (*Fī hay'at al-ʿālam*). This book, which is cited by early biobibliographers, was translated into Hebrew and into Latin. Y. T. Langermann, who edited and translated the text, says about it: 'Many of the sharp criticisms of Ptolemy which are developed in the *Doubts* can, in fact, be directed equally well at *On the Configuration*, which faithfully mirrors the astronomical theory of the *Almagest*'.[15] I have added some further observations that cast doubt on the attribution of this work to Ibn al-Haytham.[16]

To avoid so flagrant a contradiction, it is tempting to make this an early work. There is, however, no evidence to support such a conjecture – on the contrary. In fact, even in regard to much less significant matters, when Ibn al-Haytham returns to a topic he has treated before, he usually refers back to his first treatment and warns the reader that the present one now supersedes it.[17] One would therefore, *a fortiori*, expect a similar gesture here, particularly since he would be in the process of criticizing the theses defended in the first treatment. But it does not happen.

So our present knowledge of Ibn al-Haytham's astronomical work is: some people see no difficulty in attributing to him a thoroughly traditional commentary on Ptolemy, or a treatise that conforms strictly to Ptolemy, and ignore the contradiction with Ibn al-Haytham's *Doubts* and his criticisms. Others, with good reason, note the contradiction, but stop there. Much earlier, yet others had concentrated on the *Doubts* and expressed regret that Ibn al-Haytham was satisfied merely to criticize Ptolemy, without proposing another 'astronomy'. Thus the astronomer al-ʿUrḍī (d. 1266) writes:

> No one came after him (Ptolemy) to bring that art (astronomy) to completion in a correct manner; no modern scholar has added anything at all to his work or subtracted anything from it; instead, all have followed him. Some among them have raised doubts, but without contributing more than the expression of doubts, such as Ibn al-Haytham and Ibn Aflaḥ of Andalusia.[18]

[15] Y. Tzvi Langermann, *Ibn al-Haytham's On the Configuration of the World*, New York/London, 1990, p. 8.

[16] See Rashed, *Les mathématiques infinitésimales*, V, Appendix I.

[17] See for example Ibn al-Haytham, *Exhaustive Treatise on the Figures of Lunes*, in Rashed, *Les mathématiques infinitésimales*, vol. II, pp. 102–3; also vol. V, p. 267.

[18] *The Astronomical Work of Mu'ayyad al-Dīn al-ʿUrḍī: Kitāb al-Hay'ah*, edition with English and Arabic introductions by G. Saliba, Tārīkh al-ʿulūm ʿind al-ʿArab 2, Beirut, 1990, p. 214.

If we take them simply at face value, al-ʿUrḍī's words are surprising for several reasons. They seem to ignore the achievements of Thābit ibn Qurra (826–901) and all the other contributions in the following three centuries of mathematical astronomy; they also seem to place very little value on the secure observational results that astronomers obtained since the beginning of the 9th century, and they likewise seem to overlook the work on instruments. Moreover, they also seem to reflect a mistaken outlook, one that had become more extreme in our time, according to which there was an independent tradition of mathematical astronomy dedicated to criticizing errors in Ptolemy. Finally, al-ʿUrḍī's words seem to indicate that he knew no other astronomical text by Ibn al-Haytham apart from the *Doubts concerning Ptolemy*. Now, all this is very improbable, coming as it does from an astronomer like al-ʿUrḍī, the more so since his future 'boss' at Marāgha, Naṣīr al-Dīn al-Ṭūsī, knew at least Ibn al-Haytham's book *The Winding Motion*, in which the latter proposes a model of this motion that combines kinematics with some cosmology.[19] Instead, everything points to the explanation being that al-ʿUrḍī wanted to emphasize that Ibn al-Haytham had not proposed a model of the universe based jointly on the two traditions – that of the *Almagest* and that of the *Planetary Hypotheses* – a model in which a celestial kinematics and a cosmology are combined in such a way that the resulting planetary theory is coherent and capable of making predictions that are as accurate as possible; in other words, a configuration/model (*hayʾa*) like the one al-ʿUrḍī believed he had constructed in his own book.[20]

[19] According to what is reported by Naṣīr al-Dīn al-Ṭūsī, on the basis of a text by Ibn al-Haytham that is now lost (see F. J. Ragep, *Naṣīr al-Dīn al-Ṭūsī: Memoir on Astronomy – al-Tadhkira fī ʿilm al-hayʾa*, 2 vols, New York, 1993, vol. 1, pp. 215–17), the matter concerned is the deviation of the apogee and perigee of the epicycle as well as the two points on the epicycle at mean distance. Ibn al-Haytham seems to intend to construct a model using solid orbs as the mechanism for the motion. In this model, Ibn al-Haytham adds three solid orbs for the epicycles of the superior planets and five solid orbs for the inferior planets, so as to take account of the various deviations noticed by observers.

[20] Later, Ibn al-Shāṭir expressed a more qualified opinion than that of al-ʿUrḍī. This can be found in *The New Zīj* (*al-Zīj al-jadīd*, ms. Oxford, Bodleian Library, Arch. Seld. A30, fol. 2^r): 'I have noticed that eminent modern scholars, such as al-Majrīṭī, Abū al-Walīd al-Maghribī [Averroes], Ibn al-Haytham, Naṣīr al-Ṭūṣī, Muʾayyid al-Dīn al-ʿUrḍī, Quṭb al-Shīrāzī and Ibn Shukr al-Maghribī, have expressed doubts about the model of the orbs of the planets, that is, the system of Ptolemy, doubts that contest the geometrical and physical principles [he] established, and they [the scholars] have then proceeded to work to put in place principles adequate to [explaining] the motions in

(*Cont. on next page*)

In fact, al-ʿUrḍī's criticism, which in one sense misses the point, in another sense is justified. Ibn al-Haytham did indeed write an *Astronomy*, which will be discussed below. In this *Astronomy*, Ibn al-Haytham has understood that a genuine reform does not consist of constructing a model in the sense in which this was understood by al-ʿUrḍī, but first in establishing a kinematic system on a solid mathematical basis, before thinking about any kind of dynamics.

1.2. *The Configuration of the Motions of the Seven Wandering Stars*

Ibn al-Haytham's *Configuration of the Motions of Each of the Seven Wandering Stars* is a monumental achievement.[21] It deals with the 'model', or the 'structure' (*hayʾa*), that is to say, with a new astronomy or a new theory of the planets. This innovative and important book, which presents mathematical content at the cutting edge of the science of its day, has reached us in a unique manuscript, which is in a pitiful state: a substantial part of it has been cut away, the leaves are out of order, moisture has made some parts illegible and the handwriting is hard to decipher.[22]

The Configuration of the Motions of Each of the Seven Wandering Stars (henceforth *The Configuration of the Motions*) was originally organized into three books. In the first, on mathematical astronomy, Ibn al-Haytham gives the details of his planetary theory; the second he devoted to astronomical calculation or, as he writes, 'all the operations of calculation'; and the third was concerned with an astronomical instrument that was easy to manipulate and designed for precise calculation of the altitudes of the sun and the planets. Of this complete volume, only the planetary theory has come down to us. The sheer bulk of this first section is a reminder of the work's original size before so much of it was lost, and allows us to gauge the magnitude of the task Ibn al-Haytham undertook. He very probably wanted this book to encompass all parts of astronomy, just as his *Book on Optics* had done for all parts of that subject. But it likewise shows us that, at this time, a book about the model/configuration (*hayʾa*) covered not only one, but several areas of investigation: a planetary theory; a study of the procedures used in the astronomical calculations needed for compiling tables showing the parameters required for calculating positions of planets (the *zījs*); and research on astronomical instruments.

(*Cont.*) longitude and in latitude, among those that do not contest what these principles demand. They have not succeeded in this and have conceded as much in their books'.

[21] Rashed, *Les mathématiques infinitésimales*, V, Part I.

[22] Ms. St. Petersburg, 600 (formerly Kuibychev, V. I. Lenin Library).

The surviving first book is on the theory of planetary motion, which includes a prologue to the work as a whole, in which Ibn al-Haytham explains its organization and the style of his presentation. In this introduction, Ibn al-Haytham states that the style is that of demonstration, and that *The Configuration of the Motions* renders obsolete all his previous works on the same subjects. This introduction is followed by a mathematical study that takes up slightly less than half the section. It deals with fifteen propositions that feature as lemmas in the construction of the planetary theory, to which the last part of the surviving text is devoted. Note that in the first part, Ibn al-Haytham breaks new ground in the mathematics of infinitesimals since he is explicitly concerned with variations – variations of the elements of a figure as a function of other elements; variations of ratios; and variations of trigonometrical relationships. In this new research domain, Ibn al-Haytham uses infinitesimal geometry and compares finite differences. This work on variable quantities, set into motion by the needs of astronomy, made them a part of the geometry of infinitesimals.

Having completed this mathematics, Ibn al-Haytham is now in a position to construct his planetary theory. But the length of the treatment and the deep nature of the mathematics in this part of the work point to one of the motives that drive Ibn al-Haytham's astronomical research: he wants to make planetary theory even more mathematical, and much more systematically so. Here, as in the other disciplines he treats, Ibn al-Haytham takes the path opened by his predecessors from Thābit ibn Qurra on, but he advances much farther and more deeply in order to take it as far as possible. If we forget this purpose, we shall fail to understand *The Configuration of the Motions*.

But what is required for this additional mathematization to be possible within a framework that remains geometrical? For the latter to occur without running into the Ptolemaic inconsistencies that he has already censured in the *Doubts*, Ibn al-Haytham is compelled to rethink the fundamental tenets of Ptolemaic astronomy. In his eyes, then, far from being merely an instrumental or a linguistic task, this systematic mathematization was an undertaking that truly engaged with theory. That is how Ibn al-Haytham came to devise a new planetary theory that no longer concentrates on anomalies but starts by deliberately separating kinematics from cosmology.

In the *Doubts*, Ibn al-Haytham comes to the conclusion that 'the configuration (*hay'a*) Ptolemy assumes for the motions of the five planets is a false one'.[23] A few lines later, he reinforces the point: 'The order

[23] *Al-Shukūk ʿalā Baṭlamiyūs*, ed. Sabra and Shehaby, p. 34.

according to which Ptolemy organized the motions of the five planets strays from the theory'.[24] A little further on, he states:

> The configurations that Ptolemy assumed for the <motions of> the five planets are false ones; he chose them knowing that they were false, because he was unable <to propose> other ones. For the motions of the planets, there is to be found in actual bodies a true configuration that Ptolemy neither obtained nor reached.[25]

Many other similar remarks appear in various places in his writings. Under the circumstances, a mathematician of Ibn al-Haytham's stature who felt enormous respect for Ptolemy (as other comments attest) had no choice but to construct a planetary theory of his own that would rest on a solid mathematical basis and be free of the internal contradictions endemic to his predecessor. The point of Ibn al-Haytham's treatise on *The Configuration of the Motions* was precisely to bring this programme into being.

Most of the serious contradictions that Ibn al-Haytham censures set the *Almagest* against the *Planetary Hypotheses*. To characterize the irreducible inconsistencies that, in his view, vitiate Ptolemy's astronomy, one might say that they arise from the poor fit between a mathematical theory of the planets and a cosmology. Ibn al-Haytham was familiar with similar, though of course not identical, situations when, in optics, he encountered the inconsistency between geometrical optics and physical optics as understood by the philosophers. To reform optics he adopted, as it were, a kind of 'positivism' *avant la lettre*: one does not go beyond experience, and one cannot be satisfied with concepts alone when investigating natural phenomena, for one cannot acquire an understanding of the latter without mathematics. Thus, once he has assumed light is material, Ibn al-Haytham does not discuss its nature any further; rather, he restricts himself to considering its propagation and diffusion. In his optics, 'the smallest parts of light', as he calls them, retain only properties that can be treated by geometry and verified by experiment; they lack all sensible qualities except energy. In other words, he begins by insisting on making optics geometrical, or on reforming geometrical optics by leaving aside the 'why' questions that pertain to teleological physics, while remaining ready to introduce them later when he comes back to physical optics. As can be readily ascertained, this imposition of geometry led Ibn al-Haytham to study the

[24] *Ibid.*, pp. 33–4.
[25] *Ibid.*, p. 42.

propagation of light in kinematic – mechanical – terms.[26] Ibn al-Haytham adopts a similar approach in astronomy: in *The Configuration of the Motions* he deals with the apparent motions of the planets, without ever raising the question of the physical explanation of these motions in terms of dynamics. It is not the causes of celestial motions that interest Ibn al-Haytham, but only the motions themselves observed in space and time. Thus, to proceed with the systematic mathematical treatment, and to avoid the shoals that Ptolemy had encountered, he first needed to break away from any kind of cosmology. And, in fact, in this treatise Ibn al-Haytham no longer draws upon the theory of material spheres, which had appeared in his *Resolution of Doubts concerning the Winding Motion* and in the *Doubts concerning Ptolemy*. Thus the purpose of Ibn al-Haytham's *Configuration of the Motions* is clear: to construct a geometrical kinematics.

Ibn al-Haytham's second intention is implied by the first one: to avoid the difficulties found in Ptolemy's astronomy. In the *Resolution of Doubts concerning the Almagest*, he states that 'in the *Almagest* as a whole there are doubts (*aporias*) too numerous for one to list them all'.[27] All the same, in the *Doubts concerning Ptolemy* he distinguishes between doubts that can be resolved without modifying the structure of the theory and those whose elimination requires that the theory undergo radical reform.[28] One of the best examples of the latter type is the concept of the equant, exposed as an error in the *Doubts* and banished from *The Configuration of the Motions*. Ibn al-Haytham rejects the idea because one cannot, at the same time, suppose that a sphere rotates uniformly on its axis and suppose that this same rotation takes place about a line that is not a diameter of the sphere. In rejecting the equant, Ibn al-Haytham is already distancing himself very considerably from Ptolemy.

[26] R. Rashed, 'Optique géométrique et doctrine optique chez Ibn al-Haytham', *Archive for History of Exact Sciences*, 6.4, 1970, pp. 271–98; repr. *Optique et Mathématiques: Recherches sur l'histoire de la pensée scientifique en arabe*, Variorum reprints, Aldershot, 1992, II.

[27] *Fī ḥall shukūk al-Majisṭī*, ms. Istanbul, Beyazit 2304, fol. 195^{r}.

[28] *Al-Shukūk ʿalā Baṭlamiyūs*, ed. Sabra and Shehaby, p. 5: 'We shall not mention in this book all the doubts contained in his works, but shall only mention the passages that contradict one another and the mistakes that cannot be rectified; the ideas he has put in place, and the motions of the planets he has arrived at, collapse if we cannot obtain true methods or uniform models or <correcting> these passages and these errors. As for the remaining doubts, they do not impute error to the established principles and they can be resolved without any of these principles being overturned or altered'.

Not least, Ibn al-Haytham had written two books on astronomical observation and the errors to which it is subject. He was moreover well informed about the wealth of observations built up over two centuries. Accordingly, Ibn al-Haytham's third intention in writing *The Configuration of the Motions* was to construct a planetary theory that explained these observations.

These three intentions – mathematization, avoiding Ptolemy's contradictions and accounting for the observations – work together to fulfil Ibn al-Haytham's overall purpose for *The Configuration of the Motions*, that is, to set up a completely geometrical celestial kinematics. But in order to achieve this goal, he needed to find a way of measuring time. To this end, he introduced a new concept, that of 'required time', that is, a period of time measured by an arc.

A close examination of the way he organizes his exposition of planetary theory shows that Ibn al-Haytham begins by proposing simple – in effect, descriptive – models of the motions of each of the seven planets. As the exposition progresses, he makes the models more complicated and increasingly subordinates them to the discipline of mathematics. This growing mathematization leads him to regroup the motions of several planets under a single model. And it is precisely the mathematical nature of the model which makes this regrouping possible, specifically starting from Proposition 24. This step obviously has the effect of privileging a property that is common to several motions. In this way Ibn al-Haytham opens up the way to achieving his principal objective: to establish a system of celestial kinematics. He does so without as yet formulating the concept of instantaneous speed, but by using the concept of mean speed, represented by a ratio of arcs.

Here we shall explain the principal results Ibn al-Haytham obtained. A detailed commentary together with an edition of the text and a French translation of it is published elsewhere.[29]

2. THE STRUCTURE OF *THE CONFIGURATION OF THE MOTIONS*

The first extant section of *The Configuration of the Motions* divides into two parts. The first, which is mathematical and chiefly devoted to the study of variable quantities, comprises 15 propositions. The second part deals with planetary theory.

[29] Rashed, *Les mathématiques infinitésimales*, vol. V, Part I.

2.1. *Research on the variations*

The fifteen propositions with which the section begins may be separated into several groups. The first consists of the first four propositions, which deal with the variation of trigonometrical functions such as $\frac{\sin x}{x}$. Ibn al-Haytham gives rigorous proofs of the following propositions:

1. If the measures in radians of the arcs α and α_1 of a circle are such that $\alpha + \alpha_1 \leq \frac{\pi}{2}$ and $\alpha > \alpha_1$, then

$$\frac{\alpha}{\alpha_1} > \frac{\sin \alpha}{\sin \alpha_1} \text{ and } \frac{\alpha + \alpha_1}{\alpha_1} > \frac{\sin(\alpha + \alpha_1)}{\sin \alpha_1}.$$

2. If the measures in radians of the arcs α and α_1 of a circle and of the arcs β and β_1 of a different circle are such that

$$\beta + \beta_1 < \alpha + a_1 < \frac{\pi}{2} \text{ and } \frac{\alpha}{\alpha_1} = \frac{\beta}{\beta_1} = \frac{1}{k} \text{ (where } k < 1\text{)},$$

then

$$\frac{\sin \alpha}{\sin \alpha_1} < \frac{\sin \beta}{\sin \beta_1} \text{ or } \frac{\sin \alpha}{\sin . \alpha} < \frac{\sin \beta}{\sin . \beta}.$$

As a corollary to this proposition, Ibn al-Haytham proves that

$$\frac{\sin(\alpha + \alpha_1)}{\sin \alpha_1} < \frac{\sin(\beta + \beta_1)}{\sin \beta_1} \text{ or } \frac{\sin(1+k)\alpha}{\sin k\alpha} < \frac{\sin(1+k)\beta}{\sin k\beta}.$$

Ibn al-Haytham had proved this proposition in his treatise *On the Hour Lines*.[30]

3. If the measures in radians of the arcs α and α_1 of a circle and of the arcs β and β_1 of a different circle are such that

$$\beta + \beta_1 < \alpha + \alpha_1 \leq \frac{\pi}{2} \text{ and } \frac{\sin(\beta + \beta_1)}{\sin \beta_1} = \frac{\sin(\alpha + \alpha_1)}{\sin \alpha_1},$$

[30] R. Rashed, *Les mathématiques infinitésimales*, vol. V, Part II.

then

$$\beta < \alpha \text{ and } \frac{\beta}{\beta_1} < \frac{\alpha}{\alpha_1}.$$

4. If the measures in radians of the arcs α and α_1 of a circle and of the arcs β and β_1 of a different circle are such that

$$\beta + \beta_1 < \alpha + \alpha_1 \leq \frac{\pi}{2},\ \alpha_1 < \alpha,\ \beta_1 < \beta \text{ and } \frac{\sin\beta}{\sin\beta_1} \leq \frac{\sin\alpha}{\sin\alpha_1},$$

then

$$\frac{\alpha}{\alpha_1} > \frac{\beta}{\beta_1}.$$

And if

$$\frac{\sin(\beta + \beta_1)}{\sin\beta_1} \leq \frac{\sin(\alpha + \alpha_1)}{\sin\alpha_1},$$

then

$$\frac{\beta + \beta_1}{\beta_1} < \frac{\alpha + \alpha_1}{\alpha_1} \text{ and } \frac{\beta}{\beta_1} < \frac{\alpha}{\alpha_1}.$$

The second group is made up of the next three propositions (5, 6 and 7), which also deal with variable quantities and variable ratios. In the first two (5 and 6) Ibn al-Haytham considers changes in the angular position of a point on a quadrant of a circle. In Proposition 7, he examines changes in right ascension. In the course of these propositions, he compares finite differences, calls upon ideas about the geometry of infinitesimals and makes use of the sine rule (which was known to such contemporary mathematicians as Abū al-Wafā' al-Būzjānī and Ibn 'Irāq).[31]

In Propositions 5 and 6, Ibn al-Haytham considers a sphere with center ω on which positions are described with respect to a great circle ABC of diameter AC, its pole K and the great circle KC orthogonal to ABC (Fig. 120). A great circle of diameter AC cuts the arc KB in the point D. With any point, such as H, on the arc CD there is associated a great circle KH that cuts the arc CB at point P, and a circle through H parallel to (ABC)

[31] M.-Th. Debarnot, *Al-Bīrūnī: Kitāb maqālīd 'ilm al-hay'a. La Trigonométrie sphérique chez les Arabes de l'Est à la fin du Xe siècle*, Institut Français de Damas, Damascus, 1985.

which cuts the arc KC at point U; we have $\widehat{PH} = \widehat{CU}$. The arcs PH and CP are, respectively, the *inclination* (the declination if the reference circle is the equator) and the *right ascension* of the point H with respect to the circle ABC.

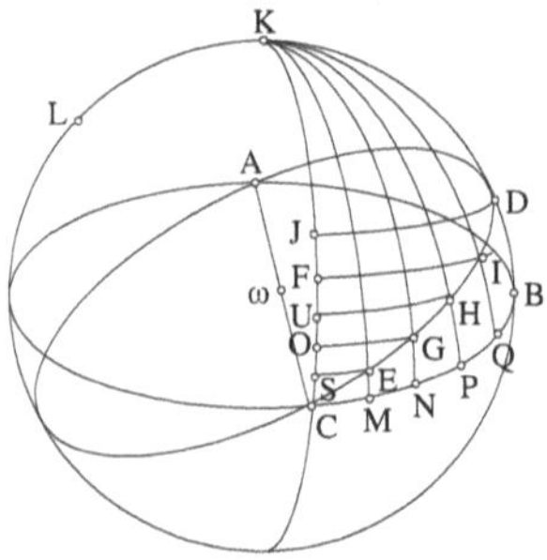

Fig. 120

First of all, Ibn al-Haytham considers how the *inclination* of arc PH varies when the point H describes the arc CD.

Let the (rectilinear) dihedral angle between the planes ABC and ADC be α, we have $B\hat{\omega}D = \alpha$, so $\widehat{BD} = \alpha$. Let us put $\widehat{CH} = x$ and $\widehat{PH} = \widehat{CU} = y$, we have $0 \leq x \leq \frac{\pi}{2}, 0 \leq y \leq \alpha$.

The proposition has two parts that can be summarized as follows (Fig. 121):

a) The arc CD is divided into n equal parts at the points with spherical abscissae x_i, $0 \leq i \leq n$, $x_0 = 0$ and $x_n = \frac{\pi}{2}$.

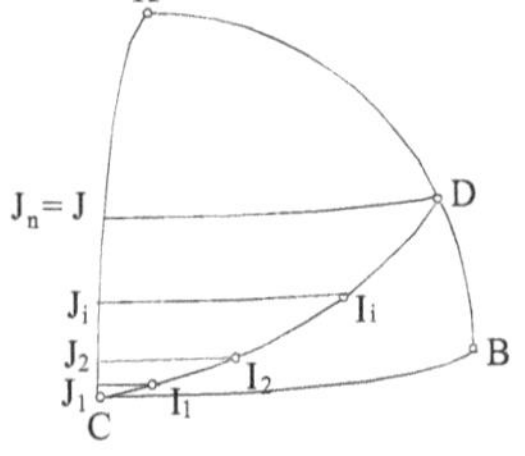

Fig. 121

For $\Delta x = x_i - x_{i-1} = \frac{\pi}{2n}$ we have $\Delta y = y_i - y_{i-1}$. We show that Δy decreases when i increases from 1 to n. In other words, y is a concave function of x.

b) We consider two equal arcs with a common endpoint, with $x_i < x_j < x_k$ and $x_j - x_i = x_k - x_j$.

We show that from (a), we have $y_j - y_i > y_k - y_j$. This result may be expressed in the form

$$\frac{x_k - x_j}{x_j - x_i} > \frac{y_k - y_j}{y_j - y_i},$$

or as

$$\frac{y_k - y_j}{x_k - x_j} < \frac{y_j - y_i}{x_j - x_i},$$

which is to say that the gradient of the graph of y as a function of x decreases as x increases.

Proposition 6 extends this result to unequal arcs, such as arcs IJ and JK, where $x_i < x_j < x_k$ and $x_j - x_i \neq x_k - x_i$.

• If the two arcs in question that have an endpoint in common are commensurable, the result follows from a) and b).

• For the case in which the same two arcs are incommensurable, Ibn al-Haytham gives a *reductio ad absurdum* argument to show that it is impossible to have

$$\frac{x_k - x_j}{x_j - x_i} \leq \frac{y_k - y_j}{y_j - y_i}.$$

We note that, after proving the required inequality holds when the magnitudes are commensurable, Ibn al-Haytham proves the general case by 'extension by continuity', giving a rigorous abductive (apagogic) proof, and by applying his extension of Lemma 1 of *Elements* X.

So we have an argument based on infinitesimals for extending by continuity an inequality for which we have, as yet, no earlier example. We also note that Ibn al-Haytham is treating arcs and angles as magnitudes to which one can apply proportion theory.

Let us now return to his discussion of the variation of the inclination and show that his results are correct:

Let us put $y = \widehat{PH}$ as a function of $\widehat{CP} = x$. We have $y = f(x)$.

In the spherical triangle CHP, the arcs PH and PC are orthogonal, so $\hat{P} = \frac{\pi}{2}$, and the angle between arcs CP and CH is the angle between their tangents, equal to $B\hat{\omega}D$, so we have $\hat{C} = \alpha$.

The relation

$$\frac{\sin \widehat{CH}}{\sin \hat{P}} = \frac{\sin \widehat{PH}}{\sin \hat{C}}$$

therefore yields

$$\sin x = \frac{\sin y}{\sin \alpha};$$

therefore y as a function of x is given by

$$\sin y = \sin \alpha \cdot \sin x, \qquad y = \text{Arc}\sin \cdot (\sin \alpha \cdot \sin x),$$

we have

$$\cos y\, dy = \sin \alpha \cdot \cos x\, dx,$$

that is

$$y'_x(x) = \frac{dy}{dx} = \frac{\sin \alpha \cdot \cos x}{\sqrt{1 - \sin^2 \alpha \cdot \sin^2 x}};$$

from which it follows that

$$y'' = -\frac{\sin \alpha \cdot \cos^2 \alpha \cdot \sin x}{\left(1 - \sin^2 \alpha \cdot \sin^2 x\right)^{\frac{3}{2}}}.$$

So for $0 < x < \frac{\pi}{2}$, we have $y' > 0$ and $y'' < 0$, $y = \widehat{PH} = f(x)$ increases from 0 to α.

But $f'(x)$ decreases over the interval $\left[0, \frac{\pi}{2}\right]$ and the function f is thus concave; therefore

$$\frac{x_m - x_k}{x_j - x_i} > \frac{y_m - y_k}{y_j - y_i}.$$

If in this expression we take:

- $x_m - x_k = x_j - x_i = \frac{\pi}{2n}$, we recover result (a).
- $x_m - x_k = x_j - x_i$, we recover the result for case (b) for equal arcs.
- $x_m - x_k \neq x_j - x_i$, we recover the result for case (c) for unequal arcs.

  If $x_j = x_k$, the arcs concerned are contiguous.

  If $x_j < x_k$, the arcs concerned are disjoint.

In the seventh proposition (Fig. 122), Ibn al-Haytham considers the right ascension $\widehat{CP}$ when the point H describes the arc CD. We put $\widehat{CH} = x$ and $\widehat{CP} = z$ for $0 \leq x \leq \frac{\pi}{2}, 0 \leq z \leq \frac{\pi}{2}$, we have:

a) as in considering the inclination, we divide the arc CD into n equal parts at points with spherical abscissa x_i. For the increment $\Delta x = x_i - x_{i-1}$ the corresponding increment in the right ascension, $\Delta z = z_i - z_{i-1}$, and using Menelaus's theorem for the arcs of great circles, we show that Δz increases when i increases from 1 to n.

b) Ibn al-Haytham next says that, as in the treatment of the inclination, one can generalize this result by considering two arcs lying on the arc CD, be they equal to one another or unequal, contiguous or disjoint, commensurable or incommensurable. Thus, for arcs I_iI_j and I_kI_m with $x_i < x_j \leq x_k < x_m$, one will have

$$\frac{x_m - x_k}{x_j - x_i} < \frac{z_m - z_k}{z_j - z_i}.$$

In other words, z is a convex function of x.

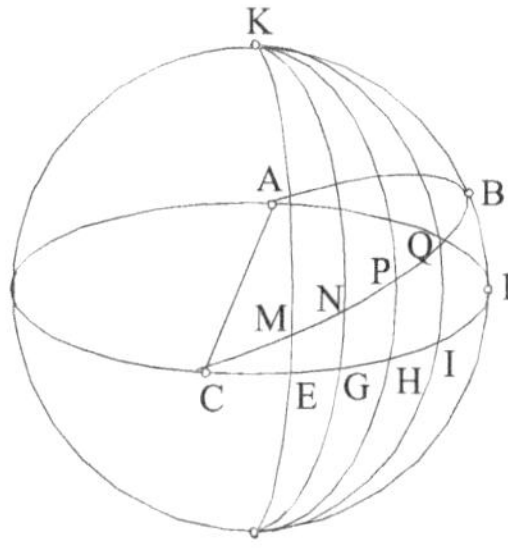

Fig. 122

Let us return to his discussion of the right ascension.

Considering $z = \widehat{CP}$ as a function of $x = \widehat{CH}$, when H describes the arc CD, $z = g(x)$.

The four circles involved are all great circles, and Menelaus's theorem yields

$$\frac{\sin \widehat{CH}}{\sin \widehat{HD}} = \frac{\sin \widehat{CP}}{\sin \widehat{PB}} \cdot \frac{\sin \widehat{KB}}{\sin \widehat{KD}}$$

$$\widehat{CH} = x,\ \widehat{HD} = \frac{\pi}{2} - x,\ \widehat{CP} = z,\ \widehat{PB} = \frac{\pi}{2} - z,\ \widehat{DB} = \alpha,\ \widehat{KD} = \frac{\pi}{2} - \alpha.$$

We therefore have

$$\frac{\sin x}{\cos x} = \frac{\sin z}{\cos z} \cdot \frac{1}{\cos \alpha},$$

which gives

$$\tan z = \cos \alpha \cdot \tan x.$$

$$z = \text{Arc} \tan (\cos \alpha \cdot \tan x) = g(x).$$

So we have

$$(1 + \tan^2 z)\, dz = \cos \alpha \cdot (1 + \tan^2 x)\, dx,$$

$$z' = g'(x) = \frac{\cos\alpha\left(1+\tan^2 x\right)}{1+\cos^2\alpha\cdot\tan^2 x} = \frac{\cos\alpha}{\cos^2 x+\cos^2\alpha\sin^2 x};$$

from which it follows that

$$z'' = \frac{\sin 2x \cos\alpha \ \sin^2\alpha}{\left(\cos^2 x+\cos^2\alpha\sin^2 x\right)^2}.$$

So for $0 < x < \frac{\pi}{2}$, we have $z' > 0$, z increases from 0 to $\frac{\pi}{2}$. We also have $z'' > 0$, $z' = g'(x)$ increases from 0 to $\frac{1}{\cos\alpha}$, hence the result Ibn al-Haytham obtained for the increment Δz.

As in the discussion of the inclination, Ibn al-Haytham indicates that his result can be extended to give an inequality involving differences of the right ascensions for unequal arcs, first in the case where these arcs are commensurable, then in the general case by using an argument of extension by continuity.

The third group is made up of Propositions 8 and 9. Ibn al-Haytham considers a circle (D, DC), that is, with center D and radius DC, a point E on DC, as well as the equal arcs AB, BH, HI such that chord $AB < EC$, and he shows that $A\hat{E}B < B\hat{E}H < H\hat{E}I$ (Fig. 123).

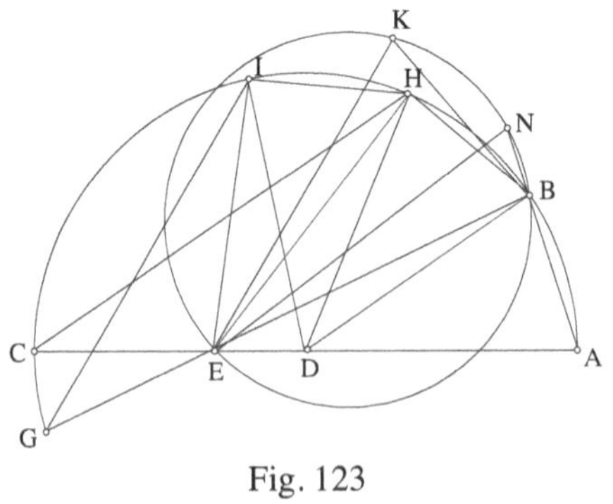

Fig. 123

If we posit $A\hat{D}B = \theta$, where $\theta \in [0, \pi]$ and $A\hat{E}B = \varphi$, we see that Ibn al-Haytham is considering how φ varies as a function of θ.

In Proposition 9 he considers the direction of its variation.

The fourth group is concerned with the variation of ratios in ever more complicated cases. This work is done in Propositions 10, 11, 12, 14 and 15. Proposition 13 is a lemma to do with technique. In this group, although Proposition 10 does not raise the complicated question of the range of the variables, Propositions 11 and 12, on the one hand, and Propositions 14 and 15, on the other, all require a long discussion, which is given in our commentary.[32]

In Proposition 10, Ibn al-Haytham considers two perpendicular planes $\mathscr{P}$ and $\mathscr{Q}$, two points A and C on their line of intersection, a semicircle of diameter AC lying in the plane $\mathscr{P}$, and a circular arc whose chord is AC, an arc smaller than a semicircle in the plane $\mathscr{Q}$ (Fig. 124).

One tries to prove that there exists a point D such that $DE \perp AC$ and $EB \perp AC$ (where B lies on the semicircle) and such that one has $\frac{DB}{DC} > k > 1$, which is the given ratio. One shows first that there exists a unique point K on AC such that $\frac{KA}{CK} = k^2$.

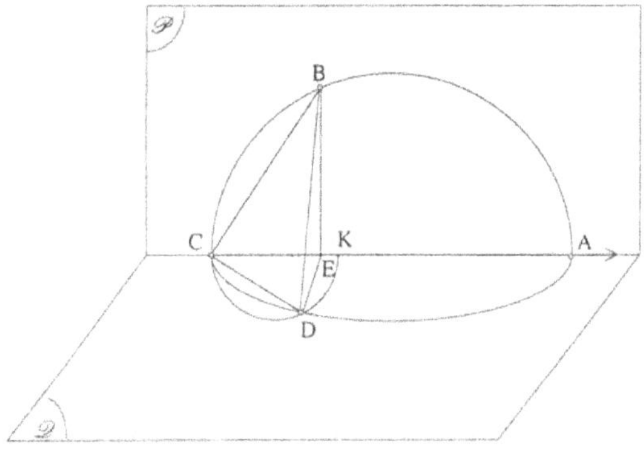

Fig. 124

One then draws a circle of diameter CK in the plane $\mathscr{Q}$ and shows that any point D on the circle yields the ratio.

[32] R. Rashed, *Les mathématiques infinitésimales*, vol. V, Part I.

In Propositions 11 and 12, we consider the meridian circle *ABC* for a given place *G*, the celestial poles *A* and *C*, a circle with center *O* that is parallel to the horizon for *G* and cuts the meridian circle in *D* and *E*, a circle of center *Q* that is parallel to the equator and cuts the meridian circle in *H*, the horizontal circle in *L* and the plane of the circle with center *Q* cuts *DE* in *X* (Fig. 125).

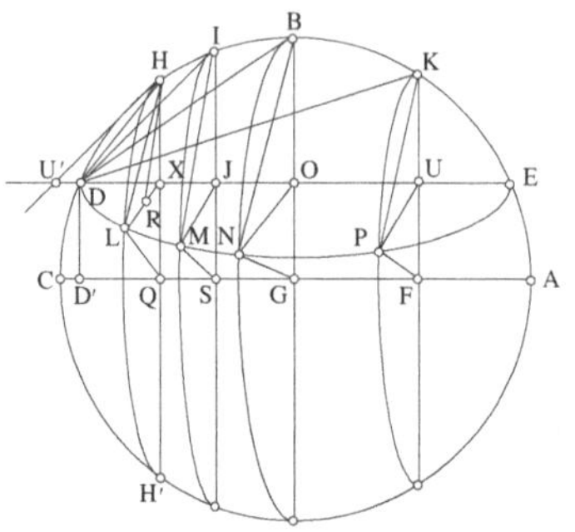

Fig. 125

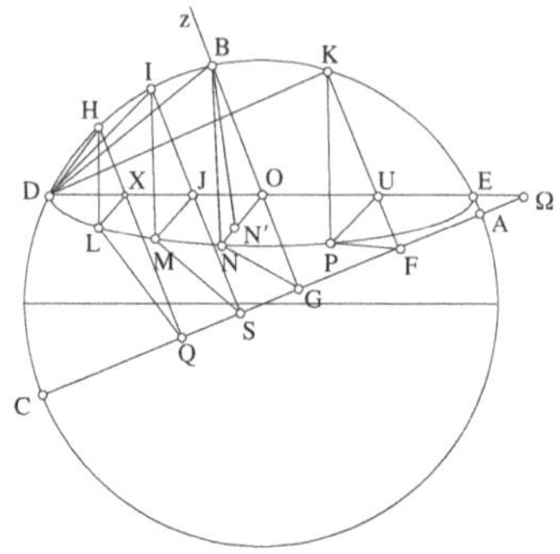

Fig. 126

Ibn al-Haytham shows that when point *X* moves along *DE* from *D* towards *E*, point *L* describes the parallel circle with center *O* and the ratio $\frac{HL}{HD}$ decreases and tends to 0.

In Proposition 12, one assumes that pole *A* is above the horizon, and that *GOz* is the vertical at point *G*; we have $D\hat{X}H = D\hat{O}z$, an angle independent of the position of *X* (Fig. 126). Ibn al-Haytham shows that when *X* moves along *DE*, the arc *HE* decreases, $\sin H\hat{D}X$ decreases and therefore $\frac{HX}{DH} = \frac{\sin H\hat{D}X}{\sin D\hat{X}H}$ also decreases from *D* to *E*.

Finally, Propositions 14 and 15 bring into play the celestial sphere for a given place, its axis, the two poles *Π* and *Π'*, the meridian and horizontal planes for that place (pole *Π* is assumed to lie on or above the horizon).

In Proposition 14, Ibn al-Haytham considers *ADB*, the meridian for an arbitrary place, and *ABC*, a horizontal circle; two circles parallel to the equator cut the meridian in *E* and *D*, the circle *ABC* in *I* and *C* and a great circle with diameter *ΠΠ'* in *I* and *K* (Fig. 127). Ibn al-Haytham proves that:

$$\text{if } \widehat{BE} < \widehat{BD} \le \frac{1}{2}\widehat{ADB}, \text{ then } \frac{\widehat{IE}}{\widehat{EB}} > \frac{\widehat{CD}}{\widehat{DB}} > \frac{\widehat{CK}}{\widehat{KI}}.$$

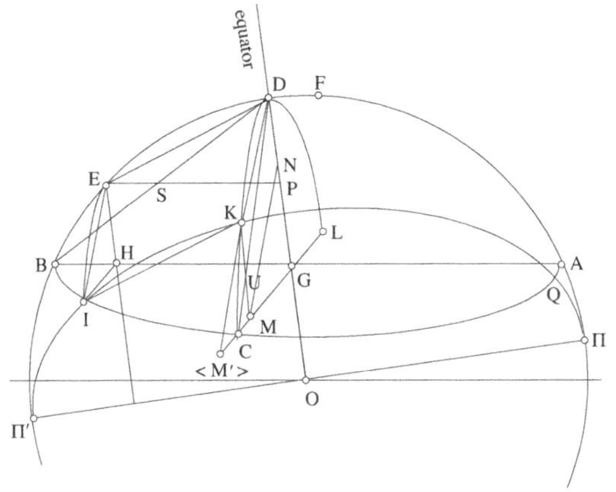

Fig. 127

We are in fact concerned with how $\frac{\widehat{IE}}{\widehat{EB}}$ varies as a function of arc *BE*; that is to say, we want to show that this ratio decreases when *E* moves from *B* towards *F* along the chord of the meridian (where *F* is the midpoint of the arc *AB*).

Proposition 15 generalizes the preceding one. These two propositions show that, using the geometrical means at his disposal, Ibn al-Haytham investigated the variation of certain trigonometrical ratios. This was a line of inquiry that he could not complete but that set in motion some new mathematical research, as will become clear from our commentary on the translation.

2.2. *The planetary theory*

Once he has proven these fifteen mathematical propositions, Ibn al-Haytham immediately goes on to consider the apparent motions of the seven planets. He deals with the apparent motion on the celestial sphere, as seen from a given place, of a planet that is carried around by the universe's diurnal rotation about its axis, in the case where the planet in question has rising and setting points on the horizon of the given place of observation (always in the northern hemisphere). From the very first propositions, Ibn al-Haytham shows, using results Ptolemy obtained for the orbs of the planets and for the different motions of the planets, that the observed trajectory of each planet's apparent motion, as seen on the celestial sphere, differs from the horary circle passing through a point of this trajectory. In other words, it is different from the circle parallel to the equator swept out by a star whose position coincides, at a given moment, with that of the planet.[33] He deals in turn with the moon, the sun, and the five planets. For the motion of the latter along the celestial orb,[34] he distinguishes direct motion, retrograde motion and the planet's stations.

From this investigation, Ibn al-Haytham draws out and defines two new concepts: 'the required time' (*al-zamān al-muḥaṣṣal*), and 'the inclination proper to the required time' (*al-mayl alladhī yakhuṣṣu al-zamān al-muḥaṣṣal*). The 'required time' corresponds to two known positions of the planet during a motion of known duration. It is measured by an arc of the horary circle, and is equal to the difference between the right ascensions of the two observed positions. The inclination proper to the 'required time' is equal to the difference of their inclinations. Note that, since the celestial sphere rotates uniformly, that physical time can be represented by an arc of the horary circle, this concept of 'required time' is essentially a geometrical one. This is precisely how Ibn al-Haytham represents physical time, which has the added advantage of allowing him to draw upon proportion theory when time is involved.

Ibn al-Haytham then shows that, in all possible cases, there exists a ratio greater than the ratio of the required time to the inclination for that time. Thanks to this property, he proves that, for each of the planets observed from a given place, the planetary position whose altitude above the local horizon is a maximum does not correspond to the point of the planet's meridian transit, unlike the situation for a star. For a planet, the

[33] In his treatise on *The Different Altitudes of the Wandering Stars*, composed before this one, Ibn al-Haytham writes as if the trajectory of this apparent motion can be identified with a horary circle (see Rashed, *Les mathématiques infinitésimales*, vol. V, Part I).

[34] In Arabic astronomy, the word *falak* designates the orb as defined in the *Almagest*, *i.e.* the spherical shell within which the planet moves. Every planet has its own orb. For example, Thābit ibn Qurra (d. 901) wrote in his *Almagest Simplified*: 'The orb in which the moon moves is the nearest orb to the earth and it is thick (*lahu sumk*). The moon moves sometimes in its upper part, sometimes in its lower part, and sometimes between them. The same happens for all the other planets' (Thābit ibn Qurra, *Œuvres d'astronomie*, ed. R. Morelon, Paris, Les Belles Lettres, 1987, p. 5, for Arabic text with French translation). This was the conventional meaning of the word in the Arabic tradition of Ptolemy, and it is also the sense in which Ibn al-Haytham employed the word in the works he wrote before *The Configurations of the Motions*. In this last book, Ibn al-Haytham used the word *falak* in a new – and unconventional – meaning, indeed so unconventional that the word 'orb' seems in places inappropriate. The right translation, as we shall see later, would be 'trajectory', 'path', or even simply 'orbit'. But, as Ibn al-Haytham himself continued to use the same word, though with a new meaning, we have no choice but to follow his example.

maximum altitude is greater than that of its meridian transit and, depending on the position of the planet in its trajectory, it will reach maximum altitude either before meridian transit (hence to the east of the meridian) or after meridian transit (to the west of the meridian).

The inquiry into the apparent motion of a planet, when it is above the horizon, ends with a discussion of the case in which the geographical latitude of the place of observation is equal (or very close) to the complement of the maximum declination of the observed axis. Ibn al-Haytham shows that, for places such as these, the planet may set in the east and then rise in the east, or rise in the west and then set in the west.

The work whose outline is sketched above presents a concept of astronomy that is new in several respects. Ibn al-Haytham sets himself the task of describing the motions of the planets exactly in accordance with the paths they draw on the celestial sphere. He is neither trying 'to save the phenomena', that is, to explain the irregularities in the assumed motion by means of artifices such as the equant (which he criticizes in his *Doubts concerning Ptolemy*), nor trying to account for the observed motions by appealing to underlying mechanisms or hidden natures. He wants to give a rigorously exact description of the observed motions in terms of geometry. The only mechanical device involved in describing the motions of the five planets is an epicycle, which is employed to account for their retrograde motions and variable speeds near apogee and perigee. Ibn al-Haytham no doubt knew that using an epicycle and a deferent was equivalent to using an eccentric, and also knew the precise conditions for this equivalence.

What Ibn al-Haytham therefore proposes – thereby driving him farther from the Ptolemaic tradition – to describe the motion in two dimensions on the celestial sphere. In his view, the motion appears to be composed of two elementary motions along great circles of the celestial sphere. The free parameters are the speeds of the elementary motions, treated as independent of one another. But for planets whose trajectory has a variable inclination to the ecliptic, Ibn al-Haytham nevertheless draws upon an epicycle to account for the variation in the inclination, thus temporarily returning to a three-dimensional model. That puts him squarely in the Ptolemaic tradition, but without appealing to the equant.

The guiding principle of Ibn al-Haytham's description is thus clear: to use Ptolemy's mechanisms as sparingly as possible. In considering the apparent motions of the planets on the celestial sphere, always with respect to the horizon, Ibn al-Haytham picks out four reference points: the rising, the meridian transit, the setting, and the maximum altitude. He shows that this last point is unique and may lie to the east or to the west of the point of meridian transit.

The new astronomy no longer aims at constructing a model of the universe, but only at describing the apparent motion of each planet, a motion composed of elementary motions, and, for the inferior planets, also of an epicycle. Ibn al-Haytham considers various properties of this apparent motion: localisation and the kinematic properties of the variations in speed. In the last part of *The Configuration of the Motions*, he considers the apparent motion of the planet on the celestial sphere during the course of a day and proves two conclusions: that the planet reaches its maximum altitude once and only once, and that it reaches any altitude less than the maximum twice, once on each side of the maximum altitude. For altitudes greater than that of meridian transit, the two points at which the planet reaches such an altitude are on the same side of the meridian. Together, these discussions add up to twenty-one propositions.

In this new astronomy, as in the old one, every observed motion is circular and uniform, or composed of circular and uniform motions. Ibn al-Haytham considers three basic motions: the diurnal motion parallel to the equator; the motion of the oblique orb relative to the axis (the line joining the two poles of the ecliptic); and the motion of the nodes of the proper orb. The observed motion of a planet is composed of these three motions plus, for the five planets only, a motion on an epicycle. For the sun, only the first two basic motions are involved. To find these, Ibn al-Haytham uses various systems of spherical coordinates: equatorial coordinates (the required time and its proper inclination), which are primary; horizon coordinates (altitude and azimuth); and ecliptic coordinates.

The use of equatorial coordinates marks a break with Hellenistic astronomy. In the latter, the motion of the orbs was measured against the ecliptic, and all coordinates were ecliptic ones (latitude and longitude). Thus, basing the analysis of the planets' motion on their apparent motions drives a change in the reference system for the data; we are now dealing with right ascension and declination. Ibn al-Haytham's book thus transports us into a different system of analysis.

Ibn al-Haytham then considers, for any given planet, the variation in the speed of the inclination, measuring it by the mean speed over an interval that is itself variable. He looks at the change in altitude of the planets between their rising and setting. Ibn al-Haytham carries out all these investigations rigorously, using the mathematical propositions he proved in the first part and relying constantly upon considerations involving infinitesimals. The geometrical proofs he brings to bear assume only that the motion of the planet is from east to west, and that it is monotonic along the north-south axis.

When the geometry is conceptualized in this way, the question of a possible motion of the earth does not arise, because we are concerned only with the motion of the planet on the celestial sphere as it appears to a terrestrial observer. In other words, we have a kind of phenomenological description of the motions of the planets, which can, however, be given only in terms of spherical geometry, infinitesimal geometry, and trigonometry. There is nothing surprising about this since Ibn al-Haytham is concerned to insure that his description employs only minimal hypotheses about the properties that characterize the motions: variation of speed and day-by-day variation of altitude.

Let us briefly summarize the various chapters of the section devoted to astronomy.

I. *The apparent motion of the planets*

In the first part of the section devoted to astronomy, Ibn al-Haytham starts from results Ptolemy proved for each of the seven planets (the three fundamental motions) and introduces definitions of the 'required time', the inclination of the motion of the planet, and the inclination of the ascending node. He investigates in turn: 1. The motion of a planet between its rising and meridian transit; 2. The motion of known duration between two points of known position.

1.1 *The apparent motion of the moon between rising and meridian transit*

Ibn al-Haytham begins by citing the results proven by Ptolemy in relation to the inclined orb of the moon, the position of this orb in relation to the circle of the ecliptic and to the nodes, that is, to the points of intersection of these two orbs. Ibn al-Haytham considers as fixed the dihedral angle between the plane of the inclined orb and the plane of the ecliptic. In fact, this angle varies little, remaining close to 5°. The orb of the moon would thus lie within the zodiac.

Ibn al-Haytham then reminds us that the motion of the moon on its orb is in the direction of the zodiacal signs (direct motion) and that each node has a uniform motion around the ecliptic in the direction opposite to that of the zodiacal signs (retrograde motion). Thus the north pole of the lunar orb, X, describes on the celestial sphere a circle centered on the pole of the ecliptic, P, and each point of the lunar orb describes a circle parallel to the ecliptic (in retrograde motion). Now, the angle between the circle of the ecliptic and the circle of the equator is constant; because the nodes move, however, the inclination of the lunar orb to the equator of the celestial sphere varies. The inclination will be equal to the arc of the great circle HX, where H is the north pole of the celestial equator.

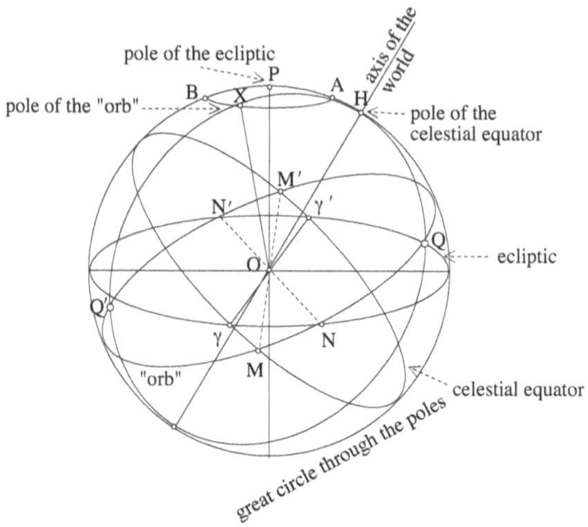

Fig. 128

Ibn al-Haytham investigates in minute detail how this arc varies as the node N makes a complete circuit around the ecliptic. In this preliminary inquiry, he ignores the precession of the equinoxes (in his terms, the retrograde motion of the equinoxes), which is very slow. He treats the planes of the celestial equator, of the ecliptic, and of the circle through the poles of the celestial sphere as if they were all fixed with respect to one another. He will later return to this procedure, to give it more detail, when he considers the maximum and minimum inclinations of the orb of each of the seven planets to the equator. He behaves as if he were deliberately constructing a simple model first, in order to increase its complexity later.

Ibn al-Haytham then defines the northernmost and southernmost points of the lunar orb with respect to the equator. These points are the mid-points of the semicircles into which the orb is divided by the diameter, that is, its line of intersection with the plane of the equator. Accordingly, they lie on the great circle HX that passes through the poles of the lunar orb and those of the equator; their inclination to the equator is equal to HX and is thus variable (Fig. 128).

Ibn al-Haytham then investigates the apparent motion of the moon, in relation to a horizon $ABCD$, between its rising at B and its meridian transit at a point N (Fig. 129, where ABC is the eastern half of the horizon circle); he considers first the case in which the motion along its orb is from north to south, then the case in which it is from south to north. He points out that his argument does not involve the horizon, and is consequently applicable to the motion of the moon between any point B east of the meridian and the point of its meridian transit. At this point, Ibn al-Haytham introduces the following three definitions:

'*Required time*': the time a fixed star takes to travel from a point *B* to a point *I* on the meridian; this is the arc *BI*. It is also the difference of the two right ascensions, $\delta(B, N)$, between the moon's initial position *B* and its final position *N*. This arc *BI* will also be called the right ascension of the motion.

Inclination of the motion of the moon: $\widehat{IN} = \Delta(B,N)$, the difference between the inclinations of the initial position *B*, and the final position *N*.

Inclination of the motion of the ascending node: $\widehat{IQ}$.

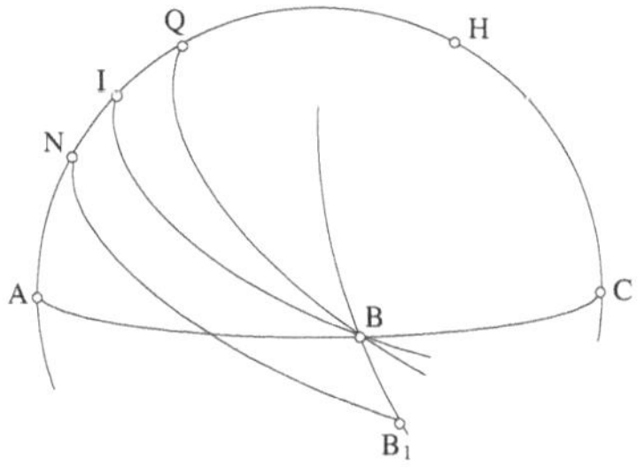

Fig. 129

This investigation of the apparent motion of the moon from its rising to its meridian transit is interrupted by a discussion of the relative positions of two circles through *B*, whose poles are respectively the pole of the equator and the pole of the ecliptic. Finally, Ibn al-Haytham considers the motion of the moon between its meridian transit and its setting, making use of the concepts he has already defined.

Note that, in this geometrico-kinematic model, Ibn al-Haytham does not introduce an epicycle since, as he says, 'the epicycle of the moon does not diverge from the plane of the orb, hence the center of the moon does not diverge from the plane of the inclined orb'.[35]

1.2. *The apparent motion of the sun between its rising and its meridian transit*

Ibn al-Haytham covers the same stages as in the previous investigation: he begins by recalling what is known about the orb of the sun (the ecliptic) and the sun's proper direct motion through the zodiacal signs. He defines

[35] R. Rashed, *Les mathématiques infinitésimales*, vol. V, p. 429, 23–25.

the points of the orb that are called equinoxes and solstices. He then deals with two examples, which are referred to a horizon *ABCD*, concerned with the apparent motion of the sun between its rising at *B* and its meridian transit. In the first case, the motion of the sun along its orb is from north to south, in relation to the equator; and in the second case, from south to north. In each example, Ibn al-Haytham finds the arcs that represent 'the required time' and the inclination of the motion of the sun.

This investigation is simpler than the one he carried out for the motion of the moon, which required that one take account of the node's motion along the ecliptic.

1.3. *The apparent motion of each of the five planets between rising and meridian transit*

Here, as in previous cases, Ibn al-Haytham begins by recalling what Ptolemy had established. He also tells us that his investigation will not take into account the motion of the node, since, he writes, it is 'a slow motion that does not become perceptible'.[36] We should remember that Ibn al-Haytham has always maintained that, unlike the case of mathematics, where reasoning is exact, in physics we always allow a certain amount of approximation. And here, the inclination of the plane of the epicycle to that of the orb is variable. Accordingly, its variation must be taken into account when investigating the motion of each of the five planets towards the meridian circle. Ibn al-Haytham does exactly this when he considers the motion of a planet between its rising at a point *B* on the horizon, and its meridian transit. He distinguishes three cases: when the planet's motion is direct; when it is retrograde; and finally when the planet is stationary. Ibn al-Haytham's investigation ends with a conclusion on the planets as a whole, concerning the 'required time' and the 'inclination of the motion'.

2. In the preceding section of the work, the two positions considered for each of the seven planets were the rising at the point *B*, and the meridian transit at point *N*; sometimes, the motion considered was from point *N* to setting. In this part of his work, Ibn al-Haytham investigates, for each of the seven planets, an apparent motion of known duration between two points *A* and *B*, whose position on the celestial sphere is known. He shows that the 'required time' and the 'inclination of the motion' are then known.

Ibn al-Haytham begins by dealing briskly with the case of the sun, which is simple because his model omits the precession of the equinoxes.

[36] *Ibid.*, p. 429, 2.

Thus if A and B are, respectively, the starting and end points of the motion, we at once have:

• 'the required time': $\delta(A, B)$, the difference of the right ascensions of the two points A and B;

• 'the inclination of the motion': $\Delta(A, B)$, the difference of the declinations of the two points A and B, that is, the difference of their inclinations to the equator.

The investigation of the motion of the moon must, however, take account of the motion of the ecliptic and the motion of the node of the orb of the moon.[37] Here, as in the case of the sun, the motion is described in equatorial coordinates, namely 'required time' and 'proper inclination'.

For each of the inferior planets (Venus and Mercury), the ecliptic coordinates – ecliptic latitude and longitude – depend on the inclination of the epicycle to the orb.[38] All the same, if at some known time the ecliptic coordinates are known, we use them to find the equatorial coordinates. Ibn al-Haytham carries out his investigation just as he had done in the case of the moon.

For the superior planets (Mars, Jupiter, and Saturn), the motion of the nodes is very slow, and insensible over the course of a day. As a result, the arc that corresponds to arc KG, which is parallel to the ecliptic in the case of the moon, is insensibly small; the point G merges with the point K and therefore lies on the horary circle AD (Fig. 130).

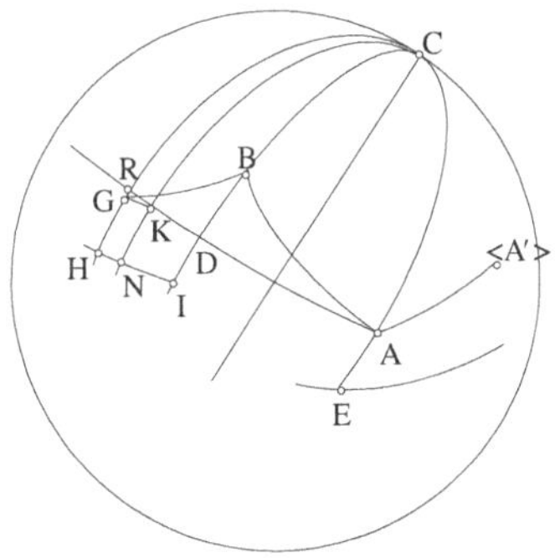

Fig. 130

Ibn al-Haytham concludes with a generalization about the five planets:

[37] See commentary on Proposition 20 in R. Rashed, *Les mathématiques infinitésimales*, vol. V, pp. 189 ff.

[38] *I.e.* the circle on which the epicycle moves (*circulus deferens*). We shall find this formulation more than once.

if the motion of the planet along its orb is direct, 'the required time' $\delta(A, B)$ is less than the known time, as happens also for the sun and the moon; and if the motion of the planet is retrograde, the 'required time' is then greater than the known time.

II. *The inclination of the planets to the equator*

Ibn al-Haytham begins by discussing first the sun, then the moon and finally the five planets. As always, he first reminds the reader of Ptolemy's results. Here Ibn al-Haytham further determines, in each case, the ecliptic coordinates of the point I, the southernmost point of the orb with respect to the equator.

In the case of the sun, the dihedral angle α between the plane of its path (the ecliptic) and the plane of the equator is constant ($\alpha = 23°27'$). This angle α is the maximum inclination of points on the ecliptic with respect to the equator, and corresponds to the solstices. The two points of maximum inclination are thus the first point of Cancer to the north of the equator, and the first point of Capricorn to the south.

In the case of the moon, the dihedral angle β between the orb of the moon and the circle of the ecliptic is constant, but the orb of the moon rotates about the axis of the ecliptic. Accordingly, the dihedral angle δ between the orb of the moon and the plane of the equator is variable; it depends on the position of the ascending node. If the ascending node is at point γ (spring equinox), we have $\delta = \alpha + \beta$. But if the descending node is at point γ, the ascending node is then at point γ', the autumn equinox, and we have $\delta = \alpha - \beta$. In either case, the positions of the northernmost and southernmost points of the inclined orb are known.

For the case in which the ascending node is not at an equinoctial point, Ibn al-Haytham embarks on a very detailed investigation using spherical trigonometry, in which he applies Menelaus's theorem four times, and shows that, if we know the position of a node on the ecliptic, we can calculate the maximum inclination of the inclined orb with respect to the equator, and find the position, with respect to the ecliptic, of the northernmost or southernmost point of the inclined orb with respect to the equator.

For the superior planets, the procedure is the same as for the moon, since the inclinations of their orbs to the plane of the ecliptic are more or less constant: for Mars, $1°51'$; for Jupiter, $1°19'$; and for Saturn, $2°30'$. On the other hand, the inclination of the orb of each of the inferior planets to the ecliptic is variable. Ibn al-Haytham accordingly devotes many pages to studying this problem.

He begins by examining the inclination as a function of the planet's position in its orb, a position for which there is a corresponding point on the eccentric. He shows that this inclination is known at any known time.

He goes on to investigate the case in which the nodes are at the equinoctial points. When related to the ecliptic, the northernmost and southernmost points of the orb relative to the equator are the solsticial points. We calculate the inclination relative to the equator as we did for the moon. Ibn al-Haytham next considers the case in which the nodes are not the equinoctial points. The positions of the northernmost and southernmost points relative to the equator are found from the northernmost and southernmost points relative to the ecliptic, and he then employs the same method as before.

Ibn al-Haytham goes on to describe – still for the inferior planets – the oscillating motion of the plane of the inclined orb about the nodal line. The motion of the nodal line is very slow, and for the purposes of this calculation, the line is accordingly assumed to be fixed. So any point *I* of the orb describes a circle with the nodes as its poles, and the point will have a to-and-fro motion along an arc of the orb. With this point *I* is associated a point *L* that represents its position in regard to the ecliptic; this point *L* will also have a to-and-fro motion along an arc of the ecliptic. In his investigation of the motion of points *I* and *L*, Ibn al-Haytham takes point *I* as lying, successively, on each of the four arcs into which the orb is divided by the nodes and the northernmost and southernmost points. He assumes that the initial position of the orb is when its inclination to the ecliptic is at a maximum, and he calls the two points in question *I* and *L*.[39] He first describes the motion of points *I* and *L*. Next he shows that the circular arc described by point *I* in a known time is known; finally, he shows that the arc of the ecliptic described by point *L* in a known time is known.

III. From Proposition 24 to the end of the book, Ibn al-Haytham proposes general models for the planets as a set, models that are constructed with the help of the mathematical propositions he has already proven. His work, which is explicitly analytical and uses infinitesimals, concerns itself with some kinematic properties of the motion. In this instance, one cannot follow Ibn al-Haytham's procedure without examining his demonstration in detail, which we do in the commentary of his text.[40] Here, however, we shall merely present a general outline.

[39] The great circle through the pole cuts the orb in *I* and the ecliptic in *L*. The points *I* and *L* have the same ecliptic longitude.

[40] See Rashed, *Les mathématiques infinitésimales*, vol. V, Part I.

In the first four propositions – 24 to 27 – Ibn al-Haytham investigates the variation of a planet's mean speed. He expresses the mean speed as the inverse ratio $\frac{\delta(X,Y)}{\Delta(X,Y)}$, where X and Y are two general known positions of a planet in its orb, $\delta(X, Y)$ is the 'required time', and $\Delta(X, Y)$ is the difference between the inclinations of points X and Y with respect to the equator. Ibn al-Haytham proves that, if we consider the four arcs into which the orb is divided by the diameter, that is, the line of intersection of the planes of the orb and the equator, and the northernmost and southernmost points of the path with respect to the equator, and if we take two positions X and Y on one of these arcs, then there always exists a ratio k such that $k > \frac{\delta(X,Y)}{\Delta(X,Y)}$.

Note that the known time is a real interval that can be measured by the motion of the planet. Ibn al-Haytham's idea of comparing 'required time', an equatorial coordinate, to this known time looks like the beginnings of a kinematic description of the motion.

In the next group of propositions, Ibn al-Haytham investigates the apparent motion of a planet above the horizon of a given place. This observed motion depends on the place and on the date of the observation. In the course of this investigation, Ibn al-Haytham uses the planet's equatorial coordinates, and consequently its position in its trajectory, the inclination of the orb to the equator, and the inclination of the equator to the horizon; that is to say, he uses the geographical latitude of the place where the observation is made. Throughout this investigation, Ibn al-Haytham assumes that the celestial sphere is inclined to the south; the observation site thus has a northern latitude. The case of the *sphaera recta*, that is to say, the case where the observer is on the terrestrial equator, appears as a special case. Ibn al-Haytham assumes that the planet's meridian transit takes place between the zenith and the southern horizon, which means that the geographical latitude of the place where the observation is made must be greater than the declination of the planet for the date in question. He also assumes that the latitude of the observation site is smaller than the complement of the declination. Ibn al-Haytham makes a detailed study of the part played by the latitude, which leads him to consider the cases in which meridian transit occurs at the zenith or north of the zenith, and finally the case of places whose latitude is equal to the complement of the maximum declination of the planet.

So, in two propositions, 28 and 29, Ibn al-Haytham investigates the altitudes of a planet above the horizon. Let us suppose that the planet rises at point B and crosses the meridian at D. The arc BD which it describes is to the east of the meridian plane. Let the altitude of the planet above the horizon be h (Fig. 131). Ibn al-Haytham shows that on arc BD there exist:

- points of altitude $h > h_D$ (the altitude of point D). Let M be one of these points;
- at least one point X on the arc BM such that $h_X = h_D$;
- at least two points with the same altitude h with $h_D < h < h_M$, one on the arc XM and the other on the arc MD.

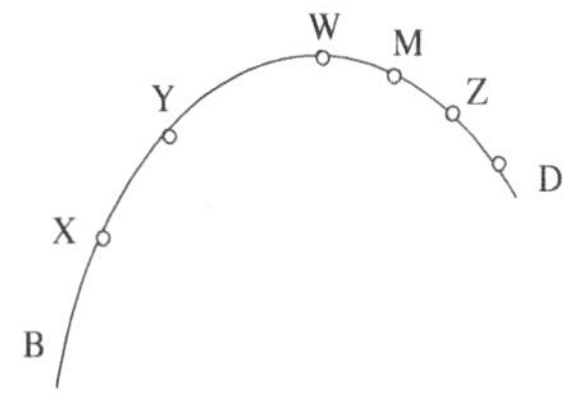

Fig. 131

He also shows that, after it crosses the meridian at D, the planet continues its motion towards the western horizon and its altitude h decreases from h_D to 0. Any altitude $h < h_D$ is thus reached at least once.

Ibn al-Haytham also shows that, if h_m is the maximum altitude, the planet will reach it only once, say at a point W; and that altitude h_D will be reached once and only once at a point $X \neq D$, such that $X \in \widehat{BW}$.

In Proposition 29, Ibn al-Haytham investigates the movement of the planet from the southernmost to the northernmost points of its trajectory. The planet crosses the meridian at G and sets at D. The arc GD that it describes is to the west of the meridian (Fig. 132).

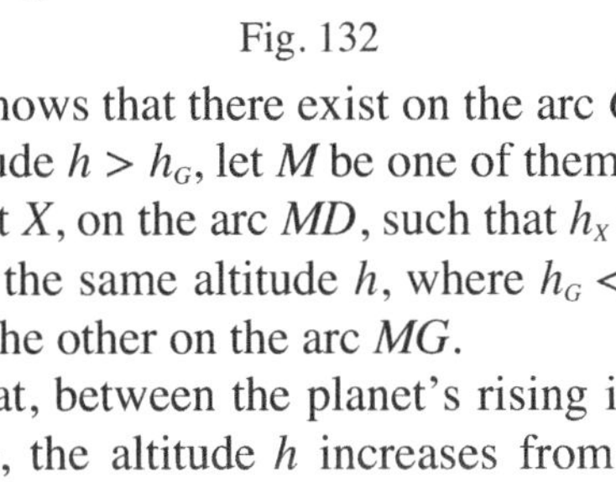

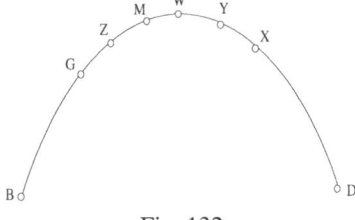

Fig. 132

Ibn al-Haytham shows that there exist on the arc GD:

- points with altitude $h > h_G$, let M be one of them;
- at least one point X, on the arc MD, such that $h_X = h_G$;
- two points with the same altitude h, where $h_G < h < h_M$, one of which is on the arc XM and the other on the arc MG.

He also shows that, between the planet's rising in the east at B and its meridian transit at G, the altitude h increases from 0 to h_G and that any altitude is reached at least once.

Later on Ibn al-Haytham returns to this investigation to calculate the altitudes reached by the planet to the west of the meridian. He shows that if h_m is the maximum altitude, the planet reaches it only once – let it be at point *W*; and that the altitude h_G, which is that of the planet's meridian transit, is reached once, and only once, at a point other than *G* – let it be at point *X* on the arc *WD*; that any altitude $h < h_G$ is reached once, and only once, at a point between *X* and *D*; and that any altitude *h* such that $h_G < h < h_M$ is reached at two points and at only two points – one on arc *GW* and the other on arc *WX*.

In Proposition 30, Ibn al-Haytham proves that the point at which maximum altitude is reached is unique; then, in Proposition 31, he returns to the investigation of altitudes to the east of the meridian. In these two propositions, Ibn al-Haytham once again introduces innovations in infinitesimal geometry. He is in fact developing an entirely original new method of working in spherical geometry: he considers infinitesimal curvilinear triangles on the sphere (triangles whose sides are not necessarily arcs of a great circle) – he constructs a sequence of such triangles whose sides tend to zero – and he handles these triangles as if they were infinitesimal rectilinear triangles. What we encounter here is in effect a geometry of infinitesimals like that used later in differential geometry.

In order to sum up some results that Ibn al-Haytham established in his investigation of the point *D* at which the planet crosses the meridian (that is, in this group of Propositions 28 to 36 where he is investigating the altitudes of a planet), let us consider the meridian plane, with pole *Z*, and the equator, whose north pole is *N* (Fig. 133).

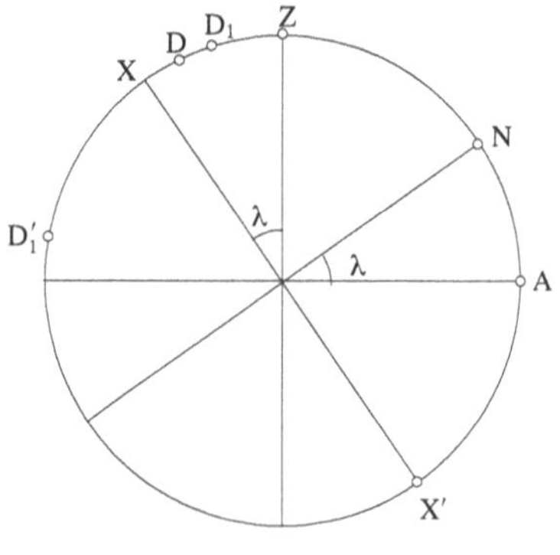

Fig. 133: $\alpha < \lambda < \frac{\pi}{2} - \alpha$

Let λ be the latitude of the place, δ the declination of the planet at meridian transit, and α the maximum value of the declination; we have arc AN = arc $XZ = \lambda$, arc $XD = \delta$, arc XD_1 = arc $ZD_1' = \alpha$.

We consider only places in the northern hemisphere, and we use the sun as our example, so $\alpha = 23°27'$.

We may summarize the investigation of the position of D, as a function of the geographical latitude λ and the date, in the form of a table. Let us assume that $\alpha < \lambda < \frac{\pi}{2} - \alpha$.

latitude	date	position of D
$\lambda = 0$ terrestrial equator	• Spring and Autumn equinoxes • Summer solstice • Winter solstice • Spring and Summer • Autumn and Winter	$D = Z = X$ $D = D_1$ to the north of Z $D = D_1'$ to the south of Z D between Z and D_1 to the north of Z D between Z and D_1 to the south of Z
$\lambda = \alpha$ tropic of Cancer	• day of the Summer solstice $\delta = \lambda$ • any other day $\delta < \lambda$	$D = D_1 = Z$ D lies on the arc ZD_1', south of Z
$0 < \lambda < \alpha$ northern tropical zone	• day of the Summer solstice $\delta = \alpha$, so $\delta > \lambda$. The declination $\delta = \lambda$ will be reached once in the Spring and once in the Summer on these two dates between these two dates for any other day of the year	D at D_1 north of Z D is at Z D lies on the arc ZD_1, north of Z D lies on the arc ZD_1', south of Z
$\lambda > \alpha$	for any day of the year	D lies south of Z

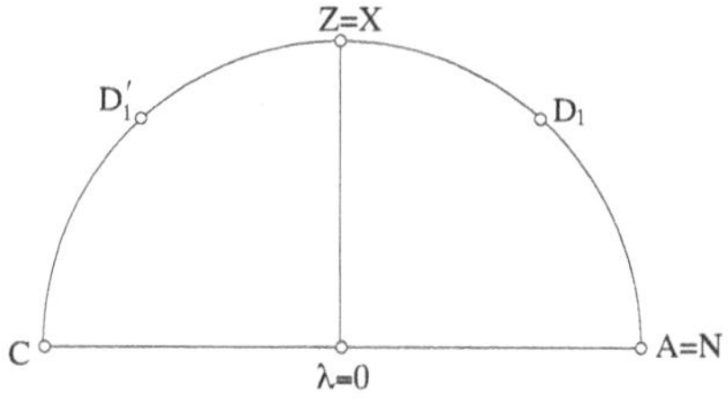

Fig. 134.1

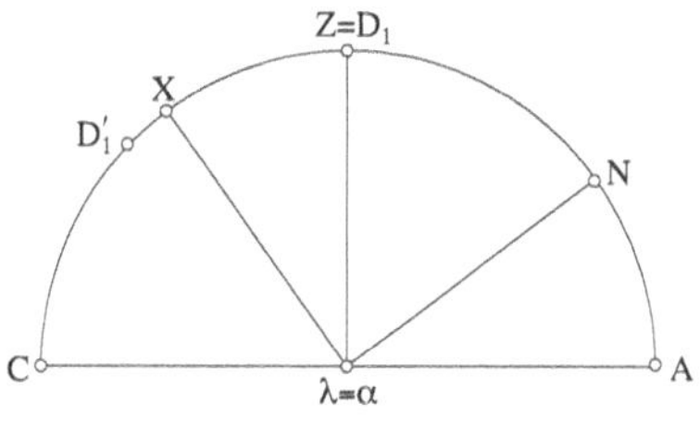

Fig. 134.2

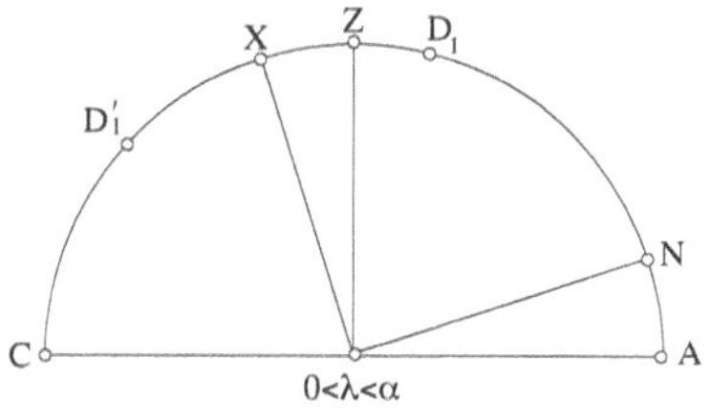

Fig. 134.3

In the case of the *sphaera recta*, whether meridian transit is north or south of the zenith Z, we can apply the method employed in Proposition 28 or Proposition 29 and show that the planet will have equal altitudes h that are pairwise equal ($h > h_D$) either to the east of the meridian, or to the west of it.

In the case where the point D, the point of meridian transit, is at the zenith Z,[41] the maximum altitude of the planet is h_D, and any altitude $h < h_D$ will be reached once, and only once, to the east, and the same will apply for the west.

So far, Ibn al-Haytham has considered places north of the equator with latitude $\lambda < \frac{\pi}{2} - \alpha$, where α is the inclination of the orb to the equator; to complete his investigation of the trajectory of a planet seen above the horizon, he considers places with northern latitude $\lambda = \frac{\pi}{2} - \alpha$ or $\lambda \cong \frac{\pi}{2} - \alpha$ and shows that in such places, and on particular dates, the planet in question may set in the east and rise in the east and that, on other dates, it may set in the west and rise in the west.

Let $BHID$ (Fig. 135) be the meridian plane for some place, BD the diameter of the horizon, EG the diameter of the equator, H the pole of the equator $\widehat{BH} = \lambda = \frac{\pi}{2} - \alpha$, $\widehat{HZ} = \alpha$. If we draw $BI \parallel EG$ and $DI' \parallel EG$, we have $\widehat{BG} = \widehat{EI} = \widehat{ED} = \widehat{GI'} = \alpha$, so the circles with diameters BI and DI' touch the horizon of the place in question at B and D respectively. The trajectories of the planet's diurnal motion therefore lie between these two circles.

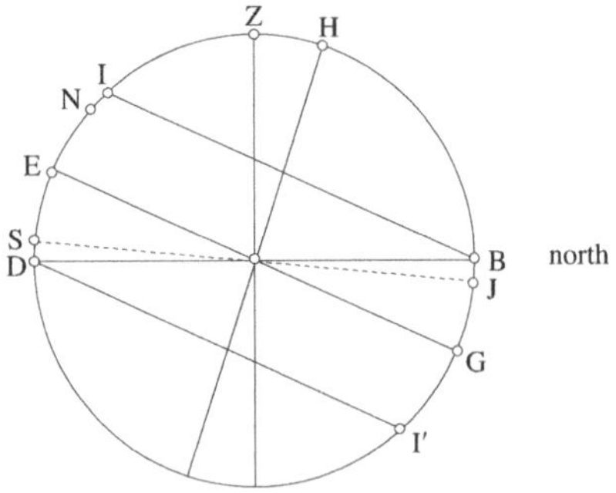

Fig. 135

[41] See Rashed, *Les mathématiques infinitésimales*, vol. V, p. 255.

We have assumed that the planet reaches point *B*, the north cardinal point of the horizon in question, *ABCD*, at the time it gets to the northernmost point of its trajectory, that is to say, at the moment when its declination is a maximum and equal to α. So thereafter the declination decreases and the trajectory of the planet moves away from the circle *BI* and begins to cut the meridian again at point *N* above the horizon.

Ibn al-Haytham then defines:

• a point *L* that belongs to this trajectory and is above the horizon *ABCD* and to the east of *B*;

• a horizon circle with diameter *JS*, of latitude $\lambda + \varepsilon$ which shares the same meridian and is such that point *L* is below the horizon.

But the points *B* and *N* are above this horizon *JS*, so when the planet moves from *B* towards *L*, it sets at a point on the eastern part of this horizon, and when it moves from *L* towards *N*, it rises at a point that is likewise on the eastern part of this horizon.

At the other extreme, the planet is assumed to be at the southernmost point of its trajectory, at the point *B* of the horary circle *BQI* (Fig. 136), and this point *B* is the south cardinal point of the horizon in question, which has latitude $\lambda = \frac{\pi}{2} - \alpha$. The declination thus increases thereafter and the trajectory of the planet diverges from the circle *BQI* and begins to cut the meridian again at a point *N* above the horizon. The method is accordingly the same as in the previous part. Ibn al-Haytham defines:

• a point *L* that lies on this trajectory and is above the horizon *ABCD* and to the west of *B*;

• a horizon circle *AJCS* at latitude $\lambda - \varepsilon$ that shares the same meridian and is such that point *L* is above it.

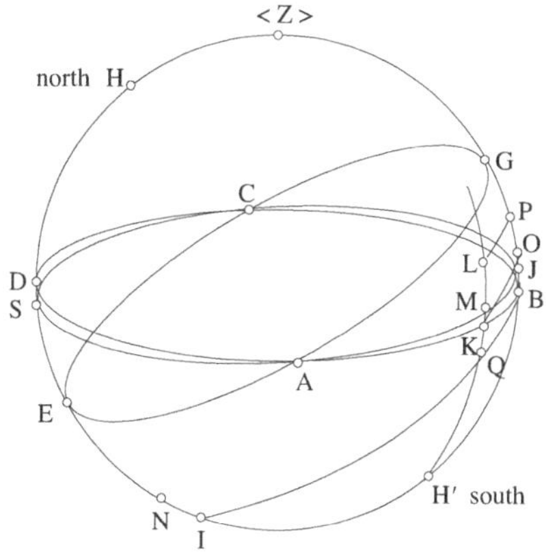

Fig. 136

When the planet moves from B towards L, it rises at a point on the western part of this horizon, and when it moves from L towards N, it sets at a point on the western part of this horizon.

Ibn al-Haytham has thus shown that on the day when a planet reaches its maximum northern declination α, there exist places in the northern hemisphere, with latitude $\lambda = \frac{\pi}{2} - \alpha + \varepsilon$, on whose horizons the planet sets and rises in the east, and that on the day when the planet reaches its maximum southern declination α, there exist places in the northern hemisphere, with latitude $\lambda = \frac{\pi}{2} - a - \varepsilon$, on whose horizons the planet rises and sets in the west.

In both cases, the points at which the planet rises and sets are very close to one another.

We have sketched the principal results that Ibn al-Haytham obtains in his *Configuration of the Motions*. Our aim was not so much to expound all the results in detail, which we have done elsewhere, but rather to give an overview of what he was trying to do in his book. All the way through *The Configuration of the Motions* he directs his efforts to constructing a descriptive phenomenological theory of the celestial motions, as they are seen by an observer on the earth. This theory, as one can easily assure oneself, does not incorporate any idea of a teleological physics, though it does not conflict with what Aristotle calls the most physical parts of mathematics, which here is geometrical optics, a subject reformed by Ibn al-Haytham himself. When Ibn al-Haytham is constructing his astronomy, his obvious concern is, as we have noted, to adopt at each stage the least possible number of hypotheses.

Thus his theory for the motion of the planets calls upon no more than observation and conceptual constructs susceptible of explaining the data, such as the eccentric circle and in some cases the epicycle. However, this theory does not aim to describe anything beyond observation and these concepts. In no way is it concerned to propose a causal explanation of the motions. In this respect, *The Configuration of the Motions* is both in the astronomical tradition that Ibn al-Haytham inherited and in a tradition that continues after Ibn al-Haytham as far as Kepler. To sum it up, in *The Configuration of the Motions*, Ibn al-Haytham's purpose is purely kinematic; more precisely, Ibn al-Haytham wanted to lay the foundations of a completely geometrical kinematic tradition.

Carrying out such a project involves first of all developing some branches of geometry required for solving new problems that arise from this kinematic treatment: Ibn al-Haytham took a huge step forward in spherical geometry as also in plane and spherical trigonometry. To get a measure of how far he has advanced beyond the Greeks, one need only compare *The Configuration of the Motions* with Chapters 9 to 16 of the first book of Ptolemy's *Almagest*; and to appreciate the distance that separates him from his contemporaries one may compare *The Configuration of the Motions* with, for example, the *Almagest* of Abū al-Wafā' al-Būzjānī. As we have seen, Ibn al-Haytham considers the changes in infinitesimal magnitudes that necessarily arise in astronomical research.

In astronomy, there are two major processes that are jointly involved in carrying through this project: freeing celestial kinematics from cosmological connections, that is, from all considerations of dynamics, in the ancient sense of the term; and to reduce physical entities to geometrical ones. The centers of the motions are geometrical points without physical significance; the centers to which speeds are referred are also geometrical points without physical significance; even more radically, all that remains of physical time is the 'required time', that is, a geometrical magnitude. In short, in this new kinematics, we are concerned with nothing that identifies celestial bodies as physical bodies. All in all, though it is not yet that of Kepler, this new kinematics is no longer that of Ptolemy nor of any of Ibn al-Haytham's predecessors; it is *sui generis*, half way between Ptolemy and Kepler. It shares two important ideas with ancient kinematics: every celestial motion is composed of elementary uniform circular motions, and the center of observation is the same as the center of the Universe. On the other hand, it has in common with modern kinematics the fact that the physical centers of motions and speeds are replaced by geometrical centers.

There remains a major question, that of the relation of this kinematics to the celestial dynamics of the day, that is to say to cosmology. The question is relevant here only if we come across evidence that Ibn al-Haytham had intended to write on cosmology once he had completed *The Configuration of the Motions*. In that case, one would expect a new cosmology to go with the new kinematics. In fact none of the titles that have come down to us, none of the manuscripts of Ibn al-Haytham's undoubtedly authentic astronomical works gives grounds for claiming that such a cosmology, based on the new kinematics, ever existed. The only cosmology text known to have been composed by Ibn al-Haytham (that is of well-attested authenticity) is earlier than *The Configuration of the Motions* since it forms part of his treatise on the winding motion. When in his *Resolution of Doubts concerning the Winding Motion*, he himself mentions this work (now lost), he writes:

> The winding motion to which Ptolemy referred, and from which arise the motions in latitude of the five planets, can only be according to the

> configuration that I demonstrated and according to the account that I gave. It is a configuration that is not subject to any impossibility or any absurdity. From this motion is generated a motion of the planet which, by the motion of its center, produces a curve imagined as if the planet were wound around on the body of the small sphere which moves the body of the planet. It is because of the winding of this curve around the body of the epicycle that this motion has been called the winding motion, and for no other reason.[42]

There is no room for doubt: in his treatise on the winding motion, Ibn al-Haytham had indeed proposed a model for the motions in latitude of the epicycles of the five planets, a model in which he considered the physical 'small spheres' that moved the celestial bodies; in other words, he had proposed a cosmology. Many other passages of the treatise confirm this.

Now, from the order of composition of Ibn al-Haytham's writings that we have already established, we know that, of these writings, the two books on the winding motion were composed before the *Doubts concerning Ptolemy*. Moreover, while in the first two books he makes use of the idea of an equant, in the last one he criticizes it, and eventually ends up completely excluding it from *The Configuration of the Motions*. Furthermore, since Ibn al-Haytham emphasizes in the introduction to *The Configuration of the Motions* that the results described in this work supersede any different ones to be found in all his other writings, we may safely conclude that *The Configuration of the Motions* was written after the *Doubts concerning Ptolemy* and, *a fortiori*, after the two books about the winding motion. Thus Ibn al-Haytham's contribution to cosmology is (as it were) local, since it relates only to a particular motion and antedates the *Doubts* and *The Configuration of the Motions*. We proved elsewhere that *The Configuration of the Motions* is also later than his treatise on *The Different Altitudes of the Wandering Stars*.[43]

Another argument in favour of this historical and conceptual sequence is drawn from the language used in *The Configuration of the Motions*. The book not only contains new concepts such as 'required time' and 'proper inclination for the required time', but also terms from ancient astronomy whose meaning has changed. For example let us consider a concept central to traditional astronomy, that of *falak*. It is well known that in traditional astronomy this term signifies 'orb'. It refers to the various solid bodies connected with a specific planet. These solid bodies move with uniform circu-

[42] *On the Resolution of Doubts relating to the Winding Motion* (*Fī ḥall shukūk ḥarakat al-iltifāf*), ms. St Petersburg, B1030/1, fols 15^{v}–16^{r}.

[43] See Rashed, *Les mathématiques infinitésimales*, vol. V, Part I.

lar motions, and the sum of these motions constitutes the apparent motion of the planet under consideration, as seen from the earth, which is at the center of the universe. In this system, a planet does not have a motion of its own, it is moved by something else, and one cannot speak of the motion of a planet along its particular orbit, but only of its apparent motion resulting from the composition of the motions of its various spheres. This same word *falak* is also used in the same context to designate the (plane) circles that are the lines on the sky that correspond to the solid bodies in question.

In fact, Ibn al-Haytham uses this term *falak* in these senses in all the works we have cited above, except in *The Different Altitudes of the Wandering Stars*, where he does not need it. On the other hand, in *The Configuration of the Motions*, the term *falak* no longer has the same meaning. In this book, it refers mainly to the apparent trajectory of a particular planet across the celestial vault, and everything else derives from the analysis of this apparent motion, without reference to solid bodies that might move the planet in question. This semantic difference, taken together with the new concepts, shows that *The Configuration of the Motions* was composed after the books we referred to earlier. This difference alone also shows that this treatise cannot be placed within a purely Ptolemaic tradition. One might almost translate the term as the 'orbit' of a planet[44] since the apparatus of the orb, in the sense in which the term was conventionally understood, no longer enters into it.

In the *Doubts*, we have seen a turning point in Ibn al-Haytham's astronomical thought. There is every indication that *The Configuration of the Motions* is the most substantial result produced by this change. The book gives us a new astronomy even though it retains a geocentric framework within which all motions are circular and uniform. We have a break with tradition despite the background of continuity.

We need to know the reasons for such a change. On this matter the available texts are silent. We may, however, offer the following hypothesis. In the absence of a theory of gravitation, the mathematician-astronomer was faced with two alternatives: either to abide by the traditional principle whereby the motion of each planet derives from a cause specific to that body, and thus to construct a cosmology of material spheres; or to accept the necessity of abandoning that route and instead to start by constructing a kinematic account, thus acknowledging the primacy of kinematics in any investigation of dynamics. In many of his astronomical writings, Ibn al-

[44] However, we do not do this in the translation, preferring to retain period usage. We simply need to alert the reader here.

Haytham had been tempted by the first alternative. But, once he had engaged upon mathematizing astronomy and had noted not only the internal contradictions in Ptolemy, but doubtless also the difficulty of constructing a self-consistent mathematical theory of material spheres using an Aristotelian physics, he turned to the second alternative, that of giving a completely geometrized kinematic account. His experience in optics perhaps helped him to take this step: here kinematics and cosmology are entirely separated to effect a reform of astronomy, just as in optics, work on the propagation of light is entirely separated from work on vision to effect a reform of optics; in the one case as in the other, Ibn al-Haytham arrived at a new idea of the science concerned.

– 2 –

FROM THE GEOMETRY OF THE GAZE TO THE MATHEMATICS OF THE PHENOMENA OF LIGHT

Let us begin by remembering one fact and evoking one metaphor. Optics, the oldest and most mature of the physical sciences because it was the first to be mathematized, was developed and transformed in its ancient and classical periods around the Mediterranean. This is the fact. As to the metaphor, it pertains to the production and circulation of scientific knowledge around the Mediterranean. The inland sea presents itself as a locus, a *topos* in the Aristotelian sense, a place of exchange between most civilizations of the ancient world, both those that encircle its shores and those that are more distant from it. To illustrate this situation, let us leave the minefield of metaphors and examine briefly the predicament of the historian before returning to the science itself. If, for example, a historian nowadays wants to study the beginnings of Hellenistic optics, he will inevitably encounter the work of Diocles from the second century B.C. His research will therefore require an examination of the Arabic version of this work, the only one that survived. If he now turns to a later period of Hellenistic optics, he will have to devote most of his efforts to studying the fundamental contribution attributed to Ptolemy (2nd century). He will now have to be content with the 12th-century Latin translation that Admiral Eugenius of Sicily made from the 9th-century Arabic version, which in turn had been translated from the Greek; indeed, both the Arabic and the Greek texts are lost. Assume now that our historian is interested only in Arabic optics, and is sufficiently cavalier to neglect the Greek sources and the Arabic translations of the latter; he will nevertheless not be able to avoid the Latin and Hebrew translations from the Arabic. We know that one of the first works in Arabic optics was written by the philosopher Abū Isḥāq al-Kindī. Of his book on optics, there remains only the Latin translation, which became an essential touchstone as much for Roger Bacon as for John Pecham and Robert Grosseteste. But if our historian studies a specific chapter of Arabic optics, such as that devoted to atmospheric phenomena, he will inevitably encounter the book of the Andalusian, Ibn Muʿādh, *De crepusculis*, which exists only in Latin and Hebrew translations.

It would be easy to multiply such examples, all of which point to the fact that this is a highly specific situation in both time and space. If indeed one compares our historian to a colleague working on later (*e.g.*, 18th-century) optics or to a historian of optics in a different cultural era (*e.g.*, China), the historian of science and mathematics in the Mediterranean cultures up to the 17th century must follow a more sinuous path. The latter must travel everywhere without reprieve; he cannot rely on a fixed point; unless he wants to run the risk of completely missing his quarry, he must resist every temptation of culture-centrism and of linear history. But the predicament that is peculiar to our historian is in fact only a reflection of the meandering constitution and diffusion of optics itself. Let us therefore take a moment to pause on the elaboration of this science, and to consider, not the research of today's historian, but the journey of the scholar of yesteryear.

In Greek and Hellenistic antiquity, research in optics is essentially divided into five chapters, which sometimes are superimposed or overlap: optics proper, that is, the geometrical study of the perception of space and of perspectival illusions; catoptrics, that is, the geometrical study of the reflection of visual rays in mirrors; burning mirrors, the study of the converging reflection of solar rays on mirrors; atmospheric phenomena, such as the halo and the rainbow; and finally, the study of vision by natural philosophers and physicians. In the chapters that simultaneously treat vision and the propagation of light, the doctrine of the 'visual ray' dominates. The latter consists of a divergent beam emitted by the eye, that is, a cone whose vertex is the eye, and whose edges consist of the visual rays that propagate rectilinearly and skim over the objects that interfere with them. According to this teaching, to see is to illuminate, and the conditions of propagation are those of vision. Each type of problem, whether propagation or vision, immediately refers back to the other. In this twofold motion of referral were grounded not only the conditions of possibility of ancient optics, but also its limits and the obstacles to its development.

On this account of the visual ray, vision is conceived as a tactile act, a palpation at a distance. This doctrine was developed in the *Optics* attributed to Euclid, found its nearly definitive formulation in the *Optics* attributed to Claudius Ptolemy, and dominated optics until the end of the 10th century and beyond. In this sense, optics is only a geometry of perception, a geometry of the gaze. Its main theme concerns the variation of the magnitude of the visible in relation to the variation of its distance from the eye as source, or better yet, the eye as headlight, and in relation to its position among other objects. In this optics, the primary concern is with the visible object, not as such, but only in relation to the question: how does the object appear to the eye? It is in some sense a geometry of appearance, in which light and vision have no distinct ontological status. In this regard, recall the postulates of Euclid's *Optics*.

I. Assume that the straight lines that emanate from the eye are carried over a space of immense magnitudes (μεγεθῶ μεγάλων).

II. That the figure bounded by these visual rays is a cone having its vertex in the eye and its base at the boundaries of the seen magnitudes.

III. And that the magnitudes on which the visual rays fall are seen; whereas those on which the visual rays do not fall, are not seen.

IV. And that the magnitudes seen under a greater angle appear larger; whereas those that are seen under a smaller angle, appear smaller, and those that are seen under equal angles, appear equal.

V. And that the magnitudes seen under rays that are more lifted up, appear higher; whereas those that are seen under rays that are more lowered, appear lower;

VI. And that likewise the magnitudes seen under rays that are more to the right, appear more to the right, whereas those that are seen under rays more to the left, appear more to the left;

VII. Finally, that the magnitudes seen under more numerous angles appear more distinctly.

In this *Optics*, the eye is the headlight, as it were, the vertex of a cone of visual rays that illuminate objects, and it is the angles and directions of these rays that determine the appearance of visible things.

On this Euclidean base, a geometrical science of optics is constituted that reaches its most accomplished form in Ptolemy's *Optics*. In the four books of this work, Ptolemy treats successively: direct vision, vision by reflection in different mirrors (plane, spherical-concave, spherical-convex, cylindrical, conic), and finally vision by refraction. Even if the doctrine remains essentially the same, the perception now becomes the result of a complex judgment, grounded in the intervention of several faculties of the soul.

Thus it is this optics-as-geometry-of-the-gaze that was transmitted into Arabic, along with most of the Greek and Hellenistic works on burning mirrors, optical meteorology, as well as the writings of philosophers and physicians. This Arabic dependence on Greek and Hellenistic optics – and, one might even say, on it alone – nevertheless did not obstruct the emergence of rather precocious, innovative research. Indeed, very soon after the massive transmission of Greek writings, the history of the discipline developed three foci: the rectification or correction of these writings, the accumulation of new results, and the renewal of its main domains. It took only two centuries to prepare what, in the end, became a genuine revolution, one that left a permanent mark on the history of optics, and arguably on the history of physics. It is this dialectical movement between a solid continuity and a profound break that I would now like to sketch.

After gaining access to Euclid's *Optics* and Anthemius of Tralles's treatise on *Burning Mirrors*, the mid-9th century philosopher and scientist al-Kindī composes several treatises on the various branches of optics. In the wake of Euclid, he assigns himself the task of 'giving an exposition of the ancients' teaching', 'developing what they had begun', and correcting the errors they made. Thus, in his book, lost in Arabic but preserved in Latin, entitled *Liber de causis diversitatum aspectus* [*De aspectibus*], he wants to demonstrate what Euclid had postulated. Indeed, one quarter of the *De aspectibus* is devoted to justifying the rectilinear propagation of light rays, using geometrical considerations on shadows and the passage of light through slits. Having thus established rectilinear propagation, al-Kindī returns to the theory of vision. He begins by bringing up and criticizing the main doctrines known since Antiquity, in the end adopting extramission. He does not, however, accept the Euclidean doctrine without seriously amending it. According to him and in contrast to Euclid's view, the visual cone is not composed of discrete rays, but presents itself as the volume of continuous radiations. The importance of this amendment derives from the idea that it undergirds: that of the ray. Al-Kindī pushes aside a purely geometric conception of the ray: rays are not geometrical straight lines, but impressions produced by three-dimensional bodies. In al-Kindī's words:

> a ray is an impression of the luminous body on opaque bodies, the name of which is derived from that of light on account of the alteration of the accidents that have occurred to bodies that receive this impression. Therefore the impression along with what is in that impression – all of that is a ray. But the body that produces the impression is a body with three dimensions: length, width, and depth. Therefore the ray does not follow straight lines between which there would be intervals.[1]

Important in itself, this critique of the concept of ray in a sense prepares the fundamental step that Ibn al-Haytham will take approximately a century and a half later: the distinction between the light and the straight line along which it propagates. Al-Kindī, however, must still explain the diversity of perception according to the various regions of the cone. On this

[1] R. Rashed, *Œuvres philosophiques et scientifiques d'al-Kindī*, vol. I: *L'optique et la catoptrique*, Leiden, E. J. Brill, 1997, pp. 459–61. Transl. here by M. H. Shank.

occasion, he distances himself from the positions of both Euclid and Ptolemy by assuming that a visual cone issues from every point of the eye.

Al-Kindī next investigates the reflection of visual rays in various types of mirrors. He devotes an entire book to burning mirrors, which situates itself both in continuity with the ancient scientists and against them. In it, al-Kindī proposes to remedy the inadequacies of Anthemius of Tralles's work, which he completes. Did not the latter indeed take as an unquestionable truth the legend according to which Archimedes had set fire to the Roman fleet, without even demonstrating that it was possible? Did he not work on the construction of a mirror in which twenty-four rays are reflected towards a single point, without rigorously determining the distance from this point to the mirror? This is the task that al-Kindī proposes to take up. To this end, he studies five types of burning mirrors: the dihedral mirror, the conical mirror, the spherical-concave mirror, the catoptric system of twenty-five mirrors, the parabolic mirror.

One of the important results of this research on the propagation and focalisation of light is that, following al-Kindī, no scientist of any renown in optics will omit the study of burning mirrors from his research program. Such is at least the case for the two most important authors: Ibn Sahl and Ibn al-Haytham. This topic is no longer a separate speciality, as it had been in Antiquity, but is now a central chapter of optics. Moreover, as we shall see, this investigation will in fact lead in the 10th century to the inauguration of a new chapter – the geometrical theory of lenses – and, with Ibn Sahl, around 980, to anaclastics.

Before Ibn Sahl, catoptricians raised questions about the geometrical properties of mirrors, and the ignition that they produced at a given distance. This is in effect the problem that Diocles, Anthemius of Tralles, and al-Kindī all set themselves. Ibn Sahl immediately modifies the question by considering not only burning mirrors, but also burning instruments, that is, those that are susceptible of causing ignition not only by reflection, but also by refraction. Ibn Sahl then successively studies the following in relation to the distance of the source (finite or infinite) and the mode of ignition (reflection or refraction): the parabolic mirror, the ellipsoidal mirror, the plano-convex lens, and the biconvex lens. In each of these sections, he undertakes a theoretical study of the curve, then details a mechanical procedure to draw it. For the plano-convex lens, for example, he begins by studying the curve as a conic section, then proceeds to the continuous drawing of an arc of hyperbola, in order to take up the investigation of the plane tangent to the surface generated by the rotation of this arc around a fixed straight line, and finally to return to the laws of refraction. If one wants to understand Ibn Sahl's investigation of lenses, however, one must first determine the state of his knowledge of refraction.

While he was investigating the fifth book of Ptolemy's *Optics*, Ibn Sahl composed another treatise, still extant, on which Ibn al-Haytham would later comment. In this *Proof that the Celestial Sphere is Not of the Utmost Transparency*, Ibn Sahl applies to the study of refraction concepts that were already present in Ptolemy. In this inquiry, however, the concept of medium occupies an important place. Ibn Sahl shows that every medium, including the celestial sphere, is characterized by a certain opacity that defines it. But Ibn Sahl's genuine discovery is that, in his treatise on *Burning Instruments*, he *characterizes* the medium by a certain ratio. It is precisely this concept of a constant ratio characteristic of the medium that constitutes the master idea of his investigation of refraction in lenses.[2]

Ibn Sahl had conceived and constituted an area of inquiry on burning instruments, and on dioptrics in addition. But since he was forced to think of other conics besides the parabola and the ellipse (notably the hyperbola) as anaclastic curves, he was very naturally led to discover Snel's law. We can therefore understand that dioptrics, when it is born with Ibn Sahl, treats only what concerns the propagation of light, independently of problems of vision. The eye thus has no place in the midst of burning instruments, no more than the subject of vision. The point of view that he adopts when analyzing the phenomena of light is thus deliberately objective. Rich in technical material, this new discipline is in fact very poor in physical content: it is fleeting, and restricted to a few energetic considerations. For example, at least in the writings that have come down to us, Ibn Sahl never tried to explain why certain rays change direction or become concentrated when they enter a new medium: it was enough for him to know how a bundle of rays parallel to the axis of a hyperbolic plano-convex lens yields a convergent bundle by refraction. To the question of why the concentration of rays produces ignition, Ibn Sahl is content to answer by defining the luminous ray by its action of igniting by postulating – as many of his successors would do long thereafter – that heating is proportional to the number of rays.

With Ibn Sahl, we are on the eve of one of the first revolutions in optics, if not physics. Scarcely a generation after Ibn Sahl, Ibn al-Haytham begins his work.

Compared to the writings of the Greek and Arabic mathematicians that precede him, even Ibn al-Haytham's optical work at first glance reveals two striking characteristics: extension and reform. Upon closer examina-

[2] See R. Rashed, *Geometry and Dioptrics in Classical Islam*, London, al-Furqān, 2005, Chap. II.

tion, one notices that the first characteristic is a material remnant of the second. Indeed, no one before Ibn al-Haytham included in his research so many domains pertinent to so many diverse traditions: philosophical, mathematical, medical. The titles of his books nicely illustrate this wide spectrum: *The Light of the Moon*, *The Light of the Stars*, *The Halo and the Rainbow*, *Spherical Burning Mirrors*, *Parabolic Burning Mirrors*, *The Burning Sphere*, *The Form of the Eclipse*, *The Quality of Shadows*, *The Discourse on Light*, as well as his magisterial book on *Optics*, translated into Latin in the 12th century, and studied and commented on in Arabic and Latin until the 17th century. Ibn al-Haytham thus broached not only the traditional themes of optical research, but also many other new ones, eventually encompassing the following domains: optics, meteorological optics, catoptrics, burning mirrors, dioptrics, the burning sphere, physical optics.

A more detailed examination reveals that, in most of his work, Ibn al-Haytham pursues a program to reform the discipline, a program that leads him to take up in turn its various problems. The founding act of this reform consists, for the first time in the history of optics, in clearly distinguishing between the conditions for the propagation of light and the conditions for the vision of objects.[3] On the one hand, it led to providing a physical foundation for the rules of propagation: its basis is a mathematically supported analogy between the motion of light and a mechanical model of the motion of a solid ball hitting an obstacle.[4] On the other hand, it led to proceeding geometrically everywhere, by observation and experimentation. Optics no longer has its former meaning of a geometry of perception. Henceforth it includes two parts: a theory of vision, to which are also associated a physiology of the eye and a psychology of perception, and a theory of light, to which are linked a geometrical optics and a physical optics. The organization of his *Optics* already reflects this new circumstance. It includes chapters devoted entirely to propagation (the third chapter of Book I and Books IV to VII); others treat vision and problems relevant to it. Among other things, this reform leads to the emergence of new problems, such as the famous problem of Alhazen in catoptrics, the investigation of the spherical lens and of the spherical diopter, not only as burning instruments, but also as optical instruments, in dioptrics; and to experi-

[3] R. Rashed, 'Optique géométrique et doctrine optique chez Ibn al-Haytham', *A.H.E.S.*, 6, 4, 1970, pp. 271–98 (repr. in *Optique et mathématiques*, II), and *id.* 'Lumière et vision: L'application des mathématiques dans l'optique d'Ibn al-Haytham', in R. Taton (ed.), *Roemer et la vitesse de la lumière*, Paris, Vrin, 1978, pp. 19–44 (repr. in *Optique et mathématiques*, IV).

[4] R. Rashed, 'Optique géométrique et doctrine optique chez Ibn al-Haytham', pp. 281 ff.

mental control, both as a research practice and as a standard of proof in optics and in physics more generally.

Let us now follow the implementation of this reform in the *Optics* and in other treatises. The *Optics* opens with a rejection and a reformulation. At the outset Ibn al-Haytham rejects out of hand all variants of the doctrine of the visual ray (extramission) and takes sides with the philosophers who defend an intromissionist doctrine of the forms of visible things. Nevertheless, there is a fundamental difference between him and such philosophers as his contemporary Avicenna: Ibn al-Haytham does not consider the forms that the eye perceives as *totalities* that emanate from the visible thing under the effect of light. Rather, he treats them as reducible to their elements: from every point of the visible object, a ray emanates towards the eye. The latter has become a simple optical instrument without *visual spirit* (πνεῦμα ὀπτικόν). The crux of the problem thus became to explain how the eye perceives the visible thing by means of these rays emitted from every point of the visible thing.

After a short introductory chapter, Ibn al-Haytham devotes the second and third chapters of his *Optics* to the foundation of this new edifice. In the second, he determines the conditions of possibility for vision, whereas in the third he turns to the conditions of the possibility of light and its propagation. These conditions, which Ibn al-Haytham presents in each case as empirical notions, that is, as the result of a regulated observation or a controlled experiment, are so many constraints on the elaboration of the theory of vision, and therefore of his new style of optics. Ibn al-Haytham enumerates six conditions of vision: the visible thing must be self-luminous or illuminated by something else; it must face the eye, that is that one must be able to draw a straight line from each of its points to the eye; the medium that separates it from the eye must be transparent, without being blocked by any opaque obstacle; the visible thing/object must be more opaque than the medium; it must have a certain volume in relation to visual acuity.[5] Ibn al-Haytham writes that these are the concepts 'without which vision cannot take place'. One cannot avoid noticing that these conditions do not hark back, as they do in ancient optics, to those of light and its propagation. The most important of the latter, established by Ibn al-Haytham, are the following: light exists independently of vision and outside of it; it moves with a very great velocity and not instantaneously; its intensity diminishes insofar as it moves away from the source; the light from a luminous source (substantial) and that from an illuminated object (secondary or accidental) are propagated to the surrounding bodies, penetrate transparent media,

[5] Ibn al-Haytham, *Kitāb al-Manāẓir*: Books I–III (*On Direct Vision*), ed. A. I. Sabra, Kuwait, 1983, p. 189.

illuminate opaque bodies, which in turn emit light; light is propagated from every point of the luminous or illuminated object, along straight lines, in transparent media, and in all directions; these virtual straight lines along which light (substantial or accidental) is propagated form with the latter the 'rays'; these lines can be parallel or cross each other, the substantial and the accidental light do not mix in either of the two cases; reflected or refracted lights propagate along straight lines in specific directions. Note that none of these concepts refers to vision. Ibn al-Haytham complements them by other concepts pertaining to color. According to him, colors exist independently of light in opaque bodies. Consequently, only light emitted by these bodies (secondary or accidental light) accompanies the colors that are then propagated according to the same principles and the same laws as light. As we have explained elsewhere, it is this doctrine of colors that compelled Ibn al-Haytham to make concessions to thc philosophical tradition, forcing him retain the language of 'forms' already emptied of its content, when he was treating only light.

Henceforth, a theory of vision will have to meet not only the six conditions of vision, but also the conditions of light and of its propagation. Ibn al-Haytham devotes the rest of Book I of his *Optics* and the next two books to the elaboration of this theory, in which he takes up the physiology of the eye and a psychology of perception, both integral to his new intromissionist theory.

Three books of the *Optics* (IV to VI) treat catoptrics. Although this domain is as ancient as the discipline itself and Ptolemy had widely studied it in his *Optics*, it had never been the object of research as extensive as Ibn al-Haytham's. Beyond these three voluminous books of his *Optics*, Ibn al-Haytham devotes to catoptrics other works that complete it, when dealing with cognate problems such as that of burning mirrors. Among other characteristics, Ibn al-Haytham's catoptric research distinguishes itself, by the introduction of physical concepts, both to explain known concepts and to understand new phenomena. It is during this research that Ibn al-Haytham asks himself new questions, notably the problem that is now associated with his name.[6]

Let us now consider the several axes of Ibn al-Haytham's catoptric research. He presents again the law of reflection, explaining it by means of the previously mentioned mechanical model. He then studies this law for various mirrors: plane, spherical, cylindrical, and conical. In each case, he

[6] The famous 'problem of Ibn al-Haytham' was magisterially analyzed by M. Nazif, *Al-Ḥasan b. al-Haytham, Buḥūhuhu wa-kushūfuhu al-baṣariyya* (*Ibn al-Haytham, His Optical Researches and Discoveries*), 2 vols, Cairo, 1942–1943, pp. 487–521.

focuses above all on determining the plane tangent to the surface of the mirror at the point of incidence in order to determine the plane that is perpendicular to the latter and includes the incident ray, the reflected ray, and the normal to the point of incidence. Here, as in his other investigations, in order to verify his results experimentally, he invents and constructs an apparatus that is inspired by the one Ptolemy had built to study reflection, but more complex[7] and suitable for all cases. Ibn al-Haytham also studies the image of an object and its position for the various mirrors. He takes up an entire class of problems: the determination of the incidence of a given reflection for the various mirrors, and inversely. He also raises for these various mirrors the problem that is now associated with his name ('Alhazen's problem'): given any two points in front of a mirror, how to determine on its surface a point such that the straight line that joins this point to one of the two given points be the 'support' of the incident ray, whereas the straight line joining this point to the other given point is the 'support' of the reflected ray. Ibn al-Haytham solves this problem, which very soon becomes more complex.[8]

Ibn al-Haytham carries out this catoptric research in other treatises, some of which post-date the *Optics*, such as the *Spherical Burning Mirrors*.[9] It is in this highly interesting work that Ibn al-Haytham discovers longitudinal spherical aberration.

The seventh and last book of Ibn al-Haytham's *Optics* is devoted to dioptrics. As he had done for catoptrics, Ibn al-Haytham inserts into this book the elements of a physical (mechanical) explanation of refraction. Moreover this book is completed by such works as his treatise on the burning sphere, and even his *Discourse on Light*, in which, following up on Ibn Sahl, he returns to the concept of the medium.

In this seventh book of the *Optics*, Ibn al-Haytham begins by relying on two qualitative laws of refraction, and on several quantitative rules, all of which are experimentally controlled by means of an apparatus that he invents and builds, as in the preceding case. The two qualitative laws, which his predecessors Ptolemy and Ibn Sahl knew, can be stated as follows: (1) the incident ray, the normal to the point of refraction, and the

[7] *Ibid.*, pp. 685–90.

[8] This is 'Ibn al-Haytham's problem'; cf. note 6.

[9] 'Al-marāyā al-muḥriqa bi-al-dā'ira', fourth treatise in Ibn al-Haytham, *Majmū' al-rasā'il*, Hyderabad, 1938–1939. See also E. Wiedemann, 'Ibn al-Haythams Schrift über die sphärischen Hohlspiegel', *Bibliotheca Mathematica*, 3rd series, 10, 1909–10, pp. 393–407, and H.J.J. Winter and W. 'Arafat, 'A Discourse on the Concave Spherical Mirror by Ibn al-Haytham', *Journal of the Royal Asiatic Society of Bengal*, 3rd series, *Science*, 16, 1950, pp. 1–6.

refracted ray are all in the same plane; the refracted ray gets closer (or farther) from the normal if light passes from the less (or more) refracting medium to the more (or less) refracting one; (2) the principle of inverse return.

Instead of following the path that Ibn Sahl blazed by discovering Snel's law, however, Ibn al-Haytham returns to the ratios of angles and establishes his quantitative rules:

- the angles of deviation vary in direct proportion to the angles of incidence: if in the medium n_1, one takes $i' > i$, one will have $d' > d$ in medium n_2 (where i is the angle of incidence, r the angle of refraction, d the angle of deviation; $d = |i - r|$);
- if the angle of incidence increases by a certain quantity, the angle of deviation increases by a smaller quantity: if $i' > i$, $d' > d$, one then has $d' - d < i' - i$;
- the angle of refraction increases in proportion to the angle of incidence: if $i' > i$, one has $r' > r$;
- if light penetrates from a less refracting medium into a more refracting one ($n_1 < n_2$), one has $d < \frac{1}{2} i$; inverting the path, one has $d < \frac{i+d}{2}$, and one will be $2i > r$;
- Ibn al-Haytham takes up again the rules that Ibn Sahl had stated in his work on *The Celestial Sphere*; he affirms that, if light penetrates at the same angle of incidence from a medium n_1 into two different media n_2 and n_3, then the angle of deviation is different for each of these media, on account of the difference of opacity. If, for example, n_3 is more opaque than n_2, then the angle of deviation will be greater in n_3 than in n_2. Inversely, if n_1 is more opaque than n_2, and n_2 than n_3, the angle of deviation will be greater in n_2 than in n_3.

Contrary to what Ibn al-Haytham believed, these quantitative rules are not all generally valid.[10] But all are verifiable within the limits of the experimental conditions that Ibn al-Haytham effectively considered in his *Optics*: the media are air, water, and glass, with angles of incidence no greater than 80°.

[10] M. Nazif, *Al-Ḥasan b. al-Haytham, Buḥūhuhu wa-kushūfuhu al-baṣariyya*, pp. 720–3, and R. Rashed, 'Le discours de la lumière d'Ibn al-Haytham (Alhazen)', *Revue d'histoire des sciences*, 21, 1968, pp. 197–224, at pp. 201–4 (repr. in *Optique et mathématiques*, V).

Ibn al-Haytham devotes a substantial part of the seventh book to investigating the image of an object by refraction, notably when the surface separating the two media is either plane or spherical. It is in the course of this investigation that he pauses at the spherical diopter and the spherical lens, thus effectively continuing Ibn Sahl's research, but with one profound modification; his study of the diopter and of the lens appears in the chapter devoted to the problem of the image, and is not separated from the problem of vision. For the diopter, Ibn al-Haytham considers two examples, according to whether the source, which is a point located at a finite distance, is located on the concave or the convex side of the diopter's spherical surface.[11]

Next Ibn al-Haytham studies the spherical lense, devoting particular interest to the image it produces of an object. He restricts himself, however, to examining one single case, when the eye and the object are on the same diameter. In other words, he studies the image, through a spherical lens, of an object placed in a specific position on the diameter that passes through the eye. His procedure brings to mind that of Ibn Sahl when he was studying the hyperbolic bi-convex lens. Ibn al-Haytham considers separately two diopters, and applies the results he has previously obtained. During this investigation of the spherical lens, Ibn al-Haytham draws on the spherical aberration of a point at a finite distance for the case of the diopter in order to study the image of a segment that is a portion of the segment defined by the spherical aberration.

In his treatise on *The Burning Sphere*, one of the high points of research in classical optics, Ibn al-Haytham details and fine-tunes several results about the spherical lens that he had already obtained in the *Optics*. Moreover, he also returns here to the question of ignition by means of this lens. It is in this treatise that we encounter the first deliberate study of spherical aberration for parallel rays that are incident on a glass sphere and undergo two refractions. During this investigation, Ibn al-Haytham uses the numerical data from Ptolemy's *Optics* for the two angles of incidence of 40° and 50°. To explain this phenomenon of the focalization of light propagated along trajectories parallel to the diameter of the sphere, he returns to the angular values instead of applying the law now known as Snel's Law.[12]

In this treatise on *The Burning Sphere*, just as in the seventh book of his *Optics* or in others of his dioptric writings, Ibn al-Haytham presents his research in a rather paradoxical fashion. Whereas he takes great care to invent, set up, and describe experimental arrangements that are very sophisticated for their day, and that allow him to determine numerical val-

[11] R. Rashed, *Geometry and Dioptrics in Classical Islam*, Chap. III, esp. pp. 162 ff.
[12] *Ibid*., pp. 170 ff.

ues, in most cases he avoids giving these values. When he does so, as in the treatise on *The Burning Sphere*, it is with parsimony and circumspection. To account for this attitude (already noted above), at least two possible explanations stand out. The first pertains to the style of scientific practice itself: quantitative description seems not yet to be a normative requirement. The second is probably linked to the latter: the experimental set-ups could only yield approximate values. It is probably for this reason that Ibn al-Haytham used the values that he borrowed from Ptolemy's *Optics*.

Experimentation, however, was already a category of proof in optics. The meaning of this discipline changed, becoming a science of luminous phenomena. It was no longer a psychological discipline, but a physical one. Neither reflection nor refraction could be considered exclusively as causes of error. In the optics of the ancients, which is an optics of the visual ray, the image is effectively a mirage: in the absence of the onlooker who is gazing, it has no objective existence, no reason for being. In Ibn al-Haytham, by contrast, the image acquires an objective status. This physical and material anchorage of optics, with the epistemological conditions for bringing it about, will be consolidated by Ibn al-Haytham's Arabic and Latin successors, and especially his 17th century readers – above all, Kepler, Descartes, and Huygens.

CONCLUSION

THE PHILOSOPHY OF MATHEMATICS

Historians of Islamic philosophy take a particular interest in what some people occasionally see fit to call *falsafa*. In their view, it is one of these doctrines of Being and the Soul that authors from Islamic culture developed, a doctrine indifferent to other kinds of knowledge and independent of all determination except for its bond with religion. On this account, these philosophers would thus be in the Aristotelian tradition of Neoplatonism, heirs of late antiquity with an Islamic hue. This historical bias insures, in appearance at least, a smooth transition from Aristotle, Plotinus, and Proclus, among others, to the philosophers of Islam from the 9th century onward. But the price of this move is high: it often (though not always) results in a pale and impoverished picture of philosophical activity and transforms the historian into an archaeologist, albeit one deprived of the latter's tools. Indeed, it is not unusual for such an historian to set his main task as the excavation of the site of Islamic philosophy, in search of the vestiges of Greek works lost in their original language but preserved in Arabic translation; or failing that, to be content with the shards of the ancient philosophers' writings, which historians of Greek philosophy have often studied with skill and competence.

To be sure, some historians have recently turned to doctrines elaborated in other fields, at the edge of the Greek heritage's wake: the philosophy of law, which jurists developed brilliantly; the profound and refined philosophy of *kalām*, *i.e.*, that of the theologian-philosophers; the Sufism of such great masters as al-Ḥallāj and Ibn ʿArabī, etc. Such studies have enriched and improved the picture, and they reflect more faithfully the philosophical activity of the time. They also enable us better to understand the place of the Greek heritage in Islamic philosophy.

But the sciences and mathematics have not yet drawn the same appreciation as law, *kalām*, linguistics, or Sufism. Even today, the essential links between sciences and philosophy, and notably between mathematics and philosophy, are left in the lurch. To be sure, the links between mathematics and philosophy are sometimes tackled in the works of the philosophers of Islam such as al-Kindī, al-Fārābī, Ibn Sīnā, etc., but in a manner that in effect is completely external. By this, I mean that, one reports their views on the links between the two domains; one attempts to connect these views to Platonic or Aristotelian doctrines; and one checks for the presumed influence of the neo-Pythagoreans. In other words, one never tries to understand the repercussions of their mathematical knowledge on their philosophies, nor even the impact, on their philosophical teachings, of their activities as scientists, which is what they were in the vast majority of cases. This lacuna is not the fault of historians of philosophy alone; historians of sciences are also responsible for it. Admittedly, to examine the links between the sciences and philosophy, one needs a particularly wide range of competencies – a linguistic knowledge much more refined than that required to work in geometry, which is syntactically elementary and lexically poor; and a knowledge of the history of philosophy itself. If to the shortfall on these prerequisites, one adds a conception of the relations between science and philosophy inherited from the ambient positivism, it is easier to understand the deep indifference of the historians of science in this domain. One should not have to repeat that the links between the sciences and philosophy form an integral part of the history of science.

Truth be told, the situation is rather paradoxical. For seven centuries, scientific and mathematical research of the most advanced kind was elaborated in the Arabic language and in the urban centers of Islam. Is it likely that philosophers, who themselves were sometimes mathematicians, physicians, etc., should have carried out their philosophical activity as recluses? Is it probable that they were indifferent to the transformations taking place under their eyes, and remained blind to the steady stream of scientific results? The list is impressive: an astronomy critical of Ptolemaic models, the reform and renewal of optics, the creation of algebra, the invention of algebraic geometry, the transformation of Diophantine analysis, the discussion of the theory of parallels, the development of projective methods, etc. In the face of such an unprecedented flowering of disciplines and successes, how can one imagine that the philosophers might have been so insensitive as to remain confined to the relatively narrow framework of the Aristotelian tradition of Neoplatonism? The apparent poverty of classical Islam's philosophy perhaps owes more to historians than to history.

And yet, to examine the links between philosophy and science or philosophy and mathematics (our limited focus here) as they transpire only in the philosophers' works, is to go down the road only one third of the way. It is also necessary to question the mathematician-philosophers and the mathematicians. But the decision to consider only mathematics demands first a justification, which is all the more necessary because this move is in no way the specific prerogative of Islamic philosophy.

No scientific discipline has contributed more to the genesis of theoretical philosophy than mathematics. None has cultivated more numerous or more ancient links with philosophy. Since antiquity, mathematics has continuously offered philosophers food for thought about the central themes of the discipline; it has supplied them with methods of exposition, argumentative procedures, sometimes even the tools appropriate to their analyses. Finally, mathematics presents itself to the philosopher as an object of study: he indeed applies himself to clarifying mathematical knowledge itself by studying its object, its methods, and by probing into the characteristics that make it apodictic. From the entire span of the history of philosophy, philosophers have constantly raised questions about the conditions of mathematical knowledge, its origins, its power of extension, the nature of the certainty it reaches, and its place among other kinds of knowledge. The philosophers of classical Islam were no exception: al-Kindī, al-Fārābī, Ibn Bājja, Ibn Sīnā, and Maimonides, also did so, among many others.

Although they are less obvious, other ties have linked mathematics and theoretical philosophy. Often they have collaborated in order to hammer out a method, even a logic, such as the encounter between Aristotle and Eudoxus about the 'axiomatic method', or al-Ṭūsī's appeal to combinatorial analysis to solve the philosophical problem of emanation from the One. Of all the forms that this link may take, however, one is particularly notable; this time, it was not philosophers, but mathematicians who created it. We refer to the doctrines that mathematicians have elaborated to justify their own practices. The conditions most propitious to these theoretical constructions typically come together in the following circumstances: when a mathematician in the forefront of contemporary research confronts an insurmountable obstacle owing to the unsuitability of available mathematical techniques for handling the newly emerging mathematical objects. Consider, for example, the several variants of the theory of parallels, notably from the time of Thābit ibn Qurra (d. 901), or the kind of *analysis situs* that Ibn al-Haytham conceived, or the doctrine of indivisibles in the 17th century.

In the main, the relations between theoretical philosophy and mathematics are established in four types of works, whose authors are: philosophers; philosopher-mathematicians such as al-Kindī, Muḥammad ibn al-Haytham (not to be mistaken for al-Ḥasan ibn al-Haytham[1]); mathematician-philosophers such as Naṣīr al-Dīn al-Ṭūsī, etc.; and mathematicians such as Thābit ibn Qurra, his grandson Ibrāhīm ibn Sinān, al-Qūhī, Ibn al-Haytham, etc. By restricting oneself to only one category or the other while examining the relations between philosophy and mathematics, one inevitably loses a dimension that is essential to this domain.

On several occasions, we have simply taken a few core samples in order to reveal the richness of the field – much more a sounding than a systematic examination of the domain.

Indeed, such a project deserves a big volume that still remains to be written. That said, the road that seemed best adapted to the task goes beyond the straightforward exposition of the views that philosophers presented about mathematics and their importance. Rather, it seeks out the themes that they broached, the intimate links that united mathematics to philosophy, and their role in the scaffolding of doctrines and systems – that is, the organisational role of mathematics. For philosopher-mathematicians, we will show among other things how they proceeded to solve philosophical problems mathematically, a fruitful approach that generated new doctrines, and even new disciplines. For mathematicians, we will highlight their attempts to solve mathematical problems philosophically, and we shall see that this approach is both necessary and profound.

To shed more light on these different situations, I deal with the following topics in succession:

1. Mathematics as conditions and sources of models for philosophical activity. Of the many philosophers who could illustrate this theme, we have selected but two: a philosopher-mathematician (al-Kindī), and a philosopher who, without being a mathematician, nevertheless was knowledgeable in mathematics (Maimonides).

2. Mathematics within philosophical synthesis. It is within the first known synthesis, that of Ibn Sīnā, that mathematics entered directly into philosophical work. Not the least result of this is the 'formal' inflection of ontology, which made possible the mathematical treatment of a

[1] See R. Rashed, *Les Mathématiques infinitésimales du IX*^{e} *au XI*^{e} *siècle*, vol. II: *Ibn al-Haytham*, London, al-Furqān, 1993, pp. 8–19; 2000, vol. III, pp. 937–41; English translation: *Ibn al-Haytham and Analytical Mathematics*. A History of Arabic Sciences and Mathematics, vol. 2, Culture and Civilization in the Middle East, London, Centre for Arab Unity Studies, Routledge, 2012, pp. 11–26, 364–81; and *Ibn al-Haytham's Theory of Conics, Geometrical Constructions and Practical Geometry*. A History of Arabic Sciences and Mathematics, vol. 3, Culture and Civilization in the Middle East, London, Centre for Arab Unity Studies, Routledge, 2013.

philosophical problem. In this rubric, naturally we consider here Ibn Sīnā, a philosopher well read in mathematics, whose contribution was continued by the mathematician Naṣīr al-Dīn al-Ṭūsī.

3. The third theme was mainly cultivated by mathematicians dealing with the problem of mathematical discovery. This concerns the *ars inveniendi* and the *ars analytica* with Thābit ibn Qurra, Ibrāhīm ibn Sinān, al-Sijzī, and Ibn al-Haytham.

Note that the point at issue in these chapters is not individual works, but a genuine tradition characterized by names and titles and extending over several centuries.

1. MATHEMATICS AS CONDITIONS AND MODELS OF PHILOSOPHICAL ACTIVITY: AL KINDĪ AND MAIMONIDES

The links between philosophy and mathematics are essential to the reconstitution of al-Kindī's system (9th century). It is precisely this dependence that the philosopher notices when he writes a book entitled *Philosophy can only be Acquired through Mathematical Discipline*[2] and when, in his epistle on *The Quantity of Aristotle's Books*,[3] he presents mathematics as propaedeutic to the teaching of philosophy. He even goes so far as to confront the philosophy student, warning him that he faces the following alternative: either to begin with the study of mathematics before tackling Aristotle's books in the order that al-Kindī specifies – only then can he hope to become a true philosopher; or to do without mathematics, in which case he becomes a mere reciter of philosophy, assuming he can memorize by heart. After listing the different categories of Aristotle's books, al-Kindī writes:

> This is the number of his books that we have already mentioned and that a perfect philosopher must know after mathematics, that is, the [areas of] mathematics I have defined by name. For if someone lacks knowledge of mathematics, *i.e.* arithmetic, geometry, astronomy and music, in order to use these books [of Aristotle] for the rest of his life, he will not be able to perfect his knowledge of them, and all his efforts will only allow him to master the <ability> to repeat them, if he can remember by heart. As for deep knowledge of them and the way to acquire it, these are completely non-existent without knowledge of mathematics.[4]

[2] Al-Nadīm, *Kitāb al-Fihrist*, ed. R. Tajaddud, Teheran, 1971, p. 316.

[3] Al-Kindī, *Rasā'il al-Kindī al-falsafiyya*, ed. M. ʿA. Abū Rīda, 2 vols, Cairo, 1369/1950, vol. I, pp. 363–84.

[4] *Ibid.*, pp. 369–70.

For al-Kindī, then, mathematics is the very foundation of philosophical instruction. Although it is not our purpose here, by exploring more deeply the role of mathematics in al-Kindī's philosophy, one could understand more rigorously the specificity of his work. Indeed, historians often illuminate it from two rather different angles. According to the first interpretation, al-Kindī appears as a Muslim representative of the Aristotelian tradition of Neoplatonism, in effect a late-antique philosopher twice over. The second interpretation sees him as a follower of philosophical theology (*kalām*), a theologian who effectively switched languages in order to speak that of Greek philosophy. But if we restore to mathematics the role that it actually played in the elaboration of his philosophy, the fundamental options that al-Kindī faced will appear before our very eyes. According to the first, which originates in his Muslim convictions, and is expressed and articulated in the tradition of philosophical theology, notably the doctrine of al-Tawḥīd (God's unity), revelation gives us the truth, which is one and rational. The second refers us to Euclid's *Elements* as a method and model: whereas the rational can be reached very concisely and almost instantaneously by revelation, it can also be reached by a collective and cumulative effort of philosophers, starting from the truths of reason, which are independent of revelation and must correspond to the criteria of geometrical proof. At the time of al-Kindī, the Aristotelian tradition of Neoplatonism furnished these truths of reason, which served as primitive notions and postulates. In philosophical theology, they were chosen to replace the truths that revelation offers, insofar as they met the requirements of geometric thought and made possible an exposition with an axiomatic appearance. It was then that the 'mathematical examination (*al-faḥṣ al-riyāḍī*)' became the instrument of metaphysics.

That is in fact the case for the epistles in theoretical philosophy, like *First Philosophy*, the *Epistle for Explaining the Finitude of the Body of the World*, etc.[5] To take the latter text as an example, al-Kindī methodically proceeds to prove the inconsistency of the concept of infinite body. He begins by defining the primitive terms *magnitude* and *homogeneous magnitudes*. He then introduces what he calls 'the certain proposition (*qaḍiya ḥaqq*)'[6] or, as he explains elsewhere, 'the first true and immediately intelligible premises (*al-muqaddimāt al-uwal al-ḥaqiyya al-maʿqūla bi-lā tawassuṭ*)',[7] or else 'the first, evident, true and immediately intelligible

[5] R. Rashed and J. Jolivet, *Œuvres philosophiques et scientifiques d'al-Kindī*. Vol. II: *Métaphysique et cosmologie*, Leiden, E. J. Brill, 1998.

[6] *Ibid.*, p. 161, *l.* 16.

[7] *Ibid.*, *Philosophie Première*, p. 29, *l.* 8.

premises',[8] *i.e.*, tautological propositions. These are expressed in terms of primitive notions, of order relations on them, of union and separation operations on them, of finite and infinite predications. The following statements illustrate such propositions: homogeneous magnitudes that are no bigger than each other are equal; or, if, to one of the equal homogeneous magnitudes, one adds a magnitude that is homogeneous with it, then they become unequal.[9] Finally, al-Kindī proceeds by *reductio ad absurdum*, by adopting a hypothesis: the part of an infinite magnitude is necessarily finite.

This is the path that al-Kindī follows in his other writings. Again following his *First Philosophy*, he uses *more geometrico* in his epistle *On the Quiddity of What Cannot be Infinite and of What is Called Infinite*, in which he seeks to demonstrate the impossibility that the world and time are infinite. Here too, al-Kindī begins by stating four premises: 1° 'Of anything from which some thing is taken away, what remains is smaller than what was before the subtraction was carried out'; 2° 'From anything, if something is taken away, if what is taken away is put back to the former, it returns to the original quantity'; 3° 'For all finite things, if they are put together, a finite thing is obtained'; 4° 'If there are two things such that one is smaller than the other, then the smaller measures the greater or measures a part of it, and if it entirely measures it, then it measures a part of it'.[10] From these premises directly inspired by Euclid's *Elements*, al-Kindī intends to establish his philosophical proposition. Assuming an infinite body from which some finite thing is taken away, he asks whether the remainder is finite or infinite. He then shows that both hypotheses lead to contradictions, and he concludes that an infinite body cannot exist. He goes on to show that the same thing goes for the accidents of the body, notably time; but time, motion, and the body mutually imply one another. He then shows that there is no infinite time *a parte ante* and that neither the body, nor motion, nor time is eternal. Therefore there is no eternal thing; the infinite is only potential, as is the case for number. However brief, these examples show how al-Kindī articulated at once mathematical principles and means, and philosophy, according to the Aristotelian tradition of Neoplatonism. Note however that al-Kindī the philosopher was also a mathematician as his works in optics[11] and mathematics[12] attest. In philosophy, he was also

[8] *Ibid.*, *On the Unicity of God*, p. 139, *l.* 1.

[9] *Ibid.*, *Epistle for Explaining the Finitude of the Body of the World*, p. 160.

[10] *Ibid.*, *On the Quiddity of What Cannot be Infinite*, p. 150.

[11] R. Rashed, *Œuvres philosophiques et scientifiques d'al-Kindī*. Vol. I: *L'optique et la catoptrique d'al-Kindī*, Leiden, E.J. Brill, 1996.

[12] R. Rashed, 'Al-Kindī's commentary on Archimedes' *The Measurement of the Circle*', *Arabic Sciences and Philosophy*, 3.1, 1993, pp. 7–53.

familiar not only with the accounts of Aristotle and of the Aristotelian and Neoplatonist tradition, but also with the commentaries of such Aristotelians as Alexander of Aphrodisias.

As to Maimonides (1135–1204), without being mathematically productive like al-Kindī, he was nevertheless informed about the subject. The philosopher obviously has enough knowledge of mathematics to try, pen in hand, to read, perhaps even to teach and to comment upon, mathematical works like Apollonius's *Conics*, *i.e.* the highest level works at the time. But his commentary never pertains to the fundamental ideas, to the properties closely studied in this work; he is interested only in the elementary techniques of proof, which for the most part were taught in the first six books of Euclid's *Elements*. To put it briefly and clearly, his commentary was simply not on the same level as the works upon which he commented. Why then did Maimonides spend so much time and energy for such a meagre result? Using Maimonides' own words, we can certainly invoke the role of mathematics in training the mind (*tarwīḍ al-dhihn*) to reach human perfection.[13] But there is more: at issue are the other relations between mathematics and philosophy. We restrict ourselves to the most important ones.

Maimonides' starting point, it must emphasized, is dogma, not philosophy: 'to elucidate, the difficulties of dogma (*mushkilāt al-sharīʿa*)', he says, 'and to make plain its hidden truths, which are far above the comprehension of the multitude'.[14] This has been one of the major tasks of philosophy since al-Kindī (see his treatise on the number of Aristotle's books), which consists in reaching the truth passed on by the Scriptures through reason and philosophical speculation. And to accomplish this task, or even simply to begin it, a perfect accord had to be assumed between the two kinds of truth, that of the Scriptures and that of reason and philosophy. This 'concordance' rests on a principle that Ibn Rushd (1126–1198) formulated as follows: 'a truth does not contradict a truth but accords with it and bears witness to it'.[15] In this respect, the means that Maimonides chose are the same as those that his predecessors used: 'the method of demonstration about which there can be no doubt (*al-ṭarīq alladhī lā rayba*

[13] Maimonides, *Dalālat al-Ḥāʾirīn (The Guide of the Perplexed)*, ed. H. Atay, Ankara University, Ilâhîyat Fakültesî Yayinlari 93, Ankara, 1972; repr. Cairo, n.d., p. 80; *The Guide of the Perplexed*, English transl. S. Pines, Chicago, University of Chicago Press, 1963, p. 75.

[14] Maimonides, *Dalālat al-Ḥāʾirīn*, ed. Atay, p. 282.

[15] *Manāhij al-adilla*, p. 32.

fīhi',[16] *i.e.* to establish by 'true demonstration (*al-burhān al-ḥaqīqī*)' the truths of dogma: the existence of God, His unity and His incorporeality. For these philosophers, however, such a demonstration could only follow a mathematical pattern. For this to be so, however, it was necessary to use a language other than that of Revelation, a language whose concepts, defined by reason alone, are endowed with a certain ontological neutrality.

'True demonstration', *i.e.* according to the model of mathematics, is the necessary way for the truths of Revelation also to attain the status of truths of reason, which is not in any way specific to a particular religion, whether revealed or not. Such is the first relation between mathematics and philosophy. As we shall see, however, these relations occur at different levels. First of all, the general approach of Maimonides consists in borrowing concepts from the Aristotelian philosophy of his predecessors, and proof and exposition techniques from mathematics; this is effectively the approach that Maimonides has used, for example, in the greater part of Book II of the *Guide*. His method thus follows that of the geometers, to whom he owes some of the techniques of proof – mainly the *reductio ad absurdum* – he uses to ground each element of his exposition. In the *Guide*, there are twenty-five such elements – twenty-five lemmas, most of which are presented as statements without proof, since Maimonides believed that his predecessors had rigorously demonstrated them. To these lemmas, he adds one postulate, and from these twenty-six propositions he deduces his 'principal theorem': GOD EXISTS, HE IS UNIQUE, AND HE IS NEITHER A BODY NOR IN A BODY. The importance of this passage derives less from the strength of the proof than from its deliberate use of a geometrical mode of exposition in metaphysics. Ever since Aristotle, the first lemmas themselves have been susceptible to logico-mathematical treatment, an approach revived by al-Kindī, and taken up by many a subsequent metaphysician, such as Ibn Zakariyā al-Rāzī, Abū al-Barakāt al-Baghdādī (d. *c.* 1164), Fakhr al-Dīn al-Rāzī (1150–1210), Naṣīr al-Dīn al-Ṭūsī (1201–1274), among others. Finally, these lemmas reappear together in commentaries on the *Guide* by al-Tabrīzī and later by Hasdai Crescas (1340–*c.* 1412). At issue is the impossibility of an infinite magnitude, and the impossibility of an infinite number of coexisting magnitudes. The third lemma states the impossibility of an infinite chain of causes and effects, whether material or not, thus banning in advance an infinite regress of causes. Three propositions follow the three lemmas. The first deals with change; four categories are subject to change: substance, quantity, quality, and place. The second concerns motion: every motion is a change and a transition from

[16] Maimonides, *Dalālat al-Ḥā'irīn*, ed. Atay, p. 187; *Guide*, Pines transl. p. 180.

potentiality to actuality. The third proposition enumerates the species of motion. The seventh lemma states: 'Everything changeable is divisible. Hence everything movable is divisible, and is necessarily a body; but everything that is indivisible is not movable; hence it will not be a body at all'.[17] The eighth lemma asserts that: 'Everything that is moved owing to accident must of necessity come to rest'.[18] The ninth, that 'every body that moves another body moves the latter only through being itself in motion when moving the other body'.[19] This is how the statement of the preliminary propositions proceeds, the fourteenth of which posits that locomotion precedes all other motions, and the twenty-fifth that 'the principles of an individual compound substance are matter and form'.

These twenty-five lemmas, a few of which we have just cited, all come from Aristotelian philosophy. They are nevertheless not homogeneous: their origins, as well as their logical complexity, distinguish them. Maimonides certainly does not ignore this heterogeneity, since he generally gives us his sources in bulk: 'the *Physics* and the commentaries on it', and 'the *Metaphysics* and the commentary on it'. The books of the *Physics* and *Metaphysics* are easy to identify: the third and eighth book of *Physics* and the tenth and eleventh of *Metaphysics*. But it is a completely different matter, which we will not take up here, when it comes to situating precisely the commentaries on the *Physics* and the commentary on the *Metaphysics*. Maimonides describes the logical complexity of the lemmas as follows: 'some <lemmas> become manifest with very little reflection and are demonstrative premises and first intelligibles or notions approaching the latter' and 'others require a number of demonstrations and premises leading up to them. However, all of them have been given demonstrations as to which no doubt is possible'.[20] In other words, there are lemmas which are so close to axioms that they become self-evident 'with very little reflection (*al-ta'ammul al-aysar*)'; and others that are so far from being axioms that proving them requires several intermediate propositions, a task which has been carried out by Aristotle, his commentators, and his successors. The twenty-five lemmas of the system belong to one of these two types.

Maimonides is fully aware that, to deserve its name, a proof must be both universal and compelling. But this is not the case for the question discussed here, if one takes into account the irreducible opposition between the two truths, revealed and philosophical, concerning the eternity of the

[17] *Dalālat al-Ḥā'irīn*, ed. Atay, p. 249; *Guide*, Pines transl. p. 236.
[18] *Dalālat al-Ḥā'irīn*, ed. Atay, p. 251; *Guide*, Pines transl. p. 236.
[19] *Dalālat al-Ḥā'irīn*, ed. Atay, p. 252; *Guide*, Pines transl. p. 236.
[20] *Dalālat al-Ḥā'irin*, ed. Atay, p. 272; *Guide*, Pines transl. p. 239, substituting 'lemma' for Pines's 'premise'.

world. For the proof to be similar to a mathematical proof, *i.e.*, truly apodictic, it should be always valid, whether one believes in the eternity of the world or not. It is therefore as a mathematician, so to speak, and against his own conviction that Maimonides introduces the eternity of the world into the system as a postulate, raising the number of the preliminary propositions to twenty-six. In this connection, he states without the slightest ambiguity:

> I shall add to the <lemmas> mentioned before, one further <lemma> that affirms as necessary the eternity of the world. Aristotle deemed it to be correct, and the most fitting to be believed; we therefore grant it conventionally (*'alā jihat al-taqrīr*) in order to show what we wanted to demonstrate.[21]

Maimonides thus introduces the eternity of the world as a postulate that is necessary for the completion of the system, and hence for the deduction of his 'theorem'. This conventional but non-arbitrary aspect of the proposition shines with its full brilliance when one knows that Maimonides does not believe in the eternity of the world. Here is what he has to say on this issue, for example:

> For according to me the correct way, which is the method of demonstration about which there can be no doubt, is to establish the existence and the oneness of the deity and the negation of corporeality through the methods of the philosophers, which methods are founded upon the doctrine of the eternity of the world. This is not because I believe in the eternity of the world, or because I concede this point to the philosophers; but because it is through this method that the demonstration becomes valid and perfect certainty is obtained with regard to these three things: I mean the existence of the deity, His oneness, and His not being a body – and all this without making a judgment upon the world's being eternal or created in time.[22]

In fact, Maimonides knew that the problem of the eternity of the universe cannot have a positive solution; some will later say that dialectical reason comes up against an antinomy, since one would have to determine the properties of things that do not yet exist.

Maimonides surely conceived the architectonics of this part of the *Guide* in the manner of a mathematical exposition, following a geometrical order. This order appears as a condition for the certainty of metaphysical

[21] *Dalālat al-Ḥā'irīn*, ed. Atay, p. 272. Pines translates the last sentence of this quotation as follows: 'We shall grant him this premise by way of a hypothesis [here Pines reads *'alā jihat al-taqdīr* 'in conformity with Ibn Tibbon's Hebrew translation'] in order that the clarification of that which we intended to make clear should be achieved' (p. 239).

[22] *Dalālat al-Ḥā'irīn*, ed. Atay, p. 187.

knowledge, namely that of God, of His existence, of His unity, and of His incorporeality. This seminal idea, already present in al-Kindī, will later reappear in Spinoza. But, as Crescas had pointed out, the major remaining problem is to know if these twenty-five propositions have effectively been demonstrated; and, whether the 'theorem' can truly be deduced from them. These two questions will continue to haunt the successors of Maimonides. Thus commentaries by al-Tabrīzī and Crescas both seem to demonstrate these propositions. Maimonides himself makes an attempt at this deduction, which we can present only in the barest outline, but with an emphasis on the spirit in which he undertook it.

According to the twenty-fifth lemma, to exist, every compound individual substance needs a mover, which suitably prepares the matter and enables it to receive the form. According to the fourteenth lemma, however, there necessarily exists another mover, which can be of a different species, prior to the latter mover. According to the third lemma, this chain of movers/mobiles is necessarily finite: the motion thus ends at the celestial sphere. The latter is endowed with locomotion, since this motion precedes every other motion for the four categories of change, according to the fourteenth lemma. But according to the seventeenth lemma, everything that moves necessarily has a mover, therefore the celestial sphere necessarily has a mover. And this mover is either outside the mobile or inside it. This is a necessary division. If the mover is outside the mobile, it is either outside the celestial sphere, or else it is not in a body; in the latter case, the mover is said to be 'separate' from the celestial sphere. If the mover is inside the latter, it must be either a force distributed throughout, or an indivisible force, like the soul in a human. One thus confronts four possibilities, three of which Maimonides will reject as impossible by means of different lemmas. The last remaining possibility is that of a non-body outside the celestial sphere, separated from it, which moves it with locomotion in space. Maimonides concludes his long chain of reasoning with the following words:

> It accordingly has been demonstrated (*faqad tabarhana*) that it is necessary that the mover of the first sphere, if the movement of the latter is regarded as eternal and perpetual, should not at all be a body or a force in the body; in this way the mover of this sphere would have no movement, either according to essence or to accident, and would not be subject to division or to change, as has been mentioned in the fifth and the seventh of the <lemmas>. Now this is the deity, may His name be sublime; I am referring to the first cause

> moving the sphere, and it is absurd that there should be two or more of them… [QED].[23]

We have just seen that, for Maimonides, there are three senses in which mathematics can be considered a condition of metaphysical knowledge. Most immediately, mathematics is an exercise of the mind. Secondly, it provides a model of construction – an architectonics – that allows one to reach certainty. Finally, it offers procedures for demonstration, notably the method of *reductio ad absurdum*. But these relations between mathematics and metaphysics are not the only ones that appear in the *Guide*. We recently drew attention to another relation that is no less important: mathematics can play the role of means of argumentation in metaphysics. The most famous and most pertinent example comes from Apollonius's *Conics*: the asymptote to an equilateral hyperbola allows one to reflect on the problem of the relations between imagining and conceiving. In his treatment of *kalām*, Maimonides indeed intends to refute the thesis according to which: 'for reason, everything that can be imagined is possible'. To do so, he wants to prove the negation of this thesis: some things exist that one cannot imagine, that is, one cannot in any way represent them to the imagination, but one can establish their existence by demonstration. In other words, for Maimonides, there is no principle that allows one to pass from imagination to metaphysical reality. He formulates his thesis thus:

> Know that there are things that a man, if he considers them with his imagination, is unable to represent to himself in any respect, but finds that it is as impossible to imagine them as it is impossible for two contraries to agree; and that afterwards the existence of the thing that is impossible to imagine is established by demonstration as true and existence manifests it as real.[24]

With these words, as we have shown elsewhere,[25] Maimonides gives a new inflection to the problem of demonstrating what one cannot conceive, a problem that the mathematician al-Sijzī had already raised in the 10th century. To illustrate this question, Maimonides draws on the very same example that his predecessor had discussed. Proposition II.14 of Apollonius's *Conics* concerns asymptotes to an equilateral hyperbola: the

[23] *Dalālat al-Ḥā'irīn*, p. 276; *Guide*, Pines transl., p. 246, substituting 'lemma' for Pines's 'premise'.

[24] *Dalālat al-Ḥā'irīn*, p. 214; *Guide*, Pines transl., p. 210.

[25] R. Rashed, 'Al-Sijzī et Maïmonide: commentaire mathématique et philosophique de la proposition II–14 des *Coniques* d'Apollonius', *Archives internationales d'histoire des sciences*, no. 119, vol. 37, 1987, pp. 263–96. English transl, 'Conceivability, Imaginability and Provability in Demonstrative Reasoning: al-Sijzī and Maimonides on II.14 of Apollonius' *Conics Sections*', *Fundamenta Scientiae*, vol. 8, no. 3/4, 1987, pp. 241–56.

curve and its asymptotes can always get nearer each other as long as they are prolonged indefinitely, and yet they never meet. Maimonides writes:

> This cannot be imagined and can in no way enter within the net of the imagination. Of these two lines, one is straight and the other curved, as has been made clear there in the above-mentioned book. Accordingly, it has been demonstrated that something that the imagination cannot imagine or apprehend and that is impossible from its point of view, can exist.[26]

The imagination to which Maimonides appeals here is the mathematical imagination: even for it, nothing guarantees the passage to metaphysical reality. But without risk of contradiction, one can assert that what is true for the mathematical imagination is *a fortiori* also true for all other forms of this faculty. Appealing to this proposition from the *Conics* seems, in Maimonides' mind, to have much greater force than citing a mere example: it is a form of argumentation that the metaphysician borrows from mathematics.

In conclusion, just like his predecessors since al-Kindī, Maimonides has found in mathematics both a model for architectonics, demonstration procedures, and means of argumentation. For him, therefore, the role of mathematics in no way reduces to that of propaedeutic to the teaching of philosophy. We now understand that when Maimonides devoted time and energy to the acquisition of even a modest amount of mathematical knowledge, it was because he thought of this, as his predecessors had, as a profoundly philosophical task: that of solving metaphysical problems mathematically.

2. MATHEMATICS IN THE PHILOSOPHICAL SYNTHESIS AND THE 'FORMAL' INFLECTION OF THE ONTOLOGY: IBN SĪNĀ AND NAṢĪR AL-DĪN AL-ṬŪSĪ

In his monumental *al-Shifā'*, as well as in his book *al-Najāt* and his *Danish-Nameh*, Ibn Sīnā (980–1037) gives a particularly prominent place to the mathematical sciences. In the *Shifā'* alone, he devotes no fewer than four books to them. To this, one should also add several independent writings in astronomy and music. It has not been adequately understood that, in all of these writings, the presence of mathematics has two concurrent significations. We have seen that al-Kindī is interested in mathematics from a double point of view, as a philosopher but also as a mathematician. Thus when he treats burning mirrors, optics, sundials,

[26] *Dalālat al-Ḥā'irīn*, ed. Atay, p. 215; *Guide*, transl. Pines, pp. 210–11.

astronomy, and when he comments on Archimedes, he does so as a mathematician. For him as a philosopher, mathematics is also a source of inspiration and a model of argumentation. Al-Kindī's approach outlived him in the writings of Muḥammad ibn al-Haytham. Ibn Sīnā belongs only in part to this tradition. His mathematical knowledge is wide-ranging even as it remains classical. He probably knew the writings of Euclid, Nicomachus of Gerasa, and Thābit ibn Qurra on the amicable numbers. But he was also familiar with elementary algebra, number theory, and some works in Diophantine analysis. He seems not to have known about contemporary research, however, witness his statements about the regular heptagon. One can therefore safely say that Ibn Sīnā had a good knowledge of mathematics, sufficient to allow him to deal with certain applications but without engaging in genuine mathematical research. In other words, it is just as incorrect to reduce Ibn Sīnā's mathematical knowledge to Euclid's *Elements* and Nicomachus of Gerasa's *Introduction to Arithmetic* as it is to make of him a 10th-century mathematical counterpart to al-Kindī. For this great logician, metaphysician, and physician, mathematics played a different role than it did for al-Kindī. It was not only a source of inspiration in certain philosophical inquiries, but also an integral part of his philosophical synthesis. Precisely herein lies the significance of his inclusion in *al-Shifā'* of four books devoted in sequence to the disciplines of the *quadrivium*. The question at hand, then, is to assess the philosophical implications of their presence in this work.

In fact, if one restricts oneself to Ibn Sīnā's theoretical statements about the status of mathematics, the nature of their objects, and the number of the disciplines that constitute it, one can conclude that he is the direct heir of tradition: the status of mathematics is defined by means of the Aristotelian theory of the classification of sciences, which is itself grounded in the famous doctrine of Being; its objects are defined by means of the theory of abstraction; as to the number of the disciplines, it is the well-known quartet transmitted by the ancient Greek tradition. At issue therefore is the middle science (*al-'ilm al-awsaṭ*) of the three disciplines that constitute theoretical philosophy, the objects of which are divided between physics, mathematics, and metaphysics (an order that the composition of the *Shifā'* follows) as a function of their materiality and changeability. Thus mathematics focuses on objects abstracted from the sensible realm and separated from physical objects, which are material and mobile. The disciplines that constitute it are those of the *quadrivium*: Arithmetic, Geometry, Astronomy, and Music. It is to this doctrine that Ibn Sīnā always returns in the *Shifā'* (not only in the *Isagoge* that opens it, but also in the *Metaphysics*) as well as in a short work devoted to the classification of the sciences, among other writings.

> The various kinds of sciences either focus on considering entities insofar as they are in motion, according to their conception and constitution, and according as they pertain to specific matters and species; or focus on considering entities insofar as they are separated from these matters, according to conception but not constitution; or again, they focus on considering entities insofar as they are separated according to both constitution and conception.
>
> The first part of these sciences is physics; the second part is pure mathematics, in which the science of numbers is famous. As to knowledge of the nature of numbers *qua* numbers, it does not pertain to this science. The third part is metaphysics. Since entities by nature fall into these three parts, these just are the theoretical philosophical sciences.
>
> Practical philosophy either pertains to the teaching of opinions, the use of which allows one to organize participation in ordinary human affairs, and <this part> is known as the organization of the city; it is called politics; or, it pertains to that which allows one to organize participation in private human affairs, and <this part> is known as the organization of the household <economics>; or it pertains to that which allows one to organize the state of an individual person in order to edify his soul: this is called ethics.[27]

There is nothing new in this conception. If therefore one does not go beyond Ibn Sīnā's Aristotelian preferences, one cannot hope to grasp the genuine role that mathematics play in the *Shifā'*. One should perhaps wonder above all if such a statement of principle corresponds to the philosopher's mathematical knowledge, and if the theoretical classification reflects an eventual *de facto* classification. To gauge and understand the gap, if any, between these two classifications, one must first take into account Ibn Sīnā's mathematical studies. We consider here only arithmetic, even though geometry gave him much food for thought (the fifth postulate, for example, as in the *Danish-Nameh*).

Starting first with biographical matters alone, we know that while Ibn Sīnā was getting his philosophical training, he was learning Indian arithmetic and algebra. Only later did he learn logic, Euclid's *Elements*, and the *Almagest*, witness the reports of such biobibliographers as al-Bayhaqī, Ibn al-ʿImād, Ibn Khallikān, al-Qifṭī, Ibn Abī Uṣaybiʿa. Thus al-Bayhaqī notes:

> When he was ten years old, he knew by heart several fundamental literary texts. His father was then studying and meditating upon a treatise by the Brothers of Purity. He too was meditating on it, and his father put him in

[27] *Al-Shifā'*, *al-Manṭiq*, 1. *al-Madkhal* (*Isagoge*), ed. G. Anawati, M. al-Khuḍayrī, F. al-Ahwānī, Cairo, 1952, p. 14.

touch with a vegetable merchant named Maḥmūd al-Massāḥ who knew Indian computation and algebra and *al-muqābala*.[28]

Citing Ibn Khallikān, Ibn al-ʿImād recalls this biographical detail in similar language: 'When he was ten years old, he had perfected the knowledge of the Glorious Qur'an, and of literature, and he knew by heart some of the foundations of religion, Indian calculation, and algebra, and *al-muqābala*'.[29] Ibn Sīnā himself writes: 'My father directed me to a man who sold vegetables and who practised Indian computation, so that he might teach me'.[30]

Since these new disciplines – Indian arithmetic and algebra – were unknown to the Alexandrians, however, they could not find a place in the traditional classification of sciences without, at the very least, altering its general framework, if not completely overturning its underlying concepts. Now in Ibn Sīnā's classification, these disciplines appear under the single rubric of 'secondary parts of arithmetic (*al-aqsām al-farʿiyya*)'.

Ibn Sīnā does not explain this concept of 'secondary parts of arithmetic'; he is satisfied merely to list its contents.

> The secondary parts of the mathematical sciences – some branches of the <science of> numbers: the science of addition and of separation of the Indian arithmetic; the science of algebra and *al-muqābala*. And the branches of the science of geometry: the science of measurement; the science of mobile ingenious devices; the science of the traction of heavy bodies; the science of weights and balances; the science of instruments specific to the techniques; the science of perspectives and mirrors; the science of hydraulics; and the branches of astronomy: the science of astronomical tables and of calendars. And the branches of music: the use of marvellous and curious instruments such as the organ and the like.[31]

We thus know only that arithmetic has, as its secondary parts, Indian computation and algebra. But the number of arithmetic disciplines to which Ibn Sīnā alludes is not limited to these last two, which he lists in his classification of sciences. Indeed we have already mentioned the book, called *al-arithmāṭīqī*, that he devotes to this science of calculation in the *Shifāʾ*. To this, one must add two disciplines: the first, which Ibn Sīnā

[28] *Tārīkh Ḥukamāʾ al-Islām*, ed. M. Kurd Ali, Damascus, 1946, p. 53.

[29] Ibn al-ʿImād, *Shadharāt al-dhahab fī akhbār man dhahab*, Beirut, n.d., vol. III, p. 234; see also Ibn Khallikān, *Wafayāt al-Aʿyān*, ed. I. ʿAbbās, Beirut, 1969, vol. II, pp. 157–8.

[30] Al-Qifṭī, *Taʾrīkh al-Ḥukamāʾ*, ed. J. Lippert, Leipzig, 1903, p. 413 and Ibn Abī Uṣaybiʿa, *ʿUyūn al-anbāʾ*, ed. N. Riḍā, Beirut, 1965, p. 437.

[31] *Parts of the Rational Sciences*, p. 112.

names, but whose status he never specified, is *al-ḥisāb*; the second is present only through its objects: integer Diophantine analysis.

The theory of numbers, *al-arithmāṭīqī*, Indian arithmetic, algebra, *al-ḥisāb*, and integer Diophantine analysis: six disciplines that overlap and are sometimes superimposed in order to cover the study of numbers. The reality is evidently much more complex than might appear from Ibn Sīnā's schematic classification of the sciences. In order to untangle the intertwinings of these disciplines and to clarify their relations, however, one must briefly recall the work of contemporary mathematicians. Indeed the latter used two different terms to distinguish, on the one hand, the arithmetic in the Hellenistic tradition and its Arabic development: the theory of numbers (*'ilm al-a'dad*); and on the other hand, the discipline designated by the phonetic transcription of *ἡ ἀριθμητική* (*al-arithmāṭīqī*) Although their connotations are not completely unrelated, each of these terms nevertheless referred to a distinct tradition. The expression 'theory of numbers (*'ilm al-a'dad*)' referred to the arithmetic books of Euclid's *Elements*, as well as to such later works as those of Thābit ibn Qurra, for example; whereas the phonetic transcription of *ἡ ἀριθμητική* (*al-arithmāṭīqī*) designated the arithmetic tradition of the neo-Pythagoreans, that is, in the sense in which Nicomachus of Gerasa understands it in his *Introduction*, a book that Ibn Qurra had translated under the title *Kitāb al-Madkhal ilā 'ilm al-'adad* (*Introduction to Arithmetic*).[32] Although not systematic, this terminological difference between the 9th and 10th centuries seems to measure the gap that separated the two disciplines at the time. In order to understand how that gap was later understood, consider what Ibn al-Haytham writes:

> The properties of numbers can be displayed in two ways: the first is induction, for if one follows the sequence of numbers one by one, and if one distinguishes them, by distinguishing and considering them, one finds all of their properties; to find the number in this way is called *al-arithmāṭīqī*. This is shown in the work of *al-arithmāṭīqī*, a work by Nicomachus of Gerasa. The other way in displaying the properties of numbers proceeds by demonstrations and deductions. All the properties of number grasped by demonstrations are contained in these three books [of Euclid] or what is reduced to them (*Sharḥ Muṣādarāt Kitāb Uqlīdis*, ms. Feyzullah 1359, fol. 213v).[33]

For this eminent mathematician each way of proceeding is a science; his remark is all the more important because Ibn al-Haytham demanded

[32] Nicomachus of Gerasa, *Kitāb al-Madkhal ilā 'ilm al-'adad*, translated by Thābit ibn Qurra, ed. W. Kutsch, Beirut, 1958.

[33] R. Rashed, 'Ibn al-Haytham et le théorème de Wilson', *Archive for History of Exact Sciences*, 22.4, 1980, pp. 305–21.

rigorous demonstrations everywhere and without restriction. And in fact, in the 10th century at least, these two traditions offered mathematicians the same conception of the object of arithmetic: an arithmetic of integers represented by line segments. But whereas, in the theory of numbers, the standard of demonstration is obligatory, in *al-arithmāṭīqī*, one can proceed by simple induction. For the scientists of the 10th century, the difference between the two traditions thus reduces to a distinction of methods and of standards of rationality.

In Ibn Sīnā, one finds precisely this conception of this connection between the two disciplines. In *al-Shifāʾ*, arithmetic appears twice: the first time in the Geometry section of *al-Shifāʾ* in which he merely summarizes Euclid's arithmetic books; as to the second time, he composes his own book of *al-arithmāṭīqī* – a text that will be read and taught for centuries – and whose genuine foundations, according to the author, are primarily found in the *Elements*. It is perhaps this picture of the relation between the two disciplines that explains why, in his *al-arithmāṭīqī*, Ibn Sīnā did not merely summarize Nicomachus, as he had done for number theory with Euclid's *Elements*. This approach thus clarifies how in this field, he had distanced himself from the neo-Pythagorean tradition. And in fact, all the ontological and cosmological considerations that burdened the concept of number have henceforth been banished from *al-arithmāṭīqī*, which is now treated as a science. The only remnant of this earlier stage is the philosophical goal common to all branches of philosophy, both theoretical and practical, namely the perfection of the soul. It is thus against the neo-Pythagoreans that Ibn Sīnā writes:

> Among those who treat the art of arithmetic, it is customary to appeal, in this place and analogous places, to developments foreign to this art, and even more foreign to the custom of those who proceed by demonstration, and closer to the discourses of rhetoricians and poets. One must abandon <this custom>.[34]

Here he even partly gives up the traditional language, drawing on that of the algebraists, in order to express the successive powers of an integer. This is how, to name the powers of an integer, philosophers came to use the terms 'square (*māl*)', 'cube (*kaʿb*)', 'square-square (*māl māl*)' to designate the successive powers of the unknown.[35]

Under these conditions, there was nothing to prevent Ibn Sīnā from integrating into his *al-Arithmāṭīqī* theorems and results obtained elsewhere,

[34] *Al-Shifāʾ*, *al-Ḥisāb* (*al-Arithmāṭīqī*), ed. ʿA. Maẓhar, Cairo, 1975, p. 60. Note that a few lines earlier, Ibn Sīnā clearly mentions them by their name, *i.e.* the Pythagoreans.

[35] *Ibid.*, p. 19.

without needing to repeat the proof, if there was one. So it was that without demonstrating it, he takes up Thābit ibn Qurra's theorem on amicable numbers, in the latter's pure Euclidean style. Ibn Sīnā likewise recalls several problems of congruence.

> If you add evenly-even numbers and one, if you get a prime number, provided that, if the last of them is added, and if the preceding one is taken away, and if the sum and the remainder are prime, then the product of the sum by the remainder, and then the whole by the last of the added numbers, yields a number that has a friend; its friend is the number obtained by adding the sum and the remainder, multiplied by the last of added numbers, and by adding the product to the first number that had a friend. These two numbers are amicable.[36]

To these two traditions, it is appropriate to add a third, to which Ibn Sīnā also alludes, namely that of integer Diophantine analysis. In the logical part of *al-Shifā'* devoted to demonstration, Ibn Sīnā indeed takes as an example the first case of Fermat's conjecture, which had already been treated by at least two 10th-century mathematicians, al-Khujandī and al-Khāzin. Ibn Sīnā writes:

> If one asks [...] if the sum of two cubic numbers is a cube, just as the sum of two square numbers was a square, one is raising a problem of arithmetic (*ḥisāb*).[37]

One can clearly see here that the word *ḥisāb* seems to designate a discipline that includes disciplines other than the Euclidean theory of numbers and *al-arithmāṭīqī*. Indeed, by *ḥisāb*, Ibn Sīnā seems to mean a science that includes all those that treat numbers, whether rational or algebraically irrational or irrational. The last paragraph of his *al-arithmāṭīqī* leaves no room for doubt about it. This is how it reads:

> That is what we wanted to say about the science of *al-arithmāṭīqī*. We have omitted some cases, the mention of which we considered to be external to the rule of this art. What remains in the science of *al-ḥisāb* is what is convenient for us in the use and determination of numbers. Finally, what remains in practice is along the lines of algebra and *al-muqābala*, of the Indian science of addition and of separation. As to the latter, however, it is best to mention them among the secondary parts.[38]

Everything points to the fact that, in *al-Arithmāṭīqī* as well as in his summary of Euclidean arithmetic books, Ibn Sīnā, like his predecessors and his contemporaries, restricts his inquiry to natural integers. As soon as he

[36] After correcting some errors in the Cairo edition, p. 28.

[37] *Al-Shifā'*, *al-Manṭiq*, 5. *al-Burhān*, ed. A. E. Afifi, Cairo, 1956, pp. 194–5.

[38] *Al-Shifā'*, *al-Arithmāṭīqī*, p. 69.

runs into problems that would force him to examine the conditions of rationality, whether to find a positive rational solution, or, more generally, to consider a class of irrational numbers, he finds himself outside of these two sciences. The term of *ḥisāb* therefore covers the whole of this arithmetic research, which takes place thanks to disciplines such as algebra, Indian computation and their analogues. These disciplines therefore take on a character that is instrumental and applied, so to speak, which sets them against ancient number theory. It is precisely by means of this instrumental and applied character that, as one can readily check, Ibn Sīnā in his classification distinguishes the whole of the 'secondary parts', defining them as such. Thus the 'secondary parts (*al-aqsām al-farʿiyya*)' of physics are medicine, astrology, physiognomy, oneiromancy, the art of divination, talismans, theurgy, and alchemy.

However, to understand how far Ibn Sīnā's own theoretical classification has moved beyond the traditional Greek and Hellenistic classifications, one must go back to his predecessor al-Fārābī (872–950). It was Steinschneider who first asked whether Ibn Sīnā's little treatise on *The Parts of the Rational Sciences* was linked to al-Fārābī's classification in his *Enumeration of the Sciences*. Steinschneider went on to deny any connection between the two works. Wiedemann confirms this opinion, and maintains that Ibn Sīnā enumerates only separate sciences, whereas al-Fārābī designates and characterises them by their interconnections; or, as he puts it, 'Ibn Sīnā essentially enumerates the individual sciences, whereas al-Farābī characterizes them in a mutually dependent representation'.[39]

This convergence makes good sense since an examination of the 'secondary parts' of arithmetic in Ibn Sīnā shows that they are none other than those disciplines that al-Fārābī brings together under the heading, 'the science of ingenious procedures', and which he defines as follows:

> the science of the manner of proceeding when one applies to physical bodies everything, the existence of which is proved, by predication and demonstration, in the aforementioned mathematical sciences; and when one brings it about and puts it into action in physical bodies.[40]

Indeed according to him, mathematical science has, as its object, lines, surfaces, solids and numbers, and it considers them as intelligible in themselves, and separated (*muntaziʿa*), that is, abstracted from physical

[39] E, Wiedemann, 'Über al-Fârâbîs Aufzählung der Wissenschaften (*Die Scientiis*)', in *Aufsätze zur arabischen Wissenschafts-Geschichte*, Hildesheim, G. Olms, 1970, p. 327. 'Ibn Sīnā zählt im wesentlichen die einzelnen Wissenschaften auf, während al-Farābi sie in zusammenhängender Darstellung charakterisiert'.

[40] *Iḥṣāʾ al-ʿUlūm*, ed. ʿU. Amīn, Cairo, 1968, p. 108.

objects. In order to discover and intentionally to make manifest the mathematical concepts present in these physical objects by means of the art, it would be necessary to construct procedures and to invent techniques and methods that allow one to overcome the obstacles presented by the materiality and sensibility of these objects. In arithmetic, al-Fārābī writes, these ingenious procedures include, among other things, 'the science known to our contemporaries by the name of algebra and *al-muqābala*, and that which is analogous to it'.[41] He also notes, however, that 'this science is common to arithmetic and geometry' and, a little later, that it

> includes the ingenious procedures for determining the numbers that one seeks to determine and to utilize, those among the rationals and the irrationals, for which Euclid gave the principles in Book X of his *Elements* (*Usṭuqusāt*), and those that are not mentioned in this book. Indeed, since the ratio of rationals to irrationals, the one to the other, is like the ratio of numbers to numbers, then every number is homologous to a certain rational or irrational magnitude. If therefore one determines the numbers that are homologous to the ratios of magnitudes, then one determines these magnitudes in a certain way. This is why one posits certain rational numbers so that they might be homologous to rational magnitudes, and certain irrational numbers so that they might be homologous to irrational magnitudes.[42]

In this crucial text, algebra is distinguished as a science in two respects: whereas it is apodictic like every other science, it nevertheless constitutes the domain of application not of one science alone, but of two simultaneously, namely arithmetic and geometry. As to its objects, it includes geometrical magnitudes as well as numbers, which can be both rational and algebraically irrational. Confronted by this new discipline that they must take into account, the new classifications of the sciences, with their aspirations to universality and exhaustiveness, must also justify one way or the other the abandonment of certain Aristotelian theses. This is how designations such as 'the science of the ingenious procedures', 'secondary parts', … came to be created, in order to carve out a non-Aristotelian zone in the midst of a classification that remains decidedly Aristotelian.

The philosophical impact of such a reworking goes far beyond, and especially much deeper than, a simple taxonomic change. If algebra is indeed common to both arithmetic and geometry without making any concessions about its status as a science, it is insofar as its very object, the 'algebraic unknown' – the 'thing (*shay'*, *res*)' – can designate indifferently a number or a geometric magnitude. What is more, since a number can also be an irrational, 'the thing' then designates a quantity that will be known

[41] *Ibid*., p. 109.
[42] *Ibid*.

only by approximation. Thus the object of the algebraists, 'the thing' must be sufficiently general to hold a variety of contents; but it must in addition exist independently of its own determinations, so that one can always improve the approximation. Aristotelian theory clearly cannot give an account of such an object's ontological status. One must therefore bring to bear on the problem a new ontology that makes it possible to discuss an object stripped of the very characteristics that alone would have made it possible to determine that of which it is the abstraction. This is an ontology that must also allow us to know an object without being in position to represent it exactly.

It is specifically since al-Fārābī that one begins to see in Islamic philosophy the development of an ontology that is sufficiently 'formal', as it were, to meet the aforementioned requirements, among others. In this ontology, 'the thing (*al-shay'*)' takes on a more general connotation than the existent. It is in this sense that al-Fārābī writes: 'One can call a thing everything that has a quiddity, whether it is exterior to the soul or whether it is [merely] conceived of in any manner whatsoever.' Whereas the 'existent is always said of every thing that has a quiddity, external to the soul, and cannot be said of a quiddity that is merely conceived'. Thus, according to him, the 'impossible (*al-mustaḥīl*)' can be called 'thing' but cannot be called 'existent'.[43] In the history of mathematics, such a tendency became more pronounced between al-Fārābī and Ibn Sīnā. Al-Karajī in particular gives algebra a more general status, and emphasizes the extension of the concept of number. Al-Bīrūnī, a contemporary of Ibn Sīnā, goes even further when he dares to write:

> The circumference of a circle stands in a given ratio to its diameter. The number of the one also stands to the number of the other in a ratio, even if it is irrational.[44]

On the plane of philosophy, Ibn Sīnā as an important metaphysician integrates al-Fārābī's concept into a doctrine that he wanted to make more systematic and that he presented in *al-Shifā'*. According to this doctrine, like the 'existent' and the necessary, the thing is given in immediate evidence or, to use Ibn Sīnā's own language, it is immediately inscribed in the soul and, along with these two other ideas, stands at the origin of all others. Whereas the existent picks out the same meaning as 'asserted (*muthbit*)' and 'achieved (*muḥaṣṣal*)', the thing according to Ibn Sīnā is that which the attribution (the statement) characterizes. Thus every existent is a thing but the reciprocal is not exactly the case, even though it is impossible

[43] *Kitāb al-Ḥurūf*, ed. M. Mahdi, Beirut, 1970, p. 128.
[44] *Al-Qānūn al-Mas'ūdī*, Hyderabad, 1954, vol. I, p. 303.

for a thing to exist neither as a concrete subject nor in the mind.[45] This is not the place for a full description of Ibn Sīnā's doctrine. Suffice it to point out that this new ontology which is neither Platonic nor Aristotelian, originated at least in part from new gains in the mathematical sciences.

If these gains led Ibn Sīnā to inflect ontology in a direction that can be called 'formal', they had a similar effect on his conception of the ontology of emanation, as we shall see later in relation to al-Ṭūsī's commentary.

The emanation from the One of the Intelligences and the celestial orbs as well as other worlds – that of nature and that of corporeal things – is one of the central doctrines in Ibn Sīnā's metaphysics. This doctrine raises a question that is at once ontological and noetic: from a unique and simple being, how can there emanate a multiplicity, which is also a complexity, that in the end includes both the matter of things and the forms of bodies and human souls? This ontological and noetic duality throws up the question as an obstacle, like a difficulty that is simultaneously logical and metaphysical and must be unravelled. One can therefore understand, at least in part, why Ibn Sīnā in his various writings never tires of returning both to this doctrine and implicitly to this question.

Studying the historical evolution of Ibn Sīnā's thought about this problem in his different writings could show us how he was able to amend his initial formulation by taking such a difficulty into account. If we consider only *al-Shifā*ʾ and *al-Ishārāt*, Ibn Sīnā presents the principles of this doctrine as well as the rules for the emanation of multiples from one simple unity. His explanation looks like a well articulated and organized exposition, but it does not count as a rigorous proof: indeed, Ibn Sīnā does not in fact give here the syntactic rules that are capable of embracing the semantics of emanation. Precisely here lies the difficulty of deriving a multiplicity from the One. Long ago, however, this derivation was already perceived as a problem and examined as such. Naṣīr al-Dīn al-Ṭūsī, the mathematician, philosopher, and commentator on Ibn Sīnā, not only grasped the difficulty, but wanted to provide the missing syntactic rules.

To understand his contribution, we must return first to Ibn Sīnā, not only to recall the elements of his doctrine, but also to grasp in his synthetic and systematic exposition, to the extent that we can, the formal principle whose presence made possible the introduction of the rules of combinatorial analysis. It was this principle that effectively allowed Ibn Sīnā to develop his exposition deductively. He had to ensure, on the one hand, the unity of Being, which is then predicable of everything univocally and, on the other hand, an irreducible difference between the First Principle and its

[45] *Al-Shifā*ʾ, *al-Ilāhiyyāt* (I), ed. G. Anawati and S. Zayed, Cairo, 1960, pp. 29 ff. and 195 ff.

creations. He then elaborated a general – in effect, a 'formal' – concept of Being. Considered as being, it is the object of no determination, not even of modalities; it is only being. It is not a genus, but a 'state' of all that is, and can be grasped only in its opposition to nonbeing, without it being the case, however, that nonbeing precedes it in time; this opposition is strictly of a rational order. On the other hand, the First Principle alone receives its existence itself.[46] It is therefore the only necessary existence: only in this case does existence coincide with essence. All other beings receive their existence from the First Principle, by emanation. This ontology and the cosmogony that accompanies it provide the three points of view from which being is considered: as being, as emanation[47] from the First Principle, and as the being of its quiddity. From the first two points of view, the emphasis lies in this being's necessity, whereas the third reveals its contingency. Briefly sketched, these are the concepts on which Ibn Sīnā will establish the following postulates:

1. There exists a First Principle, a Being that is necessary by its essence, one, indivisible by any means, and is neither a body nor in a body.
2. The totality of being emanates from the First Principle.
3. The emanation occurs neither 'according to an intention (*ʿalā sabīl qaṣd*)' nor to reach a goal, but by a necessity of the First Principle's being, that is, by its auto-intellection.
4. From the One emanates only the One.
5. There is a hierarchy of emanation, extending from entities whose being is the most perfect (*al-akmalu wujūdan*) to those whose being is the least perfect (*al-akhassu wujūdan*).

[46] Ibn Sīnā distinguishes between existence and essence for all other beings; on this point, see D. Saliba, *Sur la Métaphysique d'Avicenne*, Pau, 1926; A.-M. Goichon, *La Distinction entre existence et essence*, Paris, 1957; G. Verbecke, *Le statut de la métaphysique*, Introduction to Simone Van Riet's edition, *Avicenna Latinus, Liber de Philosophia Prima*, Louvain/Leiden, 1977.

[47] See L. Gardet, 'En l'honneur du millénaire d'Avicenne', *Revue Thomiste*, LIXe année, t. LI, no. 2, 1951, pp. 333–45; A. Hasnawi, 'Fayḍ (épanchement, émanation)', in A. Jacob (ed.), *Encyclopédie philosophique universelle*, vol. II, Paris, 1990, pp. 966–72; N. Heer, 'Al-Rāzī and al-Ṭūsī on Ibn Sīnā's Theory of Emanation', in P. Morewedge (ed.), *Neoplatonism and Islamic Philosophy*, Albany, State University of New York Press, 1992, pp. 111–25. See also in P. Morewedge (ed.), *Neoplatonism and Islamic Philosophy*, State University of New York Press, Albany, 1992: Th.-A. Druart, 'Al-Fārābī, Emanation, and Metaphysics', pp. 127–48; P. Morewedge, 'The Neoplatonic Structure of Some Islamic Mystical Doctrines', pp. 51–75; J. Owens, *The Relevance of Avicennian Neoplanism*, pp. 41–50; M. E. Marmura, 'Quiddity and Universilaty in Avicenna', pp. 77–87.

At first sight, one might suspect that several of these postulates contradict each other (2 and 4, for example) or lead to contradictory consequences. It is to avoid this initial impression that Ibn Sīnā introduces additional determinations during the course of his deduction. Thus, from 1, 2, 4 and 5, it follows that the totality of being, in addition to the First Principle, is a whole ordered by the predecessor-successor relation, which is at once logical and axiological, taking into account both the priority of being and its excellence. Indeed, if one excludes the First Principle, every being can have only one predecessor (as well as the predecessor of its predecessor and so on). Moreover every being, including the First Principle, can have only one successor (respectively, the successor, its successor …). But the philosopher Ibn Sīnā and his commentator al-Ṭūsī knew that, taken literally, this order forbids the existence of multiple beings, that is, of their independent coexistence, without some of them being logically prior to others or more perfect than they, which makes this order manifestly false, as al-Ṭūsī states.[48] It is therefore necessary to introduce additional qualifications as well as intermediate beings.

Now 1 and 2 for their part forbid that the multiplicity proceed from the 'impulses' (*nuzūʿāt*) and 'perspectives' (*jihāt*) of the First Principle, for to assume impulses and modalities in it is to deny its unicity and simplicity. Finally, 3, 4, and 5 imply that the emanation, as act of the First Principle, is not in the image of a human act, since its Author knows neither intention nor goal. Everything therefore indicates that one must introduce intermediate beings (*mutawassiṭa)*, which undoubtedly are hierarchically ordered, but which make possible an account of the multiplicity-complexity.

Let us begin properly with the First Principle and designate it by the first letter of the alphabet, *a*, as does Ibn Sīnā in his treatise *Nayrūziyya*. The First Principle 'intellects' itself by essence. In its auto-intellection, it 'intellects' the totality of being of which it is the proper principle,[49] without there being in it any obstacle to the emanation of this totality, or any rejection of it. Only in this sense can one say of the First Principle that it is the 'agent' (*fāʿil*) of the totality of being.

Having conceded this, however, one must still explain how this necessary emanation of the totality of being takes place without it being necessary to add anything at all that could contradict the Unicity of the First Principle. According to 1, 4, 5, a single being emanates from the First Principle, which necessarily belongs to the second tier of existence and

[48] Naṣīr al-Dīn al-Ṭūsī, *al-Ishārāt wa-al-Tanbihāt*, ed. S. Dunya, Cairo, 1971, p. 216.

[49] *Al-Shifāʾ*, *al-Ilāhiyyāt* (II), ed. M. Y. Musa, S. Dunya and S. Zayed, revised and introduced by I. Madkour, Cairo, 1960, p. 402, *l.* 16.

perfection. But since it emanates from a unique, pure, and simple being that is at once pure truth, pure power, pure goodness..., without any of these attributes existing in it independently in order to guarantee the unity of the First Principle, this derivative being can only be a pure Intellect. This implication respects 4, for, if this intellect were not pure, one would have to conclude that, from the One, there emanates more than one. This is the first separated Intellect, the first effect (*ma'lūl*) of the First Principle. Like Ibn Sīnā, we call it *b*.

Everything is now in place to explain the multiplicity-complexity. By essence, this pure Intellect is an effect: it is therefore contingent. But, as an emanation from the First Principle, it is necessary, since it has been 'intellected' by the latter. On this ontological duality is superimposed a noetic multiplicity: this pure Intellect knows itself and knows its own being as contingent being; that is, it knows that its essence is different from that of the First Principle, which is necessary; conversely, it knows the First Principle as necessary Being; finally, it knows the necessity of its own being as an emanation from the First Principle. I have here paraphrased what Ibn Sīnā writes himself in *al-Shifā'*.[50] He forestalls the objections of an imaginary detractor by noting that this multiplicity-complexity is not a hereditary characteristic, so to speak: it is not from the First Principle that the pure Intellect receives it, and this for two reasons. First, the contingency of its being pertains to its own essence, and not to the First Principle, which gave it the necessity of its being. Conversely, its knowledge of itself, as well as the knowledge it owes to the First Principle, is a multiplicity, which is a result of the necessity of its being originating from the First Principle. Thanks to these conditions, Ibn Sīnā can reject the accusation that he has attributed any multiplicity to the First Principle.

Next, Ibn Sīnā describes how, starting from this Pure Intellect, there emanate the other separated Intellects, the celestial Orbs, and the Souls that allow Intellects to act. Thus, from the pure Intellect *b*, there emanates, by its intellection of *a*, a second Intellect, namely *c*; and by its intellection of its essence, the Soul of the ninth celestial Orb; and by its intellection of its being as a contingent being, the body of this ninth Orb. Let us call the Soul of this Orb and its body *d*.

Ibn Sīnā continues in this fashion with his description of the emanation of Intellects, celestial Orbs with Souls and their bodies. From this point on, from every Intellect, there emanate the matter of sublunary things, the forms of bodies and human souls. Even if Ibn Sīnā's explanation has the advantage of not separating the question of the multiplicity originating from the one from the question of complexity (that is from the ontological

[50] *Ibid.*, pp. 405–6.

content of the multiplicity), his account nevertheless does not allow for a rigorous knowledge of the latter, insofar as he gives no general rule. Ibn Sīnā does nothing more than lead the elements to the Agent Intellect.

It is at this very point that al-Ṭūsī enters the picture. He will demonstrate that, from the First Principle, there effectively emanates, according to Ibn Sīnā's rules and by means of a small number of intermediaries, a multiplicity, such that every effect will have only one independently existing cause. We shall see that the price of this clear progress in the knowledge of multiplicity is the impoverishment of the ontological content: what is left of the multiplicity-complexity is in fact only the multiplicity.

Indeed, in his commentary of *al-Ishārāt*, al-Ṭūsī introduces the language and the procedures of combinations in order to follow up the emanation to the third tier of beings. At this point, he stops applying these procedures and concludes: 'if we go beyond these tiers [the first three], there may exist a denumerable multiplicity (*lā yuḥṣā ʿadaduhā*) in a single tier, and to infinity'.[51] Al-Ṭūsī's intention is therefore clear, and the procedure he applies to the first three tiers leaves no room for doubt: one must provide the proof and the means that Ibn Sīnā did not have. At this stage, however, al-Ṭūsī is still far from his goal. Indeed it is one thing to proceed by combinations for a number of objects, it is something else again to introduce a new language with its syntax. Here the language is that of combinations. Now it is precisely to the introduction of this language that al-Ṭūsī devotes his efforts in a separate treatise,[52] the title of which leaves no room for ambiguity: *On the Demonstration of the Mode of Emanation of Things in an Infinite <Number> Beginning from the First Unique Principle*. This time, we shall see al-Ṭūsī take a general approach by means of combinatorial analysis. Al-Ṭūsī's text and the results it contained did not disappear with their author. They reappeared in a late treatise completely devoted to combinatorial analysis. Thus al-Ṭūsī's solution not only set out a distinctive style of inquiry in philosophy, but also represented a most interesting contribution to the history of mathematics itself.

Al-Ṭūsī's idea is to tackle this problem with a combinatorial approach. But, for combinatorics to offer a possible solution, one must be sure that the *time* variable is neutralized. In the case of the doctrine of emanation,

[51] Ed. Dunya, vol. III, pp. 217–18.

[52] R. Rashed, 'Combinatoire et métaphysique: Ibn Sīnā, al-Ṭūsī et al-Halabi', in Roshdi Rashed and Joël Biard (eds), *Les Doctrines de la science de l'antiquité à l'âge classique*, Ancient and Classical Sciences and Philosophy, Leuven, Peeters, 1999, pp. 61–86. German translation: 'Kombinatorik und Metaphysik: Ibn Sīnā, al-Ṭūsī und Halabi', in Rüdiger Thiele (ed.), *Mathesis, Festschrift siebzigsten Geburtstag von Matthias Schramm*, Berlin, Diepholz, 2000, pp. 37–54.

this translates as follows: either to set becoming aside or, at the very least, to interpret it purely in logical terms. We have seen that Ibn Sīnā had already offered this condition. It has been correctly noted that emanation does not take place in time, and that anteriority and posteriority must be understood as essential, and not taken in a temporal sense.[53] This interpretation, which in my view is fundamental to the Avicennan system, draws on its own specific conception of the necessary, the possible, and the impossible. In a word, note that in his *al-Shifā'*,[54] Ibn Sīnā takes up this ancient problem by rejecting at the outset the ancient doctrines, which in his view are all circular: to define one of the three terms, they use one or the other of the two remaining ones. To break this circularity, Ibn Sīnā therefore seeks to restrict the definition of each term by bringing it back to the concept of existence. He then distinguishes what is considered in itself as existing necessarily, from what, also considered in itself, can exist and can also not exist. For him, necessity and contingency are inherent in the beings themselves. As to possible being, its existence as well as its non-existence depend on a cause exterior to it. Contingency therefore does not appear to be a downgraded form of necessity, but rather a different mode of existence. It may even be that possible being, while remaining such in itself, may have a necessary existence on account of the action of another being. Without going here into all of the subtleties of Ibn Sīnā's exposition, we note only that, on this specific definition of the necessary and the possible, Ibn Sīnā grounds the terms of the emanation in the nature of beings, immediately neutralising the *time* variable, as emphasized above. From these definitions, he deduces propositions, the majority of which he establishes by *reductio ad absurdum*. He shows that the necessary cannot not exist, that it cannot, by essence, have a cause, that its necessity encompasses all of its aspects, that it is one and cannot in any way allow multiplicity, that it is simple, without any composition. In each of these points, the necessary is opposed to the possible. It is therefore in the very definition of the necessary and the possible, and in the dialectic that

[53] *Al-Shifā'*, *al-Ilāhiyyāt* (II), ed. M. Y. Musa, S. Dunya and S. Zayed, revised and introduced by I. Madkour, p. 266–7. See L. Gardet, 'En l'honneur du millénaire d'Avicenne'; H. A. Davidson, *Proofs for Eternity Creation and the Existence of God in Medieval Islamic and Jewish Philosophy*, New York/Oxford, 1987; Th.-A. Druart, 'Al-Farabi and Emanationism', in John F. Wippell (ed.), *Studies in Medieval Philosophy*, Washington, The Catholic University of America Press, 1987, pp. 23–43; P. Morewedge, 'The Logic of Emanationism and Ṣūfism in the Philosophy of Ibn Sīnā (Avicenna)', Part II, *Journal of the American Oriental Society*, 92, 1972, pp. 1–18; A. Hasnawi, 'Fayḍ (épanchement, émanation)'.

[54] See especially Book 3, Chapter 4 of *Qiyās*, vol. IV, ed. Zayed, 1964.

engages both of them, that the anteriority of the First Principle and its relations with the Intelligences are forever fixed.

If therefore one can describe emanation without appealing to time, it is insofar as its proper terms are given in a logic of the necessary and the possible. The point at issue here is not whether or not this doctrine involves difficulties: it is rather that Ibn Sīnā himself had already guaranteed the conditions for introducing combinatorics into the problem.

We have stated that from a, b emanates; the latter is therefore in the first tier of effects. From a and b together emanates c, the second intellect; from b alone emanates d – namely, the celestial Orb. One therefore has, in the second tier, two elements c and d, neither of which is the cause of the other. But so far, one only has four elements: the First Cause a and three effects, b, c, and d. Al-Ṭūsī calls these four elements the *principles*. Let us now combine these four elements two by two, then three by three, and finally four by four. One successively obtains six combinations – *ab*, *ac*, *ad*, *bc*, *bd*, *cd* –, four combinations – *abc*, *abd*, *acd*, *bcd* –, and one combination of four elements – *abcd*. Tallying the combinations of these four elements one by one yields a sum of 15 elements, 12 of which belong to the third tier of effects, without the ones being intermediate beings used to derive the others. This is what al-Ṭūsī presents in his commentary of the *Ishārāt*, as well as in his treatise mentioned above. As soon as one goes beyond the third tier, however, matters quickly become complicated, and al-Ṭūsī has to introduce into his treatise the following lemma:

The number of combinations of n *elements is equal to*

$$\sum_{k=1}^{n}\binom{n}{k}.$$

To calculate this number, al-Ṭūsī uses the equation

$$\binom{n}{k}=\binom{n}{n-k}.$$

Thus, for $n = 12$, he gets 4095 elements. Note that, to deduce these numbers, he shows here the expressions of the sum by combining the letters of the alphabet.

Next, al-Ṭūsī goes back to the calculation of the number of elements of the fourth tier. He then considers the four principles with the 12 beings of the third tier; he gets 16 elements, from which he gets 65,520 effects. To obtain this number, al-Ṭūsī proceeds with the help of an expression equivalent to

$$(*)\qquad \sum_{k=0}^{m}\binom{m}{k}\binom{n}{p-k} \text{ for } 1 \leq p \leq 16, m = 4, n = 12,$$

the value of which is the binomial coefficient

$$\binom{m+n}{p}.$$

With the exception of *a*, *b*, and *ab*, none of these elements is an intermediary for the others. Thus al-Ṭūsī's response is general, and (*) gives a rule that allows one to know the multiplicity in each tier.

Having established these rules and given the example of the fourth tier, with its 65,520 elements, al-Ṭūsī is in a position to state that he has answered thc question 'of the possibility of the emanation of the denumerable multiplicity starting from the First Principle on condition that only one emanates from the One, and without the effects being successive (in a chain), which was to be proved'.

Al-Ṭūsī's success in making Ibn Sīnā's ontology speak the language of combinatorial analysis was the motor of two important evolutions: in both Ibn Sīnā's doctrine and the combinatorics. This time, the question of multiplicity is clearly kept at a certain distance from that of the complexity of the being. Al-Ṭūsī worries very little about the ontological status of each of the thousands of beings that constitute the fourth tier, for example. But there is more. Metaphysical discourse now allows us to discuss a being without giving us the means of representing it exactly to ourselves. This 'formal' (so-to-speak) evolution of ontology, which is conspicuous here, merely amplifies a tendency that was already present in Ibn Sīnā, and that we already emphasized earlier in his discussions of 'the thing (*al-shay*')'. This 'formal' movement is further accentuated by the possibility of designating beings by letters of the alphabet. Even the First Principle is no exception to the rule, since the letter *a* designates it. Here, too, al-Ṭūsī amplifies an Avicennan practice, but he inflects its meaning. In the epistle *al-Nayrūziyya*, Ibn Sīnā draws upon this symbolism, but with two variants: on the one hand, he assigns to the succession of the letters of the Arabic alphabet (following the order *abjad hawaḍ*) the value of an order of priority and of logical anteriority; on the other hand, he uses the numerical values of the letters ($a = 1$, $b = 2$, etc.). As to al-Ṭūsī, even though he implicitly keeps the order of priority by designating, as Ibn Sīnā does, the First Principle by *a*, the Intellect by *b*, he has abandoned this hierarchy in favor of the conventional value of the symbol. Their numerical value has simply disappeared. This was necessary for the letters to become the object of combinatorics. As a mathematician and a philosopher, al-Ṭūsī has recast Ibn Sīnā's doctrine of emanation in a formal direction, thus favoring a trend already present in Ibn Sīnā's ontology.

3. FROM *ARS INVENIENDI* TO *ARS ANALYTICA*

For reasons internal to the evolution of their discipline, the mathematicians of the 9th century encountered the problem of the duality of order: is the order of exposition identical to the order of discovery? This question was very naturally first raised about the very model of mathematical composition at that time and for many centuries to come, namely, Euclid's *Elements*. To this problem, Thābit ibn Qurra devotes a treatise in which he maintains that the order of exposition of the *Elements* is nothing but the logical order of the demonstrations and differs from the order of discovery. To characterise the latter, Thābit develops a 'psycho-logical' doctrine of mathematical discovery. In a sense, we are already on the terrain of the philosophy of mathematics.

This question of order will rather quickly be engulfed in a more general problematic, that of analysis and a profoundly transformed synthesis. Occasionally broached by Galen, Pappus, and Proclus, this topic in Antiquity never approached the range that characterized it in the 10th century; the development of mathematics and the conceptions of new chapters in it from the 9th century onward had broad repercussions on both the extent and the understanding of this topic. With the latter, a genuine philosophy of mathematics develops. Indeed one witnesses in succession the elaboration of a philosophical logic of mathematics, then a project of *ars inveniendi* and finally of an *ars analytica*.

It all seems to have started with Ibrāhīm ibn Sinān (909–946), who wrote a whole book completely and uniquely devoted to analysis and synthesis entitled *On the Method of Analysis and Synthesis in the Problems of Geometry*.[55] The importance of this event is obvious: henceforth, analysis and synthesis designate a domain in which the mathematician invests his energies, both as a geometer and as a logician-philosopher. Consider how Ibn Sinān discusses his enterprise and his intention:

> In this book, I have set out in exhaustive fashion a method designed for students that contains everything necessary to solve the problems of geometry. In it, I have presented in general terms the various classes of geometrical problems; I have then subdivided these classes and illustrated

[55] R. Rashed and H. Bellosta, *Ibrāhīm ibn Sinān. Logique et géométrie au Xe siècle*, Leiden, E. J. Brill, 2000, Chap. I.

> each one with an example; I have then led the student toward the road by which he can know into which of these classes he should place the problems put to him, by which he will know how to carry out the analysis of the problems – as well as the subdivisions and necessary conditions for doing so – and to carry out the synthesis of them – as well as the conditions necessary for this – and then how he will know if the problem belongs to those that are soluble only once or several times, and in general, all that it is necessary to know on the subject.
>
> I have pointed out into which kind of error geometers fall in analysis, on account of their practice – a habit that they have acquired: excessive abbreviation. I have also indicated the reason why, for geometers, there may appear to be, in propositions and problems, a difference between analysis and synthesis, and I have shown that their analysis differs from synthesis only on account of the abbreviations, and that, if they had completed their analysis as they ought, it would have been identical to synthesis; the doubt would then have been removed from the heart of those who suspect them of producing in synthesis things that they had not mentioned earlier in analysis – things, lines, surfaces, and the like, that one sees in their synthesis, without having mentioned them in analysis. I have shown this and illustrated it with examples. I have presented a method thanks to which analysis is such that it coincides with synthesis; I have issued a warning about the things that geometers tolerate in analysis, and I have shown what kind of error follows if one tolerates them.[56]

Ibn Sinān's intention is clear, and his project is well articulated: to classify geometrical problems according to different criteria in order to show how to proceed in each class, by analysis and synthesis, and to put on display the loci of errors in order to avoid them. Here is a broad outline of his classification.

1. The problems whose assumptions are completely given
    - 1.1 The true problems
    - 1.2 The impossible problems
2. The problems for which it is necessary to modify some hypotheses
    - 2.1 The problems with discussion (diorism)
    - 2.2 The indeterminate problems
        - 2.2.1 The indeterminate problems strictly speaking
        - 2.2.2 The indeterminate problems with discussion
    - 2.3 The overabundant problems
        - 2.3.1 The indeterminate problems to which an addition is made
        - 2.3.2 The problems with discussion to which an addition is made
        - 2.3.3 The true problems to which an addition is made

[56] *Ibid*., pp. 96–8.

To this is also added the modal classification of propositions.

This classification is based on several criteria: the number of solutions, the number of hypotheses, their compatibility, and their eventual independence.

Al-Samaw'al, a little more than two centuries later, takes up this classification, always starting from the number of solutions and the number of hypotheses.[57] He further refines the classification. He draws a distinction between identities and the problems that have an infinite number of solutions without being identities. He furthermore introduces the concept of undecidable problems, one for which one can demonstrate neither the existence nor the negation.[58] Unfortunately, he does not give an example. The least one can say, however, is that al-Samaw'al was able to inflect the Aristotelian concept of the necessary, the possible, and the impossible toward those of computability and semantic undecidability.

In his book, Ibn Sinān discusses other logical problems, such as the place of auxiliary constructions, the reversibility of analysis, and apagogic reasoning. Thus in Ibn Sinān's book, analysis-and-synthesis conjointly presents itself both as a discipline and as a method. The former is in effect a philosophical and pragmatic logic, insofar as it makes possible the association of an *ars inveniendi* and an *ars demonstrandi*; the latter is a procedure based on a theory of demonstration that Ibn Sinān endeavoured to elaborate.

A generation after Ibn Sinān, the mathematician al-Sijzī (last third of the 10th century) conceives of a different project, that of an *ars inveniendi* that corresponds to requirements that are at once logical and didactical requirements. Al-Sijzī begins by enumerating the methods aimed at facilitating mathematical discovery, at least seven of them. Among the latter, there is in fact a primary one, 'analysis and synthesis', and several specific methods that grant this primary one effective means of discovery. Among the latter, one finds the method of pointwise transformations and the method of ingenious procedures. All these specific methods have in common the idea of transforming and varying the figures as well as the propositions and solution procedures. When he summarizes his project, al-Sijzī writes:

[57] S. Ahmad and R. Rashed, *Al-Bāhir en Algèbre d'As-Samaw'al*, Damas, Presses de l'Université de Damas, 1972.

[58] R. Rashed, *Entre arithmétique et algèbre. Recherches sur l'histoire des mathématiques arabes*, Sciences et philosophie arabes – Études et reprises, Paris, Les Belles Lettres, 1984, p. 52. English transl.: *The Development of Arabic Mathematics: Between Arithmetic and Algebra*, Boston Studies in the Philosophy of Science 146, Dordrecht, Kluwer, 1994.

As the examination of the nature of propositions (*al-askhāl*) and of their properties in themselves is surely carried out in one of these two ways: either we imagine the necessity of their properties by making their species vary, an imagination that draws on sensation or what the senses have in common; or by positing these properties and also the lemmas that they require, successively, by geometrical necessity [...].[59]

For al-Sijzī, then, the *ars inveniendi* includes essentially only two paths. All specific methods are grouped around the first path, whereas the second is none other than 'analysis and synthesis'. Together, the following three features – the distinction, the nature of the first path, and finally the intimate relation between the two – make al-Sijzī's conception unique and reflect the novelty of his contribution.

It should also be noted that the first of the two paths splits into two, according to the two meanings of the term *shakl*. This word, which the translators of Greek mathematical writings chose to render διάγραμμα, can indifferently designate, as the Greek word does, both the figure and the proposition.

This double meaning is not too fraught with ambiguity as long as the figure graphically translates the proposition in a rather static manner, if I may say so; in other words as long as geometry essentially remains a study of figures. But everything becomes more complicated when one begins to transform the figures and to make variations on them, as is already the case in some branches of geometry at the time of al-Sijzī. The double reference then requires a clarification. Let us begin with the first meaning, that of 'figure'.

In this treatise, al-Sijzī recommends that one proceed by varying the figure in three different situations: when one makes a pointwise transformation; when one varies one element of the figure when all the others remain constant; finally, when one chooses an auxiliary construction. Now these different procedures have several elements in common. First the goal: one always seeks to attain, thanks to the transformation and the variation, the invariable properties of the figure associated with the proposition, the ones that properly characterize it. It is precisely these invariable properties that are stated in the figure as a proposition. The second element also pertains to the goal: variation and transformation are means of discovery insofar as they lead to invariable properties. This is where the imagination intervenes, a power of the soul capable of drawing on the multiplicity offered by the senses by means of the variable properties of the figures, the

[59] R. Rashed, *Les Mathématiques infinitésimales du IXe au XIe siècle*, vol. IV: *Méthodes géométriques, transformations ponctuelles et philosophie des mathématiques*, London, 2002, Appendice I, p. 818.

invariable properties, the essences of things. The third element concerns a specific role of the figure, now as a representation: the role, often mentioned by al-Sijzī, of fixing the imagination, of helping it in its task when it draws on sensation. No less important, the fourth pertains to the figure-proposition duality: there is no biunivocal relation. A variety of figures can correspond to one and the same proposition; likewise, an entire family of propositions can correspond to a single figure. In fact, al-Sijzī decided to treat this last case at great length. These new relations between figure and proposition, which al-Sijzī was the first to point out, as far as I know, require that one conceive a new chapter of the *ars inveniendi*: the analysis of figures and of their connections to propositions. This is precisely what al-Sijzī seems to have inaugurated.

A generation later, Ibn al-Haytham (d. after 1040) conceives another project: that of founding a scientific art, with its rules and vocabulary. Ibn al-Haytham begins by reminding the reader that mathematics is founded on demonstrations. By demonstration, he means 'the syllogism that necessarily indicates the truth of its proper conclusion'.[60] This syllogism in turn is composed 'of premises, the truth and validity of which the understanding recognizes without entertaining any doubts about them; and of an order and arrangement of these premises such that they compel the hearer to be convinced of their necessary consequences and to believe in the validity of what follows from their arrangement'.[61] The Art of analysis (*Ṣināʿat al-taḥlīl*) provides the method for obtaining these syllogisms, *i.e.* 'to pursue the search for their premises, to contrive to find them, and to try to find their arrangement'.[62] In this sense, the Art of analysis is an *ars demonstrandi*. It is also an *ars inveniendi*, insofar as one is led, thanks to this art, 'to undertake the hunt for the unknowns of the mathematical sciences and how to carry on seeking the premises (literally 'to hunt [*taṣayyud*] for the proofs'), which are the material of the demonstrations indicating the validity of what one discovers about the unknowns of these sciences, and the method for obtaining arrangement of these premises and the figure of the combination'.[63]

For Ibn al-Haytham, it is indeed an *Ars* (τέχνη, *ṣināʿat*) *analytica* that one must conceive and construct. As far as I know, no one before him

[60] R. Rashed, 'La philosophie mathématique d'Ibn al-Haytham. I: *L'analyse et la synthèse*', *Mélanges de l'Institut Dominicain d'Études Orientales du Caire*, 20, 1991, pp. 31–231, p. 36; *Les Mathématiques infinitésimales*, vol. IV, pp. 162 ff.

[61] R. Rashed, 'La philosophie mathématique d'Ibn al-Haytham. I: *L'analyse et la synthèse*'.

[62] *Ibid.*

[63] *Ibid.*, p. 38.

considered analysis and synthesis as an art or, more precisely, as a double art of both demonstration and discovery. In the first, the analyst (*al-muḥallil*) must know the principles (*uṣūl*) of mathematics. This knowledge must be undergirded by an 'ingenuity' and an 'intuition shaped by the art' (*ḥads ṣināʿī*). This intuition is indispensible for discovery and also proves necessary when synthesis is not strictly the inversion of analysis, but requires the discovery of supplemental properties and data. Knowledge of principles, ingenuity and intuition are so many means that the analyst must have in order to discover mathematical unknowns. The 'laws' and the 'principles' of this analytical art still remain to be known. This required knowledge is the object of a discipline that pertains to mathematical foundations and that treats the 'knowns'. It, too, has yet to be constructed. This last characteristic is peculiar to Ibn al-Haytham insofar as nobody before him, not even Ibn Sinān, thought of conceiving an analytical art founded on its own mathematical discipline. To the latter, Ibn al-Haytham devotes a second treatise, *The Knowns*,[64] which he had promised in his treatise on *Analysis and Synthesis*.[65] He himself presents this new discipline as the one that gives the analyst the 'laws' of this art and the 'foundations' that complete the discovery of properties and the grasp of premises; in other words, it touches the foundations of mathematics, the prior knowledge of which, as we stated, is indeed necessary to complete the art of analysis: these are the concepts called the 'knowns'.[66] Note that every time he treats a foundational problem, as he does in his *On the Quadrature of the Circle*,[67] Ibn al-Haytham comes back to the 'knowns'.

According to Ibn al-Haytham, a concept is said to be 'known' when it remains invariable and admits no change, whether or not a knowing subject thinks this concept. The 'knowns' designate invariable properties, independent of the knowledge that we have of them, and remain unchanged even though the other elements of the mathematical object vary. The goal of the analyst, according to Ibn al-Haytham, is precisely to find these invariable properties. Once these fixed elements have been attained, his task is complete, and one can then turn to synthesis. The *Ars inveniendi* is neither

[64] R. Rashed, 'La philosophie mathématique d'Ibn al-Haytham. II: *Les Connus*', *Mélanges de l'Institut Dominicain d'Études Orientales du Caire* (*MIDEO*), 21, 1993, pp. 87–275; *Les Mathématiques infinitésimales*, vol. IV.

[65] R. Rashed, 'La philosophie mathématique d'Ibn al-Haytham. I: *L'analyse et la synthèse*', p. 68.

[66] *Ibid.*, p. 58.

[67] R. Rashed, *Les Mathématiques infinitésimales*, vol. II, pp. 91–5; English transl. *Ibn al-Haytham and Analytical Mathematics*, pp. 101–3.

mechanical nor blind, but it is by dint of ingenuity that it leads to the 'knowns'.

In order to constitute itself, the analytical art therefore requires a mathematical discipline that must itself be constructed. The latter contains the 'laws' and the 'principles' of the former. The analytical art cannot, according to this conception, be reduced to any logic, but its own logical part is immersed in this mathematical discipline. We can thus see the limits of this art's extension.

These briefly sketched contributions indicate here several situations in which mathematicians have treated the philosophy of mathematics. Earlier we have seen other situations in which philosopher-mathematicians and mathematician-philosophers contributed to the philosophy of mathematics. These contributions are obviously part of the history of philosophy, the history of the science, and the history of mathematical thought in classical Islam. To forget these contributions is at once to impoverish the history of philosophy and to truncate the history of mathematics.

INDEX OF NAMES

INDEX OF WORKS